8th TSME-International Conference on Mechanical Engineering (TSME-ICoME 2017)

IOP Conference Series: Materials Science and Engineering
Volume 297

Bangkok, Thailand
12 – 15 December 2017

ISBN: 978-1-5108-5836-7
ISSN: 1757-8981

Printed from e-media with permission by:

Curran Associates, Inc.
57 Morehouse Lane
Red Hook, NY 12571

Some format issues inherent in the e-media version may also appear in this print version.

Printed by Curran Associates, Inc. (2018)

For permission requests, please contact the Institute of Physics
at the address below.

Institute of Physics
Dirac House, Temple Back
Bristol BS1 6BE UK

Phone: 44 1 17 929 7481
Fax: 44 1 17 920 0979

techtracking@iop.org

Additional copies of this publication are available from:

Curran Associates, Inc.
57 Morehouse Lane
Red Hook, NY 12571 USA
Phone: 845-758-0400
Fax: 845-758-2633
Email: curran@proceedings.com
Web: www.proceedings.com

Table of contents

Volume 297

8th TSME-International Conference on Mechanical Engineering (TSME-ICoME 2017)

12–15 December 2017, Bangkok, Thailand

Accepted papers received: 9 January 2018
Published online: 7 February 2018

Preface

Papers

Alternative Energy and Combustion (AEC)

Computation and Simulation Techniques (CST)

Thermal System and Fluid Mechanics (TSF)

Preface

The 8th TSME International Conference on Mechanical Engineering (TSME–ICoME 2017) was hosted by the Department of Mechanical and Aerospace Engineering, Faculty of Engineering, King Mongkut's University of Technology North Bangkok, Bangkok, Thailand. The conference was held at Arnoma Grand Hotel, Bangkok, from 12–15 December 2017. The conference is a leading conference of the Thai Society of Mechanical Engineering (TSME) and represents one of the only exclusive events in Thailand where everyone "mechanically-engineering-related" can share their ideas and establish long-lasting collaborations and networks.

The scientific program covered a wide range of topics, including Alternative Energy and Combustion (AEC), Automotive, Aerospace and Marine Engineering (AME), Applied Mechanics, Materials and Manufacturing (AMM), Biomechanics and Bioengineering (BME), Computation and Simulation Techniques (CST), Dynamic Systems, Robotics and Controls (DRC), Engineering Education (EDU), Energy Technology and Management (ETM), Industrial Entrepreneurship (IEI), and Thermal System and Fluid Mechanics (TSF).

This year conference brought together more than 200 delegates from Asia, Europe, Australia and North America to exchange knowledge by in-depth discussions and to promote international and multi-disciplinary collaborations under the theme of "Industry 4.0: Challenges and Solutions."

This volume of IOP Conference Series: Material Science and Engineering includes 72 selected papers presented in the conference. These peer-reviewed manuscripts are qualified representations of the conference topics and will raise readers' interest in these areas.

On behalf of the editing team, we would like to express our gratitude to the members of both the Conference Program Committee and the Organizing Committee, whose commitment and tireless devotion led to the success of this conference. We are also pleased to acknowledge our partners, sponsors, and participants for their contribution. Finally, we are grateful to King Mongkut's University of Technology North Bangkok for the financial support of TSME–ICoME 2017.

Assoc. Prof. Dr. Sinchai Chinvorarat
Prof. Dr. Suwat Kuntanapreeda
Conference Co-chairs of TSME–ICoME 2017

IOP Conf. Series: Materials Science and Engineering **297** (2018) 011001 doi:10.1088/1757-899X/297/1/011001

Message from the President of TSME

In 2010, TSME started to hold yearly international conferences under the name of the **International Conference on Mechanical Engineering (ICoME)** with collaboration from the American Society of Mechanical Engineers (ASME), the Japan Society of Mechanical Engineers (JSME), and the United Kingdom's Institute of Mechanical Engineers (IMechE).

The conference is a key mechanism to develop the researches from the knowledge exchanging which has contributed association networks and constituted a unique research community. The Mechanical Engineering research is an important role to enhance the growth of country especially in the industry sectors. This cooperation made a great opportunity of representing Mechanical Engineering to make the difference and success of engineering society.

In this regard, I would like to express my congratulation and appreciation to the department of Mechanical and Aerospace Engineering, Faculty of Engineering, King Mongkut's University of Technology North Bangkok for the dedication to organize the 8th TSME–ICoME where will be held at Anoma Grand, Bangkok from 12–15 December 2017. Thank you for your supporting team and all staff for hard working to make this conference possible.

I hope the conference will be success and belief that this collaboration will be further strengthened to the Mechanical Engineering in the years to come.

Asst. Prof. Dr. Chinda Charoenphonphanich
President of Thai Society of Mechanical Engineers (TSME)

IOP Conf. Series: Materials Science and Engineering **297** (2018) 011002 doi:10.1088/1757-899X/297/1/011002

Conference Program Committee

Honorary Chair

Asst. Prof. Dr. Chinda Charoenphonphanich
President, Thai Society of Mechanical Engineers (TSME)

General Conference Co-chairs

Assoc. Prof. Dr. Sinchai Chinvorarat	King Mongkut's Univ. of Technology North Bangkok, Thailand
Prof. Dr. Suwat Kuntanapreeda	King Mongkut's Univ. of Technology North Bangkok, Thailand

General Conference Secretariat

Asst. Prof. Dr. Petch Jearanaisilawong	King Mongkut's Univ. of Technology North Bangkok, Thailand

International Scientific Committee

Prof. Dr.-Ing. Heinz Peter Berg	Brandenburg University of Technology Cottbus-Senftenberg, Germany
Prof. Dr. Ashim K. Datta	Cornell University, USA
Prof. Dr. Ashwani K. Gupta	University of Maryland, USA
Prof. Dr. Katsunori Hanamura	Tokyo Institute of Technology, Japan
Prof. Dr.-Ing. Lothar Kroll	Chemnitz University of Technology, Germany
Prof. Dr. Ocktaeck Lim	University of Ulsan, Korea
Prof. Dr. Jau-Huai Lu	National Chung Hsing University, Taiwan
Prof. Dr. Naoki Maruyama	Mie University, Japan
Prof. Dr. Shinichi Nakasuka	University of Tokyo, Japan
Prof. Dr. Hironori Saitoh	Sojo University, Japan
Prof. Dr. Marco P. Schoen	Idaho State University, USA
Prof. Dr. Ning Zhu	Shizuoka Institute of Science and Technology, Japan
Assoc. Prof. Dr. Elmi Abu Bakar	Universiti Sains Malaysia, Malaysia
Assoc. Prof. Dr. Huynh Thanh Cong	Ho Chi Minh City University of Technology, Vietnam
Assoc. Prof. Dr. Seokwoo Jeon	Korea Advanced Institute of Science and Technology, Korea
Assoc. Prof. Dr. Iman K. Reksowardojo	Institute of Technology Bandung, Indonesia
Assoc. Prof. Dr. Hendri Syamsudin	Institute of Technology Bandung, Indonesia
Assoc. Prof. Dr. Le Anh Tuan	Hanoi University of Science and Technology, Vietnam
Assoc. Prof. Dr. Pham Huu Tuyen	Hanoi University of Science and Technology, Vietnam
Asst. Prof. Dr. Pham Tuan Anh	Ho Chi Minh City University of Technology, Vietnam

Local Scientific Committee

IOP Conf. Series: Materials Science and Engineering **297** (2018) 011002 doi:10.1088/1757-899X/297/1/011002

Host and Organizing Committee

Host

Department of Mechanical and Aerospace Engineering (MAE)

Faculty of Engineering

King Mongkut's University of Technology North Bangkok (KMUTNB), Bangkok, Thailand

Organizing Committee

Advisory Board

Prof. Dr.-Ing. Suchart Siengchin
President, King Mongkut's University of Technology North Bangkok (KMUTNB)

Assoc. Prof. Dr. Somrerk Chandra-Ambhon
Vice President for Research and Technology Development, KMUTNB

Assoc. Prof. Dr. Udomkiat Nontakaew
Dean, Faculty of Engineering, KMUTNB

Dr. Temsiri Sapsaman
Associate Dean for Research and Innovation, Faculty of Engineering, KMUTNB

Asst. Prof. Dr. Boonchai Watjatrakul
Head, Department of Mechanical and Aerospace Engineering,
Faculty of Engineering, KMUTNB

Co-chairs

Assoc. Prof. Dr. Sinchai Chinvorarat

Prof. Dr. Suwat Kuntanapreeda

Secretariat

Asst. Prof. Dr. Petch Jearanaisilawong

Secretariat Assistant

Asst. Prof. Dr.-Ing. Krongkaew Laohalidanond

Finance, Registration and Exhibits

Asst. Prof. Dr. Chak Chantalakhana

Asst. Prof. Dr. Patcharin Saechan

Publicity and Public Relations

Asst. Prof. Dr. Tonkid Chantrasmi

Dr. Thada Suksila

IOP Conf. Series: Materials Science and Engineering **297** (2018) 011002 doi:10.1088/1757-899X/297/1/011002

Members

IOP Conf. Series: Materials Science and Engineering **297** (2018) 011003 doi:10.1088/1757-899X/297/1/011003

Honorary Speaker

The Role of DAAD Support in Fostering Mechanical Engineering Capacity for Development

Mr. Stefan Bienefeld
Head of Development Cooperation and Transregional Programs
German Academic Exchange Service (DAAD)

Keynote Speakers

Microturbine- and Turbo-Fuel-Cell Technology for Highly Effective Generation of Electricity -- Example of an International Technology Transfer Supported by the DAAD

Univ.-Prof. Dr.-Ing. Heinz Peter Berg
Chair of Combustion Engines and Flight Propulsion,
Brandenburg University of Technology Cottbus-Senftenberg, Germany

Abstract. The development of decentralized energy supply systems and structures has received an increasingly important role in the recent years. Construction of new large-scale and base-load power plants will no longer take place in liberalized energy markets because of economic and political reasons and for the sake of acceptance. In this context, the development of flexible power plants with maximum efficiencies is inevitable. Hybridization of energy production processes is a very promising approach, because the various pros and cons of each individual process are in this case optimally combined and thus the disadvantages are decreased. One possibility is the combination of the electrochemical SOFC (Solid Oxide Fuel Cell) process with the recuperated Joule-process. The possible potential of this combination has been shown in various papers and is not the focus of this presentation. In the examples presented, the international technological transfer supported by DAAD (German Academic Exchange Service) and the cooperation with KMUTNB (King Mongkut's University of Technology North Bangkok) is shown on the field of MGT (Micro Gas Turbine), which provides the basic technology for the hybrid application of future MGT-SOFC systems. This future highly efficient energy converters are called by the network partners "Turbo-Fuel-Cell".

The international cooperation between the Department of Mechanical and Aerospace Engineering (MAE) at KMUTNB, Brandenburg University of Technology (BTU) and industrial partners is also presented in the research and education field of MGT-Technology and the planed Turbo-Fuel Cell system. This team is working closely together in the education, research and technology realization supported by the DAAD in the University-Business-Partnership Program with a focal on the MGT-Technology.

The developed MGT is an advanced turbo machinery technology in combination with foil bearing. This technology is necessary for the realization of a "Turbo-Fuel-Cell"-process (MGT-SOFC system) with a long life time and less operating costs. Foil bearings are an innovative solution for small size, high speed high temperature turbo machinery. In addition, the SOFC-technology is another important key technology. Here, the SOFC stacks are integrated instead of the combustion chamber in the MGT process. Because of the combination between this advanced turbo machinery process and embedded SOFC stacks, the system has a high dynamic response (100kW/s) and a very high effective efficiency of 68% with a future potential of nearly 80%.

The presentation also highlights the importance of this technology for the future of habitats and mobility.

IOP Conf. Series: Materials Science and Engineering **297** (2018) 011003 doi:10.1088/1757-899X/297/1/011003

Biography. Heinz Peter Berg founded the Chair Combustion Engines and Flight Propulsion and the Institute of Transport Engineering at the Brandenburg Technical University in 1997. Since then he has been a full university professor and company founder.

He has worked in various executive functions of cooperating development companies and research institutes for 25 years. He started his professional career as an academic associate and scientist at the Technical University Darmstadt in 1982. After the degree as an Dr.-Ing. of the Technical University Darmstadt, he was a founding team member of BMW Rolls Royce GmbH (now Rolls-Royce Deutschland). At BMW Rolls-Royce (from 1991), he directed the test program area, the fluid dynamic test department, the aerothermal test department and the whole section for engine component testing of all international divisions. During this period the first German turbojet engine (BR 700 family) after the second world war was designed, developed, built and certified.

In the time periode 1997-2005 Prof. Dr. Berg was Vice-President of the Brandenburg Research & Development Company and Engineering Cooperation Center GmbH. From 2005 until now he was founder and share-holder in a variety number of technology-oriented companies, which are integrated into a technology network. Since 1997 he is a full Univ.-Professor, a member of various technology organizations, an expert (e.g. DFG, SFB) and entrepreneur.

In addition to the research and development field of combustion engines and flight propulsion systems, Prof. Berg is very concerned with future technologies in the field of energy conversion and the network creation of international institutions.

Micro/nano/pico-satellite Challenge for Novel Space Development and Utilization

Prof. Dr. Shinichi Nakasuka
Department of Aeronautics and Astronautics, University of Tokyo, Japan

Abstract. University of Tokyo already developed 8 satellites, and seven of them were launched and operated successfully in space. Two CubeSat "XI-IV (2003)" and "XI-V (2005)" were primarily for space engineering education and pico-bus demonstration, but from the third satellite "PRISM" we have been challenging towards more practical applications. PRISM aimed at obtaining 30 m GSD Earth remote sensing images, which was actually achieved in 2009. Our fourth satellite "Nano-JASMINE," which is now waiting for launch, has "astrometry" mission to obtain very precise 3D map of more than 400,000 stars in space. From 2010, I lead a nationwide micro-satellite project named "Hodoyoshi Project," through which, three Earth remote sensing satellites "Hodoyoshi-1,3,4" were launched in 2014 by Russian rocket "DNEPR," which showed excellent performance of taking Earth images of 6m, 40m and 240m GSD. We are now seeking practical applications of these images for agriculture, forestry, fishery, disaster monitoring, etc. Based on the bus technologies developed in Hodoyoshi Project, in December 2014, we launched the world first 50kg class deep space probe "PROCYON," which escaped from the Earth gravitational field, and various observation and experiment were conducted successfully in deep space. Based on the obtained deep space technologies, we started development of 6U CubeSat "EQUULEUS" targeting towards Earth Lunar Lagrange Point 2, with the planned launch by NASA's SLS rocket in 2019.

In my talk, I will first show the features of micro/nano/pico-satellites, followed by the University of Tokyo's history of development and in-orbit operation results of series of micro/nano/pico-satellites. In the latter part, I will describe current and future visions of applying these satellites to various practical missions, including space science and exploration, Earth observation, entertainment, and education support program.

Biography. Prof. Nakasuka graduated from University of Tokyo in 1983 and got Ph.D in 1988. He joined IBM Research during 1988-1990, and then worked for Department of Aeronautics and Astronautics, University of Tokyo as a lecturer in 1990, as an Associate Professor in 1993, and became a Professor in 2004. He developed and launched seven micro/nano/pico-satellites successfully. He is a member of JSASS, SICE, and IAA, and the Chairperson of IFAC Aerospace Technical Committee and UNISEC-GLOBAL. He has been a member of Space Policy Committee of Japanese Cabinet Office since 2012.

IOP Conf. Series: Materials Science and Engineering **297** (2018) 011003 doi:10.1088/1757-899X/297/1/011003

Kaizen Approach to Smart Manufacturing for Connected Industries

Prof. Dr. Yasuyuki Nishioka
Department of Engineering and Design
Faculty of Engineering and Design, Hosei University, Japan
President of the Industrial Value Chain Initiative

Abstract. Manufacturing industry is facing tremendous changes in digitization. From a system integration viewpoint, approaches toward manufacturing digitization are usually on a top-down basis. It could be true that an optimized production system can be specified in the cyber world. However, factory floor's reality sometime refuses such sophisticated systems. On the other hand, Japanese production management usually make a decision based on fact-based reality. This fact-based and case-based system thinking creates valuable outcomes with respect to quality and productivity. So-called Kaizen is a typical example of a bottom-up approach for smart manufacturing.

This talk, first, shows the Industrial Value Chain Reference Architecture (IVRA), which various kinds of factories can refer to improve performance with respect to digitized manufacturing. IVRA consists of Smart Manufacturing Units (SMUs) and Portable Loading Units (PLUs). SMU has three views: asset view, management view, activity view. Kaizen approach can be realized in the activity view of an SMU where PDCA cycles perform for continuous improvement.

In order to implement IoT/ICT system in a manufacturing enterprise, it is necessary to keep interoperability among IoT devices, applications, data storages, and related tools. In order to achieve interoperability among such digital components, IVRA also addresses reference models of IVI platforms. The concept of Loosely Defined Standard (LDS) is applied in management of interoperability between IVI components, in which data commonality cannot be fixed unless improvement of the system as a whole is terminated.

Using LDS, IVRA allows the factory managers to describe scenarios of daily operations and inter-relations between actors. Then an SMU can be defined as a set of models including Actor, Activity, Thing, Information, Logic and Data. The talk also illustrates the procedure of bottom-up system improvement of a cyber-physical SMU. Factory floor managers, production engineers and chiefs of workers make some operational scenarios from their actual concerns. Then the scenarios are redesigned using data and logics in the cyber world. Assigning appropriate services provided by platform component suppliers to the data and logics, an IVI platform gives interoperability framework to those components.

Industrial Value Chain Initiative (IVI) provides a software tool, IVI modeler, to all the members for describing and sharing scenario models, IVI modeler is used by working groups to draw four kinds of diagrams. This tool performs well for facilitating common understanding and defining scenarios and terms. Models defined by the IVI modeler can play a key role in integrating the physical world and cyber world. In other words, a matching system between platform components and the practical operational scenarios is implemented by those models.

Finally, this talk discusses a platform ecosystem framework. In order to make collaboration and win-win relation among platformers, component suppliers, system integrators and factory managers, general rules of profile specification and distribution are identified. Since the platformers and component suppliers have their own data models and interfaces, IVI proposes a profile publication framework and a common dictionary for specifying the profiles in accordance with the concept of LDS. Many future possibilities of cross industrial collaboration are also discussed in the concluding remarks.

Biography. Yasuyuki Nishioka received Bachelor of Engineering from Waseda University in 1985, Master of Engineering from University of Tsukuba in 1992 and Doctor of Philosophy from University of Tokyo in 1996. He is currently a professor in the Department of Engineering and Design at Hosei University, Japan. Since 2015, Prof. Nishioka has been President of the Industrial Value Chain Initiative.

IOP Conf. Series: Materials Science and Engineering **297** (2018) 011003 doi:10.1088/1757-899X/297/1/011003

Technology Fusion in Series Production of Lightweight Structures

Dipl.-Ing. Michael Heinrich
Federal Cluster of Excellence "MERGE Technologies for Multifunctional Lightweight Structures," Chemnitz University of Technology, Germany
Director of Competence Centre for Lightweight Structures, Chemnitz, Germany

Abstract. Research in production technology is currently being driven by mega-trends such as industry 4.0, resource efficiency, circular economy, human-machine interaction and demographic change. Therefore, issues relating to economic sustainability, future working and living environments, big data and human effected production processes are at the forefront of this research. The creative scope for possible technology combinations and interactions has in-creased significantly due to the paradigm shift that is occurring in production technology with regard to holistic approaches and sustainability. Therefore, the newly developed technology variants investigated in MERGE provide an excellent basis for new research approaches. For example, the sensor integration technologies tested in MERGE open up new paths of communication to aid greater understanding of material, energy and information flows with the involvement of people across the entire life cycle of lightweight structures.

Networking digital functionalisation technologies with existing analogue production processes pursued in the MERGE Cluster aim to equip components with sensors across the board to depict the entire life cycle. A configurable information feedback between production parameters and the resulting component properties in the use phase provides the basis for detailed product and process optimisation for multifunctional lightweight structures. To expand upon this, interfaces between virtual and real-life processes will be researched, based on multi-criteria algorithms with a view to creating self-learning systems.

Biography. Michael Heinrich, became his diploma (Dipl.-Ing.) for mechanical engineering, especially for lightweight structure and polymer technology, in 2004 at the Chemnitz University of Technology (CUT), Germany. He stays at the Institute of Lightweight Structures (Head: Prof. Dr.-Ing. habil. Lothar Kroll) at the CUT and make his Ph.D.-thesis on the effects of micro injection molded composites based on carbon nano tubes for electrical contacting of piezo ceramics. Furthermore, since 2008, he is leader of Competence Centre for Lightweight Structures. Since 2015, he is research assistant at the Research Centre for Textile Semi-finished Products and Preforms (STEX) at Fraunhofer-Institute for Machine Tools and Forming Technology (IWU), Germany. He also has been in the Scientific Management Board of the Federal Cluster of Excellence "MERGE Technologies for Multifunctional Lightweight Structures" (MERGE), established at CUT, since 2017. From 2004 until now, more than 50 articles, proceedings and patents have been published.

Lessons Learned from KNACKSAT Satellite

Prof. Dr. Suwat Kuntanapreeda
Department of Mchanical and Aerospace Engineering, Faculty of Engineering,
King Mongkut's University of Technology North Bangkok, Thailand

Abstract. KNACKSAT, an acronym for KMUTNB Academic Challenge of Knowledge SATellite, is a 1U-CubeSat satellite (roughly 10 x 10 x 10 cm, 1.3 kg) developed by King Mongkut's University of Technology North Bangkok (KMUTNB). Since September 2015, the KNACKSAT project has been funded by Broadcasting and Telecommunications Research and Development Fund for the Public Interest, Office of The National Broadcasting and Telecommunications Commission (NBTC), Thailand. It's the first satellite entirely built in Thailand, entirely by a university team. The satellite will be launched into a 575-km sun-synchronous orbit in 2018. The communication between the satellite station and the ground station is based on amateur radio frequencies. The main mission of KNACKSAT include (1) developing a communication system using amateur radio frequencies, (2) taking images from space, (3) testing 3-axis attitude control algorithms by using magnetic torquers, (4) verifying a deorbit technology by a magnetic torque and (5) confirming the uses of Commercial Off-The-Shelf (COTS) components in space. But the larger objectives of the KNACKSAT project might be even more important: the educational and technology demonstration objectives. The educational objectives is to enhance the knowledge and experience of the students in satellite design and construction. On a national level, the project inspires and promotes interest in science and technology for young Thais.

The KNACKSAT project started without prior in-house experience. However, from developing all subsystems in-house, the team has gained valuable experiences that can be considered as fundamentals for development of larger satellites in Thailand in the near future.

In my talk, I will first present some relevant histories and then introduce the KNACKSAT project. Last, I will share the lessons learned from more than two years of development. The lessons are identified from both point of view of engineering and management.

Biography. Suwat Kuntanapreeda received the B. Eng. degree in Mechanical Engineering from King Mongkut's University of Technology North Bangkok (KMUTNB), Thailand in 1990, and the M.S. and Ph.D. degrees in Mechanical Engineering from Utah State University, USA in 1992 and 1995, respectively. He is currently a professor in the Department of Mechanical and Aerospace Engineering at KMUTNB.

Fuel Upgrading to Cleaner and High Value Fuels

Prof. Dr. Ashwani K. Gupta
Department of Mechanical Engineering
University of Maryland, College Park, MD 20742, USA

Abstract. Clean energy harvesting from various kinds of feedstocks is a challenge but recent research and development efforts show a novel fuel reforming path that can be used in a wide variety of industries extending from commercial to aerospace application. Fuels of different physical and chemical properties could be made available, however, they must be tailored for specific applications. In addition it is important that we effectively and efficiently use low grade fuels and reform them to cleaner fuels, such as wastes and biomass to hydrogen and methane. Carbon dioxide is a greenhouse gas that is known to contribute to global warming. Although there are other contributors to global warming but the share of CO_2 is significant. Most of the CO_2 is produced from the combustion of fossil fuels and our use of fossil fuels has been increasing due to increased activity and population. The carbon during the combustion process is converted to CO_2 with the subsequent release of thermal energy. In many parts of the world most of the electricity is produced from fossil fuels, with coal being the highest. The demand for electricity continues to grow in the developed countries with significant more in some of the developing countries. This has resulted in dramatic increase in global ambient concentrations of CO_2. Significant efforts are being pursued to curtail the emission of CO_2 or capture and store it in order to maintain the ambient concentrations to some acceptable levels.

This talk will focus on reforming wastes and biomass to clean energy with due consideration to energy sustainability, global climate change and mitigation of greenhouse gases emission to the environment. The emphasis will be on sustainability with due considerations to efficiency and fuels used in different energy using sectors. Carbon emissions continue to increase through the increase use of fossil fuels but fossil fuels offer are near carbon neutral. . Gasification and pyrolysis are important steps to reform the fuel properties so that they can be used in the current energy conversion systems for various energy and power use with due consideration to environmental impact. Gasification results from a lab scale reactor and its effect on syngas yield and characteristics will be provided using different feedstock materials.

IOP Conf. Series: Materials Science and Engineering **297** (2018) 011003 doi:10.1088/1757-899X/297/1/011003

Biography. Ashwani Gupta is a Distinguished University Professor at the University of Maryland, College Park, USA. He received his PhD from the University of Sheffield, and higher doctorate (DSc) from the University of Sheffield and also from the University of Southampton, UK. He received Honorary doctorate from the University of Wisconsin Milwaukee, King Mungkut University of Technology North Bangkok, and University of Derby, UK. He is Honorary Fellow of ASME and Fellow of AIAA, SAE, AAAS and RAeS (UK). He is co-editor of Environmental and Energy Science book series by CRC Press and Associate editor of 4 journals. He served as Director of Propulsion and Energy group at AIAA and also a member of Board of Directors at AIAA. He has received several honors and awards and several best paper awards from AIAA, ASME. At the University of Maryland he received President Kirwan award and College of engineering research award. He has published over 700 technical papers, 3 books and edited 13 books. His research interests include high intensity distributed combustion, HiTAC, swirl flows, combustion, sulfur chemistry, wastes and biomass to energy, biofuels, high speed flows and mixing, advanced diagnostics and miniature sensors, and air pollution.

Invited Speakers

Topic	**Speaker**
Minimally Invasive Surgical Training System with Force Feedback	Asst. Prof. Dr. Jackrit Suthakorn Mahidol University, Thailand
Serveillance by Aerial Gimbal	Assoc. Prof. Dr. Prasatporn Wongkamchang Navaminda Kasatriyadhiraj Royal Air Force Academy, Thailand
Waste to Fuel : The use of diesel from municipal waste plastic in BMTA buses	Assoc. Prof. Dr. Kanit Wattanavichien Chulalongkorn University, Thailand
Optimum Allocation of Heating Surfaces in Sugar Juice Evaporation System	Prof. Dr. Somchart Chantasiriwan Thammasat University, Thailand
Biomass Gasification Research for Thai Ceramic Industry	Prof. Dr. Nakorn Tippayawong Chiang Mai University, Thailand
Gas-Liquid Two-Phase Flows in Micro and Mini Channels	Prof. Dr. Somchai Wongwises King Mongkut's University of Technology Thonburi, Thailand
Numerical Heat Transfer Analysis in Heat Exchangers with Vortex Generators	Assoc. Prof. Dr. Pongjet Promvonge King Mongkut's Institute of Technology Ladkrabang, Thailand

IOP Conf. Series: Materials Science and Engineering **297** (2018) 011004 doi:10.1088/1757-899X/297/1/011004

Academic and Industrial Partners

Brandenburg University of Technology Cottbus-Senftenberg, Germany

b-tu
Brandenburgische
Technische Universität
Cottbus - Senftenberg

Chemnitz University of Technology, Germany

TECHNISCHE UNIVERSITÄT
CHEMNITZ

The Waste Incineration Research Center (WIRC), KMUTNB, Thailand

WIRC

Protecting the Environment through Excellence in R&D

Michelin Siam Co., Ltd., Thailand

MICHELIN
A better way forward

IOP Conf. Series: Materials Science and Engineering **297** (2018) 011004 doi:10.1088/1757-899X/297/1/011004

Sponsors

President Sponsors

b-tu Brandenburg University of Technology Cottbus - Senftenberg

DAAD Deutscher Akademischer Austauschdienst German Academic Exchange Service

Federal Ministry for Economic Cooperation and Development

Platinum Sponsors

TCEB THAILAND CONVENTION & EXHIBITION BUREAU

ISUZU

Gold Sponsors

LAVISION FOCUS ON IMAGING

ryts Instruments

SCG CEMENT-BUILDING MATERIALS

Silver Sponsors

SMC

สมาคมศิษย์เก่าฯ ME21 KMUTNB Alumni Assoc.

AD Training Didactic Systems Teachware

THAIRUNG

CAD-IT CONSULTANTS

SIGMA Solutions

Cradle MSC Software Company

ARTITH VENTILATOR MACHINERY www.artith.com

ADT SYSTEMS INNOVATION. DRIVEN.

KINETICS

Dynamic

Bronze Sponsors

Roller successful products

สมาคมศิษย์เก่าฯ ME24 KMUTNB Alumni Assoc.

SOMBOON GROUP

IOP Conf. Series: Materials Science and Engineering **297** (2018) 011005 doi:10.1088/1757-899X/297/1/011005

Conference Photographs

b-tu
DAAD
ISUZU

TSME-ICoME
2017
ISUZU

TSME-ICoME
2017

IOP Conf. Series: Materials Science and Engineering **297** (2018) 011006 doi:10.1088/1757-899X/297/1/011006

Peer review statement

All papers published in this volume of *IOP Conference Series: Materials Science and Engineering* have been peer reviewed through processes administered by the proceedings Editors. Reviews were conducted by expert referees to the professional and scientific standards expected of a proceedings journal published by IOP Publishing.

IOP Conf. Series: Materials Science and Engineering **297** (2018) 012001 doi:10.1088/1757-899X/297/1/012001

Production of charcoal briquettes from biomass for community use

S. Suttibak* **and W. Loengbudnark**

Department of Energy Engineering, Faculty of Technology, Udon Thani Rajabhat University, Udon Thani, 41000, Thailand

* Corresponding Author: E-mail: suntorn_su@hotmail.com, Telephone Number: 66-42-211-040, Fax Number: 66-42-221-978

Abstract. This article reports of a study on the production of charcoal briquettes from biomass for community use. Manufacture of charcoal briquettes was done using a briquette machine with a screw compressor. The aim of this research was to investigate the effects of biomass type upon the properties and performance of charcoal briquettes. The biomass samples used in this work were sugarcane bagasse (SB), cassava rhizomes (CR) and water hyacinth (WH) harvested in Udon Thani, Thailand. The char from biomass samples was produced in a 200-liter biomass incinerator. The resulting charcoal briquettes were characterized by measuring their properties and performance including moisture content, volatile matter, fixed carbon and ash contents, elemental composition, heating value, density, compressive strength and extinguishing time. The results showed that the charcoal briquettes from CR had more favorable properties and performance than charcoal briquettes from either SB or WH. The lower heating values (LHV) of the charcoal briquettes from SB, CR and WH were 26.67, 26.84 and 16.76 MJ/kg, respectively. The compressive strengths of charcoal briquettes from SB, CR and WH were 54.74, 80.84 and 40.99 kg/cm^2, respectively. The results of this research can contribute to the promotion and development of cost-effective uses of agricultural residues. Additionally, it can assist communities in achieving sustainable self-sufficiency, which is in line with our late King Bhumibol's economic sufficiency philosophy.

1. Introduction

Biomass is clean, renewable and environmentally friendly energy. It has very much less sulfur, nitrogen and ash, which cause SO_2 and NO_x formation, compared to fossil fuels [1]. Biomass is a huge energy resource since there are so many raw materials that can be used for energy such as wood slivers, agricultural residues, manure, food residues, garbage and wastewater, among others. Biomass is renewable energy as it is replenished on a shorter timescale than fossil fuels [2]. Moreover, Thailand is endowed with large biomass resources. Generating energy from biomass could not only reduce the need for imported fossil fuels, but also increase job opportunities for the population.

Thailand is an agrarian country. There are many agricultural residues produced. According to data from the Department of Agricultural Economics in 2015, Thailand's economic crops were sugarcane, rice, cassava and corn. Annual production of these crops totaled 171.7 million tonnes. This was comprised of 106.3 million tonnes of sugar cane, 28.4 million tonnes of rice, 32.4 million tonnes of cassava and 4.6 million tonnes of corn [3]. The residue to product ratios of sugarcane bagasse, sugarcane leaves and tops, rice husks, rice straw, cassava stalk, cassava rhizomes and corn stalks are 0.291, 0.302, 0.230, 0.447, 0.088, 0.490 and 0.250, respectively [4]. Thus, total crop residues amounted to 102.1

million tonnes per year. Some residues were used as fuel for power generation. For example, sugar mills burn sugarcane bagasse and rice mills burn rice husk as power plant fuels, while other materials are use as livestock feed. However, less than a half of total residues were taken. The remaining were disposed of through incineration, which aggravates air pollution. Using agricultural residues as energy is an attractive option. There are three types of biomass thermal conversion technologies. These include direct combustion, pyrolysis, and gasification [5-7]. Direct combustion is a process of burning biomass at low temperatures for long durations. The product gained from direct combustion is a solid called char that can be used as fuel. Pyrolysis is a process of converting biomass into energy under controlled temperatures (400-500°C) without oxygen. The pyrolysis product is liquid called bio-oil. Generating energy by pyrolysis has been researched and is currently economically infeasible in Thailand. Gasification, which similar to pyrolysis, is the conversion of biomass by burning it at a high temperature (800°C) in the absence of oxygen to produce a syn-gas. Manufacturing syn-gas is quite complicated and is also still under research and development. Each technology has its own advantages and disadvantages. Selection of the most appropriate technology depends on how the end products will be used. In Thailand, the most widely used technology at a community level is direct combustion owing to its simplicity, convenience and worthiness. As discussed above, direct combustion produces a solid material. Its particle size varies depending on the type of raw material used. However, to increase the efficiency of biomass fuel, making charcoal briquettes is a favored technique. At present, charcoal briquettes are made from small sized traditional charcoal. Small charcoal is ground to a powder and then mixed with binders, flour and water. All the ingredients are mixed together and then mechanically pressed into the desired shape. Moisture is removed from the briquettes either by sun drying or in a heated drier. Briquetted charcoal has better quality than the traditional charcoal. It is suitable for cooking and widely used in both noodle and grill restaurants. According to restaurant owners interviewed in the Udon Thani area, the cost of charcoal briquettes is 10 THB per kilogram and there is high demand for it.

This research aimed at sustainably introducing charcoal briquetting at the community level while using agricultural residues. Making charcoal briquettes in this setting requires simple techniques using a single machine. This will help agriculturists increase their income. Pressed charcoal briquettes must be dried to less than 8% moisture according to the community product standard. Charcoal briquette quality is competitive with traditional charcoal both in terms of its heating value and burning time. The raw materials used in this research were locally harvested sugarcane bagasse, cassava rhizomes, and water hyacinth. Charcoal briquette properties were compared with traditional charcoal. The results of this research will be shared with local communities to encourage them to sustainably use biomass in accordance with the philosophy of a sufficiency economy.

2. Experimental method and analysis

2.1. Biomass samples

The biomass studied in this research consisted of sugarcane bagasse (SB), cassava rhizomes (CR) and water hyacinth (WH). They are commonly found in Udon Thani. The samples were sun dried before being burned in a 200 liter incinerator in the absence of oxygen. All three types of samples were subjected to proximate analysis (moisture, volatile matter, ash and fixed carbon), ultimate analysis (carbon, hydrogen, nitrogen, sulfur and oxygen) and calculated heating value analysis.

2.1.1. Proximate analysis. Proximate analysis is the determination of the percentages of moisture, volatile matter, ash and fixed carbon of a substance. Moisture, volatile matter and ash determinations followed ASTM E1756-01, E872-82 and E1755-01 standards, respectively. The apparatus used consisted of a ceramic cup with a cover, metal bottom plate, desiccators, a four decimal digital scale and a high temperature incinerator. Fixed carbon was determined as the remaining portion of the sample after measuring moisture, volatile matter and ash.

2.1.2. Ultimate analysis. Ultimate analysis was used to determine the carbon, hydrogen, nitrogen, sulfur and oxygen contents of biomass samples. The oxygen content was calculated as the portion that remained after accounting for the other elements. A Leco Carbon Hydrogen Nitrogen & Sulfur analyzer, Model Carbon 628,628 S, was used. The analysis was done at the Center for Scientific and Technological Equipment, Suranaree University of Technology.

2.1.3. Heating values. Heating values were obtained by calculation. The higher heating values (HHV) were calculated using Sheng and Azavedo's equation (1) [8], while the lower heating values (LHV) were obtained using ECN's equation (2) [9], in which the results of ultimate analysis are required inputs.

$$HHV\left(\frac{MJ}{kg}\right) = -1.3675 + 0.3137C + 0.7009H + 0.0318O^{*} \quad (1)$$

where C and H are percentages of carbon and hydrogen (dry basis), respectively.

$$LHV\left(\frac{MJ}{kg}\right) = HHV - 2.442 \times 8.936\left(\frac{H}{100}\right) \quad (2)$$

2.2. Incinerating biomass

The biomass samples were burnt with a controlled amount of air in a 200 liter incinerator, as shown in figure 1. This was done to find the appropriate time to close the air inlets for each biomass sample.

Figure 1. Biomass incinerator used in this study.

2.3. Char grinding

After the biomass was converted to char, the char were ground to a smaller size, less than 0.5 cm. This promoted easy mixing with binders and a smooth surface of the desired shape. Char from cassava rhizomes is hard. Therefore it required mechanical grinding. Char from sugarcane bagasse and water hyacinth was soft enough to be ground by hand.

2.4. Mixing and pressing

Cassava starch and water were used as binders to agglomerate the ground charcoal. These materials were mixed and mechanically stirred. The machine used for this purpose was equipped with both stirring and pressing mechanisms, as shown in figure 2. The ratio of char particles: cassava starch : water were determined in previous research. The mixing ratio for cassava rhizomes, sugarcane bagasse, and water hyacinth were 5 kg: 0.5 kg: 6 liter, 2 kg : 0.1 kg : 2 liter, and 2 kg : 0.1 kg : 2 liter, respectively [10].

Figure 2. Charcoal briquette machine with a screw compressor.

2.5. Drying charcoal briquettes
According to the community product standard, the moisture content of charcoal briquettes must be less than 8%. However, the moisture content of the freshly pressed charcoal briquettes was 30-40%. So, drying was needed to remove moisture.

2.6. Charcoal briquette properties testing and analysis
The basic properties of biomass charcoal briquettes were analyzed in accordance with ASTM standards as follows.

2.6.1. Heating value. Heating value was determined following the ASTM D 5865 standard using an oxygen bomb calorimeter or calculation. For calculations, the higher heating values (HHV) and lower heating values (LHV) from ultimate analysis are required inputs. The higher heating value can be determined using equation (3) [11]. The lower heating value and hydrogen content can be determined using equation (2) [9].

$$HHV_{dry}\left(\frac{MJ}{kg}\right) = 0.3491C + 1.1783H + 0.1005S - 0.10340O - 0.0151N - 0.0211A \quad (3)$$

where C, H, S, O, N and A are the percentages of carbon, hydrogen, sulfur, oxygen, nitrogen and ash (dry basis), respectively.

2.6.2. Bulk density. Bulk density was determined using equation (4).

$$\rho = \frac{m}{V} \quad (4)$$

where m and V are mass and volume of charcoal briquettes, respectively.

2.6.3. Moisture content. Moisture content was analyzed following the ASTM D 3173 standard.

2.6.4. Volatile matter content. Volatile matter content was determined following the ASTM D 3175 standard.

2.6.5. Ash content. Ash content was analyzed following the ASTM 3174 standard.

2.6.6. Fixed carbon. Fixed carbon was determined following the ASTM 3172 standard.

IOP Conf. Series: Materials Science and Engineering **297** (2018) 012001 doi:10.1088/1757-899X/297/1/012001

2.6.7. The elemental composition. The elemental composition of charcoal briquettes was analyzed using the same method as for biomass

2.6.8. The compressive strength. The compressive strength of briquettes was determined using a Compression Testing Machine, STYE-2000. All three types of biomass charcoal briquettes with external diameter, internal diameter and length were 40, 16 and 80 mm, respectively were subjected to compressive strength testing.

2.6.9. Extinguishing time testing Extinguishing time testing is a method to determine how long the charcoal briquettes burn compared to traditional charcoal. The test was done by burning charcoal briquettes in a community gasifier stove.

2.7. Technology transfer and passing on knowledge
The innovation and technology from this research have been shared with people who interested in biomass charcoal briquette manufacturing. This was done through a new entrepreneur launch as part of a university knowledge-based project.

3. Results and discussion

3.1. Characteristic of biomass samples
Table 1 shows results of proximate analysis, ultimate analysis, and heating value calculations of three biomass types, i.e., sugarcane bagasse (SB), cassava rhizomes (CR), and water hyacinth (WH). The results show that percentages of volatile matter in cassava rhizomes and sugarcane bagasse were similar but cassava rhizomes had less ash. From the ultimate analysis, it was seen that cassava rhizomes and sugarcane bagasse had almost the same percentage of carbon. This value has a direct impact on the heating value. For water hyacinth, the percentage of carbon and heating values were less, while the ash content was greater than the other two types of biomass.

Table. 1 Characteristic of biomass samples.

Analysis	CR	SB	WH
Proximate Analysis (wt.%, dry basis)			
Moisture[a]	8.03±0.05	7.24±0.10	9.08±0.12
Volatile matter	77.63±0.56	77.70±0.26	55.53±0.38
Fixed carbon*	14.53±0.71	8.65±0.27	11.33±0.53
Ash	7.20±0.16	13.13±0.13	32.31±0.23
Ultimate Analysis (wt.%, dry basis)			
Carbon	46.63±0.11	47.05±0.02	37.11±0.12
Hydrogen	7.79±0.02	7.33±0.02	6.53±0.07
Nitrogen	0.59±0.02	0.39±0.01	1.25±0.02
Sulfur	0.07±0.01	0.29±0.07	0.37±0.01
Oxygen*	37.68±0.13	31.74±0.08	22.16±0.37
H/C molar ratio	2.00±0.01	1.87±0.00	2.11±0.02
O/C molar ratio	0.61±0.00	0.57±0.00	0.45±0.01
Molecular formula	$CH_{2.00}O_{0.61}$	$CH_{1.87}O_{0.57}$	$CH_{2.11}O_{0.45}$
Heating value by calculation (MJ/kg, dry basis)			
HHV	19.94±0.03	19.56±0.02	15.61±0.07
LHV	18.24±0.03	17.97±0.02	14.18±0.06

*Calculated by difference.
[a] wet basis.

3.2. Optimization of biomass burning

Figure 3 illustrates yield of char product obtained from burning three different types of biomass in a 200 liter incinerator with a controlled amount of air. The results showed that percentages of charcoal obtained from cassava rhizomes (CR), sugarcane bagasse (SB) and water hyacinth (WH) were 47.08, 26.08 and 58.33, respectively. The burning times before closing the air inlets were 20, 10 and 15 minutes, respectively.

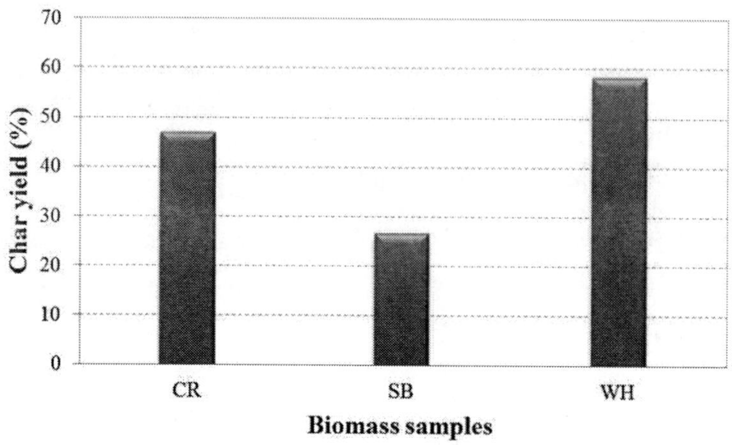

Figure 3. Yield of char product obtained after burning biomass samples.

3.3. Properties of biomass charcoal briquettes

The shape of the charcoal briquette is a hollow cylinder with a 10 mm inner diameter, a 40 mm outer diameter, and a 100-120 mm length, as shown in figure 4. Table 2 describes properties of three types of biomass charcoal briquettes. The moisture contents of all types of biomass charcoal complied with the community product standard (less than 8%). The CR charcoal briquettes gave the highest lower heating value (by calculation) at 26.84 MJ/kg followed by SB charcoal briquettes, 26.67 MJ/kg, and WH charcoal briquettes, 16.67 MJ/kg. According to the community product standard, the lower heating value must not less than 20 MJ/kg. The CR and SB charcoal briquettes met this standard. This result is consistent with the results of Phutteesakul et al. [12]. The CR charcoal briquettes had the highest density. Additionally, the CR, SB and WH briquettes could withstand applied loads of 80.84, 54.74 and 40.99 kg/cm^2, respectively. CH and SB charcoal briquettes cracked through the vertical axis, from top to bottom (figure 5). WH charcoal briquettes homogeneously cracked the whole briquette. Comparison of the three biomass charcoal briquettes revealed that CH charcoal briquettes had the best properties, followed by SB charcoal briquettes.

Figure 6 shows extinguishing times of biomass charcoal briquettes compared with traditional charcoal. The CR charcoal briquettes had the longest extinguishing time, followed by SB charcoal briquettes. The WH charcoal briquettes not only had a low extinguishing time but also a high ash content. Thus, when manufacturing WH charcoal briquettes, more charcoal particles should be added to improve its properties, especially the heating value.

Table 2. Properties of biomass charcoal briquettes.

Analysis	CR	SB	WH
Proximate analysis (wt.%, dry basis)			
Moisture[a]	5.04±0.07	6.15±0.08	4.57±0.04
Volatile matter	20.96±0.39	19.53±0.34	13.33±0.37
Fixed carbon*	70.37±0.52	66.59±0.27	56.77±0.25
Ash	8.41±0.19	13.50±0.09	29.70±0.47
Ultimate analysis (wt.%, dry basis)			
Carbon	71.01±0.23	70.23±0.10	46.72±0.75
Hydrogen	4.07±0.05	3.82±0.03	3.14±0.11
Nitrogen	0.30±0.03	0.54±0.01	1.00±0.06
Sulfur	0.02±0.01	0.01±0.01	0.36±0.03
Oxygen*	16.16±0.40	11.84±0.07	19.01±0.62
H/C molar ratio	0.69±0.01	0.65±0.01	0.81±0.03
O/C molar ratio	0.17±0.00	0.13±0.00	0.31±0.01
Molecular formula	$CH_{0.69}O_{0.17}$	$CH_{0.65}O_{0.13}$	$CH_{0.81}O_{0.31}$
Heating value by oxygen bomb calorimeter (MJ/kg)			
HHV	24.25±0.64	22.66±0.46	15.74±0.45
Heating value by calculation (MJ/kg, dry basis)			
HHV	27.73±0.13	27.50±0.03	17.44±0.39
LHV	26.84±0.12	26.67±0.03	16.76±0.37
Bulk density (kg/m^3)	1,494±10.1	1,146±23.7	696.2±6.0
Compressive strength (kg/cm^2)	80.84±5.62	54.74±1.23	40.99±1.12

*Calculated by difference.
[a] wet basis.

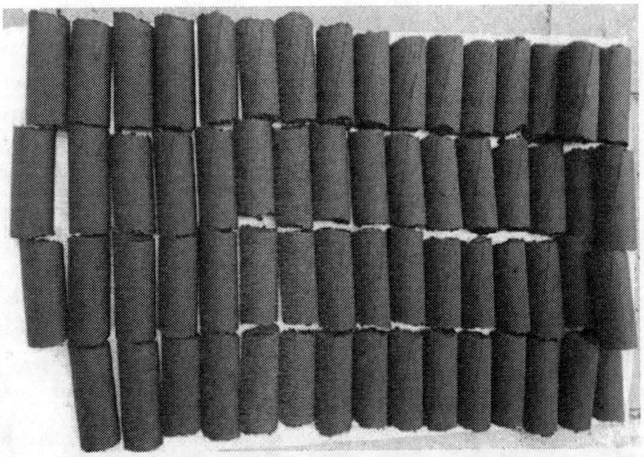

Figure 4. Finished biomass charcoal briquettes.

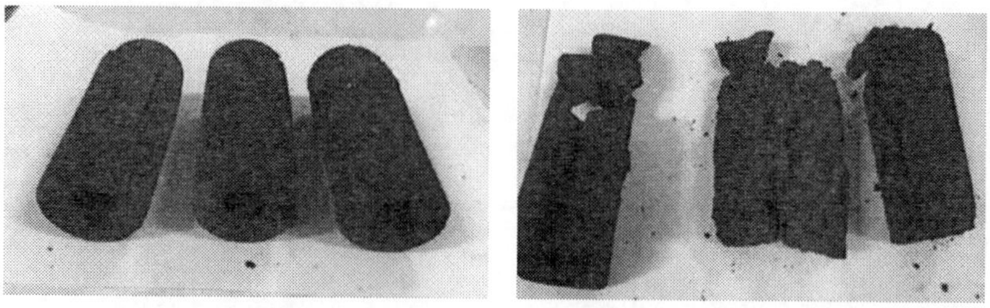

(a) CR charcoal briquette (b) SB charcoal briquette

Figure 5. Cracking characteristic of biomass charcoal briquettes after compressive strength testing.

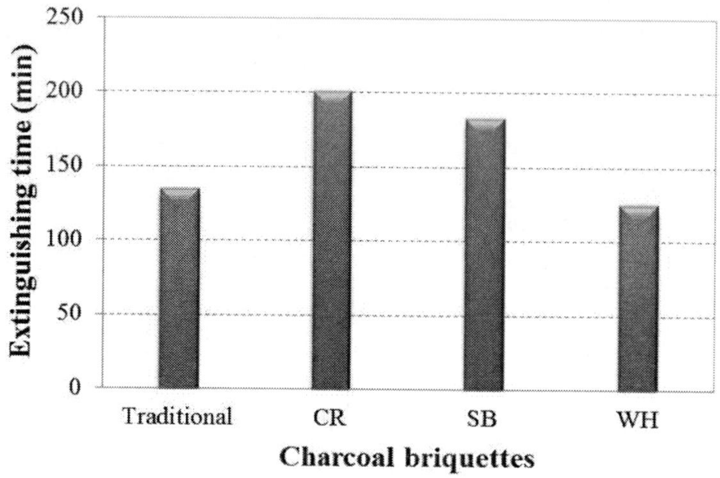

Figure 6. Extinguishing time of biomass charcoal briquettes compared with traditional charcoal.

3.4. Technology transfer and passing on knowledge

The innovation and technology from this research have been shared with people interested in manufacturing charcoal briquettes from biomass through a new entrepreneur launch as part of a university knowledge-based project. The project was held on 27-29 March 2017 at Udon Thani Rajabhat University (figure 7).

Figure 7. Technology and knowledge transfer for biomass charcoal briquetting.

4. Conclusions

Three biomass types commonly found in Udon Thani were tested for their suitability for making charcoal briquettes, sugarcane bagasse (SB), cassava rhizomes (CR), and water hyacinth (WH). Fresh biomass samples were sun dried and then burnt in a 200 liter incinerator with a controlled amount of air. The resulting char particles were mechanically pressed into a hollow-cylindrical shape. It was found that the CR charcoal briquettes had the best properties in terms of heating value, compressive strength, and extinguishing time. The innovation and technology from this research have been shared with people who are interested in manufacturing biomass charcoal briquettes to encourage them to use biomass sustainably.

Acknowledgement

Financial support granted by the National Research Council of Thailand and the Research and Development Institute, Udon Thani Rajabhat University, Thailand are gratefully acknowledged. Thanks are also extended to Prof. Jeffrey Nash, Miss Radchaneekron Jandong, Mr. Prakasit Suanboon and Mr. Montree Thonganamai for their research assistance.

References

[1] Zhang L, Xu C and Campagne P 2010 *Energy Conversion and Management* **51** pp 969 - 982
[2] Pattiya A and Suttibak S 2012 *Journal of Analytical and Applied Pyrolysis* **95**(1) pp 227-235
[3] Office of Agricultural Economics, *Acricultural statistics of Thailand 2015*, Available from http://www.oae.go.th, accessed on 14/12/2016.
[4] Pattiya A 2011 *Energy Sources, Part A: Recovery, Utilization, and Environmental Effects* **33** pp 691-701
[5] Eseltine D, Thanapal S S, Annanalai K and Ranjan D 2013 *Fuel* **113** pp 379-388
[6] Bridgwater A V 2012 *Biomass and Bioenergy* **38**(0) pp 68 - 94
[7] Van der Stelt M J C, Gerhauser H, Kiel J H A and Ptasinski K J 2011 *Biomass and Bioenergy* **35** pp 3748-3762
[8] Sheng C D and Azevedo J L T 2005 *Biomass and Bioenergy* **28**(5) pp 49- 507
[9] ECN, *Phyllis, database for biomass and waste,* Energy research Centre of the Netherlands, in Energy research Centre of the Netherlands. Available from http://www.ecn.nl/phyllis/defs.asp, accessed on 20/10/2016.
[10] Suanboon P, Thong a namai M and Jandong R 2016 *Production of charcoal from cassava rhizome and sugarcane bagasse* (Faculty of Technology, Udon Thani Rajabhat University, Thailand)
[11] Channiwala S A and Parikh P P 2002 *Fuel* **81**(8) pp 1051-1063
[12] Phutteesakul R, Suwakantakul A and Kucharrat A 2010 *J. of the Education Industry, Srinakharinwirot University* **4**(2) pp 18-28

IOP Conf. Series: Materials Science and Engineering **297** (2018) 012002 doi:10.1088/1757-899X/297/1/012002

Study on the micro direct ethanol fuel cell (Micro-DEFC) performance

Penyarat Saisirirat* **and Bordindech Joommanee**

Department of Mechanical Engineering Technology, College of Industrial Technology, King Mongkut's University of Technology North Bangkok, 1518, PrachaRat 1 Road, Wongsawang Bangsue, Bangkok, 10800, Thailand

* Corresponding Author: penyarat@gmail.com, penyaratc@kmutnb.ac.th

Abstract. The direct ethanol fuel cell (DEFC) is selected for this research. DEFC uses ethanol in the fuel cell instead of the more toxic methanol. Ethanol is more attractive than methanol by many reasons. Ethanol is a hydrogen-rich liquid and it has a higher specific energy (8.0 kWh/kg) compared to that of methanol (6.1 kWh/kg). Ethanol can be obtained in great quantity from biomass through a fermentation process from renewable resources such as sugar cane, wheat, corn, and even straw. The use of ethanol would also overcome both the storage and infrastructure challenge of hydrogen for fuel cell applications. The experimental apparatus on the micro direct ethanol fuel cell for measuring the cell performance has been set for this research. The objective is to study the micro direct ethanol fuel cell performance for applying with the portable electronic devices. The cell performance is specified in the terms of cell voltage, cell current and power of the cell at room operating temperature and 1 atm for the pressure and also includes the ethanol fuel consumption. The effect of operating temperature change on the electrical production performance is also studied. The steady-state time for collecting each data value is about 5-10 minutes. The results show that with the increase of concentrations of ethanol by volume, the reactant concentration at the reaction sites increases so the electrochemical rate also increases but when it reaches the saturated point the performance gradually drops.

1. Introduction

Nowadays an increasing number of the population is going fast and the demand of energy is very much increased. Energy from fossil fuel is one of the main energy sources to use today, but it causes environmental pollution and tends to cost more. It has a chance to expire in the near future. We cannot deny that we will use energy because energy is essential for the lives of human beings at the present and in the future. Presently, humankind realizes about the importance of using renewable energy as a clean alternative to preserve the environment. In the future, it cannot be denied that alternative energy will be utilized. Many alternatives have been found and studied, including solar energy, hydropower from dams and, not long ago, fuel cells.

A fuel cell is a device that converts the chemical energy from a fuel into electricity through a chemical reaction of positively charged hydrogen ions with oxygen or another oxidizing agent. Fuel cells are different from batteries in that they require a continuous source of fuel and oxygen or air to sustain the chemical reaction, whereas in a battery the chemicals in the battery react with each other to generate an electromotive force (emf) [1]. Fuel cells can produce electricity continuously for as long as these inputs are supplied. The first fuel cells were invented in 1838. Since then, fuel cells have been used in many other applications. Fuel cells are used for primary and backup power in commercial, industrial and residential buildings, in remote or inaccessible areas. They are also used to power small electrical appliances as well as vehicles. There are many types of fuel cells, but they all consist of an anode, a cathode, and an electrolyte that allows positively charged hydrogen ions (or protons) to move between the two sides of the fuel cell. The anode and cathode contain catalysts that cause the fuel to undergo oxidation reactions that generate positively charged hydrogen ions and electrons. The hydrogen ions are drawn through the electrolyte after the reaction. At the same time, electrons are drawn from the anode to the cathode through an external circuit, producing direct current electricity. At the cathode, hydrogen ions, electrons, and oxygen react to form water [1]. As the main difference among various fuel cell types is the electrolyte, fuel cells are classified by the type of electrolyte they use and by the difference in startup time ranging from 1 second for proton exchange membrane fuel cells (PEM fuel cells, or PEMFC) to 10 minutes for solid oxide fuel cells (SOFC). In the present, the fuel cells are divided into 9 types [2]: 1. Alkaline Fuel Cell (AFC) 2. Phosphoric Acid Fuel Cell (PAFC) 3. Molten Carbonate Fuel Cell (MCFC) 4. Solid Oxide Fuel Cell (SOFC) 5. Proton Exchange Membrane Fuel Cell (PEMFC) 6. Regenerative Fuel Cell (RFC) 7. Direct Alcohol Fuel Cells (DAFCs) that are divided into 2 types, 8. Direct Methanol Fuel Cell (DMFC) and 9. Ethanol Fuel Cell (DEFC).

For a fuel cell that uses hydrogen gas, and equipment to store hydrogen is quite expensive. Fuel cell such as a Proton Exchange Membrane Fuel Cell (PEMFC) has high performance but needs to use the purity of hydrogen and the platinum electrodes, making it rather expensive. Fuel cells that use alcohol as fuels are divided into 2 types: 1. Direct Methanol Fuel Cell (DMFC) and 2. Direct Ethanol Fuel Cell (DEFC). Previously, a lot of research focused on the study and development of Direct Methanol Fuel Cell (DMFC). There is a downside and many other problems in reaction kinetics for the slow oxidation of methanol fuel and across the water from one electrode to the one electrode (Crossover) [3, 4]. Matsuoka et al. [5] reported that Direct Methanol Fuel Cell (DMFC) electrodes have slow reaction kinetics because of obstruction of carbon monoxide (CO) catalyst platinum (Pt) at low temperatures and high surface quality of the catalysts for membrane.

Methanol is non-renewable energy described as volatile and highly flammable. Ethanol fuel, which is good choice for fuel cell is less toxic and has higher energy density than methanol. Also ethanol can be produced from biological processes in agriculture [6]. Ethanol can be obtained in great quantity from biomass through a fermentation process from renewable resources such sugar cane, wheat, corn, and even straw. Bio-generated ethanol (or bio-ethanol) is thus attractive since growing crops for biofuels absorbs much of the carbon dioxide emitted into the atmosphere from fuel used to produce the biofuels, and from burning the biofuels themselves. This is in sharp contrast to the use of fossil fuels. The use of ethanol would also overcome both the storage and infrastructure challenge of hydrogen for fuel cell applications. In a fuel cell, the oxidation of any fuel requires the use of a catalyst in order to achieve the current density required for commercially viable fuel cells, and platinum-based catalysts are some of the most efficient materials for the oxidation of small organic molecules [7].

Due to fast development in science and technology, fuel cells are developed to smaller and more compact size for usage with the portable electronic devices. Direct Ethanol Fuel Cells (DEFCs) have dimensions of about 125 mm x 115 mm x 102 mm and can produce the electric power about 102 W for longer than 3,600 hours [8]. Ethanol also remains the easier fuel to work with for widespread use by consumers. In addition, ethanol is a hydrogen-rich liquid and it has a higher specific energy (8.0 kWh/kg) compared to that of methanol (6.1 kWh/kg).

The objective of this research is to study the micro direct ethanol fuel cell performance especially on the ethanol fuel consumption rate for applying with the portable electronic devices. The cell performance is specified in the terms of cell voltage and power density at any specific current density for several operating conditions. In this work, the power is limited to 10 W, which is suitable for portable electronic devices. The experimental results will be compared with the theoretical calculations.

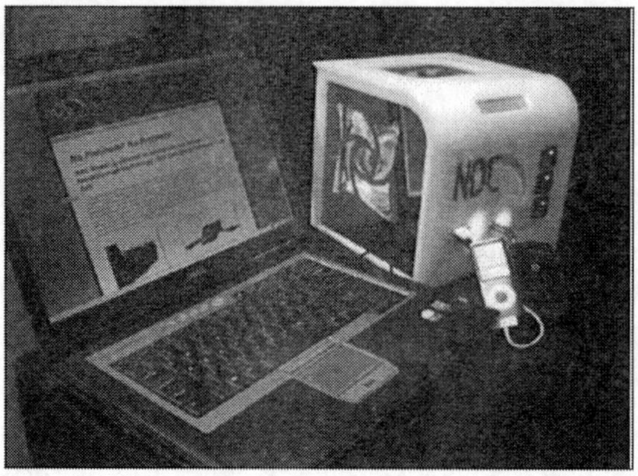

Figure 1. Direct ethanol fuel cell by horizon fuel cell technology [8].

2. Experimental apparatus

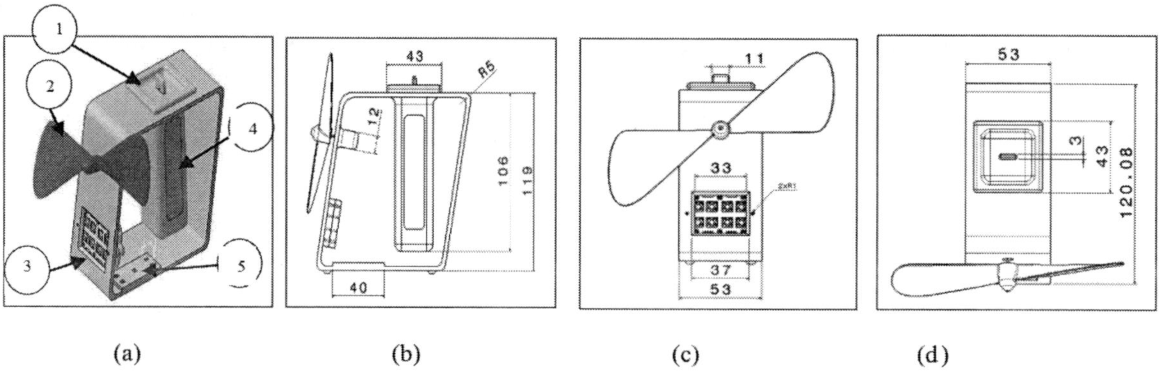

Figure 2. (a) direct ethanol fuel cell; (b) drawing front view; (c) drawing side view; (d) drawing top view (all dimensions are in millimeters)

The experimental apparatuses are composed of 1. Micro-DEFC, 2. Ethanol fuel 95% by volume, 3. Pure water, 4. 100-ml cylinder, 5. 10-ml syringe and 6. Paper measure acid-alkaline. The details on the figure 2 are: 1. Lid set (Micro-DEFC), 2. Vane set, 3. Membrane (MEA), 4. Set structure (Micro-DEFC) and 5. Cylinder solution. Bio-energy feed ethanol set demonstrate changing the ethanol fuel to electrical energy without combustion in the process. It can work continuously for many hours. The main experimental study is about the consumption of ethanol fuel for electricity production.

3. Theoretical calculations

3.1. *The density of electrical current*

Ideal gas law is an equation of state with the number of moles (N) and volume (V) [9]:

$$PV = NRT \tag{1}$$

where $P =$ Pressure (Pa) , $V =$ Fuel feed rates (mm^3/min) , $N =$ Number of moles (mole/s) $R =$ Universal gas constant = 0.082 m^3atm/kg mol k, and $T =$0°C. Lateral move (N) to get number of moles to represent the following equation to determine density of electrical current (i). The density of electrical current (i) from 1 molecule of hydrogen gas can transfer 2 electrons so flow rate of hydrogen gas can be calculated as [3]:

$$N_{H_2} = \frac{i}{2F} \tag{2}$$

where i = Density of electric current (A/m^2) and F = Faraday constant = 9.64853 x 10^4 (C/mol).

One molecule of oxygen can transfer 4 electrons, so the flow rate of oxygen is

$$N_{O_2} = \frac{i}{4F}. \tag{3}$$

When using air instead oxygen that the air flow rate can be calculated as

$$N_{air} = \frac{i}{4F} x \frac{1}{0.21}. \tag{4}$$

The density of electrical current (i) can be calculated as

$$i = N \cdot 2F \tag{5}$$

3.2. *The cell voltage*

According to the theory, the voltage of a single-cell DEFC when fed ethanol at 25 (°C) and 1 (atm) by the chemical reaction is as follows [10]:

$$Anode: C_2H_5OH + 3H_2O \rightarrow 12e^+ + 12H^+ + 2CO_2 \; (E^0_{anode} = 0.084\ V) \tag{6}$$

$$Cathode: 12e^+ + 12H^+ + 3O_2 \rightarrow 6H_2O \; (E^0_{cathode} = 1.229\ V) \tag{7}$$

$$Overall: C_2H_5OH + 3O_2 \rightarrow 3H_2O + 2CO_2 \; (E^0_{cell} = 1.145\ V) \tag{8}$$

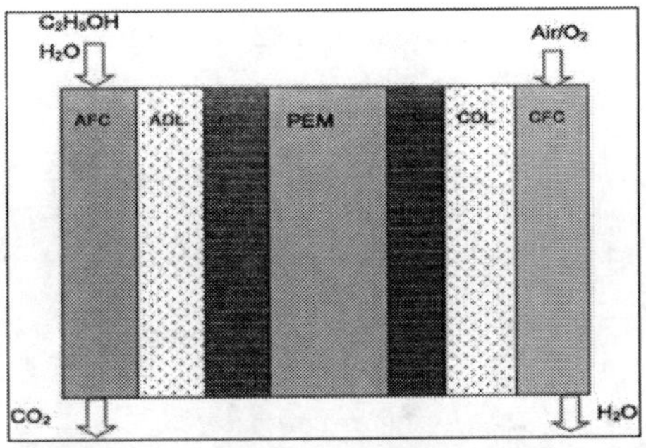

Figure 3. The system of direct ethanol fuel cell (DEFC).

In figure 3, the texts with their corresponding meanings are PEM : Membrane , CDL: Cathode gas diffusion layer, ADL : Anode gas diffusion layer, CCL : Cathode catalyst layer, ACL : Anode catalyst layer, CFC : Splice plate cathode, and AFC : Splice plate anode. The voltage (V) in a single-cell DEFC for portable electronic devices can be calculated using the following equation [10]:

$$E_{cell} = E^0_{cell} - \eta_{conc} - \eta_{ohmic} - \eta_{act} \tag{9}$$

where E_{cell} = Voltage of fuel cell (V), E^0_{cell} = Voltage of fuel cell according to the theory (V), η_{conc} = Concentration of voltage loss related to kinetics of reaction (V), η_{ohmic} = Voltage loss related to resistance (V) and η_{act} = Voltage loss related to reaction involved to kinetics (V).

The movement of substrate on anode catalyst layer and cathode catalyst layer results in the loss concentration inside the cell. The concentration of voltage loss can be calculated from equation (10) [11]:

$$\eta_{conc} = \frac{RT}{zF} ln(\frac{I_{lim}}{I_{lim} - i}). \tag{10}$$

While limitation of electric current flow follows equation (11):

$$I_{\lim} = Z_a F D_b^{eff} \frac{C_b}{L_b} \tag{11}$$

where $R=$ Universal gas constant = 8.314 (J/g mol k) , z =Number of electrons in system, D_b^{eff} = Spread efficiency coefficient of ethanol at the anode , C_b =Concentration of ethanol at the anode (molar) and L_b =The width of flow channel at the anode (mm).

The cell voltage resistance loss can be calculated from equations (12)-(15):

$$\eta_{ohmic} = iR_{total}, \tag{12}$$

$$R_{total} = \eta_{membr} + \eta_{contact}, \tag{13}$$

$$\eta_{membr} = \frac{l_m}{K_m^o} i, \tag{14}$$

$$\eta_{contat} = \frac{l_m + 2l_c}{K_s}, \tag{15}$$

while the voltage loss related to reaction involved to kinetics can be calculated from the equations (16)-(17) [12]:

$$\eta_{act,anode} = \frac{RT}{\alpha_a z_a F} \ln(\frac{i}{i_o}) \tag{16}$$

$$\eta_{act,cathone} = \frac{RT}{\alpha_c z_c} \ln(\frac{i}{i_o}) \tag{17}$$

From equation (9), we can calculate the voltage in DEFC single cell as follows:

$$\begin{aligned} E_{cell} &= E_{cell}^0 - \eta_{conc} - \eta_{ohmic} - \eta_{act} \\ &= E_{cell}^0 - (\frac{RT}{z_a F} ln(\frac{i_{lim}}{i_{lim} - i})) - (i(\frac{l_m}{K_m} + \frac{l_m + 2l_c}{K_s})_0) \\ &\quad - (\frac{RT}{\alpha_a z_a F} ln(\frac{i}{i_o}) + \frac{RT}{\alpha_c z_c F} ln(\frac{i}{i_o}) \end{aligned} \tag{18}$$

3.3. The flow rate of ethanol for electrochemical reaction

The ethanol fuel mole flow rate can calculate by equation (19) and the ethanol volume flow rate be calculated by equation (20) as shown below

$$n_i = \frac{i}{nF} \tag{19}$$

$$v_i = \frac{n_i M_i}{p_i} \tag{20}$$

where n_i = Ethanol fuel mole flow rate (mol/s), n = Number of electrons in system , v_i = Ethanol volume flow rate (cm^3/h) , M_i =Molecular (g/mol) and ρ_i =Density of ethanol (g/cm^3).

3.4. The total power and efficiency of fuel cell

Electrical power (P) can be calculated from the following equation:

$$P = VxI \tag{21}$$

Once voltage (V) is known then we can calculate the efficiency of fuel cell ($\eta_{actual,E}$) from the ratio between actual electrical potential (E_{cell}) and electrical potential from the reaction between the ethanol and oxygen (E_{cell}^o) as

$$\eta_{actual,E} = \frac{E_{cell}}{E^o_{cell}} x100\%. \quad (22)$$

The parameters used in the calculations are shown below in table 1.

Table 1. The parameters use in the calculation.

Parameters	Constants	Units	References
Design parameters			
Active area	4	cm^2	Fixed
Current density	0.175	A/cm^2	Fixed
Catalyst layer thickness, l_c	0.01	cm	
Diffusion layer thickness, l_d	0.14	cm	
Membrane thickness, l_m	0.01778	cm	
Operating parameters			
Pressure	1	atm	Fixed
Faraday constant, F	96,500	$Cmol^{-1}$	
Gas constant, R	8.314	J/molK	
Effective diffusion coefficient of ethanol in diffusio layer D^d_{EtOH}	3.916×10^{-2}	cm^2s^{-1}	[13]
Void volume fraction of anode diffusion layer , ε^d	0.4		[14]
Electronic conductivity of solid phase (PtRu/C),K_s	8.13×10^{-6}	Scm^{-1}	[13]
Protonic Conductivity of ionomer, K_m	0.1416	Scm-1	[13]
Anode transfer coefficient , α_a	0.089	cm^2	[13]
Cathode transfer coefficient , α_c	1.0	A/cm^2	[13]
Component parameters			
Oxygen density	1.141	gcm^{-3}	Fixed
Co_2 density	0.77	gcm^{-3}	Fixed
Ethanol density	0.785	gcm^{-3}	Fixed
Water density	1	gcm^{-3}	Fixed

4. Results and discussion

The goal of the DEFC study in this research was to study on the performance of the micro direct ethanol fuel cell (Micro-DEFC). It is also to study the consumption of ethanol fuel and its effects on the electrical production. The experimental performance testing studied both of the ethanol blended with pure distilled water and the ethanol blended with conventional drinking water. The effect of operating temperature change on the electrical production performance is also studied. The cell performance is specified in the terms of cell voltage and power of the cell for several operating conditions. The experimental results are shown below.

Table 2 shows the experimental results that all experiments were carried out by changing the ethanol concentration and changing the operating temperature of the fuel cell. The ethanol concentration changed from 5, 7, 9, 11, 13 and 15 ml dissolved in 60 ml of purified distilled water at operating temperature of 30, 35, 40, 45, 50 and 55 °C. The electrical production variables of the fuel cell, i.e. the electrical current, the electrical cell voltage and the electrical power were measured. Comparison of the results obtained from the theoretical calculations and the actual measurement of fuel cell performance that will be described in the next section.

Table 2. Experimental performance with ethanol blended with pure distilled water.

Ethanol concentration (95 %) (ml. blended with 60 ml. of pure distilled water)	Experimental results no.1				Experimental results no.2			
	I (mA)	V (V)	P (mW)	T (°C)	I (mA)	V (V)	P (mW)	T (°C)
5	11.2	0.95	10.6	30.8	11.2	0.96	10.5	34.6
7	12.8	0.92	12.8	30.8	15.4	0.92	14.2	34.4
9	13.1	0.90	11.9	31.2	14	0.95	13.3	34.6
11	10.3	0.92	9.58	31.2	15	0.94	14.2	34.8
13	12.1	0.89	10.8	31.2	15.3	0.91	13.9	34.8
15	11.2	0.91	10.2	30.8	15	0.92	13.8	34.4

Ethanol concentration (95 %) (ml. blended with 60 ml. of pure distilled water)	Experimental results no.3				Experimental results no.4			
	I (mA)	V (V)	P (mW)	T (°C)	I (mA)	V (V)	P (mW)	T (°C)
5	12.3	0.88	10.9	41	12.9	0.95	12.2	44.4
7	11.7	0.94	11.1	40.8	15.9	0.94	15.0	44.6
9	12.4	0.94	11.7	40.8	14.6	0.94	13.8	44.4
11	13.6	0.94	12.9	41.2	14	0.94	13.2	44.6
13	13.9	0.94	13.1	41	15.3	0.92	14.1	44.4
15	14.0	0.95	13.3	40.8	15.4	0.92	14.3	44.6

Ethanol concentration (95 %) (ml. blended with 60 ml. of pure distilled water)	Experimental results no.5				Experimental results no.6			
	I (mA)	V (V)	P (mW)	T (°C)	I (mA)	V (V)	P (mW)	T (°C)
5	13.3	0.95	12.7	51	13.9	0.96	13.4	54.6
7	14.6	0.96	14.1	50.8	16.7	0.95	16.0	54.6
9	15.3	0.97	14.8	51.2	14.8	0.95	14.1	54.4
11	18	0.96	17.3	51.2	15.7	0.95	14.9	54.6
13	18.8	0.96	18.2	50.8	16.6	0.93	15.4	54.4
15	19.3	0.98	19.0	51.2	14.6	0.94	13.7	54.6

Ethanol concentration (95 %) (ml. blended with 60 ml. of pure distilled water)	Experimental results no.7				Experimental results no.8			
	I (mA)	V (V)	P (mW)	T (°C)	I (mA)	V (V)	P (mW)	T (°C)
5	14.3	0.97	13.9	60.4	15.9	0.99	15.8	64.6
7	16.6	0.96	16.0	60.6	16.9	0.98	16.6	64.6
9	15.7	0.96	15.2	60.4	17.4	0.98	17.2	64.6
11	16.9	0.97	16.5	60.4	17	0.98	16.7	64.4
13	15.8	0.95	15.1	60.4	17.7	0.98	17.3	64.6
15	16.3	0.95	15.5	60.4	15.9	0.96	15.3	64.6

The experiments observed that when increasing the concentration of ethanol 5 ml to 15 ml dissolved in 60 ml of pure distilled water and increasing the temperature to 30 to 55 °C, the electricity production variables, such as the value of the fuel cell current, the fuel cell voltages and the power of the fuel cell were increased in accordance with the ethanol concentration and increased with temperature. Electrical charges gradually increase from low ethanol concentration and low temperatures to the highest point at a concentration of 13 ml of ethanol dissolved in 60 ml of pure distilled water at 40-45 °C. The production of electric charge of the fuel cell gradually decreases over time until the fuel has been exhausted when the ethanol concentration was increased to more than 13 ml dissolved in 60 ml purified distilled water, the electrical conductivity of the fuel cell was reduced. At 30-35 °C, for ethanol fuel blended with pure distilled water at different concentrations, the fuel cell must wait for about 5-10 minutes before it will start producing electrical charges. The experimental temperature was increased to 40-55 °C, it was found that the fuel cell started reacting faster with electricity that it took only 2-3 minutes before it can produce electricity efficiently.

Table 3. The average results of ethanol blended with pure distilled water.

Ethanol concentration (95 %) (ml. blended with 60 ml. of pure distilled water)	Average trial values			Temperature	Times
	I (mA)	V (V)	P (mW)	T (°C)	(S)
5	11.2	0.95	10.6	31-32	20
7	15.4	0.92	14.2	34-35	20
9	12.4	0.94	11.7	39-40	20
11	14.0	0.94	13.2	44-45	20
13	18.8	0.96	18.2	49-50	20
15	14.6	0.94	13.7	54-55	20

Table 3 shows the average results obtained from data in table 2. The ethanol concentration and temperature were increased. It was found that the electrical production variables of the fuel cell, such as electrical current, voltage and power of fuel cell varied by increasing ethanol concentration and temperature. The maximum electrical value is at 13 ml of ethanol, dissolved in 60 ml of pure distilled water at 40-45 °C. The electrical current is equal to 18.8 mA, the voltage is equal to 0.96 V and the power is equal to 18.2 mW. Increasing the ethanol concentration and temperature affects these electric values of the fuel cell. By increasing the ethanol concentration and temperature, the fuel cell will gradually reacts to produce electrical charge and will gradually ramp up to the highest value and then gradually decrease until fuel is completely fed. It does not affect the output voltage of the fuel cell. In the experiment, when the ethanol concentration and temperature were increased, the cell voltage values were not significantly different. Increasing the concentration of ethanol and temperature affects the electrical current and power but does not effect the electrical potential at all or minimally. Table 4 is the theoretical calculation of the value of electrical currents, fuel cell voltages and power of the fuel cell for comparison these results with the experimental results as shown above and the last column of the table 4 is the fuel cell efficiency calculation results. Based on the calculation, the value of electrical current and power are even higher. On the other hand, the fuel cell electrical potential and the fuel cell efficiency decrease visibly. At the ethanol concentration of 5 ml dissolved in 60 ml of pure distilled water at a working temperature of 31-32 °C has the highest efficiency of 92.524% and the maximum voltage of 1.0594 V. At the concentration of 15 ml of ethanol dissolved in 60 ml of pure distilled water and at the working temperature of 54-55 °C, the maximum current is 695.214 mA and the maximum output power is 618.495 mW.

Table 4. The fuel cell efficiency calculation results when blended ethanol with pure distilled water.

Ethanol concentration (95 %) (ml. blended with 60 ml. of pure distilled water)	Average trial values			Temperature	Fuel cell efficiency
	I (mA)	V (V)	P (mW)	T (°C)	(%)
5	231.738	1.0594	245.503	31-32	92.524
7	324.433	1.0258	332.804	34-35	89.591
9	417.128	0.9917	413.699	39-40	86.618
11	509.822	0.9577	488.275	44-45	83.645
13	602.518	0.9236	556.541	49-50	80.671
15	695.214	0.8896	618.495	54-55	77.698

Next, the experiment was repeated but conventional drinking water, instead of pure distilled water, was used to blend with the ethanol. The experiment steps remain the same. When the ethanol concentration is increased from 5 ml to 15 ml dissolved in 60 ml of conventional drinking water and the operating temperature of the cell is in the range of 30-65 °C, the production of electricity such as electrical current, the fuel cell voltage and the power of the fuel cell increases with the ethanol concentration and increases with the temperature. The fuel cell electrical values are gradually increased from low ethanol concentration and lower temperatures to the highest point at a concentration of 13 ml of ethanol dissolved in 60 ml of conventional drinking water at 49-50 °C, the electric charge of the fuel cell will gradually increase. It continues to decrease over time until all the fuel is gone, when the ethanol concentration is increased to more than 13 ml, the electrical output of fuel cell is reduced. The experimental results are shown in table 5.

Table 6 shows the average results obtained from data in table 5. The experiment was conducted to change the blendture of ethanol from pure distilled water to conventional drinking water. It was found that the production of electrical values of the fuel cell, such as electrical currents, the fuel cell voltages and the power of the fuel cell varied by increasing ethanol concentration and temperature. The maximum electrical values were at 13 ml of ethanol dissolved in 60 ml of conventional drinking water at 49-50 °C. The electrical current was equal to 19.6 mA, the fuel cell voltage was equal to 0.91 V and the fuel cell power was equal to 18.0 mW. Increasing the ethanol concentration and temperature affects the electrical values of the fuel cell. By increasing the ethanol concentration and temperature, the fuel cell gradually reacted to produce the electrical value and gradually increased to the maximum value and then gradually decreased until the fuel is completely fed. However, it did not affect the production of the electric potential of the fuel cell. In the experiment, when the ethanol concentration and temperature were increased, the fuel cell voltages were not significantly different. Increasing the ethanol concentration and temperature affects the value of fuel cell electrical currents and power, but does not affect the voltage significantly.

Table 5. Experimental performance of ethanol blended with conventional drinking water.

Ethanol concentration (95 %) (ml. blended with 60 ml. of conventional drinking water)	Experimental results no.1				Experimental results no.2			
	I (mA)	V (V)	P (mW)	T (°C)	I (mA)	V (V)	P (mW)	T (°C)
5	6.5	0.86	5.6	30.4	15.2	0.83	12.6	34.6
7	7.3	0.86	6.3	30.8	15.6	0.9	14.1	34.6
9	7.8	0.87	6.8	30.2	16.0	0.93	15.0	34.6
11	8.5	0.87	7.4	30.4	17.5	0.91	16.0	34.2
13	10.2	0.86	8.7	30.4	18.2	0.91	16.6	34.6
15	14.5	0.87	12.6	30.6	17.4	0.93	16.3	34.6

Ethanol concentration (95 %) (ml. blended with 60 ml. of conventional drinking water)	Experimental results no.3				Experimental results no.4			
	I (mA)	V (V)	P (mW)	T (°C)	I (mA)	V (V)	P (mW)	T (°C)
5	17.8	0.89	15.6	40.6	15.9	0.9	14.3	44.4
7	17.4	0.91	15.8	40.4	17.5	0.9	15.7	44.4
9	18.2	0.90	16.4	40.8	18.5	0.9	16.7	44.4
11	18.1	0.92	16.7	40.4	18.4	0.90	16.7	44.4
13	18.5	0.90	16.7	40.4	18.4	0.90	16.6	44.4
15	19.9	0.90	18.0	40.4	17.3	0.91	15.8	44.4

Ethanol concentration (95 %) (ml. blended with 60 ml. of conventional drinking water)	Experimental results no.5				Experimental results no.6			
	I (mA)	V (V)	P (mW)	T (°C)	I (mA)	V (V)	P (mW)	T (°C)
5	16.0	0.91	14.7	50.4	16	0.91	14.6	54.6
7	16.0	0.93	14.9	50.4	16.0	0.92	14.8	54.4
9	16.7	0.93	15.6	50.8	17.6	0.92	16.2	54.6
11	18.4	0.93	17.2	50.4	19.62	0.92	18.1	54.6
13	19.6	0.91	18.0	50.4	21.9	0.93	20.6	54.2
15	17.5	0.92	16.1	50.6	17.9	0.92	16.5	54.4

Ethanol concentration (95 %) (ml. blended with 60 ml. of conventional drinking water)	Experimental results no.7				Experimental results no.8			
	I (mA)	V (V)	P (mW)	T (°C)	I (mA)	V (V)	P (mW)	T (°C)
5	16.7	0.91	15.2	60.4	16.7	0.91	15.2	64.4
7	17.3	0.92	15.9	60.4	17.8	0.92	16.4	64.6
9	17.2	0.92	15.9	60.4	18.8	0.92	17.4	64.6
11	20.2	0.92	18.5	60.4	20.4	0.92	18.7	64.6
13	23.0	0.92	21.2	60.4	25.0	0.92	23.2	64.4
15	19.0	0.92	17.3	60.8	21.1	0.92	19.5	64.2

Table 6. The average results of ethanol blended with conventional drinking water.

Ethanol concentration (95 %) (ml. blended with 60 ml. of conventional drinking water)	Average trial values			Temperature	Times
	I (mA)	V (V)	P (mW)	T (°C)	(S)
5	6.5	0.86	5.6	30-31	20
7	15.6	0.90	14.1	34-35	20
9	18.2	0.90	16.4	39-40	20
11	18.4	0.90	16.6	44-45	20
13	19.6	0.91	17.8	49-50	20
15	17.9	0.92	16.5	54-55	20

From the experimental results, we can see that the electrical production of the fuel cell are not much different, in terms of the electrical currents, the electrical potential and the power of the fuel cell, and whether ethanol blended with pure distilled water or ethanol blended with conventional drinking water was used as fuel. Therefore, it can be concluded that ethanol can be used in combination with conventional drinking water instead of pure distilled water for electricity production in a direct ethanol fuel cell for reducing the operating costs. To increase the amount of electricity produced and the power of the cell, the fuel cells can be connected as a series of stacks according to the amount of electricity or power required.

5. Conclusion

Direct Ethanol Fuel Cells or DEFCs are a subcategory of proton-exchange fuel cells in which ethanol is used as the fuel. Their main advantage is the ease of transport of ethanol, which is an energy-dense yet reasonably stable liquid at all environmental conditions. DEFC is a new fuel cell that is similar to the PEMFC fuel cells, where both are used as polymer-type membrane electrolytes. However, in this fuel cell itself, it is an anode catalyst, extracting hydrogen from liquid ethanol. A more efficient version of a direct fuel cell would play a key role in the theoretical use of ethanol as a general energy transport medium, in the hypothesized ethanol economy. The goal of the DEFC study in this research was to study on the performance of the micro direct ethanol fuel cell (Micro-DEFC) for usage in portable electronic devices. It is also to study the consumption of ethanol fuel, which affects the electric current production. The effect of operating temperature change on the electric production performance is also studied too. The cell performance is specified in the terms of the cell voltage, the cell electrical current and power of the cell for several operating conditions. The results show that with the increase of concentrations of ethanol by volume, the reactant concentration at the reaction sites increases so the electro-chemical rate also increase but when it reaches the saturated point the performance gradually drops. From the averaged data, that the maximum current production, maximum cell voltage and maximum power of the fuel cell are 18.8 mA, 0.96 V and 18.2 mW, respectively. They are occurred at the ratio of ethanol equal to 13 ml in 60 ml of pure distilled water and at the cell operating temperature of 49-50 °C and atmospheric pressure. From the calculation results show that with increase of ethanol concentrations and the operating temperature of the fuel cell that the electrical current production and electrical power production also increase. On the other hand, the electrical potential of the fuel cell and the efficiency of the fuel cell are noticeably reduced. When changing the substance blended with ethanol from pure distilled water to conventional drinking waterit was found that the production of electrical values of the fuel cell, such as electrical currents, the fuel cell voltages and the power of the fuel cell varied with increasing ethanol concentration and temperature. The maximum electrical values were at 13 ml of ethanol dissolved in 60 ml of conventional drinking water at 49-50 °C. The electrical current was equal to 19.6 mA, the fuel cell voltage was equal to 0.91 V and the fuel cell power was equal to 18.0 mW. By increasing the ethanol concentration and temperature, the fuel cell gradually reacted to produce the

IOP Conf. Series: Materials Science and Engineering **297** (2018) 012002 doi:10.1088/1757-899X/297/1/012002

electrical value and gradually increased to the maximum value and then gradually decreased until the fuel is completely fed. However, it did not affect the production of the electric potential of the fuel cell. In the experiment, when the ethanol concentration and temperature were increased, the fuel cell voltages were not significantly different. Increasing the ethanol concentration and temperature affects the values of fuel cell electrical currents and power, but does not effect on the voltage.

The production of electricity, electrical potential and electrical power when using ethanol blended with pure distilled water or when using ethanol blended with conventional drinking water shows little difference. Therefore, it can be concluded that ethanol can be used in combination with conventional drinking water instead of pure distilled water for the production of electrical values in a fuel cell to save the operating costs. To increase the amount of electricity produced and the power of the cell, the fuel cell can be connected as a series of stacks according to the amount of electricity or power required.

Acknowledgments

This research was supported by College of Industrial Technology, King Mongkut's University of Technology North Bangkok.

References

[1] Larminie J and Dicks A 2003 *Fuel Cell Systems Explained* (New York: Wiley)
[2] Prommes P 2015 *Princess of Naradhiwas University Journal.* **7** pp 157-170
[3] Wongyao N, Therthianwong A and Therthianwong S 2011 *Energy. Convers. Manage.* **52** pp 2676-2681
[4] Xu Q, Zhao TS, Yang WW and Chen R 2011 *Int. J. Hydrogen. Energy.* **36** pp 830-838
[5] Matsuoka K , Iriyama Y, Abe T, Matsuoka M and Oguni Z 2005 *J. Power Sources.* **150** pp 27-31
[6] Roelofs KS, Hirth T and Schiestel T 2011 *Mater. Sci. Eng. B.* **176** pp 727-735
[7] Song SQ, Zhou WJ and Zhau ZH 2005 *Int. J. Hydrogen. Energy.* **30** pp995-1001
[8] NDcPower. *EOS vs Conventional Fuel Cell Technology* (Horizon Fuel Cell Technologies Pte Ltd) [1 February 2012] [http://www. ndcpower.com/green-power.php; 2011. 2007]
[9] Hoonsom M 2013 *PEM Fuel Cells and Chemical Chemistry Analysis* (*Master Thesis*) (Department of Chemical Technology, Faculty of Science, Chulalongkorn University)
[10] Abdullah S and Kamarudin SK 2015 *Int. J. Hydrogen. Energy.* **40** pp 11943 – 11948
[11] Andreadis GM , Podias AKM and Tsiakaras PE 2008 *J. Power Sources.* **181** pp 214-227
[12] Zhu H and Kee RJ 2003 *J. Power Sources.* **117** pp 61-74
[13] Jeng KT and Chen CW 2002 *J. Power Sources.* **112** pp 367-375
[14] Suresh NS and Jayanti S 2011 *Int. J. Hydrogen. Energy.* **36** pp 14648-14658

Research of interaction between technological and material parameters during densification of sunflower hulls

Peter Križan*, Miloš Matúš,Juraj Beniak and Ľubomír Šooš

Institute of Manufacturing Systems, Environmental Technology and Quality Management, Faculty of Mechanical Engineering, STU in Bratislava, Nám. Slobody 17, 812 31 Bratislava, Slovakia

* Corresponding Author: peter.krizan@stuba.sk

Abstract. During the biomass densification can be recognized various technological variables and also material parameters which significantly influences the final solid biofuels (pellets) quality. In this paper, we will present the research findings concerning relationships between technological and material variables during densification of sunflower hulls. Sunflower hulls as an unused source is a typical product of agricultural industry in Slovakia and belongs to the group of herbaceous biomass. The main goal of presented experimental research is to determine the impact of compression pressure, compression temperature and material particle size distribution on final biofuels quality. Experimental research described in this paper was realized by single-axis densification, which was represented by experimental pressing stand. The impact of mentioned investigated variables on the final briquettes density and briquettes dilatation was determined. Mutual interactions of these variables on final briquettes quality are showing the importance of mentioned variables during the densification process. Impact of raw material particle size distribution on final biofuels quality was also proven by experimental research on semi-production pelleting plant.

1. Introduction

The pelletization of biomass is a mass and energy densification process for materials that possess low bulk densities. The quality of biofuels (final product of densification) is determined by the end-user's requirements on the heating system and the handling properties [1, 2]. In general, the biofuels quality depends on the properties of the raw material (feedstock) – in terms of biomass type, moisture content and particle size; and quality management of the manufacturing process – in terms of operating conditions, technological variables, pelletizer type and binding agent [3, 4]. Low-quality pellets can cause operational problems in combustion systems, including undesired effects in the equipment such as slugging, fouling or corrosion, and may originate substantial amounts of gaseous and particulate matter (PM) emissions [1,2,4]. There are known many projects and applications where we can see energy recovery of different types of raw materials. With the development of biomass energy recovery technologies and with the gradual decrease of the quantities of usable wood biomass, increases the requirements for the processing of other types of biomass, e.g. agricultural - herbaceous biomass[5]. Sunflower hulls as an unused source is a typical product of agricultural industry in Slovakia belongs to the group of herbaceous biomass. Sunflower as herbaceous plant which is widely and usually cultivated on Slovakia has a high potential of usage in energy sector. In case of sunflower hulls densification (in general in case of herbaceous biomass) there are some important differences with comparing the wooden biomass. There are some problems come from the material (sunflower hulls) composition and also from material structure. These parameters influencing also lonely treatment, densification and also the combustion of final biofuels from sunflower hulls [3].

The main goal of presented experimental research is to determine the impact of technological variables and impact of mechanical treatment of raw feedstock (sunflower hulls) on final biofuels quality. Each type of raw feedstock has its own specific properties, chemical composition and behavior during mechanical treatment and densification. Each small change in the properties of the raw feedstock can affect the final biofuels quality [5]. The variety of raw feedstock properties requires different approach and technological conditions [6]. Due to the design of an economically efficient method of biomass treatment process, it is necessary to know and determine the optimal parameters for densification with respect to the type of raw feedstock [5, 6]. On the other hand, the applicable technical standards for the quality of biofuels are valid and must be complied with. According to the valid EN standards [7,8] the quality of biofuels is evaluated from mechanical indicators of quality (bulk density, mechanical durability, particle density, etc.) points of view and from thermal-chemical indicators of quality (moisture content, heating value, etc.) points of view. According to the main goalit was important to determine not only the impact of technological variables and raw material treatment variables on final biofuels quality but also on biofuels dilatation. The experimental findings presented here are significant from biofuels production point of view and also from densification machines constructions point of view [5].

The solution of this project comes directly from the requirement of practice, where the company demands to increase the quality of the produced pellets.

2. Materials and Methods

2.1. Monitored dependent variables

On the basis of our knowledge and experiences from executed measurements we can identify the impact of technological variables (compression pressure, compression temperature) and the impact of raw feedstock parameters (type of raw feedstock, moisture content and particle size) on biofuels quality [5, 6]. During this experimental study only the impact of compression pressure, compression temperature and particle size distribution on final biofuels quality was determined. All mentioned variables influences the conditions in pressing chamber during densification and thus influences the final biofuels quality and also stabilization time. Biofuels stabilization time is the time interval during which dilatation occurs [9]; it is also the time interval during which the biofuel stabilizes. Biofuel stabilization time takes approximately 24 hours according to the type of the raw feedstock and densification technology used but it can also take longer. Standard DIN 52182 [10] describes the conditions and process for detecting biofuels density after stabilization time. Biofuel dilatation is an effect during which the biofuel dimensions and weight are changing (diameter, length and weight). These changes come from the raw feedstock properties and also external variables of the densification technology. Dilatation directly influences the biofuel quality [9].

With a suitable combination of the mentioned variables and parameters we can significantly impact the densification process, mainly from the final biofuels quality point of view [5]. We will be able to find the optimal conditions for each specific input setting (type of raw feedstock and its parameters).

2.2. Experimental methodology and process

According to the main goal it was important to divide the experimental research into the two phases:

1.) experimental research with using of experimental pressing stand (see figure 1)

2.) semi-operation tests with using of production pelleting plant (see figure 2)

Experiment research was carried out in laboratory conditions at temperature 23°C, with and also without additional pressing temperature. The main goal of first phase was to determine the impact of compression pressure and compression temperature on final biofuels quality and on biofuels dilatation. These results were obtained without influence of additional pressing temperature [11, 12] because additional pressing temperature influences acting radial pressures and friction conditions in pressing chamber [12, 13]. Experiment was realized on available experimental pressing stand (see figure 1) which uses a vertical single axis pressing. Pressing stand is inserted to the hydraulic press which is a source for pressing pressure in various levels. According to the designed experimental plan within the range from 63 MPa to 191 MPa of compression pressure was experimental research executed. The raw material was fed into the pressing chamber and compression with a hydraulic piston was commenced. Due to an aim of this research pressing chamber with 140 mm length and ø 20 mm diameter was used in pressing stand during experiment. Also the influence of compression temperature was determined, when the 100°C additional temperature was used.

To maintain the experimental plan and ensure correct results, 7 briquettes with ø 20 mm diameter had to be pressed for each setting. This is the minimum number of values for processing using known statistical and mathematical methods [5]. Holding time in pressing chamber was chosen according to our previous experiences, it´s mean the compression pressure was applied on each briquette for 7 seconds. The quality of the briquettes is defined by a number of quality indicators, but in this case (briquettes with a diameter greater than ø 8 mm) we used as the evaluation parameter the briquettes density. The briquettes density was calculated on the basis of the briquettes weight and volume ratio. For each briquette, its weight, length and diameter were measured. The length and diameter of the briquettes were determined according to EN ISO 17829 [14] and the volume of the briquettes was then calculated. For each setting, the average value of the density obtained was calculated. After the briquettes has been stabilized, i.e. after the expansion of the briquettes has stabilized, the briquettes have been reweighed and measured. In this way, it is generally possible to define the dilatation of the briquettes and to determine the influence of the said technological parameters on the briquettes dilatation [9].

Figure 1. Hydraulic press with experimental pressing stand.

Figure 2. Pelletizing press KAHL 33-390.

The main goal of second phase was to determine the impact of particle size distribution on final biofuels quality during semi-operational tests. For the pellets production a vertical pelletizing machine KAHL 33-390 with flat round matrix was used (see figure 2). During the pelletizing process, a matrix with ø 6 mm holes was used. The constant operating parameters for the pelletizing press were the pressing pressure (12 MPa), the pressing temperature (100 °C) and the circumferential speed of rolls (2.2 $m.s^{-1}$). The interaction between the raw feedstock moisture content, the compression pressure, and the compression temperature during the pelletizing of the biomass is of major importance [12, 13]. If the moisture content of the pressed material is very low, or very high for that matter, the particles become inconsistently arranged, and the resulting pellets becomes unstable [11,15]. Past research showed that moisture has an effect on the plasticization of the lignin. In the present study the initial moisture content of sunflower hulls was low (9.24%), so that additional water was added to the

IOP Conf. Series: Materials Science and Engineering **297** (2018) 012003 doi:10.1088/1757-899X/297/1/012003

biomass to increase the raw feedstock moisture content before pelletizing. Adequate compression temperature ensures that the lignin plasticizes in the pressed material, and the optimal compression pressure provides densification of raw material [12, 13]. A certain moisture level causes a higher temperature in the densification process and thus a higher quality of pellets [12].

Semi-operational tests for pelletizing of sunflower hulls of both prepared particle size distribution (sample 1 and sample 2) were carried out without technical and technological complications. The pelleting process was pretty continuous. During constant feeding of raw feedstock the water flow was increased respectively reduced until the pellets with the highest mechanical quality were reached. The diameter of the pellets produced was ø 6 mm. Subsequently, pellets were subjected to laboratory testing and the determination of mechanical quality indicators and some chemical-thermal quality indicators according to the technical EN standards [7, 8].

2.3. Raw sunflower hulls properties

In this study the sunflower hulls were used. Sunflower hulls as an unused source is a typical product of agricultural industry in Slovakia belongs to the group of herbaceous biomass. Sunflower as herbaceous plant which is widely and usually cultivated on Slovakia has a high potential of usage in energy sector. Initial bulk density of used sunflower hulls according to EN ISO17828 [16] was measured. Bulk density of sunflower hulls is 0.1614 kg.dm^{-3}.

Initially, the particle size distribution was analysed with the aid of a Retsch Vibrating Sieve Equipment AS 200. Other different particle size distribution were processed by a hammer mill Stoza ŠV 5 with ø 4 mm sieve. For pellets production, untreated sunflower hulls (sample 1) and sunflower hulls treated by single level shredding (sample 2) were used. Figures 3 and 4 show the prepared samples before and after shredding. Figure 5 shows the raw material particle size distribution of both samples used.

Moisture content of sunflower hulls before disintegration and pelletizing was measured with the aid of a Kern MRS 120-3 balance. This measurement consisted in heating the raw feedstock (gravimetric method of moisture content measuring) [17] at 105 ± 2°C until a constant weight was achieved. Due to the complex process of moisture content change, only one moisture content level was used in this study. Sunflower hulls with 9.24% of moisture content was used for disintegration, for pelletizing and for whole experimental research.

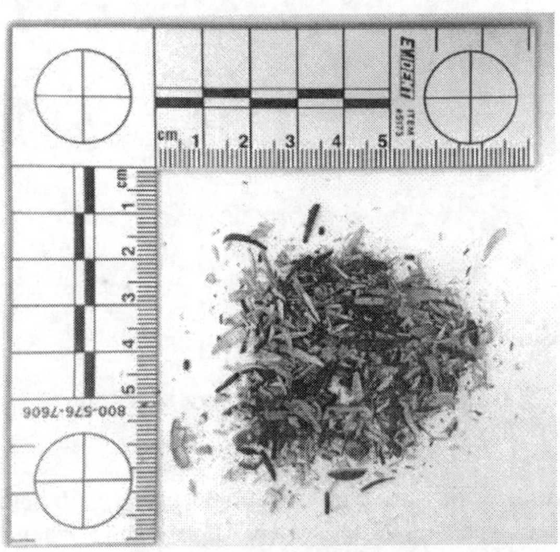

Figure 3. Untreated sunflower hulls - sample 1.

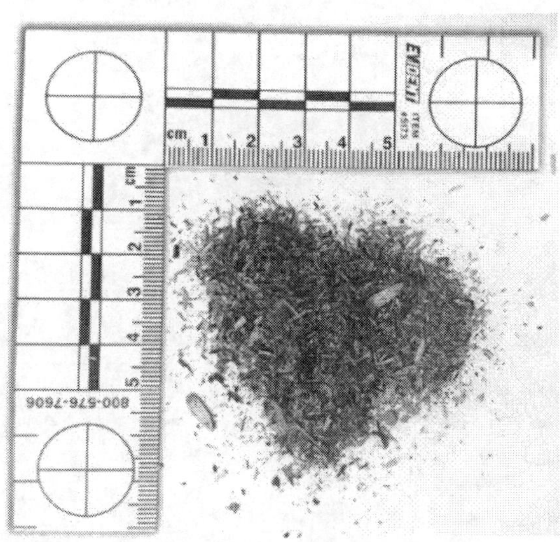

Figure 4. Treated sunflower hulls - sample 2.

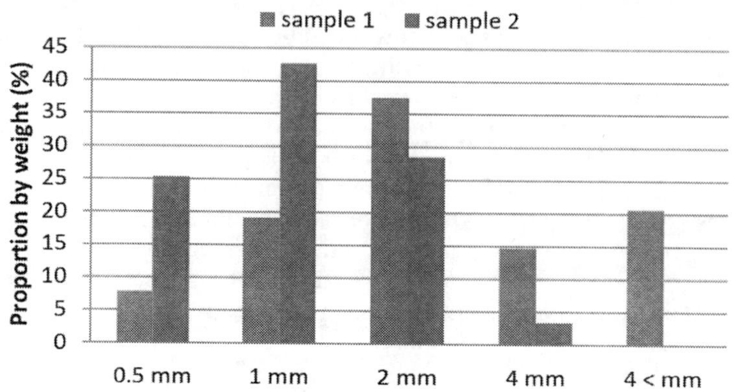

Figure 5. Raw feedstock particle size distribution of both samples studied.

3. Results and Discussion

All the results in this paper, showing the interrelation between the different variables, were depicted graphically, as seen in the figures below. The results of this experimental research are applicable to the choice of dimensions of densification machines, and gave a comprehensive overview of the behavior of different variables during the sunflower hulls densification process.

3.1. Effect of compression pressure and temperature on briquettes quality and dilatation

During the following experimental research only the untreated sunflower hulls (sample 1) were used. Briquettes with ø 20 mm diameter for these purposes on the experimental pressing stand were produced and can be seen in figures 6 and 8. The interdependence of briquettes density from compression pressure can be seen in figures 7 and 9. These figures describe results related to production also with and without additional compression temperature. In general, as the compression pressure increases so too does the briquette density.

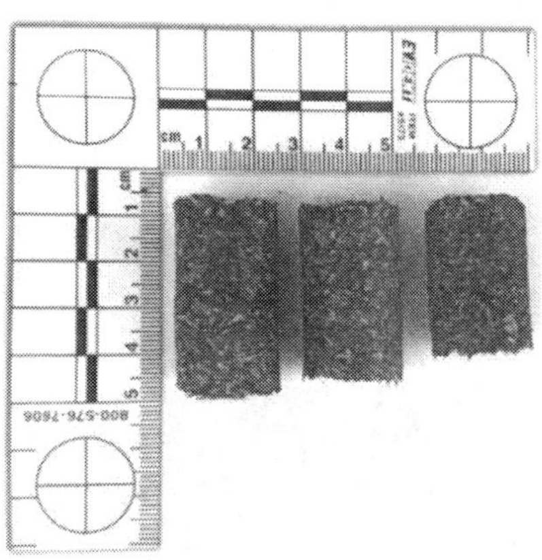

Figure 6. Briquettes produced at T = 23°C.

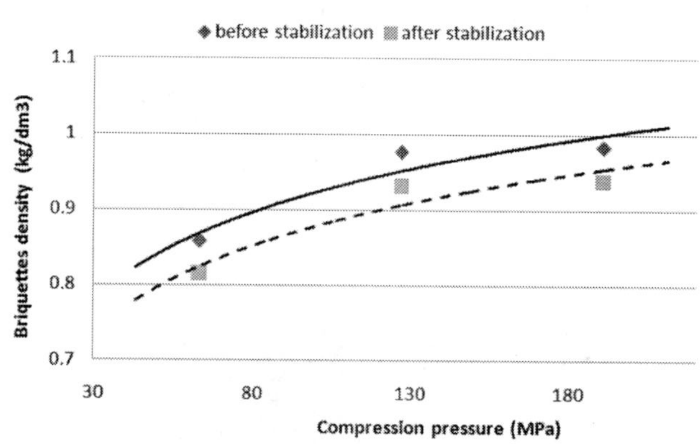

Figure 7. Briquettes density, related to compression pressure at T = 23°C.

These interdependences also highlights the importance of briquette stabilization time. This effect can significantly distort the results of the experiment. The dilatation effect is produced when the compression pressure is removed, and, as a result of the relationship of the interactions between the compression temperature, moisture content, and input particle size [9,18]. Two of these interdependencies can be seen in figure 7 and also in figure 9. One is valid for densities evaluated immediately after releasing the briquettes from the pressing chamber, *i.e.*, before stabilization, and the second for densities evaluated after the stabilization time. During the pressing without any additional temperature (see figure 7), the briquette density changed and dilatation negatively influenced the briquette density. Average density difference of 4.82 % was determined.

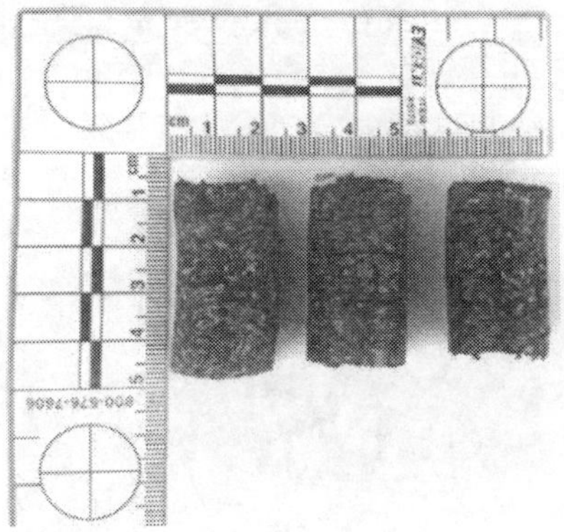

Figure 8. Briquettes produced at T = 100°C.

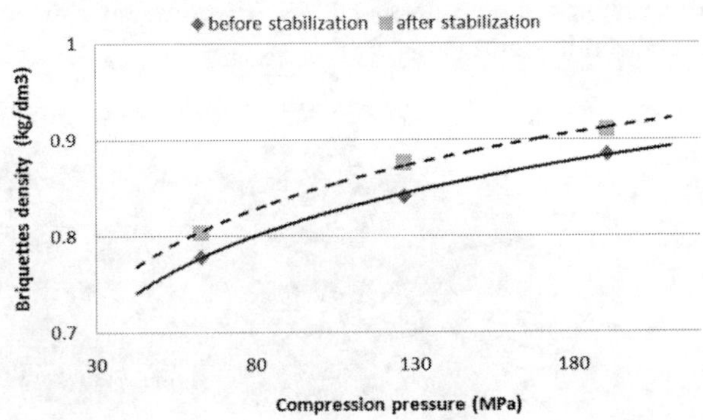

Figure 9. Briquettes density, related to compression pressure at T = 100°C.

Similar measurement was done also with additional compression temperature of 100 °C. The briquette density changed but dilatation positively influenced the briquette density. Average density difference of 3.39 % was determined. During densification, lignin is releasing from the biomass which acts as a natural glue. If the briquette isn't created in optimally adjusted conditions, dilatation of the briquette can also significantly influence the absorption of moisture from surroundings. In practice, the stabilization time can be decreased by cooling. In figure 10 the graphical dependence of differeing density before and after stabilization on the compression pressure and compression temperature can be observed. This figure clearly proves the impact of compression pressure and temperature on briquette dilatation and means that with increasing of compression pressure and also temperature, briquettes dilatation is decreasing.

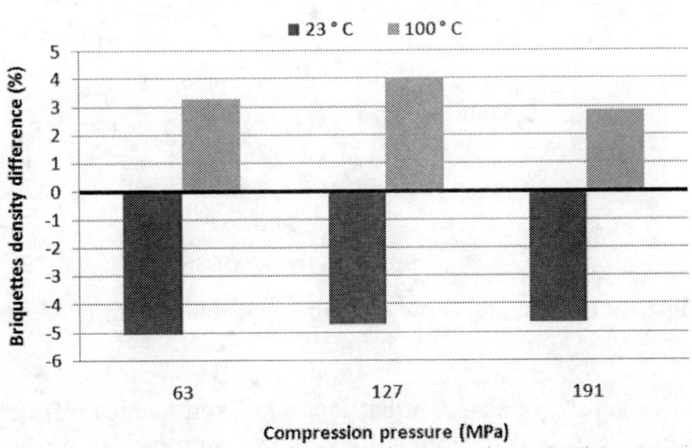

Figure 10. Graphical density difference dependent on compression pressure at both levels of compression temperatures.

The positive impact of compression temperature on the final briquette dilatation was proven. In general (in case of wooden) the compression temperature has a positive impact on the lignin component of the compressed feedstock [18, 19]. Lignin, cellulose and hemicelluloses content are

IOP Conf. Series: Materials Science and Engineering **297** (2018) 012003 doi:10.1088/1757-899X/297/1/012003

very important during densification and influences creation of binding mechanisms between material fibers and particles. The compression temperature influences lignin plasticization [19, 20]. With increased compression temperature during densification, the lignin becomes more and more liquid, plastic. Temperature significantly affects the binding mechanism forming between the solid particles of the compressed material [18, 20]. A more effective binding of the cellular structure of the compressed feedstock and the required plasticization of lignin is possible only under induced compression temperatures [18, 19]. Binding of raw feedstock particles in sunflower hulls briquettes can be seen in figure 11.

Figure 11. Binding of raw feedstock particles in sunflower hulls briquettes.

The results presented in figure 12 were obtained as a result of densification at different compression temperature levels (23 and 100 °C) and the displayed interdependencies are valid for case after briquettes stabilization. This figure shows the relationship between briquette densities the compression pressure at different levels of compression temperatures during densification of sunflower hulls.

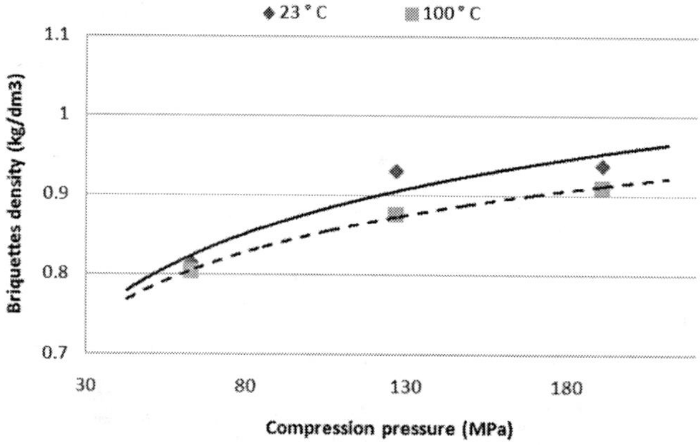

Figure 12. Binding of raw feedstock particles in sunflower hulls briquettes.

Results displayed in figure 12 are shown that in case of sunflower hulls densification were higher densities achieved during densification without additional compression temperature. This fact can be attributed to the nature and chemical composition of sunflower hulls. With low contents of lignin and hemicelluloses in the raw feedstock, the additional temperature during densification has no significant influence on the final briquettes density [5,15,19]. Average values of densities achieved during densification without additional compression temperature were higher about 3.57 % in comparing with densities achieved during densification at 100 °C of compression temperature.

3.2. Effect of particle size distribution on pellets quality

The main aim of following part of research was to determine the effect of particle size distribution during semi-operational tests with using of production pelleting plant. As was mentioned above two different particle size distribution were processed by a hammer mill. For pellets producing, untreated sunflower hulls (sample 1) and sunflower hulls treated by single level shredding (sample 2) were used. Pellets with ø 6 mm diameter from both samples representing different particle size distribution (see figure 5) were produced and can be seen in figures 13 and 14.

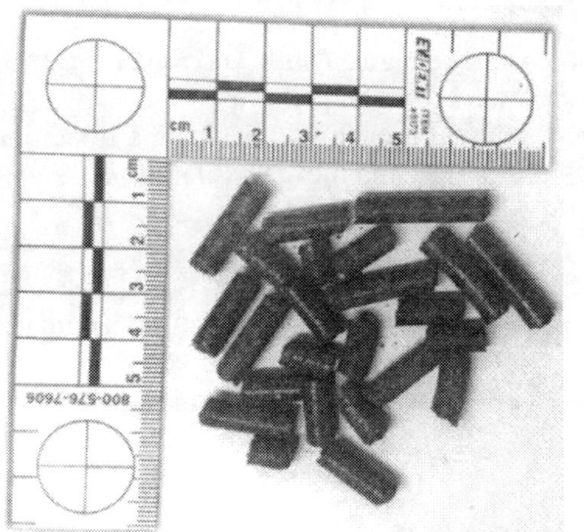

Figure 13. Sunflower hulls pellets - sample 1.

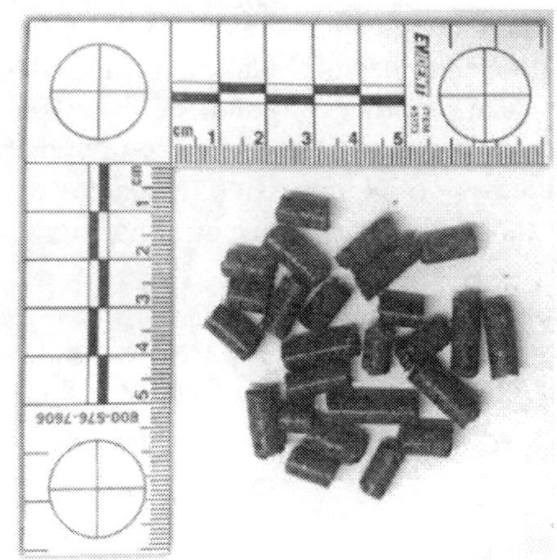

Figure 14. Sunflower hulls pellets - sample 2.

In general the particle size of the input sawdust has an important effect on the densification process because larger particles increase the energy needed for densification. Moreover, pellets densified from large particles have lower homogeneity and strength [11]. A large portion of fine particles allows for better raw material densification. In general (case of wooden), with increasing of particle size, the bond strength between particles decreases, causing them to crumble [18]. In terms of densification, it is very important that the binding forces are created between particles. The strength of the resultant bonds increases with decreasing of particle size [6,19]. It is known that the size of bindings between particles and the particle size distribution affects the physical properties of the pellets. When pelletizing, especially without additives, the surfaces of the raw material particles must contact with the greatest possible area [20]. The size of this contacting area increases as the particle size becomes smaller and higher compression pressures are applied.

Table 1. Properties of pellets from sunflower hulls.

Parameter	Pellets (obtained from company)	Pellets (sample 1)	Pellets (sample 2)
Moisture content (%)	8.61	5.60	5.99
Heating value ($MJ.kg^{-1}$)	-	19.07	19.06
Bulk density ($kg.dm^{-3}$)	0.422	0.611	0.552
Particle density ($kg.dm^{-3}$)	0.856	1.237	1.231
Mechanical durability (%)	81.63	87.08	79.26
Pellets abrasion (%)	32.79	12.63	20.51
Hardness (N)	197.5	190.3	207

Table 1 presents the physical properties of the produced pellets from both samples. It is seen that with decreasing of the particle size, the physical properties of pellets changed. The resulting pellets are uniform but of medium quality. What we can attribute to the properties of the raw feedstock - mainly to its chemical composition (low content of lignin, hemicelluloses, etc.). According to presented results we can see that in case of sunflower hulls densification the higher pellets quality was reached from the untreated sample. It means that positive influence of particle size reduction is not in this case significant.

4. Conclusion

The main conclusions that can be withdrawn from this study are as follows:

- With a suitable combination of the influencing variables can be significantly (positively) affected the densification process of sunflower hulls, mainly from the final biofuels quality point of view.
- Increasing of compression pressure positively influences the biofuels final density and dilatation.
- Increasing of compression temperature positively influences only the biofuels dilatation.
- Reducing of particle size does not significantly affect the final pellets quality.

Acknowledgment

The paper is a part of the research done within the project VEGA 1/0420/16 – "Determination and research of parameters influence on final briquettes quality at waste biomass densification". The authors would like to thank to the Ministry of Education of Slovak Republic and to the Slovak Academy of Sciences.

References

[1] Fernandes U and Costa M 2012 *Fuel Process. Technol.* **103** pp 51-56
[2] Garcia Maraver A, Zamorano M, Fernandes U, Rabacal M and Costa M 2014 *Fuel* **119** pp 141-152
[3] Lisý M, Baláš M, Špiláček M and Skála Z 2015 *A.Polytechnica* **55** pp 401-406
[4] Lisý M, Baláš M, Špiláček M and Skála Z 2014 *A.Polytechnica* **54** pp 42-51
[5] Križan P 2015 *The densification process of wood waste* (Berlin De Gruyter open)
[6] Križan P, Šooš Ľ, Matúš M, Beniak J and Svátek M 2015 *Wood research* **60**(2) pp 301-316
[7] EN ISO 17225-22014 (European Committee for Standardization Brussels Belgium)
[8] EN ISO 17225-62014 (European Committee for StandardizationBrussels Belgium)
[9] DIN 52182 1976 (Deutsches Institut für Normung Berlin Germany)
[10] Križan P, Matúš M, Beniak J and Kováčová M 2014 *A. Polytechnica* **54** pp 135-41
[11] Kaliyan N and Morey Vance R 2009 *Biomass and Bioenergy J.* **33**(3) pp 337-359
[12] Nielsen NPK, Gardner DJ and Poulsen T 2009 *Wood and Fiber Science* **41** 4 pp 414-415
[13] Y Li and H Liu 2000 *Biomass and Bioenergy J.* **19** pp 177-186
[14] EN ISO 17829 2016 (European Committee for Standardization Brussels Belgium)
[15] Križan P, Matuš M, Šooš Ľ and Beniak J 2015 *J. Energies* **8**(7) pp 6382-6398
[16] EN ISO 17828 2016 (European Committee for Standardization Brussels Belgium)
[17] EN ISO 18134-12016 (European Committee for Standardization Brussels Belgium)
[18] Serrano C, Monedero E, Lapuerta M and Portero H 2011 *Fuel Processing & Technology* **92** pp 699-706
[19] Mani S, Tabil L G and Sokhansanj S 2006 *Biomass Bioenergy J.* **30** pp 648–654
[20] Thoreson C P, WebsterK E, DarrM J and Kapler E J 2014 *J. Energies* **7**(6) pp 4019-4032

IOP Conf. Series: Materials Science and Engineering **297** (2018) 012004 doi:10.1088/1757-899X/297/1/012004

The Turbo-Fuel-Cell 1.0 – family concept

Compact Micro Gas Turbine (MGT) – Solid Oxide Fuel Cell (SOFC) energy converters in the 100-500 kW electrical power range for the future.

H. P. Berg[1,*], A. Himmelberg[1], M. Lehmann[1], R. Dückershoff[2] and M. Neumann[3]

[1] Brandenburg University of Technology (BTU), Department of Combustion Engines and Flight Propulsion, Siemens-Halske-Ring 14, 03046 Cottbus, Germany,
[2] University of Applied Sciences Mittelhessen (THM), Friedberg, Department of Mechanical Engineering, Mechatronics and Material Technology, Wilhelm-Leuschner-Str. 13, 61169 Friedberg, Germany
[3] Professor Dr. Berg and Kiessling GmbH, Burger Chaussee 25, 03044 Cottbus, Germany

* Corresponding Author: peter.berg@b-tu.de

Abstract. The "Turbo-Fuel-Cell-Technology" has been described as a MGT-SOFC hybrid system consisting of a recuperated micro gas turbine (MGT) process with an embedded solid oxide fuel cell (SOFC) subsystem. SOFC stacks are connected to "SOFC stack grapes" and are equipped with the so called HEXAR-Module. This module is composed of a high-temperature heat exchanger (HEX), an afterburner (A) and a steam reformer (R). The MGT-concept is based on a generator driven directly by the turbomachine and a recuperator, which returns the exhaust heat to the pressurized compressor outlet air. This provides the necessary base for a highly effective, pure MGT process and the "MGT-SOFC-high-efficiency process". This paper describes the concept and the thermodynamic background of a highly effective and compact design of the "Turbo-Fuel-Cell 1.0-Family" in the electrical performance class from 100 to 500 kW. The technological state of the system is shown and a rating of the system with comparative parameters is discussed. It becomes visible that all necessary basic technologies should be available and that the technology (for stationary applications) can have the "entry into services (E.I.S.)" in the next 10 years. The MGT-SOFC performance map under different operation conditions is discussed. This article also provides an overview of the research on MGT-SOFC-Systems and the scenario of an energy supply network and a mobile energy conversion of the future introduction.

1. Introduction

In the redesign of our future habitats, research is looking for novel solutions for mobility and energy supply. These must have high fuel flexibility in order to be able to be used in a wide variety of ways and to enable an infrastructural change (for example, the transition from conventional fuels to fuels for energy storage from renewable sources).

In the present article, the "Turbo-Fuel-Cell 1.0 (TFC1.0)-Family" concept is suggested and discussed physically. The basis for compactness results from a multilayer pressure vessel (DE patent application number: 10 2017 107 003.6 / MLC = multi-layer container) with integrated hybrid micro-gas turbine SOFC process. The "TFC1.0-Family" System is designed for the 100 to 500 kW electrical energy range with a very high overall efficiency $\eta_{MGT-SOFC}^{real}$. Furthermore, the system can also be scaled to smaller capacities.

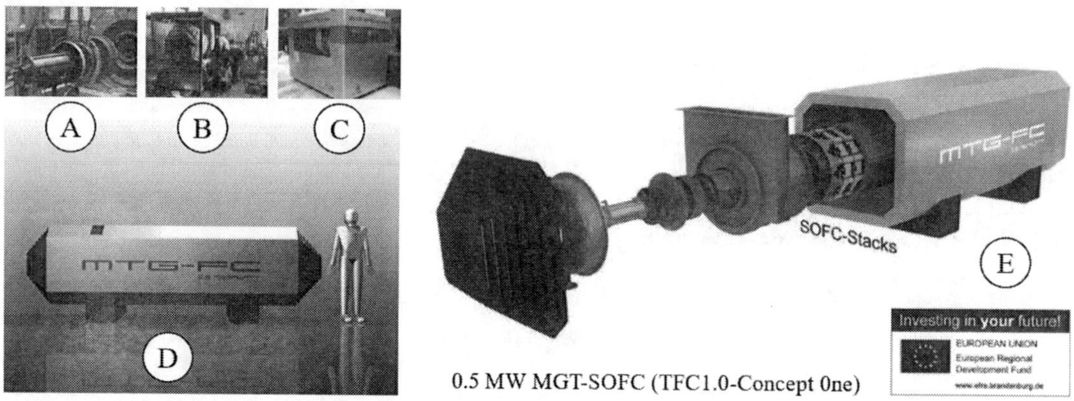

Figure 1. View of the first stationary TFC1.0-Concept (from 2014) based on the MGT-100 technology developed by BTU-CS. Motivation for the MAXEFF-research programme sponsored by the European Union (European Regional Development Fund)

In the years 2010-2014, research and development projects at the BTU-CS (Brandenburg University of Technology Cottbus-Senftenberg) were carried out in cooperation with industry partners for an air-bearing micro gas turbine (figure 1, A and B) with a high efficiency potential. The result was a 100 kW CHP with high fuel flexibility (figure 1, C). This basic machine was chosen for the concept of a first highly compact micro gas turbine SOFC system (figure 1, D and E, MTG-FC 500 from the year 2014). As the "TFC1.0 Concept One", it was the source of the "TFC1.0-Family (100 to 500 kW)" based on MLC technology. In addition, smaller performance classes were analysed, which also use the MLC basic principle and turbomachinery with air bearings.

Own studies also show a lower application range down to 20 KW (automotive application / TFC-M1), while studies of Siemens-Westinghouse on pilot plants clearly demonstrate the application areas over 150 kW [2, 3]. The first "TFC1.0-270" system (example in this publication) will be based on an air-bearing micro gas turbine (MGT), a technology from B+K (Prof. Dr. Berg and Kiessling GmbH, Germany) which is currently in the industrialization process.

This machine concept (figure 2) has the necessary technology (air bearing, recuperator, turbomachine and generator). Furthermore, there are already industrialized SOFC stacks on the market.

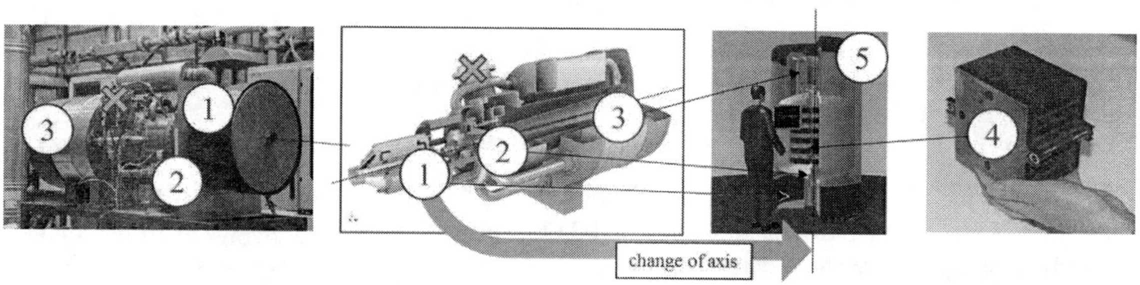

Figure 2. Overview about the combination of MGT and SOFC and an illustration of the change in the rotational axis of the MGT

For the "TFC1.0-270 (250-300 kW)" this MGT basic technology (figure 2: left side /without combustion chamber, yellow cross) is to be used. The components generator (1), turbo set (2) and recuperator (3) are arranged vertically (figure 2: "change in the rotational axis") in combination with the MLC (5). Conventional SOFC stacks (4) are to be integrated as stack grape in the MLC.

The "Turbo-Fuel-Cell 1.0" will be realized with an embedded SOFC system (instead of the combustion chamber). As SOFC technology, the planar technology is used. This has the advantage of a relatively high power density [4], which contributes to the compactness of the system. However, the sealing in the fuel cell stub must be ensured by a careful selection of materials and by means of sensible preloading. Another advantage is the high degree of industrialization of the planar SOFC technology. The resulting high availability at a comparatively low price level represents advantageous boundary conditions for the realization of the "TFC1.0-Family Series".

Therefore, planar SOFC technologies are currently under investigation in a BTU-CS research program (EU-funded, acronym "MAXEFF"). In this case, an examination of the pressure charge (in the MGT pressure range) is carried out, in particular taking into account the structural integration and design concept. In addition to the planar SOFC technology, tubular SOFC technologies have been used in connection with a hybridized gas turbine process [2, 5, 6].

The monolith concept and the Sulzer-Hexis technology (combined tube stack module) are also to be mentioned. The monolith concept (MSOFC) promises the highest power density (8 kW/kg, without charging) and could therefore be of great interest to the mobile sector [6; 4]. However, the monolith concept with the above-mentioned high conduction densities does not yet exist in industrialization.

2. Examples of some future applications

For the Turbo-Fuel-Cell (TFC) technology, there will be a large number of applications in the stationary and mobile sectors. In both cases a TFC1.0-270 can be the right system for the entry into market (E.I.S.) in 2027.

2.1. *Use as immobile application*

As a decentralized energy supplier, a TFC system can be used in the highly effective conversion of chemically stored energy into electricity (gas to power) in the future network with renewable energies (figure.1 (D) and figure 2 (5) a sketch of a stationary application).

A practical advantage is the relatively fast availability for the "Smart Grids", since one does not have to wait for a new energy carrier (e.g. hydrogen) including infrastructure, but can also use fuels of the same market. TFC-converters could be used further in the modification of the infrastructure (e.g. hydrogen economy). This hybrid system can provide electrical energy faster than a single MGT or SOFC system. Müller [7] has made an investigation on a Capstone C60 MGT with SOFC and found a response of 100 kW/s.

2.2. *Use as mobile application*

As an energy converter for the mobile sector, an application in the rail, automotive and marine sector can be conceived as shown in figure 3 and 4. The TFC for rail applications (figure 3) is very similar to the stationary situation (figure 1 and 2). At present, rail vehicles with diesel-electric drive systems are mainly used on non-electrified track connection. A TFC system can eliminate the NO_X and particulate problems with low CO_2 emissions. A TFC system represents a zero emission system. In image of figure 3 this application area is exemplarily visualized.

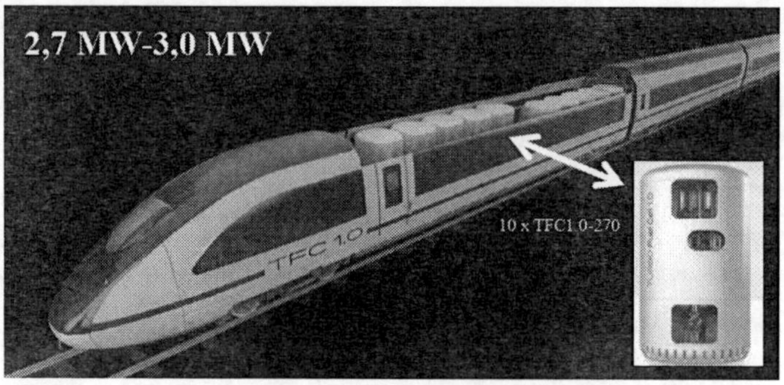

Figure 3. Example of a train application

Ten TFC1.0-270 packs can provide 2.7 to 3 MW emission-free electrical power. The fuel could initially be diesel or natural gas with the option of a subsequent conversion to other chemical energy carriers, e.g. "Power to gas" - fuel / hydrogen H_2

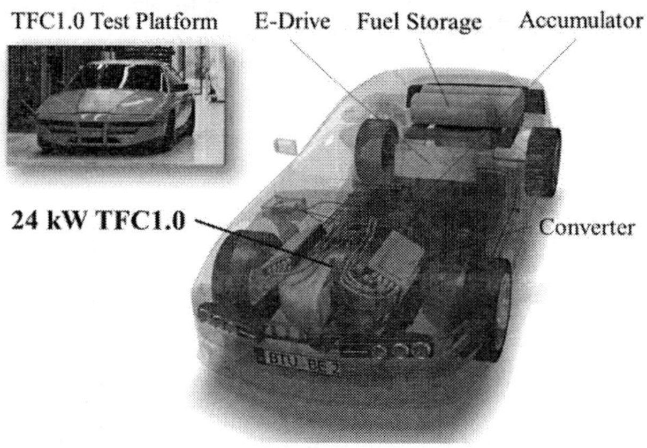

Figure 4. Example of an automotive application with a TFC1.0-24M (24 kW)

A further future use is the automotive sector (figure 4, e.g. range extender for the E-mobility) and also in the field of aviation technology (APU systems, figure 5). Here, the monolith concept (MSOFC) technology may be of interest instead of the state of the art SOFC technology. A power weight ratio above 8 kW/kg can be achieved by turbocharging and the use of MLC technology. In combination with buffer batteries, a powerful actuation without a range problem can be represented. With the use of e.g. carbon-fibre-reinforced polymers or other fibres, the MLC technology allows a very compact and light design for the pressure containing outer skin. The image in figure 4 on the upper left shows the driving test bed of BTU-CS for automotive applications.

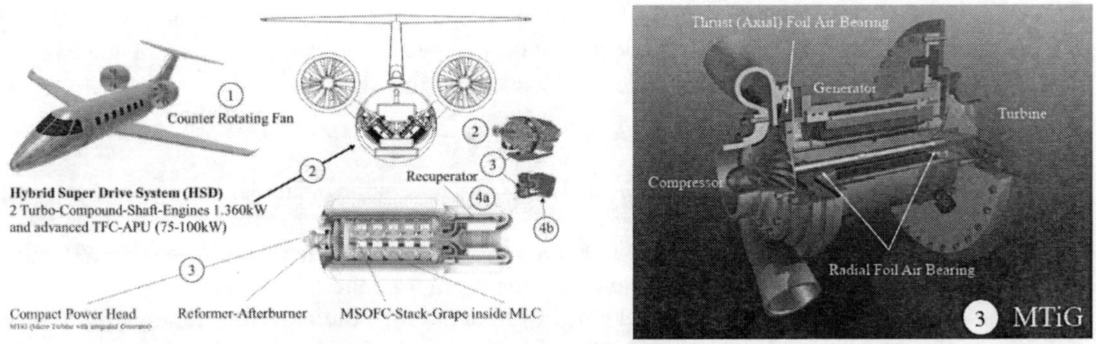

Figure 5. Example of a TFC-APU aerospace application (left) with a highly compact turbo generator (3) and main details of the "MTiG" (right)

Figure 5 shows an example of an aircraft concept (MTOW 9,300 kg, $Ma_{Cruise} = 0.75$, $h_{Cruise} = 11$ km) with a hybrid main drive [8, 9]. With an advanced fan technology (CRF = Counter Rotating Fan), two hybrid turbo compound engines are realizing the take-off power of 1,360 kW. The parallel-hybrid shaft-power system represents a combination between the turbo compound and an electric drive. The power coupling between the rotary piston engine (core engine) and the air-bearing power head (MTiG = micro turbine with integrated generator) are electrically coupled. The MTiG represents a derivative from the existing TFC powerhead. Furthermore, the drive system uses an MTiG based APU, which uses an advanced recuperator (oval tube bundle heat exchanger [7]. This APU could be replaced by a TFC system. In this case, the MTiG, the oval tube recuperator and the MSOFC would be used in conjunction with the aforementioned main engine. The result would have been an aviation-technical "TFC-AERO-System". With it, the desired overall high power weights of the entire drive system of about 2-3 kW/kg can be realized in the future.

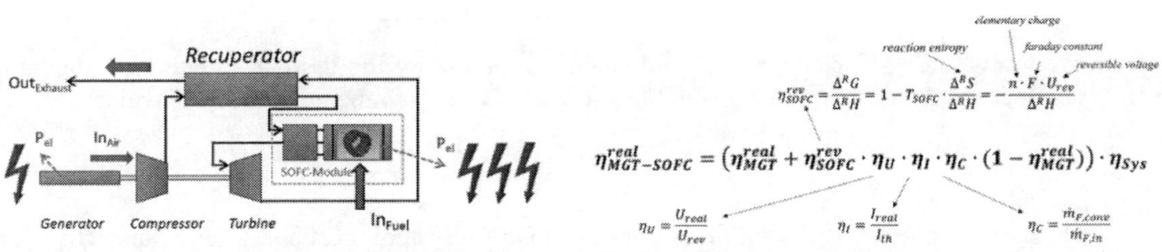

Figure 6. Sketch of the MGT-SOFC process (left) an the fundamental equations (right)

3. Thermal cycle of the MGT-SOFC process and the TFC-design concept

The cycle of the TFC (MGT-SOFC-process) is sketched schematically in figure 6. The basic principle is sufficiently known from the literature and has already been described in numerous publications, e.g. [6, 8, 9]. For this reason, a brief description in connection with the MLC (multi-layer container) is given below and the design of the TFC system is presented.

3.1. *Basics of the MGT-SOFC process*

According to [1] the efficiency chain of the MGT-SOFC process is shown in figure. 6. In this case, the efficiency η_{MGT}^{real} of the MGT base (depreciated with the heat and pressure losses of the MGT-SOFC system) is approximated. η_{SOFC}^{rev} represents the efficiency of the ideal fuel cell.

$$\eta_{SOFC}^{rev} = -\frac{n \cdot F \cdot U_{rev}}{\Delta^R H} = -\frac{n \cdot e \cdot N_A \cdot U_{rev}}{\Delta^R H} \tag{1}$$

(n = charge number, F = Faraday constant, U_{rev} = reversible voltage, e = elementary charge, N_A = Avogadro constant)

The Gibbs enthalpy $\Delta^R G$ is the enthalpy of formation $\Delta^R H$ minus the reaction entropy $\Delta^R S$ multiplied with the operating temperature T_{SOFC} (idealized assumption = isotherm) of the fuel cell.
η_{SOFC}^{rev} is further reduced by the electrochemical efficiency η_U, the current efficiency η_i and the fuel conversion Efficiency η_C. The electrical work

$$\Delta W_{electrical\_SOFC} = \Delta^R H_{High\ Syn\ Gas} \cdot \eta_{SOFC}^{rev} \cdot \eta_U \cdot \eta_I \cdot \eta_C \tag{2}$$

is delivered by the SOFC (figure.7). The "High Syn Gas" is supplied to the SOFC from the reformer. It is generated by a steam reforming process (in the example CH_4 + H_2O steam) using the heat energy $\Delta Q_{Reformer}$. In addition to the waste heat from the SOFC ($\Delta^R Q + \Delta^I Q$), further heat $\Delta^F Q$ is produced by the oxidation of the "Low Syn Gas" in the afterburner.
In the MGT, the heat energy

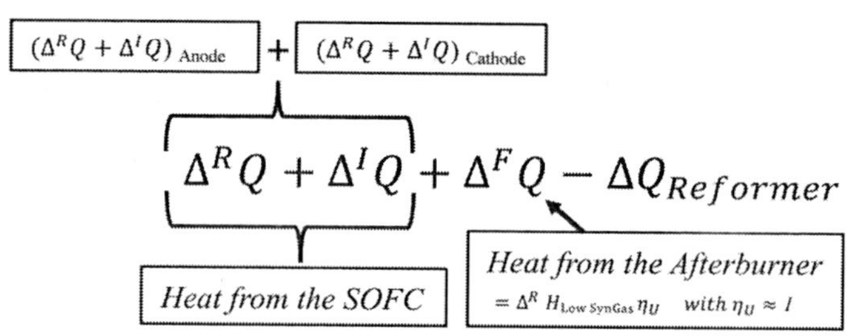

is converted with the efficiency η_{MGT}^{real} into electrical energy by the thermal circuit with the generator (η_{MGT}^{real} includes in the definition of this publication all losses up to the generator). The electrical work

$$\Delta W_{electrical\_Generator\_MGT} = \eta_{MGT}^{real} \left(\Delta^R Q + \Delta^I Q + \Delta^F Q - \Delta Q_{Reformer}\right) \tag{3}$$

is delivered by the MGT-Generator.

The Energy delivered by the whole TFC (MGT-SOFC)-System must be further reduced by the losses of the electrical converters, controllers and regulating units with the system efficiency η_{Sys}. The electrical work

$$\Delta W_{Turbo\ Fuel\ Cell} = \eta_{Sys} \cdot \left(\Delta W_{electrical_{SOFC}} + \Delta W_{electrical\_Generator\_MGT}\right) \tag{4}$$

is delivered by the overall TFC (MGT-SOFC) -System.

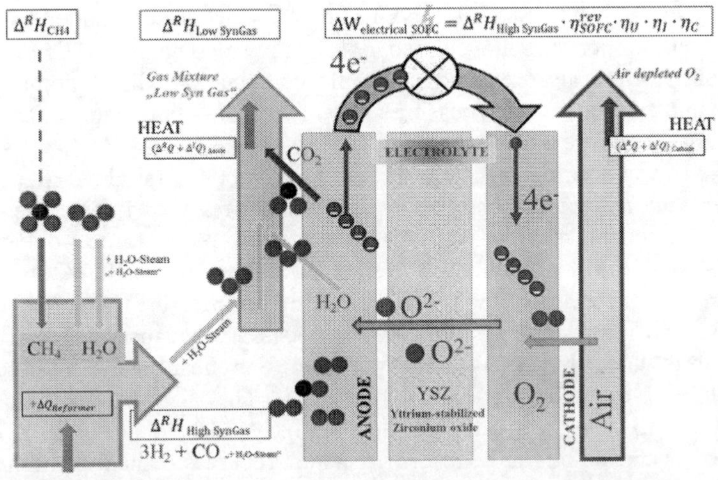

Figure 7. Sketch of the energy situation in a SOFC-Cell

The TFC1.0-270 will operate in its first application with natural gas. Therefore in this article the results are discussed with the fuel CH_4 with the formation enthalpy $\Delta^R H_{CH_4}$. The combination of the SOFC with the MGT results in a revaluation of the efficiency, because the anergy of the SOFC is available as exergy for the thermal MGT-cycle. The overall efficiency of the TFC can therefore be described in the first approximation by $\eta^{real}_{MGT-SOFC}$. With the electrical work of the MGT-SOFC System:

$$\Delta W_{Turbo\ Fuel\ Cell\_CH4} \approx \Delta^R H_{CH4}\ \eta^{real}_{MGT-SOFC} = \Delta^R H_{CH4} \left(\eta^{real}_{MGT} + \eta^{real}_{SOFC} \cdot (1 - \eta^{real}_{MGT})\right) \cdot \eta_{Sys} \quad (5)$$

For the simplest explanation, the MGT-SOFC can be described by the process cycle shown in figure 6. E.g. the TFC1.0-270 is based on the MGT basic machine concept shown in figure. 2 with air bearing without combustion chamber. Simplified the combustion chamber is replaced by the MLC-HEXAR-SOFC system which is described below.

3.2. *The Multi-Layer Container (MLC) with an integrated SOFC-Stack-Grape*

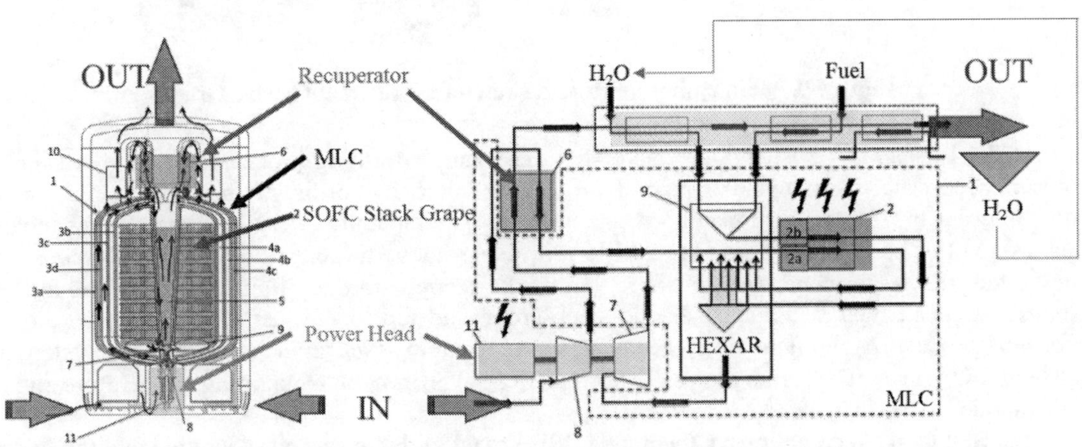

Figure 8. Sketch of the MLC = Multi-Layer Container
(patent application number: 10 2017 107 003.6) with integrated SOFC-Stack-Grape (left) and description of the functionality on a simplified TFC(MGT-SOFC)-cycle (right)

The FC module is typically composed of SOFC stacks (SOFC = Solid Oxide Fuel Cell). In the TFC, they are located in a stack cluster (Stack-Grape, figure 9), which is surrounded by the MLC.
However, another high-temperature fuel cell technology may also be used. The FC module (SOFC post-combustion reactor-reformer heat exchanger) replaces "the so-called" combustion chamber of a recuperated MGT and is installed in a pressure vessel (pressure dependent on the pressure ratio of the compressor of the MGT / figure 8, pressure vessel (-1-)).

The air is sucked in and compressed by the compressor (-8-) (slightly higher than the operating pressure of the high-temperature fuel cell due to component pressure losses). The compressed air flows through the multi-skinned pressure vessel (-1-) through the recuperator (-6-) via the multi-skinned pressure vessel (-1-) with the channels 4a, 4b. Before the entry into the SOFC stack grape, another heat exchange takes place (in the range of the SOFC operating temperature). The heat exchange takes place on the duct wall 3b and in the pipe system. This is necessary because, in most operating points, the exhaust gas heat behind the turbine (-7-) is not sufficient to preheat the air in the recuperator (6) (to the operating temperature of the high-temperature fuel cells). This air flows to the cathodes (-2a-) of the high-temperature fuel cells. On the anode side, the "High Syn Gas" occurs (-2b-). The process of "High Syn Gas" generation is carried out in the reformer, which receives its heat from the post-combustion of the anode ("Low Syn Gas") and cathode exhaust gas mixture.

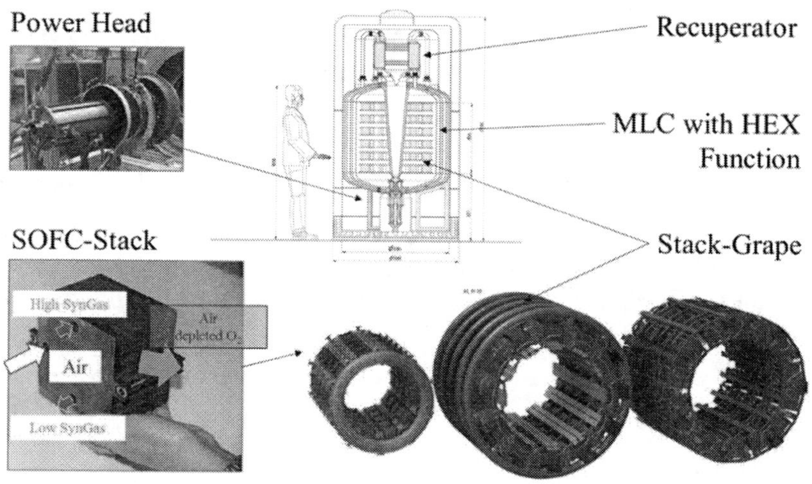

Figure 9. Main components and sketch of 3 different Stack-Grapes

In "TFC1.0-Family" systems, the water for the steam reforming process (with the concept of a superimposed water-steam cycle) is passed through the outer shell of the multi-skinned pressure vessel (-4c-) in order to "trap residual heat loss" and to keep the temperature very low (e.g. allows the use of plastics). In this case, the water is brought to a pressure above the operating pressure, preheated and evaporated in a steam generator (in the region of the recuperator, see figure 10) with the aid of the exhaust gas heat, then superheated in the stack-grape and fed to the methane stream for the steam reforming process. At the same time, the steam serves as an inert gas in the sheathed tube system of the high and low syngas. The "High Syngas" reformed in (-9-) is brought to operating temperature and flows to the anode.

In the SOFC, the oxygen ions at the anode (-2b-) react with the carbon monoxide and the hydrogen of the fuel gas (syngas from the reforming) to water and carbon dioxide (figure 7). The oxygen is fed to the SOFC stack on the cathode side (-2a-) by the heated air. There, each oxygen molecule O_2 is divided into $2O^{2-}$-ions with the aid of four electrons $4e^-$. These diffuse through the YSZ-electrolyte to the anode to oxidize the "High Syn Gas". Between the anode and the cathode, an electron potential difference result. The resulting direct current is converted in an inverter (not shown here) to the desired demand. There is no complete conversion in the SOFC stack. In order to use the remaining portion of the combustible components in the anode exhaust gas, this is fed to a post-oxidation (post-combustion). The residual oxygen in the cathode exhaust gas serves as an oxidation partner. The resulting hot exhaust gases flows from the Afterburner-Reformer module (AR see figure 10) to the turbine. Here the expansion takes place. The turbine (7) drives the generator (11) and the compressor (8) via a common shaft. The turbine exhaust gas delivers heat energy to the inner vessel wall and in the recuperator (6) to the compressed air. The remaining heat energy is used in heat exchangers for preheating the steam, the fuel gas and can be used for heating purposes.

The MLC (1) forms through its concentric channels the main supply lines of the circular process. It also provides a compact containment for the process. In the case of steam reforming with a superimposed water-steam circuit, the concentric channel located outside serves for water pre-heating (4c) by means of residual heat losses.

The MLC design allows a very compact design of an MGT-SOFC small-scale power station by means of channels (4) layered around a central axis (including the heat-insulating layers, 3). In the MLC design, the channels are arranged in such a way that waste heat is always transported back through the fluid flow direction.

The pressure vessel is formed in the TFC1.0 family by the outer layer which is at a lower advantageous temperature level (approximate ambient temperature). With TFC1.0, the exhaust gas removal takes place centrally and helps in recuperation of heat. The inner pressure receiving passage is therefore metallic. It also serves as a diffuser between the turbine outlet and the recuperator due to its conical shape.

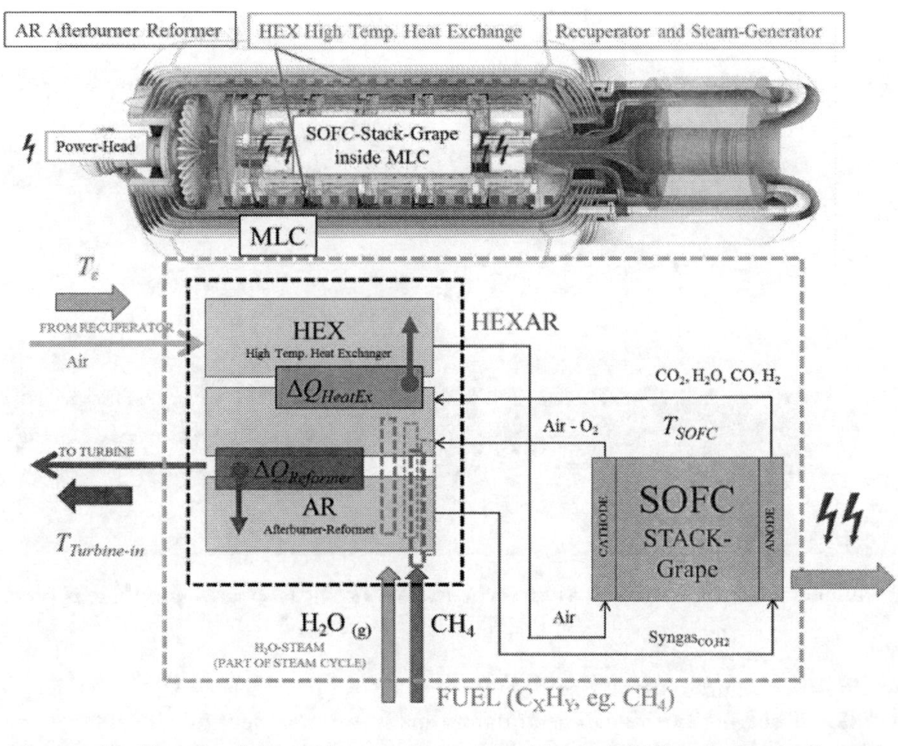

Figure 10. Sketch of the main components inside the MLC and circuit diagram

The MLC concept is based on annular layers (insulation, channel wall, channel, channel wall, insulation, channel wall, channel, channel wall, etc.). Only the cool channel wall near the surroundings must absorb the pressure in the outer area. Thus, the pressure vessel can be made more thin-walled and made of a more cost-effective material. This also affects the heat capacity favourably. Due to the radial integration of the process-related guide channels, these can be designed very thin-walled since no large pressure differences act. Furthermore, these may also have smaller leakages in the case of the process air guidance, since only relatively small mass flow components (advantageously: in the direction of the hotter inner ring / in the "SOFC" direction) flow due to the low pressure difference. Due to the drastic reduction of the heat capacity of the system, the time behaviour (heating process, load change, etc.) is favourably influenced.

4. Turbo technology of the TFC1.0-system

In the TFC process the MGT is important as a heat engine and for the air supply of the fuel cell. In addition, the system becomes more powerful due to the pressure boosting. In TFC-power units up to 500 kW the combination of small size single stage radial turbine and compressor pair are well suite option. At a speed range from 30,000-130,000 rpm, the size and weight of the rotor components are lying in the range of 60 N at 90,000 rpm. The temperature of the rotor components is roughly between the compressor outlet temperature and the turbine inlet temperature. Due to the fact that the ion exchange in SOFC process is directly related to the surface area, oil contamination in the process medium will result in (unburn) particle deposition on the Cathode surface in the stacks which leads to efficiency degradation [2]. Therefore, the MGT for a hybrid system has to be completely oil-free to ensure the long term efficiency. Together with the high-speed, high-temperature and oil-free process gas requirement, the foil bearing is one of the promising solutions for the power unit in MGT-SOFC system.

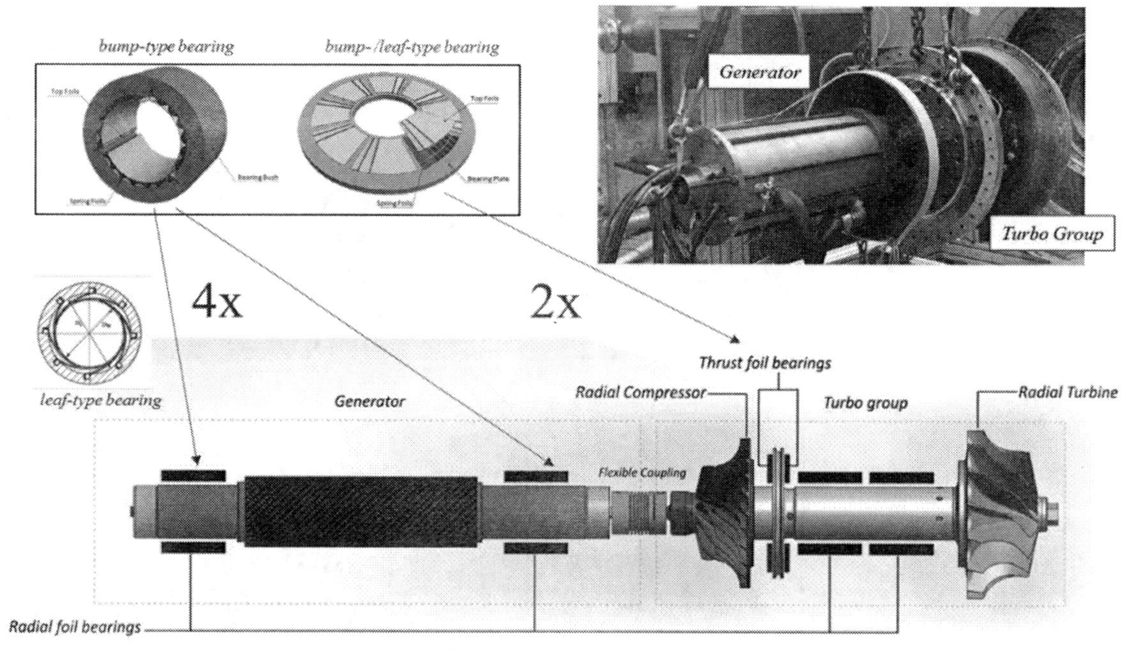

Figure 11. General configuration of bump type radial and thrust bearing and a leaf type bearing in a typical TFC1.0 power head

Foil bearings are hydrodynamic, self-acting gas bearings which generally consist of a top foil, compliant foil and bearing bushing. The foils are made from a sheet of nickel base super-alloy. The assembly of these components forms a hydrodynamic wedge contour to the rotating shaft. The hydrodynamic lubrication film pressure is formed between the opposing surfaces due to the high relative speeds [3].

The top foil is supported by a series of compliant structures which has a spring like function usually made of corrugated foil. The compliant foil acts as an elastic foundation that permits the top foil to locally deflects in response to changes in the hydrodynamic pressure and accommodates misalignment, shock load, thermal distortion, centrifugal growth of the shaft and engine housing. These effects limit the use of rigid air bearings and represent the most important advantage of compliant foil bearings. The small sliding motions between contact surfaces of top foil, compliant foil and bushing caused by shaft motion during the operation generates Coulomb damping. Additionally viscous damping mechanisms can be realized by air chambers between the gas foil bearing Components. These viscous and Coulomb damping effects increase the dynamic stability of the rotor bearing system. Further research in development of innovative coating technologies reduced friction and increased the thermal resistance above 300°C in operation with a number of start stop cycles up to 40,000 [3, 4, 5, 6].

5. Characteristics and performance maps of the Turbo-Fuel-Cell1.0 (example: TFC1.0-270)

SOFC fuel cells require an operating temperature T_{SOFC} of 700°C to 900°C (TFC1.0-270, T_{SOFC} = 740°C to 880°C) which is generally above the outlet temperature of the recuperator (TFC1.0-270, $T_{\varepsilon_{Design\_Point}} = 580°C$) and below the turbine inlet temperature (TFC1.0-270, $T_{Turbine_{in\_Design\_Point}} = 888°C$).

To close the cycle in the heat supply of the recuperated micro turbine process the heat input

$$\Delta Q_{Input} = \Delta Q_{HeatEx} + \Delta Q_{SOFC-Turbine} = \Delta^R Q + \Delta^I Q + \Delta^F Q - \Delta Q_{Reformer} \tag{6}$$

is required (see Chapter 3).

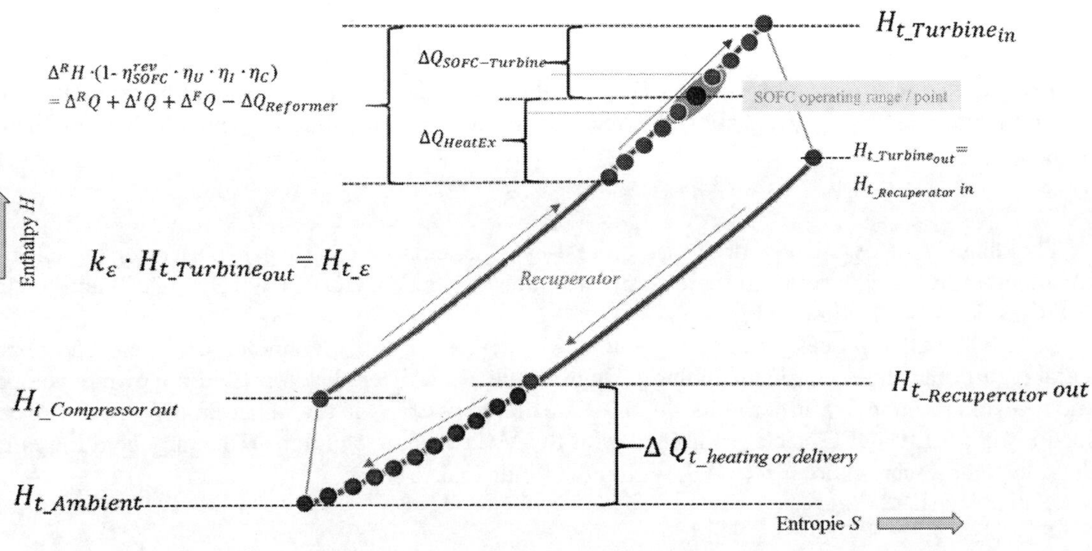

Figure 12. Illustrated thermodynamic cycle of the MGT-SOFC process

Only the heat quantity ΔQ_{Input} of the SOFC can be used for the thermal cycle of the MGT. It is determined using the SOFC efficiency chain ($\eta_{SOFC}^{rev} \cdot \eta_U \cdot \eta_I \cdot \eta_C$). The heat quantity is the not in electricity converted chemical energy and can be described by using the formation enthalpy $\Delta^R H$ of the fuel.

$$\Delta^R H \cdot (1 - \eta_{SOFC}^{rev} \cdot \eta_U \cdot \eta_I \cdot \eta_C) = \Delta Q_{HeatEx} + \Delta Q_{SOFC-Turbine} = \Delta Q_{Input} \tag{7}$$

In this equation ΔQ_{HeatEx} is the amount of heat required to increase the recuperator outlet temperature T_ε to the operating temperature of the fuel cell T_{SOFC} and $\Delta Q_{SOFC-Turbine}$ is the amount of heat required to increase the operating temperature of the fuel cell T_{SOFC} to the turbine inlet temperature $T_{Turbine\_in}$. In the TFC1.0 the operating temperature of SOFC T_{SOFC} does not match the exit temperature of the recuperator T_{ε}, so that the needed amount of heat ΔQ_{HeatEx} has to be exchanged by a the high-temperature heat exchanger HEX which is part of inner layer of the MLC (see figure 10).

In order to design the thermal arrangement between the MLC-HEX-function, the SOFC-Grape-channels and the afterburner-reformer (AR, see figure 10) the size X was introduced in [1] as the ratio between the exchanged heat quantity Q_{HeatEx} and the required amount of heat ΔQ_{Input} to close the MGT-cycle.

$$X = \left(\frac{\Delta Q_{HeatEx}}{\Delta Q_{Input}}\right) \tag{8}$$

In [1] it is shown that X can be approximated.

$$X \cong \left(\frac{T_{SOFC}-T_{\varepsilon}}{T_{Turbine_{in}}-T_{\varepsilon}}\right)_p = \left(\frac{T_{SOFC}-k_{\varepsilon}\cdot T_{Turbine_{out}}}{T_{Turbine_{in}}-k_{\varepsilon}\cdot T_{Turbine_{out}}}\right)_p = \frac{\theta-k_{\varepsilon}\pi^{\phi}}{1-k_{\varepsilon}\pi^{\phi}} \tag{9}$$

The quality of the recuperator is represented by the factor $k_{\varepsilon} < 1$. $\pi_{Turbine}$ is the pressure ratio of the turbine and θ is the ratio of the SOFC operating temperature T_{SOFC} and the turbine inlet temperature $T_{Turbine_{in}}$

$$\theta = \frac{T_{SOFC}}{T_{Turbine_{in}}} \tag{10}$$

For ϕ the polytrophic efficiency of the turbine η_T and the adiabatic exponent γ of the exhaust gas is used

$$\phi = \frac{1-n}{n} = \frac{1-\gamma}{\gamma}\,\eta_T \tag{11}$$

The targeted operating temperatures T_{SOFC} of the SOFC-Stack-Grape is generally below the turbine inlet temperature $T_{Turbine_{in}}$ and above the recuperator outlet temperature T_{ε}. The X-values are therefore in the range:

$$0 \leq X \leq 1 \tag{12}$$

The diagrams shown below are based on real-cycle calculations. CH_4 gas is used for fuel. The CH_4 is converted by a steam reforming process in a synthesis gas (H_2, CO / "High Syn Gas") and fed to the SOFC as described in figure 10.

For MGT SOFC process is the design point of interest, which is technologically feasible. Herein a suitable turbomachinery basic technology (figure 1 and 2) has been developed during own development and research programs. Furthermore, the SOFC technology exists in an industrialized version. It can be seen in figure 13 (SOFC stacks combined with the MGT basic technology), that an operating range at the selected pressure ratio of $\pi = 4.5$, within the limits makes sense.

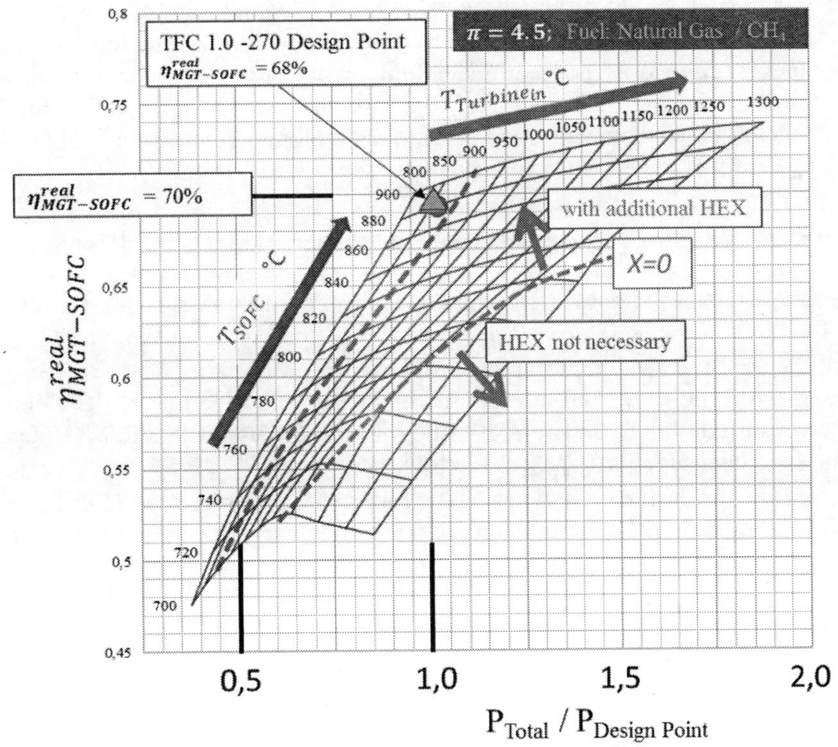

Figure 13. Performance maps of a calculated TFC1.0-System at $\pi = 4.5$ (e.g. TFC1.0 $P_{Design\ Point} = 270\ kW$

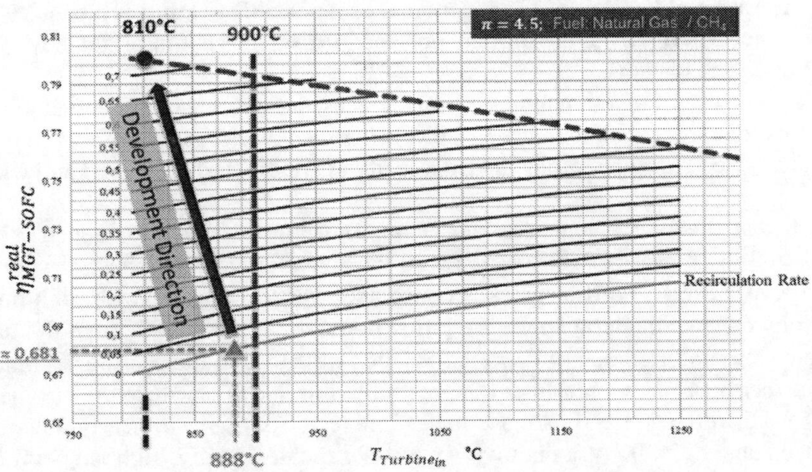

Figure 14. Influence of the recirculation rate according to the turbine inlet temperature and the efficiency

It is seen that the overall efficiency $\eta^{real}_{MGT-SOFC}$ increases with increasing turbine inlet temperature $T_{Turbine_{in}}$ while the SOFC operating temperature T_{SOFC} is constant. In this case, X decreases. It is also

IOP Conf. Series: Materials Science and Engineering **297** (2018) 012004 doi:10.1088/1757-899X/297/1/012004

physically interesting and logical that the efficiency will decreases after reaching the X = 0 line. In this case, the temperature of the SOFC is always equal to the recuperator outlet temperature (coupled).

For realising an operation on the right site of the green dotted line (border line $T_{Turbine\_in} = 900°C$), a higher technological effort is needed (turbine cooling, material). So, it is better to design the TFC1.0-processes on the left side of the border line, because the use of an uncooled turbine is possible. Furthermore it fits to the development line of future technologies (planed for a "TFC 2.0-Family"). In this case, the proportion of recirculation is to be increased, as the turbine inlet temperature drops (figure 14).

This potential of recirculation has the strongest impact, as it directly affects the efficiency $\eta_{SOFC}^{real} = \eta_{SOFC}^{rev} \cdot \eta_U \cdot \eta_I \cdot \eta_F$. An increase in the recirculation (fuel gas recirculation) leads to an increase of η_F. This increases the efficiency η_{SOFC}^{real} of the SOFC. Unfortunately, this results in a lower heat generation. At too low heat generation ΔQ_{Input}, there is the possibility that the MGT process no longer can be closed. Figure 14 shows the influence of the recirculation ratio. One can clearly see the major impact on the overall efficiency. Furthermore, it can be seen that the line of development goes in the direction of lower turbine inlet temperatures. Above the red dotted line the MGT-cycle is no longer possible because the heat supply ΔQ_{Input} is too low.

In the publication [1] also further variations can be found (e.g. pressure ratio of $4 < \pi < 5$) In the case of the TFC-process for natural gas conversion (under the boundary condition of the selected SOFC with $T_{SOFC} \approx 860°C$) the selected pressure ratio and turbine inlet temperature can be considered as logical.

6. Conclusion and forecast

State of the art hybrid MGT-SOFC systems are illustrated in a variety of publications [7; 8; 9; 3]. The individual components (turbine parts, heat exchangers, afterburner chamber, pressure vessel, etc.) are always connected to pipe lines. These must be insulated in order to minimize heat loss. A particularly high requirement is placed on the pressure vessel in which the high-temperature fuel cells are located. This container must be able to withstand the high operating pressure at high operating temperatures (700°C-900°C) and comply with the statutory container guidelines. The same applies to the connecting pipelines. As a result, the plants are relatively large and heavy, the heat losses (reduction of the exergy potential) are high and the material costs (high temperature resistant steels, more insulation, more material use, etc.) are higher. The MLC concept makes it possible to build compact MGT-SOFC systems in the future. The described multi-skinned pressure vessel is an important key component for the realization of compact systems.

Furthermore, the use of aerodynamic air bearing is essential for the implementation of the described TFC-family. Oil-free process air is indispensable; otherwise the SOFC modules very quickly degenerate and become unusable. This technology is also already available, industrialised and will be continuously improved.

It is shown that the components for the technology are available to industrialize a MGT-SOFC process in about 8 to 10 years (TFC1.0-270) with the achievement of highest efficiencies ($\eta > 60\% - 65\%$). To achieve these high efficiencies it is not necessary to operate with high pressure ratios. The turbo technology can use moderate pressure ratios in the area of $\pi \approx 4.5$. Also the turbine can operate in a reasonable temperature range of $T_{Turbine} = 800 - 900°C$, so that no special blade cooling system is required. Furthermore, the increasing of the recirculation rate will reduce the turbine inlet temperature (from about 890°C to 810°C) and increase the efficiency (from 68% to nearly 80%). Additionally, the in [1] introduced heat ratio "X" is an innovative tool for the design of a high temperature heat exchanger (HEX) inside the Multilayer Container (MLC). It has been shown that a practical range of $0.6 \leq X \leq 1$ can be considered.

Due to the great advantages of the flexible hybrid technology, the MLC design and the modular design of the basic technologies, a large number of applications, application areas and performance classes within the TFC product family are possible and feasible as described.

Acknowledgement and future cooperation
The realization of a compact MGT-SOFC system is a major challenge for research, development and engineering. It has an important future potential for the world of tomorrow.

We would like to thank the European Union and the Investment Bank of the State of Brandenburg (ILB) for the MAXEFF-project. Furthermore we like to thank our students, employees, cooperating colleagues and industrial partners.

At this point, we would also like to point out that we are looking forward to a future national and international cooperation with other institutions and industrial partners.

Together, the research, development and marketing of the described technology can be realized at an international level and provides an importance for future generations.

References
[1] Berg H P, Dückershoff R, Lehmann M and Prechavut N 2017 *S. ETC2017-266*
[2] Hassmann K 2001 *Fuel Cells.* **1**. pp 78-84
[3] George R A 2000 *Journal of Power Sources* **86** pp 134-139
[4] Kurzweil P. 2013 *Brennstoffzellentechnik - Grundlagen, Komponenten, Systeme, Anwendungen.* 2 (Wiesbaden : Springer-Verlag)
[5] Zhou L, et al. 2008 *Electrochimica Acta* **53**(16) pp 5195-5198
[6] Zhang X, et al. 2010 *Journal of Power Sources* **195**(3) pp 685-702
[7] Mueller F, Gaynor R, Auld A E, Brouwer J, Jabbari F and Samuelsen G S 2008 *Journal of Power Sources* **176** pp 229-239
[8] Berg H P, Malenky U, Antoshkiv O and Mattke R 2015 *Deutscher Luft- und Raumfahrtkongress.* 2015.
[9] Berg H P, Himmelberg A, Malenky U, Meincke M and Soontornpasatch T 2016 *Deutscher Luft- und Raumfahrtkongress*
[10] Ahlinder S, Berg H P and Reile E 2005 *Numerical Heat Transfer*
[11] Berg H P and Krienke C 2015 *Magdeburger Maschinenbau-Tage 12*
[12] Calisea F, et al. 2006 *Gas Turbine System* **31**(15) pp 3278-3299
[13] Prechavut N, Berg H P 2015 *Deutscher Luft- und Raumfahrtkongress*
[14] Dellacorte C 2011 *Tribology Transactions.* **4** pp 674-684
[15] Howard S, Dellacorte C, Valco M J, Prahl J M and Heshmat H 2001 *Tribology transactions.* **4** pp 657-663
[16] Heshmat H, Hryniewicz P, Walton li J F, Jahanmir S and Dellacorte C 2005 *Tribology Int..* **11-12** pp 1059-1075
[17] Sammes N M, Du Y and Bove R 2005 *Journal of Power Sources* **145**(2) pp 428-434

IOP Conf. Series: Materials Science and Engineering **297** (2018) 012005 doi:10.1088/1757-899X/297/1/012005

Performance and emission analysis of single cylinder SI engine using bioethanol-gasoline blend produced from Salvinia Molesta

Priyank Gupta[1,*], Partha Protim Das[1], M. Mubarak[2] and A. Shaija[2]

[1] Mechanical Engineering Department, Sikkim Manipal Institute of Technology, Majitar, 737136, India

[2] Mechanical Engineering Department, NIT Calicut, Calicut, 673601, India

*Corresponding Author: priyankmee0032@ymail.com

Abstract. Rapid depletion of world's crude oil reserve, rising global energy demand and concerns about greenhouse gases emission have led to the high-level interest in biofuels. The biofuel, bioethanol is found as an alternative fuel for SI engines as it has similar properties those of gasoline. Higher areal productivity with fast growth rate of microalgae and aquatic weeds makes them promising alternative feedstocks for bioethanol production. In this study, bioethanol produced from S.molesta (aquatic weed) using combined pre-treatment and hydrolysis followed by fermentation with yeast was used to make bioethanol- gasoline blend. The quantity of bioethanol produced from S.molesta was 99.12% pure. The physical properties such as density and heating value of bioethanol were 792.2 kg/m^3 and 26.12 MJ/kg, respectively. In this work, the effects of bioethanol–gasoline (E5) fuel blends on the performance and combustion characteristics of a spark ignition (SI) engine were investigated. In the experiments, a single-cylinder, four-stroke SI engine was used. The tests were performed using electric dynamometer while running the engine at the speed (3200 rpm), and seven different load (0, 0.5, 1, 1.5, 2, 2.5 and 3 kW). The results obtained from the use of bioethanol–gasoline fuel blends were compared to those of gasoline fuel. The test results showed an increase of 0.3% in brake thermal efficiency for E5. From the emission analysis, reduced emissions of 39 ppm unburned hydrocarbon, 1.55% carbon monoxide and 2% smoke opacity, respectively was observed with E5 at full load. An increase in CO2 by 0.17% and NOx by 86.7 ppm was observed for E5 at full load.

1. Introduction

The depletion of fossil energy resources, escalating prices of petroleum reserves with increased greenhouse gas emissions have forced researchers to use alternative fuels for energy needs[1]. Bioethanol is considered as a promising alternative fuel for SI engines due to its higher octane number and lower emissions of carbon monoxide, particulates, unburned hydrocarbon and carbon dioxide [2, 3]. The bioethanol can be produced from sugar bearing feedstocks like sugar cane, sweet potato, potato and lignocellulosic feedstocks like wood, grass, microalgae and aquatic weeds (freshwater and sea weed) [4]. The production of bioethanol from microalgae and aquatic weeds like S.molesta are quite promising due to its higher biomass productivity and lesser land requirement for its cultivation [5]. S.molesta (Kariba weed) is an aquatic weed found abundantly as a free floating plant on the fresh water bodies like ponds, lakes, dams causing the problems like clogging of hydroelectric dams, restricts irrigation and makes the water unsuitable for drinking purposes [6].

Characteristics of bioethanol such as octane number and specific gravity make it significant substitute of gasoline for modern spark-ignition (SI) engines. Bioethanol have cetane number lesser than diesel which does not provide sufficient self-ignition quality to use it in the compression-ignition engines. However, it can also be added with diesel to form bioethanol-diesel blends to reduce exhaust emissions for compression-ignition engines [7, 8]. Ethanol–gasoline blends are represented by an 'EX', where E stand for type of blend (bioethanol-gasoline) and X stands for percentage of bioethanol in the blends such as E5, E15, or E90. Bioethanol also have some disadvantages like less energy density as compared of gasoline. However, it is considered as high energy efficient fuel. One of the differentiable characteristics of bioethanol from other fuels is that it contains high oxygen content, so that called as oxygenated fuel. The existence of high oxygen content in bioethanol leads the better combustion of the blends with improved performance characteristics in SI engines [9, 10].

Numerous studies have been done over past recent years on the effect of the bioethanol-gasoline blends on performance analysis from spark-ignition engines. It has been seen that some emissions such as CO, UBHC and smoke can be decreased by using bioethanol-gasoline blends but not others such as CO_2, NOx. A/F ratio is one of the most emissions affecting property and emissions of CO, UBHC are strongly depending on A/F ratio [11]. As A/F ratio increases, CO and UBHC emissions decrease. Regarding decreased emissions, HC and CO emissions were observed to reduce with increased bioethanol quantity in bioethanol-gasoline blends [12]. This happens because of the presence of high oxygen content which acts as an oxidizing agent. HC emissions is also lead by the flame quenching which happens at the combustion chamber wall and the absorption of fuel vapor into oil layers on the cylinder wall during intake and compression, followed by desorption of fuel vapor into the cylinder during expansion and exhaust [13]. Uncontrolled increased emission of NOx has been reported while increasing bioethanol in blends, but these are not significant and consistent in many of the cases [8]. It may be because of higher oxygen availability in blends at the higher temperature but NOx emissions should be decreased theoretically with increased bioethanol content in blends because of the reasons discussed.

In this study, bioethanol produced from S.molesta (aquatic weed) using combined pre-treatment and hydrolysis followed by fermentation with yeast was used to make bioethanol-gasoline blend (E5). The effects of bioethanol–gasoline (E5) fuel blends on the performance and combustion characteristics of a spark ignition (SI) engine were investigated. In the experiments, a single-cylinder, four-stroke SI engine was used. The tests were performed using electrical dynamometer while running the engine at the speed (3200 rpm), and seven different load (0, 0.5, 1, 1.5, 2, 2.5 and 3 kW). The results obtained from the use of bioethanol–gasoline fuel blends were compared to those of gasoline fuel.

2. Fuel Specification

Fuel properties such as, calorific value and density of bioethanol, E5 fuels were measured and compared with gasoline and are depicted in table 1. Calorific value of bioethanol, 26.12 MJ/kg, measured by using Bomb calorimeter was found a little less than that of ASTM standard, which is 27 MJ/kg. This is because of its impurity mainly contains water.

Table 1. Characteristics of test fuels

Properties	Bioethanol (99.12% pure)	E5	Gasoline
Chemical formula	C_2H_5OH	-	C_4-C_{12}
Calorific value (MJ/kg)	26.12	43.20	44.45
Density (kg/m^3)	792.23	742.38	740.12

3. Performance evaluation of SRIRAM HONDA GK 200 engine

A petrol engine selected for the constant speed load testing of E5 fuel was a single cylinder, constant speed, four stroke SRIRAM HONDA engine, the specifications of which are shown in table 2. A constant speed of 3200 rpm was selected, which is little higher than the rated speed in order to check the performance characteristics at higher loads as compared to rated power. All the readings were taken from zero to full load using electrical dynamometer. Three readings were taken for all parameters at each load and average values were considered for reducing the error. Alternator efficiency was assumed as 80% for the calculation of brake power. The schematic view of the experimental test rig is shown in figure 1.

Table 2. Technical details of testing rig

Model	SRIRAM HONDA GK 200
BHP	2 HP @ 3000 rpm
SFC at rated power	600 g/kWh
Bore	67 mm
Stroke length	56 mm
No. of stroke	4
No. of Cylinder	1
Cooling system	Air cooled
Dynamometer type	Electrical

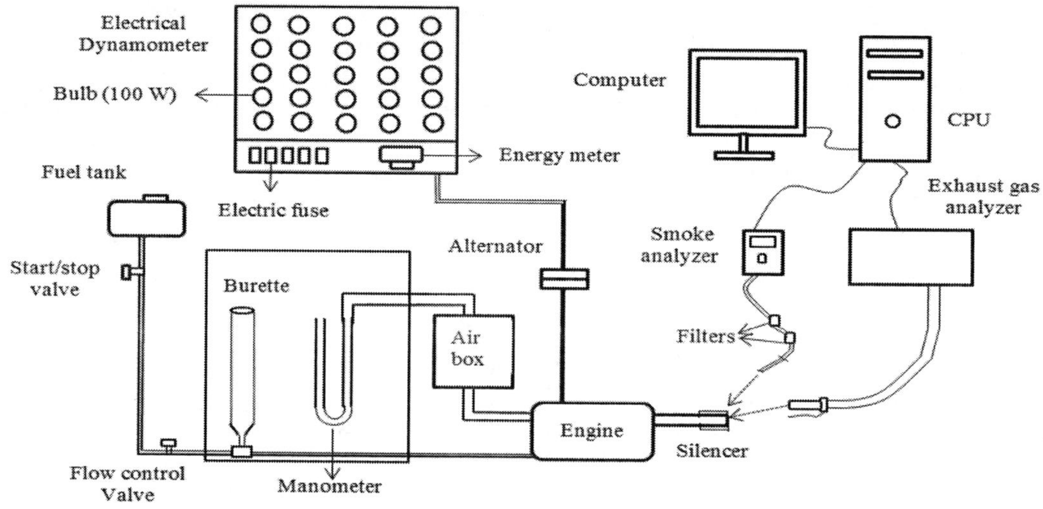

Figure 1. Engine test rig with performance and emission analysis unit

3.1 Performance analysis

The total fuel consumption, brake specific fuel consumption, brake thermal efficiency and A/F ratio were calculated and plotted against BP. The schematic view of test rig is shown in figure 1. Engine performance characteristics were observed with Gasoline alone, and E5.

3.2 Emission analysis

Relative volumes of CO, HC, O_2, NOx and CO_2 in exhaust gases from the single cylinder, four stroke SI engine were measured using AVL DiGas 444 exhaust gas analyzer with RS232 cable. The temperature of the catalyst has been measured with K type thermocouple and data are stored using data logger. Continuous observations were taken at every load and final measurements were calculated by taking averages of minimum 4 observations. The AVL 437C smoke meter was used to measure the opacity of exhausted gases. The opacity is the extinction of light between light source and receiver. Measurements have been taken according to the same test procedure, explained above. Exhaust gas analyzer and smoke meter range, accuracy and uncertainties details are shown in table 3.

Table 3. Details of Exhaust gas analyzer and smoke meter range, accuracy and uncertainties

Model of gas analyzer	AVL DiGas 444 Exhaust Gas Analyzer		
measured quantity	Range	Accuracy	Uncertainties (%)
CO	0-10%	±0.02%	±0.2
CO2	0-20%	±0.3%	±0.15
HC	0-10000 ppm	±20 ppm	±0.2
NOx	0-5000 ppm	±10 ppm	±0.2
	AVL 437 Smoke meter		
Smoke	NA	0.01%	±1.0

4. Results and discussion

4.1 Performance analysis of the engine using gasoline and E5

4.1.1 Brake thermal efficiency (η_{bth}). Brake thermal efficiency is defined as brake power of a heat engine as a function of thermal input from the fuel. It was calculated by using equation (1).

$$\eta_{bth} = \frac{B.P}{m_f \times C.V} \quad (1)$$

where, *B.P* is brake power, m_f is mass flow rate of fuel, and *C.V* is calorific value of fuel used. It was noticed that the brake thermal efficiency of the engine higher using E5 as compared to that using gasoline as shown in figure 2. This may be because of higher effect of the low calorific value of E5 than its TFC. There is not any significant improvement of brake thermal efficiency for gasoline and E5 has been seen at low BP till 0.6 kW. The maximum brake thermal efficiencies, 14.04 and 14.18% were observed for gasoline and E5 at the BP of 2.3925 and 2.4023 kW respectively. The maximum improvement of 2.44% of brake thermal efficiency was observed at the BP of 1.978 and 1.974 kW for gasoline and E5.

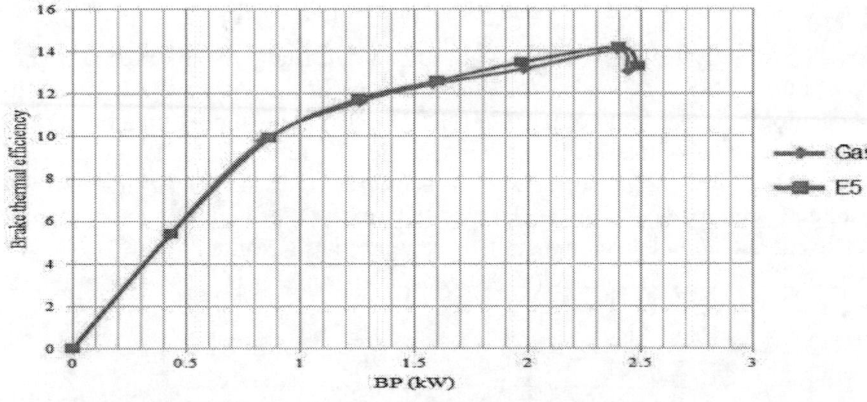

Figure 2. Variation of brake thermal efficiency with brake power

4.1.2 Brake specific fuel consumption (BSFC). It is the fuel flow rate per unit brake power output. It measures how efficiently an engine is using fuel supplied to produce work. It was calculated using equation (2).

$$BSFC = \frac{m_f}{B.P} \quad (2)$$

where, m_f is mass flow rate of fuel used and *B.P* is brake power.

Figure 3 shows the comparison for BSFC of gasoline and E5 for different loading. BSFC first decreased, reach at lowest point and then increased with increasing load for both gasoline and bioethanol. This is because of reduction of heat loss due to higher opening of throttle at higher loadings. The theoretical A/F ratio of gasoline is 1.6 times higher than bioethanol [52]. It has been seen that BSFC for E5 is little higher than gasoline. This is because of higher TFC of E5 for producing same output power as compared to gasoline. It was seen that after BP of 1.8 kW, there is not any significant difference in BSFC for gasoline and E5. The lowest BSFC, 0.576 and 0.587 kg/kWh for gasoline and E5 were observed at the BP of 2.3925 and 2.4023 kW, respectively.

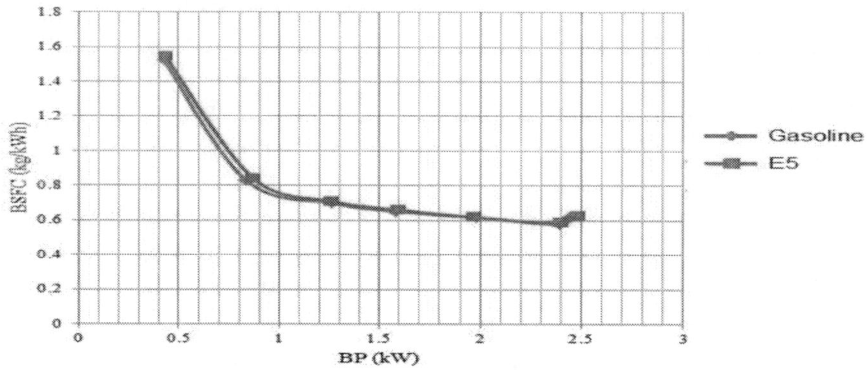

Figure 3. Variation of brake specific fuel consumption with brake power

4.1.3 A/F ratio. A/F ratio is useful to know the engine operating conditions. As load increases, A/F ratio decreases because of higher fuel consumption to produce required power output. It was calculated using equation (3).

$$A/_F = \frac{m_a}{m_f} \tag{3}$$

where, m_a is mass flow rate of air and m_f is mass flow rate of fuel used.

Engines runs with slightly lean A/F mixture at no load and rich A/F mixtures at higher loads. A/F ratios of 14.989 and 15.556 were noted for gasoline and E5, respectively at no load conditions. Theoretically, stoichiometric A/F ratio of gasoline (14.7) is 1.6 times higher than bioethanol [8]. Hence by adding bioethanol with gasoline reduces stoichiometric A/F ratio of blends and provide actual lean A/F ratio. The variation of A/F for gasoline and E5 is shown in figure 4. The A/F ratio for E5 was higher than that of gasoline for all loads. This is because of the presence of high oxygen content in E5, which leads to more efficient and complete combustion as compared to gasoline.

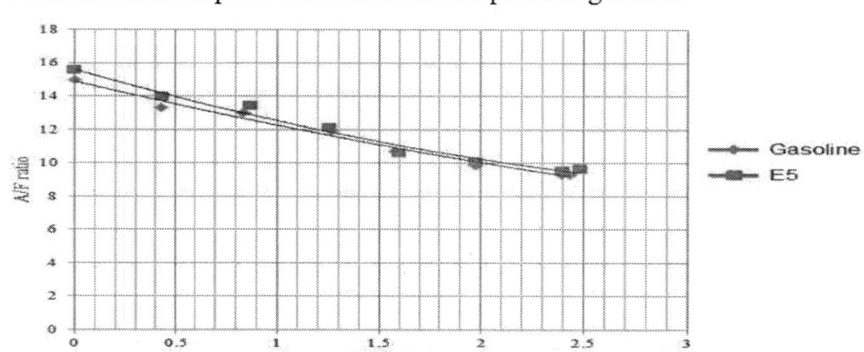

Figure 4. Variation of A/F ratio with brake power

4.2 Emission analysis of the engine with Gasoline and E5

4.2.1 Unburnt hydrocarbon. (UHBC). UBHC emissions in SI engine may happen due to flame quenching or incomplete combustion (partial burning or complete misfire). It strongly depends on A/F ratio. As A/F ratio decreases, UBHC emission increases as shown in figure 5. A little reduction in UBHC emission using E5 as fuel was observed compared to gasoline. This may be because of more efficient combustion of E5 due to presence of high oxygen content and higher A/F ratio as compared to gasoline. Significant reductions of UBHC for E5 by 8.80, 10.17, 25.25, 23.83, 8.34, 19.66, 43.67, 39.54 ppm as compared to gasoline were recorded.

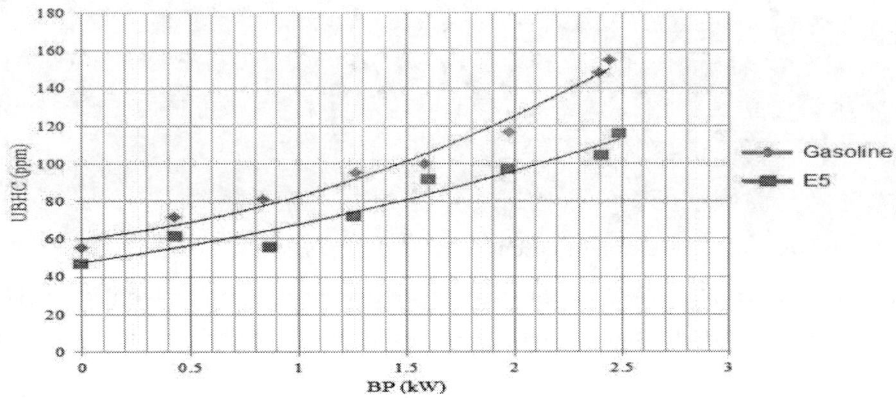

Figure 5. UBHC vs BP

4.2.2 Carbon dioxide (CO_2). CO_2 is the product of complete combustion of fuel in the combustion chamber of the engine. The basic trend of CO_2 emissions is that it decreased with increased load on the engine. This might be due to relatively incomplete combustion of fuels and lower A/F ratios achieved at higher loads. Higher emissions of CO_2 for E5 were observed as compared of gasoline as shown in figure 6. This is because of the more efficient combustion of E5 due to high oxygen content and higher A/F ratio achieved at every load as compared to gasoline. Little increment of CO_2 emissions by 0.215, 0.97, 0.64, 0.76, 0.20, 0.70, 0.175, 0.55% as compared to gasoline were recorded.

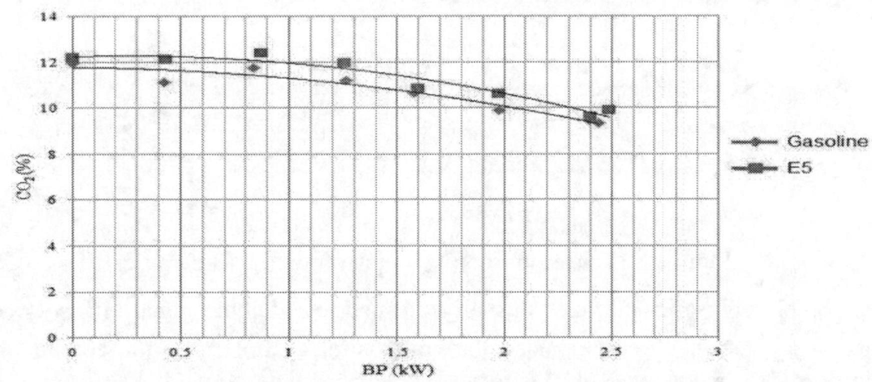

Figure 6. Variation in carbon dioxide emission with brake power

4.2.3 Carbon monoxide (CO). CO is the main constituent as a result of incomplete combustion of fuels in engines and strongly depends on A/F ratio. As A/F ratio decreases, CO emission increases because of incomplete combustion. A significant reduction in the percentage of CO emissions using E5 as compared to gasoline is shown in figure 7. This is because of the higher A/F ratio achieved and efficient combustion of E5 due to high oxygen presence as compared to gasoline. At no load condition, 4.23%

and 3.0% of CO emissions were recorded for gasoline and E5, respectively. Significant reductions of CO emissions by 1.23, 1.96, 1.54, 1.72, 1.10, 0.748, 1.55, and 1.54 % as compared to gasoline were recorded.

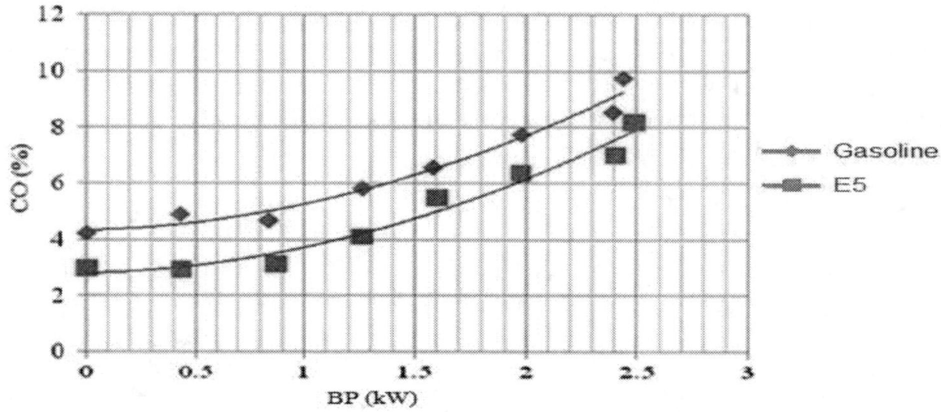

Figure 7. Variation of carbon monoxide emission with brake power

4.2.4 Oxides of nitrogen (NOx). NO_x in the exhaust are due to high oxygen content in the fuel or high temperatures inside the combustion chamber. As load increases, NOx emission increases because of higher temperature at higher loads. Figure 8 shows the comparison of NOx emissions for gasoline and E5. A significant increment in NOx emission for E5 was observed which increased with load. This might be because of the high oxygen content present in E5 as compared to gasoline. Increased NOx emissions for E5 by 18.34, 11.25, 30.25, 59.67, 65.67, 92.67, 86.67, 68.37 ppm as compared to gasoline were recorded.

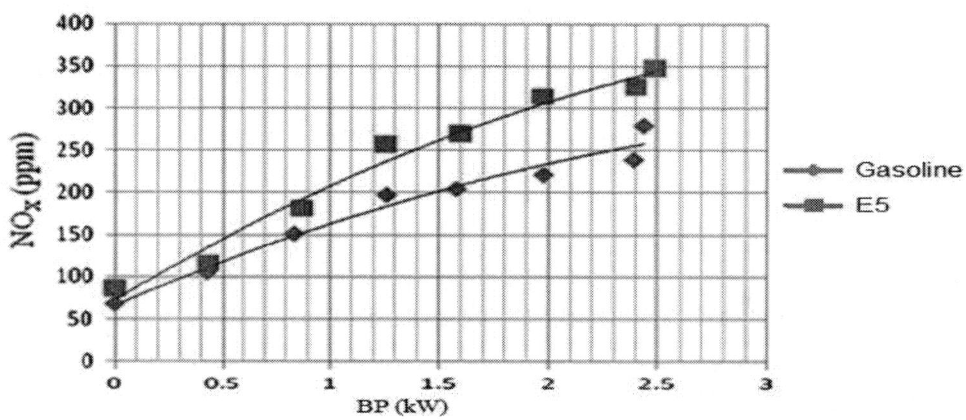

Figure 8. Variation of NOx emission with brake power

4.2.5 Smoke. Smoke is solid particles, usually formed by dehydrogenation, polymerisation and agglomeration reactions which occur inside the combustion chamber. In the combustion process of different hydrocarbons, acetylene (C_2H_2) is formed as intermediate product. These acetylene molecules after simultaneous polymerisation dehydration produce carbon particles. Figure 9 shows the difference in the smoke emission for E5 and gasoline at different loads. A little reduction in smoke emission was observed for E5 as compared to gasoline.

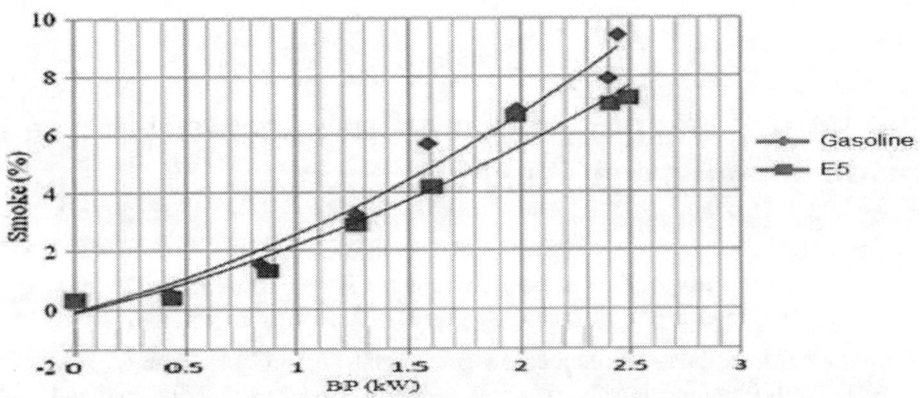

Figure 9. Variation of smoke emission with brake power

5. Conclusion

Salvinia molesta can be used as an alternative feedstock for the production of bioethanol which can be used to make bioethanol-gasoline blends as a fuel for SI engines. The performance and emission analysis of a single cylinder SI engine was analyzed using E5 and compared with that using gasoline. The test results showed an increase of 0.3% in brake thermal efficiency for E5. From the emission analysis, reduced emissions of 39 ppm unburned hydrocarbon, 1.55% carbon monoxide and 2% smoke opacity, respectively were observed with E5 at full load. An increase in CO_2 of 0.17% by volume and increase in NO_x of 86.7 ppm were observed for E5 at full load. It is observed that for E5 blend there is no significant improvement in performance analysis but shows a weighty improvement in emission analysis as compare to that of using gasoline.

References

[1] Balat M 2011 *Energy Convers. Manage.* **52** pp 858-875
[2] Hsieh W, Wu T and Lin T 2002 *Atmospheric Environment* **36** pp 403-410
[3] Al-Hasan M 2003 *Energy Conversion and Management* **44** pp 1547-1561
[4] Mubarak M, Shaija A, Suchithra T. V 2013 *23rd National Conf. on I. C. Engine and Combustion* (SVNIT, Surat, India, 13-16, December 2013)
[5] Singh J and Gu S 2010 *Renew and Sust En rev* **14** pp 1367-1378
[6] Brennan L and Owende P 2010 *Renew and Sust En rev* **14** pp 557-577
[7] Bielaczyc P, Woodburn J, Klimkiewicz D, Pajdowski P and Szczotka A 2013 *Fuel Proc. Tech.* **107** pp 50–63
[8] Schifter I, Diaz L, Rodriguez R, Gómez J P and Gonzalez U 2011 *Fuel* **90** pp 3586–3592
[9] Siddegowda K B and Venkatesh J 2013 *Inter. J. of Innov Res in Sci, Eng and Tech.* **2** (9) pp 2319-8753
[10] Pal A 2014 *Int. J. of Thermal Technologies* E-ISSN 2277 – 4114
[11] Kumar J, Ansari N A, Verma V and Kumar S 2013 *American Journal of Engineering Research* **02** (07) pp 191-201
[12] Canakci M, Ozsezen A N, Alptekin E and Eyidogan M 2013 *Renewable Energy* **52** pp 111-117
[13] Eyidogan M, Ozsezen A N, Canakci M and Turkcan A 2010 *Fuel* **89** pp 2713–2720

Ignition improvement by injector arrangement in a multi-fuel combustor for micro gas turbine

O. Antoshkiv[1,*], T. Poojitganont[1], S. Jeansirisomboon[2] and H. P. Berg[1]

[1] BTU Cottbus - Senftenberg, Siemens-Halske-Ring 14, Cottbus, 03046, Germany
[2] Gulf Group Company, 67 Wireless Road, Bangkok, 10330, Thailand

* Corresponding Author: antoshki@b-tu.de

Abstract. The novel combustor design also has an impact on the ignitor arrangement. The conventional ignitor system cannot guarantee optimal ignition performance in the usual radial position. The difficult ignitability of gaseous fuels was the main challenge for the ignitor system improvement. One way to improve the ignition performance significantly is a torch ignitor system in which the gaseous fuel is directly mixed with a large amount of the combustor air. To reach this goal, the ignition process was investigated in detail. The micro gas turbine (MGT) ignition was optimised considering three main procedures: torch ignitor operation, burner ignition and flame propagation between the neighbour injectors. A successful final result of the chain of ignition procedures depends on multiple aspects of the combustor design. Performed development work shows an important step towards designing modern high-efficiency low-emission combustors.

1. Introduction

The decentralization of electric energy generation requires development of modern powerful and environmental friendly co-generators. The micro gas turbine (MGT) is a power system with low maintenance costs. Furthermore, a gas turbine can operate by using different fuels. The limited fossil energy sources and increasingly stringent emission regulations demand today an environmental friendly combustion technology, working on alternative fuels. This leads to a new combustion chamber architecture. In comparison to the conventional combustor, the multi-fuel combustion chamber operates under a higher swirl number and longer dilution zone. Due to this fact, the combustor prepares much homogeneous air-fuel mixture, which causes lower NOx-emissions.

Combustion is the basic of technology concerning operation of internal combustion engines. However, it is considered to be the main source of pollution in the natural environment that can threaten inhabitants. Since the pre-industrialized era, Green House Gas (GHS) concentration is continuously observed, and it is found that in 2012 CO_2 was about 40% higher than in the mid-1800s. Among human activities that produce greenhouse gases, the use of energy is by far the largest source of emission (around 83%). Within energy sector, oxidation of CO_2 during combustion dominates the total GHG emissions [1].

A gas turbine is an energy converter that change chemical energy contained in fuel to mechanical energy. In the energy sector, gas turbine usually cooperates with generator in an effort to convert mechanical energy to electrical energy. Main components of general gas turbine are compressor, combustor, and turbine. First of all, air is compressed into high pressure condition by multi stages of compressor, following that air and fuel well mix in a combustion zone and initially ignited by spark igniters. Gas turbine burner can be considered as a decisive part of combustor placed in a combustion chamber, where compressed air and fuel are mixed. It can be classified as premixed or non-premixed burner types. An ideal burner should guarantee complete combustion, flame stability, low emissions and high reliability in a wide range of operation condition [2], even if in an unexpected situation. Figure 1 (left) illustrates burner configurations for specific types of fuels, which is designed by BTU Cottbus-Senftenberg's researchers for a micro gas turbine (MGT) [3].

2. Previous research work

One crucial demanded characteristics of gas turbine combustor is ability for fast and reliable flame transfer from operated burner to others. This flame transport can be described a light-across phenomenon, which is a part of a light-around phase that flame is distributed to all burners in a combustion chamber. In principle, flame propagation is influenced not only by a volumetric expansion, but also buoyancy force and burner aerodynamics [4-5]. Almost of practical combustors consist typically of multi-burners placed side-by-side around its periphery similar to combustor demonstrated in figure 1 (right).

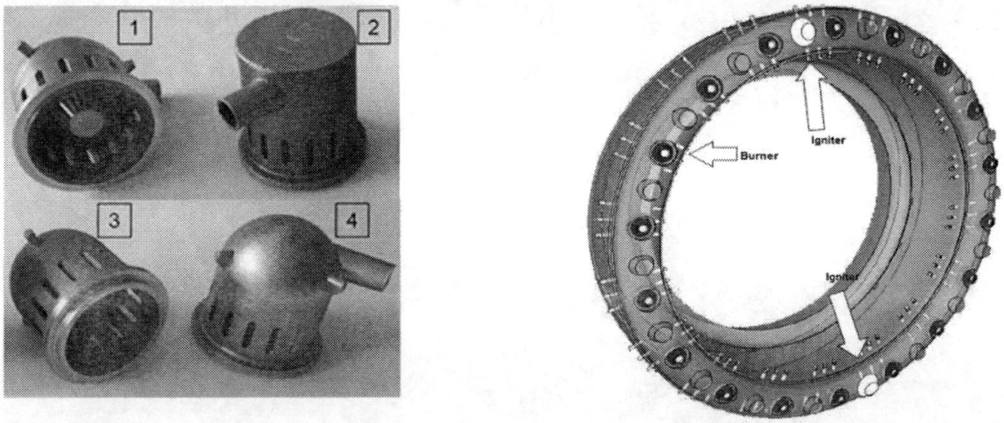

Figure 1. MGT Kerosene/Diesel Injector (1), Methane/Natural Gas Burner (2), LNG Injector (3), INSBRuk™ Burner (4) and Turbomeca Vesta combustor 18 main burners and 2 ignitors [4,6-7].

The non-reacting direct numerical simulation of twin swirling jet at Re = 5000 with 3 swirl levers (S = 0.68, 1.08 and 1.42) was investigated in order to study the interaction between swirling jet burners. This study showed that twin swirl jets strengthen and augment turbulence and Reynolds stress, compared with an isolated one because of the aerodynamic interaction of consecutive burners. In addition, the higher swirl number leads to the larger intensive interaction between burners as shown in figure 2 [8]. In consistent, Kao Y.H. et al. [9] proposed that multi-burner aerodynamic profile is different from a single burner as it is also concerned with adjacent burner interaction even if for a cold-flow (non-reacting). Moreover, not only the inter-swirler space influenced on multi-burner aerodynamics and flow field profiles, but it is also affected by the distance of burners to the walls [9].

IOP Conf. Series: Materials Science and Engineering **297** (2018) 012006 doi:10.1088/1757-899X/297/1/012006

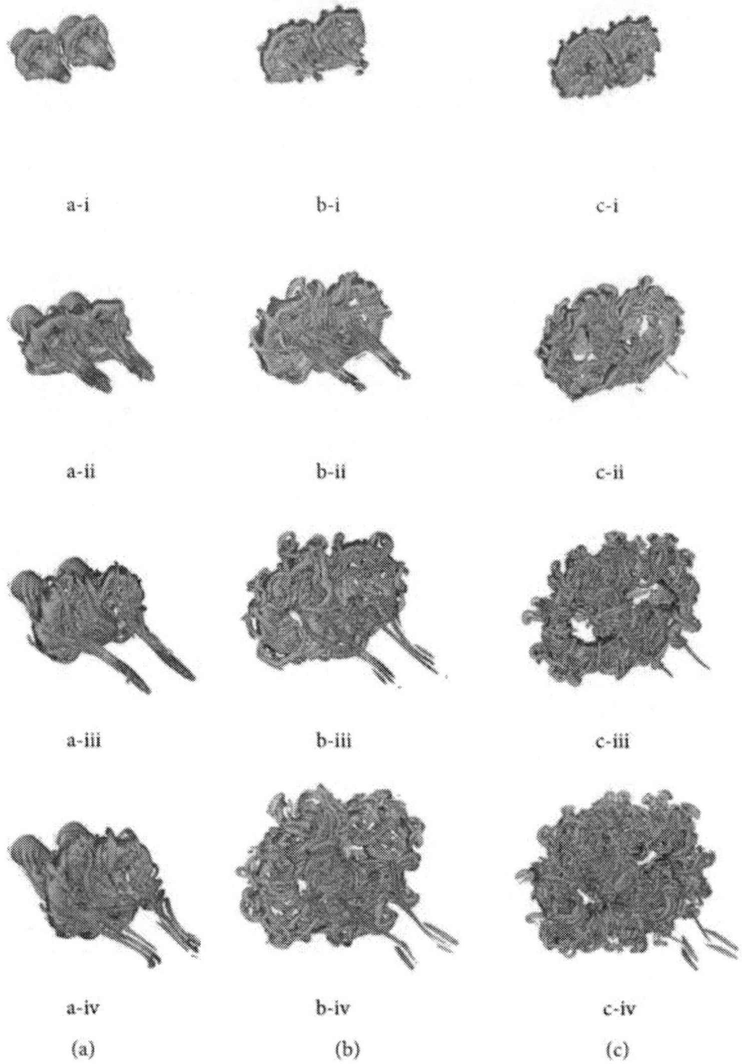

Figure 2. The twin swirling jet burner interaction with different swirl number *s=0.68 (a), s= 1.08 (b)* and *s = 1.42 (c)* from t = 2.5 ms to 10 ms (from I to IV correspondingly) [8].

A large eddy simulation of a complete gas turbine chamber was conducted in 2013. It is important that thermal expansion played a great role on the flame propagation to neighbour burners rather than turbulence because the flame front moved at faster speed than turbulent flame speed [6, 10].

Not only, interconnection of consecutive burner aerodynamics and thermal expansion influence on flame propagation, but distance also affects flame dynamics. Barré, D. et al. studied methane/air non-premixed combustion and addressed that the ignition sequence strongly depends on the spacing of burners. If spacing increases, axial propagation becomes dominant and successive time to a full ignition also rises in figure 3 [7].

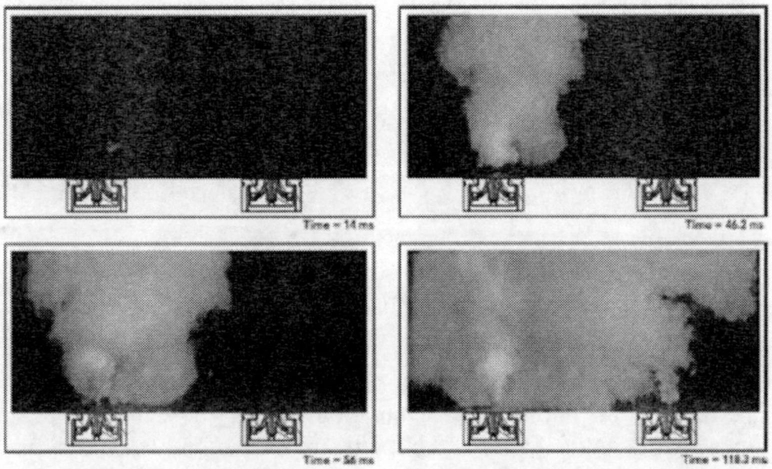

Figure 3. Fast visualization of flame dynamics for large distance between burners [7].

It is obvious that increasing space between burners leads to a lower flame propagation possibility. In consequence, burner angle was implemented in order to improve light-across ignitability performance and ease the difficulty of large distance light-across. Figure 4 demonstrates the typical burner arrangement in the MGT combustor developed at the BTU Cottbus - Senftenberg.

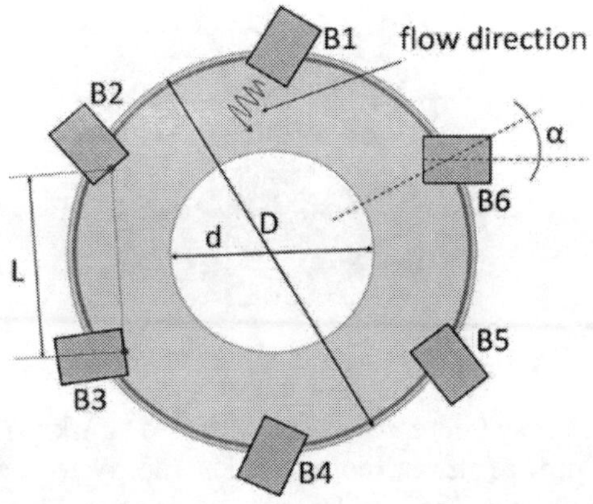

Figure 4: MGT Burners arrangement [11].

3. Experimental Configuration

3.1. *Test hardware*

Experimental burner has a configuration showed in figure 5 and mounted around the periphery of MGT combustion chamber in operation (figure 6). In the case of this experiment, a combustion liner is designed in order to place burner inside. Gas is supplied circumferentially through 8 millimetres diameter inlet at the bottom of the cup and distributed through 15 small holes (figure 5).

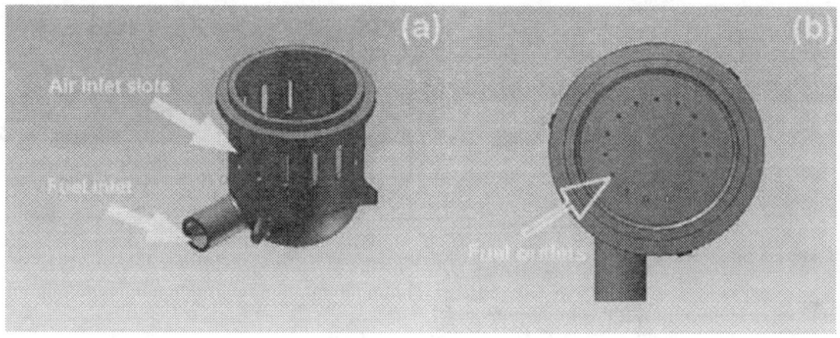

Figure 5. Burner configuration isometric view (a) and top view (b).

Whereas surrounding air inside the combustor frame flows through 15 swirling vane slots on the symmetric circumference of the burner surface, creating swirling flow in order to stabilize flame.

In this experiment two identically aligned burners are placed on aluminium profiles, which can be linearly adjusted to change the distance between burners, whereas the limit of the nearest and farthest spaces is defined by the diameter of flame tube and the dimensions of combustion liners. In addition, angle orientation is adapted by an adjustable aluminium connector with an angle lock screws.

Figure 6. Position of burners in real MGT combustor.

Air and fuel are mixing at atmosphere condition. A fixed spark ignitor operates at a voltage range between 5-10 kV near the centre of the burner provides energy to create a spark kernel. In addition, overall fuel mass flow rates are controlled by a Bronkhorst flow meter model F-103D with the operation flow rate 1.5 g/sec and temperature 20 ^{0}C at 1 bar. After that, a fuel flow divides to distribute fuel equally to all burners. Similarly, air is supplied by an air compressor unit with a maximum capacity of 60 g/s at operating atmospheric pressure. According to the schematic (figure 7), compressed air flow rate is adjusted by a by-pass control valve and compressed air passes through an air distributor.

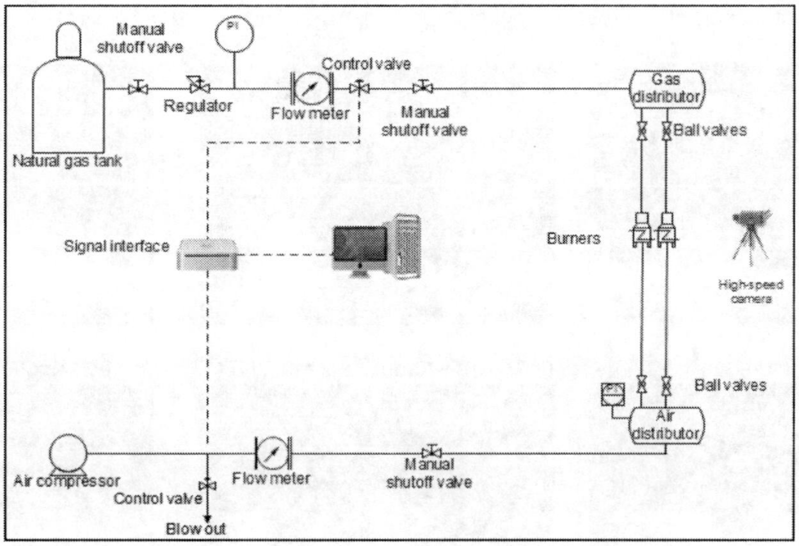

Figure 7. Experimental schematic set-up.

In order to observe flame propagation among variation of equivalence ratios, all valves are controlled by in-house software programmed in LabVIEW. Pressure gauges are used to monitor both air and fuel pressure locally without data interface. The uncertainty of the data during experiment could be claimed that it is lower than 5% as usual.

3.2. *High-speed imaging system*

General components arrangement at the test rig is shown in figure 8. The combustor chamber walls consist of quartz glasses with sealing between their edges at aluminium profiles. On the back side, a black screen is placed in order to facilitate observation and picture taking. High-speed camera system consists of IRO connected with phantom high speed camera. IRO and high-speed controllers control the trigger and interface with the Davis software [12].

Figure 8. High-speed camera system components.

4. Discussion of results

All of propane-air mixture combustion experiments were conducted at the atmospheric condition. Temperature and pressure are around 25 °C at humidity 51% and 1004.11 mbar with a constant air mass flow rate at 6 g/s. The mass flow rate shown in the pictures is averaged over the numbers of experiments ($\dot{m}_{avg}$). The experimental conditions are shown in table 1 with the increased step of distance until reaching 70 mm with at least 3 experiments at each condition. Flame colour is blue. Flame enlargements in both axial and horizontal directions according to rising amount of fuel mass flow (figure 9).

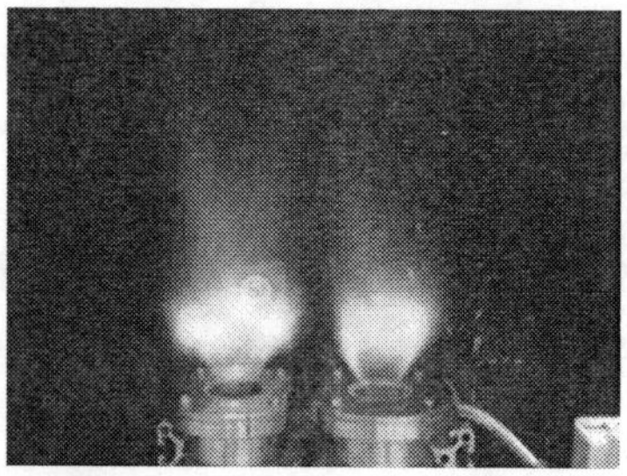

θ=0°,Δd=20mm,$\dot{m}_{avg}$=0.268g/s

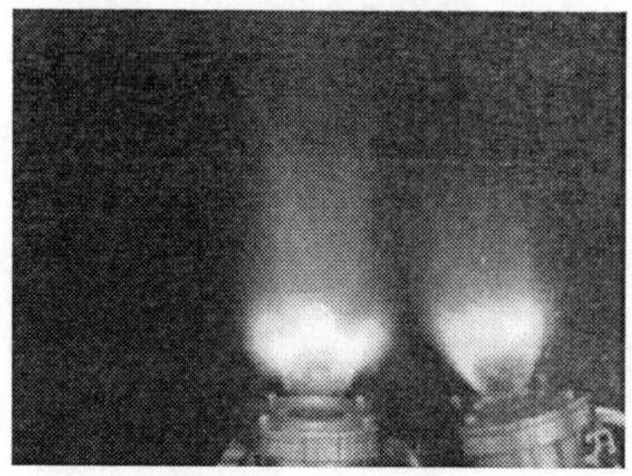

θ=15°,Δd=40mm, $\dot{m}_{avg}$=0.277g/s

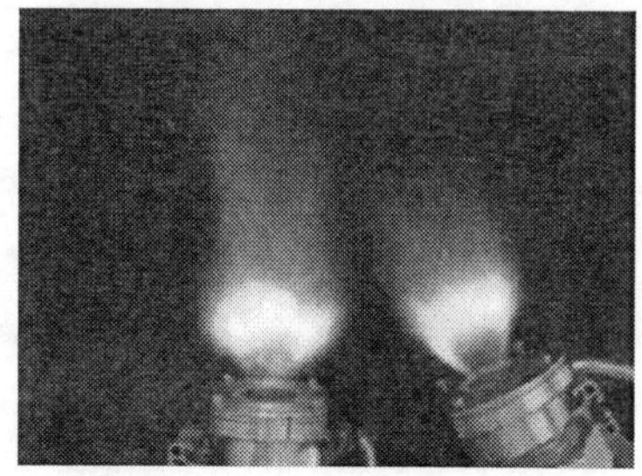

θ=30°,Δd=60mm,$\dot{m}_{avg}$=0.283g/s

θ=0°,Δd=30mm,$\dot{m}_{avg}$=0.300g/s

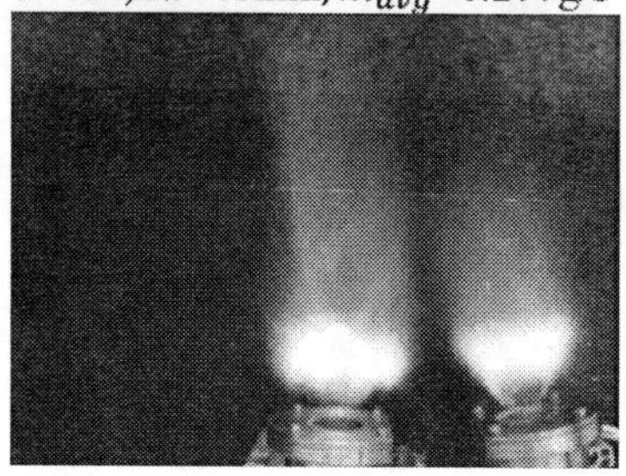

θ=15°,Δd=50mm,$\dot{m}_{avg}$=0.317g/s

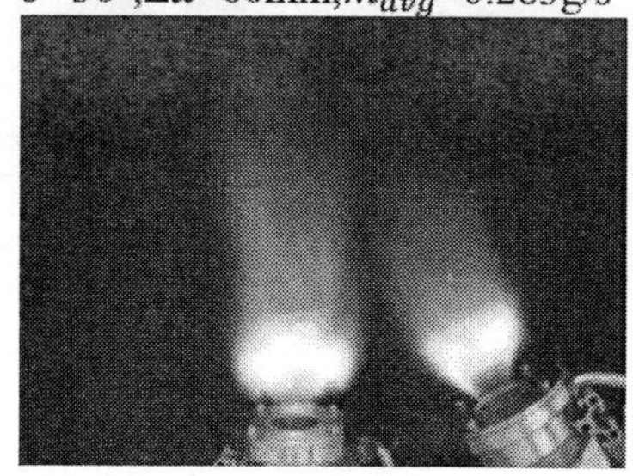

θ=30°,Δd=70mm,$\dot{m}_{avg}$=0.307g/s

Figure 9. Flame feature at different fuel mass flow rates.

Required fuel mass flow rate provoked flame propagation at various burner spaces and tilted angles as illustrated in figure 10. It is obvious that all tilted angles have excellent linear relation of distance and fuel mass flow for all interested ranges with minimal experiment deviation as an error bar. Among all angles, 75° required less fuel mass flow for the flame jump. The larger tilted angle, the smaller amount of fuel is necessary to achieve flame propagation.

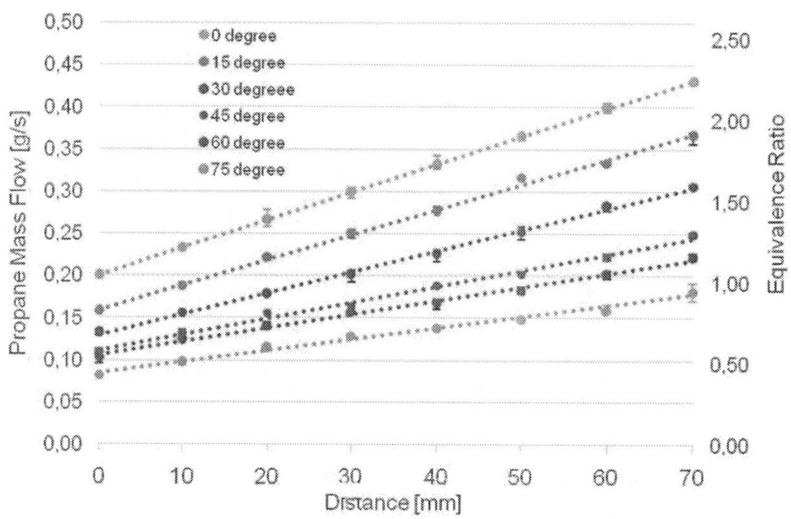

Figure 10. Mass flow rate and distance relation of all tilted angles with error bar.

In order to evaluate flame propagation capability between consecutives burners [2], dimensionless term, called ignitability unit (I_{ig}), is introduced in order to compare the capability of flame propagation among particular conditions.

$$I_{ig} = \frac{1}{\emptyset} = \frac{AFR_{LACR}}{AFR_{stoich}} \tag{1}$$

Data in figure 10 are shown to explain the relation of ignitability unit as a function of distance and burner angles which is illustrated in figure 11. Curves were fitted using the least-squares method of polynomial order 2 for the points approximation. The approximated mathematical functions are shown in table 1. Additionally, experiment pictures of all angles (15°, 30°, 45°, 60°, and 75°) are also available in the appendix.

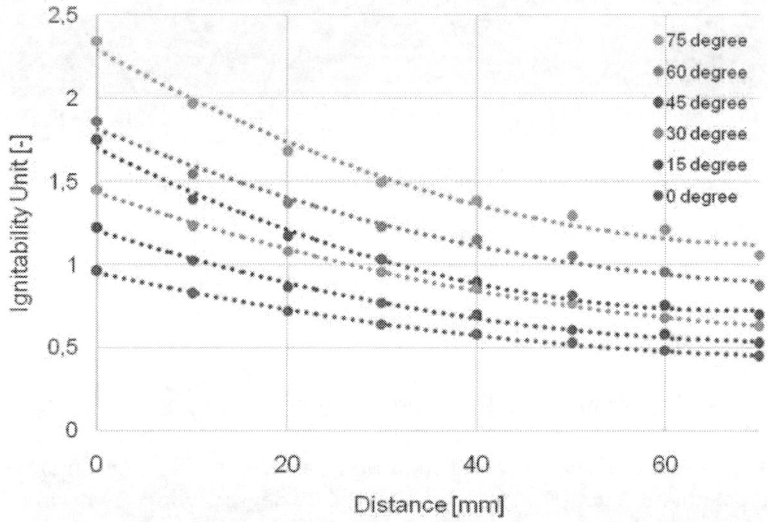

Figure 11. Ignitability unit and distance relation at all tilted angles.

It is noticeable that 75° achieves ignitability units above unity for all interested distance. In contrast with 0° position, it was not able to surplus unity for all lengths. While the others have both ignitability higher and lower unity.

Table 1. Approximated formulas of ignitability unit by the least-squares method.

Tilted angle	$I_{ig} = f(d); I_{ig} = ignitability; \Delta d = distance$
0°	$I_{ig} = 0.00008\Delta d^2 - 0.0126\Delta d + 0.9492$
15°	$I_{ig} = 0.0001\Delta d^2 - 0.018\Delta d + 1.2014$
30°	$I_{ig} = 0.0001\Delta d^2 - 0.0193\Delta d + 1.4322$
45°	$I_{ig} = 0.0002\Delta d^2 - 0.0292\Delta d + 1.704$
60°	$I_{ig} = 0.0001\Delta d^2 - 0.0235\Delta d + 1.8143$
75°	$I_{ig} = 0.0002\Delta d^2 - 0.032\Delta d + 2.2931$

In order to illustrate the relation of all parameters, surface plot was implemented to explain ignitability as a function of tilted angle and distance (figure 12). It is visible that the larger distance of consecutive burners, the lower ignitability at all tilted degrees is required. Moreover, larger burner angle leads to higher ignitability at specific distance, compared with smaller one. Among all angles, the highest ignitability is clearly at the 75° and descended to 0°.

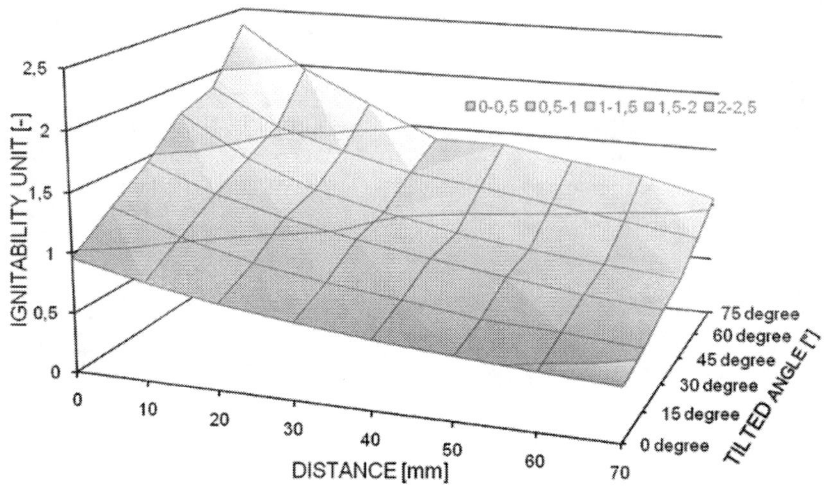

Figure 12. Ignitability unit as a function of distance and tilted angle in surface plot.

5. Experimental Investigation of Flame Propagation

5.1. Conditions of performed experiments

In order to investigate the dynamics of flame propagation is enough to observe the tracking of OH radicals using OH* light. The high-speed imaging system was used for this aim. To avoid the heat effect on high-speed system and for safety operating condition for high-speed imagining investigation, the fuel mass flow was adjusted at 0.3 g/s ($\emptyset < 1$). Tilted angle 0° degree, which is norm of gas turbines, and 45 degree, which is intermediate angle, are of interest. The experimental conditions and settings of the high-speed camera were identical during all performed tests.

During the experiments, air mass flow rate has been fixed at 6 g/s, whilst burners distance and tilted angles are varied in the range of 0-70 mm and 0°-7°, respectively. Fuel mass flow rates of each specific cases providing successive flame propagation were defined. It is understandable that larger distances required higher fuel flow rates to maintain flame jump to another burner. At distance 70 mm of tilted angle 0°, the largest overall fuel mass flow of 0.864 g/s was needed.

5.2. Example of Tilted Angle 0° and Distance 15 mm

The important experimental conditions and the high-speed camera parameters are illustrated in table 2.

Table 2. Experimental conditions of tilted angle 0° with distance 15 mm.

Experimental Parameters	Values
$\dot{m}_{air}, [g/s]$	*0.6*
$\dot{m}_{C_3H_8}, [g/s]$	*0.5*
$T, [°C]$	*27*
$P, [atm]$	*1*
RH%, [%]	*44*
Gain, [-]	*40*
Gate, [*ns*]	200 000
Exposure Time, [μs]	229.805

Figure 13 shows the recorded distribution of OH* during light-across. Until the reactive particles reach second burner, it moves toward downstream of the second burner in an axial direction into the recirculation zone. After it arrives the region near the outlet of the consecutive burner, higher light emission was observed and the concentration of OH* increased significantly, following shortly by a fully develop of the combustion reaction (Pic.103). Finally, the steady state of reaction was achieved with lower OH* distribution and with the smaller region of OH* territory, compared to the previous stage; however, it showed quite steady location without considerable movement.

Large amount of gray scale high-speed pictures are exported by DaVis and post-processing to measure light intensity by the software called ImageJ, where the quantity of light intensity is normalized by its maximum value over the observed time period was conducted in order to present a better comparison of the light intensity degree at particular times.

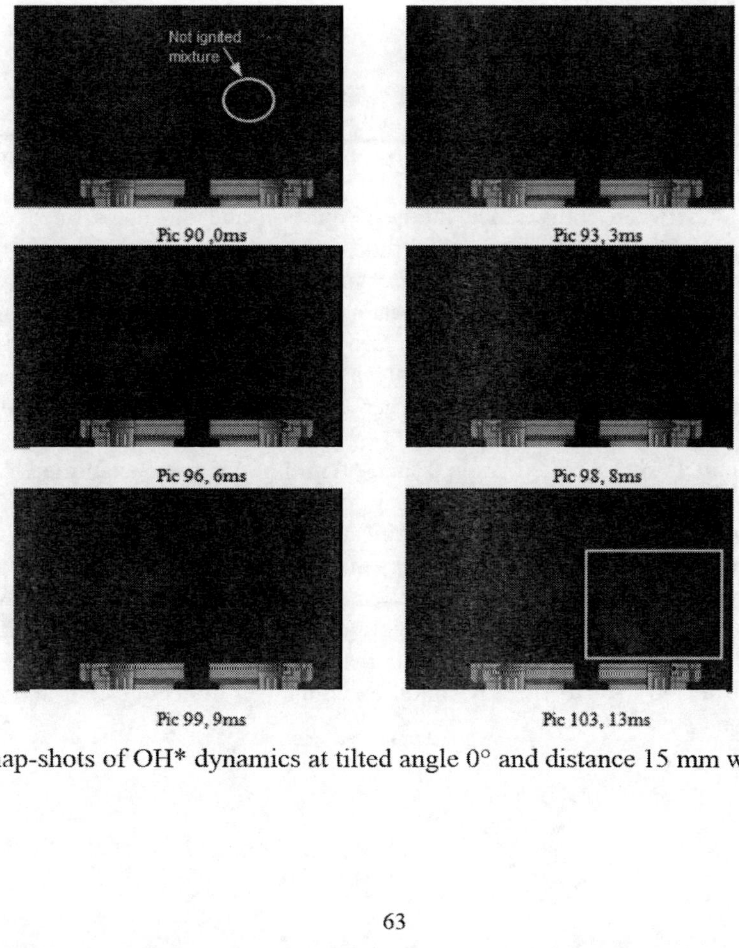

Figure 13. Snap-shots of OH* dynamics at tilted angle 0° and distance 15 mm with 1K resolution.

IOP Conf. Series: Materials Science and Engineering **297** (2018) 012006 doi:10.1088/1757-899X/297/1/012006

Normalized-light intensity (figure 14) over time was observed and is demonstrated in the box in Pic. 103 of figure 13. The figure shows OH* distribution during obvious stages of development of flame propagation. In the first step, light intensity (OH*) of second burner is not indicated. A formed cloud starts travelling to downstream (Pic. 90-98). In other words, higher ignition energy is generated, which is enough to ignite second burner. Finally, complete ignition of the consecutive burner is noticeable with the medium light intensity fluctuation.

The plot in figure 14 shows the time development of normalized-light intensity at 0° tilted angle and distance 15mm achieved after data acquisition.

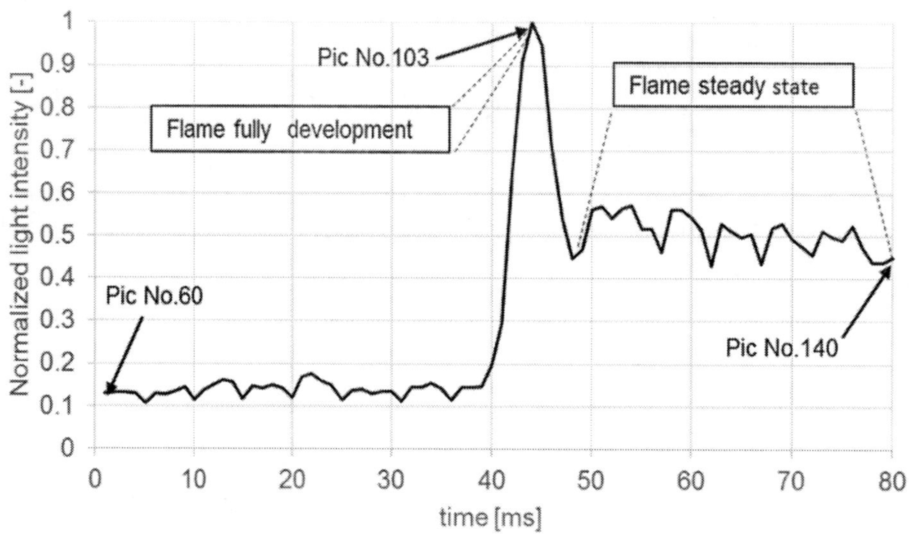

Figure 14. Time development of normalized-light intensity at 0° tilted angle and distance 15mm.

6. Conclusions

Test rig with twin burners has been implemented to study the correlation of the distance between burners and their tilted angles in order to get a better understanding of the condition driving a light–across ignition phenomenon. The methodology applied for the experiments were also contributed rationally in detail.

The relationships between equivalence ratios of ignition and distances of all tilted angles show apparently linear trends (figure 10). The Ignitability unit, which is a reciprocal of global equivalence ratio, was introduced as a parameter to evaluate and compare flame propagation potential among experimental cases. Tilted angle 75° degree performed the highest ignitability above unity at all interested distance, compared with the others (figure 12). In other words, the greater tilted angle, the larger distance between burners can be achieved at a particular ignitability unit. It was found that a tilted burner angle compensates a difficulty of a larger distance between burners and facilitates leaner combustion.

Flame behaviour at 15 mm of tilted angle 0° and 60 mm of 45° were examined by LaVision's high-speed imaging system equipped with OH* filter in order to eliminate interfered wave lengths. OH* radical behaviour of both cases were almost the same. OH* appeared rarely in the region of an unignited burner at a lower equivalence ratio. Increasing the equivalence ratio, OH* accumulated and augmented to be a cloud at the downstream region. At this state, the cloud moved upstream to an unignited burner (figure 13). Until it reached a lower area near the outlet of burner, OH* quantity enlarged substantially over a territory and the highest light intensity emitted at this state. Finally, OH* concentration decreased with shrinking in dispersed areas to a steady condition with lower light emissions and a minor fluctuation beyond this timeframe.

The results provide a hint and highlight an approach to increase gas turbine ignitability and mitigate a flame propagation trouble of consecutive burners for lean and ultra-lean combustions. Possibility to reduce burner units around combustion chamber is accessible as a result in lower weight per unit of power output, cost effectiveness and less maintenance expenses. In addition, the data can be used for validating computational simulations of combustion flame propagation as a benchmark and offering more insights of the light-across phenomenon.

Other aspects of properties such as local equivalence ratios, local temperature, element chemical compositions, etc. are also beneficial for better understanding the nature of excited radicals, promoting flame propagation across sectors as well.

References

[1] International Energy Agency 2013 *CO2 Emission from Fuel Combustion Highlights* (Paris: IEA)
[2] Lefebvre A H and Ballal D R 2010 *Gas Turbine Combustion* (Phiadelphia: Tayloy & Francis)
[3] Antoshkiv O, Berg H P and Mundstock J 2013 *Conf. 26. Deutscher Flammentag* Duisburg-Essen
[4] Boileau M, Staffelbach G, Cuenot B, Poinsot T and Bérat C 2008 *Combustion and Flame* **154** 2
[5] Bach E Kariuki J J Dawson R and Mastorakos E 2013 *Conf. 51st AIAA Aero. Sci. Meeting* Grapevine (Dallas/Ft. Worth Region)
[6] Bourgouin J F, Durox D, Schuller T, Beaunier J and Candel S 2013 *Combustion and Flame* **160** p 1398
[7] Barrè D, Esclapez L, Cordier M, Riber E, Cuenot B, Staffelbach G, Renou B, Vandel A, Gicquel L Y and Cabot G 2014 *Combustion and Flame* **161** p 2387
[8] Xu W, Gui N, Ge L and Yan J 2014 *J. of Computational Engineering* **2014**
[9] Kao Y, Tambe SB, Jeng S 2014 *Proc. ASME. Turbo Expo vol 4A* (Düsseldorf)
[10] Warnatz J, Maas U and Dibble R W 1996 *Combustion* (Hiedelberge: Springer)
[11] Antoshkiv O Berg H P Jobusch C and Poojitganont T 2013 *Proc. XXI Int. Sym. on Air Breathing Engines vol 2* p 1217
[12] LaVision GmbH 2014 *DaVis 8.2 software* (Göttingen: LaVision GmbH)

Hydrothermal carbonization of rice husk for fuel upgrading

N. Suteerawattananonda[1,*], N. Kongkaew[2] and S. Patumsawad[1]

[1] Department of Mechanical and Aerospace Engineering, King Mongkut's University of Technology North Bangkok, 1518 Pracharat I Rd. Wongsawang Bang-Sue, Bangkok 10800, Thailand
[2] Department of Materials Handling and Logistics Engineering, King Mongkut's University of Technology North Bangkok, 1518 Pracharat I Rd. Wongsawang Bang-Sue, Bangkok 10800, Thailand

[*] Corresponding Author: s.natayos@gmail.com, +66874551133

Abstract. The biomass is popularly used as renewable energy. In Thailand rice is the most consume agricultural products. Agricultural residues from rice husk can be an energy resource. However, alkali and alkali earth materials (AAEMs) in biomass ash are the causes of corrosion and erosion problem in the heat exchanger equipment, while the acidity of ash affects the slagging agglomeration problem. Reduction of alkali and alkali earth materials can minimize the problem. In order to challenge the reduction of alkali and alkali earth materials in biomass ash, hydrothermal carbonization process was selected. Thai rice husk was used as sample to compare the result of treatment. The rice husk was heated under the condition of different temperature ranged from 180°C to 250°C, at operate pressure ranges from 12 bar to 42 bar with residence holding reaction time 1 hour. The results of proximate analysis show that the percentage by mass of fixed carbon are increased 2 times, but volatile matter is decreased by 40% and ash content is decreased by 11% due to the increment of temperature. Meanwhile, the X-Ray fluorescence (XRF) analysis results show the decreasing of alkali and alkali earth materials are reduced.

1. Introduction

The renewable energy was invented to fulfill the energy consumption demand due to population which is growing up day by day. The biomass is the renewable energy with the advantages, the biomass is green energy and biomass can be found in everywhere. While the combustion produces CO_2, agriculture biomass consumes CO_2. The CO_2 transfers in cyclic and it is not added to the atmosphere, with the CO_2 cyclic the biomass were classified to the green energy. The biomass can be found in everywhere and this is the advantage over the fossil fuel. Because of the fossil need the proper environment to discover, so its take more value than biomass.

Combustion is the oldest and easiest method to convert the biomass to energy. However, use of biomass as an energy resource in combustion process has some disadvantages due to high moisture content, uncertainty of heating value, and high ash (alkali metal) content resulting in fouling and corrosion problem in the boiler.

In 2016, Thailand consumes 2,862 kilotons of rice husk to produce energy which is 5.78% of total biomass used as an energy resource. [1] Generally, rice husk has high alkaline in composition. The case of low effective when use in the direct combustion process. Hydrothermal is one of treatment processes to

IOP Conf. Series: Materials Science and Engineering **297** (2018) 012007 doi:10.1088/1757-899X/297/1/012007

improve quality of solid fuels. The study of hydrothermal carbonization applied to coconut fiber and eucalyptus leaves in two different temperature range, 150°C-300°C and 300°C-375°C. Hydro-char fuel properties were significantly improved compare to feedstock biomass [2]. The acidity component in tropical woody biomass ash was minimized by using the hydrothermal carbonization pretreatment in mild subcritical condition [3, 4].

The objective of this paper is to study the effect of temperature to properties of Thai rice husk treated under hydrothermal process.

2. Material and Methods

2.1. Material

Rice husk is the agricultural residue products from rice production obtained in Thailand. The 1cm sizes of rice husk (figure 1.) were dried in large batch at 95°C with 24hr residence time.

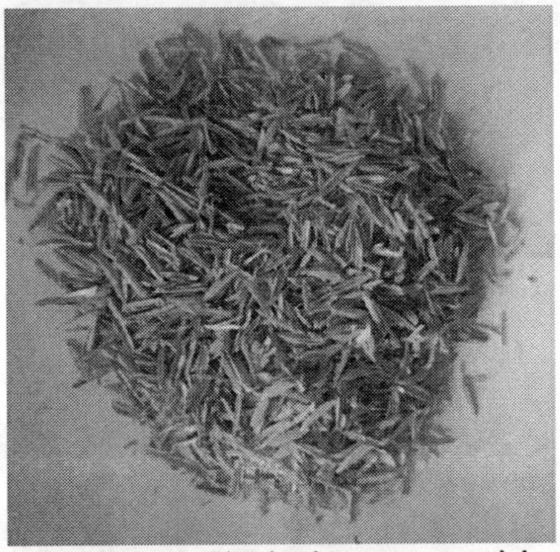

Figure 1. Rice husk as raw material

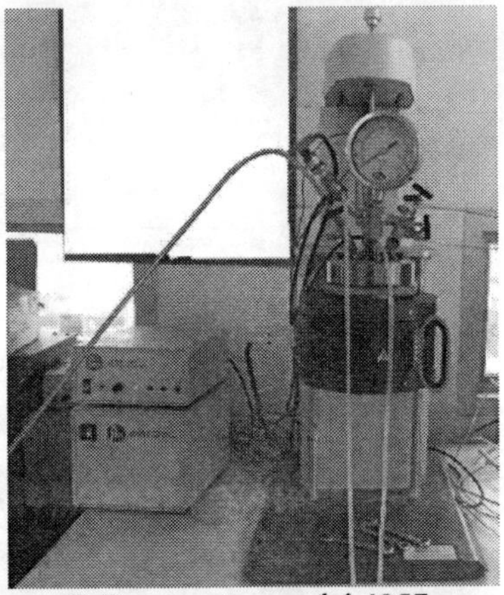

Figure 2. Parr reactor model 4857

2.2. Methods

The hydrothermal carbonization process was performed via Parr reactor model 4857 (figure 2.), under the variety operating temperature 180, 200, 220 and 250. The 50g (dry basis) of raw material was put and 500ml deionized water was added, with the purpose of control the pressure range from 12 bar to 42 bar in the reactor vessel. Then, seal the reactor and purge residual air by using the Nitrogen gas. Then heating the reactor with electric coil to the desired temperature and maintains residence time for 1 hour. After the reactor cools, the bio-gas was removed, the bio-char and bio-oil were separated by a filter. The bio-char was dried before analyze process.

3. Result and Discussion

3.1. Proximate and ultimate analysis

Proximate analysis was done under the ASTM D7582. CHN analyzer (LECO Corporation, 628 Series: CHN) was used in the ultimate analysis. Proximate analysis of bio-char result from table 1 shows the decrease of moisture content from 8.75% to 3.26%,. Also, the volatile matter content is reduced from 67.44 to 23.23%. In the opposite, the fixed carbon is increased from 12.81% to 25.91% as shown in Figure 3. The ultimate analysis shows that the percentage of carbon is slightly up from 37.76% to 40.27%. The hydrogen is slightly down from 5.12% to 4.57% and the oxygen is also decreased from 37.09% to 28.23%. The effect from the thermal decomposition of hemicellulose which the most sensitive components of three main components is first decomposition compound related to the lower of H, O and moisture when the temperature increase. In the opposite the retained lignin and cellulose fraction generates the non-condensable gases such as H_2, CO, CO_2 and CH_4 from char that relate to the increment of carbon content

[3]. The lower in volatile matter and higher carbon at higher temperature of hydrothermal carbonization are satisfied to the study of conversion sewage sludge into hydro-chars at 200°C hydrothermal carbonization process was studied by Chao He [5]. The heating value increased at higher temperature from the study of D. Kalderis [6] also satisfy with this study.

However, the hydrothermal carbonization pre-treatment causes the weight loss in biomass which related to loss in energy, the proximate analysis results were weighted by mass loss as shown in figure 3. The ash component in entire samples remains the same as raw material. While the increment of hydrothermal temperature the volatile matter and oxygen content trends go down. In the opposite, the fixed-carbon trends are raised. Whereas the carbon content form ultimate analysis weighted by mass loss cannot see the significant different.

Table 1. The result of proximate analysis and ultimate analysis

Temperature	Proximate (%, d.b.)				Ultimate (%, d.b.)				Mass loss (%)
	Moisture (w.b., %)	Ash	Volatile Matter	Fixed-Carbon	%C	%H	%N	%O	
Raw Rice husk	8.75	19.75	67.44	12.81	39.01	5.73	0.21	35.29	0
180 c	5.191	22.99	63.32	13.68	39.57	5.10	0.31	32.02	23.60
200 c	4.23	25.34	61.35	13.30	41.23	4.76	0.36	28.30	26.58
220 c	4.19	26.48	57.52	16.00	42.03	4.77	0.45	26.25	31.40
250 c	3.26	31.99	42.09	25.91	46.64	3.96	0.69	16.70	44.80

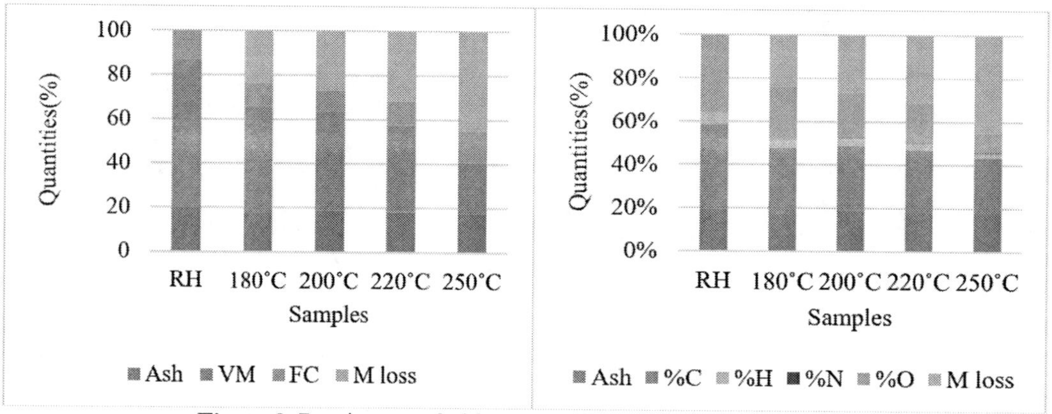

Figure 3. Proximate and ultimate analysis weighted by mass loss

3.2. Energy

The energy loss was concerned with purpose of fuel uses. Heating value trends are increased due to the increase of temperature, but the mass yield trend goes down (figure 3.). The correlation between mass yield and heating value were specified in term of energy loss. The highest heating value of solid is 19.96 MJ/kg at 250°C, but the mass yield is the lowest at 55.19%. However, the energy loss is maximum at 250°C but minimum at 180°C

Table 2. Energy and mass yield

Temperature	HHV (d.b., MJ/kg)	Mass yield (%)	Energy loss (%)
Raw material	16.77	100	0
180°c	15.92	76.40	4.60
200°c	15.69	73.42	5.25
220°c	17.03	68.60	5.08
250°c	19.96	55.19	5.75

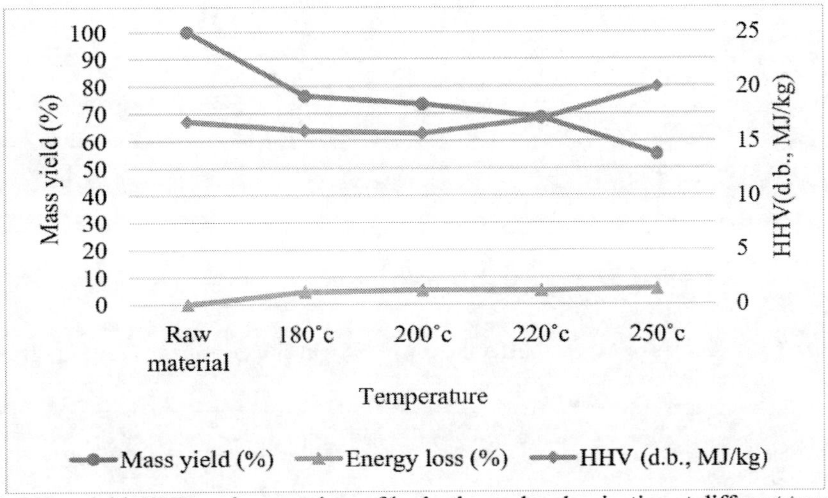

Figure 4. Mass yield, HHV and energy loss of hydrothermal carbonization at different temperature.

3.3. Ash component analysis

The use of X-Ray fluorescence (XRF) analysis with an X-Ray fluorescence spectrometer (Bruker, model S8-Tiger) to obtain the ash component in term of oxide compound. The chart of CaO, K_2O and P_2O_5 and residence temperature trend are decreasing (figure 5-8.). The significant downward trend of alkaline at 180°C indicated that the alkaline was pulled out at lower temperature. Therefore, at the higher temperature the alkaline quantities are remained approximately. These results correlated with the study of C.A. Cuvilas [3] in tropical woody biomass treated by hydrothermal carbonization at 175°C-350°C. Whereas the change of temperature affect the reduction of alkaline in biomass ash but slightly affect to the reduction of metal as in the figure 9. Shown the fairly constant quantities of Fe_2O_3. The summary of ash compound quantities in term of slagging index revealed that there are not significant change at temperature above 200°C (table 3.).

Table 3. Ash component form XRF analysis

Sample	Quantities (%)						
	MgO	CaO	Na_2O	K_2O	P_2O_5	TiO_2	Fe_2O_3
Raw material	0.443	0.470	N/A	2.17	0.683	N/A	5.99E^{-4}
180°c	0.133	0.376	N/A	0.652	0.205	N/A	4.26E^{-4}
200°c	0.135	0.167	N/A	0.453	0.119	N/A	3.79E^{-4}
220°c	0.136	0.184	N/A	0.507	0.17	N/A	4.86E^{-4}
250°c	0.128	0.123	N/A	0.556	0.177	N/A	4.58E^{-4}

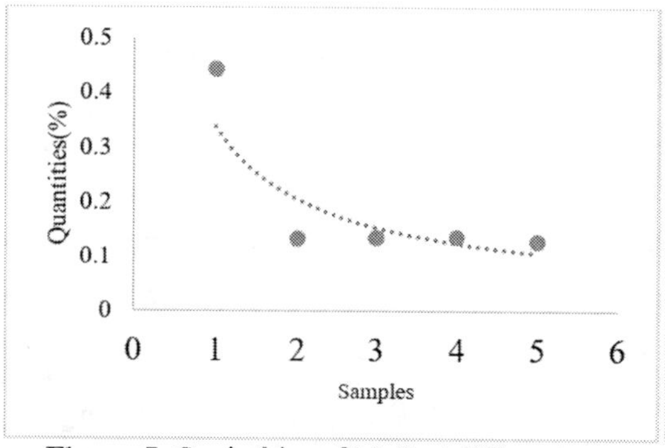

Figure 5. Quaintities of MgO at different temperature.

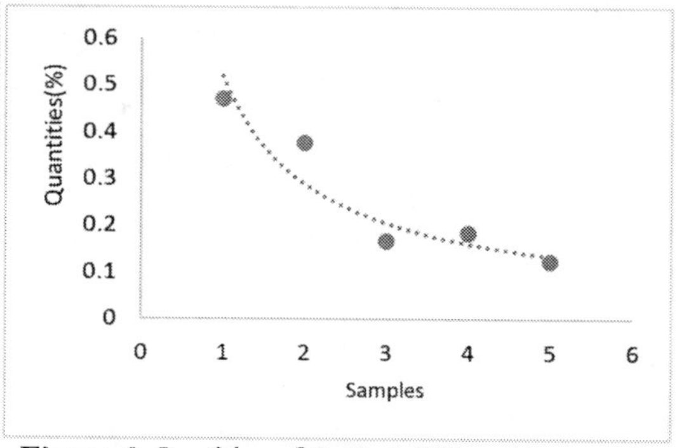

Figure 6. Quatities of CaO at different temperature.

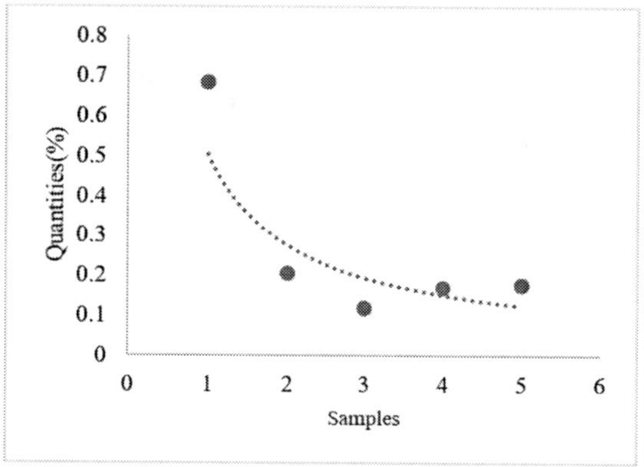

Figure 7. Quantities of K_2O at different temperature.

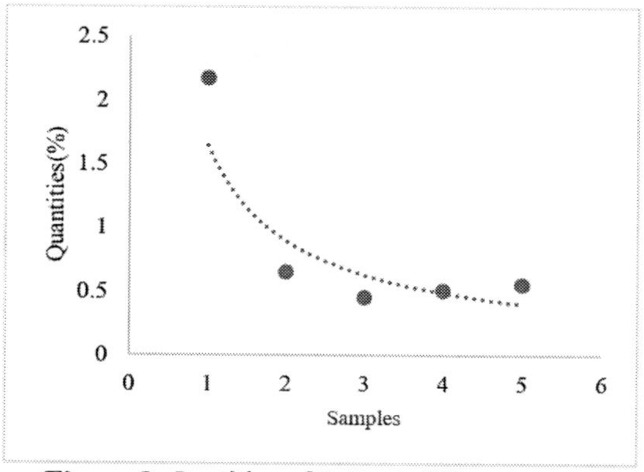

Figure 8. Quatities of P_2O_5 at different temperature.

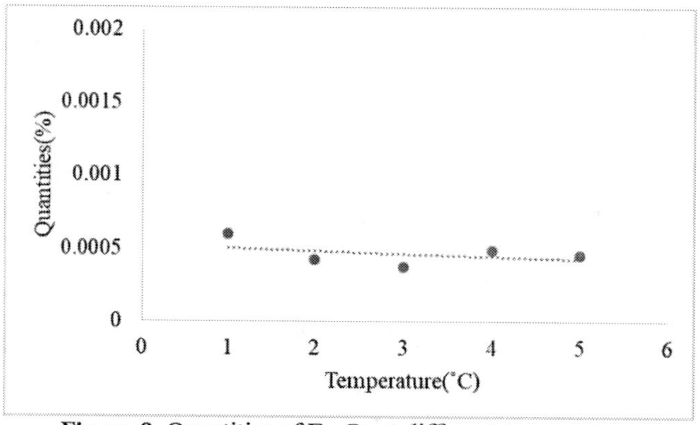

Figure 9. Quantities of Fe_2O_3 at different temperature.

3.4. Fuel classification

The H/C and O/C plot in figure 10. show the steadily decrease due to the increment of temperature. The treated rice husk properties trends go toward to the coal and anthracite. The data indicate that the treated rice husk fuel properties were upgraded.

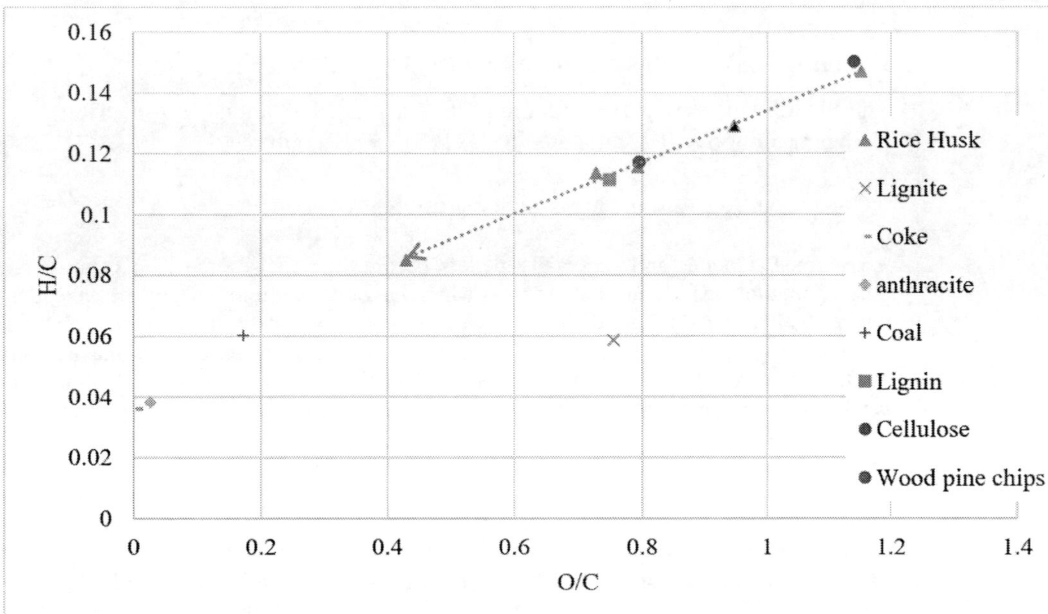

Figure 10. The O/C and H/C classification plot

4. Conclusion

The change of hydrothermal carbonization temperature in these study indicate that the alkaline component in ash were treated at low temperature of the process. Then the higher temperature process does the slightly effect on the alkaline component in ash. However, the combustion properties are getting better due to the increase of temperature. As the fuel uses, hydrothermal carbonization at 180°C is the lowest percentage energy loss of energy.

References

[1] Department of Alternative Energy Development and Efficiency 2016 *Thailand Table of Energy Balance*
[2] Zhengang Lui and Balasubramanian R 2012 *Procedia Environmental Sciences* **16** pp 159 – 166
[3] Cuvilas C A, Kantarelis E and Yang W 2015 *Energies* **8** pp 2165-2175
[4] Zhao P, Shen Y, Ge S, Chen Z and Yoshikawa K 2014 *Applied Energy* **131** pp 345–367
[5] Chao He, Giannis A and Wang J Y 2013 *Applied Energy* **111** pp 257–266
[6] Kalderis D, Kotti M S, Méndez A and Gascó G 2014 *Solid Earth* **5** pp 477–48
[7] Febrero L, Granada E, Regueiro A and Miguez J L 2015 *Energies* **8** pp 9794-9816

IOP Conf. Series: Materials Science and Engineering **297** (2018) 012008 doi:10.1088/1757-899X/297/1/012008

Development of the compaction machine for the production of new shapes of pressed biofuels

Ľubomír Šooš *, Miloš Matúš, Juraj Beniak and Peter Križan

The Slovak University of Technology, Faculty of Mechanical Engineering, Námestie slobody 17, Bratislava, 812 31, Slovak Republic

* Corresponding Author: lubomir.soos@stuba.sk

Abstract. Briquettes and especially pellets became the fuel of the 21st century. These are pressed biofuels made from the biomass which have the required heat, shape, size, density and mechanical properties. Today, these pressed biofuels are made in the form of a block, cylinder, n-angle octagonal, either without or with the holes. Several analyses confirm that neither a block, nor the cylinder is the optimal shape for the production of pressed biofuels, both in terms of the production, storage, automated transport in the combustion process and the optimum combustion process. For this reason, we began to analyse different shape, size, density and mechanical properties of briquettes and pellets. In the first part of this article, the biofuel is described from these points of view. The result of this analysis is the new optimized spheroid shape of the pressed biofuels. The goal of the second part of the article is the construction design of a new compacting machine for manufacturing of the optimized shape of the compacted piece. The task is demanding due to the fact that in comparison to the production of cylindrical or square-shaped compacted pieces, the manufacturing of ´quasi-spherical´ compacted pieces is discontinuous. Furthermore, unlike the standard types of compaction presses which compact the material between the two cylinders, it is necessary to hold the compacted piece for certain time under high pressure and at the high temperature. In this way, the lignin contained in compacted raw material becomes plastic and no further binding material needs to be added. The kinematics of a new compactor was therefore divided into two stages- ´the stage of compacting´ and ´the stage of load bearing capacity. This article describes an innovative and patent protected principle of compactor construction. The prototype of a designed machine has already been produced in our department. The first test results of this machine production as described in the conclusion of the paper confirm that kinematics and compactor construction were both correct.

1. Introduction

Prehistoric man uses nature as a source of raw materials and energy. Firstly this was wood, then crude oil, natural gas and other materials that mankind was, over a period of time, capable of processing. Compacting technologies have been in operation for over 150 years. The first patent for the production of wood briquettes comes from the year 1864, [1]. Wood briquettes have been used in the USA for 75 years. One of the best fuel logs (briquettes) was manufactured and sold as the "pres-to-log" in 1933. In recent times there has occurred a worldwide 'boom' in using refined material, mostly in the form of pellets. The main advantage of pellets is the possibility of fully-automated delivery in the process of combustion, along with very precise regulation of the combustion process, even for low-performance furnaces. Until the present, pellets throughout the world are produced just in a cylindrical shape. But often the question is asked: Is the cylindrical pellet, with its dimensions, optimal from the standpoint of size and shape? The goal of the presented paper is to search out an answer to this question.

IOP Conf. Series: Materials Science and Engineering **297** (2018) 012008 doi:10.1088/1757-899X/297/1/012008

2. Types of refined bio-fuels

Among the best-known technologies of compaction, we class briquetting, pelleting and compaction. The resultant products of these technologies are briquettes, pellets and granules (figure 1).

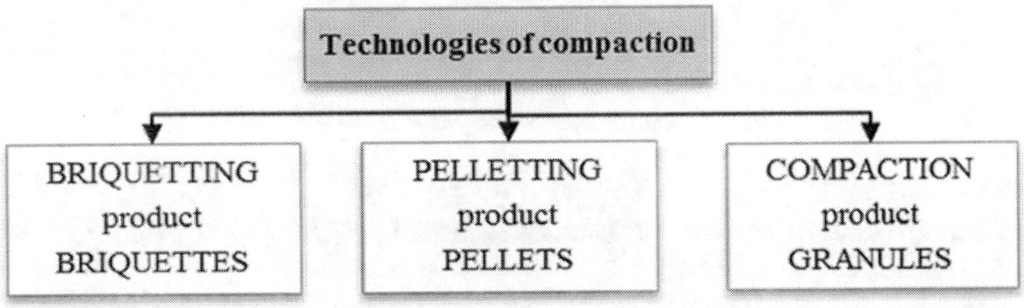

Figure 1. Technologies of compaction.

2.1 Briquettes

Briquettes in the past were the most widespread type of. A great advantage of these refined fuels, in comparison with classic fuels, is that according to need we can in their production change all three decisive parameters: shape, size and density and thereby optimize expenses for their production or requirements for shipping or storage, but also the requirement from the standpoint of their optimal energy recovery.

On the other hand, however the production of a briquette press falls dramatically (up to 30%) as a result of the openings.

Briquettes are manufactured on mechanical, hydraulic or screw briquette presses. The highest quality and strength of briquettes is achieved on screw briquette presses. But with this method there is greater wear on the working equipment – the screw press. The molds produced are generally in the form of cylinders, blocks or n-angles. In conformity with DIN 51731 the characteristic size is greater than 30 milimeters.

Briquettes are suitable for kitchen stoves, brick ovens, level and central heating and for fireplaces. Burning time of briquettes is from 30 – 60 minutes, depending on their shape and size. Larger pressed fuels catch fire less easily, burn worse, and their combustion tends to be less ideal. Briquettes with larger diameters are also made with holes (figure 2).

a) cylindrical briquettes without holes b) octagonal briquettes with holes c) block briquettes

Figure 2. Actual shapes of briquettes

2.2 Pellets

The percent share of pellets on the refined fuels market is increasing dramatically. Currently pellets are made exclusively in a cylindrical shape, with diameters of 6-25 millimetres and lengths The percentage of pellets on the refined fuel market is increasing dramatically. Currently, pellets are made in a cylindrical shape, with diameters of 6-25 millimeters and lengths of pellets equals to two to five times the diameter (figure 3).

Figure 3. Pellets

Greater sized pellets are used only for combustion in large furnaces and cement-making. The greatest advantage of pellets is that, due to the size and homogeneity of the fuel, a fully-automated combustion process is possible. The disadvantages of pellets are the high ratio of the fuel's surface to its volume. With this is related the burning of volatile materials even at low temperatures (above 200°C), the very rapid burning process (10 – 20 minutes), and also high wear on the functioning parts (die, roller ...) of the pelleting mills. Another disadvantage is the high demand associated with the production technology. The material must be of high quality, homogeneous, able to disintegrate into very small particles, and to have optimal humidity. The high investment costs associated with the manufacture of the fuel and with its usage are also disadvantageous. The input power for a unit of volume output "$P_{j,w}$"[W/(kg h^{-1})] is lower for pellet presses than for briquettes. This has to do with the fact that the ratio of the area of the press openings to the total area of the die on which the press pressure acts is less than one. Only in special furnaces is it energy-effective to recover pellets. The very high demands on investments also apply to the user of the fuel. The construction of an entire heating system, including the storehouse, the feeders, the furnace, and the monitoring and management system, is a great barrier to a more rapid increase in users of this fuel in Slovakia.

2.3 Granules

Compacting is the technology for compressing waste whose tractability and humidity fit the parameters required for pressing between two counter-revolving smooth or grooved rollers. The results of compacting are granules or agglomerates of a slab shape which, after slicing into granules, are usable in further processing, primarily in the chemical and metallurgy industries. Granules are produced in various sizes and shapes, from cylindrical to spherical. Granules have a very convenient shape as regards the requirements for automated feeding. The ratio of the surface to volume is also advantageous with respect to the combustion process and wear on functioning parts.

The classic compacting technology is not very adapted to the compression of biomass. This is because the fundamental binder in biomass is lignin. In order for lignin to work as a binder, two basic conditions must be met. At the moment of compression, the lignin must be in a plastic state (temperature around 100 °C) and at the same time the pressed fuel must be subjected to very high pressure for a determined period of time. In compacting, the main problem is to ensure the required cooling of the molding under great pressure. So to achieve the required strength of the pressed materials, a binder is added.

Figure 4. Granules

3. Optimization of shape, size and density of pressed fuels

The quality and success of using refined biofuel (figure 5) depends in particular on:

1) the material properties of the compacted material (kind of material, size of the compacted particles, humidity, ...),
2) compaction technology used, (size and shape of the pressed fuels, working pressure, speed of compaction, temperature, endurance, energy consumption, wear on functioning parts, ...),
3) combustion conditions (sort, type and size of the combustion equipment, ...),
4) requirements for feeding in the combustion process (size, shape, ...),
5) requirements for storage (minimization of volume and space requirements).

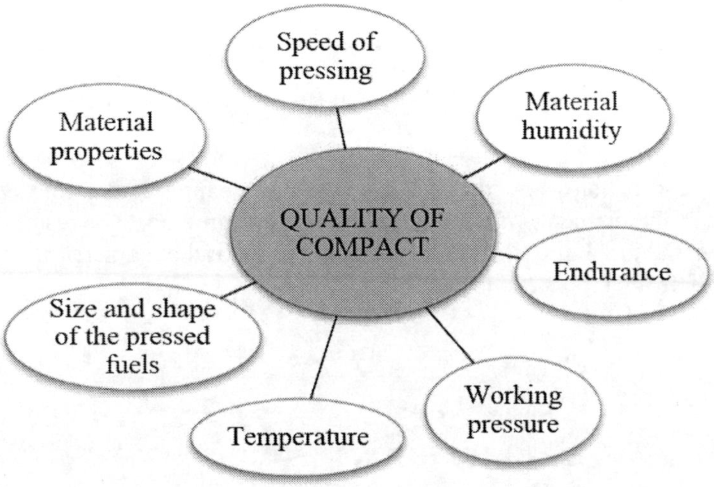

Figure 5. Factors influencing pressed fuel properties

Requirements for dimensional and mechanical properties, chemical composition and emissions are defined by DIN 51731, or Ö-NORM 7135 standards. Another mark of alternative pressed fuels is the possibility to influence their:

1) shape (block, ball, n-angle, cylinder, with or without openings),
2) size (from 6 millimeters to several dozen centimeters),
3) density (of the pressed fuel itself minimally > 1 kg dm^{-3}, pouring density),
4) humidity (maximum 15%),

strength (abrasion < 3%, mechanical strength under bending and pressure).

3.1 Surface to volume ratio optimization

It is interesting to compare the ratio of the surface area to the volume (S/V) of the pressed fuel. This is the function of the shape and size of the pressed fuel. With a higher ratio S/V (small pressed fuels) it is possible to more accurately manage the process of combustion, but the fuel burns more rapidly (10 – 20 min.), which is influenced, in addition to the large surface, by the combustion of the volatile materials at lower temperatures (over 200 °C). But disadvantages include the higher interface areas between the processed raw material and the pressing machines, therefore more rapid wear on the functioning areas (die, roller, piston,...) of the compaction machine. Other disadvantages are the increased requirements on the quality and uniformity of the processed particle of the processed raw material, and requirements related to the production technology. The material must be homogenous, disintegrated into very small particles, must have optimal humidity, and the optimal temperature must be assured during compacting. Unsuitable raw material or incorrect parameters during compaction will cause stoppage and complicated cleaning of the clogged die. High investment costs associated with the production of the fuel and with its usage are also problematic. The high demands on investments for the user of the fuel are too not negligible.

The cylinder is the most frequently produced shape of pressed fuels. Accordingly, we will compare how the ratio of the surface "S_c" to the volume "V_c" changes with a change of the diameter "*d*" or the height "*h*". The volume of the cylinder is defined by the relation

$$V_c = \pi r^2 h \quad (m^3) \tag{1}$$

and the roller surface

$$S_c = 2\pi r^2 + 2\pi r h \quad (m^2) \tag{2}$$

then the ratio of the surface to the volume is defined by the relation

$$\frac{S_c}{V_c} = \frac{2(\mathrm{r+h})}{rh} \quad (m^{-1}) \tag{3}$$

The first example illustrates how the "S_c/V_c" ratio will change with a constant height "*h=50*" mm and with a change of the diameter "*d*" in the range 10-25 millimeters. The result of the analysis is clear from figure 6. From the graph it is clear that with an increase of the diameter the surface to volume ratio will change by 55% from 440 to 200 (m^{-1}).

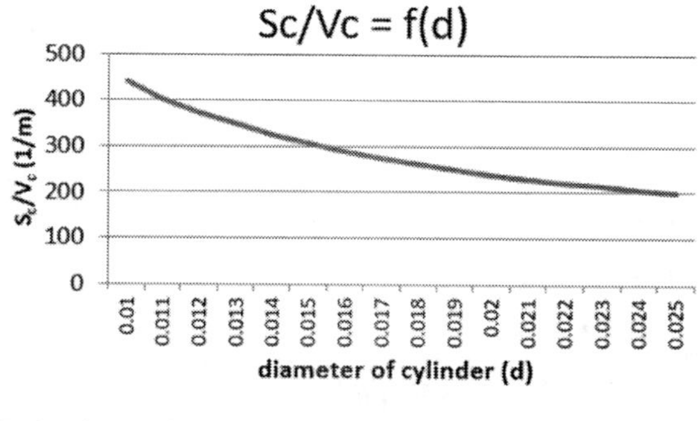

Figure 6. Change in S_c/V_c in dependence on change of the diameter "d" at constant height "h"

With the second example, with an increase of the diameter we proportionally also increased the height of the pressed fuel "*h*", in the ratio "*h=3d*" (figure 7). Pellets with a diameter d=10 mm and height h=30 mm correspond to the smallest diameter, and on the other hand the largest diameter briquettes with a diameter d=100 mm and height h=300 mm correspond to the largest diameter. From the figure it

IOP Conf. Series: Materials Science and Engineering **297** (2018) 012008 doi:10.1088/1757-899X/297/1/012008

is clear that with an increase in the diameter "d" and the height "h" the surface to height ratio fell by one order.

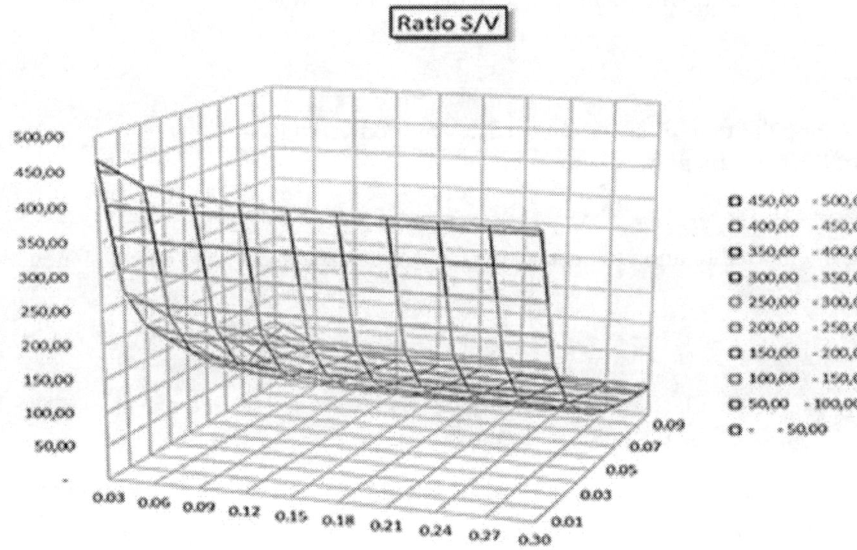

Figure 7. Change in S_c/V_c in dependence on change of the diameter "*d*" at constant height "*h*"

3.2 Density optimization

In conformity with international standards, the density of refined pressed fuels must be in the range of 1÷1,4 kg dm^{-3}. So it is generally valid that with an increase in the density of a pressed fuel the quality increases from the standpoint of resistance to interruption during transport, storage and in the automated feeding in the process of combustion. For the stated reason refined fuels must also fulfil requirements for abrasion and mechanical strength. Resistance against atmospheric moisture increases proportionally with increasing density of the pressed fuels. It is for this reason that with higher pressing power the compacted material overheats, the lignin melts down uniformly in the inner structure of the pressed fuel, and a large part of it flows to the surface which as a result becomes shiny and without pores. This causes an increase of the mechanical strength of the pressed fuel and at the same time prevents a leaking of humidity into the content of the compacted pressed fuel. With a negation of high density (>1.2 kg/dm^3) there occurs higher strain and wear on the compacting machine. Very high density can have an unfavorable influence on the burning process – imperfect combustion.

3.3. Shape optimization

From the standpoint of pressed fuel shape, the ball is optimal (figure 9a). The monitored ratio of pressed fuel surface to the unit volume is minimal. From this it follows a low wearing of the functional areas and a favorable burning method. The ball also has an optimal shape as regards automated transport of the pressed fuels. The problem however is production of such a pressed fuel. In classic compaction machines, maintaining the shape of the pressed fuel when cooling under pressure.

The changing ratio of the surface area to volume is not merely dependent on the change in the size of the particular geometric figure but also on the change if its shape. Let us compare, for instance, changes in the surface area of a cylinder and sphere in dependence on the change of their sizes.

The surface area of a sphere is calculated according to the formula:

$$S_S = 4\pi r^2 \tag{4}$$

and the volume according to

$$V_s = \frac{4}{3}\pi r^3 \tag{5}$$

The ratio of the surface and volume may be calculated as follows:

$$\frac{S_S}{V_S} = \frac{3}{r} \tag{6}$$

In order to compare the ratio of changing surfaces of monitored figures, let us assume that the volumes are of the same magnitude, i.e.

$$V_c = V_s \tag{7}$$

Subsequently, the mutual relation between the ratios of surface areas may be calculated as

$$S_s = S_c \frac{3h}{2(r+h)} \tag{8}$$

The diagram in figure 8 illustrates the change in their respective, mutual sizes. The basis for the comparison is the ratio of the diameter $'r'$ to the height $'h'$.

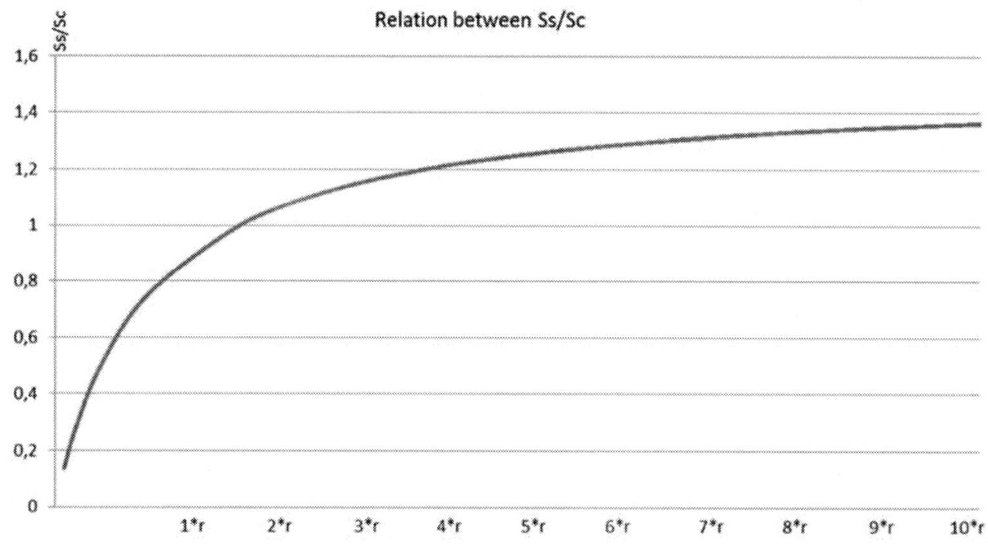

Figure 8. Relation between the ratios of surface sphere and cylinder "S_s/S_c"

Assuming that "h < 2r", the surface area of the sphere $'S_s'$ must be lower than that of the cylinder $'S_c'$. The equilibrium of the surfaces occurs at "h=2r". Upon further increase of "h", the surface area of the sphere $'S_s'$ will be higher than the surface of the cylinder $'S_c'$.

However, the particular surface areas of machine presses in the process of sphere manufacturing are higher than those created in the pressing of cylinders with higher length, (h>2r).

The intersection of two cylinders has a somewhat greater ration of pressed fuel surface to volume (figure 9b). The pressed fuel is also suitable for automated transport. Moreover, the sharp edges at the intersection of the cylinders provide better burning of the pressed fuel. For this pressed fuel shape it also possible to solve the maintenance phase problem.

In direct connection with the goals of the task, in 2011 a testing prototype of a compacting machine based on the principle of pressing between two rotating rollers was designed (figure 10). At present the

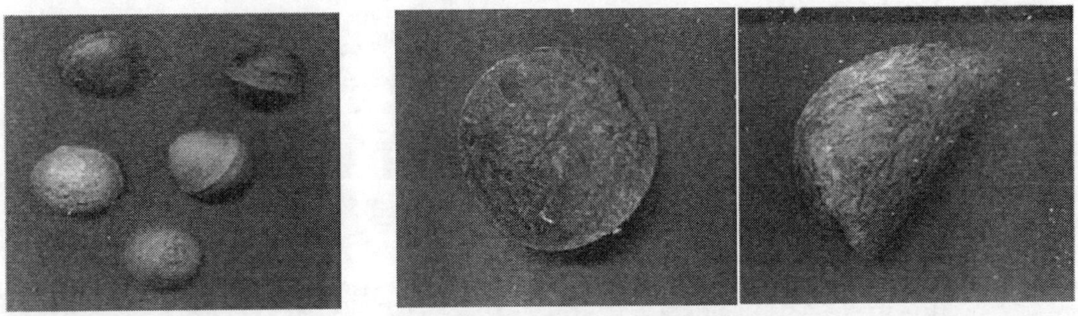

a) round pressed fuel b) intersection of two cylinders

Figure 9. New pressed fuel shapes [4], [5].

prototype is in the finishing stage, where it will have a driven exterior annulus, [5], [6]. Compaction occurs between the rollers on the inner side. The compaction machine consists of a pressing ring (1) with axial half-cylindrical recesses along its inner circumference and a second pressing disk (4) with a pressing roller groove on its circumference. The axial axes of the half-cylindrical recesses of the press ring (1) and the die roll (4) are rotated by 90 °. The output of the press roll (4) follows a calibration segment (3) encircling the axially polar recesses, which ensure that the pressings are cooled under pressure. The shape of the pressed fuel and the construction of the machine also address the maintenance problem by a so-called calibration zone (3). The heat needed during the pressing is obtained by friction. Finished fuel is selected by a breaker (2).

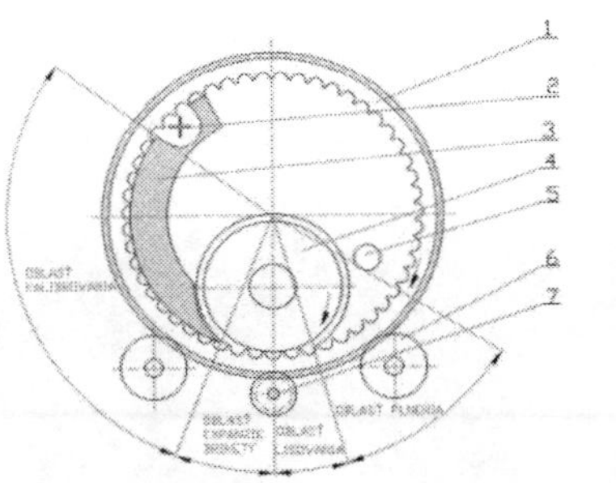

1-press ring, 2-breaker, 3-calibration segment, 4-an internal press roll, 5-filling hole, 6-supporting pulleys, 7 - pinion.

a) patent, [5] b) 3D model c) function model

Figure 10. The new compaction machine

We are currently working on the design of the multidimensional adjustable compacting machine, which presupposes and involves, on the one hand, the scalar change of the ratio between the radius of the pressing disk and the ring, and on the other hand, the change in the number of shape sequences of the ring and the disk.

4. Conclusion

The focus of the presented paper is optimization of the shape and dimensions of pressed fuels. The pressed fuels analysed were considered from the standpoint of the optimal ratio of the surface to the volume, the optimal density and optimal shape such that the standards required for automated feeding, and optimal combustion processes were fulfilled. The result of the optimization is the design of a new principle for a compaction machine which is characterized by:

1) sufficient pressure maintenance during the compaction of the pressed fuels, which assures the needed function of lignin as a binder,

2) minimal machine input with respect to a unit of output,
3) less sensitivity of the pressing technology to the type of material, particles and humidity, the possibility of variable changes to the machine's output,
4) minimal wearing of the functioning areas of the compacting machine.

References

[1] Strehler A 2000 *Ecological engineering* **16** pp 25-40
[2] Drzymala Z and Hryniewicz H 2000 *Problemy podwyzsania trwalosci brykieciarek walcowych stosowanych do kawalkowania drobnoziarnistych odpadow przemyslowych* (Akademia Górniczno-Hutniczna w Krakowie) pp 9
[3] Ľ. Žák 2011 *Viacparametrická optimalizácia rozmeru a tvaru výliskov z organického odpadu* Diploma work, KVT SjF STU Bratislava
[4] Šooš Ľ and Korgo S 2002 *Proc. 5th Conf. on Environment and Mineral (part. 2)* pp 577-583
[5] Šooš Ľ 2010 *Kontinuálny spôsob lisovania biomasy do optimálnych výliskov a zhutňovací stroj* Patent number: SK 287505. – Date of acquisition: 14. 10. 2010
[6] Gros P 2012 *Progresívme technológie na výrobu tuhých biopalív*, Práca kandidátskeho minima, KVT SjF STU Bratislava p 65

Experimental investigation of solid rocket motors for small sounding rockets

Thada Suksila

Advanced Propulsions and Rockets Lab
Department of Mechanical and Aerospace Engineering
Faculty of Engineering, King Mongkut's University of Technology North Bangkok,
1518 Pracharat Sai 1 Rd., Wongsawang, Bangsue, Bangkok 10800, Thailand

* Corresponding Author: thada.s@eng.kmutnb.ac.th, Mobile: 0963519519,
Fax.: 025869541

Abstract. Experimentation and research of solid rocket motors are important subjects for aerospace engineering students. However, many institutes in Thailand rarely include experiments on solid rocket motors in research projects of aerospace engineering students, mainly because of the complexity of mixing the explosive propellants. This paper focuses on the design and construction of a solid rocket motor for total impulse in the class I-J that can be utilised as a small sounding rocket by researchers in the near future. Initially, the test stands intended for measuring the pressure in the combustion chamber and the thrust of the solid rocket motor were designed and constructed. The basic design of the propellant configuration was evaluated. Several formulas and ratios of solid propellants were compared for achieving the maximum thrust. The convenience of manufacturing and casting of the fabricated solid rocket motors were a critical consideration. The motor structural analysis such as the combustion chamber wall thickness was also discussed. Several types of nozzles were compared and evaluated for ensuring the maximum thrust of the solid rocket motors during the experiments. The theory of heat transfer analysis in the combustion chamber was discussed and compared with the experimental data.

1. Introduction

Solid rocket motor is a rocket engine that uses the combustion process to break the chemical energy bond of the fuel and oxidiser to produce the thrust for the rocket. The fuel and oxidiser are mixed together to form a well-combined mixture whereas the binder binds the fuel and oxidiser. Many types of fuel and oxidisers can be used in a solid rocket motor [1]. As the solid rocket motor contains this explosive material, there might be some concerns regarding the safety regulations and laws depending on the geographic location, which require several documents to be presented for purchase, handling, and storage of such materials [2]. However, the study of solid rocket motor has become a fundamental subject in the field of rocket propulsion for aerospace engineering students. For that reason, many aerospace engineering institutes around the world are interested in the theoretical and experimental study of sounding rockets, as they help students to directly experience, develop, and apply their knowledge of rocket propulsion.

In 2006, T. John et al. demonstrated a launch vehicle for a small satellite using a sounding rocket [3]. In mid-2013, S. Singh designed and constructed a solid rocket motor that researchers could use as the main propulsion of a sounding rocket [4]. A. Okninski et al. developed the Polish small sounding rocket as a reusable CanSat launcher in 2015 [5]. In addition, sounding rockets for picosatellite launchers and rocket competitions have been studied and reported [6, 7].

This paper focuses on the development of the I-J class of solid rocket motors that use commercially available materials and chemicals for fuel and oxidiser as an alternative to avoid high cost, complex machining, and procurement challenges. It is a preliminary step for further development of solid rocket motors for Thai research students, which can enable them to design and construct rockets that have an altitude of 1-5 km in the near future.

2. Rocket Motor Principle

A basic rocket motor comprises three main parts: head cap, combustion chamber, and nozzle. The principal of rocket motor operation is shown in figure 1. Initially, the ignition fuse is ignited causing the propellant inside the combustion chamber to burn at a rapid rate. This combustion produces hot gas, which increases the pressure in the combustion chamber. The hot gas flows through the exit at the nozzle. Usually, the nozzle contains converging and diverging parts. The converging part has an exit area smaller than the entrance, which increases the velocity of the exhaust gas. The connecting part between the converging and diverging parts is called the throat. At this point, the velocity of the hot gas reaches the speed of sound, Mach (M) = 1. Beyond the throat, the exit area increases further to bring the hot gases to supersonic speed, as explained in the 'de Laval nozzle' [1, 2]. However, the pressure decreases as the gas flows through the nozzle.

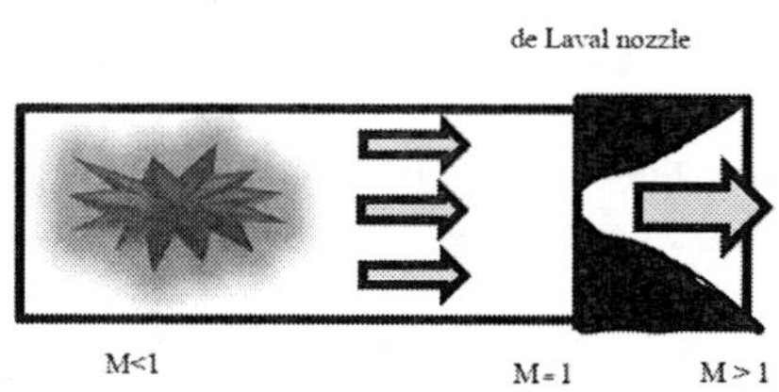

Figure 1. Principal of rocket motor operation.

The thrust is generated because of the increase in velocity of the hot gas and the pressure difference between the ambient and exit pressure at the nozzle. Thrust is given by the equation

$$F = \dot{m}V_e + (P_e - P_a)A_e, \tag{1}$$

where F is the thrust (N), $\dot{m}$ is the mass flow rate (kg/s), V_e is the exit velocity of the hot gas (m/s), P_e, P_a are the exit pressure at the nozzle and the ambient pressure respectively, and A_e is the exit area of the nozzle.

2.1. Rocket propellant

Rocket propellant is a well-combined mixture of fuel and oxidiser. The oxidiser provides oxygen for the fuel to burn. In some cases, a binder can be used for increasing the flexibility of the mixture in the casting process. The ratio of the fuel and oxidiser can be altered or the combustion chamber parameters can be changed to get different burn rates. The fuel and oxidiser are commercial grade and available with local suppliers. In this study, the appropriate ratio of fuel and oxidiser were obtained as follows.

IOP Conf. Series: Materials Science and Engineering **297** (2018) 012009 doi:10.1088/1757-899X/297/1/012009

2.1.1. Black powder. Traditionally, the black powder is prepared by dry mixing 15% charcoal powder, 10% sulphur powder, and 75% potassium nitrate by weight, and 1 g dextrin is added per 100 g of the mixture. The ingredients were mixed together for nearly 2 min in a plastic container to prevent electrostatic discharge. The black powder can be used as a fuse ignitor, as it burns rapidly. The maximum combustion temperature is approximately 1400 °C [8].

2.1.2. White mixed powder. In this study, white powder consisted of 65% potassium nitrate and 35% sugar powder by weight mixed well together. The final mixed white powder can absorb the moisture in the air; so, it must be kept sealed in an airtight container if meant for later use. The combustion temperature of the white mixed powder was approximately 1200–1300 °C.

2.1.3. Rocket Candy or R-Candy. This propellant used in our study contained a combination of sugar syrup, potassium nitrate, and sugar powder in the ratio of 65%, 18%, and 17% by weight, respectively. In this research, the rocket candy was prepared using an electric pot, as can be seen in figure 2. Initially, the sugar syrup was heated to boiling temperature. Then, sugar powder and potassium nitrate were mixed in it and the syrup was heated until it started boiling again, after which the electric pot was turned off and the hot mixture was poured into the moulds. The combustion temperature of R-Candy was approximately 1347 °C [9].

2.2. Material and propellant selection

The propellant selection criteria for cost and safe handling are commercially available. Based on the criteria, the white mixed powder was selected as the first propellant to test in rocket motor class I and the second propellant was prepared with the R-Candy for rocket motor class J. The segments used were PVC 5 and PVC 13.5 pipes, which contained the white mixed powder and R-Candy, respectively. The end burning with white mixed powder and tubular grain configurations with R-Candy were prepared. The end burning was quite easy to prepare in the PVC pipe; however, the tubular grain configuration was quite difficult to prepare in the casting and moulding process, as there was a risk of the R-Candy solidifying if the process took longer than 10 min. The wooden base was 1-in thick, and the rod used to cast and mould was ¾-in in diameter; the moulds and the PVC can be seen in figure 3.

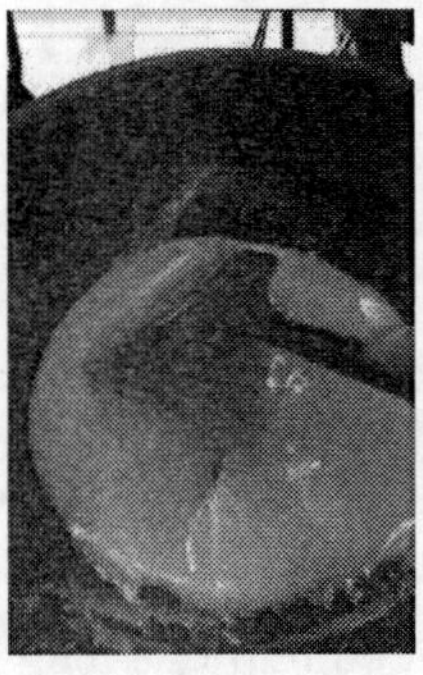

Figure 2. Rocket Candy heating in the PVC propellant electric pot.

Figure 3. Casting and moulding the propellants into segments.

IOP Conf. Series: Materials Science and Engineering **297** (2018) 012009 doi:10.1088/1757-899X/297/1/012009

3. Rocket Motor Design

The rocket motor consists of head cap, combustion chamber, nozzle, igniter charge, and propellant grain segment, as can be seen in figure 4.

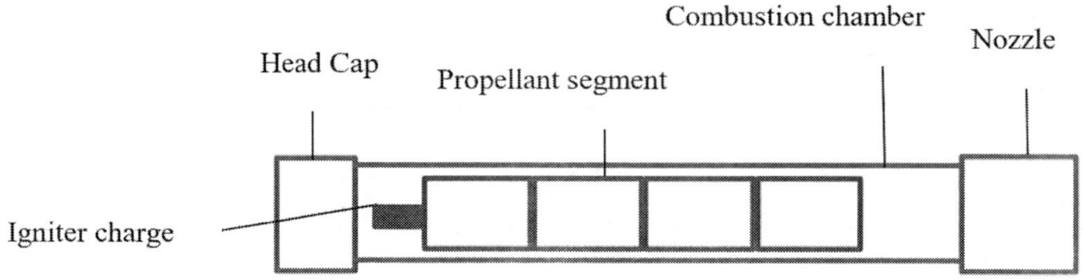

Figure 4. Basic rocket motor components.

3.1. Propellant grain segment using white powder

Four types of rocket motor nozzles were used in this study. The first was a straight connector of length ¾ in between the internal thread and outer thread. The second was the reduction connector from ¾ to ½ in, which acted as a divergent nozzle. Next was a straight internal thread connector with a washer as a throat inside the connector. Last was a straight internal thread connector without a washer. The completely assembled rocket motors are shown in figure 5. The length of the ¾-in PVC 5 tube was 5 in and it was filled with white mixed powder propellant with total propellant weights of 82.8, 85.5, 76.0, and 77.1 g, respectively. Then, the ¾-in head cap was connected to the PVC 5 tube and nozzle. The fuse ignitor was connected with the electrical wire and taped to the nozzle to avoid falling off.

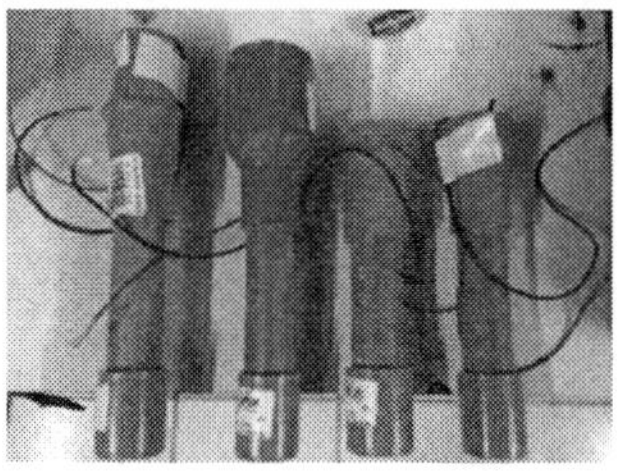

1 2 3 4

Figure 5. Rocket motors assembled with white powder propellant.

5 4 3 2 1

Figure 6. Rocket motors assembled with candy propellant.

3.2. Propellant grain segment using Rocket Candy

For this study, R-Candy with tubular grain segment was selected with PVC 13.5 as the propellant grain segment. The molten R-Candy was cast into the prepared segments. The total lengths of the propellant grain segments were 2.5 in, 5 in, 10 in, and 15 in. The propellant grain segments of length 2.5 in were selected, as the propellant could be conveniently cast and moulded into this length. Then, each 2.5 in segment was connected with electrical tape together until the required lengths of 5 in, 10 in, and 15 in were obtained. Next, the grain segments were slid into the PVC 13.5 shells of 3-in diameters. The fuse ignitor was placed at the top of the propellant segment so that it could burn rapidly and then, the head cap, outer shell, and nozzle were assembled together. The rocket numbers one to five were prepared with 2.5 in, 5 in, 10 in, and two 15 in propeller segments, respectively, as shown in figure 6.

3.3. Nozzle

Nozzle is the part that produces thrust in a rocket motor; the exhaust gases rapidly flow with increasing velocity as they pass through the throat and exit to the atmosphere. In the case of R-Candy propellant, the nozzle was cast with concrete and the washer was placed at the throat to sustain the high temperature and pressure, as can be seen in figure 7. The mixing ratio of concrete and water was 1:0.35 by weight [10]. The ratios of the divergent part of the nozzle can be seen in figure 8. The throat diameter was 1.5 cm and the exit nozzle diameter was 9.8 cm. For simplicity, the converging part of the nozzle was not included in the measurements.

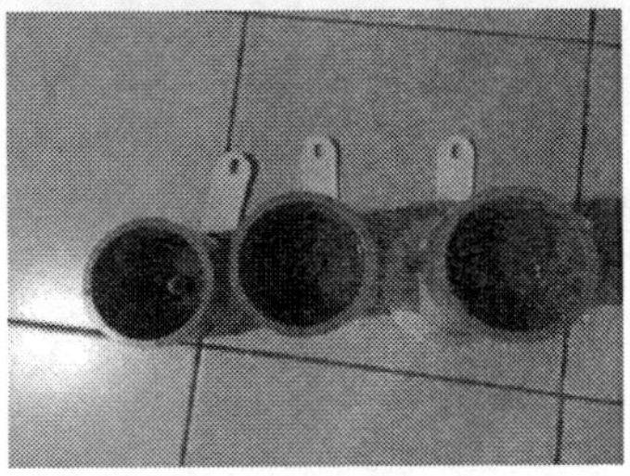

Figure 7. Nozzles manufactured using concrete and 2-in PVC connectors.

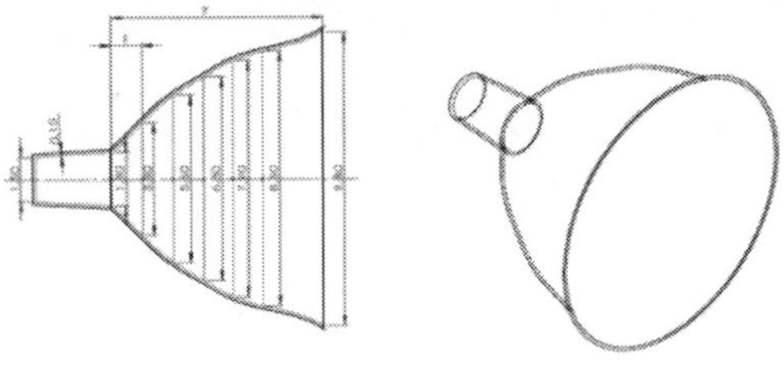

Figure 8. Dimensions of the rocket nozzle.

3.4. Ignition fuse and ignition system

Electrical ignitor switches are required for safety purposes. There are two types of switches—pre-fire and fire switches—as can be seen in figure 9. The wire from the switch was connected to the rocket motor and the other side of the switch was connected to the battery of 12 V and 8 Ah.

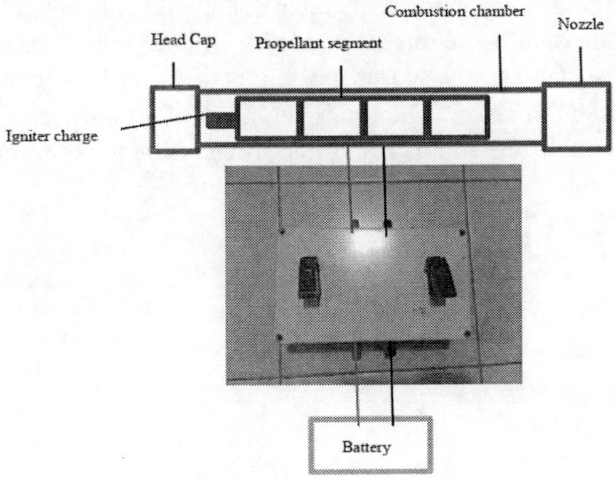

Figure 9. Electrical ignition switches.

3.5. Heat transfer analysis in combustion chamber

In this study, only conduction and convection were considered inside the combustion chamber and the radiation was neglected for simplicity. The 1D heat conduction is the transfer of heat through a material (solid, liquid, or gas) and can be expressed as:

$$q'' = -k\frac{dT}{dx} = \frac{k}{L}(T_{WH} - T_{WC}), \tag{2}$$

where q'' is the heat flux, k is the thermal conductivity of the material, L is the thickness of the material, and T_{WH} and T_{WC} are the hot wall and cold wall temperatures. The 1D heat convection is the transfer of heat from a surface (liquid or solid) to a fluid in motion and is described by:

$$q'' = h(T_G - T_W), \tag{3}$$

where h is the convection coefficient, and T_G and T_W are combustion gas temperature and wall temperature, respectively. The heat flows during conduction and convection are shown in figure 10 and figure 11, respectively. For safety reasons, the combustion chamber temperature was assumed to be approximately 2000–2500 °C.

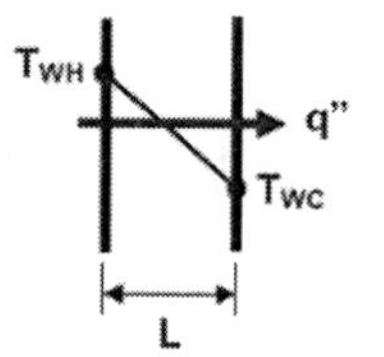

Figure 10. 1D heat conduction.

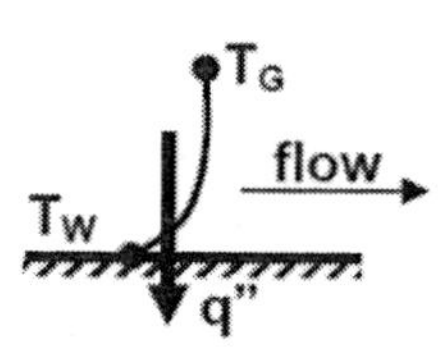

Figure 11. 1D heat convection.

3.6. Static rocket motor test stand

The static rocket motor test stands were made with square steel and connected together with screw. The width, length, and height of the stand were measured as shown in figure 12, and were 50 cm, 50 cm, and 150 cm, respectively. This test stand could be transported via a sedan. Moreover, the test stand could be used in the vertical or horizontal position. The total weight of the stand was around 30 kg. The rocket motor was tightly secured into this test stand. The thrust force of the rocket motor was recorded using a digital scale and a camera recorder, as can be seen in figure 13.

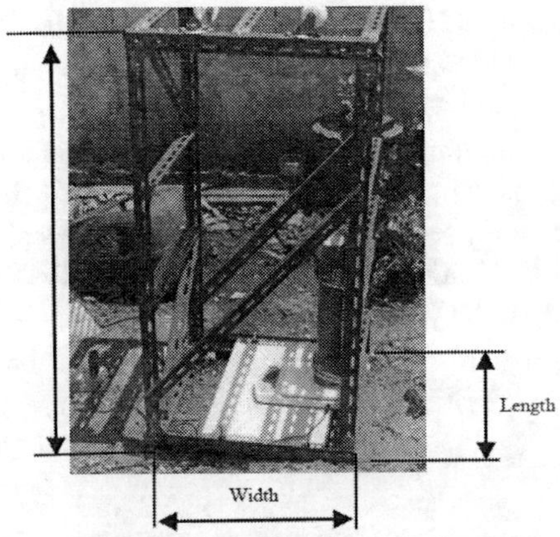

Figure 12. Rocket static test stand dimensions.

Figure 13. Static rocket testing and VDO recorder.

3.7. Safety regulations

Safety was our first priority in this lab and we followed the safety regulations for an amateur rocket [11]. In addition, a screen was used to cover the structure to prevent the debris from scattering except on one side, which was kept open to release the exhaust gases. The surrounding area was cleared and prepared for the test every time.

4. Testing and Results

The rocket motor numbers one to four, which used white mixed powder, provided the total impulse of approximately 100 N·s, which is similar to that of a class G rocket. The end burning time was determined as approximately 30 s using the camera recorder. The initial propellant weight was approximately 78–82 g and the average thrust was around 3.2–3.6 N. The total impulse was calculated as approximately 98–110 N·s and the maximum thrust was measured by a digital weight as approximately 83–103 N. However, the maximum thrust of the rocket motor number one could not be measured, as the combustion chamber melted and developed a hole of approximately 2-mm diameter because its burning time was longer than that of the other parts. Based on the results, the rocket motor number two provided the maximum thrust using the divergent nozzle. The heat flux was approximately 0.35–0.44 MW, assuming that the thermal conductivity of PVC is 0.19 W/(m K) and its melting point is approximately 160 °C [12-13]. The results for the white mixed powder rocket motor can be seen in table 1, assuming that the specific impulse (I_{sp}) of the white mixed powder and R-Candy [1-2] is approximately 130 s. The total impulse (I) can be calculated as:

$$I = (Wp) \cdot (Isp), \tag{4}$$

where I is the total impulse, W_p is the propellant weight, and I_{sp} is the specific impulse.

Table 1. Testing results for the white mixed powder rocket motors.

Rocket number	Burning time (s)	Propellant weight (g)	Average thrust (N)	Total impulse (N·s)	Maximum thrust (N)
1	32	82.8	3.29	105.59	-
2	30	85.5	3.63	109.04	103.63
3	27	76.5	3.61	97.56	93.61
4	28	77.5	3.53	98.70	83.53

The R-Candy rocket motor numbers one to four were categorised as rocket classes G, H, and I because the total impulse ranged within 80–160, 160–320, and 320–640 N·s [14]. The sensitivity of the digital scale was not very accurate because the maximum thrust varied during the test, as shown in table 2. However, the rocket numbers 4 and 5 misfired and exploded. There are several reasons for this, which include the casting process of R-Candy, alignment of the propellant segments, and expansion area ratio of the nozzle. The test stand was placed in the horizontal position, as can be seen in figure 14. It can be observed in tables 1 and 2 that the mass of the propellant increases with the total impulse, as shown in equation (3). The pressure gauge scale was connected but the scale had a wide range from 0–10000 psi because of which it could not detect and record the data for white mixed powder and R-Candy. The digital thrust via data acquisition was developed to record the thrust and pressure simultaneously; however, there were problems in the signal from the data acquisition to the computer, which resulted in unclear data.

Table 2. Testing results for R-Candy rocket motors.

Rocket number	Burning time (s)	Propellant weight (g)	Average thrust (N)	Total impulse (N.s)	Maximum thrust (N)
1	17	107.1	8.03	136.58	>18.03
2	30	223.2	9.49	284.65	>39.49
3	10	452.7	57.73	577.33	>50.00

Figure 14. Static rocket test stand firing in the horizontal position.

5. Conclusions and Recommendations

It was observed in this research that the white mixed powder could be used for small class G rockets; the walls in the combustion chamber made of PVC 5 might not be able to sustain such high temperature and long combustion duration of more than approximately 32 s for this class of rockets. The divergent nozzle is the most suitable, as it provides the maximum thrust. For rockets of higher class, the walls need to be thicker, such as of PVC 13.5, to prevent melting. However, an outer pipe should be used to cover the propellant segment in the interest of safety. The classification of R-Candy propellant rocket motors from classes G to I was successfully performed and recorded; however, the class J rocket motors misfired and exploded.

In conclusion, the rocket motors were successfully tested and classified, and this study of the rocket motor can be used for determining the flight of rockets in the near future.

Acknowledgement

This research was funded by King Mongkut's University of Technology, North Bangkok (Grant no. KMUTNB-NEW-59-34). The authors would like to thank Dr Phongsatorn Saisutjarit for his initial support.

IOP Conf. Series: Materials Science and Engineering **297** (2018) 012009 doi:10.1088/1757-899X/297/1/012009

References

[1] Sutton G P and Biblarz O 2001 *Rocket Propulsion Elements* (New Jersey: John Wiley & Sons)
[2] Larson W J, Henry G N and Humble R W 1995 *Space Propulsion analysis and Design* (New York: McGraw-Hill)
[3] John T, Lloyd J D and Stephen D H, 2006 *AIAA 4th Responsive Space Conf. RS4-2006-4004* (Los Angeles, CA) pp 1-15
[4] Singh S 2013 *Advances in Aerospace Science and Applications* **3** pp 199-208
[5] Okninski A, Marciniak B, Bartkowiak B, Kaniewski D, Matyszewski J, Kindracki J and Wolanski P 2015 *Acta Astronautica* **108** pp 46-56
[6] Goll M S and Dunham D W *A Sounding Rocket Launch Technique* (Texas Rocket Company and Kinetx inc.)
[7] Cortes E, Garnica S, Khalighi S, Yiu W L, Liu B, Pathan A, Ramos J, Suppanade N D and Wells J D 2013 *Experimental Sounding Rocket: Sonic Eagle* (8th Intercollegiate Rocket Engineering Competition, Green River, UT)
[8] Brown M E and Rugunanan R A 1989 *Propellants, Explosives, Pyrotechnics* **14** pp 69-75
[9] https://en.wikipedia.org/wiki/Rocket_candy
[10] https://en.wikipedia.org/wiki/Water%E2%80%93cement_ratio
[11] http://www.gramlich.net/projects/rocket/safety.html
[12] http://engineeringtoolbox.com/thermal-conductivity-plastics-d_1786.html
[13] https://en.m.wikipedia.org/wiki/Polyvinyl_chloride
[14] https://en.m.wikipedia.org/wiki/Model_rocket_motor_classification

Development of a hardware-in-loop attitude control simulator for a CubeSat satellite

Wittawat Tapsawat*, Teerawat Sangpet and Suwat Kuntanapreeda

Department of Mechanical and Aerospace Engineering, King Mongkut's University of Technology North Bangkok (KMUTNB), Bangkok 10800, Thailand

* Corresponding Author: s5901001810018@email.kmutnb.ac.th

Abstract. Attitude control is an important part in satellite on-orbit operation. It greatly affects the performance of satellites. Testing of an attitude determination and control subsystem (ADCS) is very challenging since it might require attitude dynamics and space environment in the orbit. This paper develops a low-cost hardware-in-loop (HIL) simulator for testing an ADCS of a CubeSat satellite. The simulator consists of a numerical simulation part, a hardware part, and a HIL interface hardware unit. The numerical simulation part includes orbital dynamics, attitude dynamics and Earth's magnetic field. The hardware part is the real ADCS board of the satellite. The simulation part outputs satellite's angular velocity and geomagnetic field information to the HIL interface hardware. Then, based on this information, the HIL interface hardware generates I2C signals mimicking the signals of the on-board rate-gyros and magnetometers and consequently outputs the signals to the ADCS board. The ADCS board reads the rate-gyro and magnetometer signals, calculates control signals, and drives the attitude actuators which are three magnetic torquers (MTQs). The responses of the MTQs sensed by a separated magnetometer are feedback to the numerical simulation part completing the HIL simulation loop. Experimental studies are conducted to demonstrate the feasibility and effectiveness of the simulator.

1. Introduction

Attitude control is a mission-critical part in satellites [1–2]. Thus, it is necessary to be tested to ensure reliable operation in space. However, entirely numerical simulations might not be adequate for testing the flight software/hardware and obtaining any mal-operations.

Hardware-In-the-Loop (HIL) simulation is a real-time simulation technique consisting of a numerical simulation part and a real hardware part. The HIL simulation is nowadays more and more used in variant fields for research and education. The main purposes of the HIL simulation can be grouped into two categories: testing real components [3, 4] and simulating complex systems [5–7].

Examples of HIL simulation applied to space-related systems can be found in [8–10]. In [8], Park et al. developed an integrated orbit and attitude HIL simulator. The simulator can be used to evaluate and verify orbit and attitude synchronous control algorithms, scenarios and relevant hardware technologies for autonomous satellite formation flying. In [9], Corpino and Stesina developed a HIL simulation system used for the verification of functional requirements of a CubeSat. It was shown that the developed system effectively supported the assembly-integration-verification (AIV) process. In [10], Benninghoff et al. used HIL simulation for testing and verifying guidance, navigation and control algorithms for on-orbit servicing of spacecraft. The simulation can effectively simulate variant safe rendezvous test cases.

This work develops a low-cost HIL simulation system used for functional testing of the attitude determination and control subsystem (ADCS) of a CubeSat satellite named KNACKSAT (KMUTNB Academic Challenge of Knowledge SATellite) [11]. KNACKSAT is fully developed in-house in Thailand (see figure 1). The satellite is scheduled to be launched into a 575-km sun-synchronous orbit in 2018. The satellite uses amateur radio frequencies for the communication between the satellite station and ground stations. The KNACKSAT project is funded by Broadcasting and Telecommunications Research and Development Fund for the Public Interest, Office of the National Broadcasting and Telecommunications Commission (NBTC), Thailand.

The objectives of the KNACKSAT project can be categorized into educational objectives and technology demonstration objectives. In term of the educational objectives, the project will enhance the knowledge and experiences in satellite design and construction. These can be considered as fundamentals for development of larger satellites in Thailand in the near future. Apart from these, the project provokes the interests in science and technology of young Thais. For the technological objectives, the main missions of KNACKSAT include (1) developing a communication system using armature radio frequencies, (2) taking images from space, (3) testing 3-axis attitude control algorisms by using magnetic torquers, (4) verifying a deorbit technology by a magnetic torque and (5) confirming the uses of Commercial Off-The-Shelf (COTS) components in space.

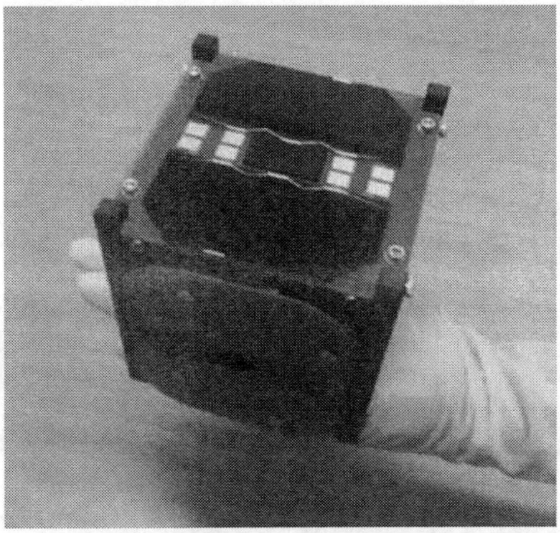

Figure 1. KNACKSAT Satellite.

KNACKSAT comprises of six subsystems: Structure Subsystem (STR), Communication Subsystem (COMM), Electrical Power Subsystem (EPS), Command and Data Handling Subsystem (CDH), Attitude Determination and Control Subsystem (ADCS), and Camera Subsystem or Payload (CAM).

Two main functions of the ADCS subsystem are de-tumbling during satellite deployment and attitude pointing. The subsystem consists of three magnetic torquers (MTQ), two three-axis magnetometers, two three-axis rate-gyros and six two-axis sun sensors. The on-board CPU is a 32 bit-microcontroller. The subsystem is summarized in figure 2. More details of the project can be found at http://www.knacksat.space.

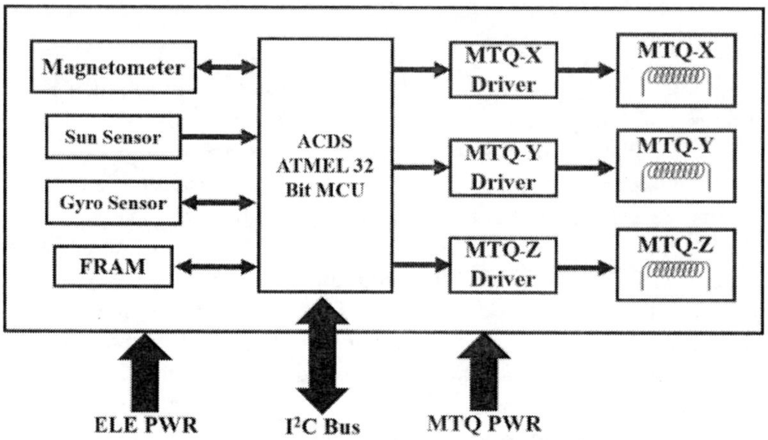

Figure 2. ADCS subsystem diagram [11].

The rest of the paper is organized as follows. The description of the developed HIL system is presented in the next section. Experimental setup and results are given in Section 3. The conclusions of this paper are drawn in Section 4.

2. System Description

The main utilization of the developed HIL simulation system is testing the flight software/hardware of the ADCS subsystem of KNACKSAT. The architecture of the system is shown in figure 3. The system consists of a numerical simulation part (indicated by blue-color block), a hardware part (indicated by red-color blocks), and a HIL interface unit (indicated by green-color blocks). The hardware part comprises the ADCS board and three MTQs of KNACKSAT. The numerical simulation part is based on the PROPAT package [12]. PROPAT is a small set of functions in Matlab to simulate and propagate orbit and attitude of an Earth's satellite. The package includes an analytical orbit propagator, allied to a numerical attitude propagator. Also available are functions to compute the position of the Sun relative to Earth and the geomagnetic field in a given satellite position. PROPAT is freely available at http://www.dem.inpe.br/~val/projetos/propat/.

As shown in figure 3, the numerical simulation part outputs simulated satellite's angular velocity and geomagnetic field information to the ProMini boards through the USB-to-Serial convertor. The ProMini boards representing magnetometers and rate-gyros convert the simulation information to be the corresponding I²C sensor signals, mimicking the signals from the on-board sensors. Then, the signals are fed to the ADCS board. The CPU on the ADCS board reads the signals in the same way as reading from the on-board sensors. By using the information from the read signals, the CPU calculates the control signals based on a predefined control algorithm and then send the control signals to drive the MTQs. The actual magnetic field produced by the MTQs is measured by the magnetometer and then feedback the measured values to the numerical simulation through the DUE board to close the simulation loop. Note that the numerical simulation part uses these measured values to calculate the control torque acting on the satellite.

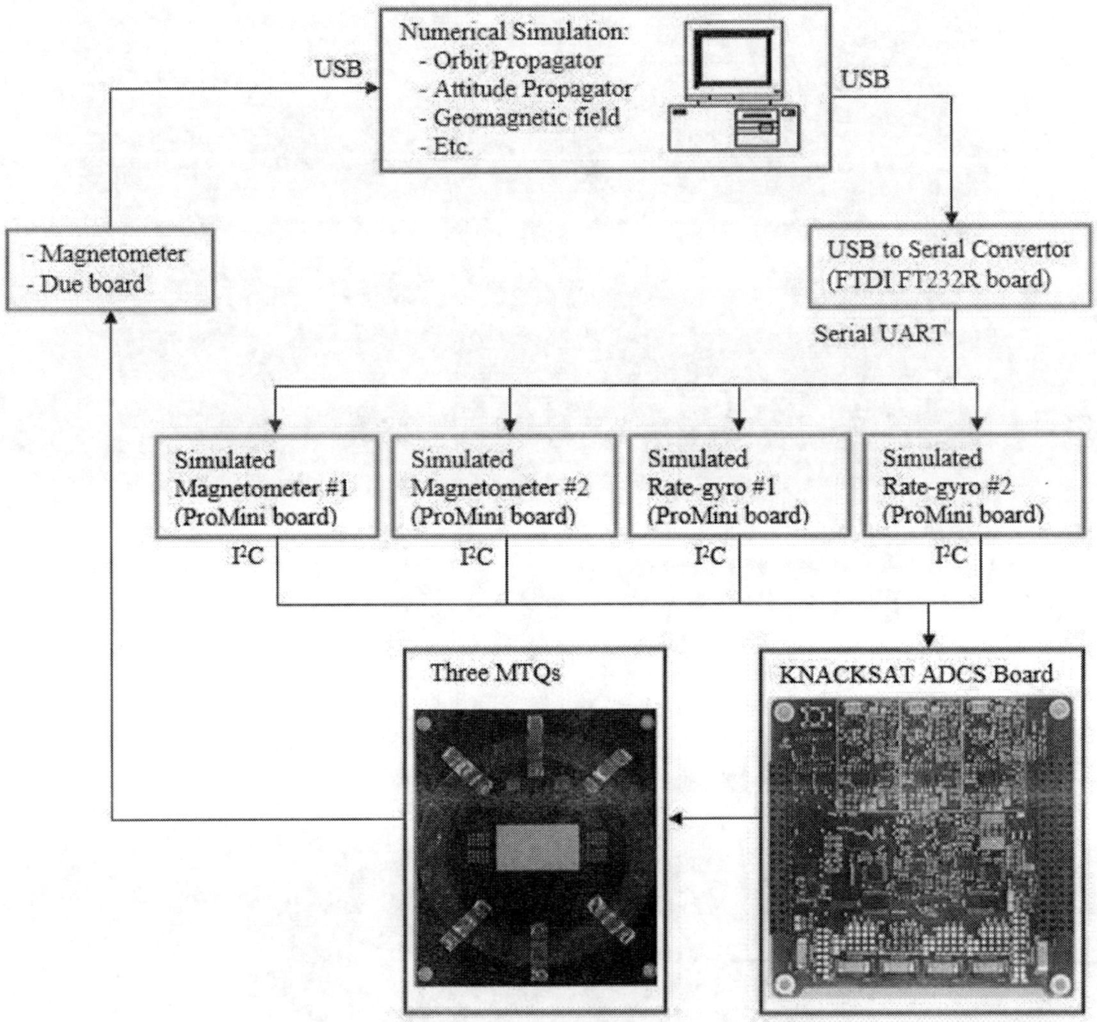

Figure 3. HIL architecture.

Figures 4 and 5 display photos of the developed HIL simulator. In figure 4, as indicted by the four red-color circles, the magnetometers and the rate-gyros are not installed on the ADCS board. The cables of the I^2C signals from the ProMini boards (i.e., simulated magnetometers and rate-gyros) are soldered to the ADCS board substituting the real magnetometers and rate-gyros. In figure 5, the three MTQs are mounted orthogonally in a cube. This mounting configuration is identical to the one on KNACKSAT. The additional magnetometer is installed in the center of the cube, measuring the magnetic field actually produced by the MTQs when they are driven by the ADCS board.

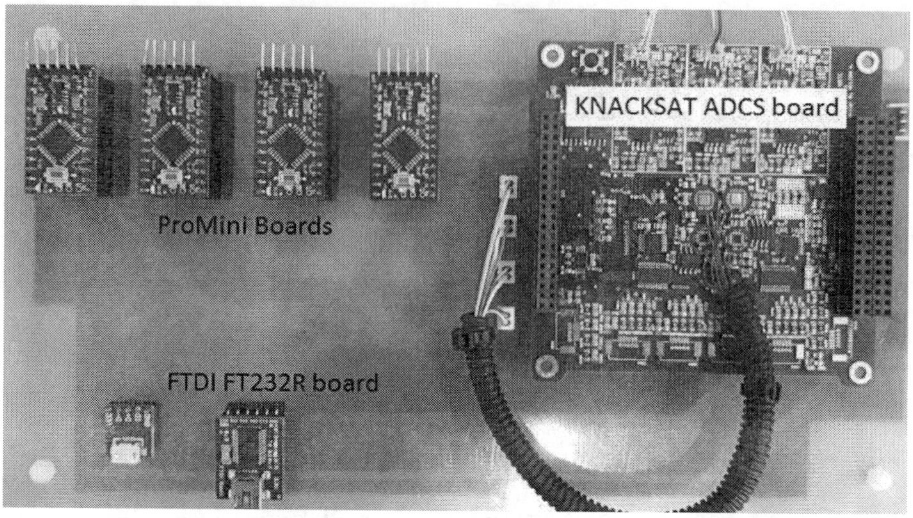

Figure 4. Photo of ADCS board connected with simulated sensors.

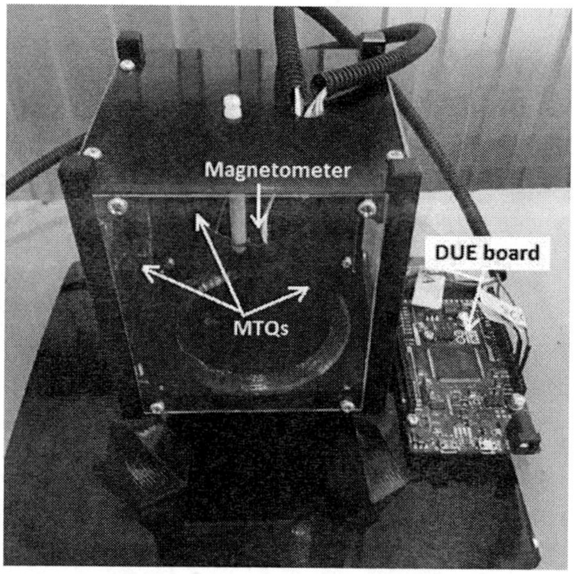

Figure 5. Photo of MTQs with a magnetometer and a DUE board.

3. Experimental Setup and Results

The developed HIL simulator was utilized to verify a detumbling control algorithm on the ADCS board. Satellites are usually tumbling when they are deployed from the laucher. Many detumbling algorithms are avaiable from the literature. The most well-known is the B-dot algorithm [13–15]:

$$\boldsymbol{m} = -k\dot{\boldsymbol{B}} \tag{1}$$

where k is a positive gain, $\boldsymbol{m} = [m_x, m_y, m_z]^T$ is the control dipole moment vector generated by the MTQs and $\boldsymbol{B} = [B_x, B_y, B_z]^T$ is the geomagnetic field vector measured by an on-board magnetometer. However, the algorithm needs differentiation of the geomagnetic field. The measurement noise in the magnetometer can cause undesired effects during differentiation. An alternative algorithm to avoid the differentiation is

$$m = -k(\boldsymbol{B} \times \boldsymbol{\omega}) \tag{2}$$

where $\boldsymbol{\omega} = [\omega_x, \omega_y, \omega_z]^T$ is the angular velocity of the satellite. The angular velocity can be measured direcly using an on-board rate-gyro.

The detumbling control algorithm (2) was used here. The numerical solver was the Dormand–Prince method with the step size of 0.1 sec. The control sampling time on the ADCS board was set to be 0.5 sec. The orbit was a circular sun-synchronous orbit with the altitude and the inclination of 575 km and 97.7 deg, respectively. The satellite was assumed to be a rigid body with the inertia matrix:

$$\mathbf{I} = \begin{bmatrix} 0.003340 & 0 & 0 \\ 0 & 0.002274 & 0 \\ 0 & 0 & 0.003342 \end{bmatrix} \text{km-m}^2.$$

There are two experiments were conducted here. In the first experiment, the measurement signals (i.e., magnetometer and rate-gyro signals) were assumed to be disturbed by measurement noises. In the other one, the signals were disturbed by both measurement noises and offsets. The initial angular velocity $\boldsymbol{\omega}_0 = [0.5, 0.5, 0.5]^T$ rad/sec and the gain $k = 6000$ were choosen in all experiments. The gain k was tuned in numerical simulations such that the tumbling converged within approximately one orbital period (i.e., 5760 seconds) and the magnitude of control dipole moment was less than 0.1 A/m^2.

Experiment 1 The zero-mean normal distrubution noises with the standard devaiation of 3 μT and 0.02 rad/sec were added to the simulated magantometer and rate-gyro signals, respectively. These values were choosen cooresponding to setting three standard deviations (3-sigma) equaling to 10% of the maximum operation values of the signals.

The results are shown in figures 6–9. As shown in figure 6, the angular velocities converge within one orbital period. The results for the last 60 sec are shown in figure 7. The control dipole moments are shown in figures 8 and 9. The results indicate that the detumbling algorithm on the ADCS sucessfully detumbling the satellite and the developed HIL simulator is effectively utilized to verify the ADCS hardware and software.

Note that the control dipole moments in figures 8 and 9 are not the values directly calculated from the control algorithm (2). They were the actual values generated by the MTQs. By using these values in the simulation, more realistic results could be obtained since it automatically included transient behaviour, nonlinearity and uncertainty of the MTQs. Our studies also found that the time constant of the MTQs is approximately 0.46 msec, which is considered to be very small comparing to the control sampling period. Thus, the effect of the delay due to the transient behaviour of the MTQs cannot be observed in the simulation. In addition, the control dipole moments were also set to zeros (see figure 9) while the ADCS board was reading the simulated magnetometers. This process is required in real systems to avoid the magnetic fields generated by the MTQs to disturb the magnetometers. The time required to read the magnetometers is approximately 0.2 sec.

Experiment 2 In addition to the measurement noises as in the previous experiment, the offsets 10 μT and 0.05 rad/sec were added to the magantometer and rate-gyro signals, respectively. These values are approximately equal to 10% of the maximum operation values of the signals. The results are shown in figures 10 and 11. As shown in figure 10, the angular velocities still converge within one orbit. However, by comparing to the results of the previous one, there are more oscillations in the angular velocities due to the effects of the offsets.

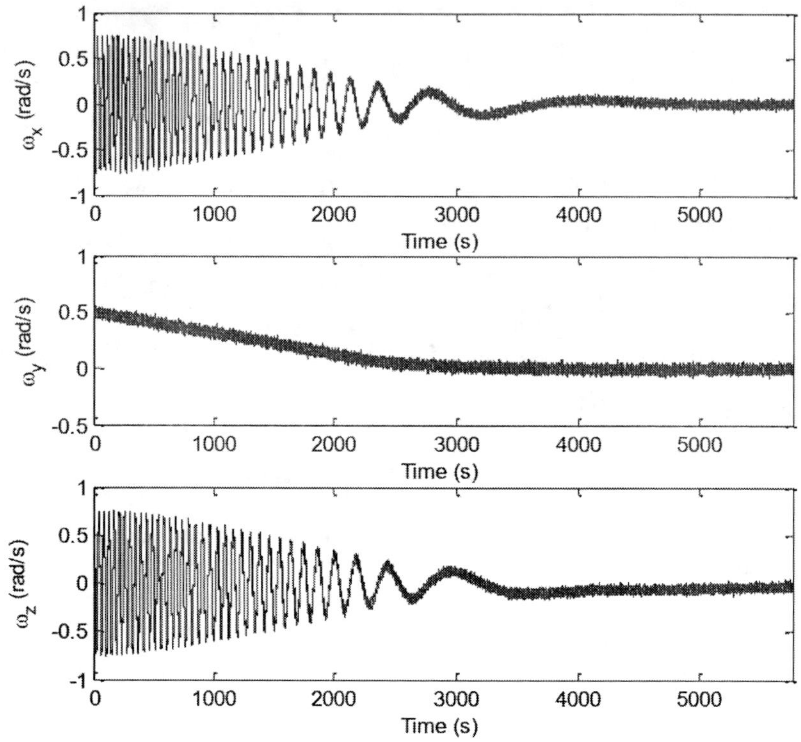

Figure 6. Satellite angular velocities (Experiment 1).

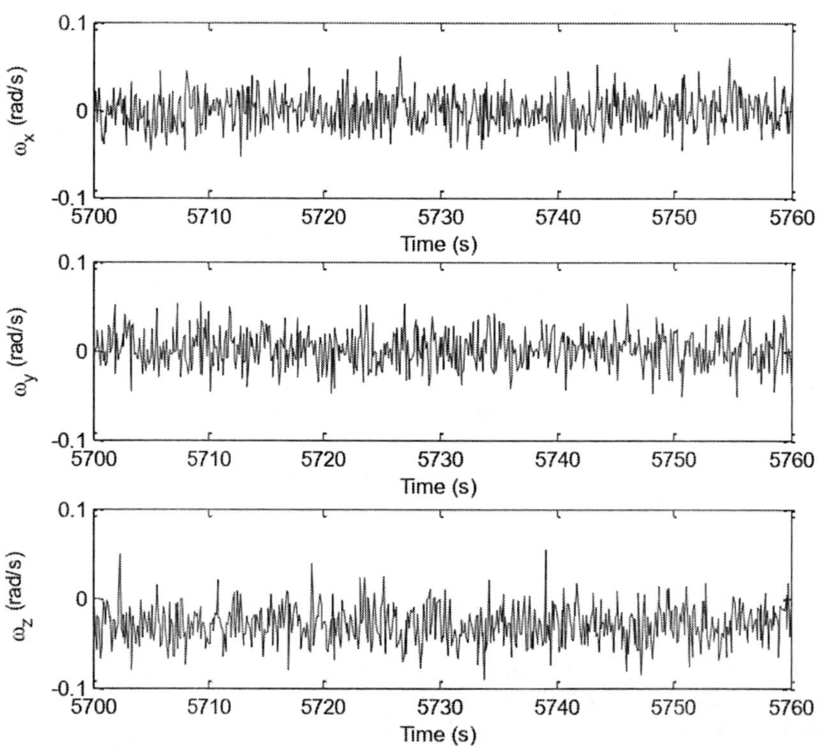

Figure 7. Zoomed satellite angular velocities (Experiment 1).

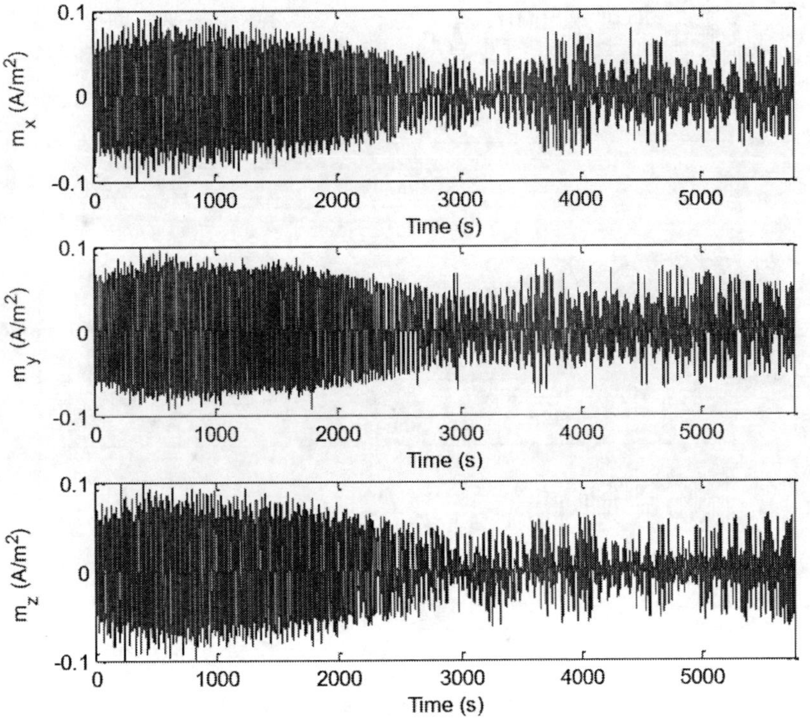

Figure 8. Control dipole moments (Experiment 1).

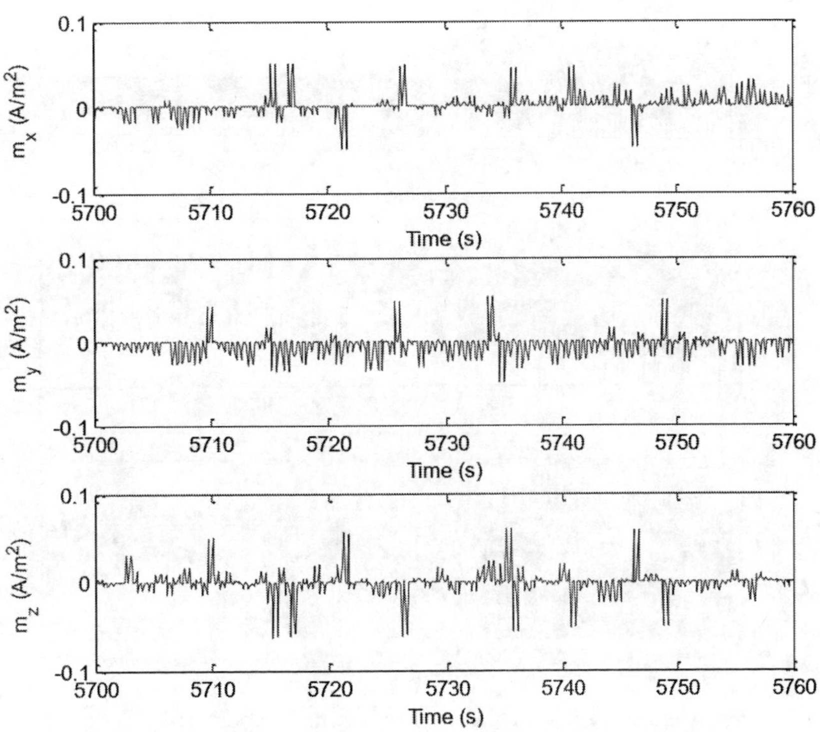

Figure 9. Zoomed control dipole moments (Experiment 1).

IOP Conf. Series: Materials Science and Engineering **297** (2018) 012010 doi:10.1088/1757-899X/297/1/012010

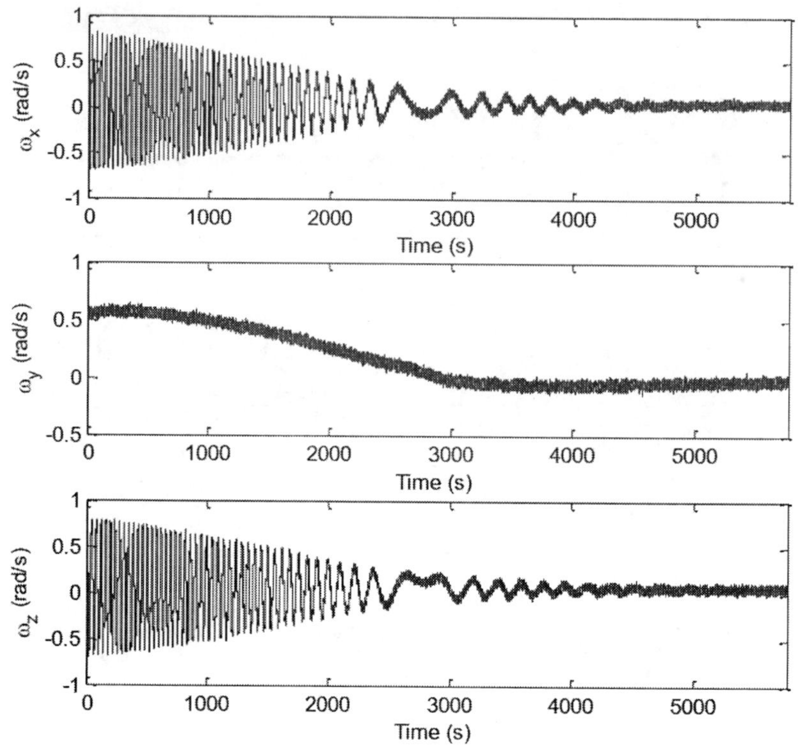

Figure 10. Satellite angular velocities (Experiment 2).

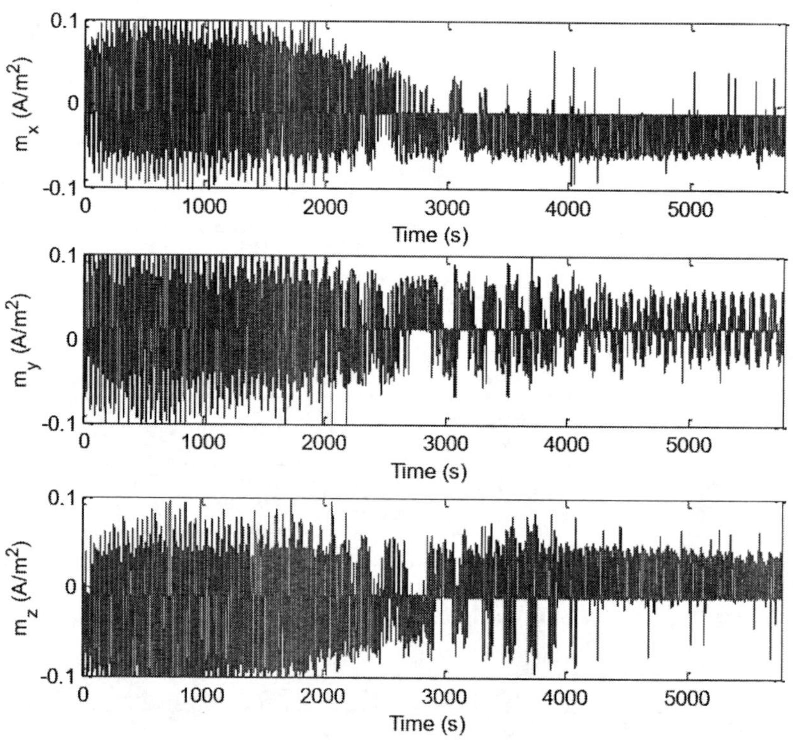

Figure 11. Control dipole moments (Experiment 2).

4. Conclusions

This paper has presented a development of a low-cost Hardware-In-the-Loop (HIL) simulator for testing the ADCS subsystem of KNACKSAT satellite. The simulator consists of a numerical simulation part, a hardware part, and a HIL interface hardware unit. The numerical simulation part is based on the PROPAT package. The hardware part is the ADCS board of KNACKSAT satellite. The developed HIL simulator was utilized to verify a detumbling control algorithm on the ADCS board. Experimental results demonstrated the simulator was effectively utilized to test the ADCS subsystem.

Acknowledgement

This research was financially supported by King Mongkut's University of Technology North Bangkok (KMUTNB-AED-G60-16).

References

[1] Larson W J and Wertz J R 1992 *Space Mission Analysis and Design* (California: Microcosm)
[2] Wertz J R 1978 *Spacecraft Attitude Determination and Control* (Dordrecht: Kluwer)
[3] Yoo C, Kang B Y and Park B 2010 *Proc. Int. Conf. Control, Automation and Systems (Gyeonggi-do)* (New York: IEEE) pp 1729–1732
[4] Kwon J J, Hong T W, Park K, Heo S J, Lee K W, Kim W W, Koo H C and Park B K 2014 *Int. J. Automotive Tech.* **15**(3) pp 429–439
[5] Herrmann S, Kaehler M, Souffrant R, Rachholz R, Zierath J, Kluess D, Mittelmeier W, Woernle C and Bader R 2012 *Computer Methods and Programs in Biomedicine* **105** pp 109–119
[6] Allotta B, Conti R, Meli E, Pugi L and Ridolfi A 2014 *Mechanica* **49** pp 615–644
[7] Bracco G, Giorcelli E, Mattiazzo G, Orlando V and Raffero M 2015 *Mechatronics* **25** pp 11–17
[8] Park H-E, Sang-Young Park S-Y, Kim S-W, Park C 2013 *Advances in Space Research* **52** pp 2052–2066
[9] Corpino S and Stesina F 2014 *IEEE Trans. Aerosapce and Electronics Systems* **50**(4) pp 2807–2817
[10] Benninghoff H, Rems F and Boge T 2014 *Acta Astronautica* **102** pp 67–80
[11] Worrakul N, Laohalidanond K, Saisutjarit P, Kuntanapreeda S and Inamori T 2017 *Proc. Asian Conference on Defence Technology(Phuket)* (New York: IEEE) pp 36–41
[12] Carrara V 2015 *Proc. Int. Symp. on Dynamic Problems of Mechanics (Rio Grande do Norte)* (Rio de Janeiro: ABCM) pp 1-12
[13] Stickler A and Alfriend K T 1976 *Journal of Spacecraft and Rockets* **13** pp 282–287
[14] Avanzini G and Giulietti F 2012 *Journal of Guidance, Control, and Dynamics* **35**(4) pp 1326–1334
[15] Lovera M 2015 *Proc. American Control Conference (Chicago)* (New York: IEEE) pp 1867–1872

IOP Conf. Series: Materials Science and Engineering **297** (2018) 012011 doi:10.1088/1757-899X/297/1/012011

On verifying magnetic dipole moment of a magnetic torquer by experiments

Aekjira Kuyyakanont[1,*], Suwat Kuntanapreeda[1] and Nisai H. Fuengwarodsakul[2]

[1] Department of Mechanical and Aerospace Engineering, Faculty of Engineering, King Mongkut's University of Technology North Bangkok, Bangkok 10800, Thailand
[2] Department of Electrical and Software Systems Engineering, The Sirindhorn International Thai-German Graduate School of Engineering, King Mongkut's University of Technology North Bangkok, Bangkok 10800, Thailand

*Corresponding Author: aekjira25757@gmail.com

Abstract. Magnetic torquers are used for the attitude control of small satellites, such as CubeSats with Low Earth Orbit (LEO). During the design of magnetic torquers, it is necessary to confirm if its magnetic dipole moment is enough to control the satellite attitude. The magnetic dipole moment can affect the detumbling time and the satellite rotation time. In addition, it is also necessary to understand how to design the magnetic torquer for operation in a CubeSat under the space environment at LEO. This paper reports an investigation of the magnetic dipole moment and the magnetic field generated by a circular air-coil magnetic torquer using experimental measurements. The experiment testbed was built on an air-bearing under a magnetic field generated by a Helmholtz coil. This paper also describes the procedure to determine and verify the magnetic dipole moment value of the designed circular air-core magnetic torquer. The experimental results are compared with the design calculations. According to the comparison results, the designed magnetic torquer reaches the required magnetic dipole moment. This designed magnetic torquer will be applied to the attitude control systems of a 1U CubeSat satellite in the project "KNACKSAT."

1. Introduction

In recent years, satellite technologies have been developed greatly. In many satellite projects, the most important priority is to control the movement of the satellite for pointing any instrument or equipment it carries in a desired direction. Currently, there are many techniques for attitude control of a satellite such as thrusters, spin stabilization, momentum wheels, control moment gyros, solar sails, gravity-gradient stabilization, magnetic torquers and pure passive attitude control. All of these attitude control devices are called "actuators." Several factors such as satellite dimensions, satellite operating time, satellite mission, must be considered for choosing an appropriate actuator type for a satellite.

This report focuses on a type of actuators which will be used in "KNACKSAT" (figure 1), a CubeSat satellite project in King Mongkut's University of Technology North Bangkok, Thailand [1]. In this case, three magnetic torquers for 3-axis attitude control have been selected for the CubeSat. The CubeSat category of interest is a CubeSat 1U which weighs 1 kg and has dimensions of 10 x 10 x 10 cm approximately, as shown in figure 1. Magnetic torquers are coils or permanent magnets which provide a moment against the local magnetic field. In the case of CubeSat 1U, magnetic torquers are appropriate for this satellite because their sizes can be minimized in the manufacturing process and they require a

IOP Conf. Series: Materials Science and Engineering **297** (2018) 012011 doi:10.1088/1757-899X/297/1/012011

low power supply. Magnetic torquer hardware is not complicated, because a magnetic torquer is a simple electromagnetic coil. However, one problem is to generate sufficient torque for the satellite mission. This report presents a methodology for measuring the torque of a magnetic torquer by placing it on an air-bearing inside a generated magnetic field provided by a Helmholtz coil. The torque produced by the magnetic torquer can be directly measured from the force acting on a gram gauge.

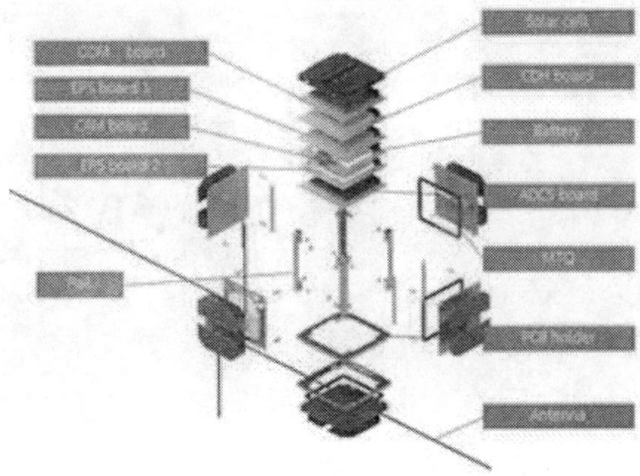

Figure 1. KNACKSAT [1]

2. Magnetic torquer

A magnetic torquer (MTQ) is one of the actuators which are widely used for attitude control in low Earth orbit satellites because MTQs are lightweight, reliable and energy-efficient [2-5]. The torque produced by an MTQ is required only to accelerate or decelerate the change in a satellite's attitude by minute amounts. The MTQs are used in attitude determination and control subsystems (ADCS) for attitude control, detumbling, and stabilization. The MTQs are classified into two types, permanent magnets and electromagnetic coils, as shown in figure 2. Both types of the actuators produce a magnetic dipole moment which interacts with the Earth's magnetic field to generate external torques on the satellite.

(http://n-avionics.com)

a. electromagnetic coils

(https://www.magnetsource.com)

b. permanent magnets

Figure 2. Magnetic torquer types

2.1. Magnetic field generation

This paper focuses on the electromagnetic coils MTQ. The MTQ produces mechanical torque to the satellite by generating its own magnetic field which interacts with the Earth's magnetic field. The torquer is built in form of a copper winding. Once a current flows through the copper wire, a magnetic field is created, as shown in figure 3.

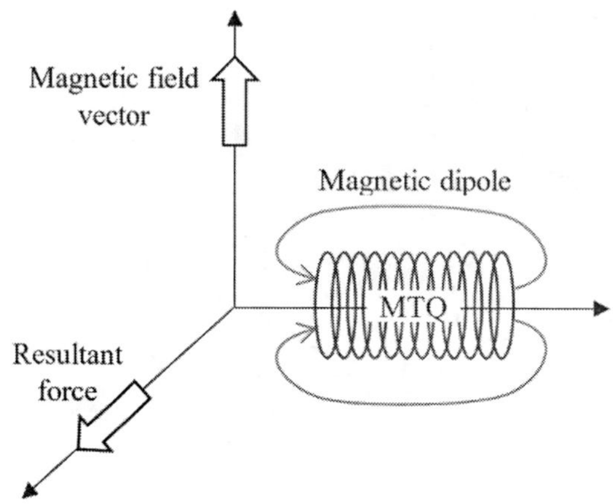

Figure 3. MTQ operating concept

The torque about an arbitrary point generated by a magnetic dipole in a uniform magnetic field is given as

$$\vec{T} = \vec{M} \times \vec{B} \tag{1}$$

where

$\vec{M}$ = magnetic dipole moment [Am^2]
$\vec{B}$ = flux density of the magnetic field [Tesla]
$\vec{T}$ = torque generated by the magnetic dipole [Nm]

2.2. Design of MTQ

First of all, the restricting parameters in the design such as magnetic dipole moment, power loss due to coil resistance, voltage, mass, and also dimension parameters have to be determined. To be able to control the angular momentum of the satellite, the actuator must have enough control capability to overcome on-orbit disturbances such as external torque due to aerodynamic drag, solar radiation pressure and the Earth's magnetic field as well.

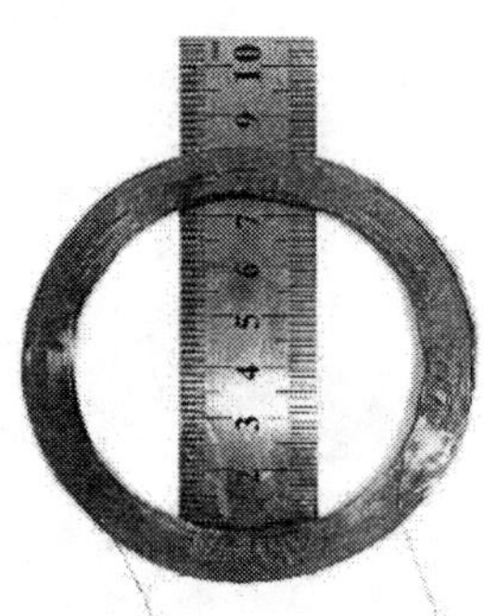

Figure 4. MTQ prototypes used in KNACKSAT

Three MTQ circular coils were designed for a CubeSat 1U satellite named KNACKSAT, as shown in figure 4. The design specification of the MTQs under the satellite conditions must be taken into account. Due to the construction constraints, the concept of coreless electromagnetic coils for 3-axis attitude control is chosen. To create torque, two parameters need to considered, the magnetic dipole moment ($\vec{M}$) and the Earth's magnetic field ($\vec{B}$) from equation (1). The magnetic dipole moment is calculated from the equation below.

$$M = ANI \tag{2}$$

where

M = magnetic dipole moment [Am^2]
A = cross section area of the MTQ [m^2]
N = number of turns of the MTQ
I = current [A]

The MTQ is required to produce a magnetic dipole moment of 0.15 Am^2 at a maximum power input of 1 W. Table 1 summarizes the specification of the designed MTQs.

Table 1. Specification of the MTQs

Parameters	Value
Magnetic dipole moment, Am^2	0.15
Outer and Inner diameter, mm	82 and 72
Thickness, mm	3
Cross section area, mm^2	5,281
Number of turn	370
Copper wire No.	AWG34
Coil resistance, Ohm	~38
Coil inductance, mH	~18

3. MTQ verification

3.1. Test principle

This experiment for verifying the designed MTQs involves the torque measurement methodology. The experiment is a direct method to measure a magnetic dipole moment of the MTQ by applying the methodology of the cross product of two magnetic fields interaction, as seen in the equation (1). In the experiment, the value of a magnetic field (B) is created by the Helmholtz coil, and the produced force of the MTQ can be measured using a gram gauge. We can calculate the torque (T) of the MTQ from an equation below where r is a distance from origin of the MTQ to the trailing edge of the gram gauge shaft.

$$\vec{T} = \vec{F} \times \vec{r} \tag{3}$$

This experimental test-bed is set on an air-bearing to ensure that, when the test-bed is rolling, there is no effect of friction. The test-bed is shown in figure 5.

Figure 5. MTQ torque measuring test-bed

3.2. *Test equipment*

3.2.1. Helmholtz coil. A Helmholtz coil is a device for generating a region of nearly uniform magnetic field [6]. It consists of two identical coils placed symmetrically along a common axis with equal currents flowing in the same direction.

For generating a uniform magnetic field, the coils are parallel to each other and separated by a distance equal to the coil radius, R. The magnetic field lines for a Helmholtz coil are shown in figure 6. Figure 7 shows our implemented Helmholtz coil. The specification of the coil is summarizes in table 2.

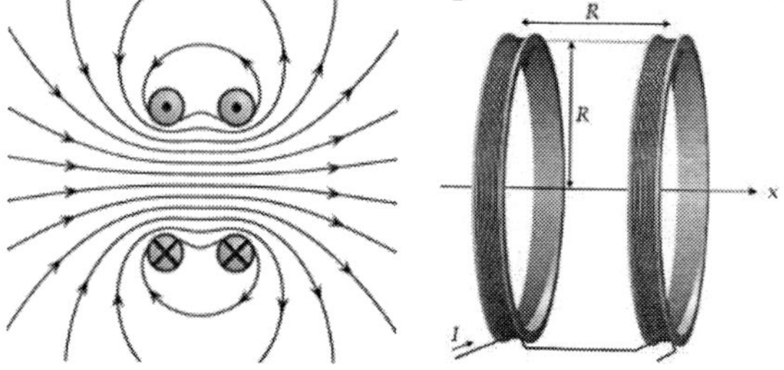

(https://en.wikipedia.org/wiki/Helmholtz_coil)

Figure 6. Magnetic field lines of Helmholtz coil

Table 2. Specification of the Helmholtz coil

Parameters	Value	
Magnetic field, M	3.2382	mT
Maximum error at closed to coils	5.4	%
Coil size	20	AWG
Number of turn in each coil, N	150	turns
Diameter of AWG20	0.8128	mm
Maximum current for chassis wiring	11	A
Weight of a coil	1.7541	kg
Coil Radius, R	40	cm
Current, I	9	A
Voltage, V	226	v

IOP Conf. Series: Materials Science and Engineering **297** (2018) 012011 doi:10.1088/1757-899X/297/1/012011

Figure 7. Helmholtz coil

3.2.2. Air-bearing. Air-bearing (figure 8) is a bearing that uses a thin film of pressurized air to provide an exceedingly low friction load-bearing interface between surfaces. The two surfaces do not touch. As they are contact-free, air bearings avoid the traditional bearing-related problems of friction, wear, particulates, and lubricant handling, and offer distinct advantages in precision positioning, such as lacking backlash and static friction, as well as in high-speed applications.

Figure 8. Air-bearing and air compressor

3.2.3. Gram gauge. A gram gauge (figure 9) offers a very quick and easy method of measuring very low force between 0 and 0.1 N. The gram gauge scale has a resolution of 0.005N.

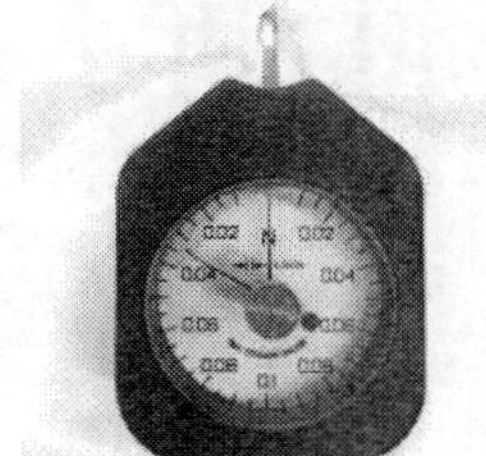

Figure 9. Gram gauge

3.2.4 Electronic hardware. The electronic hardware is needed for controlling the input current of the MTQ. To avoid friction caused by cable connections, the input current is controlled via a wireless communication system, as shown in figure 10. The system consists of a microcontroller, a current sensor, a coil driver, a Bluetooth module and a Lithium polymer battery.

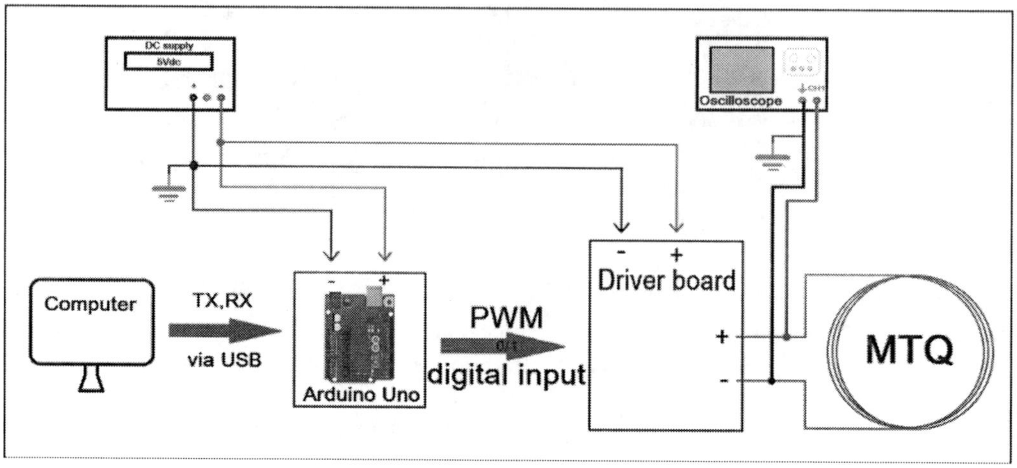

Figure 10. Electronic hardware diagram

4. Results

As mentioned in section 2.2, the maximum magnetic dipole moment of the MTQ should be 0.15 Am^2. In this experiment, the created magnetic fields are stronger than the Earth's magnetic field at LEO, as shown in figure 11, so that the produced force is sufficiently large for the measurement range of the gram gauge. Therefore, the magnetic dipole moment in the experiment is more than 0.15 Am^2, as shown in table 3.

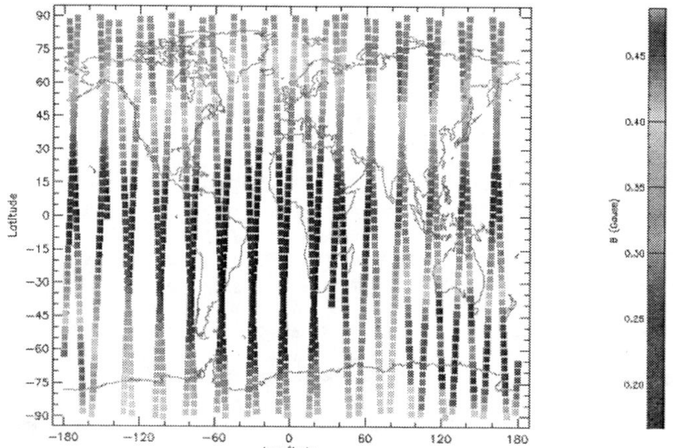

Figure 11. World map of magnetic field at 600-800 km by using 'Tsyganenko 96' model (The magnetic field plot was taken from SPENVIS: http://www.spenvis.ome.be)

Table 3. Experimental result data

Parameters	Values
Force (F), N	0.019
Distance (R), m	3.6×10^{-2}
Voltage supply to the MTQ, V	5
Current supply to the MTQ (i), A	0.142
Current supply to the Helmholtz coil, A	7.73
Magnetic field of the Helmholtz coil (B), mT	2.58
Magnetic dipole moment (M), Am^2	0.2651

5. Conclusion

In this experiment, the magnetic field generated by the Helmholtz coil was stronger than the Earth's magnetic field, so that the generated force will be large enough for the measurement range of the gram gauge. From the relationship between magnetic dipole moment and input current of the MTQ in equation (2) we can plot a graph, as shown in figure 12.

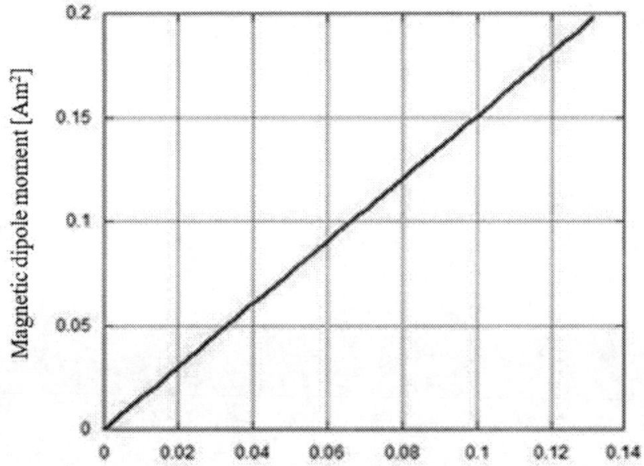

Figure 12. Magnetic dipole moment vs. current from the calculation

In the experiment, only the maximum torque generated by the MTQ was measured. In this case, the maximum torque will be obtained when we supply a maximum voltage (5V) to the MTQ. From the calculation at the input current of 0.1 A, we get a magnetic dipole moment value of 0.19 Am^2. For the experiment, the input current of the MTQ coil was not precisely regulated because it was supplied by a voltage-controlled source at 5V. Therefore, the actual current depends on the coil resistance. From the experiment, a magnetic dipole moment of 0.265 Am^2 is measured at the input current of 0.142 A. Note that the input current is larger than expected, since the actual coil resistance is about 35.21 Ω, which is less than the calculated value of 38 Ω. When recalculating the theoretical magnetic dipole moment value with the actual current, the calculated magnetic dipole moment is 0.264 Am^2. Therefore, there is a very small deviation of 0.343% between the design calculation and the experiment. From the proportional calculation, the built MTQ coil is able to generate a magnetic dipole moment of 0.186 Am^2 at the input current of 0.1 A, which exceeds the required value of 0.15 Am^2. This result demonstrates a good accuracy of the MTQ design process and it confirms that the designed MTQ can deliver the expected magnetic dipole moment to fulfill the required task.

Acknowledgment
This work was financially supported by the National Broadcasting Telecommunications Commission (NBTC), Thailand. We would like to thank Miss Chanakarn Nin-inchan and Mr. Watcharapong Narongsak for preparing the experimental equipment and testing.

References

[1] Worrakul N, Laohalidanond K, Saisutjarit P, Kuntanapreeda S and Inamori T 2017 *Proc. Asian Conference on Defence Technology(Phuket)* (New York: IEEE) pp 36–41
[2] Michael E, Polites, Charles A. Gibson, 1971 *The Space Congress Proceedings* **1**
[3] Fakhari Mehrjardi M and Mirshams M 2010 *Contemporary Engineering Sciences* **3**(5) pp 227 – 236.
[4] Slavinskisa A, Kulua E and Virua J 2014 *Proc. of the Estonian Academy of Sciences* 63 pp 242–249
[5] Lee J and Jobanputra A Ng R 2002 *Canadian Aeronautics and Space Journal* **48**(1)
[6] Hurtado-Velasco R and Gonzalez-Llorente J 2016 *Applied Mathematical Modelling* **40** pp 9835–9847.

IOP Conf. Series: Materials Science and Engineering **297** (2018) 012012 doi:10.1088/1757-899X/297/1/012012

Viscous investigation of a flapping foil propulsor

Attapol Posri, Surasak Phoemsapthawee* and Nonthipat Thaweewat

Department of Maritime Engineering, Faculty of International Maritime Studies, Kasetsart University, Thailand

* Corresponding Author: surasak.pho@ku.th

Abstract. Inspired by how fishes propel themselves, a flapping-foil device is invented as an alternative propulsion system for ships and boats. The performance of such propulsor has been formerly investigated using a potential flow code. The simulation results have shown that the device has high propulsive efficiency over a wide range of operation. However, the potential flow gives good results only when flow separation is not present. In case of high flapping frequency, the flow separation can occur over a short instant due to fluid viscosity and high angle of attack. This may cause a reduction of propulsive efficiency. A commercial CFD code based on Lattice Boltzmann Method, XFlow, is then employed in order to investigate the viscous effect over the propulsive performance of the flapping foil. The viscous results agree well with the potential flow results, confirming the high efficiency of the propulsor. As expected, viscous results show lower efficiency in high flapping frequency zone.

1. Introduction

Flying and swimming creatures have transformed their organs into flapping foils, for example, wings, fins or tails [1], to propel themselves. After million years of evolution, the hydrodynamic optimization of such flapping foil propulsor is obtained through natural selection [2]. Undoubtedly, this biomimetic propulsor interests inventors to use it as a propulsion system. Several numerical and experimental studies have been performed to clarify the flapping foil mechanism [3, 4,5]. This illuminates the way to use a flapping foil as a ship propulsion system. A study based on Boundary Element Method [6] has concluded that a flapping foil can be used as a marine propulsive device since its performance is comparable to a conventional marine propeller. However, the efficiency of the flapping foil used as an energy harvesting device has been reported to be lower than that of a rotary turbine [7].

In order to explore the parametric space of flapping foil mechanism, especially the kinematics giving optimal propulsive efficiency, researches using two-dimensional or three-dimensional Boundary Element Method have been performed to study effects of foil motions, for example angle of attack profile [8], or geometry [9]. Moreover, by attaching a flapping foil beneath ship hull either vertically or horizontally, the ship response in waves can be improved [10, 11]. Significant trust as well as anti-rolling improvement has been reported. It can simply be said that the overall ship performance is enhanced.

In an effort to use a flapping foil as a ship propulsive device, the efficiency of flapping foil has been calculated using potential flow theory and presented similarly to that of screw propeller series [6]. The transition between the two regimes, e.g., flapping foil and marine propeller, has been done by introducing an equivalent advance ratio. Another pattern in matching the propulsive foil and prime mover power has been presented [12]. A numerical study based on potential flow method [13] has presented a complete openwater characteristics of flapping foil. Such information is necessary for matching and optimizing the ship propulsion system.

The linear theory and inviscid numerical method, i.e., Boundary Element Method (BEM) is mentioned as a useful tool for hydrodynamic force prediction at high Reynolds numbers [9]. However, these inviscid methods can sometimes lead to disagreement of results in comparison with those of experiments or viscous solver at a certain range of dimensionless number describing flapping frequency such as reduced frequency, Strouhal number or advance number.

Due to a lack of boundary layer model in potential method, the results of inviscid simulation may slightly deviate from reality of physical flow. In the present study, the hydrodynamic force coefficients and propulsive efficiency obtained from inviscid model [13] will be compared to that of viscous solver. The results are presented as functions of equivalent advance number in order to be comparable to marine propeller.

2. Flapping Foil Propulsor

In this section the mechanism of flapping foil propulsor as well as the openwater characteristics is defined. In order to investigate the viscous effect over the openwater characteristics reported in [13], a flapping foil with NACA0012 section and an aspect ratio of 5 is simulated. The nomenclature in this paper is defined following the work in [13].

2.1. *Flapping Motion*

The flapping mechanism can be described by two sinusoidal motions: the heave and the pitch motions as presented in equation (1) and (2) respectively.

$$h = h_0 \sin(2\pi f t + \frac{\pi}{2}) \tag{1}$$

$$\theta = \theta_0 \sin(2\pi f t + \pi) \tag{2}$$

where h is the heave position, h_0 is the heave amplitude, θ is the pitch angle, θ_0 is the pitch amplitude or in the other hand the maximum pitch angle, f is the motion frequency. The heave and the pitch motions are defined at the middle of the chord as illustrated in figure 1.

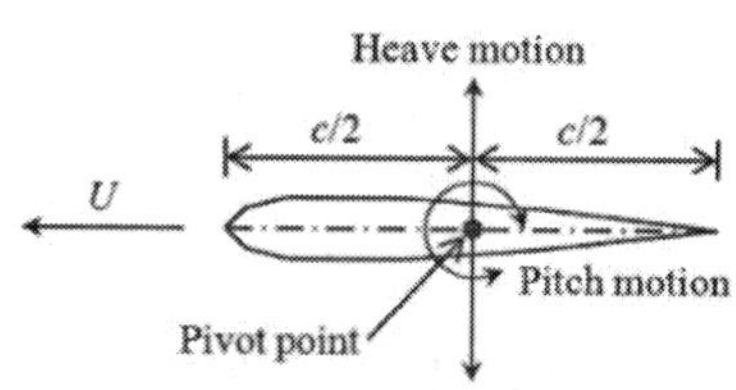

Figure 1. Definition of the flapping motion (reproduced from [13]).

2.2. *Openwater Characteristics*

Openwater characteristics are the performance of a propulsor in a uniform flow. Following the work [13], the openwater characteristics of a flapping foil propulsor consist of four non-dimensional parameters: the thrust coefficient K_T, the heave-force coefficient K_H, the pitch-moment coefficient K_P and the propulsive efficiency η. The definitions of all force coefficients are described as follow:

$$K_T = \frac{\overline{T}}{\rho f^2 D_{eq}{}^4} \tag{3}$$

$$K_H = \frac{H_{\max}}{\rho f^2 D_{eq}{}^4} \tag{4}$$

$$K_P = \frac{P_{\max}}{\rho f^2 D_{eq}^{\;5}} \tag{5}$$

where $\overline{T}$ is the average thrust, $H_{\max}$ is the maximum heave force, $P_{\max}$ is the maximum pitch moment, ρ is the fluid density. Since the flapping foil propulsor has been represented as a screw propeller, the reference length is defined as an equivalent diameter $D_{eq} = \sqrt{8h_0 b/\pi}$ where b is the foil span.

The propulsive efficiency η of a flapping foil propulsor is the ratio between the propulsive power and the input power used to drive the flapping motion over a flapping period:

$$\eta = \frac{U\overline{T}}{f\int_0^{1/f}\left(H\dot{h} + P\dot{\theta}\right)dt} \tag{6}$$

where U is the advance velocity, H is the heave force, $\dot{h}$ is the heave velocity, P is the pitch moment, $\dot{\theta}$ is the pitch angular velocity.

All of these openwater parameters are presented as functions of a dimensionless number describing the flapping frequency which is represented in this paper as an equivalent advance number J_{eq}.

$$J_{eq} = \frac{U}{fD_{eq}} \tag{7}$$

3. Simulation Tool Investigation

A commercial CFD code based on Lattice Boltzmann Method, XFlow 2016, is used to simulate the flow around flapping foil since the solver is effectively capable to work with moving object and reduce time in meshing process due to its automatic mesh generation (AMG) [14]. The solver has characteristics of Wall-Modeled Large Eddy Simulation which is applied for turbulence modeling and near-wall treatment [15, 16].

A small number of works has been done to clarify the influence of boundary domain. In order to optimize the computational expense while still possessing acceptable accuracy of the numerical results, the domain size, as well as resolutions of near wall and wake, is examined, in the same way as meshing sensitivity of conventional CFD is performed. Simulations of a simple lifting problem, a foil with an incident angle, are then performed. The simulation tool investigation is broken into two scenarios: the resolution convergence problem and the domain size problem.

The initial domain is imposed in term of chord length. The boundary condition is setup as depicted in figure 2. The size of initial domain is defined similarly to the work in [17]. The simulations employ a mirror technique since the flow is symmetric at the mid span plane. A NACA 0012 foil is then placed at the free-slip wall. The angle of attack of the foil is set to be 5 degrees. The simulations are performed with a Reynold number of 1 million.

IOP Conf. Series: Materials Science and Engineering **297** (2018) 012012 doi:10.1088/1757-899X/297/1/012012

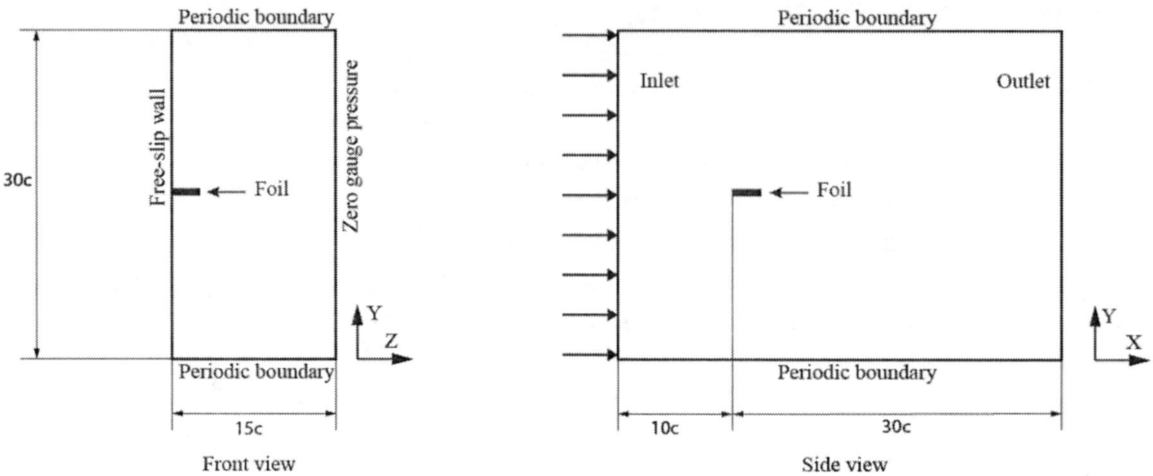

Figure 2. Initial computational domain.

3.1. *Resolution Convergence*

The resolution convergence is investigated as a function of the lattice size over the chord length. The resolutions of wall and wake region are set to be equal. The far-field resolution is always set to equal to the chord length. The adaptive resolution scheme is employed to vary the lattice size from the wall and wake regions to the far-field conditions. The investigation results are shown in figure 3 and as the computation times are shown in table 1.

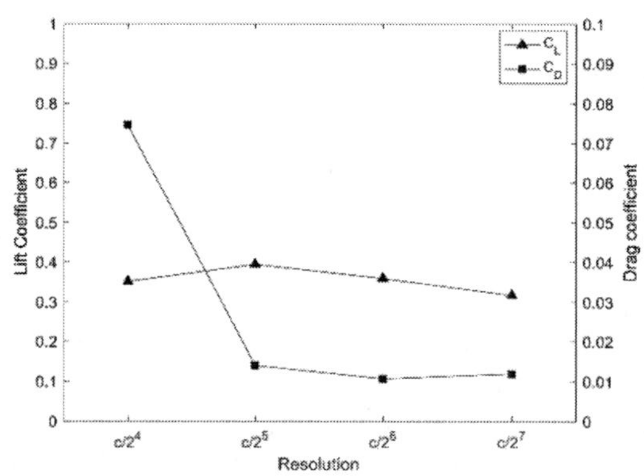

Figure 3. Lift and drag coefficients against simulation resolution.

Table 1. Simulation time in function of resolution.

Resolution	Time
$c/2^4$	8 mins.
$c/2^5$	38 mins.
$c/2^6$	3 hrs. 31 mins.
$c/2^7$	>25hrs.

According to figure 3, the resolution of $c/2^5$ is chosen to be the optimized resolution since both lift and drag coefficients do not change significantly with higher resolutions. Table 1 also indicates that the computation time increases exponentially as a function of resolution.

3.2. *Simulation Domain*

Although, XFlow has been used to simulate a flapping foil [18], the domain size investigation has not been presented. A proper domain size used to simulate the NASA trapezoidal wing [17] has been suggested in which the wall effect is negligible on the basis of a small blockage ratio. However, the domain size for foil is examined in this section.

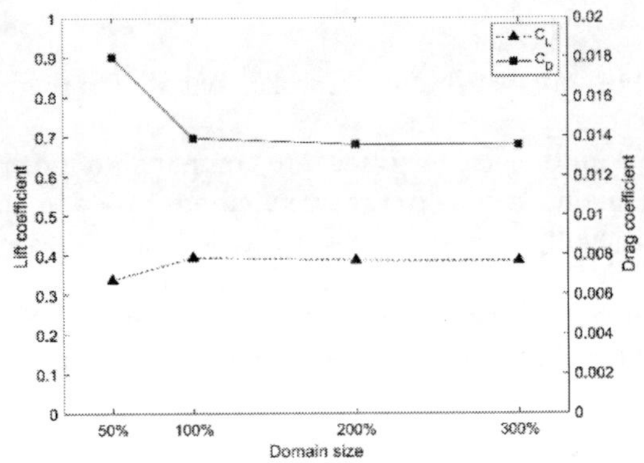

Figure 4. Lift and drag coefficient against domain size with respect to the initial domain size.

Table 2. Simulation time in function of domain size with respect to the initial domain size.

Domain size	Time
50%	47 mins.
100%	42 mins.
200%	44 mins.
300%	45 mins.

Figure 4 show the influence of the domain size in percentage of the initial domain. Although, the domain size of 100% of the initial one seems to be sufficient to accurately simulate the foil, the domain size of 200% is selected to simulate the flapping foil since the time-consuming unnoticeably changes as the domain is varied. Also, the flow physics of flapping foil is considerably different from that of a stationary wing due to its kinematics that creates an obstruction area.

4. Results and Discussion

Whilst the presence of a von Karman street coincides with drag generation, the presence of a reverse von Karman street coincides with thrust generation as illustrated in [6]. These phenomena agree very well with the change of fluid momentum. An example of vorticity distribution in the wake zone of the XFlow simulations illustrating a reverse von Karman street is present in figure 5. However, the reverse Von Karman cannot be directly implied from figure 5 since unfortunately XFlow presents vorticity in absolute value.

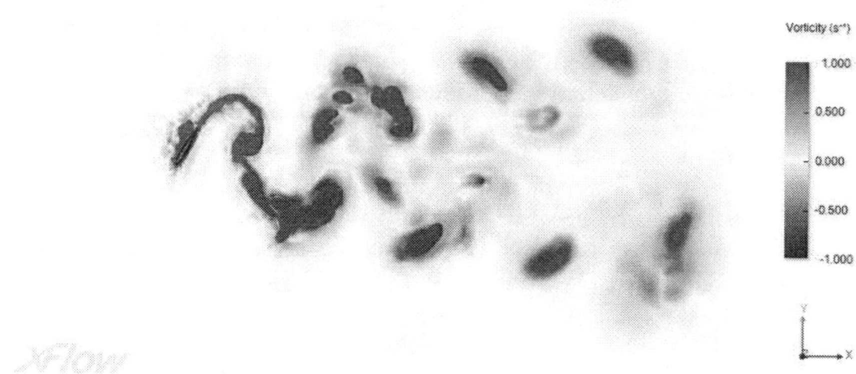

Figure 5. Example of vorticiy distribution in wake zone illustrating a reverse von Karman street.

The simulation results with the viscous effect are compared to those of potential flow theory. It can be seen from figure 6 to 9 that the results of the two methods give a good agreement for both propulsive efficiency and thrust coefficient in spite of noticeable deviations especially for kinematics with pitch amplitude of 15 degrees. At this small pitch, the viscous solver tends to give lower thrust as well as propulsive efficiency for all heave-to-chord ratio in comparison with those of BEM. For moderate pitch, e.g. θ_0 = 30° and 45°, the LBM seemingly gives a slightly higher propulsive efficiency, except for the very high heave-to-chord ratio (h_0/c = 4).

In addition, for each pitch amplitude, the difference of simulation results from both method is higher with a lower advance number. This is expectable since a lower advance number, or in other word a higher flapping frequency, corresponds to a higher angle of attack. The viscous effect undoubtedly has more influence in this zone.

Figure 10 to 13 present the comparison of heave-force and pitch-moment coefficients between the two methods. The numerical results agree very well especially for high heave-to-chord ratio. Although the deviation for the case of 15° pitch amplitude is visually noticeable, the ratio between the two results is not much different compared to the other cases of pitch amplitude. The deviation are more clearly seen on the pitch-moment coefficient with low heave-to-chord ratio and high pitch amplitude.

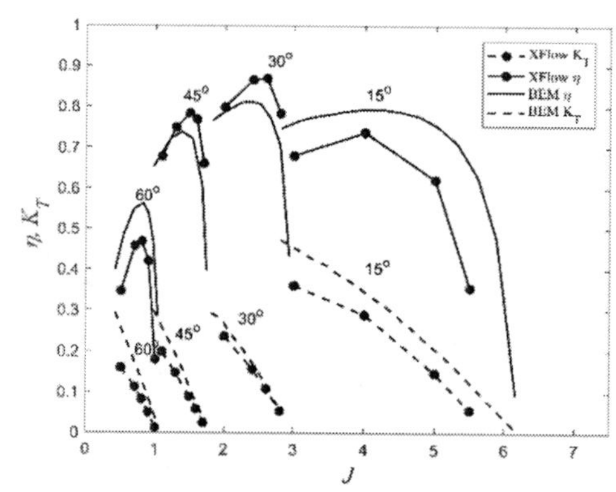

Figure 6. K_T and η in function of J_{eq}; h_0/c = 1; θ_0 = 15°, 30°, 45° and 60°.

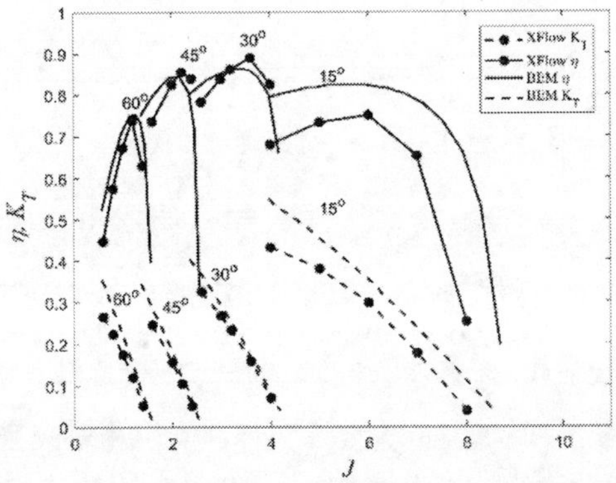

Figure 7. K_T and η in function of J_{eq}; $h_0/c = 2$; $\theta_0 = 15°$, 30°, 45° and 60°.

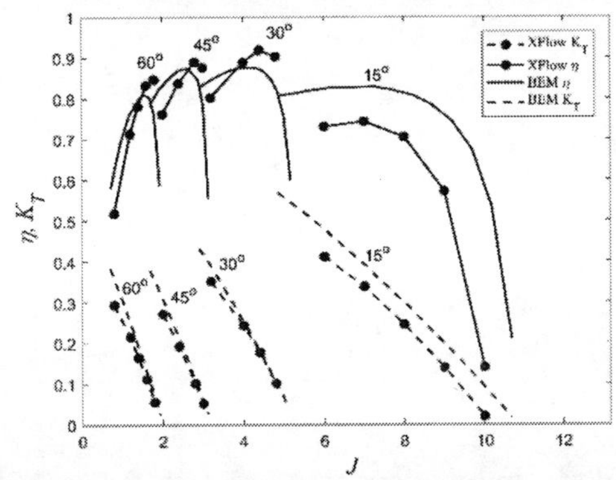

Figure 8. K_T and η in function of J_{eq}; $h_0/c = 3$; $\theta_0 = 15°$, 30°, 45° and 60°.

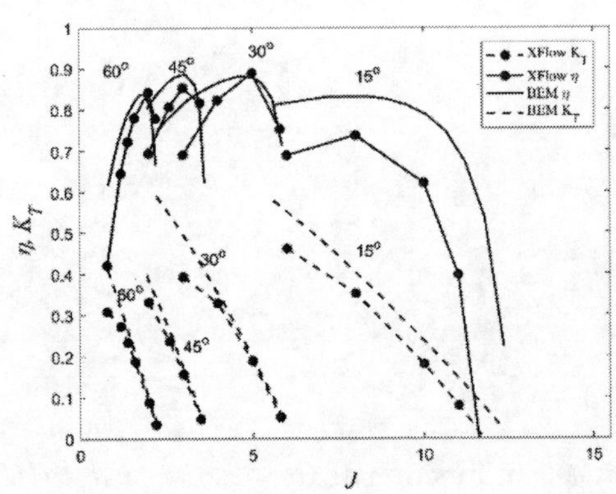

Figure 9. K_T and η in function of J_{eq}; $h_0/c = 4$; $\theta_0 = 15°$, 30°, 45° and 60°.

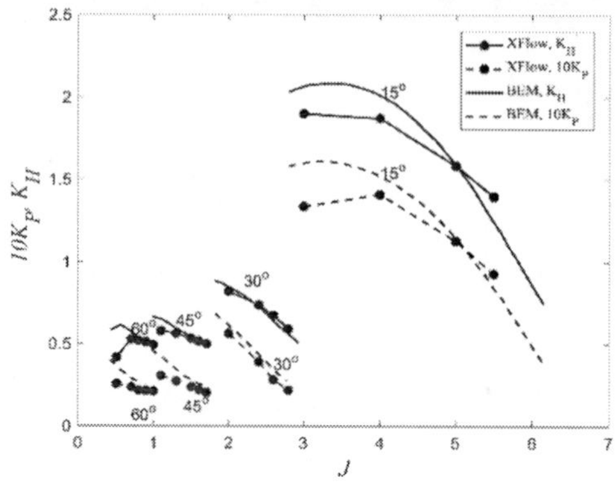

Figure 10. K_H and K_P in function of J_{eq}; $h_0/c = 1$; $\theta_0 = 15°, 30°, 45°$ and $60°$.

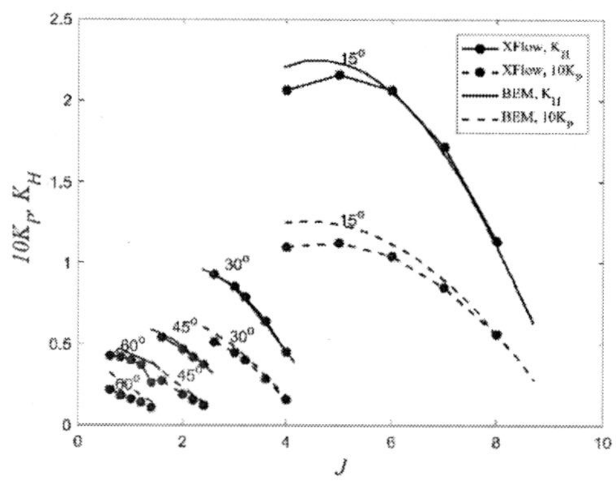

Figure 11. K_H and K_P in function of J_{eq}; $h_0/c = 2$; $\theta_0 = 15°, 30°, 45°$ and $60°$.

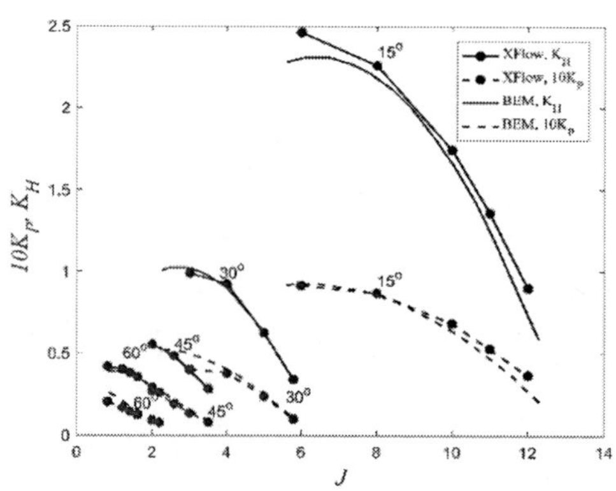

Figure 12. K_H and K_P in function of J_{eq}; $h_0/c = 3$; $\theta_0 = 15°, 30°, 45°$ and $60°$.

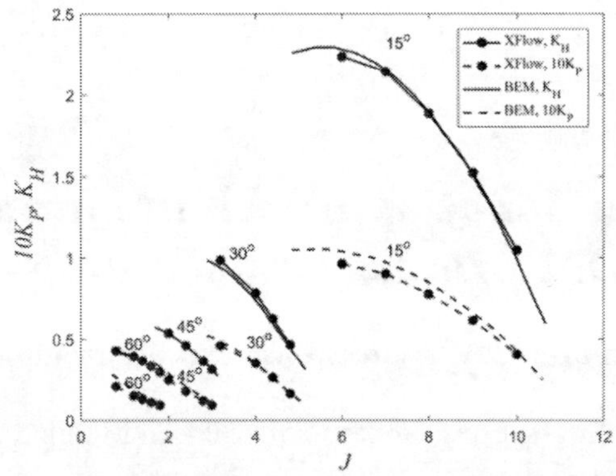

Figure 13. K_H and K_P in function of J_{eq} ; $h_0/c = 4$; $\theta_0 = 15°, 30°, 45°$ and $60°$.

5. Conclusion and Perspective

The viscous results agree well with the potential flow results. As expected, viscous results show lower efficiency in high flapping frequency zone. According to the viscous results, the possibility to use a flapping foil as a propulsion system is confirmed. The results show that the efficiency of a flapping foil propulsor can outperform a conventional screw propeller. The maximum efficiency is 90% for the viscous results and 87% for the inviscid results. For the case of 15° pitch amplitude, despite the lower efficiency compared with the other pitch amplitude cases, the propulsor can give an efficiency higher than 70% for a very wide range of equivalent advance ratios. Such propulsor can therefore be used like a controllable pitch propeller since it gives high efficiency for different sailing conditions i.e., different speeds and loads. However, from the point of view of mechanical complexity, the flapping foil seems to overcome the controllable pitch propellers.

References

[1] Rozhdestvensky K V and Ryzhov V A 2003 *Prog. Aerosp. Sci.* **39** pp 585-633
[2] Eloy C 2012 *J. Fluids Struct.* **7** pp 205-224
[3] Anderson J M, Streitlien K, Barrett D S and Triantafyllou M S 1998 *J. Fluid Mech.* **360** pp 41-72
[4] Blondeaux P, Fornarelli F, Guglielmini L, Triantafyllou M S and Verzicco R 2005 *Physics of Fluids* **17** 113601
[5] Read, Douglas A, Hover F S and Triantafyllou M S 2003 *J. Fluids Struct.* **17.1** pp 163-183.
[6] Floc'h F, Phoemsapthawee S, Laurens J M and Leroux J B 2012 *Ocean Eng.* **39** pp 56-61
[7] Xu G D, Xu W H and Dai J 2017 *Appl. Ocean Eng.* **63** pp 242-50
[8] Hover F S, Haugsdal O, Triantafyllou M S 2004 *J. Fluids Struct.* **19** pp 37-47
[9] Mantia M L and Dabnichki P 2011 *Appl. Math. Model.* **35** pp 979-990
[10] Belibassakis K A and Filippas E S 2015 *Appl. Ocean Res.* **52** pp 1-11
[11] Belibassakis K A and Politis G K 2013 *Ocean Eng.* **72** pp 227-40
[12] Politis G K and Tsarsitalidis V T 2014 *Ocean Eng.* **84** pp 98-123
[13] Phoemsapthawee S 2017 *Proc. of SIMMOD 2017* (Pattaya) pp 199-205
[14] Johnson A A 2006 *25th Army Science Conf.*
[15] XFlow 2016 *Computational fluid dynamics Theory guide* (Next Limit Technologies)
[16] XFlow 2016 *Computational fluid dynamics Validation guide* (Next Limit Technologies)
[17] Holman D M, Brionnaud R M, Martinez FJ and Mier-Torrecilla M 2012 *30th AIAA Applied Aerodynamics Conf.* (New Orleans)
[18] Sun X, Zhao L and Jiao Z 2016 *Proc. of 2016 IEEE Chinese Guidance, Navigation and Control Conf.* (Nanjing)

Application of grey-fuzzy approach in parametric optimization of EDM process in machining of MDN 300 steel

Partha Protim Das[1], P. Gupta[1], S Das[1], B. B. Pradhan[1] and S. Chakraborty[2]

[1] Mechanical Engineering Department, Sikkim Manipal Institute of Technology, Majitar, 737136, India
[2] Production Engineering Department, Jadavpur University, Jadavpur, 700032, India

*Corresponding Author: parthaprotimdas@ymail.com

Abstract. Maraging steel (MDN 300) find its application in many industries as it exhibits high hardness which are very difficult to machine material. Electro discharge machining (EDM) is an extensively popular machining process which can be used in machining of such materials. Optimization of response parameters are essential for effective machining of these materials. Past researchers have already used Taguchi for obtaining the optimal responses of EDM process for this material with responses such as material removal rate (MRR), tool wear rate (TWR), relative wear ratio (RWR), and surface roughness (SR) considering discharge current, pulse on time, pulse off time, arc gap, and duty cycle as process parameters. In this paper, grey relation analysis (GRA) with fuzzy logic is applied to this multi objective optimization problem to check the responses by an implementation of the derived parametric setting. It was found that the parametric setting derived by the proposed method results in better a response than those reported by the past researchers. Obtained results are also verified using the technique for order of preference by similarity to ideal solution (TOPSIS). The predicted result also shows that there is a significant improvement in comparison to the results of past researchers.

1. Introduction

Electro-discharge machining (EDM) emerges to be an important manufacturing process for machining hard materials [1] and its applications are widely accepted in many manufacturing industries [2]. For optimal machining performance measures in EDM, it is an important task to select proper combination of machining parameters as slight changes in a single parameter adversely affect the process. While controlling the process parameters it is capable of producing required dimensional accuracy and quality surface finish [3]. It is very necessary to select the appropriate machining parameters such as discharge current, pulse on time, pulse off time, arc gap, and duty cycle [4] which adversely affects the machining performance measures such as MRR, TWR, RWR and SR [2]. A considerable amount of work has been reported by the past researchers on the performance measurement of EDM process on the basis of MRR, TWR, RWR, and SR for various types of steels. Researchers have experimentally studied the response quality of EDM process and concluded that these qualities can be optimized by proper control of process parameters [5]. Thus, to exploit the full potential of the EDM process the control parameters are to be properly selected to obtain the optimum values of response parameters. Existing well known approaches such as grey relational analysis (GRA), analytic network process (ANP), preference ranking organization method for enrichment evaluation (PROMETHEE), VIKOR method etc. can be applied in this direction.

Deng [6] first introduced Grey systems. It is a powerful tool that deals with poor, unknown and vague data [7]. In recent years, a grey system has been effectively used for solving many multiple conflicting criteria in various fields of manufacturing [8, 9]. Fuzzy sets were first introduced by Zadeh [10] which can successfully deal with improper, uncertain and vague data. Fuzzy logic aided with GRA can further improve its performance in solving multi-objective optimization problems. Many researchers have effectively employed grey fuzzy logic in optimizing multi-objective problems [7, 9]. Soepangkat et al. [11] applied integrated fuzzy-logic based GRA in optimizing wire EDM process while machining AISI D2 steel for optimal responses of surface roughness and layer thickness. Krishnamoorthy et al. [12] improved the quality of the drilled holes in carbon fibre reinforced plastic (CFRP) composite materials by incorporating the optimal combination of drilling parameters using grey relational analysis aided with fuzzy logic.

An attempt has been made earlier by the past researchers in obtaining the parametric setting of EDM process on MDN 300 steel with copper as an electrode material. Nikalje et al. [13] studied the influence of EDM process considering discharge current, pulse on time, pulse off time, arc gap, and duty cycle as process parameters. They applied Taguchi method in obtaining the optimal parametric combination which optimizing performance measures MRR, TWR, RWR, and SR. Taguchi method is a single response optimization technique which deals with optimizing a single response while it does not take into account the effects of other performance measures. In this study emphasis has been made in optimizing the process parameters of this EDM process in machining of MDN 300 steel. GRA aided with fuzzy logic has been adopted and is applied to problems to further enhance the results obtained by past researchers.

2. Methodology

2.1. Design of experiments

The experiments are designed as per Taguchi's L_9 orthogonal array of experiments with MRR, TWR, RWR and SR as responses. The three level variations for each of discharge current, pulse on time and pulse off time is chosen for this experimentation are shown in table 1. The experimental results for all the nine experiments are shown in table 2.

Table 1. Process parameters of EDM process

Sl. No.	EDM parameter	Unit	Level 1	Level 2	Level 3
1	Discharge current	A	10	15	20
2	Pulse on time	μs	25	45	65
3	Pulse off time	μs	24	36	48

Table 2. Experimental Data

	Discharge current	Pulse on time	Pulse off time	MRR (mm^3/min)	TWR (mm^3/min)	RWR	SR (μm)
1	1	1	1	15.76	4.29	27.33	5.62
2	1	2	2	25.92	5.22	20.15	6.6
3	1	3	3	30.02	4.03	13.37	7.71
4	2	1	2	28.65	7.97	27.76	6.48
5	2	2	3	39.87	8.73	21.88	7
6	2	3	1	29.99	7.14	23.83	7.32
7	3	1	3	30.57	9.48	30.9	5.92
8	3	2	1	41.31	10.46	25.33	6.96
9	3	3	2	51.38	13	25.38	8.15

2.2 Grey relational analysis

In grey system, the data in the decision matrix are needed to be normalized (data pre-processing) in a range between 0 and 1 to make the data dimensionless and comparable. The following expressions are utilized for data pre-processing depending on the type of the considered criterion, i.e. equation (1) for larger-the-better and equation (2) for smaller-the-better type [8, 9].

$$x_i^*(k) = \frac{x_i(k) - \min x_i(k)}{\max x_i(k) - \min x_i(k)} \quad i = 1,2,\ldots,m \text{ and } k = 1,2,\ldots,n \tag{1}$$

$$x_i^*(k) = \frac{\max x_i(k) - x_i(k)}{\max x_i(k) - \min x_i(k)} \tag{2}$$

where $x_i(k)$ and $x_i^*(k)$ are the observed and normalized data respectively for i^{th} alternative and k^{th} criterion. After normalization, the grey relational coefficient (GRC) is calculated using equation (3).

$$\xi_i(k) = \frac{\Delta_{\min} + \zeta\Delta_{\max}}{\Delta_{0i}(k) + \zeta\Delta_{\max}} \tag{3}$$

where $\Delta_{0i}(k)$ is the difference between $x_i^0(k)$ and $x_i^*(k)$ ($x_i^0(k)$ is the ideal sequence). The distinguishing coefficient ζ lies between 0 and 1, usually considered as 0.5. $\Delta_{\min} = \forall j^{\min} \in i \forall k^{\min} \left\| x_0(k) - x_j(k) \right\|$ is the smallest value of Δ_{0i}; and $\Delta_{\max} = \forall j^{\max} \in i \forall k^{\max} \left\| x_0(k) - x_j(k) \right\|$ is the largest value of Δ_{0i}. A higher value of GRC for an alternative indicates that it is closer to the optimal solution with respect to a particular criterion. Grey relational grade (GRG) for an alternative is computed by averaging the GRC values corresponding to each criterion using equation (4).

$$\gamma_i = \frac{1}{n}\sum_{k=1}^{n} \xi_i(k) \tag{4}$$

where n is the number of criterion. A higher value of GRG indicates that the corresponding alternative is the best choice for the said application.

2.3 Fuzzy logic in grey relational analysis

Fuzzy set deals with imprecise and inadequate information in an efficient way to arrive at a logical conclusion for any decision making problem. Fuzzy set theory states that, in a universe of discourse X, a fuzzy subset $\tilde{A}$ of X is defined by a membership function $f_{\tilde{A}}(x)$, which takes each element x in X to a real number R in the interval of [0,1]. The function value $f_{\tilde{A}}(x)$ represents the grade of membership of x in $\tilde{A}$. The larger the value of $f_{\tilde{A}}(x)$ is, the stronger is the grade of membership for x in $\tilde{A}$.

In GRA, the use of lower-the-better and higher-the-better characteristics results in some uncertainty in the derived results which can be effectively controlled using fuzzy logic. A fuzzy logic unit consists of a fuzzifier, fuzzy membership functions, a fuzzy rule base, an inference engine and a defuzzifier. In the fuzzy logic, the membership functions are the inputs to the fuzzifier in order to fuzzify the GRC values. The inference engine performs a fuzzy reasoning of the developed fuzzy rules to generate a fuzzy value. The defuzzifier finally converts the fuzzy value into an understandable value known as grey fuzzy reasoning grade (GFRG). A fuzzy rule base comprising a set of if-then control rules is developed to represent the inference relationship between the input and output. A set of such fuzzy rules is provided as below:

Rule 1: if x_1 is a_1 and x_2 is b_1 and x_3 is c_1 and x_4 is d_1, then output (G) is e_1, else
Rule 2: if x_1 is a_2 and x_2 is b_2 and x_3 is c_2 and x_4 is d_2, then output (G) is e_2, else
Rule n: if x_1 is a_n and x_2 is b_n and x_3 is c_n and x_4 is d_n, then output (G) is e_n. (5)

IOP Conf. Series: Materials Science and Engineering **297** (2018) 012013 doi:10.1088/1757-899X/297/1/012013

where A_i, B_i, C_i and D_i are the fuzzy subsets defined by the corresponding membership functions, i.e. μ_{Ai}, μ_{Bi}, μ_{Ci} and μ_{Di} respectively. The inference engine performs fuzzy reasoning on fuzzy rules while taking max-min inference to generate a fuzzy value, $\mu_{c_0}(G)$.

$$\mu_{C_0}(G)=(\mu_{A_1}(x_1)\wedge\mu_{B_1}(x_2)\wedge\mu_{C_1}(x_3)\wedge\mu_{D_1}(x_4)\wedge\mu_{E_1}(G))\vee \dots\dots\dots$$
$$(\mu_{A_2}(x_1)\wedge\mu_{B_2}(x_2)\wedge\mu_{C_2}(x_3)\wedge\mu_{D_2}(x_4)\wedge\mu_{E_2}(G))\vee \dots\dots\dots$$
$$(\mu_{A_n}(x_1)\wedge\mu_{B_n}(x_2)\wedge\mu_{C_n}(x_3)\wedge\mu_{D_n}(x_4)\wedge\mu_{E_n}(G)) \quad (6)$$

where $\wedge$ and $\vee$ are the minimum and maximum operation, respectively. Finally, a centric fuzzification method is utilized to transform the fuzzy multi-response output, $\mu_{c_0}(G)$ into a crisp value of GFRG (G_0).

$$G_0=\frac{\sum G\mu_{C_0}(G)}{\sum \mu_{C_0}(G)} \quad (7)$$

The GFRG values are then arranged in descending order. The alternative with the maximum value of GFRG signifies it to the best choice with respect to a set of criteria/attributes.

3. Results and discussion

3.1. Grey-fuzzy analysis

The pre-processed data of experimental results is done using equation (1) and (2) where MRR is of 'higher-the-better' characteristics and TWR, RWR and SR is of 'lower-the-better' characteristics. GRC and GRG values are calculated using equation (3) and (4) and the results for each of the combination of parameters is given in table 3. In order to obtain an improved quality in the performances and to decrease the vagueness in the data, grey-fuzzy logic method is additionally used for computing the GFRG. In this paper, four inputs (GRC) and one output (GFRG) fuzzy-logic system are used. Mamdani inference engine is used which performs fuzzy reasoning with fuzzy rules for generating a fuzzy value. In total 9 fuzzy rules is developed based on 'if–then' control rule that shows inference relationship between the input GRC and output GFRG. In fuzzy logic, rules are usually developed based on the expert's opinions. But it is not always feasible to derive those rules from the knowledge of the human expert as when the number of inputs to a fuzzy system is more. Several methods have already been proposed for generating fuzzy rules automatically from numerical/experimental data [14, 15]. Generation of fuzzy rules from the numerical data mainly consists of two steps, i.e. a) to divide the data pattern into corresponding fuzzy subsets and b) to define a rule for each fuzzy subset. Segmentation of the data pattern and fuzzy rule generation are always interdependent. One of such rule is given below.

If MRR= Lowest, TWR= Highest, RWR = Lowest and SR= Highest, then GFRG = Very High.

Table 3. Normalized data, GRC, GRG and GFRG

Exp. No.	Normalized value				Grey relational coefficient				GRG	GFRG	TOPSIS Score
	MRR	TWR	RWR	SR	MRR	TWR	RWR	SR			
1	0	0.971	0.2037	1	0.3333	0.9452	0.3857	1	0.6661	0.680	0.4843
2	0.2852	0.8673	0.6132	0.6126	0.4116	0.7903	0.5638	0.5635	0.5823	0.567	0.5719
3	0.4003	1	1	0.1739	0.4547	1	1	0.3770	0.7079	0.696	0.6617
4	0.3619	0.5608	0.1791	0.6601	0.4393	0.5323	0.3785	0.5953	0.4864	0.493	0.4295
5	0.6769	0.476	0.5145	0.4545	0.6074	0.4883	0.5074	0.4783	0.5203	0.510	0.5555
6	0.3995	0.6533	0.4033	0.3281	0.4543	0.5905	0.4559	0.4266	0.4819	0.494	0.4987
7	0.4158	0.3924	0	0.8814	0.4612	0.4514	0.3333	0.8083	0.5136	0.510	0.3746
8	0.7173	0.2832	0.3177	0.4704	0.6388	0.4109	0.4229	0.4856	0.4896	0.506	0.4669
9	1	0	0.3149	0	1	0.3333	0.4219	0.3333	0.5221	0.510	0.4632

GRC values for MRR, TWR, RWR and SR are the inputs to the fuzzy logic system. The linguistic membership function for instance lowest (LT), low (L), medium (M), high (H) and highest (HT) are used to represent input variables of GRC. Likewise the output GFRG is being represented by the membership functions such as lowest (LT), very low (VL), low (L), medium low (ML), medium (M), medium high (MH), high (H), very high (VH), highest (HT). In this study triangular shaped membership function to define these membership functions are shown in figure 1 and 2. The rule-based fuzzy-logic reasoning is shown in figure 3. Maximum–minimum compositional operation by tracking the fuzzy reasoning yields a fuzzy output. At last, the defuzzifier converts the fuzzy predicted values into a crisp GRFG value by using MATLAB (R2013a) fuzzy toolbox. This GFRG values are tabulated in table 3. It can be seen from the table that experiment number 3 is having the highest GFRG value, signifying it to be the most preferred. A validation test with respect to technique for order preference by similarity to ideal solution (TOPSIS) reveals that the experiment number 3 actually gives the optimal parametric setting for the considered EDM process. It is a popular method that seeks to identify the best solution with the shortest distance to positive ideal solution and the longest distance from negative ideal solution. The computed TOPSIS scores are provided in table 3.

Table 4 and figure 4 shows the response table and corresponding graph for GFRG. It is obtained by calculating the average value of each input machining parameter at its corresponding level. The max–min column indicates that discharge current is the most significant factor among the three input parameters. In order to obtain the best responses, the optimal combination of the parameters as depicted from the table shows that discharge current must maintained at level 1 and while pulse on time and pulse off time at level 3 respectively.

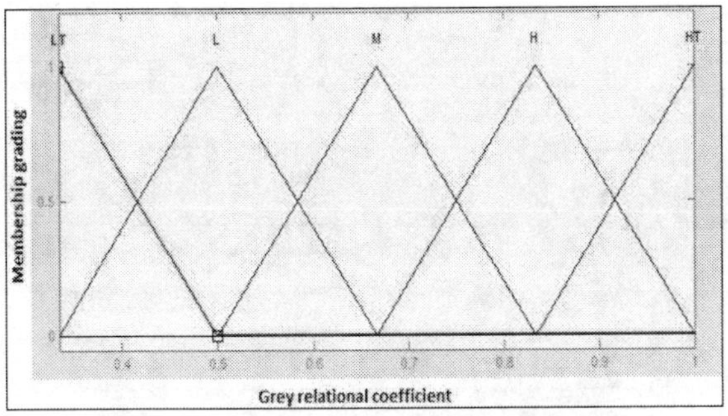

Figure 1. Input membership function

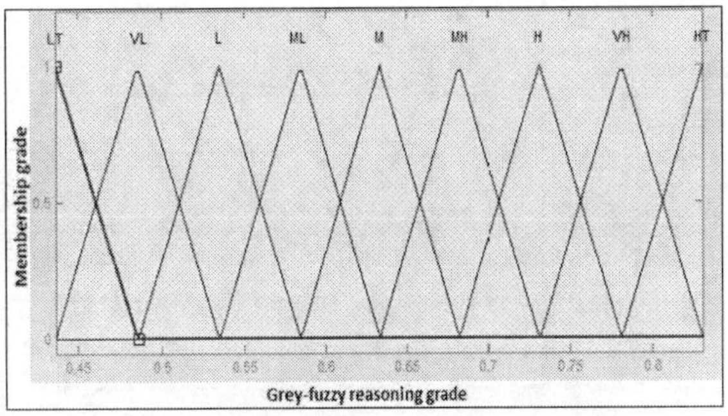

Figure 2. Output membership function

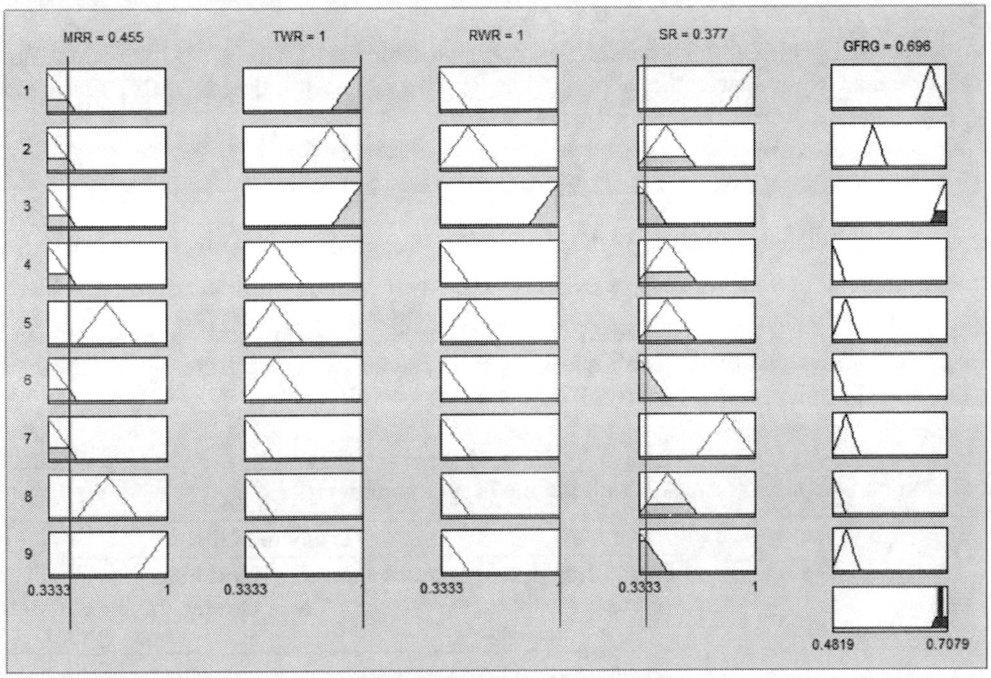

Figure 3. Rule viewer

Table 4. Response table for GFRG

EDM parameter	Level 1	Level 2	Level 3	Max-Min	Rank
Discharge current	**0.6477**	0.499	0.5087	0.1487	1
Pulse on time	0.5610	0.5277	**0.5667**	0.039	3
Pulse off time	0.5600	0.5233	**0.5720**	0.0487	2

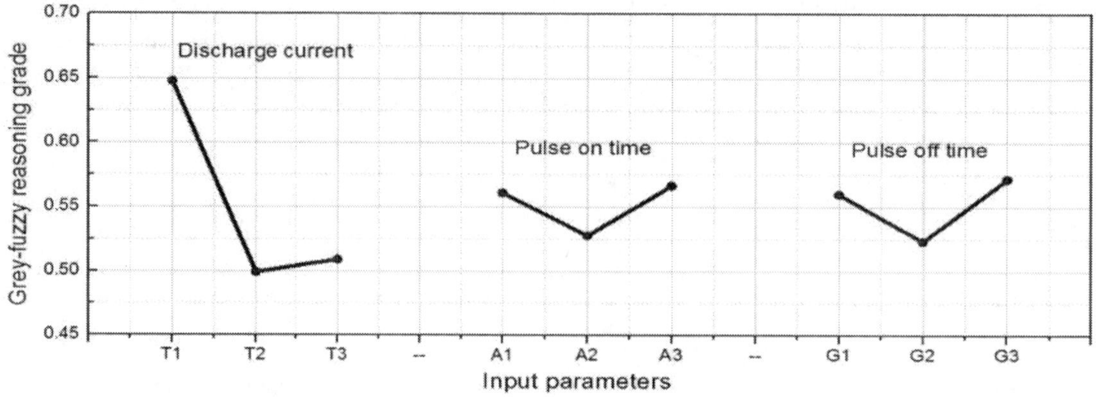

Figure 4. Response graph

3.2. Predicted GFRG

Optimal levels of machining input parameters obtained from GFRG are A1 B3 C3 which are different from the obtained parameter settings by past researchers. The predicted GFRG for the parametric combination can be estimated using the formulae.

$$G_p = G_m + \sum_{i=1}^{N} (\overline{G}_i - G_m) \tag{8}$$

where, G_p is the predicted GFRG, G_m is the mean GFRG for all the 9 experiments, $\overline{G}_i$ is the mean GFRG of the corresponding optimal i^{th} response and N is the total number of input parameters.

As shown in table 5 the predicted GFRG for both the previous and obtained input parametric combinations signifies that there is an improvement in the GFRG value from 0.6884 to 0.7064, which equals to 0.018, i.e., an improvement by 2.61%.

Table 5. Comparison table for initial and obtained input parameters

	Predicted GFRG	
Levels	Machining parameters previously obtained A3 B3 C2	Optimum machining parameters by our method A1 B3 C3
	Discharge current= 15A, Pulse on time= 45µs and Pulse off time= 24µs.	Discharge current= 10A, Pulse on time= 65µs and Pulse off time= 48µs.
GFRG	0.6884	0.7064
Improvement in GFRG	-	0.018
% improvement	-	2.61%

So as to fully justify the superiority of derived parametric setting A1B3C3 over A3B3C2, the following regression equations (showing only the main effects) are developed. Based on these equations, a comparison within the various responses at two different parametric settings is provided in table 6. It is interesting to observe that for MRR there is a substantial improvement from 30.399 mm3/min to 32.323 mm3/min with the setting A1B3C3. Similarly, for the remaining three responses it was found that their values are also marginally better with respect to the setting A1B3C3. Hence, there is also an improvement in the GFRG value by 6.45% for the proposed parametric setting.

IOP Conf. Series: Materials Science and Engineering **297** (2018) 012013 doi:10.1088/1757-899X/297/1/012013

MRR = – 13.5 + 1.72 x Discharge current + 0.303 x Pulse on time + 0.186 x Pulse off time (9)
TWR = – 2.97 + 0.647 x Discharge current + 0.0202 x Pulse on time + 0.0049 x Pulse off time (10)
EWR = 27.6 + 0.692 x Discharge current – 0.195 x Pulse on time – 0.144 x Pulse off time (11)
SR = 8.01 + 0.0367 x Discharge current – 0.0330 x Pulse on time + 0.0101 x Pulse off time (12)

Table 6. Comparison of predicted responses at two different settings

Levels	Predicted GFRG	
	Machining parameters previously obtained A3 B3 C2	Optimum machining parameters by our method A1 B3 C3
MRR (in mm^3/min)	30.399	32.323
TWR (in mm^3/min)	7.7616	5.0482
EWR	25.749	14.933
SR (in μm)	6.8679	6.0668
GFRG	0.594	0.695
% improvement	-	6.45%

4. Conclusion

From the above analysis it was found that a discharge current of 10 A, a pulse on time of 65 μs and a pulse off time of 48 μs are the optimal combination. It is also verified by the predicted results that the parametric combination found by the adopted approach significantly increases the output performance. Therefore, it is concluded that the optimization procedure proposed in this present paper significantly improved the machining of MDN 300 steel by EDM process.

References

[1] Abu Zaid OA 1997 *J. Mater Process Technol.* **68**(1) pp 27–32
[2] Ho K H and Newman S T 2003 *Int. J. Mach. Tool. Manufac.* **43** pp 1287–1300
[3] Ekmekci B, Elkoca O and Erden A 2005 *Metall. Mater. Trans.* **36B** pp 117–124
[4] Puertas I and Luis CJ 2003 *J. Mater. Process. Technol.* **143–144** pp 521–526
[5] Lin Y C and Lee H S 2009 *Int. J. Adv. Manuf. Technol.* **42** (11–12) pp 1052–1064
[6] Deng J 1989 *J. Grey. Syst.* **1** (1) pp 1–24
[7] Pattnaik S, Karunakar D B and Jha P K 2013 *Int. J. Adv. Manuf. Tech.* **67** (5–8) pp 1577–1587
[8] Hsiao Y F, Tarng Y S and Huang W J 2008 *Mater. Manuf. Processes* **23** pp 51–58
[9] Singh P N, Raghukandan K and Pai B C 2004 *J. of Mat. Proc. Tech.* **155-156** pp 1658-1661
[10] Zadeh L 1965 Fuzzy sets. *Inf. Control* **8** pp 338–353
[11] Soepangkat B O P and Pramujati B 2013 *Appl. Mech. Mater.* **393** pp 21–28
[12] Krishnamoorthy A, Boopathy S R, Palanikumar K and Davim J P 2012 *Measurement* **45** pp 1286–1296
[13] Nikalje A M, Kumar A and Sai Srinath K V 2013 *Int. J. of Adv. Manuf. Tech.* **69**(1–4) pp 41–49
[14] Wang L X and Mendel J M. 1992 *IEEE Trans. on Systems, Man and Cybernetics* **22** (6) pp 1414-1427
[15] Sugeno M and Yasukawa T 1993 *IEEE Trans. on Fuzzy Systems* **1** (1) pp 7-31.

IOP Conf. Series: Materials Science and Engineering **297** (2018) 012014 doi:10.1088/1757-899X/297/1/012014

Development and comparison of processing maps of Mg-3Sn-1Ca alloy from data obtained in tension versus compression

K.P. Rao[1,*], K. Suresh[2], Y.V.R.K. Prasad[3] and N. Hort[4]

1 Department of Mechanical and Biomedical Engineering, City University of Hong Kong, Tat Chee Avenue, Kowloon, Hong Kong
2 Department of Physics, Bharathiar University, Coimbatore 641046, Tamilnadu, India (formerly with City University of Hong Kong)
3 processingmaps.com, No. 2/B, Vinayaka Nagar, Hebbal, Bengaluru 560024, India (formerly with City University of Hong Kong)
4 Helmholtz-Zentrum Geesthacht, Institute of Materials Research, Magnesium Innovation Centre, Max-Planck-Straße 1, Geesthacht 21502, Germany

[*] Corresponding Author: mekprao@cityu.edu.hk, +852 3442 8409, +852 3442 0172

Abstract. The hot workability of extruded Mg-3Sn-1Ca alloy has been evaluated by developing processing maps with flow stress data from compression and tensile tests with a view to find the effect of the applied state-of-stress. The processing maps developed at a strain of 0.2 are essentially similar irrespective of the mode of deformation – compression or tension, and exhibit three domains in the temperature ranges: (1) 350 – 425 °C, and (2) 450 – 550 °C and (3) 400 – 500 °C, the first two occurring at lower strain rates and the third occurring at higher strain rates. In all the three domains, dynamic recrystallization occurs and is caused by non-basal slip and controlled by lattice self-diffusion in the first and second domains and grain boundary self-diffusion in the third domain. The state-of-stress imposed on the specimen (compression or tension) does not have any significant effect on the processing maps.

1. Introduction

Magnesium alloys have assumed technological importance in recent years due to their light weight and are being used for several applications in automobiles and aerospace. Two important limitations for use of these alloys are their lower corrosion resistance and creep strength. In recent years, newer alloys are being developed to combat these limitations and Mg-Sn-Ca system has been explored for this purpose. It has been observed that addition of Sn in magnesium improves the latter's corrosion resistance [1] and tensile properties [2]. Ternary addition of Ca improves the higher temperature creep strength by forming stable intermetallic CaMgSn particles in Mg-Sn matrix [3, 4]. It is observed that the ratio of Sn to Ca is important for microstructural control to optimize the strength and corrosion properties [5]. A ratio of 3:1 binds entire Ca in forming CaMgSn particles while the remaining Sn stays in solid solution. On the other hand, higher calcium content promotes the formation of Mg_2Ca particles [2], the presence of which is detrimental to the corrosion properties. Thus, Mg-3Sn-1Ca composition gives acceptable combination of corrosion resistance and creep strength. The hot workability of the as-cast alloy has been evaluated using processing map [6], which revealed that the alloy may be extruded at a much higher temperature (500 °C) than other conventional Mg alloys (300 °C). This processing map, the principles of which were discussed earlier [7-9], was developed on the basis of the temperature and strain rate dependence of flow stress obtained in constant strain rate isothermal compression tests. Uniaxial compression testing is preferred for this purpose since large strains may be imposed on the sample without the onset of complications like necking as in tension or a large non-uniformity of strain from center to surface as in a torsion bar.

It is useful to examine whether the state-of-stress in these testing techniques has any effect on the characterization of hot deformation by processing maps. In this direction, the processing maps developed on AISI 403L stainless steel on the basis of compression versus torsion data [10], have been compared and the results indicated that either techniques may be used to predict the workability domain but the instability regime is wider in the map made with torsion data than that made with compression data. This is understandable since shear stresses accentuate flow instability by adiabatic shear bands or flow localization. The aim of the present investigation is to compare the characteristics of hot deformation of high temperature extruded TX31 alloy as evaluated by processing maps developed from uniaxial tensile versus compression test data.

A processing map depicts the conditions of temperature and strain rate or speed that are best suited for performing industrial metal working operations, without resorting to trial and error attempts on shop-floor. The relevance of such a map is particularly important for successful conversion of cast alloys into their wrought form so to improve their mechanical properties through appropriate control of the evolution of microstructure during thermomechanical or hot working processes. The principles and procedures for developing processing maps [7-9] are briefly summarized here. Briefly, the work-piece undergoing hot deformation is considered to be a dissipator of power and the strain rate sensitivity (m) of flow stress is the factor that partitions power between deformation heat and microstructural changes. The efficiency of power dissipation compared to an ideal linear dissipator, occurring through microstructural changes during deformation as a function of temperature and strain rate is given by:

$$\eta = 2m/(m+1) \tag{1}$$

The variation of efficiency of power dissipation with temperature and strain rate represents a *power dissipation map* which is generally viewed as an iso-efficiency contour map. Further, the extremum principles of irreversible thermodynamics as applied to continuum mechanics of large plastic flow [11] are explored to define a criterion for the onset of flow instability given by the equation for the instability parameter ξ:

$$\xi(\dot{\varepsilon}) = \frac{\partial \ln[m/(m+1)]}{\partial \ln \dot{\varepsilon}} + m \leq 0 \tag{2}$$

The variation of the instability parameter as a function of temperature and strain rate represents an *instability map* which delineates regimes of instability where ξ is negative. A superimposition of the instability map on the power dissipation map gives a *processing map* which reveals *domains* (efficiency contours converging towards a peak efficiency) where individual microstructural processes occur and the limiting conditions for the *regimes* (bounded by a contour for $\xi = 0$) of flow instability. Processing maps help in identifying temperature – strain rate windows for hot working where the intrinsic workability of the material is maximum (e.g. dynamic recrystallization (DRX) or superplasticity) and also in avoiding the regimes of flow instabilities (e.g. adiabatic shear bands or flow localization).

2. Experimental

The nominal composition (wt.%) of the starting material Mg-3Sn-1Ca was extruded at 500 °C at a ram speed of 3 mm s^{-1} using a round shear die of 12 mm diameter with an extrusion ratio of 17:1. The average grain diameter was about 9 μm. For compression testing, cylindrical specimens with 10 mm diameter and 15 mm height were machined with the compression axis parallel to the extrusion direction. Uniaxial compression tests were conducted in the temperature range of 300 – 500 °C and true strain rate range of 0.0003 – 1 s^{-1}. Constant true strain rates were achieved by an exponential decay of actuator speed in the servo-hydraulic machine. Details of the test set-up and experimental procedure are given earlier [12]. For high temperature tensile tests, cylindrical specimens with a gage length of 30 mm and gage diameter of 6 mm were used and constant true strain rate had been achieved by ramping up the speed of the actuator exponentially. The load – stroke data were converted into true stress – true strain curves using standard equations. Processing maps were developed using the procedure described earlier using the flow stress data and different temperatures and strain rates at a constant strain. The deformed specimens were sectioned parallel to the compression or tensile axis as the case may be, polished and etched with an aqueous solution containing 10% picric acid. The microstructures were recorded close to the fracture end in the case of tensile specimens while in compression specimens they were examined in the middle of the specimen.

3. Results and Discussion

3.1. Stress-strain behavior

The true stress – true strain curves obtained on TX31 deformed in compression and tension at a temperature of 300 °C and at different strain rates are shown in figure 1. The compression curves were showing flow softening at higher strain rates and lower temperatures and a near steady state flow at lower strain rates and higher temperatures. The tensile curves on the other hand showed steady state flow only for the strain rate of 0.0003 s^{-1} and exhibited normal shapes for higher strain rates, namely reaching an ultimate tensile stress before dropping towards fracture when necking occurs.

The curves at higher temperatures are shown in figure 2 with 500 °C as an example, for compression and tension. The compression curves are of flow softening type at higher strain rates and steady-state type at lower strain rates similar to those obtained at 300 °C (figure 1). The tensile curves (figure 2) exhibited nearly steady state flow for all the strain rates except that for a strain rate of 1 s^{-1}, which started to neck at strains higher than about 0.2. From the stress – strain curves given in figures 1 and 2, it is clear that for the purpose of comparison, the flow stress data up to a strain of 0.2 at different temperature and strain rates may be valid for developing processing maps for compression *versus* tension. At higher strains, the data will acquire complications due to the three-dimensional state-of-stress in necked region in tension rendering it unacceptable for comparison with compression.

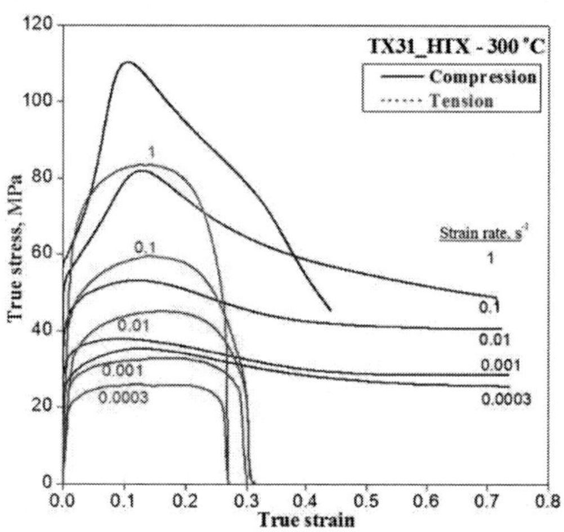

Figure 1. Stress-strain curves of TX31 alloy obtained at a temperature of 300 °C in compression *versus* tension at different strain rates.

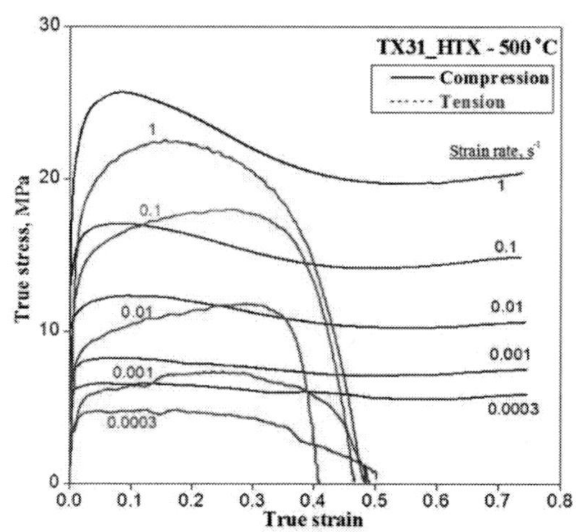

Figure 2. Stress-strain curves of TX31 alloy obtained at a temperature of 500 °C in compression *versus* tension at different strain rates.

3.2. Processing maps

The processing maps obtained at a strain of 0.2 from flow stress obtained from compression and tension tests are shown in figure 3 and figure 4, respectively. The numbers on the contours represent efficiency of power dissipation in percent. The maps exhibit three domains in each case – two in the lower strain rate range and the third at higher strain rate range. For compression map (figure 3), the lower strain rate domain (#1) occurs over a wide range of temperature of 340 – 420 °C and extend over the strain rate range of 0.0003 – 0.002 s^{-1} with a high peak efficiency of power dissipation at 28%. The higher temperature domain (#2), however, occurs in temperature and strain rate ranges given by 450 – 500 °C and 0.0003 – 0.01 s^{-1} with a peak efficiency of about 29%. In comparison, the map for tension shown in figure 4 also exhibited three domains in a similar temperature range although the upper strain rates have moved to higher values in the case of Domain #1 and Domain #2. The peak efficiency values were also higher in these two domains. It is possible that the true strain rates in tension will be lower than the nominal values, which may be the reason for the higher values exhibited by the processing maps for tension.

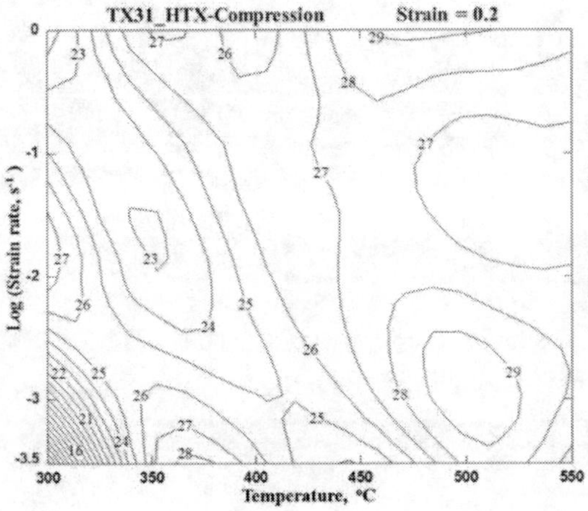

Figure 3. Processing maps of TX31 alloy at a strain of 0.2 for compression.

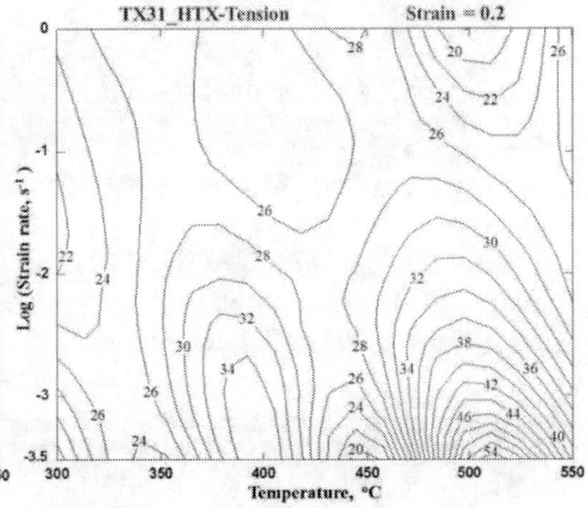

Figure 4. Processing maps of TX31 alloy at a strain of 0.2 for tension.

The microstructures obtained in Domain #1 at 500 °C/0.0003 s^{-1} on the compression and tension specimens, shown in figure 5(a) and 5(b) are similar. The microstructures obtained on specimens deformed under compression and tension in Domain #2 are shown in figure 6(a) and 6(b) and those in Domain #3 are shown in figure 7(a) and 7(b) and these are also similar. The microstructures in figure 5 to figure 7 clearly show that dynamic recrystallization has occurred in these domains. It is likely that Domain #1 and Domain #2 will merge into a single domain at large strain since the temperature range for these domains is favorable for the occurrence of non-basal slip including prismatic and second order pyramidal slip [13].

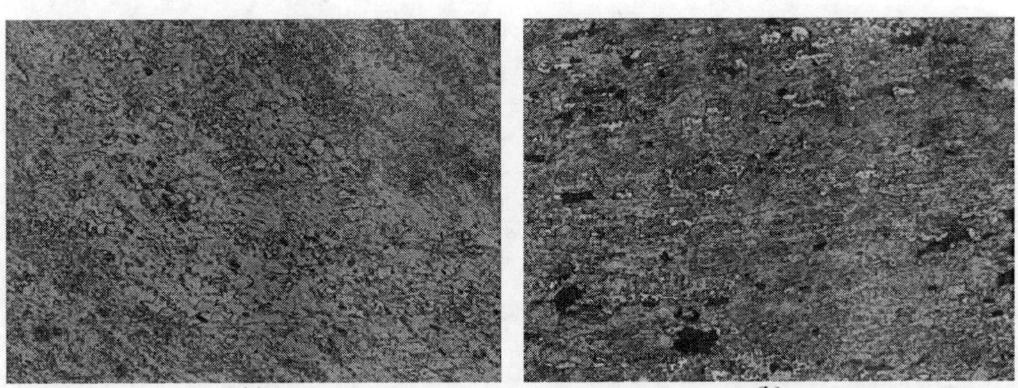

(a) (b)

Figure 5. Microstructures obtained on TX31 alloy processed at 400 °C/0.0003 s^{-1} (Domain #1): (a) Compression, and (b) Tension.

The higher strain rate domain is not fully developed at the strain selected for the comparison, i.e., 0.2, but it generally represents DRX by basal+prismatic slip and recovery by climb controlled by grain boundary self-diffusion [13]. However, all of the three domains are essentially similar for both tension and compression and represent DRX process. Thus, the results show that the hot working mechanisms have not changed irrespective of deformation imposed either by compression or tension.

(a) (b)

Figure 6. Microstructures obtained on TX31 alloy processed at 500 °C/0.0003 s^{-1} (Domain #2): (a) Compression, and (b) Tension.

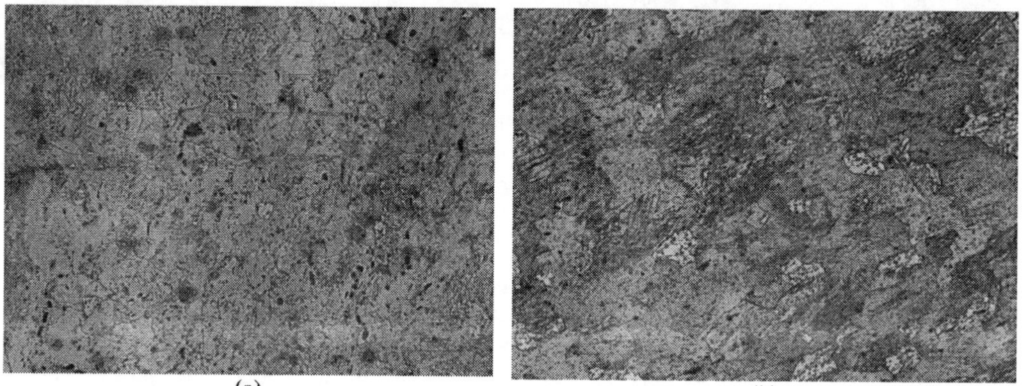

(a) (b)

Figure 7. Microstructures obtained on TX31_HTX alloy processed at 450 °C/1.0 s^{-1} (Domain #3): (a) Compression, and (b) Tension.

4. Conclusions

The hot working characteristics of high temperature extruded TX31 alloy deformed in compression has been compared with those evaluated on the basis of tensile data. The following conclusions are drawn:

(1) The shapes of true stress – true strain curves are similar and exhibited a steady state flow in both modes of deformation until a strain of about 0.2, beyond which the occurrence of necking in tension has restricted the comparison.

(2) The processing maps developed at a strain of 0.2 with flow stress data at different temperatures and strain rates, are essentially similar irrespective of the mode of deformation – compression or tension.

(3) The processing maps exhibit three domains in the temperature ranges: (1) 350 – 425 °C, and (2) 450 – 550 °C and (3) 400 – 500 °C, the first two occurring at lower strain rates and the third occurring at higher strain rates.

(4) All the three domains represent dynamic recrystallization process caused by non-basal slip and controlled by lattice self-diffusion in the first and second domains and grain boundary self-diffusion in the third domain (higher strain rates).

(5) The state-of-stress imposed on the specimen (compression or tension) does not have any significant effect on the processing maps.

Acknowledgement
The work presented in this paper has been supported by a strategic research grant from the City University of Hong Kong (Project #7002013).

References
[1] Abu Leil T, Rao K P, Hort N, Blawert C and Kainer K U 2006 *Magnesium Technology 2006* ed Luo A A, Neelameggham N R and Beals R S (TMS: Warrendale) p 281
[2] Liu H, Chen Y, Tang Y, Wei S and Niu G 2007 *J. Alloys and Compounds* **440** p 122
[3] Hort N, Huang Y, Abu Leil T, Maier P and Kainer K U 2006 *Advanced Eng. Mater.* **8** p 359
[4] Abu Leil T, Huang Y, Dieringa H, Hort N, Kainer K U, Bursik J, Jiraskova Y and Rao K P 2007 *Mater. Sci. Forum* **546-549** 69
[5] Abi Leil T, Hort N, Dieringa H, Blawert C, Huang Y, Kainer K U and Rao K P 2006 *Magnesium Technology in the Global Age* ed Pekguleryuz M O and Mackenzie L W F (CIM: Montreal) p 739
[6] Rao K P, Prasad Y V R K, Hort N and Kainer K U 2008 *J. Mater. Processing Technol.* **201** p 359
[7] Prasad Y V R K, Rao K P and Sasidhara S 2015 *Hot working guide – A compendium of processing maps*, (ASM International: Materials Park)
[8] Prasad Y V R K and Seshacharyulu T 1998 *Int. Mater. Reviews* **43** p 243
[9] Prasad Y V R K 2003 *J. Mater. Eng. and Performance* **12** p 638
[10] Venugopal S, Prasad Y V R K and Mannan S L 1996 *Mater. Letters* **26** p 161
[11] Ziegler H 1965 Some extremum principles in irreversible thermodynamics with applications to continuum mechanics *Progress in Solid Mechanics (*vol 4) ed Sneddon I N and Hill R (New York: Wiley) p 91
[12] Prasad Y V R K and Rao K P 2005 *Mater. Sci. and Eng.* **A391** p 141
[13] Prasad Y V R K, Rao K P, Hort N and Kainer K U 2009 *Mater. Sci. and Eng.* **A502** p 25

IOP Conf. Series: Materials Science and Engineering **297** (2018) 012015 doi:10.1088/1757-899X/297/1/012015

Compressive behavior of Sulcata Tortoise's carapace at high rate of deformation

Nadda Jongpairojcosit[1,2], Chinnawit Glunrawd[1] and Petch Jearanaisilawong[1,*]

[1] Department of Mechanical and Aerospace Engineering, Faculty of Engineering King Mongkut's University of Technology North Bangkok, 1518 Pracharat 1 Rd, Wongsawang, Bangsue, Bangkok 10800, Thailand
[2] Defence Technology Institute (Public Organization), Ministry of Defence Office of the Permanent Secretary of Defence (Chaengwattana) 4th Floor, 47/433 Moo 3, Ban Mai, Pak Kret, Nonthaburi 11120, Thailand

* Corresponding Author: petch.j@eng.kmutnb.ac.th

Abstract. This paper presents the dynamic compressive response of tortoise carapace at high rate of deformation. Disk specimens are cut from carapace and compressed using Split-Hopkinson Pressure Bar (SHPB) technique. The steel bar system together with a copper pulse shaper generate an incident wave that can achieve a constant rate of deformation within the specimens. The results show increasing compressive modulus and compressive strength compared to the quasi-static data of carapace. The strain waves on the incident and the transmission bars from finite element analysis based on the experimental setup agree with the test data.

1. Introduction

Sulcata tortoise or African spurred tortoise is a desert dwelling tortoise. It has a large shell, which the length from head to tail can reach 80 cm when fully grown. The shell is its natural shield to prevent its body from attack. The dome-shape of the shell known as carapace is the main part of the protection. The carapace structure has four distinctive layers, which are keratin scutes, dorsal cortex, cancellous interior and ventral cortex. The keratin scutes made of fibrous structural protein cover the outer shell surface. The other layers are bone-like materials; the dorsal and ventral cortices are dense solids, whereas the cancellous interior is a closed-cell structure. The mechanical responses under quasi-static loading and microscopic structures of turtle shells have been adapted in bio-inspired armor designs [1]-[2]. However, the discussion on the dynamic behavior of tortoise's carapace is still limited.

Split-Hopkinson pressure bar (SHPB) technique is utilized to investigate mechanical behavior of materials under high strain rate. The test system consists of striker, incident and transmission bars of the same diameter and material. These bars are axially aligned with free movement in the longitudinal direction. The sample is placed between the incident and transmission bars. The striker launched by compressed gas hits the incident bar that generates the incident wave travelling along the incident bar. Some part of the incident wave passes through the sample and the transmission bar, while the other part reflects back along the incident bar. The strain waves on the incident and the transmission bars are recorded by oscilloscope for material behavior analysis. In the case of soft materials, pulse shaper technique is implemented to increase the rise time of incident wave to achieve dynamic equilibrium. Figure 1 presents a schematic of the SHPB system.

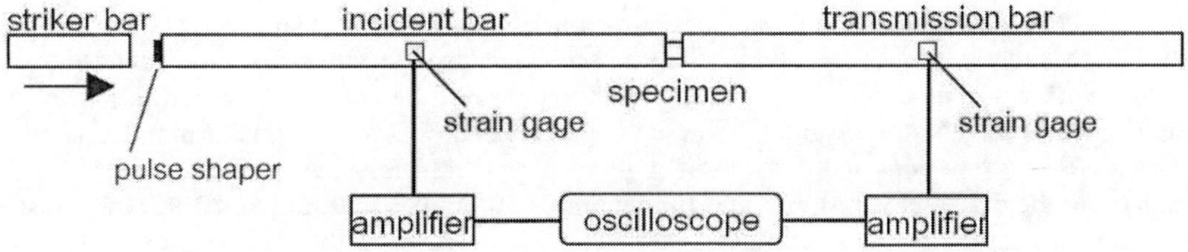

Figure 1. Split-Hopkinson pressure bar configuration in compression test.

With the assumption of one-dimensional wave propagation, the strain rate in the specimen $\dot{\varepsilon}_s$ as a function of time, t, can be calculated as [3]

$$\dot{\varepsilon}_s(t) = \frac{c_b}{l_s}(-\varepsilon_i(t) + \varepsilon_r(t) + \varepsilon_t(t)) \tag{1}$$

where c_b is the elastic wave speed of the bars, l_s is the thickness of the specimen, ε is the strain wave and the subscript i, r and t stand for incident, reflected and transmitted wave, respectively.

The strain on specimen ε_s can be derived from the time integral of Equation (1) as shown in equation (2) and the stress on specimen σ_s is given by equation (3).

$$\varepsilon_s(t) = \frac{c_b}{l_s}\int_0^t(-\varepsilon_i(t) + \varepsilon_r(t) + \varepsilon_t(t))dt \tag{2}$$

$$\sigma_s(t) = \frac{A_b E_b}{2A_s}(\varepsilon_i(t) + \varepsilon_r(t) + \varepsilon_t(t)) \tag{3}$$

where A_b and E_b are the cross-sectional area and elastic modulus of the bars, respectively. A_s is the cross-sectional area of the specimen.

The SHPB result is only valid when the specimen attains dynamic equilibrium. This condition is determined by force balance in equation (4). The forces are derived from equation (5) - (6).

$$F_1(t) = F_2(t) \tag{4}$$

$$F_1(t) = A_b E_b(\varepsilon_i(t) + \varepsilon_r(t)) \tag{5}$$

$$F_2(t) = A_b E_b \varepsilon_t(t) \tag{6}$$

where F_1 and F_2 the forces at the interfaces between the incident bar and the specimen and the transmission bar and the specimen, respectively.

The aim of this study is to examine the dynamic compression behavior of tortoise's carapace using a compressive SHPB system with steel bars and a copper pulse shaper. The finite element analysis is carried out to explore the impact behavior of the carapace. The stress-strain result at high strain rate can be used to define dynamic plastic behavior of tortoise's carapace that sheds light into the underlying energy absorption mechanism of the carapace layers.

2. Experiment and method

Circular disk samples are prepared from a dried carapace. The disk surfaces and edges are polished with sandpaper. The keratin scutes are totally removed because of its wavy surface. The dorsal and ventral cortices are slightly ground to obtain flat surfaces; the disk samples resemble bone-like sandwich structure with a porous core. The curvature and thickness of the carapace are major constraints of the sample size. Average diameter and thickness of specimens are 15.8 mm and 3.3 mm, respectively.

Our SHPB system consists of the striker, incident and transmission bars made of AISI 4340 alloy steel having the diameter of 20 mm and the length of 300 mm, 2000 mm and 1500 mm, respectively. The firing system is driven by compressed air and the velocity of the striker is controlled by the air pressure. The striker speed is measured by two infrared sensors attached at the end of the striker bar launching barrel. The strain gauges are attached on the incident and transmission bars for strain wave acquisition. The data is captured through a digital oscilloscope that has a sampling rate of $2.5x10^6$ per second. Since the tortoise carapace is a biological and porous material that has a low wave speed, a pulse shaper is required to increase rise time to attain dynamic equilibrium. Vecchio and Jiang [4] proposed that the shape of the incident wave should be tailored to conform to the shape of the transmitted pulse to achieve a constant strain rate. C1100 copper disks with a diameter of 12 mm and a thickness ranging from 1.0, 1.5 to 2.0 mm are used as pulse shapers in this study. After preliminary measurements, the 2.0 mm pulse shaper with 4 bars firing pressure provide a striking velocity of 14 m/s and an incident wave that results in dynamic equilibrium at a constant strain rate. Figure 2 shows the positions of the pulse shaper and the specimen. The system is calibrated before the experiment to confirm the coaxial impact at the bar interfaces.

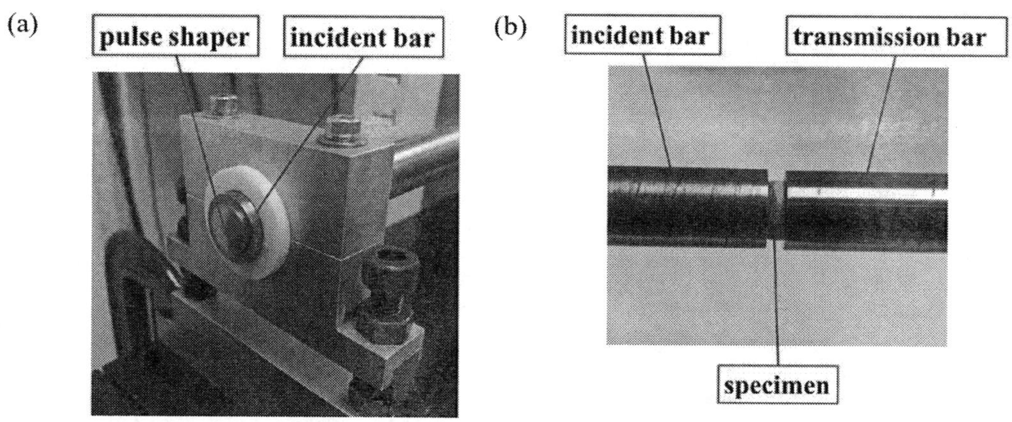

Figure2. (a) copper pulse shaper (b) tortoise carapace specimen.

To gain an insight into the impact behavior of the carapace under compression, a numerical analysis of the specimen under SHPB system is performed. The finite element model is created in ANSYS using 3D hexahedral elements and frictionless conditions on the contact surfaces. To take advantage of the structural symmetry, only a quarter of the bars' cross-section is modeled. Moreover, the specimen sandwich structure is simplified as a solid disk. The dimensions of bars, pulse shaper and specimen are replicates of the experimental values. The completed FE model consists of 45,942 elements. The AISI 4340 steel bar and copper pulse shaper are modeled as rate-dependent elastic-plastic materials with standard parameters from software's library. The elastic behavior of the carapace is taken from [5]. Due to the limitations of rate dependent plastic data, the plastic behavior of the carapace is modeled after those of human skull [6]. Table 1 lists the material parameters in the finite element model. The strain wave signals on the incident and transmission bars from the analysis are compared with the experimental data.

Table 1. Material parameters in finite element model

Steel AISI 4340			Johnson Cook Strength					
Density (kg/m^3)	Bulk modulus (GPa)	Shear Modulus (GPa)	Initial yield stress (MPa)	Hardening constant (MPa)	Hardening exponent	Strain rate constant	Thermal softening exponent	Melting Temp. (°C)
7830	159	81.8	792	510	0.26	0.014	1.03	1520

Copper		Mie-Gruneisen equation of state		
Density (kg/m^3)	Shear Modulus (GPa)	Gruneisen coefficient	C1 (m/s)	S1
8900	46.4	2	3958	1.497

Carapace		Mie-Gruneisen equation of state [6]		
Density (kg/m^3)	Shear Modulus (MPa)	Gruneisen coefficient	C1 (m/s)	S1
1470	154	1	1850	0.94

3. Results and discussion

Figure 3 presents the strain waves on incident and transmission bars. These data are derived from the voltage signal of the strain gauges using an in-house MATLAB code. The positive and negative values represent the characteristics of tension and compression in the specimen, respectively. When the striker bar hits the incident bar, it creates a compressive incident wave. Instead of a square wave, the steel bars together with the copper pulse shaper modify the incident wave and increase the rise time as shown in figure 3.

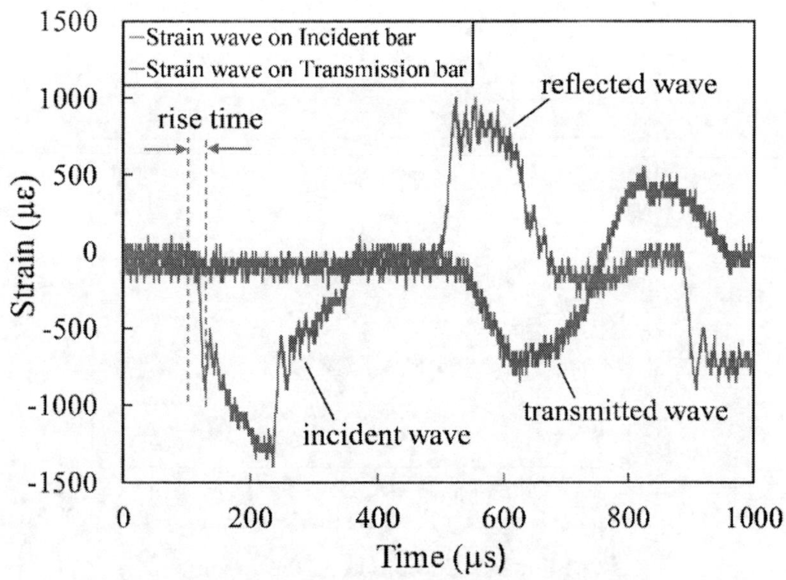

Figure 3. Strain wave signals from SHPB testing.

This pulse shaper induces a constant strain rate on SHPB testing of the carapace. It also matches with the report from Bekker et al. [6]. The incident wave travels along the bar to the interface between the incident bar and the specimen. At this point, the wave separates into reflected and transmitted waves. The reflected wave is in tension while the transmitted wave is in compression. The reflected wave is measured from the strain gage on the incident bar, and the transmitted wave is measured from that on the transmission bar.

The dynamic equilibrium is investigated by comparing forces on both ends of the specimen. From equation (5) - (6), these forces are calculated from the strains of incident, reflected and transmitted waves. Figure 4 shows that the system attained dynamic equilibrium between 25-80 microseconds. The graph also illustrates that, after the force equilibrium, the F_2 calculated from the transmitted wave is lower than F_1. This characteristic supports the fact that the wave through the soft and porous material has a low transmitted pulse and difficult to attain the dynamic equilibrium.

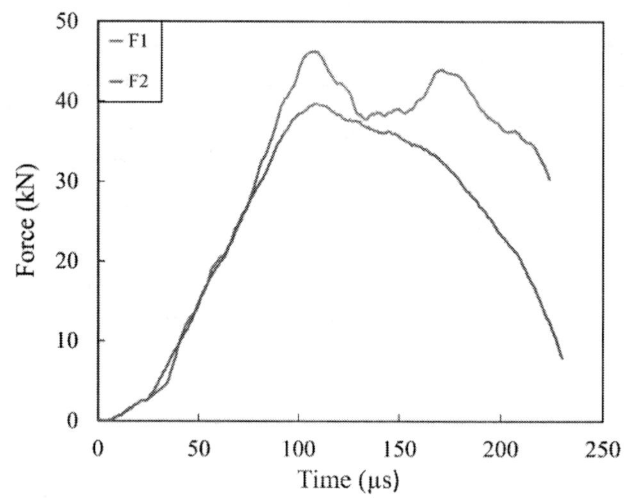

Figure 4. Forces on front (F_1) and back (F_2) surfaces of the specimen.

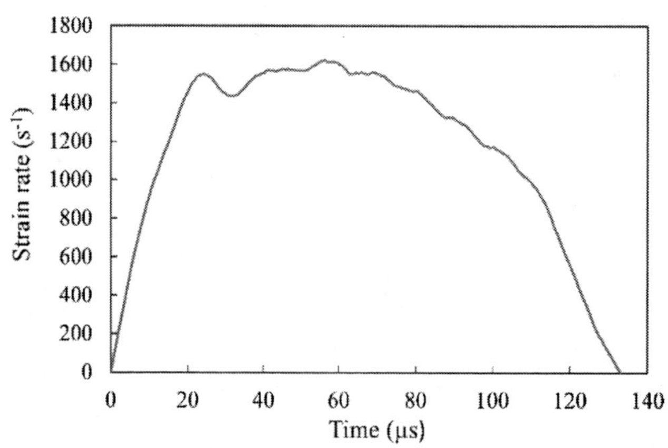

Figure 5. Strain rate on the specimen.

Figure 5 presents the strain rate on the specimen over time. The average strain rate during dynamic equilibrium (between 25-80 microseconds) is 1500 s^{-1}, which is classified as high strain rate deformation (10^2-10^4 s^{-1}).

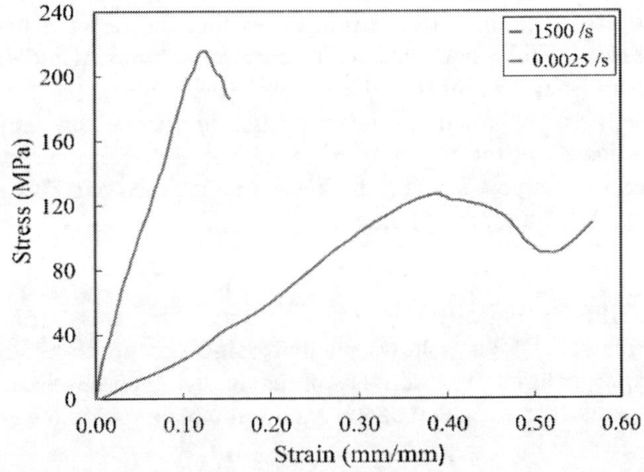

Figure 6. Dynamic and quasi-static stress-strain curve of tortoise carapace.

The stress-strain data of quasi-static compression test was taken from the literature [5]. The cube samples of tortoise carapace were compressed along the through-thickness direction at a strain rate of 0.0025 s^{-1}. The engineering stress-strain data on the specimen at 1500 s^{-1} is compared with the quasi-static data in figure 6. The graph shows the strain rate hardening characteristic of the material where compressive strength increases from 126 MPa to 215 MPa at high strain rate. In quasi-static compression, the curve shows a characteristic of sandwich structure with a porous core where the linear elastic behaviour is followed by a softening response due to the collapse of porous cells. At a strain of 0.5, the collapsed porous cells densify and the structure exhibits a hardening response. In comparison, the response of carapace in the SHPB test exhibits the characteristic of a brittle material with a small strain-to-failure. This is because at high rate of deformation, the rapid increase of stress in the structure lead to fracture of solid layers prior to the collapse and densification of the porous cores.

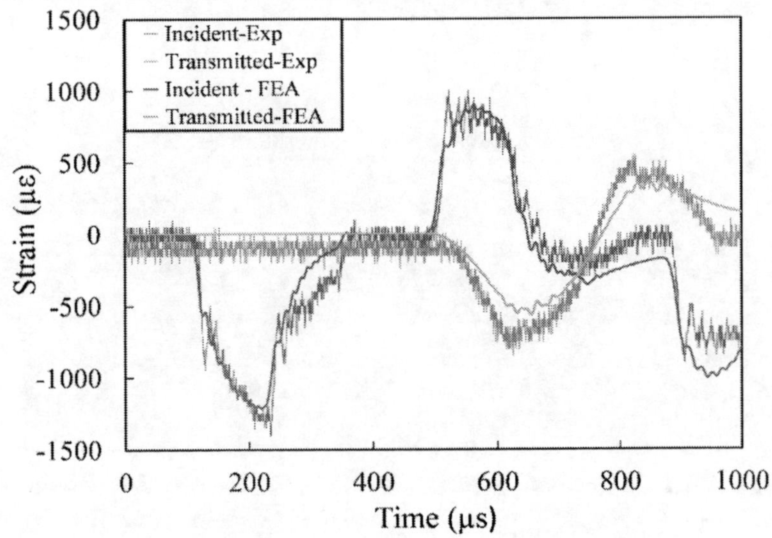

Figure 7. Comparing strain wave signals.

The normal strain in the longitudinal direction of the bar at the same position as the strain gauges are plotted in figure 7. The graph shows that the simulation result has the same trend as the test data with slightly lower magnitude of the strain. This demonstrates that the defined materials can capture the phenomenon of the experiment. With good agreements between numerical and experimental results, this simulation model demonstrates an insight into the dynamic behavior of the carapace under high strain rate regime. In particular, the mechanical response at high strain rate that cannot be observed in the experiments can be investigated in the simulation. In addition, the experimental parameters such as dimension of pulse shaper, impact velocity and thickness of specimen can be validated in this model to achieve the desired strain rate for future experiments.

4. Conclusions

This paper presents the SHPB testing of the tortoise carapace. The stress-strain data of this bone-like sandwich structure at strain rate $1500 s^{-1}$ shows an increasing compressive strength compared to the quasi-static data of the same material. The simulation results using compatible material properties and simplified model of the tortoise carapace can capture the same response as the experiment.

References

[1] Achrai B and Wagner D 2017 *J. Mech. Behav. Biomed. Mater.* **73** p 50
[2] An B and Wagner D 2017 *J. Mech. Behav. Biomed. Mater.* **71** p 54
[3] Weinong W C and Song B 2011 *Split Hopkinson (Kolsky) Bar Design, Testing and Applications* (Springer)
[4] Vecchio S and Jiang F 2007 *Metall. Mater. Tran.* **38A** p 2655
[5] Jongpairojcosit N and Jearanaisilawong P 2017 *J. Mech. Behav. Biomed. Mater.* **72** p 261
[6] Jean A, Nyein K, Zheng Q, Moore F, Joannopoulos D and Radovitzky R 2014 *Proc. Natl. Acad. Sci. U.S.A.* vol. 111 p 15310
[7] Bekker A, Cloete J, Chinsamy-Turan A, Nurick N and Kok S 2015 *Mater. Sci. Eng. C.* **46** p 443

Nanoindentation tests of Sulcata Tortoise's carapace

Nadda Jongpairojcosit[1,2] and Petch Jearanaisilawong[1,*]

[1] Department of Mechanical and Aerospace Engineering, Faculty of Engineering King Mongkut's University of Technology North Bangkok, 1518 Pracharat 1 Rd, Wongsawang, Bangsue, Bangkok 10800, Thailand
[2] Defence Technology Institute (Public Organization), Ministry of Defence Office of the Permanent Secretary of Defence (Chaengwattana) 4th Floor, 47/433 Moo 3, Ban Mai, Pak Kret, Nonthaburi 11120, Thailand

* Corresponding Author: petch.j@eng.kmutnb.ac.th

Abstract. This paper illustrates the dependence of tortoise carapace's hardness and elastic modulus on peak-load of nanoindentation tests. Sulcata Tortoise carapace is a sandwich-like structure consisting of four submicron layers: keratin scutes, dorsal cortex, cancellous interior (porous layer) and ventral cortex. Nanoindentation test is a viable option for measuring their mechanical properties due to the size of the layers. However, preliminary results of nanoindentation tests on a carapace sample vary according to the indentation peak load. As the indentation peak load increases, both measured hardness and elastic modulus decrease. A systematic approach to determine an optimal peak load for nanoindentation test is presented.

1. Introduction

Turtle shells are natural shields optimized by years of evolution. Its protective mechanism inspires numerous studies [1-3]. African spurred tortoise or Sulcata tortoise is in the same species as turtles; therefore, its shell has the same structure as turtles and terrapins. Its bony shell is composed of a dome-shape carapace and a flat plastron. Its shell length can reach 80 cm and a weight of over 100 kg. Figure 1 shows a cross-section of the tortoise 's carapace with an average thickness of 6 mm. It is consisted of four distinct layers: keratin scutes, dorsal cortex, cancellous interior and ventral cortex. Keratin scutes is the outermost layer made of Keratin, a structural protein that provides flexibility and elasticity to the shell. Beneath the keratin scutes are the bone layers. Dorsal and ventral cortices have dense or solid structures while cancellous interior has a closed-cell cellular structure. The overall Sulcata tortoise carapace resembles a sandwich-like structure with a porous core.

Nanoindentation test is a standard technique for measuring mechanical properties of both organic and inorganic materials with non-homogeneous microstructure [4-6]. Hardness and elastic modulus of the microstructures can be calculated from the load-displacement data using Oliver and Pharr method [7]. Figure 2 shows a representative load-displacement curve of elastic-plastic loading and elastic unloading with parameters for hardness and elastic modulus calculation [7]. The hardness H and the elastic modulus E can be derived by the following equation:

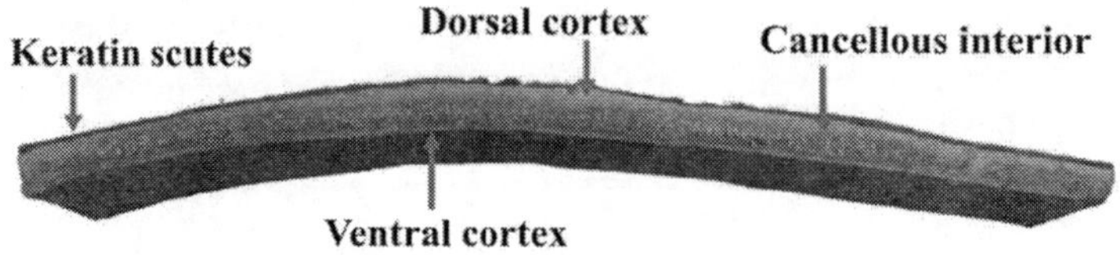

Figure 1. A cross-sectional view of carapace with four distinct layers.

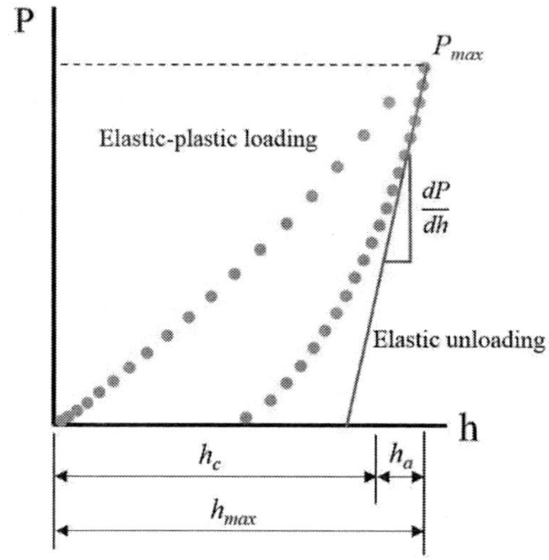

Figure 2. Analysis of load-displacement curve to determine hardness and elastic modulus.

$$H = \frac{P_{\max}}{A} = \frac{P_{\max}}{24.5h_c^2} \tag{1}$$

where $P_{\max}$ is the peak load, A is the area function that depends on the indenter. For a Berkovich diamond indenter, $A = 24.5h_c^2$. The contact depth h_c is derived from

$$h_c = h_{\max} - 0.7\frac{P_{\max}}{dP/dh} \tag{2}$$

where $h_{\max}$ is the maximum depth, and $\frac{dP}{dh}$ is the slope of load-displacement curve during unloading. The reduced modulus E^* and the material modulus E are given by equation (3) and (4), respectively.

$$E^* = \frac{dP}{dh}\frac{1}{2}\frac{\sqrt{\pi}}{\sqrt{A}} \tag{3}$$

$$E = \frac{1-\nu_s^2}{\frac{1}{E^*} - \frac{1-\nu_i^2}{E_i}} \tag{4}$$

where ν is the Poisson's ratio and subscript s and i are the values of sample and indenter, respectively.

Some studies [5-6] reported that hardness and elastic modulus were influenced by the peak loads. This characteristic of nanoindentation test can lead to erroneous measurements of material parameters. The non-homogeneous structure of turtle carapace further complicates the selection of correct peak loads.

Previous works that performed nanoindentation tests on turtle carapace did not clearly justify their choices of peak load or indentation depth. Rather, the selected values appeared to depend on the availability of test equipment and on the dimensions of the samples. For example, Achrai and Wagner [2] examined the red-eared slider turtle's carapace using the indent depth between 1.8-2.0 micron. Balani et al. [8] chose a load range of 0.1-2 mN with a maximum depth 50-300 nm to test freshwater snapping turtle's shell. Recommended peak load or indentation depth for testing tortoise shell has not been reported.

In addition to the peak load dependence, the nanoindentation test on turtle shell shows a time-dependent response during unloading because the layers exhibit creep behavior [2, 8]. This creep leads to an increase in the indentation depth after reaching the peak load. Figure 3 illustrates the load-displacement characteristic of creep behavior in nanoindentation test. The peak force must be maintained for a sufficiently long period to mitigate the effect of creep in the measurement of maximum depth.

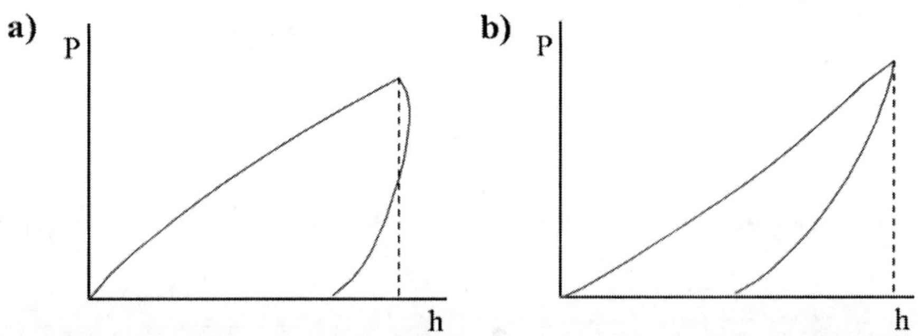

Figure 3. Load-displacement curve: a) creep during unloading b) without creep.

This paper aims to address the correlation between peak load and mechanical properties of tortoise carapace and to determine the appropriate peak load range for testing a sandwich-like structure. With an appropriate peak load, the material properties measured from nanoindentation test are validated and suitable for future reference. In addition, the systematic peak load selection procedure from this work can be followed when performing nanoindentation tests on other biological materials.

2. Experiment and method

2.1. *Nanoindentation test*

A Sulcata tortoise's shell was donated for this study. The shell was sun dried before cutting into strip specimens. The sides of the strips were polished using 1000 grit silicon carbide sandpaper. The hardness and elastic moduli of the carapace's four layers were examined by IBIS Nanoindentation system with a Berkovich diamond indenter. The indenter has a Young's modulus and a Poisson's ratio of 1000 GPa and 0.07, respectively. A fused silica was used for indenter tip's area function calibration [9].

In this study, the peak loads were selected from the capability of test equipment. The IBIS Nanoindentation system has two loading ranges: mid-range and high-range. The mid-range load has a peak load between 0-50 mN and an indentation depth between 0-2 micron. The high-range can supply a peak load between 0-500 mN and an indentation depth between 0-20 micron. Five values of peak loads were targeted: 10, 30, 100,150 and 200 mN with a holding time of 10 sec at maximum load. Approximately 9-18 indents were performed for each testing condition. The values of Poisson's ratio of each layer were taken from Achrai and Wagner [2]: the Poisson 's ratio of keratin scutes layer was 0.4, while the other layers were assigned to be 0.3.

The load and displacement response were continuously measured during loading and unloading steps. The onset of the elastic unloading curve is fitted by a tangent line for the calculation of elastic modulus and hardness of the material. However, the dorsal cortex and the cancellous interior exhibit creep response when unloading after a peak load of 200 mN. The creep behavior induces a negative slope on the load-displacement curve $\frac{dP}{dh}$; therefore, the elastic modulus and the hardness cannot be determined.

3. Results and discussion

Figure 4 and figure 5 show distributions of hardness and elastic modulus versus peak loads, respectively. The average values of hardness for keratin scutes and ventral cortex tend to decrease from low to high peak load. For dorsal cortex and cancellous interior layer, the hardness slightly increases from 10 mN to 30 mN peak load. However, the average hardness values of low peak loads (10 and 30 mN) are higher than those of high peak loads (100 and 150 mN).

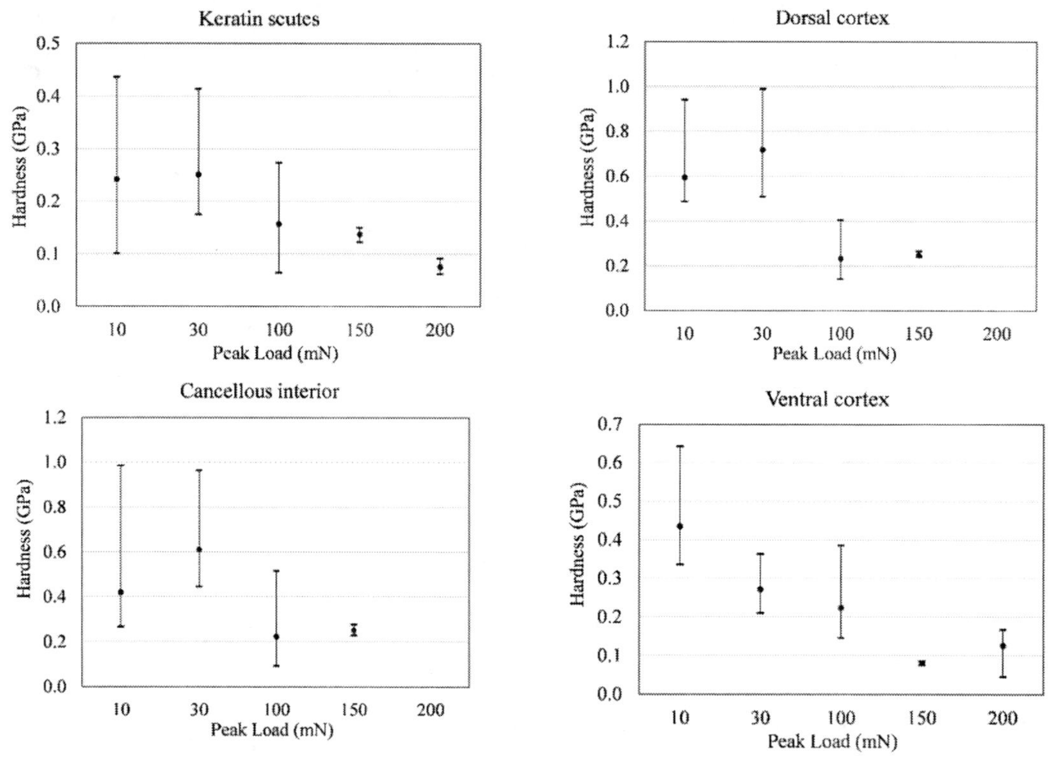

Figure 4. Hardness versus peak load of carapace layers.

The elastic moduli in figure 5 show the same characteristics for all layers of tortoise 's carapace. When increasing the peak load, the elastic modulus decreases. In addition, both hardness and elastic modulus at high peak load show less variations than those at lower ones. The discrepancies at low peak loads are influenced by the local effects such as surface roughness, porosity and variations of material properties, especially nonhomogeneity in biomaterials [10].

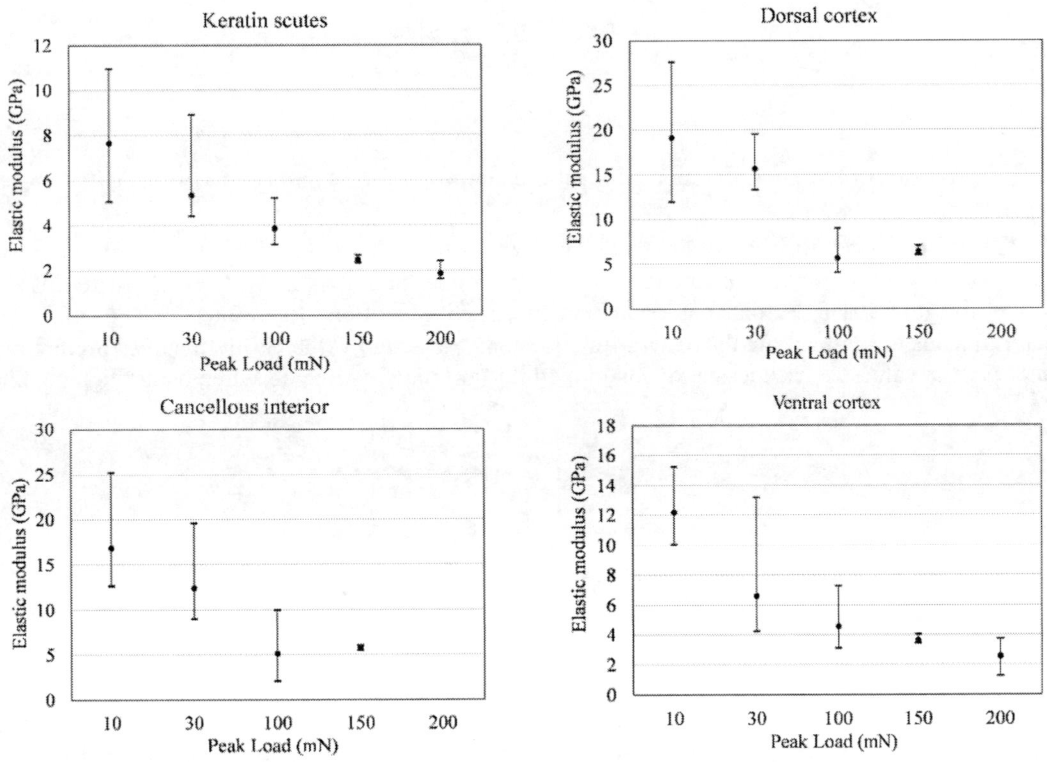

Figure 5. Elastic Modulus versus peak load of carapace layers.

However, using high peak load induces creep in the layers. When the material creeps, the indentation depth continues to sink during unloading. The negative slope of the unloading curve results in a negative modulus, which is not thermodynamically admissible. High peak load also leads to cracking of brittle layers as shown in figure 6.

Figure 6. Indention on dorsal cortex with micro-crack at peak load of 200 mN.

The high values of hardness and modulus at low peak loads are related to the indenter shape [9], [10]; the indenter tip is idealized as a sharp cone in the calculation while in the experiment the tip is round. Figure 7 illustrates the difference between area function calculated from a round tip (A) and a sharp tip (A_i) with the same contact depth (h_c). The calculated area function A_i is less than the actual area function A. From Equation (1) and (3), the area function is inversely proportionated to the hardness and elastic modulus. Therefore, the calculated hardness and modulus are overestimated for low peak loads.

IOP Conf. Series: Materials Science and Engineering **297** (2018) 012016 doi:10.1088/1757-899X/297/1/012016

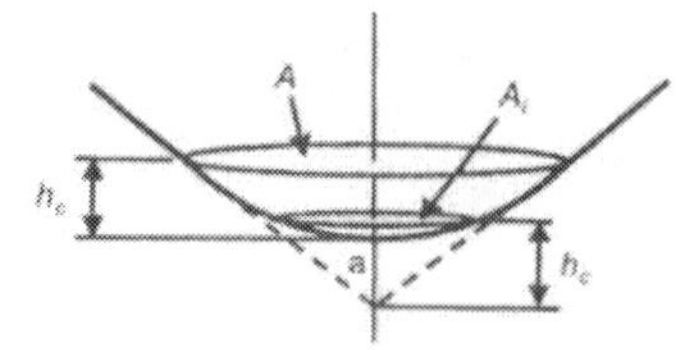

Figure 7. Area function of ideal indentation tip (A_i) and round tip (A) [9].

On the other hand, the material behavior influences the calculation of hardness and modulus at high peak loads [9]. In particular, soft materials pile-up around the indenter while brittle materials sink-in as shown in figure 8. The tortoise carapace is brittle since it is a bone-like material. The sink-in of brittle material leads to a larger calculated area function than the actual value. With this over-prediction of area function, the values of hardness and elastic modulus are underestimated when using high peak loads.

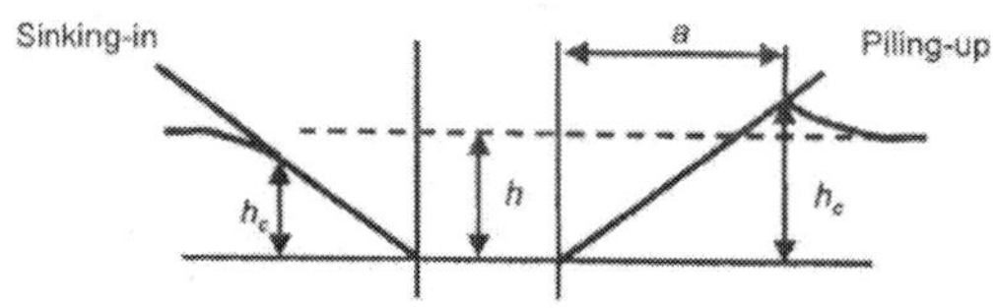

Figure 8. Material pile-up and sink-in [9].

Figure 4 and figure 5 highlight that in the mid-range (10, 30 mN) the measured data at a given peak load have large variations, except the hardness of ventral cortex. In the high-range (100, 150, 200 mN), the results have smaller variances in each peak load, which indicate an increase in data repeatability. However, at 200 mN, creeps and micro-crack occur on dorsal cortex and cancellous interior layers that result in indeterminable hardness and elastic modulus. Besides, some indentations at high peak loads have maximum depth more than the certified depth range of the indentation system. Because of these reasons, hardness and elastic modulus of tortoise carapace must be calculated from the peak load between 100 mN and 200 mN as illustrated in figure 9.

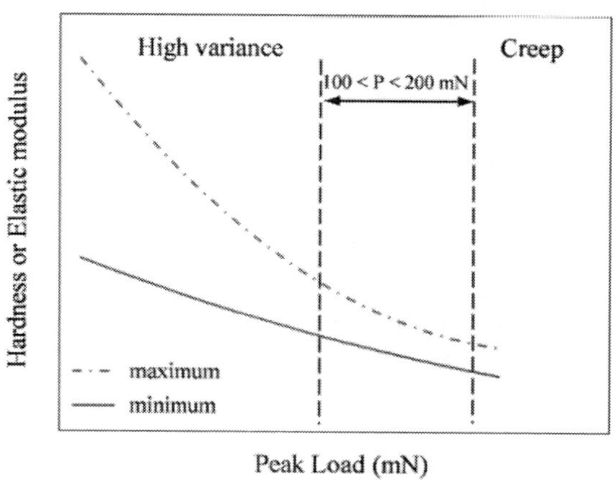

Figure 9. Peak load range for nanoindentation of Sulcata tortoise's carapace.

Table 1 presents values of hardness and elastic modulus of carapace 's layers from optimal peak loads. From the data, the hardnesses of the carapace's four layers are in the same range, except the minimum value of ventral cortex. The elastic modulus of keratin scutes is the smallest value and that of dorsal cortex is the largest.

Table 1. Mechanical properties of tortoise carapace from nanoindentation tests at peak load $100 < P < 200$ mN.

	Keratin scutes	Dorsal cortex	Cancellous interior	Ventral cortex
Hardness (GPa)	0.16-0.20	0.23-0.25	0.22-0.25	0.08-0.22
Elastic modulus (GPa)	2.5-3.9	5.5-6.5	5.2-5.8	3.6-4.6

4. Conclusions

Mechanical properties measured by nanoindentation test show the peak load dependence characteristic. The assumption in the calculation, effect of microscopic measurement and material inconsistency lead to incorrect determination of mechanical properties. The results reveal that the suitable peak load for measuring tortoise's carapace is $100 < P < 200$ mN.

Acknowledgements

The authors would like to thank Dr. C. Yinharnmingmongkol for donated tortoise's shell. The support from Faculty of Dentistry, Chulalongkorn University and technical advice from laboratory's officers are gratefully acknowledged.

References

[1] Achrai B, Bar-On B and Wagner D 2014 *J. Mech. Behav. Biomed. Mater.* **30** p 223
[2] Achrai B and Wagner D 2013 *Acta Biomater.* **9** p 5890
[3] Damiens R, Rhee H, Hwang Y, Park J, Hammi Y, Lim H and Horstemeyer F 2012 *J. Mech. Behav. Biomed. Mater.* **6** p 106
[4] Faingold A, Cohen R and Wagner D 2012 *J. Mech. Behav. Biomed. Mater.* **9** p 198
[5] Gong J, Miao H, Peng Z and Qi L 2003 *Mater. Sci. Eng. A.* **354** p 140
[6] Jang K 2006 *J. Alloys. Compd.* **426** p 312
[7] Oliver C and Pharr M 1992 *J. Mater. Res.*7**(6)** p 1567
[8] Balani K, Patel R, Keshri K, Lahiri D and Agarwal A 2011 *J. Mech. Behav. Biomed. Mater.* **4** p 1440
[9] Fischer-Cripps A C 2005 *The IBIS Handbook of Nanoindentation* (Fischer-Cripps Laboratories Pty Ltd)
[10] Menčík J 2012 Uncertainties and Errors in Nanoindentation *Nanoindentation in Materials Science* ed J Nemecek chapter 3 pp 53–86

Development of feedback-speed-control system of fixed-abrasive tool for mat-surface fabrication

K. Yanagihara* and R. Kita
National Institute of Technology, Ariake College, Higashi Hagio-Cho 150, Omuta-Shi, 836-8585, Japan

* Corresponding Author: Kiyoshi@ariake-nct.ac.jp

Abstract: This study deals with the new method to fabricate a mat-surface by using fixed-abrasive tool. Mat-surface is a surface with microscopic irregularities whose dimensions are close to the wavelengths of visible light (400-700 nanometers). In order to develop the new method to fabricate mat-surface without pre-masking and large scale back up facility, utilization of fixed-abrasive tool is discussed. The discussion clarifies that abrasives in shot blasting are given kinetic energy along to only plunge-direction while excluding traverse-direction. If the relative motion between tool and work in fixed-abrasive process can be realized as that in blasting, mat-surface will be accomplished with fixed-abrasive process. To realize the proposed idea, new surface-fabrication system to which is adopted feedback-speed-control of abrasive wheel has been designed. The system consists of micro-computer unit (MPU), work-speed sensor, fixed-abrasive wheel, and wheel driving unit. The system can control relative speed between work and wheel in optimum range to produce mat-surface. Finally experiment to verify the developed system is carried out. The results of experiments show that the developed system is effective and it can produce the surface from grinding to mat-surface seamlessly.

1. Introduction

Mat-surface is a surface with microscopic irregularities whose dimensions are close to the wavelengths of visible light (400-700 nano-meters) [1].

In common industrial techniques, mat-surface is produced by using chemical etching, electrical or laser abrasion, loose abrasive lapping, or shot blasting. These conventional processes, however, demand pre-masking of work-piece and relatively large scale back up facility.

Therefore, in order to develop the new method to fabricate mat-surface without pre-masking and large scale back up facility, new tool which is utilizing fixed-abrasive is developed in this study.

2. Fundamental idea for Tool design

2.1. Mechanism of mat-surface fabrication by means of abrasive process

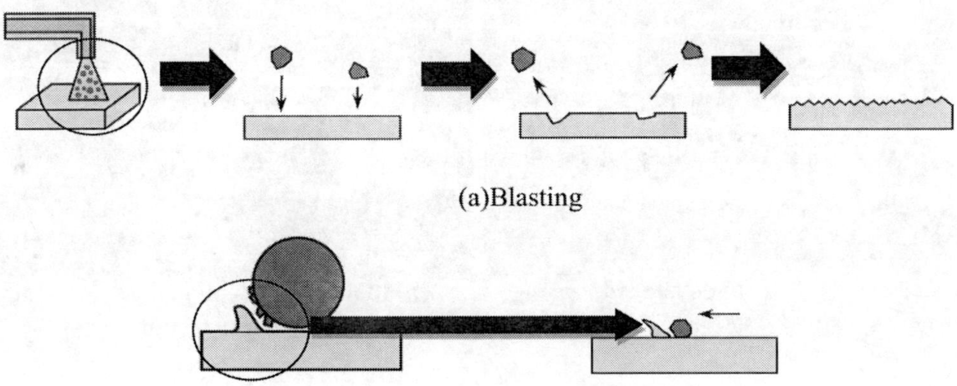

(a)Blasting

(b)Grinding

Figure 1. Comparison of surface-generate mechanism

One of the means to fabricate mat-surface is blasting. Figure 1(a) illustrates how blasting process surface of work-piece. In blasting process, hard abrasive-grains whose diameter is from μm to mm are given kinetic energy with compressed air. Then the abrasives with high kinetic energy smash work-piece, and can produce some small dimples on the surface of the work-piece as mat-surface. In this process, regarding each abrasive grain as a tool, the kinematics of tool has only movement of plunge-direction to work-piece without that of traverse-direction.

On the other hand, in grinding process with fixed-abrasive tool, abrasive grains on the tool-surface are given kinetic energy of traverse-direction against work-piece with rotational movement of spindle of grinding machine as shown in figure 1(b). When each abrasive grain touches work-piece surface tangentially, each grain work as a cutting edge. It must scratch and cut the surface of work-piece with a little volume.

Therefore, in designing new fixed-abrasive tool that can produce mat-surface, realization of the movement of abrasive grains which can be plunged into surface of work-piece is necessary without any grain movements of traverse-direction.

2.2. Driven type mat-surface fabrication tool

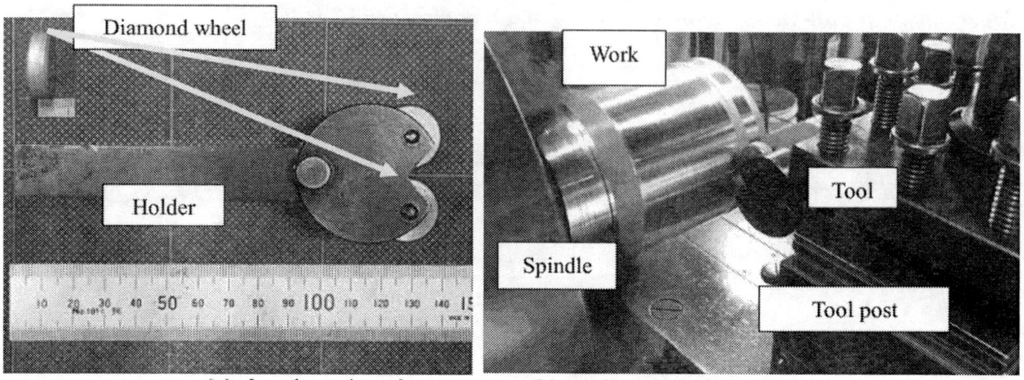

(a) developed tool (b) tooling set-up on lathe

Figure 2. Driven type of mat-surface-fabrication-tool

Table 1. Tool specifications and experimental conditions

Fixed-Abrasive Wheel	Abrasive grain	Diamond
	Mesh size	#400
	Particle size(μm)	44
	Tool width(mm)	20
Processing conditions	Work-piece speed(m/min)	Less than 497
	Feed rate(mm/min)	6.64

Figure 2 (a) shows the designed fixed-abrasive tool for mat-surface fabrication. The tool has two diamond wheels whose specifications are shown in table 1. The two wheels are just contacted to work-piece as shown in figure 1 (b) when it is used. The wheels are driven by rotation of work-piece with relative-speed zero between each wheel and work-piece, while excluding traverse movement of abrasive grains against surface of work-piece to avoid that each abrasive on wheel does not work as a cutting edge.

(a) Before (b) After

Figure 3. Surface of work-piece

3. Characteristics of Mat-surface fabrication with fixed-abrasive tool.

3.1. Proof experiment of tool-design concept

At the beginning, to confirm whether our discussed idea is right or not, the developed tool is examined with lathe, as shown in figure 2 (b). The experimental conditions are also shown in table 1. The material of work-piece is aluminum alloy in this experiment. Figure 3 shows the photo of surface of work-piece before and after experiments. In figure 3 (a): the work surface before experiment, some regular marks which are generated by cutter of lath are observed. On the other hand, in figure 3 (b): after processed with the developed tool, the cutter marks are disappeared, and mat-surface is generated.

The topography of the obtained mat-surface is measured, as shown in figure 4. Random dimples are clearly observed over processed surface. This proves that the idea to produce mat-surface by applying fixed-abrasive is right, and that the developed tool is applicable.

Processing with fixed-abrasive tool has an advantage in the control of its processing area without any masking, and can be done by installing the tool on machine tool since machine tool provides axis feeding to the installed tools.

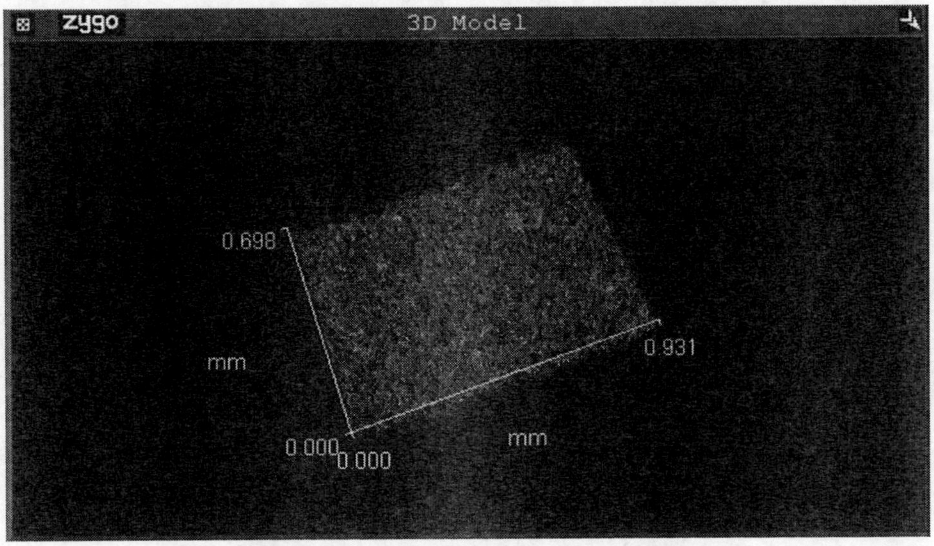

Figure 4. Topography of obtained surface

3.2. Control-ability of tool positioning

Figure 5 shows that some spots of mat-surface is selectively fabricated on the surface of work-piece with the developed tool which is installed on machine tool. It demonstrates that the tool can produce a mat-surface wherever a designer wants to fabricate it without any masking.

Figure 5. Spots of mat-surface

Table 2. Surface roughness Ra at several surface speeds

Work surface speed (m/min)	Ra(μm)
-	1.68 (pre-processed)
23	2.52
73	2.64
152	2.09
284	2.14
497	3.02

3.3. Speed of work-piece

In order to realize the processing condition that is "zero" of the relative speed between the work-piece and the abrasive wheel while traversing wheel of the whole tool-body, the two wheels of the tool are driven by the work-piece rotation. The drive force is generated by friction of the contact area between the tool-wheel and the surface of the work. If any efficiency of the process is considered, it is desired that the processing time is as short as possible. Table 2 shows surface-roughness of obtained mat-surface in different processing-speed; spindle speed is one of the conditions of lath-process. Roughness of the work-piece is 1.68 μm Ra before the experiment. From 23 to 284 m/min of processing speed, mat-surfaces of about 2μm Ra are averagely obtained. In 497 m/min of processing speed, however, surface roughness is rougher than those in the other conditions since some scratches are seen on the processed work-surface. The driving force of the developed tool basically depends on friction force between work-piece speed and grits on wheel. Therefore this demonstrates the limitation of processing speed, and at least below 284 m/min is desirable for stable processing for the developed tool.

3.4 Normal force in processing

Figure 6 shows normal force of tool in processing mat-surface. In the figure, processing-forces of the developed tool are compared with those of a knurling tool since processing mechanism of the developed way is similar to that of knurling way. In the experiment, two kinds of materials of work-piece, S35C steel and aluminum alloy, are experimented. It results that normal force of the developed tool is 1 0times lower than that of knurling tool in processing of both the steel and aluminum. The reason of the result is the difference of size of cutting edge of each. The size of one of the edges on surface of knurling-tool is mm order. It is usually used with initial indentation of edges in mm range. Otherwise the size of it on surface of fixed-abrasive is μm order. Thus the initial indentation of edge of our proposed tool is μm range. Therefore, friction force of the developed tool between tool and work is apparently estimated less than that of knurling tool.

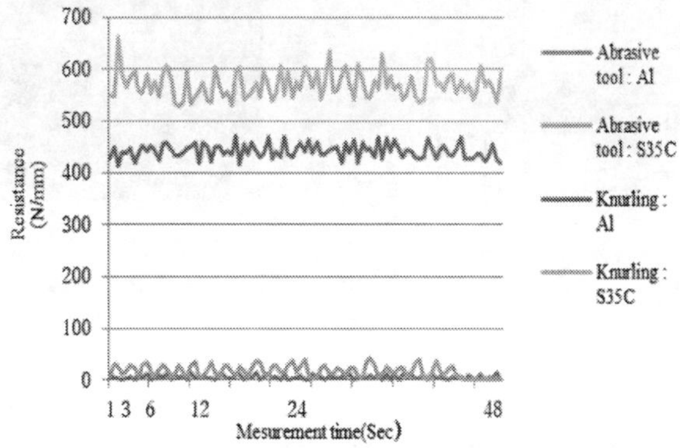

Figure 6 Comparison of normal forces

4. Development of feedback-speed-control system

4.1. System outline

In former chapter, the limit of driven-type mat-surface fabrication appeared in higher range of the work-speed since driving force of driven-type tool depends on the friction force between the tool and the work. To apply the developed mat-surface fabrication method by means of fixed-abrasive tool for wider situation with higher productivity, not passive but active control of the wheel driving is necessarily to track high speed rotation of work-piece.

Figure 7 shows active wheel-driving with feedback control system for mat-surface fabrication. The system consists of a stepping motor to drive an abrasive wheel, a rotary encoder to detect work-piece speed (revolution), and a 8-bit micro-computer to integrate the feedback system.

The rotary encoder is coupled to a contact wheel. The wheel is contacted to work-piece before processing is started. The rotary encoder is driven and produces pulses in proportion to the revolution of the spindle. The outputted pulse is input to the microcomputer. The microcomputer calculates optimum revolution of the abrasive wheel to provide relative speed of zero between the work-piece and the abrasive wheel. After the calculation, the microcomputer sends the calculated pulses which have the optimum frequency to realize the relative speed zero, to stepping-motor-driver-unit, as shown in figure 8.

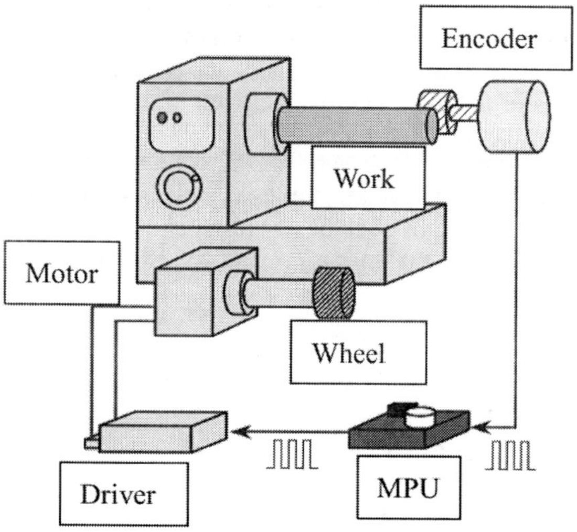

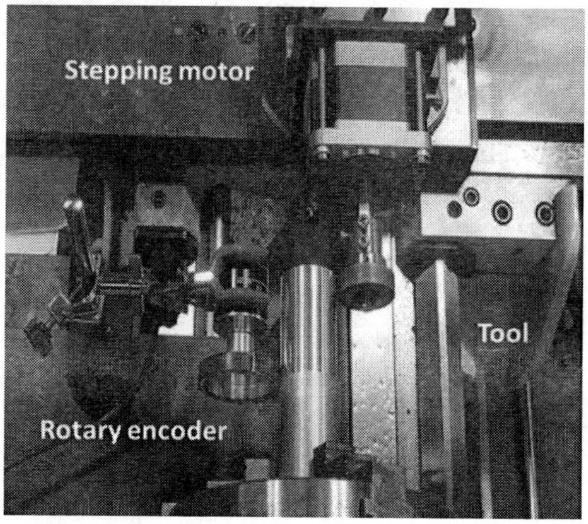

Figure 7. Work-speed feedback control for mat-surface fabrication

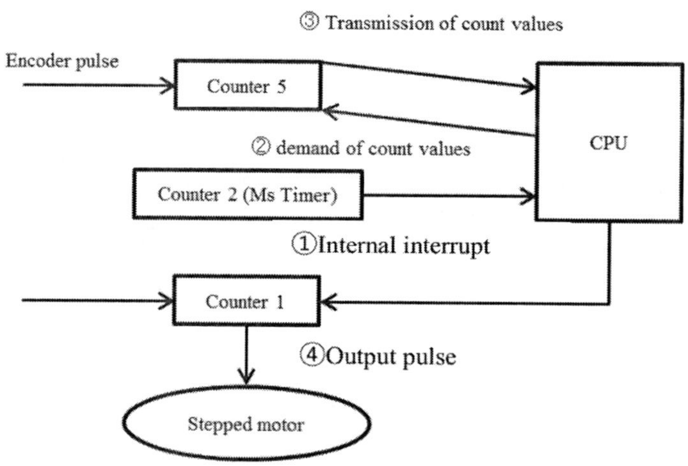

Figure 8. Control sequence among microcomputer and other devices

(a) feedback control (b) passive driven type

Figure 9. Comparison of microscope images of obtained surfaces

4.2. experiment to verify the developed system

The developed system is verified under the experimental conditions as shown in table 2. Figure 9 is micro-scope images of obtained surface. Figure 9 (a), (b) is the surface which is generated with the proposed feedback controlled, the passive driven-type controlled tool, respectively. Also, Figure 10 shows the roughness curves and the surface roughness Ra of each surface. They are almost the same surface. Therefore, it is verified whether the developed system realizes relative speed zero between work-piece and abrasive wheel, thus it can produce mat-surface as the driven-type tool which is passive controlled.

4.3. Constant relative-speed control finishing

The developed feedback control system can produce not only relative speed of zero but also keeping of constant relative speed. Figure 11 is the obtained surface when transition of relative speed between aluminum-work and tool are produced intendedly. At the beginning of the experiment, relative speed is existing to produce grinding state. Then, feed-back control is started to produce mat-surface in processing of grinding. The figure shows both grooved area that means grinding state and mat-surface that means state of relative speed zero.

The result demonstrates that the developed system can produce the surface from grinding to mat-surface seamlessly.

IOP Conf. Series: Materials Science and Engineering **297** (2018) 012017 doi:10.1088/1757-899X/297/1/012017

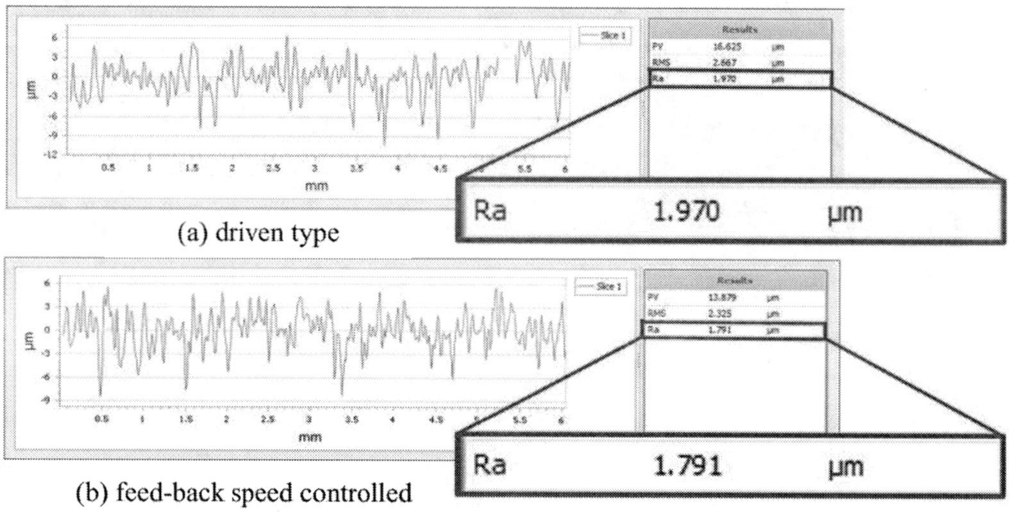

(a) driven type

(b) feed-back speed controlled

Figure 10. Comparison of roughness curves

Grinding zone | Mat-surface zone

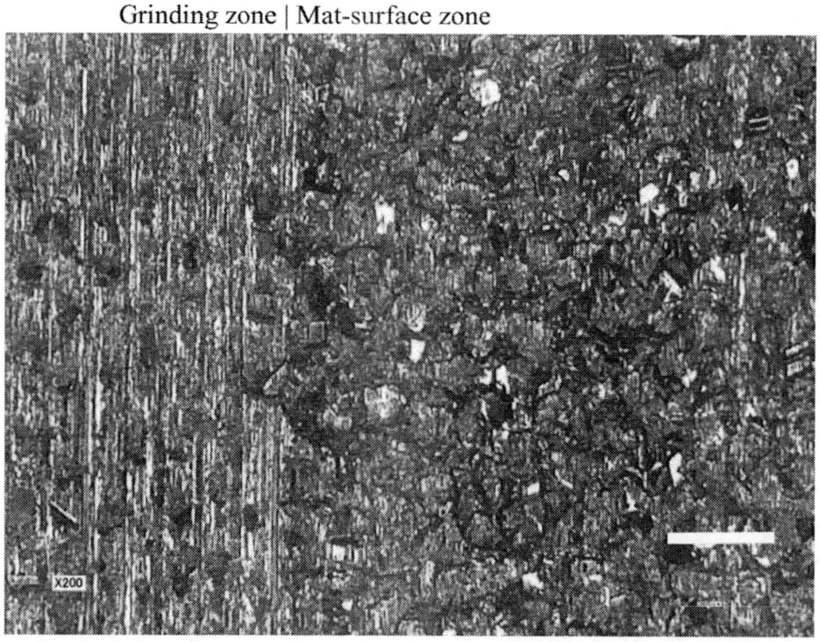

Figure 11. Surface when transition of relative speed between work and tool are occurred

5. Conclusion

In order to fabricate mat-surface without pre-masking and large-scale back-up facility, utilization of fixed-abrasive tool is discussed and new method to fabricate mat-surface with fixed-abrasive tool is proposed. The proposed method is experimented with new designed tool. The obtained results are following:

(1) To fabricate mat-surface with fixed-abrasive tool, it is important that realization of the function as abrasive grains which can be plunged into surface of work-piece is necessary without any grain movements of traverse-direction against work-piece.
(2) According to (1), new fixed-abrasive tool for mat-surface-fabrication is designed. The tool has two abrasive wheels, and the wheels are driven by movement of work-piece to prevent relative motion of traverse-direction.
(3) normal force of the developed tool is 10times lower than that of knurling tool in processing of both the steel and aluminum.
(4) Feed-back active-speed-control system is developed to produce mat-surface without any dependency of the friction force between surface condition on the tool and the work piece, and it is verified that the system is effective.
(5) the developed system can produce the surface from grinding to mat-surface seamlessly in case of processing aluminum work-piece.

Acknowledgement

Great supports of the Technical Support Center for Education and Research of N.I.T, Ariake college is gratefully acknowledged.

References

[1] Industrial Art Institute, Mat-surface Finishing (in Japanese).
[2] Kita R and Yanagihara K 2014 *Proc. of 46th student member meeting of graduation study of JSME Kyushu branch* (in Japanese).
[3] Kita R and Yanagihara K 2015 *Proc. of 2015 annual meeting of JSME, S1340103* (in Japanese).
[4] Kita R and Yanagihara K 2017 *Proc. of 48th student member meeting of graduation study of JSME Kyushu branch* (in Japanese).
[5] PAT.P 2015-021663, Method and tool design for selective fabrication of mat-surface.
[6] PAT.P 2016-147369, Method and tool design for selective fabrication of mat-surface.

Roughness and compressive strength of FDM 3D printed specimens affected by acetone vapour treatment

Juraj Beniak*, Peter Križan, Ľubomír Šooš and Miloš Matúš

Slovak University of Technology in Bratislava, Faculty of Mechanical Engineering, Nám. Slobody 17, 812 31 Bratislava, Slovak Republic

* Corresponding Author: e-mail: juraj.beniak@stuba.sk; tel. +421 905 593 953

Abstract. Rapid Prototyping technologies are the fastest growing technologies in the manufacturing of components and parts. There are many techniques which can be used with different materials and different purposes of produced part. Gradually, Rapid Prototyping systems have grown into Additive Manufacturing, because technology expansion brings faster production, improved manufactured components, and expanded palette of used materials. So now this techniques are also used for regular production of special parts, where is usual change of part design, where is necessary to produce variety of different designs and shapes. The following article deals with Fused Deposition Modelling (FDM) technology, the core of which is the manufacture models and components from thermoplastic polymers by deposition single fibres of semi-molten plastic material layer by layer. The article focuses on the results of research for testing of manufactured specimens by FDM technology. Components are modified by acetone vapour for surface smoothing. The purpose is to point out how the additional specimen treatment influence the strength properties. Presented paper shows realized experiments and measurements of compressive force on specimens and surface roughness which are influenced by acetone vapour treatment.

1. Introduction

This paper deals with one of Additive manufacturing technologies – Fused Deposition Modeling (FDM) also known as fused filament fabrication (FFF), which is focused on producing of parts and models from thermoplastic materials, such as ABS, PLA, Nylon and others.

If we want to optimize production process, first of all we need to know how the system and the device behave with regards to different settings and with expectation of necessary properties of produced components. Following paper presents research on properties of produced specimens as model strength, which is necessary to predict the material properties of produced parts. The paper especially deals with compressive strength and roughness properties of FDM 3D printed specimens. There are also published papers which deal with similar problem and research how the chemical vapor affects the surface roughness of 3D printed parts and what are the mechanical properties of exposed parts [1, 2, 3]. But the difference is between the methodology how the vapor smoothing is in the process, what time and what tools are used.

For purpose of this paper we had to produce specimens for strength measurement. There have been used of FDM device that utilizes thermoplastic material for components production.

Fused Deposition Modeling (FDM) usually uses two types of materials – one for modeling and one for support. First, the model material is used to build the model. Second, the support material is used to build a support structure on the areas where the modeling material will overhang the rest of model [4].

This technique works on a principle similar to a fuse-gun [5]. The material is unspooled from the spool to the fuse-head, where it is melted and deposited on the working table (figure 1). After the completion of the model, the support materials are either broken away or dissolved in a special bath.

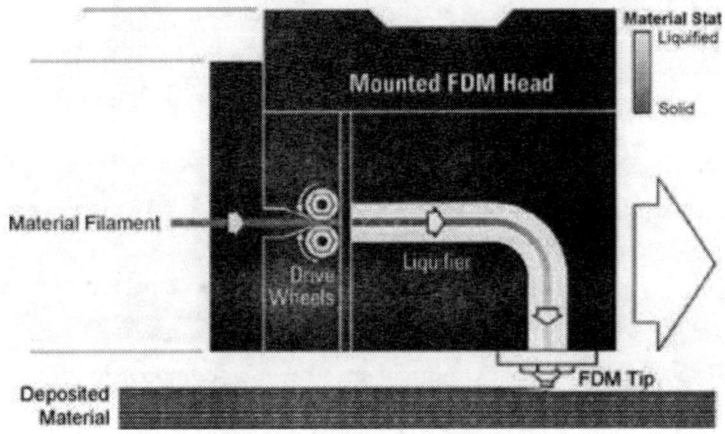

Figure 1. Scheme of FDM process [7]

The rapid prototyping use many different techniques which could be used with many different materials. Basically there can be used for different types of plastic materials, resins, and plaster. If we need the parts for final use from metallic materials, we can use different metallic materials for part production. Of course, the final properties such as parts are different, depending on chosen material, selected technology and also machine settings.

The standard materials used for plastic part modeling are ABS (Acrylonitrile Butadiene Styrene), PC (Polycarbonate), PLA (Polylactid acid), ASA (Acrylonitrile Styrene Acrylate), PETG (Polyethylene terephthalate), Nylon, HIPS (High Impact Polystyrene) or some special composite materials, for example Laywood, a composite of polymer and wood particles. All of these materials have some advantages and disadvantages. They have different processing requirements such as required temperature but also very important material properties as tensile strength, flexural strength or strength in compression. Selection of material for produced model depends mostly from available device type, because not all FDM devices are able to process all available materials. The other reason is how to choose the model material, when the device can process more types of materials that provide the final use of produced model and the material properties [6].

As is written above, it is necessary to take into consideration the ecological dimension. All mentioned plastic materials such as ABS, PC or Laywood are not environmental friendly. They practically do not decompose in the environment. But the PLA plastic material is environmentally friendly. This material is natural polymer, produced from corn, potatoes or sugar-beet. It is biodegradable, and in the environment will degrade and decompose in a few years in contrast to other polymers.

There are many of researches published in this field, which could be part of optimizing system, included in the database [7, 8, 9, 10]. All of these results are very important to proper setting of 3D printing device to reach the best result.

2. Experiment preparation

Presented research focuses on discovering strength properties of produced parts under compression load. The measurements are made of specimens from Dimension SST FDM device, produced from ABS plastic material. Printing process uses standard temperature from model material (ABS) 280°C, 3D printer chamber temperature 75°C, layer thickness 0,25mm. All specimens are produced from one spool

IOP Conf. Series: Materials Science and Engineering **297** (2018) 012018 doi:10.1088/1757-899X/297/1/012018

of material to ensure the same properties of raw material. The shape and dimension of specimens (figure 2.) are prepared in order of ASTM D695 standard, which determines conditions for plastic material testing within compression load. The diameter of cylindrical specimen is 12,7mm and the height is 25,4mm (2x12,7mm).

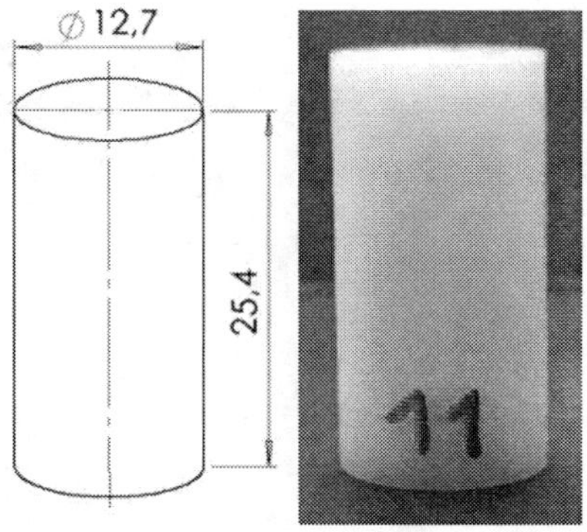

Figure 2. Scheme and picture of cylindrical specimen for compression testing of plastic materials

This research also focuses on specimens which are post-processed for surface smoothing with acetone for better surface design and roughness. This is realized by vapor smoothing. This method is done by placing specimens in a closed container with a small amount of acetone (100ml) in the bottom. This closed container is placed into a tank with hot water at 100°C. This is for heating to make acetone evaporate and cause specimen surface etching (figure 3.). Such surface will get fused and cause smoother surface.

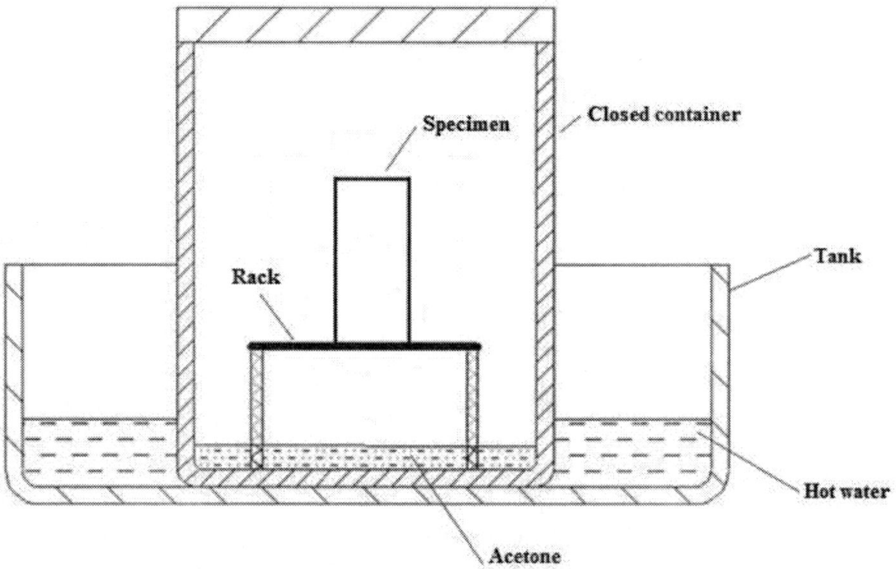

Figure 3. Scheme for specimen preparation – vapor smoothing

Research also addresses how long the specimen is influenced by acetone vapor. The required output of this research is to specify the best time to reach the best surface quality without negative or with the best positive influence on specimen compressive strength.

3. Measurement

Measured and evaluated factors are compressive strength *Fmax* (maximum compressive force when the specimen collapse) and surface roughness *Ra* (arithmetic average roughness). Because produced parts are usually mounted in different configurations, the condition of outer dimensions after surface smoothing are important. This research also accounts for this necessity and there are measured diameters of produced specimens.

4.1. Compressive force measurement and evaluation

Mechanical press is used for the measurement of compressive force. Specimens are placed between fixed and moving parts of press. Recording device records all movements and applied forces during the testing period. The final maximum compressive force, when the specimen is broken, is recorded as illustrated in figure 5. When the specimen is pressing, the whole material volume is compressed and squeezed. After the critical compression the specimen could be usually deformed by two ways. Squeezed cylinder could break at the outer circumference (figure 4 – left) or could collapse and slide down to the side (figure 4 – right).

Loading in compressive testing is realized within a constant speed press of 10 mm.min^{-1}. This speed is recommended by ASTM D695.

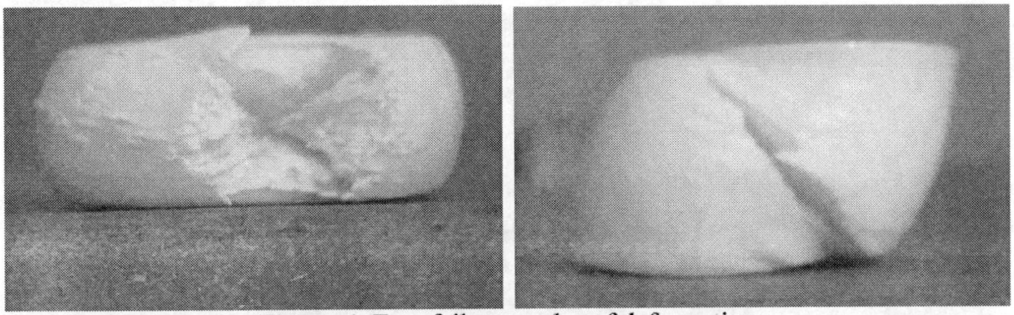

Figure 4. Two failure modes of deformation.
Symmetrical deformation (left). Specimen collapsing to the side (right)

First measurement is made with clean specimens, which are not influenced by acetone. Measured values of maximum pressure force *Fmax* for four repeating measurements for each specimen type are listed in table 1 and table 2. Average values from four repeated measured values are also in the tables. Difference Δ (N) between "clean" specimen and others specimens which have been influenced by acetone within 5 minutes, 7,5 minutes and 10 minutes are observed to analyse how the maximum force changes compared to the maximum compressive force of the clear specimen. This difference is also displayed in percentage Δ (%).

Table 1. Measured values of maximum compressive force.

	$Fmax_1$ (N)	$Fmax_2$ (N)	$Fmax_3$ (N)	$Fmax_4$ (N)	***Fmax* (N)**	Δ (N)	Δ (%)
Clean	11520	12010	12350	11825	**11926,25**		
5min	14000	14200	14950	14650	**14450**	2523,75	21,16
7,5min	14100	14050	13650	13720	**13880**	1953,75	16,38
10min	12240	13100	13300	13200	**12960**	1033,75	8,67

IOP Conf. Series: Materials Science and Engineering **297** (2018) 012018 doi:10.1088/1757-899X/297/1/012018

Table 2. Measured values of maximum compressive strength in MPa.

	$Rmax_1$ (MPa)	$Rmax_2$ (MPa)	$Rmax_3$ (MPa)	$Rmax_4$ (MPa)	***Rmax* (MPa)**	Δ (MPa)	Δ (%)
Clean	90,99	94,86	97,541	93,40	**94,19**		
5min	110,57	112,15	118,08	115,71	**114,13**	19,93	21,16
7,5min	111,36	110,97	107,81	108,36	**109,63**	15,43	16,38
10min	96,67	103,47	105,04	104,26	**102,36**	8,16	8,67

Measured values are also graphically illustrated on figure 5. The clear (not affected) specimen have the lowest value on compressive force compared to others specimens which are influenced by acetone. The best result is reached when the specimen is exposed to 5 minutes to acetone vapor. Mechanical properties of ABS plastic can be improved to about 21% higher level by this procedure.

On the other hand, higher exposure at 7,5 minutes and 10 minutes result in lower values. This is probably caused by higher volume of acetone affecting the ABS plastic material and probably degrade the internal structure of polymer strings. There could be deduced that if the time is too long, the ABS material will degrade and lose strength properties and will be useless. This assumption is based on different researches [11, 12].

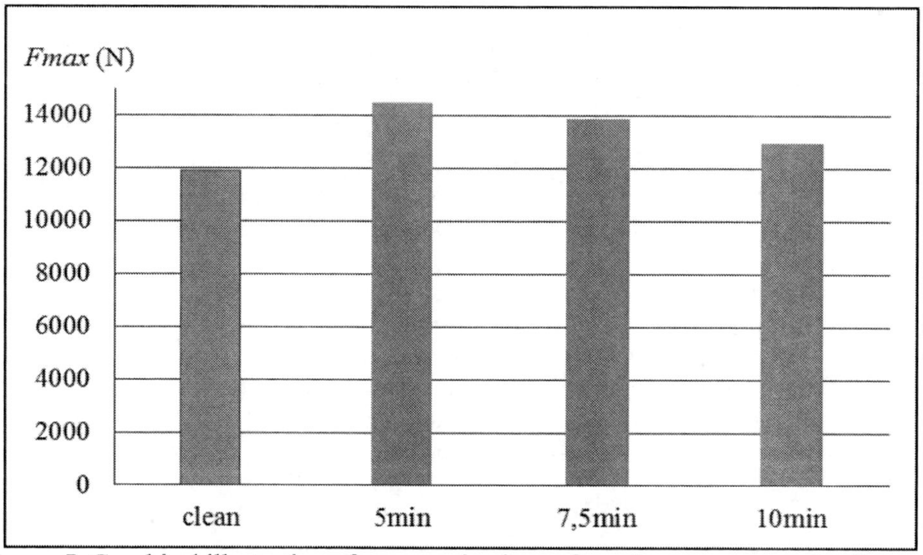

Figure 5. Graphical illustration of measured values of maximum compressive force

4.2. Roughness measurement and evaluation

Process of vapor processing of produced FDM parts have the primary purpose to smooth part surface. This has to be evaluated by measuring surface roughness to recognize what is the best time to reach the best part surface. The same specimens are used in this experiment; the roughness is measured before the specimens have been placed to the press for compression force measurement.

The change of surface is also visible by visual comparison (figure 6).

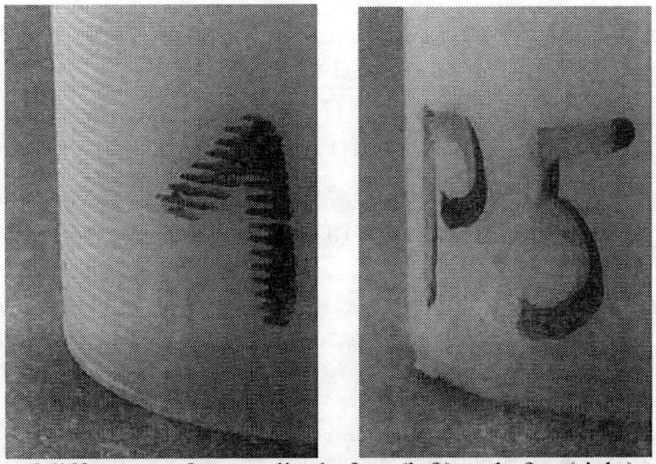

Figure 6. Presentation of different surface quality before (left) and after (right) vapor smoothing 5 min

The official measurement is realized by a special device assigned for this purpose – roughness tester Surtronic 3+. The device and measuring configuration are displayed in figure 7. This device can measure a wide spectrum of parameters, for example *Ra* - arithmetic average roughness, *Rq* - root-mean-square roughness, *Rz* - sum of maximum peak and maximum valley of a profile, *Rmax* - value of maximum peak of the profile and others.

Arithmetic average roughness *Ra* is the most common and most used in mechanical engineering, so it will be used as a comparing roughness parameter for this measurement. The specimen has to be fixed on the same table as the device to prevent any movements and distortions. The measurements have been made on different sides around the cylindrical specimen to eliminate accidental errors and irregular surface smoothing.

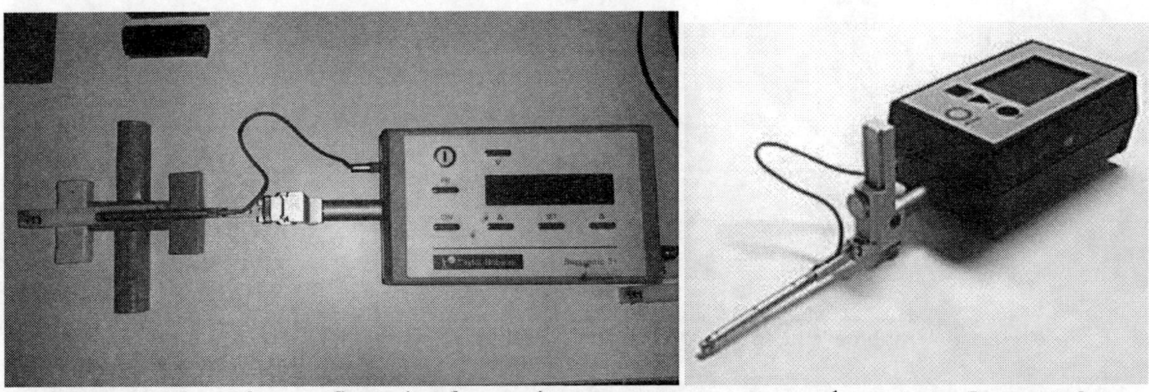

Figure 7. Measuring configuration for roughness measurement – roughness tester Surtronic 3+

Measured values of Arithmetic average roughness *Ra* for five repeating measurements for each specimen type are placed in table 3. Average values from five repeated measured values are also in the table. Difference Δ (μm) between "clean" specimen and others specimens which have been influenced by acetone within 5 minutes, 7,5 minutes and 10 minutes are observed to see how the *Ra* changed compared to a clear specimen. This difference is also displayed in percentage Δ (%), when the value *Ra* of clean specimen is set as 100%, and others values are deviations from 100%. For example in the case of specimen 5 minutes influenced by acetone, the value of 2,36 μm is just 13,95% from the value measured on a clean specimen (16,92 μm).

Table 3. Measured values of Arithmetic average roughness *Ra*.

	Ra_1 (μm)	Ra_2 (μm)	Ra_3 (μm)	Ra_4 (μm)	Ra_5 (μm)	***Ra* (μm)**	Δ (μm)	Δ (%)
Clean	16	17,8	16	17,2	17,6	**16,92**	-	100%
5min	2,2	2,8	3	1,8	2	**2,36**	-14,56	-86,05
7,5min	0,4	0,4	0,4	0,6	0,4	**0,44**	-16,48	-97,4
10min	0,6	0,4	0,2	0,4	0,4	**0,4**	-16,52	-97,64

Measured values are also graphically illustrated on figure 8. There is clear to see that clear (not affected) specimen have the highest roughness *Ra* compared to others specimens which are influenced by acetone. The best result and the best roughness is reached when the specimen is exposed to 10 minutes of acetone vapor. So it is possible to reach Arithmetic average roughness *Ra=0,4* μm, which is only 2,36% from the initial value of 16,92 μm. So we can get good and smooth surface by acetone vapor procedure. Almost the same roughness is also attained for a 7,5 min exposure. Relative change and jump are observed between a clean specimen and a 5 minute exposure. In the future it could be interesting to make experiments also with the time between 0 and 5 minutes to see how the surface is changing. More than 10 minutes is counterproductive because the etched surface or the top layers of cylindrical specimens are flowing down due to gravity.

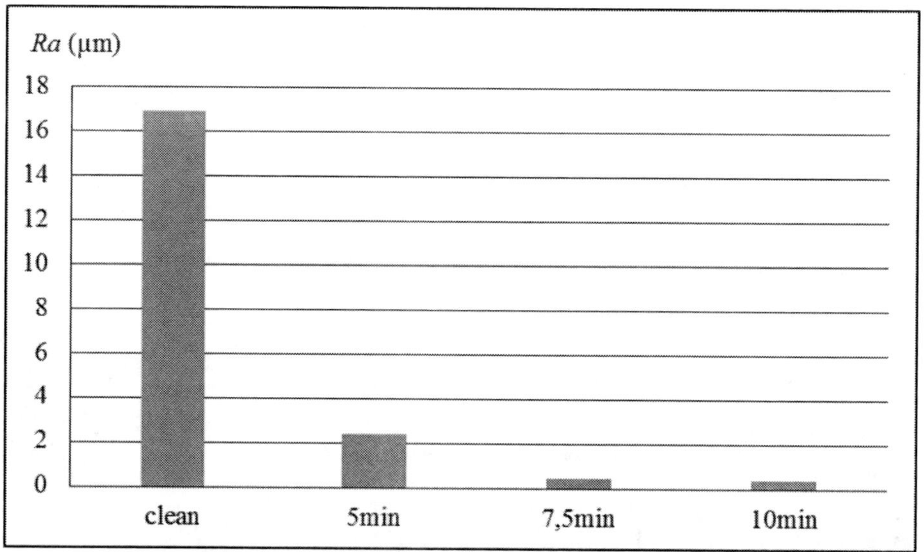

Figure 8. Graphical illustration of measured values of Arithmetic average roughness *Ra*

5. Conclusion

Presented paper shows that there is big influence of surface smoothing process for ABS parts produced on FDM device. Smoothing process rapidly influencing the compressive force and this process also reinforced the part and make it stronger compare to not influenced specimens. It is significant that the compressive strength of specimen can be increased by more than 21 percent when the part is exposed to 5 minutes of acetone vapor. On the other hand, if this time is longer, the strength of specimens is decreasing.

Smoothing process is made for primary depression of surface roughness. This process rapidly decreases roughness from Ra value of 16,92 µm down to 0,4 µm when the specimen is influenced by 10 minutes of acetone vapour without significant changes in the diameter of a cylindrical specimen.

Acknowledgement

The paper is a part of the research done within the project VEGA 1/0420/16 – Determination and research of parameters influence on final briquettes quality at waste biomass densification. The authors would like to thank to the Ministry of Education of Slovak Republic and to the Slovak Academy of Sciences.

References

[1] Anoop K Sood, Raj K Ohdar and Siba S. Mahapatra 2012, *J. of Advanced Research* **3**(1) pp 81-90
[2] Garg A, Bhattacharya A & Batish 2017 *The Int. J. of Advanced Manufacturing Technology* **89** p 2175
[3] Rupinder Singh, Sunpreet Singh, Iqwinder Preet Singh, Francesco Fabbrocino and Fernando Fraternali 2017 *Composites Part B: Engineering* **111** pp 228-234
[4] Chua C K, Leong K F and Lim C S 2003 *Rapid Prototyping, Principles and Applications* (World Scientific Co. Pte. Ltd)
[5] DUC PHAM OBE FREngAND STEFAN DIMOV, Rapid Prototyping, A time compression tool, Manufacturing Engineering Centre, Cardiff University.
[6] Beniak J, Križan P, Matúš M and Svátek M, 2015 *15th Int. Multidisciplinary Scientific Geoconference : Conf. Proc.* (Albena, Bulharsko) pp 18.-24
[7] Lee S, Kim S G, Kim H J and Ahn S H 2007 *J. of Materials Processing Technology* **187-188** (1) pp 627-630
[8] Mendonsa C, Naveen KV, Upadhyaya P and Shenoy V D, *Influence of FDM Process Parameters on Build Time Using Taguchi and ANOVA Approach*, International Journal of Science and Research (IJSR), ISSN (Online): 2319-7064
[9] Gurrala P and Regalla S 2011 *Applied Mechanics and Materials* **110-116**
[10] Svetlík J and Demeč P 2012 *Applied Mechanics and Materials : CECOL 2012: 3rd Central European Conf. on Logistics* (Trnava, Slovakia, 28-30 November 2012). 309 pp 75-79
[11] Arostegui A, Sarrionandia M, Aurrekoetxea J and Urrutibeascoa I 2006 *Polymer Degradation and Stability* **91**(1) pp 2768-2774
[12] Tiganis, B.E & Burn, S & Davis, Paul & Hill, A.J. 2002). *Thermal degradation of acrylonitrile–butadiene–styrene (ABS) blends*. Polymer Degradation and Stability. 76. 425-434. DOI: 10.1016/S0141-3910(02)00045-9.

IOP Conf. Series: Materials Science and Engineering **297** (2018) 012019 doi:10.1088/1757-899X/297/1/012019

Material selection and assembly method of battery pack for compact electric vehicle

N. Lewchalermwong[1], M. Masomtob[2,*], V. Lailuck[2] and C. Charoenphonphanich[1]

[1] Department of International College, King Mongkut's Institute of Technology Ladkrabang, Bangkok 10520, Thailand
[2] Materials for Energy Research Unit, National Metal and Materials Technology Center (MTEC), National Science and Technology Development Agency (NSTDA), Pathum Thani 12120, Thailand

* Corresponding Author: manopm@mtec.or.th

Abstract. Battery packs become the key component in electric vehicles (EVs). The main costs of which are battery cells and assembling processes. The battery cell is indeed priced from battery manufacturers while the assembling cost is dependent on battery pack designs. Battery pack designers need overall cost as cheap as possible, but it still requires high performance and more safety. Material selection and assembly method as well as component design are very important to determine the cost-effectiveness of battery modules and battery packs. Therefore, this work presents Decision Matrix, which can aid in the decision-making process of component materials and assembly methods for a battery module design and a battery pack design. The aim of this study is to take the advantage of incorporating Architecture Analysis method into decision matrix methods by capturing best practices for conducting design architecture analysis in full account of key design components critical to ensure efficient and effective development of the designs. The methodology also considers the impacts of choice-alternatives along multiple dimensions. Various alternatives for materials and assembly techniques of battery pack are evaluated, and some sample costs are presented. Due to many components in the battery pack, only seven components which are positive busbar and Z busbar are represented in this paper for using decision matrix methods.

1. Introduction

The Lithium-ion battery industry has additional significance well beyond its value chain. The battery pack system has various performance requirements related to battery performance, durability, vibration, thermal properties, safety and others, as shown in Figure 1 with related trade-offs among these requirements. Therefore, the design process involves solving the problem of these trade-offs and determining suitable specifications for the system components to achieve a well-balanced overall system with minimum cost.

2. Methods

This paper is established in order to study about material selection and assembly methods of battery pack for small electric vehicles by using DECISION MATRIX methods. The research used "Battery Pack Development for a Compact Electric Vehicle" as a case study, which is a cooperative project between National Science and Technology Development Agency (NSTDA) and Thailand Automotive Institute (TAI) in Thailand.

The researchers and engineers have studied various strategic decision-making tools to analyse and evaluate decision options, which are materials alternatives and assembly technique alternatives. These are the main factors for producing battery pack to cover efficiency and effectiveness at the lowest production cost. Decision matrix was used as a decision-making tool for selection of material and assembly technique for the battery module and the battery pack helping to examine competing alternatives based on multiple criteria that involve various ratings and weights by mathematical model. By working through a series of decision-process steps, criteria can be established for the assessment and comparison of different possible alternatives and then compare choices. This method allows us to list and weight various decision criteria deemed important for a given business situation or problem alternative. The criteria used for the selection are main variables impacts on production consist of material cost, performance, complexity as well as difficulty in assembling process, time spent and safety. Decision based on logic by considering all possible alternatives including all information about upcoming result in the future in order to apply for selection.

3. Methodology

This research studies each component of the new conceptual battery pack and provides the information regarding the material selection and the manufacturing method selection. In each component, the design criteria are specified and discussed with researchers and engineers who are the designer of the battery pack. For the material selection, the material options are filtered to a small number of alternatives. Then the decision matrix is used as the decision aid tool in order to choose the appropriate alternative. The design criteria of the new concept battery are used as the selection criteria in the decision matrix. For the manufacturing method selection, the manufacturing methods that are related to the concerned component are listed and discussed. The brief information of the related manufacturing method is also provided. After that, the selection of the manufacturing method is based on the gathered information.

4. Scope of Work

This work used "Battery Pack Development for Compact Electric Vehicle" as a case study, which is a cooperative project between National Science and Technology Development Agency (NSTDA) and Thailand Automotive Institute (TAI). The battery pack in the project uses 18650 battery cells, which accordingly have 18 mm of diameter and 65 mm of length while the number 0 means cylindrical shape.

Figure 1 shows one battery sub-module including 22 Lithium-ion battery cells that connected in parallel connection to increase capacity. At the positive side, each positive terminal of the battery cells is connected to the positive busbar via fuse wire. At the negative side, each negative terminal of the battery cells is welded with the negative busbar by spot welding. In Figure 2, one half module demonstrates eight battery sub-modules in serial connection to up voltage.

Due to many components in the battery pack, only 4 components which are positive busbar, negative busbar, Z busbar and upper battery holder are represented in this paper. To select materials such as nickel, copper, aluminium and ABS, DECISION MATRIX method is used in decision in term of cost, safety, function, performance and manufacturing. Some component is analysed via Finite Element Method (FEM) software to check safety and performance.

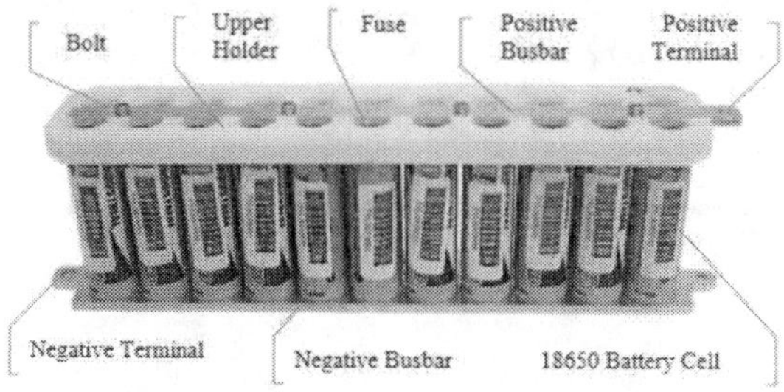

Figure 1. Twenty Two Battery Cells in Parallel for One Battery Sub-Module

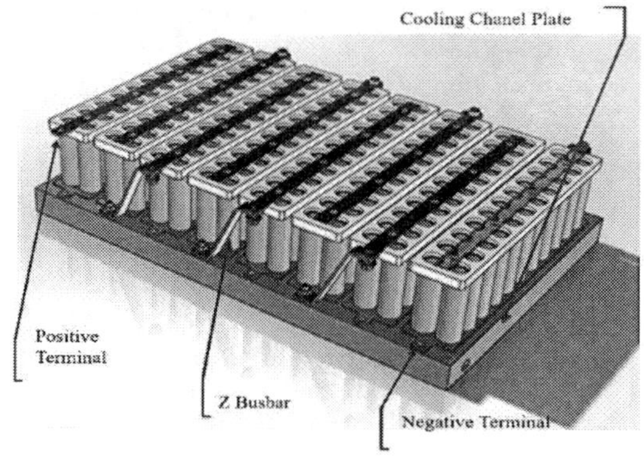

Figure 2. Eight Battery Sub-Modules in Serial Connection for half Module

4.1 Architecture analysis method

Application of product architecture design methodology to system design.

The application of architecture analysis and core competencies to identify design items and sequence of design through empirical knowledge of experienced engineers. The three underlining techniques in this approach are:

- Clarifying the relationship between performance attributes and specification of Battery Pack.
- Extraction of battery pack design items.
- Setting an efficient sequence for process of battery pack design.

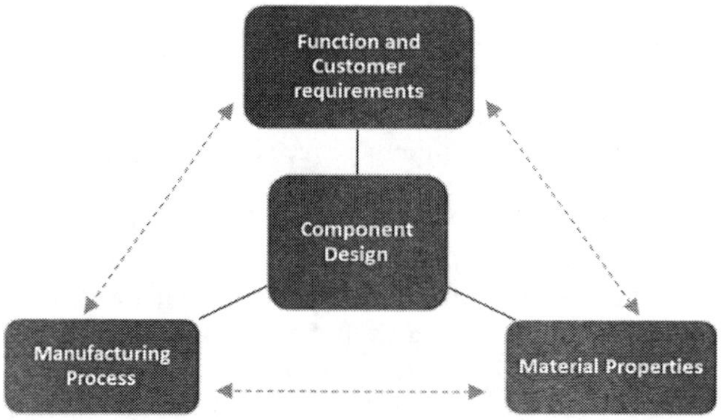

Figure 3. Factors that should be considered in component design

IOP Conf. Series: Materials Science and Engineering **297** (2018) 012019 doi:10.1088/1757-899X/297/1/012019

4.2 Decision Matrix

The decision matrix is a decision aid tool that qualitatively evaluates and prioritizes a list of abstract solutions. It also provides a system which relationships between sets of information are identified, analyzed, and rated. The decision matrix is also known as alternatives evaluation matrix, criteria-based matrix, COWS decision matrix, decision grid, opportunity analysis, problem selection matrix, Pugh matrix, weighted criteria matrix, and etc. The decision matrix is frequently used when only one option can be implemented. In addition, it can be used as a method to rank all the alternatives. Moreover, it can also help reduce the number of options in the list by filtering out the unsuitable alternatives. The procedure of the decision matrix is provided following.

- Identify alternatives - The alternatives can be product features, service features, process steps, or potential solutions.
- Identify selection criteria - The criteria may result from an affinity diagram or a brainstorming activity in the team. The customer needs are also included if possible. The list of criteria should be discussed and refined within the team.
- Assign a relative weight to each criterion - All criteria are weighted depending on the relative importance of each criterion to the concerned situation. The criterion that is more important than others is assigned the higher weight factor.

There are several options for weighting scales as shown in the following examples.

1 - 10 : 1 = the least priority, 10 = the most important

Percentage: Each criterion is assigned its weighting percentage. The summation of all weighting percentage is equal to 100 percent.

- Design scoring system - There are three options for scoring system.

Method 1 - A scoring range is established for all criteria. There are also many options for scoring range. Examples are given following.

- 1, 2, 3 : 1 = low, 2 = medium, 3 = high
- 1, 2, 3, 4, 5 : 1 = little to 5 = great
- 1, 4, 9 : 1 = low, 4 = moderate, 9 = high

Method 2 - Each alternative is ranked according to its performance for the concerned criterion. Number 1 means that it is the least desirable, while the best performance alternative in each criterion gets the highest number.

Method 3 (Pugh matrix) - One alternative is chosen as a baseline. The baseline is given the 0-score in all criteria. Then other alternatives are evaluated by comparing with the baseline. The better performance alternatives get the positive scores such as +1, +2, or +3. Meanwhile the worse alternatives get the minus scores such as -1, -2, or -3. A three-point scale (-1, 0, 1), a five-point scale (-2, -1, 0, 1, 2) or a seven-point scale (-3, -2, -1, 0, 1, 2, 3) can be used.

- Rate the alternatives - The scores of all alternatives for each criterion may come from the average values of the scores given by individual team members, or the scores may come from the consensus decision of the team.
- Total the scores - All the scores in each criterion are multiplied by their weighting factors. Then the overall score of each alternative can be obtained from the summation of the multiplied scores.

The decision matrix tables are generally formed with the list of all criteria and the list of all alternatives. The list of all criteria is located on the first column, while the list of all alternatives is placed on the first row. The second column usually stores the weighting factors.

Therefore, the decision matrix is a L-shaped matrix with 2-dimension variation.

5. Result

This research provides the feasibility study of the new concept battery production. A single sub-module of the new concept battery consists of seven components which are battery cell, positive busbar, negative busbar, cooling plate, battery mount, male electrical connector, and female electrical connector. The battery pack can comprise many battery cells which can be electrically connected in either series or parallel layout. This research studies each component of the new concept battery, and the information regarding the material selection and the manufacturing method selection is gathered and provided in this research.

Moreover, the decision matrix is utilized as the decision aid tool in order to choose the appropriate material. Meanwhile, the selection of the manufacturing method is based on the gathered information.

5.1 Positive Bus Bar and Z bar

In the material selection process for the electrical connectors, this research provides four aspects of consideration which are material compatibility aspect, electrical conductivity aspect, density aspect, and material cost aspect. All four aspects are discussed in the following subsection. Aluminium and copper are the two most popular materials that are used to produce electrical connectors, and there are also many discussions concerning the selection between theses two materials.

(a) Material Selection

Aluminium conductors are a viable alternative for bus bar applications. Using Aluminium as conductor doubles the conductivity per mass in comparison to Copper and saves more than 30% of the costs. For Data Centre applications, Aluminium Busbar is the perfect alternative to Copper, as they offer reduced cost whilst maintaining the mandatory reliability and sustainability criteria.

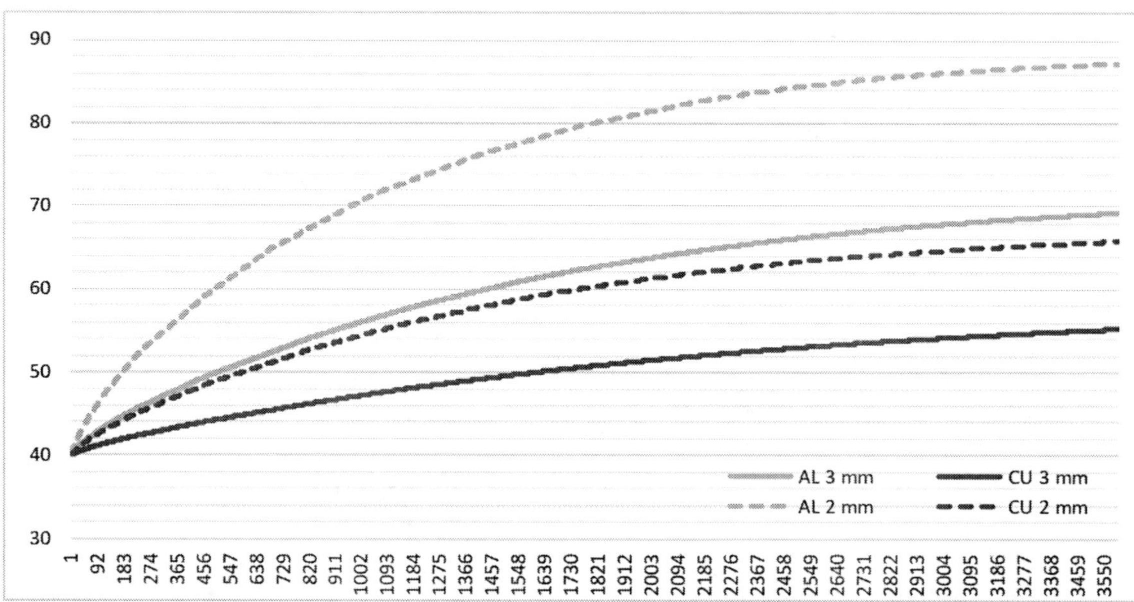

Figure 4. Comparison of temperature increasing for each material at 40 ° C by FEM software.

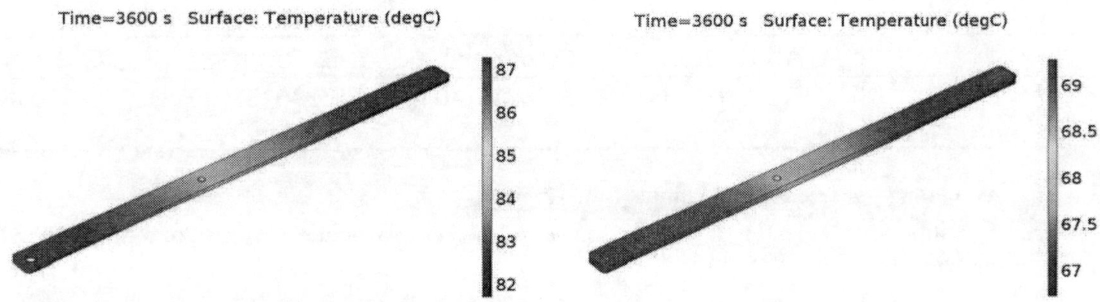

Figure 5. Thermal expansion of AL 2 mm and 3 mm

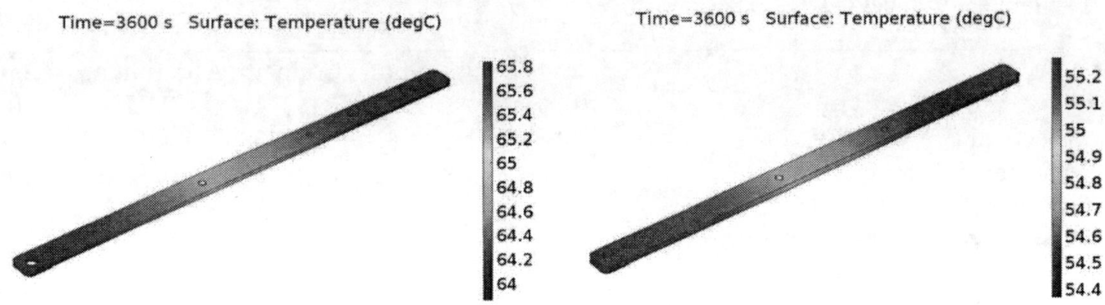

Figure 6. Thermal expansion of Cu 2 mm and 3 mm

This research increase the thickness of the positive bus bar and Z bar from 2 millimeters to 3 millimeters by using an aluminum material. The purpose is to reduce the resistance and temperature of positive bus bar and Z bar, reduce costs and reduce weight compared to copper. The results showed that Z busbar 3 mm of aluminum had lower electrical resistance.

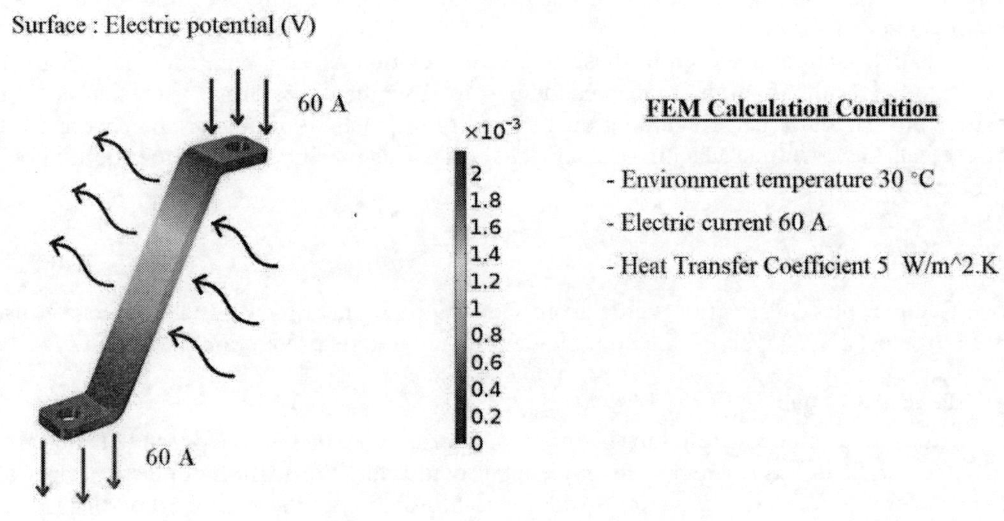

Figure 7. Condition for FEM calculation

IOP Conf. Series: Materials Science and Engineering **297** (2018) 012019 doi:10.1088/1757-899X/297/1/012019

Table 1. Electric resistance of Z Busbar

	AL 3 mm	*CU 3 mm*	*AL 2 mm*	*CU 2 mm*
Electric resistance	6.10997E-05	3.58811E-05	9.29615E-05	5.45909E-05

(b) Weighting parameters and comparison of result

All consideration aspects lead to a comparison of results. The evaluation scores are multiplied with the weighting percentage.

Table 2. Comparison of evaluation scores with weighting percentage

Consideration aspect Weighting	*percentage*	*Evaluation Score*		*Weighting calculation*	
		Aluminium	Copper	Aluminium	Copper
Thermal conductivity	30%	2.5	1	0.75	0.3
Electical conductivity	20%	3	5	0.6	1
Mechanical strength	10%	3.5	5	0.35	0.5
Material cost	40%	5	1	2	0.4
Total scores 100%	100%	14	12	3.7	2.2

(c) Manufacturing method

Regarding the manufacturing method of positive bus bar and Z bar, the information of Laser Cutting Machine (Laser CNC) is provided. According to the gathered information, Laser Cutting Machine (Laser CNC) / shearing machine technique is chosen to be the appropriate methods. This is mainly because this method has the ability to cut aluminium without distorting the material shape with short production time.

Concerning positive bus bar and Z bar, this research provides four aspects of consideration for the material selection, which are thermal conductivity aspect, mechanical strength aspect, density aspect and material cost aspect. The information regarding each aspect is supplied. In addition, the thermal simulation and conductivity simulation were also conducted in the FEM software. Then each material alternative is evaluated and scored in these four consideration aspects with the decision matrix tool. According to the result, aluminium receives the best total scores; therefore, it is chosen as the material for positive bus bar and Z bar. Regarding the manufacturing method of positive bus bar and Z bar, this research selects Laser Cutting Machine (Laser CNC) / shearing machine process to produce positive bus bar and Z bar.

5.2 Upper Battery Mount

In this topic, plastic selection guide from Quadrant Engineering Plastic Products is used as a guideline to select plastic type for the upper battery mount and its production method.

(a) Material Selection

Regarding to limit resource material for battery mount prototype, there are 2 possible material which are ABS and acrylic in this research. The guideline provides the information of plastic properties that should be concerned for plastic selection. Additional information of the plastic properties from other resources is also supplied.

Regarding the cooling connector of the new concept battery, the design criteria are provided following:

- **Thermal resistance** The heat deflection temperature value (HDT 1.8 MPa standard) and the maximum continuous service temperature value must be more than 50°C due to the battery design constraint.
- **Chemical resistance** The cooling connector material must be resistant to the coolant in the cooling system. However, the cooling fluid in the new concept battery is not yet specified, and the choosing of the cooling fluid is also beyond the scope of this thesis.

The prices of raw plastics can vary according to the different factors such as ingredients, additives, time, manufacturers, and quantities.

(b) Manufacturing method

There are several manufacturing methods of plastic parts such as injection moulding, die casting, sand casting, and machining. However, the injection moulding is concerned as the most popular plastic fabrication method. Plastic components with thin-walled are generally produced by this method. Complex shapes with good dimensional accuracy can be expected from the injection moulding. However, the cost of the tool and equipment is relatively high.

Table 3. Recommended manufacturing methods for different part shapes by Quadrant Engineering Plastic Products

Plastic shape	*Manufacturing method*
Long lengths	Extrusion
Smaller sections	
Rod, plate and tube	
Large stock shapes (heavy sections)	Casting
Rod, plate and tube	
Near net shapes	
Custom cast parts	
Various shapes in advanced engineering materials	Compression moulding
Rod, disc, plate and tube	
Plastic shape	Manufacturing method
Small shapes and thin walls in advanced engineering materials	Injection moulding
High volumes (more than 10000 parts)	

Concerning the battery mount, the plastic selection guide from Quadrant Engineering Plastic Products is used as a guideline. The information regarding the selection factors is further provided, and the design criteria of battery mount is also discussed. These design criteria are used as the input variable into the plastic material database, which results in more than thousand of usable material options. Further evaluation is conducted by using the decision matrix tool to choose the suitable material alternative. Seventy-two plastic alternatives with available price and properties information are evaluated further with the decision matrix method. The selection criteria used in this evaluation are material price, material strength, heat deflection temperature, thermal expansion and water absorption, which are based on the plastic selection guide. According to the result from the decision matrix tool, ABS receives the best total score, which is chosen as the material Regarding the manufacturing method, brief information of injection moulding is provided as the suitable manufacturing method. For the battery mount, there is another possibility to produce this component. Firstly, the battery mount can be manufactured by plastic CNC milling process of ABS which the number of plastic choices is limited.

5.3 Negative bus bar

Regarding to the negative terminal of battery cell is Nickel based material which has to be attached with negative busbar. In this case, compatibility issue is the main parameter that need to be considered. Consequently, Nickel is the best optional material for producing negative busbar. However, nickel cost is very high and nickel processing is quite difficult. In order to reduce production cost, this research study alternative material that have similar property as nickel.

(a) Material Selection

This research design to use 2 components for producing negative busbar and nickel plate is fixed material for the first component. copper and aluminium are chosen as optional material for the second component. Therefore, this research uses the selection criteria for second component of negative busbar same as Positive Bus Bar and Z bar.

(b) Manufacturing method

For lower battery mount and negative busbar, there are 2 possible optional jointing technique which are spot welding and laser welding technique.

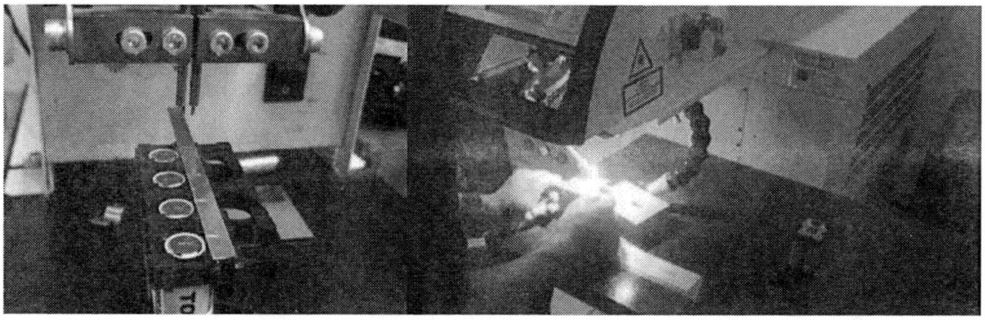

Figure 8. Spot welding and Laser welding technique

Table 4. Welding result for Negative busbar

Material	*Cycle time*	*Voltage*	*Result*	*Cost (Baht/kg)*
(a) Ni+Cu	5 sec	6	Attached well	223.95
(b) Al+Cu	5 sec	8	Can bear pressure only X and Y	42.7
(c) Ni+Al	5 sec	8	Attached well	193.45

(a) (b) (c)

Figure 9. Sample result of spot welding for negative bus bar

Concerning negative busbar, the information is provided in four aspects of consideration for the material selection, which are material compatibility aspect, electrical conductivity aspect, density aspect and material cost aspect. However, the material compatibility issue dominates the material selection of the negative busbar. Negative nickel busbar is attached to the negative terminal of battery cell, while negative aluminium-busbar is attached to negative nickel-busbar. According to the gathered information, the nickel and aluminium are chosen as the negative-busbar material. Regarding the manufacturing method of the negative busbar, brief information of Laser Cutting Machine (Laser CNC)/ shearing machine is provided.

6. Conclusion

This research is expected to able evaluate various alternatives for materials and assembly techniques of battery pack, and some sample costs are presented. A methodology that choose the most proper alternative on the basis of this analysis is presented. The methodology also consider the impacts of choice-alternatives along multiple dimensions. This research studies each component of the new concept battery, and the information regarding the material selection and the manufacturing method selection is gathered and provided in this research. Moreover, the decision matrix is utilized as the decision aid tool in order to choose the appropriate material. Meanwhile, the selection of the manufacturing method is based on the gathered information. This research is expected to distribute knowledge of battery pack development for small electric cars leading to increased business opportunities. And it can be used as a guide for automotive industry in familiar field for the maximum benefit to consumers and manufacturer as well.

References

[1] Burton C. Suedel, Jongbum Kim, and Cynthia J. Banks. Comparison of the Direct Scoring Method and Multi-Criteria Decision Analysis for Dredged Material Management Decision Making.
[2] Marcy Lowe, Saori Tokuoka, Tali Trigg and Gary Gereffi 2010 Lithium-ion Batteries for Electric Vehicles: THE U.S. VALUE CHAIN, October 5, 2010.
[3] Lisa Li, Henry Kuang, Hui Wang, Sam Yang, Assembly System Configurator for Lithium-Ion Battery Manufacturing. 2017 The regents of the university of michigan, 2017.
[4] James A. Milke, Fire Hazard Assessment of Lithium Ion Battery Energy Storage Systems, University of Maryland, College Park, USA
[5] Mahmoud M. Farag 1997 *Materials Selection for engineering design* (Prentice Hall Europe)
[6] Mikell P. Groover 2010 *Fundamentals of modern manufacturing. 4th ed.* (John Wiley & Sons, Inc.USA)
[7] AV Plastics: About injection moulding. What is injection moulding?. URL: http://www.avplastics.co.uk/what-is-injection-moulding [Effective date: 05.02.2016]

[8] Berkmanns J and Faerber M 2008 *Laser cutting. LASERLINE Technical. In: BOC, The Linde Group* pp 3-18
[9] Blue Sea Systems: Electrical Conductivity of Materials. URL: https://www.bluesea.com/resources/108/Electrical_Conductivity_of_Materials [Effective date: 08.07.2015]
[10] CustomPartNet: Material Selector. URL: http://www.custompartnet.com/materials/ [Effective date: 12.02.2016]
[11] CustomPartNet: Injection Molding. URL: http://www.custompartnet.com/wu/InjectionMolding [Effective date: 05.02.2016]
[12] Quadrant Engineering Plastic Products 2011 *Engineering Plastics. Product Guide for Design Engineers. In: The Quadrant group of companies* pp 6-14

IOP Conf. Series: Materials Science and Engineering **297** (2018) 012020 doi:10.1088/1757-899X/297/1/012020

Experimental investigations of density-induced segregation in binary granular mixtures under vertical vibrations

Tanapon Yachai, Itthichai Preechawuttipong* and Pawarut Jongchansitto

Department of Mechanical Engineering, Faculty of Engineering, Chiang Mai University, 239 Huai Kaew Rd., Muang Chiang Mai District, Chiang Mai 50200, Thailand

* Corresponding Author: itthichai.p@cmu.ac.th

Abstract. The objective of this work is aimed to investigate experimentally the influences of density ratio on segregation of cylindrical granular mixtures under vertical vibrations. Polytetrafluoroethylene, Polyamide, Polyethylene terephthalate, Balsa woods and Hinoki woods were used as the cylindrical particles, with 10 mm in diameter and 60 mm in length. The granular sample is composed of 1000 particles with two different materials with each type of 500 particles. All particles are periodically filled into the rectangular container with size 300 x 300 x 60 mm, as the initial condition. Density ratios are 2.73, 3.36, 5.27, and 13.63. The experimental operated by vertical shaker for 10 minute to achieve a steady state condition with same amplitude at 5 mm. The dimensionless vibration accelerations changed by frequency for 5 experimental, experiment the dimensionless vibration accelerations = 1, 2, 3, 4, and 5. The beginning state and the end state collected data by shooting the image. The segregation is considered by average height or the segregation coefficient. The result showed that for the density ratios of 2.73 and 3.36 for all dimensionless acceleration values, the segregation does not occur. On the other hand, for density ratio more than 5.27, the segregation occur by the way that the higher density particles move downward to the bottom of the sample, while the lower density particles move up to the top of the container. The segregation occurs when the dimensionless acceleration is more than 2 and the segregation is clearer when the dimensionless acceleration increases.

1. Introduction

Granular materials are omnipresent that can be found in our everyday lives and in nature. For example, sand, soil, pebble, floating dust in the air, and many layers of rock that overlaps below the crust, etc. These materials generally exhibit complex behaviors, i.e. sometimes behave like fluid and sometimes behave like solid. Such complex behaviors of granular materials are caused by the fact that these materials are constituted from grains or powders with a variety of size, shape, density, friction, and other physical properties. Several behaviors of granular media can be found in many industries, for instance a compaction of sand for a casting in the metal industry, a flow of grains inside the silo in the food industry, and a mixing in the manufacturing process of medicines in the pharmaceutical industry, etc. The other one interesting behavior is the segregation of materials. One of the well-known segregation phenomena is the Brazil Nut, effect [1-3]. This phenomenon provides the upward movement of the larger particles under vibrated granular bed. In practice, there are several factors inducing segregation, such as size, density, surface roughness [1-5]. Therefore, it can be seen that the segregation phenomena due to such factors are quite important in many industrial and engineering fields.

In many industries, the vibration is employed for several applications such as mixing, surface scrubbing, transport, and segregation. Particular in the segregation, it is used for a qualification of the product, e.g. the product that does not have a required weight can be separated by means of vibration. It can be said that the density and the vibration are interesting factors to be investigated. Nevertheless, previous experimental and simulation studies regarding the density-induced segregation have been mostly performed with the granular mixtures composed of different particle sizes: Hong *et al.* [6] proposed a segregation diagram for the crossover from Brazil nut effect to reverse Brazil nut effect for the binary mixtures by means of molecular dynamics simulations. The similar segregation diagram for different vibration amplitudes was also proposed by Ciamarra *et al.* [7]. It can be seen that an unclear effect between density and size has been encountered. In the present work, we therefore aim at investigating the influence of density-induced segregation in binary granular mixture with same particle size subjected to different vertical vibrations.

2. Experimental setup

The cylindrical particles were made of Polytetrafluoroethylene, Polyamide, Polyethylene terephthalate, Ultra-high-molecular-weight polyethylene, Balsa woods and Hinoki woods. The particle diameter and length were about 10 mm and 60 mm [11], respectively. The tested configurations are presented in table 1. The granular mixtures consisted of 1000 particles, which were mixed between two species of density. Each species was composed of 500 particles. The particles were then periodically filled into the container, as shown in figure 1.

Table 1. Tested configurations.

Density ratio	Materials	
	Dense (Density in g/cm^3)	Loose (Density in g/cm^3)
2.73	100	23.56
3.36	150	34.64
5.27	200	23.76
13.63	250	27.9

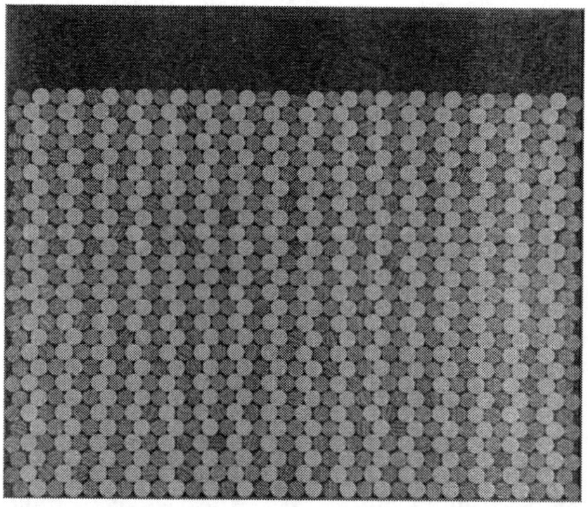

Figure 1. Example of a periodical filling into the container at an initial stage.

A vibrational testing machine designed for the present study is shown in figure 2. Note that this machine was completely installed on a standard base for the vibrational testing. The testing machine was consisted of an aluminum container covered by an acrylic sheet. The inside dimensions of the container are 300x300x60 mm. The container was connected to a scotch yoke mechanism used for converting the rotational movement into the linear movement, i.e. a vertical sine wave. The vibration was provided by an electric motor, which was connected to an electronic inverter. This electronic inverter was employed to systematically adjust the vibrational frequency for specifying the dimensionless vibrational accelerations (Γ).

The degree of vibration use the dimensionless vibrational acceleration (Γ) was defined as $\Gamma = r\omega^2/g$, where g is the gravitational force, r is the vibrational amplitude, $\omega = 2\pi f$ is the vibrational frequency. In this study, the vibrational amplitude was fixed at 5 mm, while the vibrational frequencies were progressively changed in order to obtain the values of Γ = 1-5. The container was subjected to a vertical vibration under sinusoidal waves.

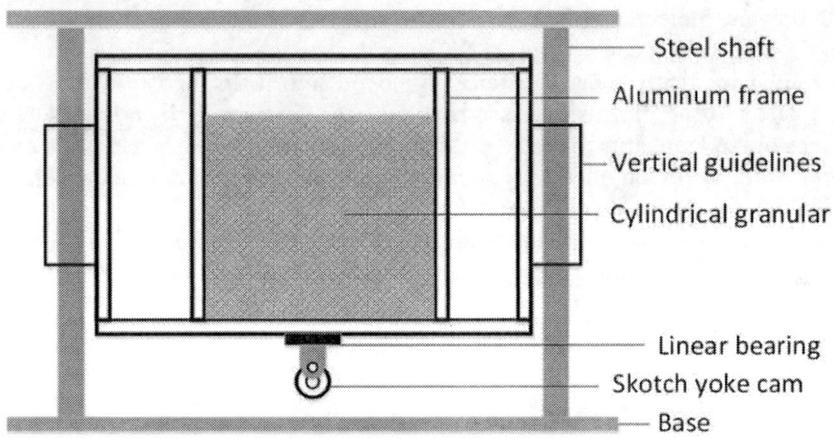

Figure 2. Drawing of a vibrational testing machine.

The degree of vibration use the dimensionless vibrational acceleration (Γ) was defined as $\Gamma = r\omega^2/g$, where g is the gravitational force, r is the vibrational amplitude, $\omega = 2\pi f$ is the vibrational frequency. In this study, the vibrational amplitude was fixed at 5 mm, while the vibrational frequencies were progressively changed in order to obtain the values of Γ = 1-5. The container was subjected to a vertical vibration under sinusoidal waves.

This experiment was taken a photo at initial stage and final stage at 10 minutes. All photos were used open-source software named "ImageJ" [12] for finding vertical positions of each centroid particle. Giving the z_i is the height of particle i with the distance from the bottom container to the vertical final position and N_t is the total number of the all particles inside the container, the average height of particle can be determined by

$$z_t = (1/N_t)\sum_{i=1}^{N_t} z_i \tag{1}$$

Segregation indicator used in this study is the segregation coefficient (H). This coefficient is employed to statistically characterize the degree of segregation of the binary mixture in a static stage, which is given by

$$H = 2\left(\frac{z_a - z_b}{z_a + z_b}\right) \tag{2}$$

where z_a is the average height of lower density particles and z_b is the average height of higher density particles. The factor 2 is a multiplication factor arising from the equal solid fraction of each particle species which limits the average height of particles in such species to $0.25 < z_a < 0.75$ and thus $-1 < H < 1$. It must be noted that in each density ratios at the initial stage, the segregation coefficient approaches to zero, as shown in figure 1. This implies that a homogeneous system is observed. Note that the higher density particles move downward to the bottom is initially observed when $H > 0$, while the higher density particles move up to the top of the container found when $H < 0$.

3. Results

Experimental results of segregation in binary granular mixture under vertical vibration are presented in this section. The results are separated into 2 parts: 1) effect of the dimensionless vibrational acceleration on the segregation coefficient and 2) effect of density ratio on the segregation coefficient.

3.1. *The effect of the dimensionless vibrational acceleration*

In general, the particles commonly arranged to have the highest density. The densest packing is the hexagonal packing or honeycomb [13]. Figure 3 shows the hexagonal packing at an equilibrium system. This packing prevents potentially the particle movement. Nevertheless, the equilibrium system created from the hexagonal packing can be broken by using the vertical vibration, i.e. this vibration practically produces a void space that the particles can be moved into it. In this study, the level of the vertical vibration is related to the dimensionless vibrational acceleration as mentioned above. It means that different values of the Γ are able to create different amount of the void space. As a result, it can be seen that the segregation coefficient increases when the dimensionless vibrational acceleration is increased, as presented in figure 4. More clearly, the packing zone for $\Gamma = 2$ is larger than that for $\Gamma = 5$, as illustrated in figure 5. This is a reason why the segregation coefficient for $\Gamma = 2$ is higher than that for $\Gamma = 5$. This result was also found by previous studies as in refs. [7, 10].

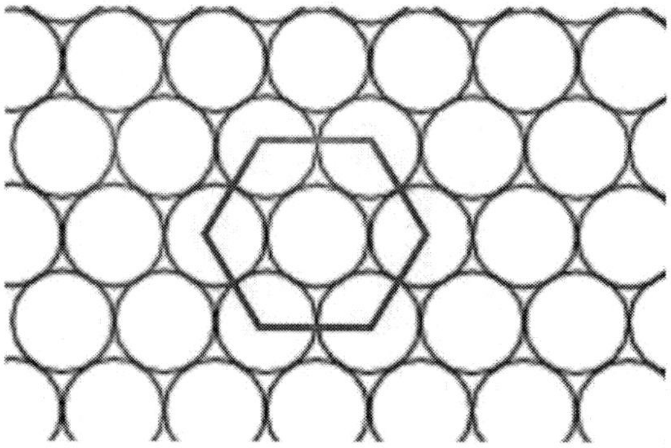

Figure 3. Hexagonal packing.

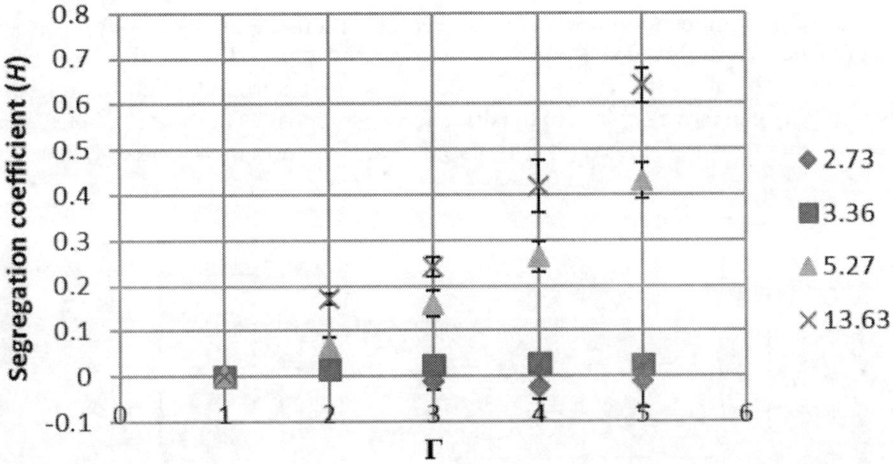

Figure 4. The segregation coefficient as a function of the dimensionless vibrational acceleration at the final stage.

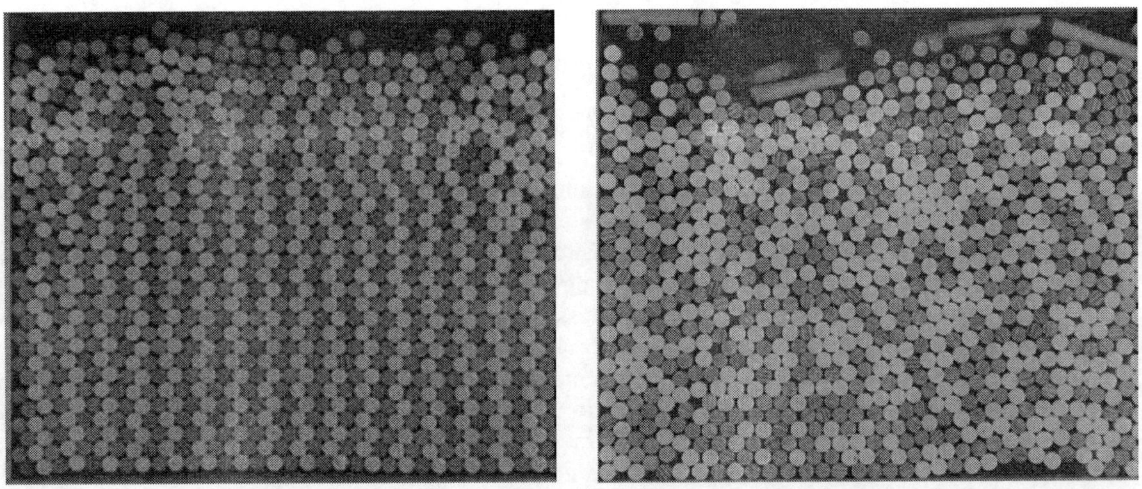

a) b)

Figure 5. Examples of the destroying packing of particles under different values of Γ at 180 seconds: a) Γ=2 and b) Γ=5.

3.2. *The effect of the density ratio*

The results of the segregation in binary mixtures due to different density ratios are shown in figure 6. It is clearly seen that the segregation coefficient increases with the density ratio. It must be noted that the particles whose have a higher density move downward to the bottom of the container. This can be explained by the momentum theory, i.e. the under the same external applied loading, particles whose have a lower density can move upward in the direction that opposites the gravity, better than the particles whose have a higher density. This result was also observed by previous works [4, 7, 14].

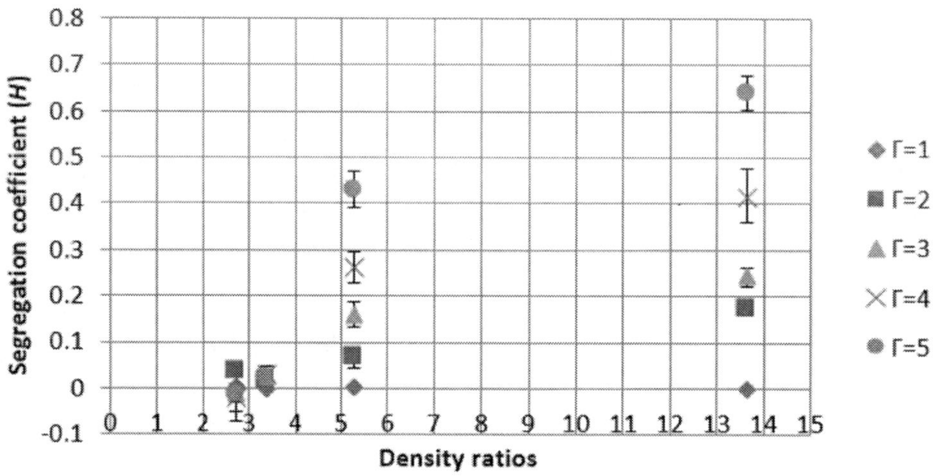

Figure 6. The segregation coefficient as a function of the density ration at the final stage at 10 minutes

4. Conclusion

The influence of density ratios on segregation phenomena in binary granular mixtures under vertical vibrations is experimentally investigated in this study. The ratios between particles whose have a higher density and particles whose have a lower density are 2.73, 3.36, 5.27, and 13.63, respectively. The particles are periodically placed into the rectangular container. The samples are then shaken in the vertical direction with five different dimensionless vibrational accelerations (Γ=1-5). The experimental results show that the segregation coefficient increases when the dimensionless vibration acceleration is increased, i.e. the segregation phenomenon occurs when the dimensionless vibrational acceleration is higher than 2 for the density ratio that segregation. This segregation is clearer when the dimensionless vibration accelerations are increased. Furthermore, the segregation coefficient increases when the density ratio is increased. The segregation does not occur for the density ratios of 2.73 and 3.36. It is observed when the density ratio is higher than 5.27.

References

[1] Jain A, Metzger M J and Glasser B J 2013 *Powder Technol.* **237** pp 543–553
[2] Liao C -C, Hsiau S -S and Wu C -S 2014 *Powder Technol.* **253** pp 561-567
[3] Metzger M J, Remy B and Glasser B J 2011 *Powder Technol.* **205** pp 42-51
[4] Sanders D A, Swift M R, Bowley R M and King P J 2004 *Phys. Rev. Lett.* **93**(20) 208002-1 – 208002-4
[5] Ulrich S, Schröter M and Swinney H L 2007 *Phys. Rev. E* **76** 042301-1–042301-3
[6] Hong D C, Quinn P V and Luding S 2000 *Phys. Rev. Lett.* **86** pp 3423-3426
[7] Ciamarra M P, De Vizia M D, Fierro A, Tarzia M, Coniglio A and Nicodemi M 2006 *Phys. Rev. Lett.* **96** 058001-1 – 058001-4.
[8] Di Renzo A, Di Maio F P, Girimonte R and Formisani B 2008 *Powder Technol.* **184** pp 214–223
[9] Lim E W C 2016 *Powder. Technol.* **27** pp 2478–2488
[10] Liffman K, Muniandy K, Rhodes M, Gutteridge D and Metcalfe G 2001 *Granul. Matter.* **3** pp 205-214

[11] Calvetti F, Combe G and Lanier J 1997 *Mech. Cohes-Frict. Mat.* **2** pp 121–163
[12] Rasband W S, *ImageJ*, (U.S. National Institutes of Health, Bethesda, Maryland, USA, http://imagej.nih.gov/ij/, 1997-2016)
[13] Andreotti B, Forterre Y and Pouliquen O 2013 *Granular Media Between Fluid and Solid* (New York: Cambridge University Press.)
[14] Lim E W C 2016 *Adv. Powder Technol.* **27** pp 2478-2488.

Numerical modelling of the influence of stromal cells on tumor growth and angiogenesis

Nobuyuki Sakiyama and Katsuya Nagayama*

Department of Mechanical Information Science and Technology, Kyushu Institute of Technology, 680-4 Kawazu, Iizuka, Fukuoka 820-8502, Japan

* Corresponding Author: nagayama@mse.kyutech.ac.jp

Abstract. According to the statistics provided by the Ministry of Health, Labor and Welfare the death of one in 3.5 Japanese people is attributed to tumor highlighting the need for active research on malignant tumors. Early detection can be cited as a countermeasure against malignant tumors, but it is often difficult to observe the growth process, and thorough understanding of the phenomena will aid in more efficient detection of such tumors. A malnourished benign tumor may create new blood vessels from existing ones and proliferate abnormally by absorbing nutrients from these newly created blood vessels to become malignant. Different factors influence the shape of tumors and shape is an important factor in evaluating their malignancy. Because interstitial cells greatly influence tumor growth, investigating the influence of stromal cells on tumor growth will help in developing a better understanding of the phenomenon.

1. Introduction

The study of malignant tumors has garnered much attention globally owing to the high mortality rates associated with them. Statistics provided by the Ministry of Health, Labor and Welfare indicate that the deaths of one in 3.5 Japanese people can be attributed to tumors of various types. A malnourished benign tumor may create new blood vessels from the existing ones and proliferate abnormally by absorbing nutrients from these newly created blood vessels to become malignant. Early detection is often cited as a countermeasure against malignant tumors [1], but a lack of clear understanding of the phenomena makes it is difficult to detect and observe their growth. Different factors influence the shape of tumors, and shape is an important factor in evaluating their malignancy. This work studies malignant tumors and the related angiogenesis using numerical analysis based on a particle model [2, 3]. In this method, calculation points or points under active consideration are assumed to move across the mass while storing relevant quantities at mass points. One particle is represented as one cell. Because the shape is maintained by balancing between particles, the direction of cell division and migration is not limited.

Stromal cells such as fibroblasts are distributed around the cancer cells as shown in figure 1. Stromal cells have great influence on tumor growth as shown in figure 2. However, thus far, we have performed tumor growth analysis without introducing stromal cells. Therefore, by introducing stromal cells into analysis, we investigate the influence of stromal cells and use it to elucidate the phenomenon of tumor growth.

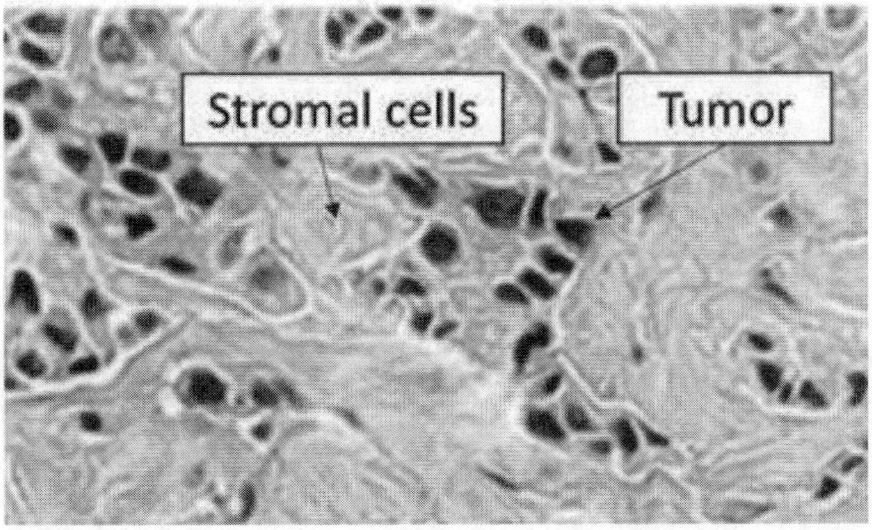

Figure 1. Distribution of stromal cells and fibroblasts in invasive breast cancer in humans.

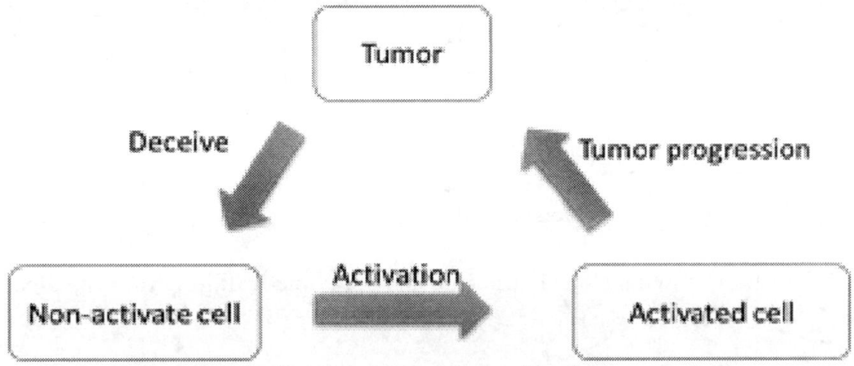

Figure 2. Relation between stromal and tumor, artery.

2. Calculation Model

For analysis, particles are first placed in a 2D region, and cancer cell group particles representing blood vessels are arranged around this region. In the next step, the amount of nutrient diffusion between the blood vessel and the cell is calculated. Equation (1) represents diffusion in cells taking into account the nutrient consumption. The left-hand side of equation (1) represents the temporal variation of the quantity of nutrition, C. The diffusion coefficient is represented by D, whereas α is the nutrient consumption coefficient.

$$\frac{\partial C}{\partial t} = D\left(\frac{\partial^2 C}{\partial x^2} + \frac{\partial^2 C}{\partial y^2} + \frac{\partial^2 C}{\partial z^2}\right) - \alpha \tag{1}$$

The time required for a cancer cell to divide is dependent on the availability of nutrients to that cell. Cancer cells with access to high nutrient concentrations divide actively while the group of cells with reduced nutrient concentration becomes dormant. The cell division time T can be obtained from equation (2). In this equation, T_{min} represents the shortest cell division time, Kc is the relaxation coefficient of the nutrition term, and C is the nutrient concentration.

$$T = \frac{T_{min}}{1+Kc}\left(1 + \frac{Kc}{c^2}\right) \tag{2}$$

Stromal cells are modelled as particles that can proliferate, migrate, etc. The nutrient consumption or diffusion and the frequency of cell division in stromal cells are modelled similar to that in cancer group cells. In stromal cells, the nutrient consumption coefficient α is set to 0.1 and the shortest time for cell division is taken as ten times as that for cancer group cells. Thus, for stromal cell particles, equation (1) and equation (2) are modified to equation (3) and equation (4) respectively.

$$\frac{\partial c}{\partial t} = D\left(\frac{\partial^2 c}{\partial x^2} + \frac{\partial^2 c}{\partial y^2} + \frac{\partial^2 c}{\partial z^2}\right) - 0.1\alpha \tag{3}$$

$$T = \frac{10T_{min}}{1+Kc}\left(1 + \frac{Kc}{c^2}\right) \tag{4}$$

Figure 3 represents the flow of calculation in the analytical model.

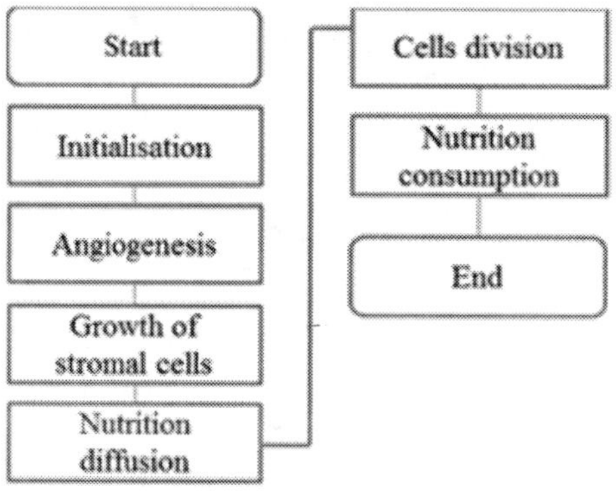

Figure 3. Flowchart showing steps in the analysis of stromal cell growth using a particle-based method.

3. Analysis Condition

First, a 2D analysis of tumor growth in the early stages of development was performed figure 4 shows the initial condition, where a single cancer cell particle was placed alongside one among 16 blood vessel particles (artery) in the region of analysis. The minimum vessel diameter was set to an initial value of 50 μm. In addition, the nutrient diffusion distance by the blood vessel was set to 200 μm. The size of one cancer cell group particle was set to be 50 μm and the minimum cell division cycle was 18 hours. Stromal cells may be supplied from blood vessels or may be pre-existing [4]. Therefore, analysis was conducted in two cases. In case 1, inactivated stromal cells are generated only in the vicinity of blood vessels, and only pre-existing but inactivated stromal cells around the cancer cells are activated. In case 2, activated stromal cells are generated around cancer cells. For comparison, we also analyzed cases without stromal cells and case 1-2 fused.

The next analysis of early-stage tumor growth and angiogenesis was performed in a three-dimensional region. Figure 5 represents the initial state, where one cancer cell particle was placed in the vicinity of the center of the region where a blood vessel was present. The initial conditions and parameter values were the same as those used in the 2D analysis. Two analysis cases were considered here, the conditions for the occurrence of stroma being the same as those in the 2D analysis.

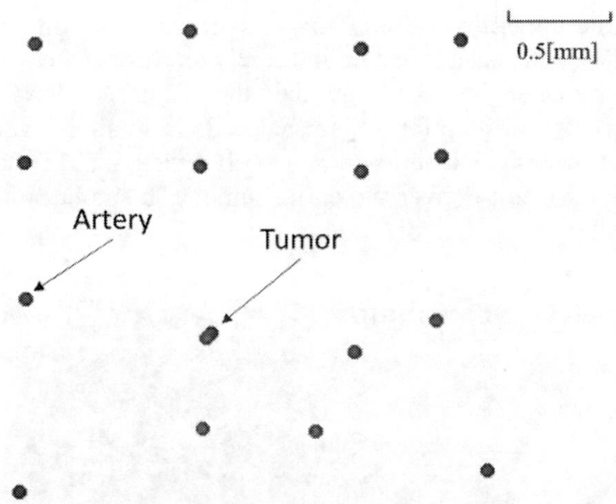

Figure 4. Initial conditions for the 2D modeling of early-stage tumor growth.

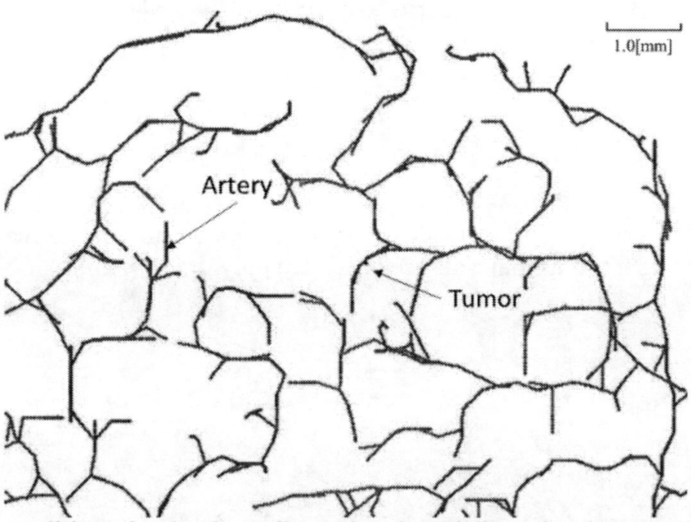

Figure 5. Initial conditions for the three-dimensional modeling of early-stage tumor growth and angiogenesis.

4. Results and Discussion

The influence of stromal cell growth on tumor development and shape was examined in detail using the methods and cases described in earlier sections. Figure 6 shows the distribution of cancer cells and stromal cells for case 1 in which stromal cells were generated only around blood vessels. In this case, the tumor was irregular in shape. This is because the inactive stromal cells around the periphery of the blood vessels push the cancer cells outward, making the tumor boundaries irregular. Figure 7 shows the distribution of cancer cells and stromal cells for case 2 in which stromal cells were generated only from the periphery of the cancer. In this case, the stromal cells were distributed so as to envelop cancer cells. Moreover, the size of the tumor as a whole was smaller than that in case 1. This is because cancer cells consume nutrients at a higher rate than stromal cells and compete with each other for nutrition. Those cells to which nutrients do not reach thus become incapable of division. Figure 8 shows the result of case without stromal cells. Figure 9 shows the result of case 1-2 fused. When stromal cells were present, the tumor became larger than in the absence of stromal cells.

The results of the three-dimensional analysis are presented in figure 10 and figure 11. The results obtained are similar to the case 2 results in the 2D analysis, where stromal cells cover the tumor. The size of the tumor as a whole became larger than the condition where stromal cells did not develop. This is because the nutrient consumption of stromal cells is smaller than that of cancer cells, and nutrition tends to reach the cancer cells leading to more cell division and consequently, larger tumors. Also, the blood vessels vascularized to cover the entire tumor. This is because the pressure inside the tumor is high, and neovascularization cannot be done.

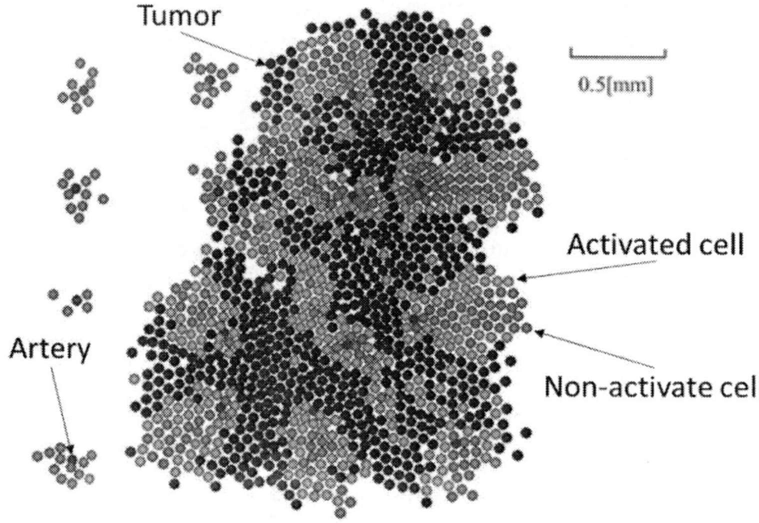

Figure 6. 2D representation of the distribution of cancer cells and stromal cells around arteries in case 1, wherein stromal cells are generated only in the vicinity of blood vessels.

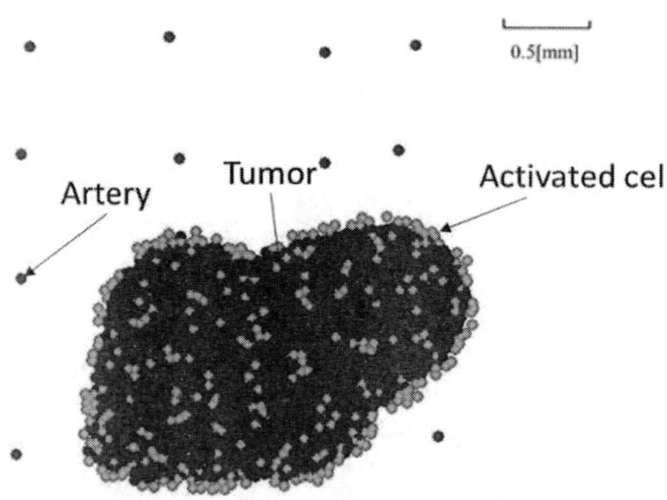

Figure 7. 2D representation of the distribution of cancer cells and stromal cells around the tumor in case 2, where stromal cells are generated only on the periphery of the tumor.

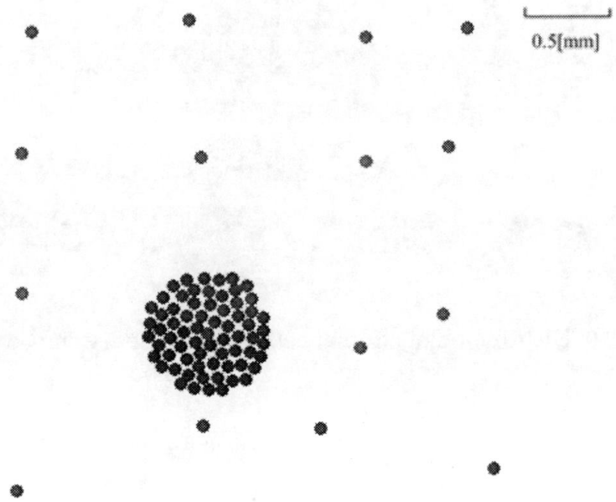

Figure 8. Case without stromal cells.

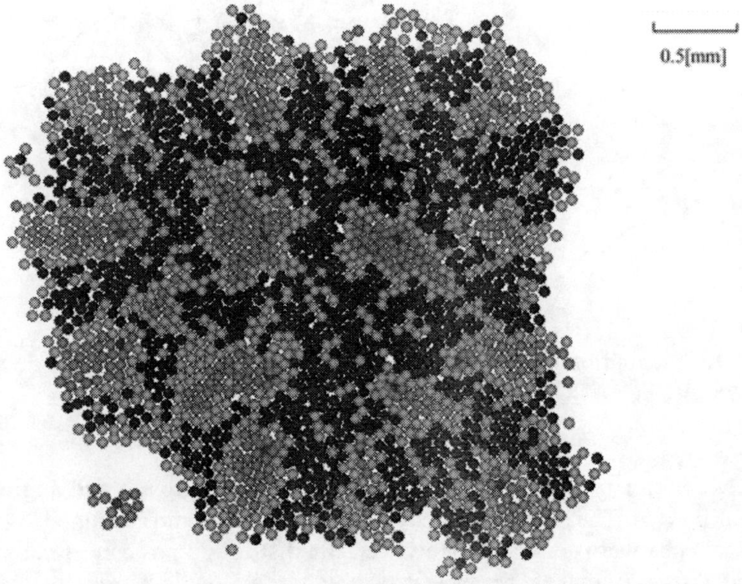

Figure 9. Case 1-2 fused.

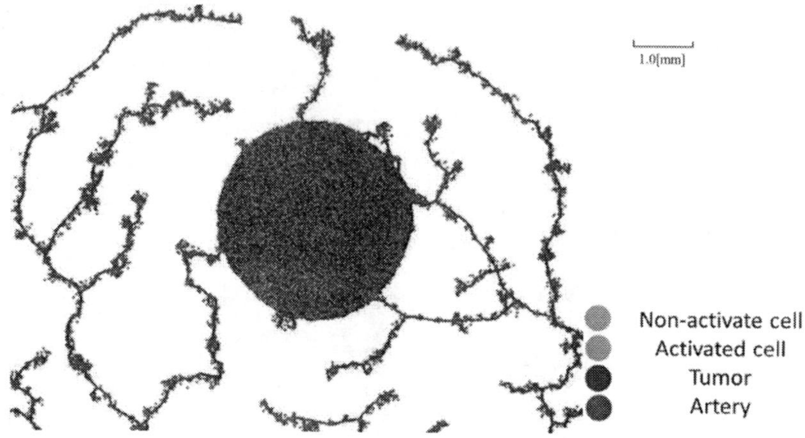

Figure 10. Distribution of stromal cells near the artery and around the tumor in the three-dimensional model

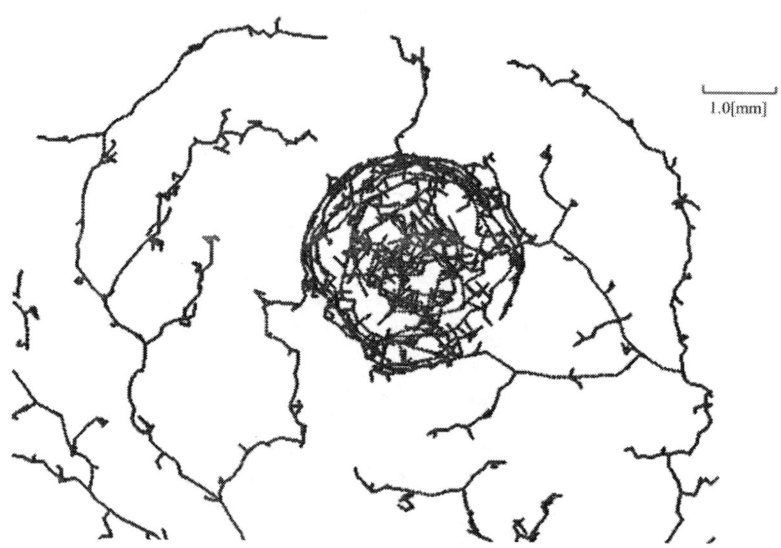

Figure 11. Distribution of newly generated blood vessels in the tumor depicted in the 3D model

5. Conclusion

In this study, we performed a numerical simulation of tumor growth and angiogenesis using a particle model and real images. The shape and size of the malignant tumor changed significantly depending on how the stromal cells were generated. However, the distribution of cancer cells and stromal cells could be approximated to real images, regardless of how stromal cells were generated. A three-dimensional numerical model was also developed, which can effectively predict tumor growth and angiogenesis.

The next steps would be the validation of the model with experimental data and refinement of the existing model so as to make it suitable for predicting the possible malignancy of tumors.

Reference

[1] Matsubara S 1992 *Cancer Biology*
[2] Nagayama K, Tomita H and Miura I 2011 *Theoretical and Applied Mechanics Japan* **60** pp.353-357
[3] Nagayama K, Oshiumi Y and Yumita M 2012 *Spring World Congress on Engineering and Technology* (SCET2012) pp 352-355
[4] Ishii G and Ochiai A 2008 *The origin of fibroblast recruited into cancer-induced stromal tissue*

Numerical simulation of the hair formation –modeling of hair cycle

Narumichi Kajihara* **and Katsuya Nagayama**

Kyushu Institute of Technology, 680-4 Kawazu, Iizuka, Fukuoka, 820-8502, Japan

* Corresponding Author: q677109n@mail.kyutech.jp

Abstract. In the recent years, the fields of study of anti-aging, health and beauty, cosmetics, and hair diseases have attracted significant attention. In particular, human hair is considered to be an important aspect with regard to an attractive appearance. To this end, many workers have sought to understand the formation mechanism of the hair root. However, observing growth in the hair root is difficult, and a detailed mechanism of the process has not yet been elucidated. Hair repeats growth, retraction, and pause cycles (hair cycle) in a repetitive process. In the growth phase, hair is formed through processes of cell proliferation and differentiation (keratinization). During the retraction phase, hair growth stops, and during the resting period, hair fall occurs and new hair grows. This hair cycle is believed to affect the elongation rate, thickness, strength, and shape of hair. Therefore, in this study, we introduce a particle model as a new method to elucidate the unknown process of hair formation, and to model the hair formation process accompanying the proliferation and differentiation of the hair root cells in all three dimensions. In addition, to the growth period, the retraction and the resting periods are introduced to realize the hair cycle using this model.

1. Introduction

In recent years, interest in the field of beauty care has increased. In particular, hair has been very important in determining the human appearance. The growth process of human hair has remained a mystery for quite some time, and with the upsurge of the beauty care fields, understanding this phenomenon has become important. Therefore, this study developed method for analyzing the formative process of the hair root using a particle model.

Hair repeats the processes of growth and retraction, with a pause interval in a repetitive cycle, which is also called the hair cycle. In the growth phase, hair is formed through processes of cell proliferation and differentiation (keratinization). Hair growth stops in the regressive period, and during the resting period, old hair falls out and new hair replaces it. The elongation rate, thickness, strength, and shape of hair are believed to be affected by the hair cycle. Therefore, many studies have dealt with observation and experimentation of hair growth. However, observation of the hair formation process that occurs internally is difficult, so the relation between the hair growth cycle and the process of hair formation is still unknown.

Therefore, in this study, we introduce a particle model as a new method to elucidate the process of hair formation, and to model the hair formation process that accompanies the proliferation and differentiation of cells in the three-dimensional hair root. The model attempts to describe the growth, regression, and resting phases of the hair cycle. This model may be useful for elucidating the phenomenon of hair formation, which prior analysis methods have failed to explain. The goal of this work is to contribute to the field of medical care and beauty by establishing the basis for this hair formation. [1, 2]

2. Object of analysis and analytical model

At first, in this research we use a method called particle method. In this method, one cell is regarded as one particle, and the equilibrium between the particles is solved to express the particle movement. As a feature, it is possible to represent a method involving deformation of cells, it is also possible to express movement and deformation at the time of cell extinction.

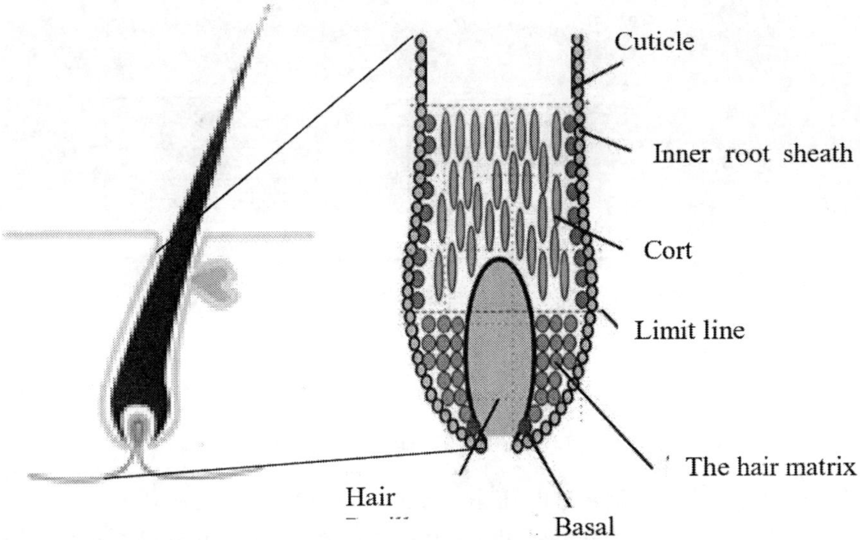

Figure 1. Cross section of hair and analytical model

Figure. 1 depicts the cross section of a hair strand and the proposed analytical model. The subject of analysis in this study is the root of the part below the cornered cells shown in the figure. The hair dermal papilla and basal cells divide from this region, resulting in hair matrix cells, cuticles, and the cortex. Further, analysis was performed on the part of the root sheath located outside, the dermal papilla by arranging the cells in an initial state and then extending them over a constant period [5].

In this study, we analyzed the process of hair formation using a particle model. This model treats a calculation point as a moving particle, which is then analyzed by tracking the movement of the particle. We introduced this method to model each particle as one cell.

In this analysis, we considered the volume and spring forces to be acting between the particles. The volumetric force is generated between all particles within a region where the force is applied. When particles enter the inside of their own region, the repulsive force causes the particles to separate, and an attractive force is exerted by the attractive force. In addition, the spring force works only between two connected particles, and maintains particles at a stable distance. The relational expressions for the volumetric force and spring forces assumed in this analysis are denoted by equations (1) and (2), respectively, where dr0 denotes the stable distance, ddr denotes the interparticle distance, k is an interparticle coeffient, k' is an elastic coefficient, r is the basic particle diameter, and a is a cell variation coefficient .[1,2]

$$\vec{F} = \left\{ k \cdot \left(1 - \frac{ddr}{dr0} \right) \cdot (a/r) \right\} \frac{\overrightarrow{ddr}}{dd0} \tag{1}$$

IOP Conf. Series: Materials Science and Engineering **297** (2018) 012022 doi:10.1088/1757-899X/297/1/012022

$$\vec{f} = \left\{ -k'\left(1 - \frac{ddr}{1.0 \times dr0}\right) \cdot \left[1 - \left(1 - \frac{ddr}{1.0 \times dr0}\right)^4\right]\right\} \frac{\overrightarrow{ddr}}{dr0} \tag{2}$$

$$\vec{x}' = \vec{x} + (\alpha \cdot \sum \vec{F} + \beta \cdot \sum \vec{f}) \tag{3}$$

The analysis protocol for this work is shown in figure 2. The initial setting and placement of each particle at the beginning of the analysis were determined. Subsequently, the cell particle was divided based on the function of the basal and hair matrix cells. At this time, the particles that exceeded the limit became cuticle or cortex, and when going up further it will kerge. For each particle, the volume and spring forces were calculated using equations (1) and (2), respectively. Finally the particles moved according to equation (3). The data is the output and the process returned to cell division. This cycle was repeated to analyze the model.

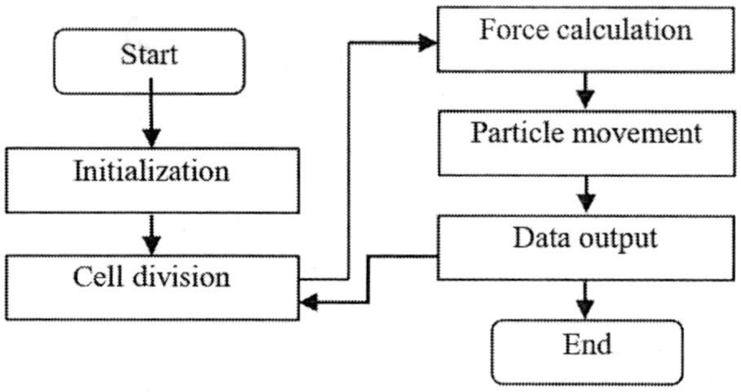

Figure 2. Analysis flowchart

Figure 3 is a schematic diagram of the hair cycle. [3] The hair cycle is a representation of the periodic growth process of hair, and it is divided into three periods of the growth period, retraction period, and rest period. The growth period is the period from the synthesis of hair to the growth of hair. Approximately one month is required to synthesize hair, and, it takes 1-6 years for hair to grow to the maximum length and width. In the retraction phase, the color pigment falls, and the root sheath begins to degenerate, this takes two to three weeks. In the telogen phase, hair growth stops and hair fall occurs, which requires a period of approximately 3-9 months. After the quiescent period, new hair is generated and the cycle returns to the growth period.

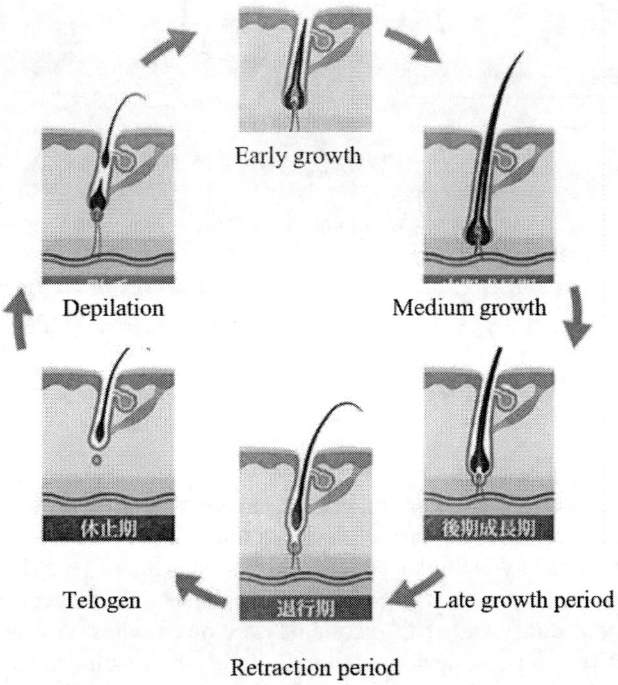

Figure 3. Hair Cycle

3. Analysis conditions

In the three-dimensional formation process model of hair, the initial arrangement comprises the inner root sheath, the hair papilla and hair matrix cells as shown in Figure 4. Further, no cells existed in the interior. Cell particles in the early stage of development have a spherical shape with a diameter of 10 μm. From this point, the cells divide and keratinize beyond the limiting line. With regard to the inner root sheath, it has an initial spherical shape of 3 μm. The quantity of inner root sheaths increases by division, which causes it to grow upward.

The frequency of basal cells formation is about one per day, and the frequency of division of the birth cells is believed to cause cell division at a rate of about three per day. Immediately after dicing, cell migration is unfettered because of the force of volume preservation. However, as the changes in cell shape become slow, the spring force develops between neighboring particles and the shape of the hair is determined by solidification of the cell group.

Each figure below has a brief caption describing it and, if necessary, a key to interpret the various lines and symbols on the figure.

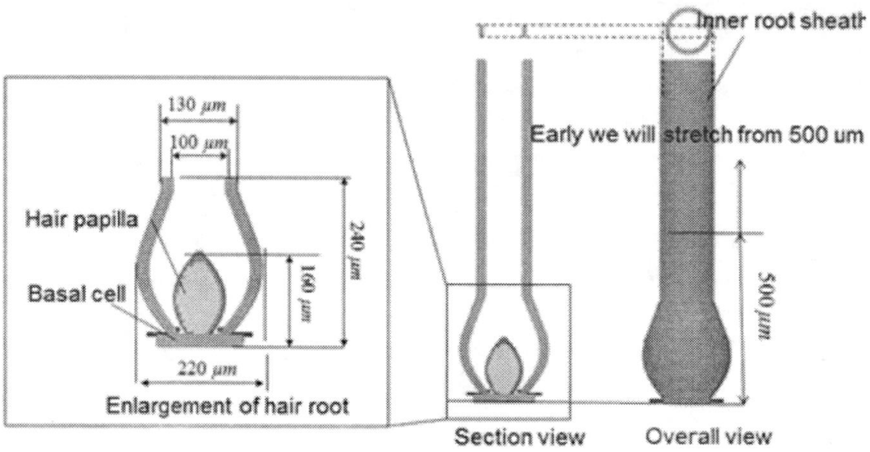

Figure 4. Initial Shape

In the modeling of the hair cycle, the growth period expresses how the hair is actually stretched by the elongation of the inner root sheath and, cell division of the hair matrix cells and the basal cells. In the retraction phase, hair retraction is expressed by the phenomenon of particles and expresses how the dermal papilla is retracting. The resting period shall be expressed by stopping the division of how the growth of hair stops.

4. Analysis Result

The results of the modeling of the hair growth process are shown in figure 5. In the growth period, internal growth was observed while the inner root sheath extended, and we confirmed that growth stopped after a certain period in the retraction phase. In the telogen phase, the hair cycle was reproduced by stopping the growth of hair, which was considered to be a state of hair sloughing.

The actual modelled phenomenon was compared to images and research results. We found that the increase of the period was consistent with results of previous studies. As shown in figure 3, we found that the growth, retraction, and rest periods coincided with the results shown in figure 5. The process model results agreed with, each period of the cycle shown in figure 5. The growth period was 1-6 years, the retraction period was 2-3 weeks, and the telogen period was 3 months, but the expression of that period was also possible. Actually, in the animation of the process, the growth period was longer than the retraction and quiescence periods, so the growth period was shortened and displayed. However, calculating the results based on the actual phenomenon was possible. In addition, although a sectional view is displayed in figure 5, since the analysis was performed in three-dimensions, a stereoscopic analysis was also possible.

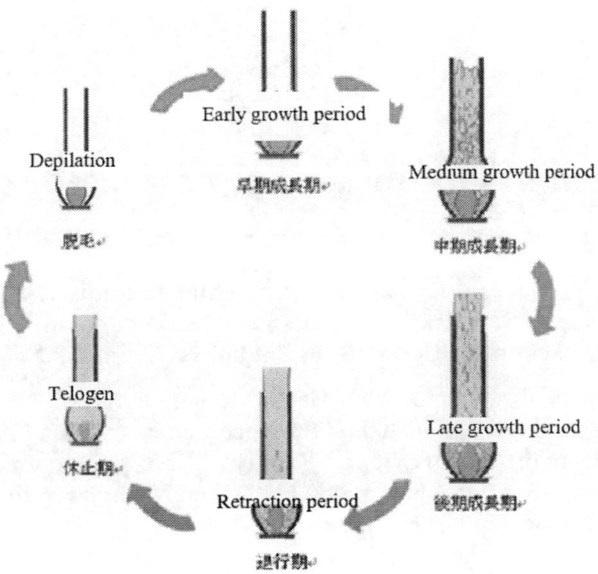

Figure 5. Analysis results of hair cycle

5. Conclusion

In this study, numerical simulation of hair formation at the hair root using a particle model was performed. By following the state of the hair growth from the proliferation of the basal cells and the hair matrix cells to extraction and contraction of the inner root sheath, fabricating the hair growth process similar to the real phenomenon was possible. In addition, we also constructed a model using the spring and volume forces, and used the model to simulate the actual hair growth cycle. From these results, we think that we can construct a hair process analysis tool with shrinking of the root sheath part. As a result of this research, it can be said that human hair grows at a rate of 5 μm per day, and its value can be quantitatively evaluated by a numerical analysis of this study. From the viewpoint of comparison with actual phenomena, hair growth is considered to be qualitatively due to large individual difference.

Future work will focus on the hair papilla portion to better approximate the actual hair growth process, and this will be utilized for analysis of hair formation mechanism, beauty diagnosis systems, etc. As part of our ongoing research, the process of hair thinning is also being considered. This interrupted phenomenon occurs when growth in early stages of growth is interrupted and enters the retraction period or the telogen period, resulting in hair thinning. [4] Therefore, it is a simulation that can change hair thickness simultaneously with growth retraction. In the future, we will analyze such types of hair growth phenomena.

Reference

[1] Shogo Matsuoka, Katsuya Nagayama and Hiroyuki Taguchi 2013 *Proc. of the 66th General Meeting of the Kyushu Branch of the Japan Society of Mechanical Engineers Proceedings* **307** pp 89-90
[2] Katsuya Nagayama, Shogo Matsuoka, Naoko Morisaki and Hiroyuki Taguchi 2015 *Open Journal of Regenerative Medicine* pp 1-5
[3] http://bio-ceutical.com/html/page3.html
[4] Anti Aging Series 1 *Practiceof hair loss/hair removal hair growth* p 93 and p 121
[5] Junya Toguchida and Kyushi Nakazawa *Regenerative medicine and tissue stem cell which is key to elucidation of disease* p 29

IOP Conf. Series: Materials Science and Engineering **297** (2018) 012023 doi:10.1088/1757-899X/297/1/012023

The development of lab-on-a-chip fabricated from two molds

A. Pramuanjaroenkij[1,*], J. Bunta[1], J. Thiangpadung[1], S. Sansaradee[1], P. Kamsopa[1], S. Sodsai[1], S. Vichainsan[1], K. Wongpanit[2], T. Maturos[3], T. Lomas[3], A. Tuantranont[3], B. Cetin[4], S. Phankhoksoong[1] and A. Tongkratoke[1]

[1]Department of Mechanical and Manufacturing Engineering, Kasetsart University, Chalermphrakiat Sakon Nakhon Province Campus, Sakon Nakhon 47000, Thailand
[2]Department of Animal Science, Faculty of Natural Resources and Agro-Industry, Kasetsart University, Chalermphrakiat Sakon Nakhon Province Campus, Sakon Nakhon 47000,Thailand
[3]Nanoelectronics and MEMS Laboratory, National Electronics and Computer Technology Center, 112 Thailand Science Park, Phahonyothin Road, Khlong Nueng, Khlong Luang, PathumThani 12120,Thailand
[4]Microfluidics and Lab-on-a-chip Research Group, Mechanical Engineering Department İhsan Doğramacı Bilkent University, Ankara 06800, Turkey

* Corresponding Author: E-mail anchasa@gmail.com

Abstract. Development of diagnostic technique of microfluidic or lab-on-a-chip (LOCs) is currently of great interest for researchers and inventors for their many advantages. It can be used as a real laboratory was many ways to help to the diagnosis faster. This research aims to develop Polydimethylsiloxane (PDMS) lab-on-a-chip (LOCs) which were produced from different molds; the silicon wafer mold and the stainless mold to investigate the flow of the biological sample as the flow in nanochannels. In addition, this research proposes a means to leakage and the blockage of the channel flow. The experimental results were found that the LOCs casted from the silicon wafer mold sandwiched by both the plasma cleaner machine and H shaped acrylic sheets showed leakages around the electrode areas because the first new electrodes were too thick, the proper thickness of the nickel electrode was at 0.05 millimeters. The LOCs casted from the stainless mold were inserted by the nickel electrodes produced by the from the prototype shaped electroplating process; this LOCs using nickel plated electrodes 2 times to make a groove on the nickel electrode backsides when pouring the PDMS into the LOCs casted from the stainless mold. It was found that PDMS was able to flow under the nickel electrode and the PDMS sheet could stick with the glass slide smoothly. In conclusion, it was possible to develop these LOC designs and new electrode fabrications continually under helps from Micro-Electro-Mechanical system, Thailand National Electronics and Computer Technology Center, since causes of the LOC problems were found, and demonstrated the feasibility of developing the LOCs for chemical detection and disease diagnostics.

1. Introduction

Lab-on-a-chip or LOC devices are miniature laboratories built on a thin glass or plastic chip of several centimeters in dimensions. The polydimethylsiloxane (PDMS) LOC is one of the simple LOC devices [1, 2]. The PDMS LOC chamber is made from polydimethylsiloxane by solven casting and drilling. These small devices can duplicate the specialized functions as their room-sized counterparts in clinical diagnoses. The advantages of these devices include significantly reduced reagent consumption, short analysis time, automation, and portability [1, 2]. Dielectrophoresis (DEP) is an electrokinetic movement of the neutral particles induced by polarization in non-uniform electric field [2, 3]. When the dielectrophoresis is positive, the particle moves towards the locations with the greatest electric field. On the other hand, if the dielectrophoresis is negative, the particles are pushed from the locations with the greatest electric field [2, 4-5].

Yang et al. [6] introduced a microfluidic device using simple microchannels for continuous real-time blood plasma separation using defibrinated chicken blood as biological samples. Their device as shown in Figure 1 was claimed to separate blood plasma almost 100% when the blood was infused 30 minutes through the device. We noted that their device was designed to separate blood plasma from an initial blood sample of up to 45% inlet hematocrit (volume percentage of cells).

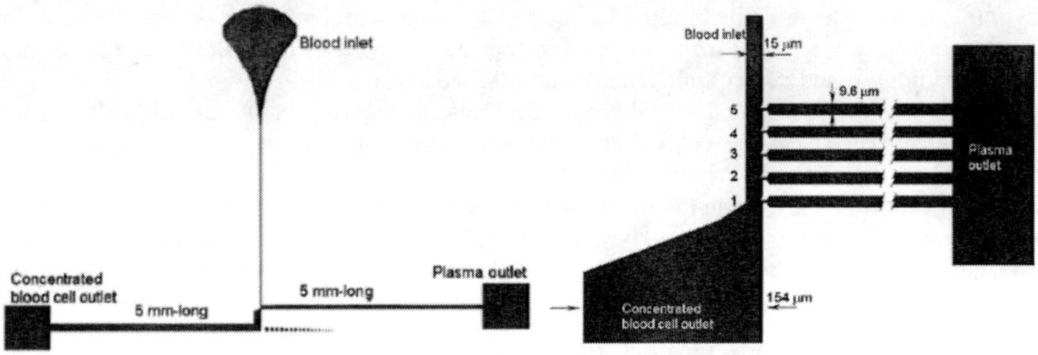

Figure 1. Schematic of the simple-microchannel microfluidic device [6].

Li and Cetin [7] introduced a LOC design of a microfluidic channel with a pair of simple electrodes to perform a continuous separation of particles based on electrical properties using alternating current dielectrophoresis (AC-DEP). Later, Cetin et al. [8] presented a novel, simple LOC device for continuous separation of particles by their sizes based on AC-DEP. They applied the non-uniform electrical field generated by means of embedded asymmetric electrodes inside the LOC device. Cetin and Li [9] also presented the LOC device (Figure 2) which could be used to continuously separate particles and cells based on the AC-DEP properties. Their results showed that white blood cells (WBCs) and 10 mm latex particles could be separated due to their electrical properties even they were in the similar size range.

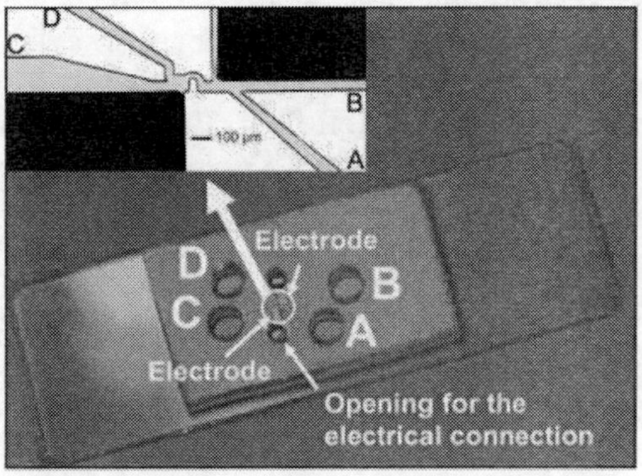

Figure 2. The final device from Cetin and Li [9].

Kerhoas and Sollier [10] published their critical review on micro-scale blood plasma separation: from acoustophoresis to egg-beaters Lab Chip; providing information about plasma analyses and target applications, existing solutions for micro-scale blood plasma extraction and the approaches to validating specific applications, since lab-on-a-chip devices were continuously developed to clinical markets. The review about DEP in microfluidics technology was published by Cetin and Li [11]; they summarized that DEP could be utilized either by DC-field or AC-field and concluded possible future research directions on DEP research as (i) replacing the bench-top instruments with the microfluidics technology for clinical application, (ii) developing the microfluidics technology to hand-held, point-of-care testing devices (Both the sample preparation and the chemical/biological analysis in one device), (iii) utilizing mechanical micromachining (milling, drilling), microinjection molding techniques [12-13] for LOC fabrication and using the polymer-based conductive materials as electrodes which may lead to inexpensive and massive fabrication of DEP-based microfluidic systems, (iv) integrating the electrorotation analysis with DEP-based systems would outcome robust and practical DEP based clinical instruments, (v) extending the proposed systems to operate with high-conductivity buffer solutions and (vi) manipulating CNTs and nanoparticles in the LOC devices for the development of the bionano/ nanotechnology-based devices and nanomaterial-based sensors.

Zeinali et al. [14] presented two alternative fabrication techniques of a microfluidic device with 3D sidewall electrodes (figure 3); (1) both the mold and the electrodes were fabricated using high precision machining and (2) the mold with tilted sidewalls was fabricated by using high precision machining and the electrodes was deposited on the sidewall by using sputtering together with a shadow mask fabricated by electric discharge machining. They found both fabrication processes were assessed as highly repeatable and robust. They claimed that both methods were complementary with respect to the channel height.

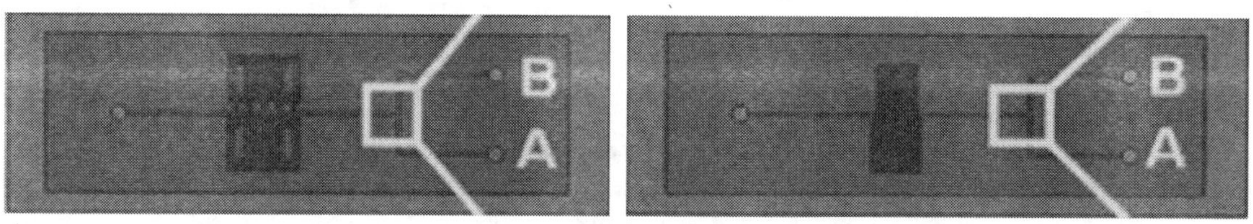

Figure 3. Alternative fabrication techniques of a micro- fluidic device with sidewall electrodes [14].

Tripathi et al. [15] published their research on a simple and efficient passive microfluidic device for plasma separation from pure blood; undiluted blood in the flow rate range of 0.3 to 0.5 ml/min. Ulum et al. [16] presented their study on the wicking and separation characteristics of blood plasma in a cotton thread matrix functioning as a microfluidic thread-based analytical device (μTAD) as shown in Figure 4 using ethylene-diaminetetraacetic acid (EDTA) anticoagulant solution. In this work, the blood of healthy Indonesian thin tailed chicken was selected to be separated. They found that the separation depended on the synergy of three factors, cotton fiber, EDTA anticoagulant and blood platelets (the platelets induced the formation of a fibrin-filter via a partial coagulation process in the EDTA-treated μTAD.

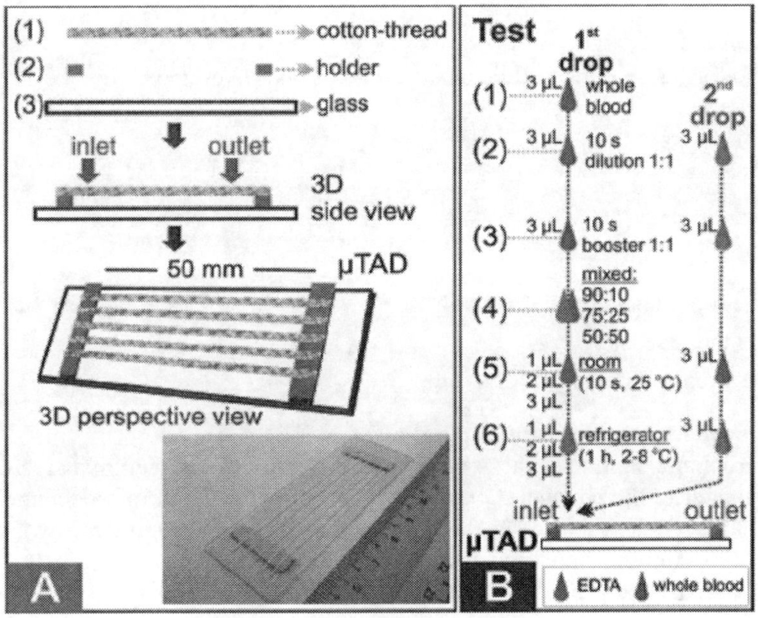

Figure 4. The microfluidic thread-based analytical device [16].

Pramuanjaroenkij et al. [2] presented the reusable Lab-on-a-Chip or LOC fabricated and developed to reduce its investment cost and to ease its operations by applying the LOC introduced by Cetin et al. [8]. In this work, chicken and chicken blood were chosen as the samples to implement two-phase flows with different red-blood-cell shapes. The chicken blood flow represented the viscous two-phase flow with the oval shape particles while the chicken blood flow represented the inadhesive samples with the round particles. Since it was hard to characterize the particle flow in the inadhesive flows, the result pictures were analyzed and presented in terms of color intensities per unit area by using a computer program called "ImageJ". The chicken-blood-flow results were validated with the hematology results which were the hematocrits to find their relationships between the hematocrits and the RBC flows. The normal-health-condition chicken with the standard hematocrits higher than 28% showed the average-different-color intensities per unit area between cathode and anode at 39.485 pixels per unit area while the lower standard hematocrit samples, the hematocrits were lower than 28%, showed the average-different-color intensities at 14.641 pixels per unit area, the lower intensity the lower hematocrit. So the LOC coupled with "Image J" exhibited their capabilities to investigate the chicken blood conditions, especially, this coupled technique consumed less time than the traditional hematology process [2]. Pramuanjaroenkij et al. [17] experimentally studied two types of lab-on-a-chips (LOCs); (1) the reusable Polydimethyl-siloxane (PDMS) LOC; casted on the silicon wafer mold, to improve its performance and (2) the new PDMS LOC casted on the stainless steel mold imported. They solved leakage problem and channel hydrophobic behaviour of the reusable LOC which was sandwiched by H-acrylic plates and flowing the bovine serum albumin (BSA) solution, respectively. They found that the sample flow without the electrical potential (0 volt) was slower than the flow with the 1-volt electrical potential but the flow was stopped when the electrical potential at 5 volts was applied, therefore, the new LOC could prove to be used with the chicken blood with the 1-volt electrical potential.

This research aims to develop Polydimethylsiloxane (PDMS) lab-on-a-chip (LOCs) which were produced from different molds; the silicon wafer mold and the stainless mold to investigate the flow of the biological sample as the flow in nanochannels. In addition, this research proposes a means to leakage and the blockage of the channel flow.

2. Basic Principles of Microfluidics

2.1. Reynolds number

The flow of a fluid through a microfluidic channel can be characterized by the Reynolds number, defined as equation (1)

$$R_e = \frac{LV_{avg}\rho}{\mu} \tag{1}$$

where L is the most relevant length scale.
μ is the viscosity,
ρ is the fluid density.
V_{avg} is the average velocity of the flow.

For many microchannels, L is equal to 4/AP where A is the cross sectional area of the channel and P is the wetted perimeter of the channel. Due to the small dimensions of microchannels, the Re is usually much less than 100, often less than 1. In this low Reynolds number regime, flow is completely laminar and no turbulence occurs-the transition to turbulent flow generally occurs in the range of Reynolds number 2000. Laminar flow provides a means by which molecules can be transported in a relatively predictable manner through microchannels. Note, however, that even at Reynolds numbers below 100, it is possible to have momentum-based phenomena such as flow separation.

2.2. Poiseuille's Law

In such a laminar flow of viscous and incompressible fluid, the pressure drop and the flow rate, as well as the effective resistance might be obtained by using the Poiseuille equation (2).

$$\Delta P = \frac{8\mu LQ}{\pi r^4} \tag{2}$$

where ΔP is the pressure drop.
Q is the volumic flow rate.
L is the length of the channel.
r is the radius of the channel.

3. Experiment Study

To produce the silicon-wafer casted PDMS LOC, the silicon wafer was patterned by using the negative photo-resist (SU-8 25, MicroChem Co., Newton, MA) technique. The dielectrophoretic chamber was made from the PDMS prepared by mixing the precursors sylgard with a curing agent at a ratio of 10:1 by volume. The prepolymer mixture was degassed at 20-50 mTorr at room temperature in desiccators pumped with a mechanical vacuum pump for 10 minutes to remove any air bubbles in the mixture. The PDMS mixtures were gradually poured onto the patterned silicon wafer or a mold. After the PDMS was cured at 100 °C for 30 minutes on the mold, the molded polymer samples were peeled off and punched into a hole in order to create a chamber. The microelectrodes were inserted into the PDMS electrode chambers manually under the microscope and, then, all components as the LOC were assembled by using acrylic plates to sandwich the components. In each LOC device, there were two inlet reservoirs, two exit reservoirs, channels with 25 micrometers (in z-direction) in height, the main channel with 100 micrometers in width, and two small reservoirs punched on top of the electrodes for the external electrical connections [2, 18-19] as shown on figure 5.

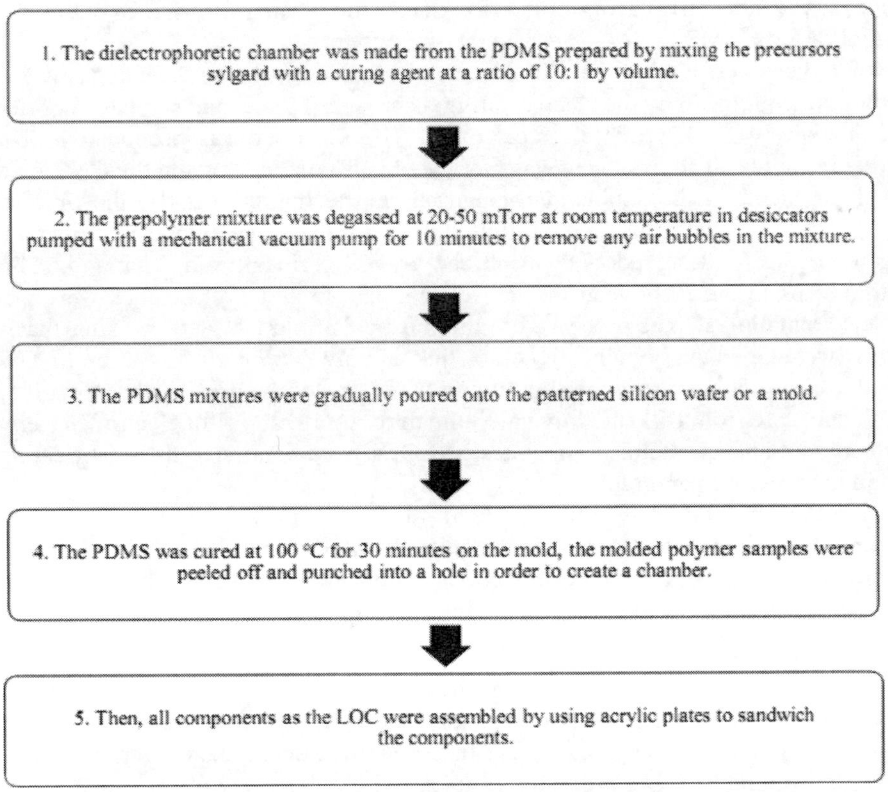

(5a)

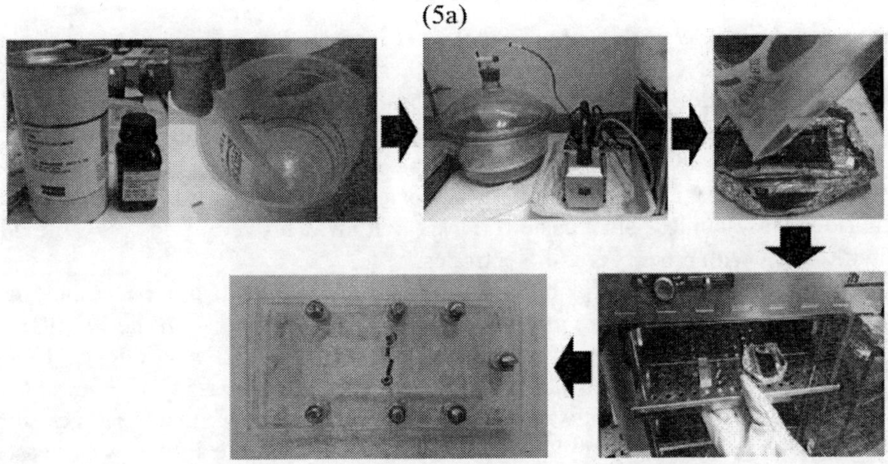

(5b)

Figure 5. (5a)The flow diagrams and (5b) the fabrication step for the silicon-wafer casted PDMS LOC.

The stainless-steel casted PDMS LOC was casted on the mold imported from Bilkent University, Turkey, there were total of four screws on both sides of the mold; two screws for each side, to hold the electrodes on their positions, control the PDMS flow and to prevent PDMS leakage. The stainless-steel casted PDMS LOC was fabricated at Nanoelectronics and MEMS Laboratory, Thailand National Electronics and Computer Technology Center by following steps; (i) washing the stainless steel mold with Isopropanol, (ii) placing electrodes on the stainless steel mold, (iii) mixing PDMS and water with ratio of 10:1 and pouring the mixture on the stainless steel mold, (iv) placing the stainless steel mold inside the vacuum chamber to get rid of air bubbles, (v) placing the mold on the hotplate with constant temperature, (vi) taking the PDMS attached with the electrodes out of the mold and placing the glass slide on the PDMS and (vii) placing both parts inside the plasma cleaner for 10 minutes, called the stainless-steel casted PDMS LOC [17] as shown on figure 6.

In this work, there were total of five electrode sets; the first set was for the silicon-wafer casted PDMS LOC and they were prepared by the electroplating technique. The second set was imported with Bilkent University, Turkey, with the stainless steel mold. The third set was produced in Thailand by the electroplating technique. The fourth set was fabricated in Thailand by using the CNC machine. And the fifth set was produced by the double-step electroplating technique. Firstly, the first and third set of electrodes made from nickel by using the traditional electroplating technique as shown on Figure 7. The fourth electrodes were designed and machined by CNC locally in Thailand. There were two electroplating steps in the fifth-electrode fabrication. The first-step electrodes were produced by the traditional electroplating step as processed in the third-electrode fabrication. After that, some parts of the electrodes were taped to allow the nickel to stick on only these parts.Then, the first electrodes were placed on the electrode positions of the silicon-wafer casted LOCs. In the meantime, the second electrodes which were from Bilkent University, the third, fourth and fifth electrodes were placed on the stainless steel mold and the stainless-steel casted LOCs were continued their fabricating processes as mentioned in the previous paragraph.

Finally, the chicken blood was tested inside all LOCs; the silicon-wafer casted LOCs with the first electrodes and the stainless-steel casted LOCs with the second, third, fourth and fifth electrodes. All LOCs were investigated with the chicken blood to find their leakage problems by sandwiching the PDMS and glass-slide parts inside the plasma cleaner.

4. Results and Discussion

The chicken blood was chosen as the biological sample, taken from the chicken by the professional caretaker and prepared by mixing the blood with the ethylenediaminetetraacetic acid (EDTA) as the anticoagulant solution, we already found that the suitable anticoagulant-sample ratio for the chicken sample at 1:1 by volume from our previous work [2, 17].

Since the second new electrodes (figure 8(b)) were ready to be used; they were stainless steel plates machined by the micro-CNC machine, we placed these stainless steel electrodes on the mold. Other stainless-steel casted PDMS processes were performed and the stainless-steel casted PDMS was sandwiched with the glass slide inside the plasma cleaner. After we injected the chicken blood sample into the stainless-steel casted LOC reservoir, we found that there was no leakage and we could connect the electrodes with the external electrical circuit to perform electric fields under the 1-voltage electrical-circuit condition. This stainless-steel casted PDMS LOC was proved to be workable and ready to be compared with LOCs with other types of electrodes.

The nickel electrodes for the silicon-wafer LOCs; obtained from the electroplating technique as shown on the left of figure 8(a), were placed on the electrode positions on the PDMS casted from the silicon wafer, the glass slide was placed on top of the first PDMS and electrodes and the silicon-wafer casted LOCs were sandwiched by the plasma cleaner as shown on the right of Figure 8(a). The chicken blood was tested inside the silicon-wafer casted LOCs, we found that leakages surrounding the electrodes and noticed that the electroplated electrodes were thick, the silicon-wafer casted LOCs with the first electrodes were not functioned.

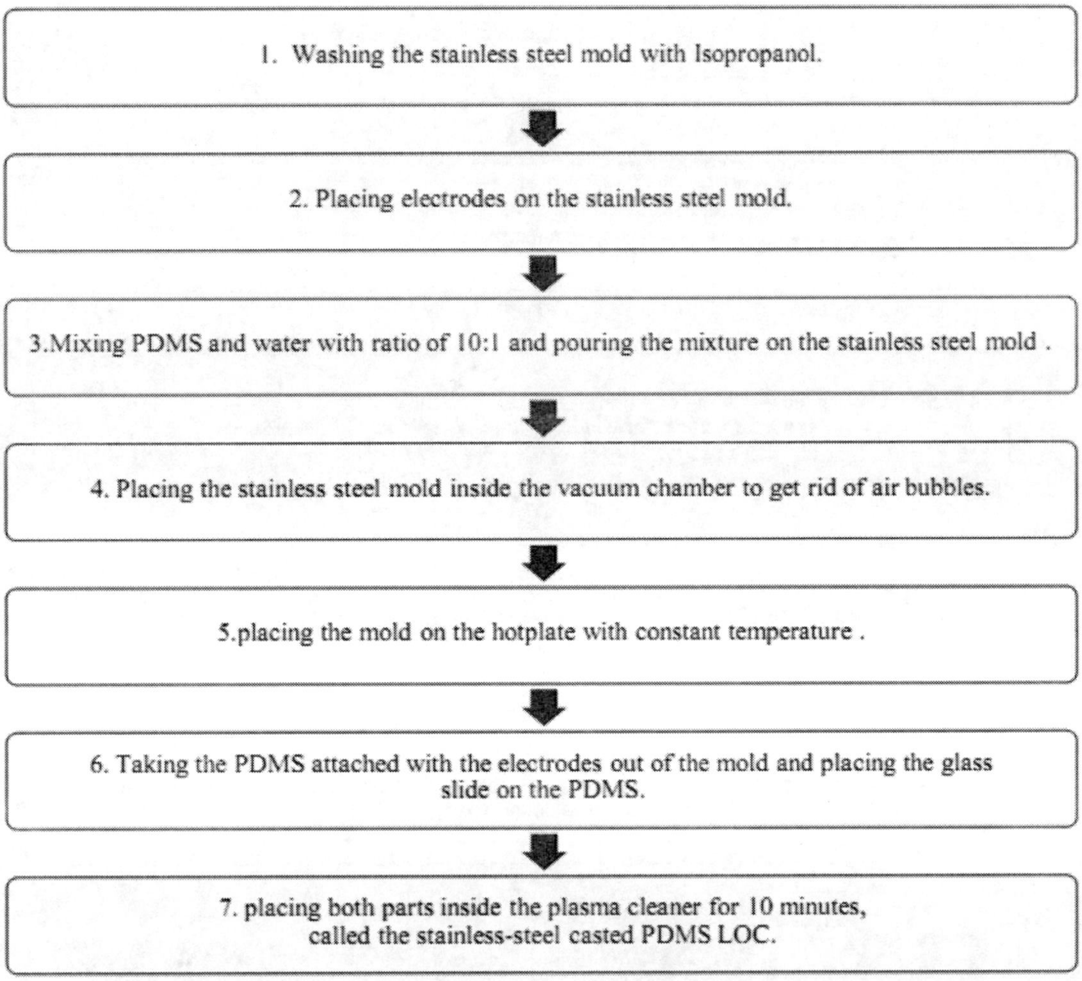

(6a)

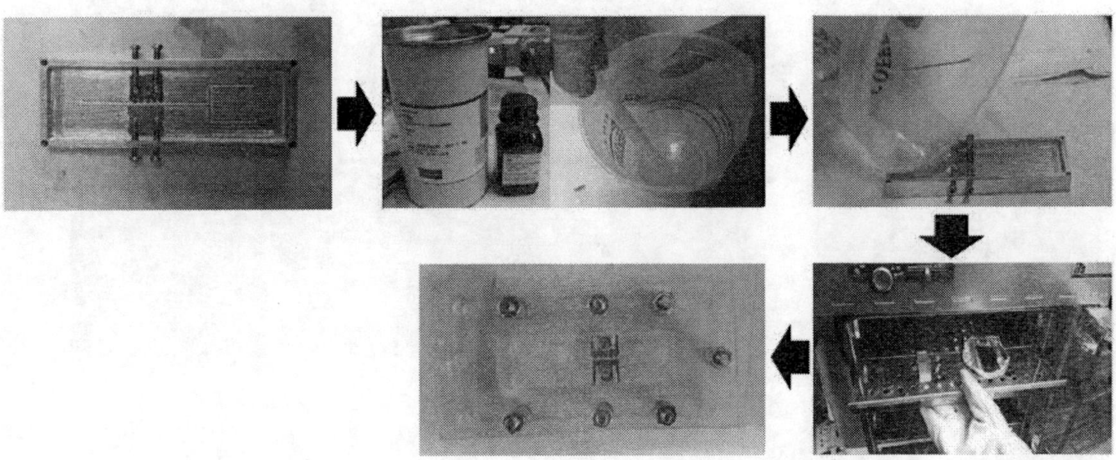

(6b)

Figure 6. (6a) The flow diagrams and (6b) the fabrication step for the stainless-steel casted PDMS LOC.

IOP Conf. Series: Materials Science and Engineering **297** (2018) 012023 doi:10.1088/1757-899X/297/1/012023

1. Placing a dry film on a stainless plate.

2. The top part of the dry film was peeled out and the left dry film was smoothening with the plate.

3. Placed a paper which was holed in electrode shapes on the dry film before the photolithography process was performed by using UV light for 25 seconds..

4. The plate was washed to get rid of the undried parts. On the other hands, the dried parts were harden by the UV light as the electrode shapes and they could not be washed.

5. The patterned plate was submerged into potassium dichromate solution for 10 minutes and then it was connected with wires.

(7a)

(7b)

1. Placing a dry film on a stainless plate.

↓

2. The top part of the dry film was peeled out and the left dry film was smoothening with the plate.

↓

3. Placed a paper which was holed in electrode shapes on the dry film before the photolithography process was performed by using UV light for 25 seconds..

↓

4. The plate was washed to get rid of the untried parts. On the other hands, the dried parts were hardened by the UV light as the electrode shapes and they could not be washed.

↓

5. The patterned plate was submerged into potassium dichromate solution for 10 minutes and then it was connected with wires.

↓

6. The electroplating process by applying 0.5 mA current for 17 hours. The electroplated plate was cleaned again in the sodium hydroxide solution. The plate with the nickel electrodes on top was dried and the electrodes were peeled out.

(7c)

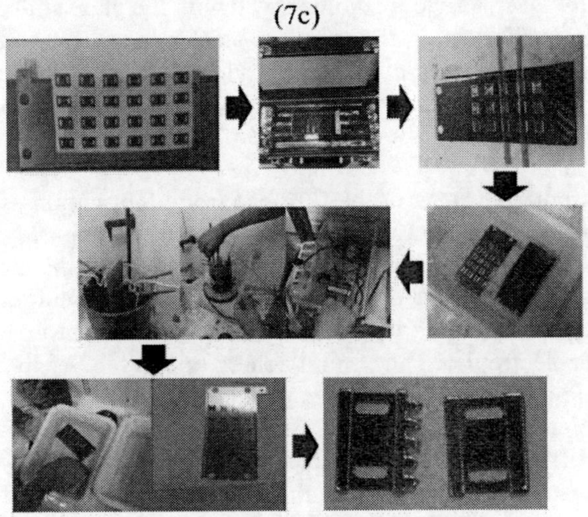

(7d)

Figure 7. (7a) The flow diagrams and (7b) the fabrication step for the electroplating technique of the first electrodes, (7c) the flow diagrams and (7d) the fabrication step for the electroplating technique of the third [20].

The electroplating processes were started with placing a dry film on a stainless plate. Then the top part of the dry film was peeled out and the left dry film was smoothen with the plate. We placed a paper which was holed in electrode shapes on the dry film before the photolithography process was performed by using UV light for 25 seconds. After that the plate was washed to get rid of the undried parts. On the other hands, the dried parts were harden by the UV light as the electrode shapes and they could not be washed. The patterned plate was submerged into potassium dichromate solution for 10 minutes and then it was connected with wires. The anode side was also connected with a nickel plate and the cathode side was connected with the stainless steel plate, the nickel dichloride solution was used as the electrolyte solution as shown in figure 6. We induced the electroplating process by applying 0.5 mA current for 17 hours. The electroplated plate was cleaned again in the sodium hydroxide solution. The plate with the nickel electrodes on top was dried and the electrodes were peeled out. We performed these electroplating processes for the first, third and fifth electrodes.

The third nickel electrodes for the stainless-steel LOCs; obtained from the electroplating technique as shown on the left of figure 8(c), were placed on the electrode positions on the stainless steel mold, the PDMS was poured on the mold and the PDMS was casted. We found that, during the PDMS pouring, the electroplated electrodes were floating and moving from their positions because the electrodes were light and not held in their positions. After the glass slide was sandwiched by the plasma cleaner with the PDMS part including the electrodes (The right of figure 8(c)), the chicken blood was tested inside the stainless-steel casted LOCs, we found that there was no leakage but we could not connect the electrodes with the external electrical circuit to perform electric fields because we could not find the certain positions to attached wires. Therefore, the stainless-steel LOCs with the third electrode was unworkable because of the electrode positions. We also discovered that the silicon wafer mold was broken at the channel part, therefore the casted channel did not function as shown in figure 9.

For the fourth electrodes, we designed them as shown on the top left of figure 8(d) and the stainless steel plate were machined according to our design. These machined electrodes were fabricated in Thailand. After the fourth electrodes were delivered, other stainless-steel casted PDMS LOC processes were performed as shown on figure 8(d). The chicken blood sample was investigated in these stainless-steel casted LOCs, we found that there were leakages under the electrode areas even we could connect the electrodes with the external electrical circuit. When we compared the second electrodes with the fourth electrodes, we noticed that there was PDMS attached on patterns on the second electrode backsides and these small PDMS surfaces adjoined well with the glass slide, these small PDMS surfaces played an important role in fabricating the stainless-steel casted LOCs. The backsides of the fourth electrodes were attempted to be milled but the electrodes were broken before the milling process was done. If one could fine a micro-milling machine, this pattern work could be done. Therefore, the last stainless-steel casted LOCs with the fourth electrodes were not workable because of leakage parts underneath the electrodes.

There were two electroplating steps in the fifth-electrode fabrication as shown on figure 8(e). The first-step electrodes were produced by the traditional electroplating step as processed in the third-electrode fabrication. After that, two edge areas along the length of the electrodes were taped and the whole electrodes were submerged into the potassium dichromate solution. The tape strips prevented these areas to touch the potassium dichromate solution, when we performed the second electroplating technique, the nickel was electroplated and could adhere on the tape-strip areas while nickel could not stick on made the middle area of the electrodes (the areas with no tape) which was covered by the potassium dichromate solution. The double electroplating technique made a groove on the backside of the fifth electrodes. The fifth electrodes were placed on the stainless mold, when the PDMS was poured on the stainless mold, theses electrodes were in the electrode positions and the PDMS could attach on the back of the fifth electrodes. The PDMS with the double-electroplated electrodes was sandwiched with the glass slides by the plasma cleaner to produce the fifth LOC. The fifth LOC was also experimentally investigated with the chicken blood and found that the chicken blood could flow through the LOC from one end to other ends without any leakages (figure 10).

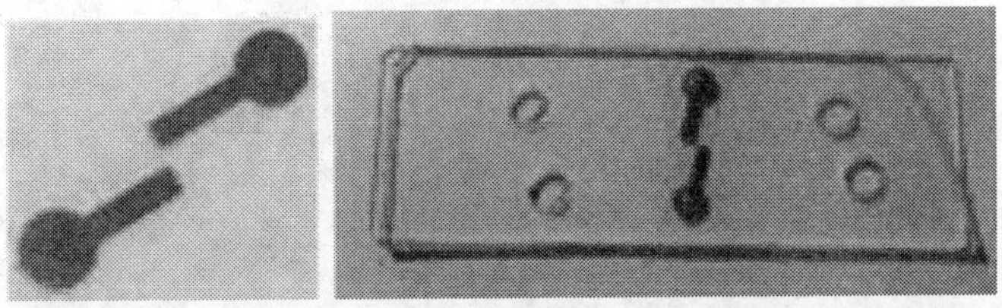

(8a)

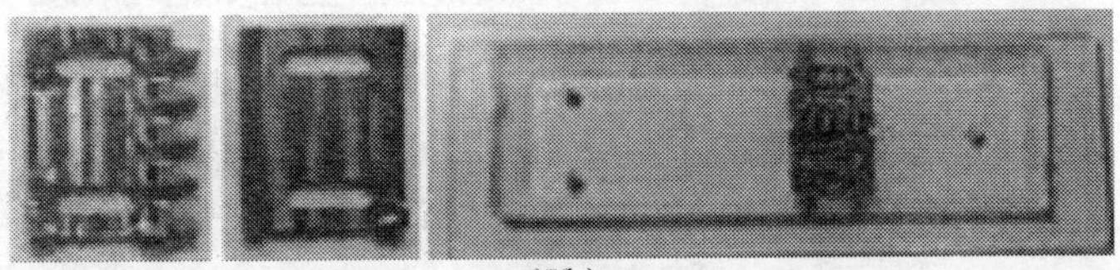

(8b)

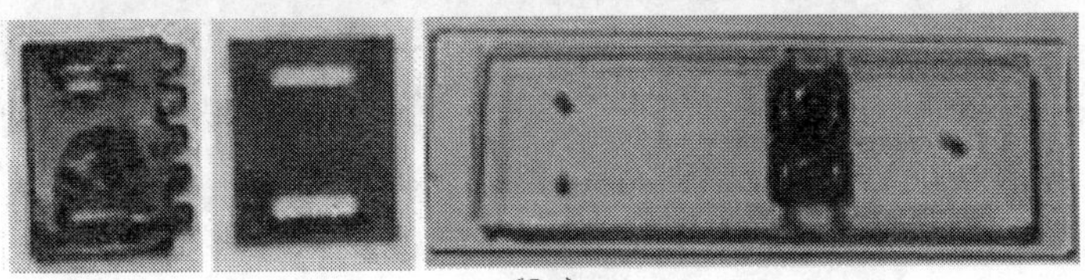

(8c)

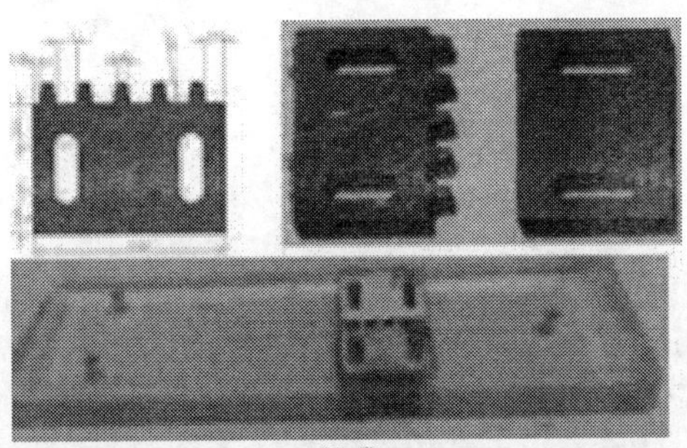

(8d)

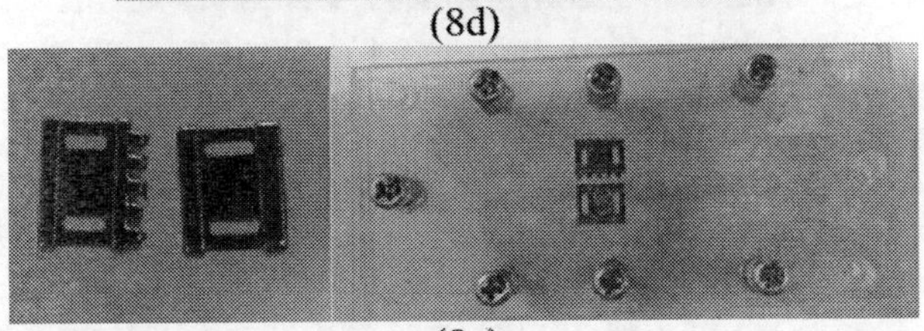

(8e)

Figure 8. (8a)The first, (8b) second, (8c) third, (8d) fourth, and (8e) fifth electrodes.

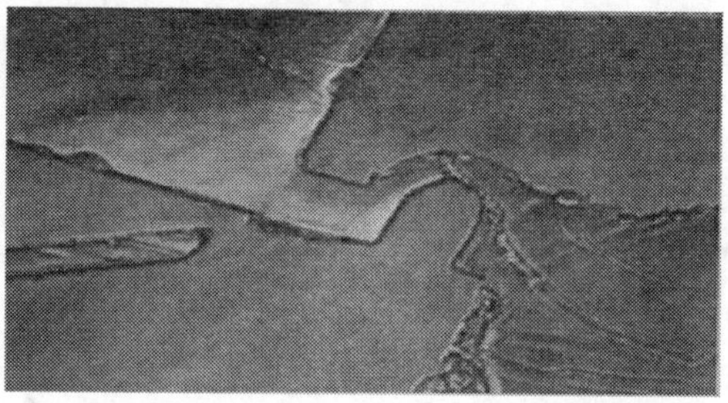

Figure 9. The broken channel on the first-electrode LOC.

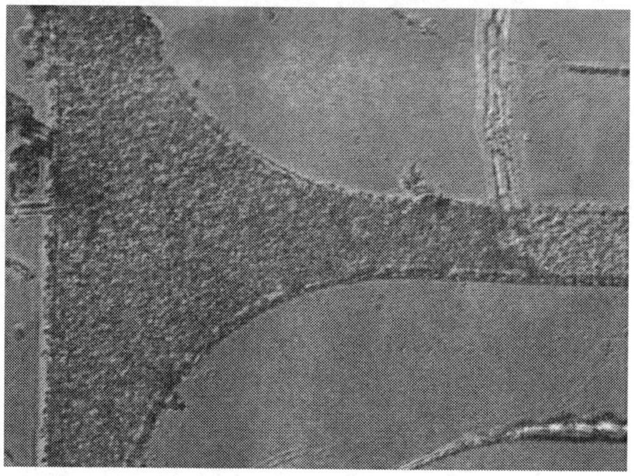

Figure 10. The chicken blood flow inside the workable fifth-electrode LOC.

5. Conclusion

Five new electrodes were fabricated and investigated to apply on two types of LOCs; the silicon-wafer casted PDMS LOC and the stainless-steel casted PDMS LOC. The five sets of the new electrodes consisted of the first sets produced from the nickel electroplating processes for the silicon-wafer casted PDMS LOC, the second sets machined from stainless steel plates at Bilkent University using the micro-CNC machine, the third sets produced from the nickel electroplating processes for the stainless-steel casted PDMS LOC, the fourth sets machined from stainless steel plates in Thailand using the traditional CNC machine the fifth sets also produced from the double-electroplating processes. All five PDMS LOCs were experimentally tested with the chicken blood as biological samples. The silicon-wafer casted PDMS LOC with the first electrodes could not be used with the samples because of its leakages around the electrodes. The stainless-steel casted PDMS LOC with the second electrodes were workable while the stainless-steel casted PDMS LOCs with the third and fourth electrodes were malfunctioned because the third set was misplaced during the PDMS casting process and the fourth electrode surfaces could not adjoin the glass slide surface as close as the second electrode surfaces did. After the problems were indicated, electrode fabrication could be improved. Since we had only one pair of the workable electrodes, the biological sample investigation could not be performed continuously. We have investigated the new technique or the double electroplating process to produce the workable electrodes as in the fifth electrode fabrication. This new technique could help us to produce numerous amount of workable electrodes and we could apply biological samples with these workable LOCs to find health condition-sample flow relationships. These workable LOCs could contribute to animal science with lower cost and less time consumption.

Acknowledgments
The research described in this paper was financially supported by Thermo-Fluid, Material Chemistry and Technology Research Group, Faculty of Science and Engineering and Office of Campus, Kasetsart University, Chalermphrakiat Sakon Nakhon Province Campus, and Kasetsart University Research and Development Institute, Kasetsart University.

References
[1] Li D 2010 *Microfluidics Based Microsystems Fundamentals and Applications,* Kakaç S, Kosoy B, Li D and Pramuanjaroenkij A eds. (Springer, Netherland) pp 377–397
[2] Pramuanjaroenkij A, Wongpanit, K, Phonong G, Chaiburi B and Kakaç S 2013 *Proc. Int. Conf. on Micro/Nanoscale Heat and Mass Transfer of the ASME (*Hong Kong, China) 4
[3] Pohl H A 1978, *Dielectrophoretic (*Cambridge United Kingdom)
[4] Fu L M, Lee G B, Lin Y H and Yang R J 2004 *IEEE/ASME Trans. on Mechatronics* **9**(2) pp 377–383
[5] Wang X -B, Hughes M P, Huang Y, Becker F F and Gascoyne P R C 1995 *Biochimica et Biophysica Acta* **1243** pp 185–194
[6] Yang S, Ündar A, Zahn J D 2006, *Lab Chip* **6** pp 871–880
[7] Cetin B, Li D 2009 *Electrophoresis* **30** pp 3124–3133
[8] Cetin B, Kang Y, Wu Z and Li D 2009 *Electrophoresis* **30** pp 766–772
[9] Cetin, B., Li, D., 2010, Lab-on-a-chip Device for Continuous Particle and Cell Separation Based on Electrical Properties via Alternating Current Dielectrophoresis, *Electrophoresis*, **31**, pp 3035–3043.
[10] Kerhoas M K, Sollier E 2013 *Lab Chip* **13** pp 3323–3346
[11] Kang Y, Cetin B, Wu Z and Li D 2009 *Electrochemical Acta* **54** pp 1715–1720
[12] Cetin B and Li D 2010 *Electrophoresis* **32** pp 2410–2427
[13] Jung W C, Heo Y M, Yoon G S, Shin K H, Chang S H, Kim G H and Cho M W 2007*Sensors* **7** pp 1643–1654
[14] Zeinali S, Cetin B, Oliaei S N B and Karpat Y 2015 *Electrophoresis* **36** pp 1432–1442
[15] Tripathi S, Kumar Y V B, Agrawal A, Prabhakar A and Joshi S S 2016 *Scientific Reports 6* **26749** pp 1-15
[16] Ulum M F, Maylina L, Noviana D and Wicaksono D H V 2016 *Lab Chip* **16** pp 1492–1504
[17] Pramuanjaroenkij A, Jankwan J, Noykuan P, Khammai A, Wongpanit K, Maturos T, Lomas T, Tuantranont A and Cetin B 2016 *Proc. Int. Conf. on Mechanical Engineering of the TSME* (Chiang Mai, Thailand)
[18] Pramuanjaroenkij A, Napinij U, Teingtit W, Boonthueng S, Maturos T, Lomas T, Tuantranont A and Kakaç S 2011 *Proc. Int. Conf. on Mechanical Engineering of the TSME* (Krabi, Thailand) 2
[19] Pramuanjaroenkij A, Thanomsit K, Chaiburi B, Phonong G, Khanaphol A, Phutthasawong W, Maturos T, Lomas T and Tuantranont A 2012 *Proc. Int. Conf. on Mechanical Engineering* (Chiang Rai, Thailand)
[20] Bunta J, Thiangpadung J, Sansaradee S 2015 *The Comparison Study on Lab-on-a-chip Fabrication from Different Molds, Senior Project*, (Department of Mechanical and Manufacturing Engineering, Faculty of Science and Engineering, Kasetsart University Chalermphrakiat Sakon Nakhon Province Campus, Thailand)

Mass diffusion coefficient measurement for vitreous humor using FEM and MRI

Komsan Rattanakijsuntorn[1,*], Anita Penkova[2] and Satwindar S. Sadhal[2]

[1]Faculty of Engineering, Department of Mechanical Engineering, Ubon Ratchathani University, 85 Sathonlamark Road, Warinchamrap, Ubon Ratchathani 34190, Thailand
[2]Aerospace & Mechanical Engineering Department, University of Southern California, Los Angeles, CA 90089-1453, United States

* Corresponding Author: komsan.r@ubu.ac.th

Abstract. In early studies, the 'contour method' for determining the diffusion coefficient of the vitreous humor was developed. This technique relied on careful injection of an MRI contrast agent (surrogate drug) into the vitreous humor of fresh bovine eyes, and tracking the contours of the contrast agent in time. In addition, an analytical solution was developed for the theoretical contours built on point source model for the injected surrogate drug. The match between theoretical and experimental contours as a least square fit, while floating the diffusion coefficient, led to the value of the diffusion coefficient. This method had its limitation that the initial injection of the surrogate had to be spherical or ellipsoidal because of the analytical result based on the point-source model. With a new finite element model for the analysis in this study, the technique is much less restrictive and handles irregular shapes of the initial bolus. The fresh bovine eyes were used for drug diffusion study in the vitreous and three contrast agents of different molecular masses: gadolinium-diethylenetriaminepentaacetic acid (Gd-DTPA, 938 Da), non-ionic gadoteridol (Prohance, 559 Da), and bovine albumin conjugated with gadolinium (Galbumin, 74 kDa) were used as drug surrogates to visualize the diffusion process by MRI. The 3D finite element model was developed to determine the diffusion coefficients of these surrogates with the images from MRI. This method can be used for other types of bioporous media provided the concentration profile can be visualized (by methods such as MRI or fluorescence).

1. Introduction

During the past decades, many researchers have attempted to develop ocular drug delivery systems. Intravitreal drug delivery has become a popular method of treatment of many retinal diseases, commonly including AMD, Diabetic Retinopathy, and Retinal Vein Occlusions. While various drugs have been successfully developed, delivering drugs to the retina requires knowledge of the relevant fluid mechanics and transport phenomena. Low drug concentrations are insufficient to treat the retinal disease and high concentrations can carry risks of side effects [1-3]. Therefore, it is crucial to know the drug distribution within the eye following delivery by intravitreal injection.

With the help from the advanced computer software nowadays, ocular drug transport models have been developed to simulate and predict drug distribution in the eye [4-7]. These models require the values of parameters that are used in the governing equations, and one of the most common uses of parameter is drug diffusivity. Therefore, the aim of this work is to develop a technique that delivers the values of diffusion coefficient of drugs in the vitreous from MRI images.

In early study by Penkova et al [8], the 'contour method' for determining the diffusion coefficient of the vitreous humor was developed. This technique relied on careful injection of an MRI contrast agent (surrogate drug) into the vitreous humor of fresh bovine eyes, and tracking the contours of the contrast agent in time. In addition, an analytical solution was developed for the theoretical contours built on point source model for the injected surrogate drug. The match between theoretical and experimental contours as a least square fit, while floating the diffusion coefficient, led to the value of the diffusion coefficient. This method had its limitation that the initial injection of the surrogate had to be spherical or ellipsoidal because of the analytical result based on the point-source model. With a new finite element model for the analysis in this study, the technique is much less restrictive and handles irregular shapes of the initial bolus.

2. Methods and materials

2.1. Apparatus

Whole bovine eyes were prepared for MRI visualization with the same method as described in the previous study [8]. Three gadolinium-based contrast agents: gadolinium-diethylenetriamine pentaacetic acid (Gd-DTPA, 938 Da), nonionic gadoteridol (Prohance, 559 Da), and bovine albumin conjugated with gadolinium (Galbumin, 74 kDa) were used as drug surrogates to model diffusive transport of different molecular weight drugs. Experiments were conducted by injecting 30 μL of the contrast agents in the vitreous of the whole eye. With the concentration measurement from the MRI signal at regular intervals for 2-3 hours, the concentration contours were constructed and matched with the theoretical contours from the finite element analysis. The best fit at different times gave fairly consistent values of the diffusion coefficient.

2.2. Governing equations and model development

A three-dimensional finite element model of the vitreous was developed to help in analyzing MRI images and calculate diffusion coefficients by comparing simulated data to MRI experimental data. To model the drug distribution in the vitreous, the mass transport equation was applied:

$$\frac{\partial c}{\partial t} + v \cdot \nabla c - D\nabla^2 c + q = 0 \tag{1}$$

where c represents the concentration of the drug, D is the diffusion coefficient of the drug in the vitreous, v is the velocity of fluid in the vitreous, and q is the release rate of the drug. Since the experiment was conducted by excluding convection in the system, and we assumed that the drug is neither generated nor degraded within the vitreous, the terms involving with v and q were set to zero. The mass transport equation can be rewritten as:

$$\frac{\partial c}{\partial t} - D\nabla^2 c = 0 \tag{2}$$

For the purpose of modelling, the domain of interest (of the vitreous) in which drugs diffused through was simply defined as a sphere with a radius of 10 mm as illustrated in figure 1(a). The geometry created for the domain of interest, were converted into a total of 105,276 finite element tetrahedral meshes. Since the rate of change of concentration over time near the center of the domain is larger than the outer, mesh size was varied by having finer meshes near the center and coarser meshes near the outer boundary as shown in figure 1(b).

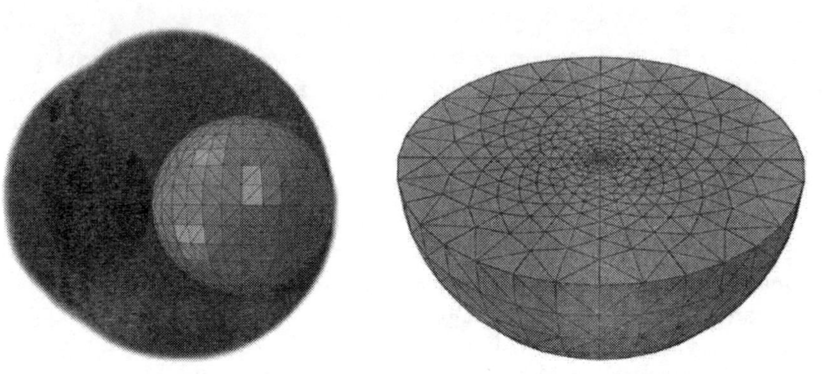

(a) (b)

Figure 1. The domain of interest and the finite element tetrahedral meshes.

2.3. Boundary and initial conditions

The vitreous humor in this study is considered to be a homogeneous substance. The lens is avascular and is considered to be impermeable to fluid flow, so we assumed that the drug surrogate does not penetrate the lens. Similarly, RPE is considered as a major barrier for the retinal delivery of hydrophilic drugs, the same assumption that we made for the lens also applied for the RPE. Since the outer boundary of the domain of interest reaches the regions of lens and RPE membrane, we simplified the boundary condition at the outer surface as zero flux condition. This kind of boundary condition would much simplify the finite element equation by eliminating the flux term as detailed in appendix A. The concentration distribution profiles of the MRI images were directly projected to the nodes in the finite element model as the initial condition for each time point. With this method, we can handle the situation that the concentration distribution profile was not symmetric.

2.4. Solution methodology

The concentration distribution in the vitreous was solved using the finite element method involving linear algebraic equations of the form,

$$[K_C]\left\{\frac{dC}{dt}\right\} + D[K_D]\{C\} = 0, \tag{3}$$

where $[K_C]$ and $[K_D]$ are the global system matrices developed from the Galerkin finite element method in appendix A. The solution employed the 3-D mesh of a total of 105,276 elements, and four nodal tetrahedral elements were used in the 3-D continuum domain. For the time discretization, we applied the well-known theta-method, which results in the equation

$$[K_C]\left\{\frac{C^{n+1} - C^n}{\Delta t}\right\} + D[K_D]\{\theta C^{n+1} + (1-\theta)C^n\} = 0 \tag{4}$$

here C^n is the concentration distribution at a specific time point, C^{n+1} is the concentration distribution at the later time point, and the parameter θ is related to the applied numerical method. It is worth emphasizing that for $\theta = 0.5$, the method yields the Crank-Nicolson implicit method which has higher accuracy and unconditional stable for the time discretization [9]. Substituting $\theta = 0.5$ in equation (4), then it can be rewritten as,

$$([K_C] + 0.5\Delta t D[K_D])\{C^{n+1}\} = ([K_C] - 0.5\Delta t D[K_D])\{C^n\} \tag{5}$$

In this equation, the concentration profile C^n was known by projecting the concentration values of a specific time point from MRI images onto the finite element domain. The unknown parameter, *D*, was floated in order to get the calculated C^{n+1} which represents the concentration at the later time step. The value of *D* between two consecutive time points can be obtained by least square fitting of C^{n+1} and the concentration profile of the later time point from MRI images.

2.5. Model validation

In order to verify the finite element diffusion model developed in the previous section, the solutions obtained from the model was compared with the exact solutions. The exact solution for a spherical bolus of radius r_0 in a spherical shell of radius *R* was derived in the previous study [8] and shown in equation (6).

$$c(r,t) = c_0 \left[\frac{r_0^3}{R^3} + 2 \sum_{n=1}^{\infty} \frac{\sin(\lambda_n r_0) - \lambda_n r_0 \cos(\lambda_n r_0)}{\lambda_n R \, sin^2(\lambda_n R)} e^{-D\lambda_n^2 t} \frac{\sin(\lambda_n r)}{\lambda_n r} \right] \tag{6}$$

where c_0 is the initial concentration of the drug deposited, *r* is the radial distance from the point of injection, *D* is the diffusion coefficient, *t* is the time, *c*(*r*,*t*) is the concentration, and λ_n is the set of constants satisfying the transcendental equation.

$$\lambda_n R = \tan(\lambda_n R) \tag{7}$$

For the model validation, we use the following parameters: $r_0 = 2$ mm, $R = 10$ mm, and $c_0 = 0.05$ in both finite element model and the exact solution, to compare the concentration values. Figure 2 shows the comparison of concentration distribution profiles from the developed model and the exact solution under the same initial and boundary conditions. The finite element solutions along the center line at a specific time (t=40 min) have a strong agreement with the exact solutions.

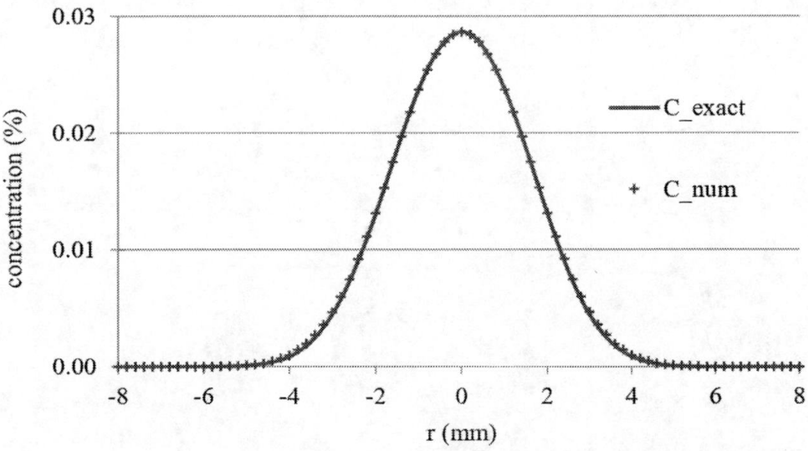

Figure 2. The comparison of concentration distribution profiles from the finite element model and the exact solution.

3. Results

Following the method of diffusion coefficient measurement by using MRI developed earlier, contours of constant concentration were established for different drug surrogates in each eye, and the comparison of these with the corresponding finite element calculation was shown in figure 3 where the (x, y, z) coordinate system in relation to the eye is shown in figure 4. While several time points for each eye were recorded and used in the analysis, only a selected example for each contrast agent is shown for illustrative purposes. The diffusion coefficients values of each drug surrogates acquired from the finite element method were presented in figure 5. We have made a comparison of previously reported D values for Gd-DTPA from the analytical method in the previous study [8] with the current method (finite element), and found that they were very consistent. The diffusion coefficient was $(3.040 \pm 0.274) \times 10^{-6}$ cm^2/s from the analytical method, and $(3.069 \pm 0.237) \times 10^{-6}$ cm^2/s from the finite element method. By comparing the diffusion coefficients between Gd-DTPA (molecular weight 938 Da), Prohance (molecular weight 559 Da), and Galbumin (molecular weight around 74k Da), we found that Galbumin had the lowest diffusivity, while Gd-DTPA had the highest diffusivity in the vitreous.

(A)
(B)
(C)

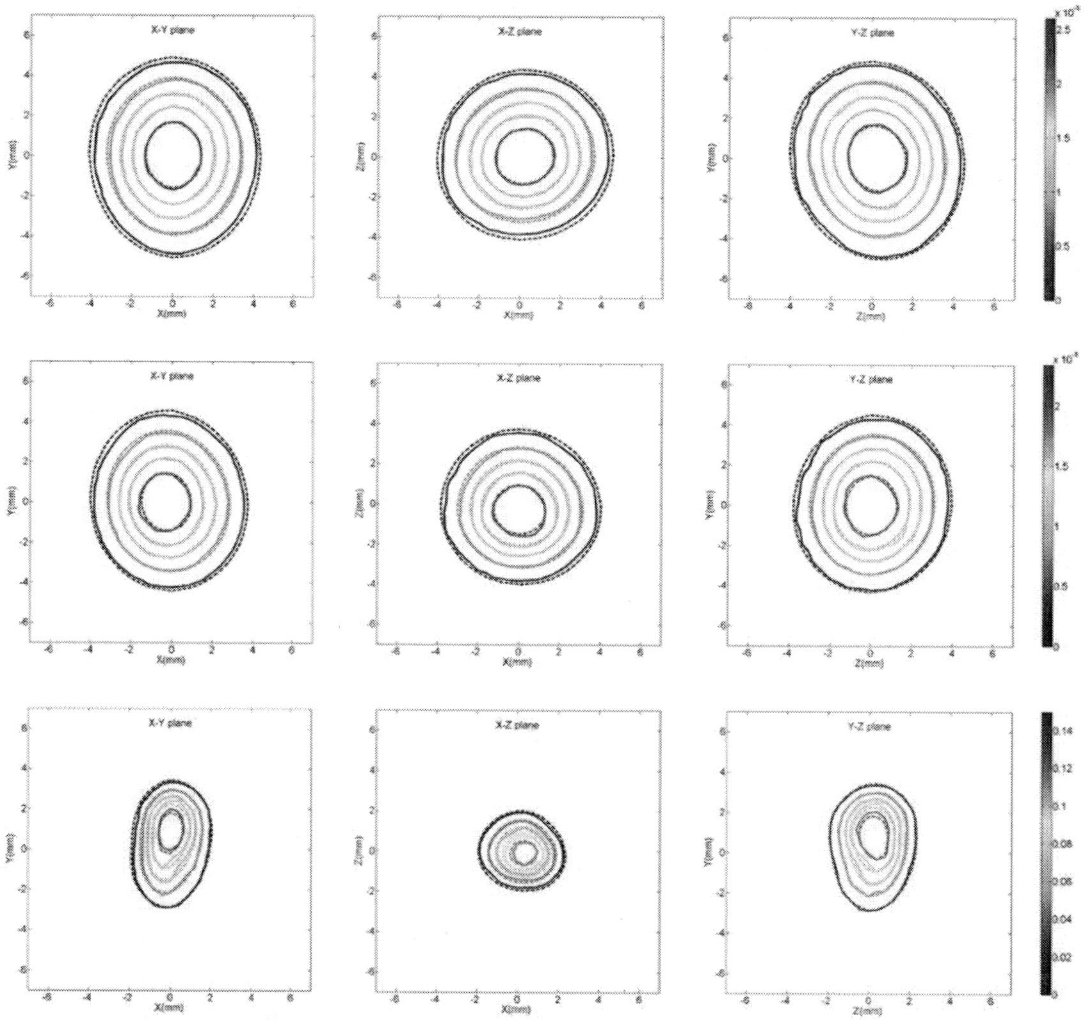

Figure 3. Contours of constant concentration in (A) Gd-DTPA at t=85 min, (B) Prohance at t=88 min, and (C) Galbumin at t=185 min (—— : measurement, ---- : finite element).

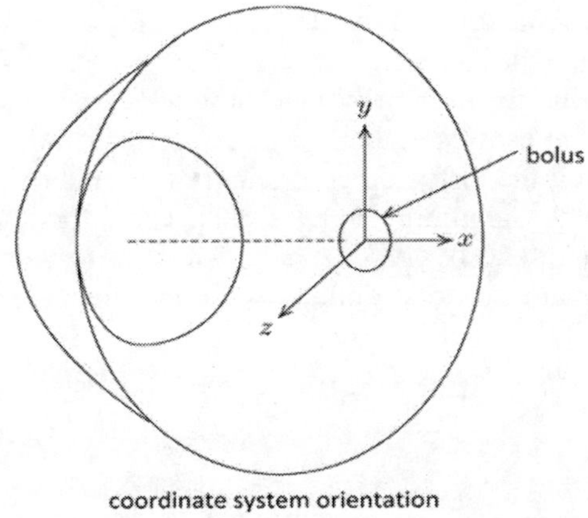

Figure 4. Coordinate system orientation with respect to the eye.
The origin located at the center of the bolus.

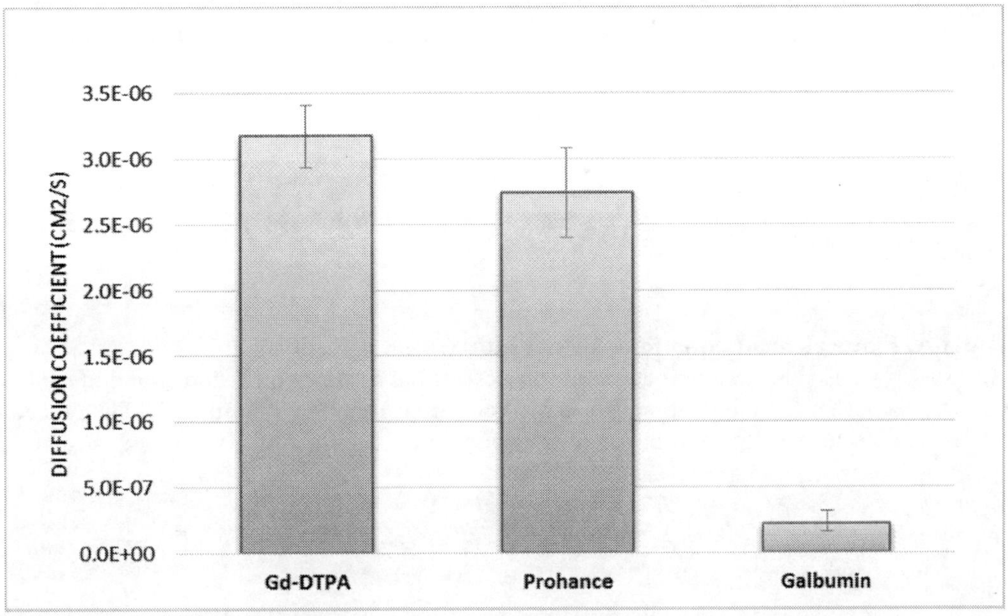

3.069±0.237 $\times 10^{-6}$ cm²/s	2.739±0.340 $\times 10^{-6}$ cm²/s	2.271±0.209 $\times 10^{-7}$ cm²/s

Figure 5. Diffusion coefficients of drug surrogates in bovine eyes.

4. Discussion

Early investigations of diffusion coefficients of different MRI contrast agents had been conducted by Gillis et al [10]. They measured diffusivity of a negatively charged contrast agent Gd-DTPA and the non-ionic gadoteridol, Prohance, in control cartilages using MRI. The diffusion coefficients for Gd-DTPA and Prohance were reported by Gillis et al as $(1.84 \pm 0.12) \times 10^{-6}$ cm^2/s and $(1.55 \pm 0.22) \times 10^{-6}$ cm^2/s respectively. In our study, we found that the diffusion coefficient in the vitreous of a smaller molecule, Prohance was $(2.739 \pm 0.340) \times 10^{-6}$ cm^2/s which was lower than the diffusion coefficient of a larger molecule, Gd-DTPA. This implied that the effect of negative net charge on Gd-DTPA is likely to enhance the diffusivity in the vitreous.

Galbumin used in this experiment is a bovine albumin conjugated with gadolinium which has negative charge. Molokhia et al estimated the free aqueous diffusion coefficient of Galbumin in water at 37 °C as 8×10^{-7} cm^2/s using the diffusion coefficient of bovine serum albumin (BSA), viscosity of water, and molecular weights of Galbumin and BSA [11]. The diffusion coefficient of Galbumin in this study was found to be $(2.271 \pm 0.209) \times 10^{-7}$ cm^2/s which is in the same order as the one estimated by Molokhia. Our lower value can also be attributed to the fact that we conducted experiments at room temperature. To compare with the smaller negative charge Gd-DTPA, Galbumin has a lower diffusion coefficient as expected.

5. Conclusion

The contour method recently developed is a very useful technique for the diffusion coefficient measurement in vitreous humor. With a new finite element model for the analysis, it can handle arbitrary shapes of the concentration contours, and we are able to show a comparison of diffusion coefficients of Gd-DTPA with Prohance and Galbumin. It has been found that both molecular weight and net charge of drugs affect the diffusion process in the vitreous humor. This method does not only work for the measurement in the vitreous humor, but also can be applied for other types of tissues in which the concentration profile can be visualized by MRI or fluorescence. The fact that we can now handle irregular shapes of the contours, the technique can be opened up to measure parameters that include convective transport effects as well.

Acknowledgement

The work in this paper is supported by NIH (National Institute of Health) under award number 5R01EY026599-02.

Appendix A. Finite element equations for the intravitreal diffusion

This section provides a detailed derivation of 'weak form' of the drug diffusion equation using Galerkin method. Basic concept and application of this method can be found in literature [9]. The procedure starts by considering the suitable form of the mass transport equation which is given by equation (2),

$$\frac{\partial C}{\partial t} - D\left(\frac{\partial^2 C}{\partial x^2} + \frac{\partial^2 C}{\partial y^2} + \frac{\partial^2 C}{\partial z^2}\right) = 0 \qquad \text{(A1)}$$

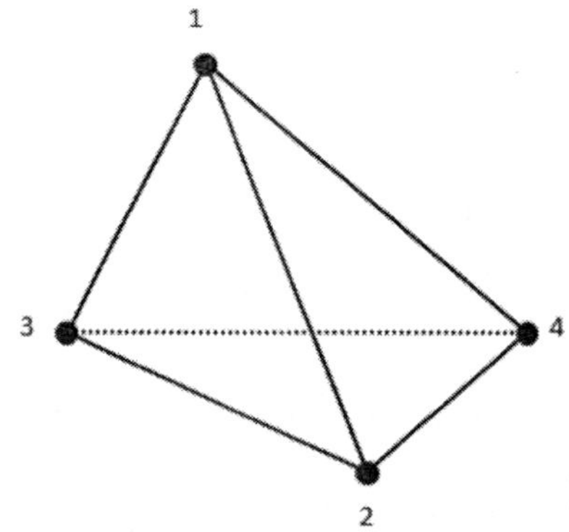

Figure A1. Linear tetrahedral element.

IOP Conf. Series: Materials Science and Engineering **297** (2018) 012024 doi:10.1088/1757-899X/297/1/012024

A domain V is divided into linear tetrahedral finite elements connected at nodes as shown in figure A1. Shape functions Ni are used for interpolation of concentration inside a finite element,

$$C = [N]\{C\}$$

where

$$[N] = [N_1 \; N_2 \; N_3 \; N_4]$$

$$\{C\} = \begin{Bmatrix} C_1 \\ C_2 \\ C_3 \\ C_4 \end{Bmatrix}$$

Here {C} is a vector of concentrations at nodes and [N] is a matrix of shape functions. Using the Galerkin method, we can write equation (A1) in the following form:

$$\int_V N_i \left[\frac{\partial C}{\partial t} - D\left\{\frac{\partial}{\partial x}\left(\frac{\partial C}{\partial x}\right) + \frac{\partial}{\partial y}\left(\frac{\partial C}{\partial y}\right) + \frac{\partial}{\partial z}\left(\frac{\partial C}{\partial z}\right)\right\}\right] dV = 0 \tag{A2}$$

Applying the divergence theorem to the terms in the bracket {}, we arrive at the relations:

$$\int_V N_i \frac{\partial C}{\partial t} dV + D \int_V \left[\frac{\partial N_i}{\partial x}\frac{\partial C}{\partial x} + \frac{\partial N_i}{\partial y}\frac{\partial C}{\partial y} + \frac{\partial N_i}{\partial z}\frac{\partial C}{\partial z}\right] dV$$

$$- D \int_S \left[N_i \frac{\partial C}{\partial x} n_x + N_i \frac{\partial C}{\partial y} n_y + N_i \frac{\partial C}{\partial z} n_z\right] dS = 0 \tag{A3}$$

where n_x , n_y , and n_z are outer normal components to the surface of the body. Since we applied the zero flux boundary condition at the surface of the body, the last terms in the integral of S is zero. Now the equation (A3) can be simplified as,

$$\left(\int_V \{N\}[N] dV\right)\left\{\frac{dC}{dt}\right\} + D\left(\int_V \left\{\frac{\partial N}{\partial x}\right\}\left[\frac{\partial N}{\partial x}\right] dV + \int_V \left\{\frac{\partial N}{\partial y}\right\}\left[\frac{\partial N}{\partial y}\right] dV + \int_V \left\{\frac{\partial N}{\partial z}\right\}\left[\frac{\partial N}{\partial z}\right] dV\right)\{C\} = 0$$

Let's denote

$$[K_C] = \int_V \{N\}[N] dV$$

$$[K_D] = \int_V \left\{\frac{\partial N}{\partial x}\right\}\left[\frac{\partial N}{\partial x}\right] dV + \int_V \left\{\frac{\partial N}{\partial y}\right\}\left[\frac{\partial N}{\partial y}\right] dV + \int_V \left\{\frac{\partial N}{\partial z}\right\}\left[\frac{\partial N}{\partial z}\right] dV$$

The discretized finite element equation for the mass transport equation can be rewritten as in equation (3),

IOP Conf. Series: Materials Science and Engineering **297** (2018) 012024 doi:10.1088/1757-899X/297/1/012024

$$[K_C]\left\{\frac{dC}{dt}\right\} + D[K_D]\{C\} = 0 \tag{A4}$$

References

[1] Pflugfelder S C, Hernandez E, Fliesler S J, Alvarez J, Pflugfelder M E and Forster R K 1987 *Arch. Ophthalmol.* **105** pp 831–837

[2] Ozkiris A and Erkilic K 2005 *Can. J. Ophthalmol.* **40** pp 63–68

[3] Bhavsar A, Ip M and Glassman A 2007 *Am. J. Ophthalmol.* **144** pp 454–456

[4] Stay M, Xu J, Randolph T and Barocas V 2003 *Pharm Res.* **20**(1) pp 96–102

[5] Balachandran R and Barocas V 200 *Pharma. Res.* **25** pp 2685–2696.

[6] Kathawate J and Acharya S 2008 *Int.J. Heat Mass Transfer* **51** 5598–5609.

[7] Haghjou N, Abdekhodaie M J, Cheng Y L and Saadatmand M 2011 *World Acad. Sci. Eng. Technol.* **77** p 706

[8] Penkova A, Rattanakijsuntorn K, Sadhal S S, Tang Y, Moats R and Hughes P M 2014 *Int.J. Heat Mass Transfer* **70** pp 504-514

[9] Lewis R W, Nithiarasu P and Seetharamu K N 2004 *Fundamentals of the Finite Element Method for Heat and Fluid Flow* (NJ: Wiley)

[10] Gillis A, Gray M and Burstein D 2002 *Magnetic Resonance in Medicine.* **48**(6) pp 1068-1071

[11] Molokhia S A, Jeong E K, Higuchi W E and Li S K 2009 *Exp. Eye Res.* **88**(3) pp 418–425

Improved design of a cone-shaped rotating disk for shear force loading in a cell culture plate

Boonrit Keawprachum[1], Nuttapol Limjeerajarus[2], Chalida Nakalekha Limjeerajarus[3] and Pornsak Srisungsitthisunti [1,*]

[1] Production Engineering Program, Faculty of Engineering, King Mongkut's University of Technology North Bangkok 1518 Pracharat I Rd., Wongsawang, Bangsue, Bangkok 10800, THAILAND
[2] Research Center for Advanced Energy Technology, Faculty of Engineering, Thai–Nichi Institute of Technology 1771/1 Pattanakarn, Suan Luang, Bangkok 10250, THAILAND
[3] Department of Physiology, Faculty of Dentistry, Chulalongkorn University, 34 Henri Dunant Rd, Pathumwan, Bangkok 10330, THAILAND

*Corresponding Author: Email: pornsaks@kmutnb.ac.th, Tel.: +66-2913-2500-24 Ext. 8218, Fax.: +66-2587-0029

Abstract. In our previous study, a cone-shaped rotating disk had been designed and proposed for generating shear force on the cell in a cell culture plate. This study aims to improve the design of the rotating disk that could provide a better uniformity of shear stress distribution. The top of the cone was designed to be trimmed off to obtain a flat head area. The effect of tilt angle (θ) was numerically studied using computational fluid dynamics (CFD) technique in ANSYS-Fluent software. The results revealed that for 500 rpm, the new designed rotating disk with a height of cone-shaped top to the plate bottom h = 1 mm and $\theta = 25°$ provided the best uniformity of 0.820 which was better than that of the previously designed.

1. Introduction

Mechanical force loading like shear force can affect dental cell growth and regeneration. Shear stress, produced by torsion and shear between the teeth and alveolar bone, is the most common type of mechanical stress induced on periodontal ligament (PDL) cells [1]. PDL is also regarded as a porous tissue surrounded by a massive vascular network. When pressure builds up between the teeth and alveolar bone, pressurized interstitial fluid produces fluid shear stress at the PDL cell membrane [2]. Therefore, shear stress was thought to be the most appropriate biomechanical stress for PDL tissue engineering [1]. In previous studies, the proper shear stress required to apply to PDL cells was in the range of 0.1-6 Pa [3, 4].

Applying shear force onto the cells can be done in many ways. Force loading via angular velocity of the medium in a cell culture plate is one of the appropriate techniques as the loading apparatus does not directly contact the cells. However, the local angular velocity of the medium varies proportionally with the radius length of the rotating disk. Thus, the shear force exerted on the cell is not uniformly distributed throughout the bottom of the culture plate. To effectively study the effects of shear force on cell regeneration and its gene expression, the rotating disk must be carefully designed so as to obtain the shear force uniform distribution.

However, the cells are at the bottom of the culture plate and the rotating disk rotates at the surface of the medium. Therefore, this problem becomes three dimensional velocity profiles of which the angular velocity at the bottom of the culture plate is not linearly proportional to the radius of the rotating disk. Hence, the numerical analysis is needed to estimate the shear stress distribution at the bottom.

In our previous work, a cone-shaped rotating disk design with different heights of cone-shaped tip to the bottom of the cell culture plate, tilt angles, medium volumes had been numerically investigated. The simulation results suggested that for 5 cm^3 medium, the height of 2.47 mm and the tilt angle of 15° should be used in order to obtain the best uniformity of shear stress distribution at the plate bottom of 0.817 at 500 rpm [5]. Nevertheless, considering the geometry of the cone shape, the uniformity of the shear stress distribution can be improved by reducing the difference in radius of the top and the bottom of the cone. This can be done by trimming off the top of the cone by which the radius of the top will be increased at a closer distance to the bottom of the culture plate, as compared with the ordinary cone shape.

In this study, the height of the cone-shaped top (flat head) to the bottom of the cell culture plate (h) was fixed at 1 mm. This was because the smaller height provided the higher shear stress at the same rotational speed, and most importantly, the required volumes of the medium and the costly growth factor would become less. However, the height lower than 1 mm would be improper as the cells may be peeled off from the scaffold. Thus, the objective of this research was to numerically study the effect of the tilt angle of a top-flat cone-shaped rotating disk (new design) on the shear stress distribution, so that an optimum tilt angle under a given condition, i.e., h = 1 mm and the disk diameter of 34 mm, can be suggested with higher uniformity.

2. Model Development

The top of the cone was trimmed off to get a flat head (figure 1). The fluid model used to analyze the shear stress distribution was developed based on the finite volume method using ANSYS FLUENT software. The governing equations used in this work included the steady-state mass conservation equation and the steady-state Navier-Stoke's equations which are coupled and can be expressed in the general form as follows.

$$\nabla \cdot (\rho\phi\vec{V}) = \nabla \cdot (\Gamma_\phi \nabla\phi) + S_\phi \tag{1}$$

where ρ denotes to the fluid density, ϕ is the transported quantity (mass and momentum), $\vec{V}$ is the velocity vector, Γ_ϕ is the transportation quantity diffusivity, and S_ϕ is the source term.

The fluid model was discretized into 291,132 computational f polyhedral cells. The density and viscosity of the medium fluid were 1012.95 $kg{\bullet}m^{-3}$ and 0.00282 $kg{\bullet}m^{-1}{\bullet}s^{-1}$, respectively.

The assumptions conditions followed the previous work [5] which following: (i) the system operated under the steady state and isothermal condition, (ii) the fluid velocity at all fluid-solid boundaries was equal to that of the solid boundary (no-slip condition), (iii) fluid medium was isotropic and homogeneous, and (iv) the cell height at the bottom of the cell culture plate was negligible.

The SIMPLE (Semi-Implicit Method for Pressure Linked Equations) algorithm was used to solve the Navier-Stokes equations iteratively. The calculation in double digit precision was used to get highly accurate simulation results. To prevent the oscillation of the solution, the second-order upwind discretization scheme was selected. The under-relaxation factors of the pressure, density, body forces, and momentum were set at 0.27, 1, 1, and 0.55, respectively. The solutions were iterated until the specified convergence criterion of 10^{-5} was achieved.

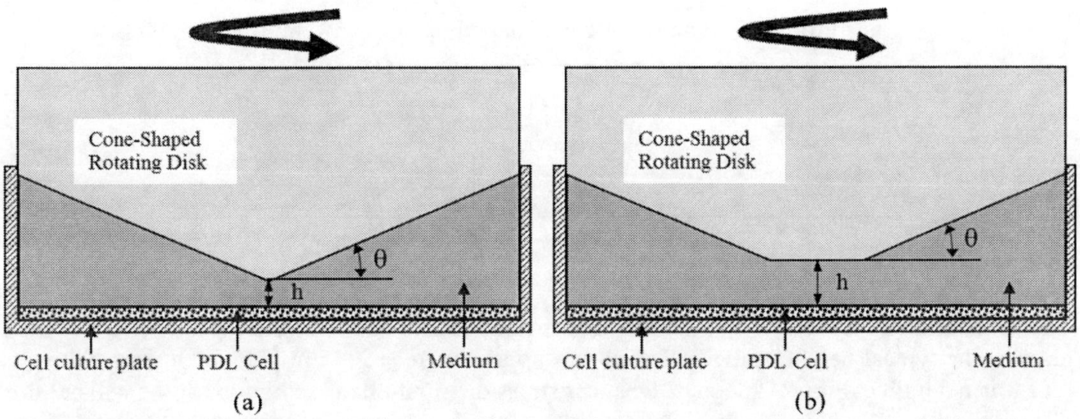

Figure 1 Schematic diagrams of a) the previous and b) the new cone-shaped rotating disk design, respectively

In this study, the diameter of the rotating disk was set at 34 mm according to the standard size of cell culture plate (34 mm diameter ×10 mm depth) and the height *h* was set at 1 mm. The tilt angle was set at 15, 20, 25 and 30 degrees. These criteria were selected based on the fact that the top of the rotating disk must not touch the bottom of the plate, and the surface of the respective medium must touch the cone bottom but must not overflow when the rotating disk is dipped into the medium. Table 1 presents the conditions of the tilt angles and their related medium volumes used in this simulation. The rotational speeds were varied from 100 to 500 rpm. In order to correctly and systematically analyze how uniform the stress distribution was in each case, an index called "uniformity" was introduced. The uniformity can be calculated using the equations (2) and (3). The higher value of uniformity means the shear stress is more evenly distributed in the entire bottom area.

The uniformity indexer presents how a specified field variable varies over a surface, where a value of 1 indicates the highest uniformity. The uniformity index can be weighted by area or mass: the area-weighted uniformity index captures the variation of the quantity (for example, the species concentration), whereas the mass-weighted uniformity index captures the variation of the flux (for example, the species flux) [6].

The area-weighted uniformity index γ_a of a specified variable φ is calculated using the following equation:

$$\gamma_a = 1 - \frac{\sum_{i=1}^{n}\left[\left(\left|\varphi_i - \bar{\varphi}_a\right|\right)A_i\right]}{2\left|\bar{\varphi}_a\right|\sum_{i=1}^{n} A_i} \tag{2}$$

where i is the facet index of a surface with n facets, and $\bar{\varphi}_a$ is the average value of the field variable over the surface:

$$\bar{\varphi}_a = \frac{\sum_{i=1}^{n}\varphi_i A_i}{\sum_{i=1}^{n} A_i} \tag{3}$$

Table 1 The simulation conditions of the tilt angles and their respective medium volumes

h (mm)	Tilt angle of cone-shaped (θ)	Medium volume (cm^3)
1	15	2.8
	20	3.7
	25	4.8
	30	5.9

3 Results and Discussion

The investigation focused on the shear stress distribution and its uniformity at the bottom of the cell culture plate as the typical height of the PDL cell was approximately 500 nm [7] which was relatively small and thus it is negligible. Figure 2 depicts the shear stress distribution at the bottom of the cell culture plate with different tilt angles. The results indicated that at the same rotational speed (500 rpm), the shear stress distribution seemed to be more uniform as the tilt angle increased. Thus, we further investigated the area-averaged shear stress and its uniformity at various rotational speeds. Figures 3 and 4 display the area-averaged shear stress and the uniformity index simulated with different tilt angles in the rotational speed range of 100-500 rpm. Obviously, the area-averaged shear stress was directly proportional to the rotational speed. For the effect of tilt angle, the average shear stress increased with the decreasing tilt angle decreased. This was because with the narrower tilt angle, the overall surface of the cone-shaped rotating disk rotated closer to the plate bottom, resulting in greater average angular velocity. Nonetheless, the increasing rate of the average shear stress slowed down as the tilt angle became lesser.

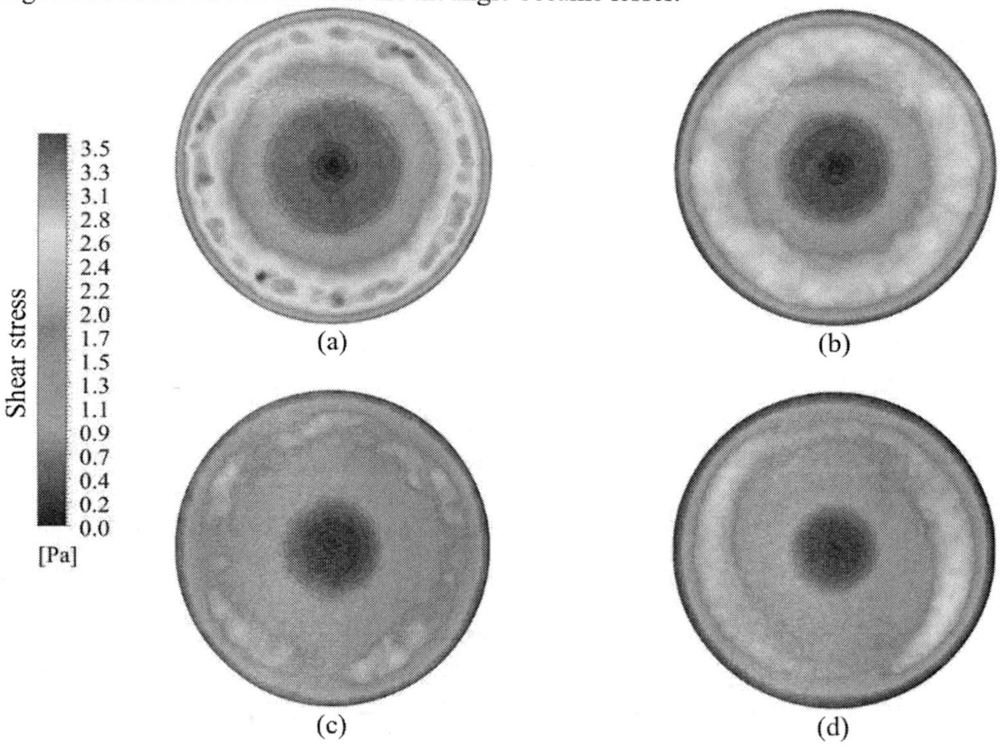

Figure 2 Shear stress distribution at the bottom of the cell culture plate at 500 rpm with the tilt angles of a) 15°, b) 20°, c) 25° and d) 30°

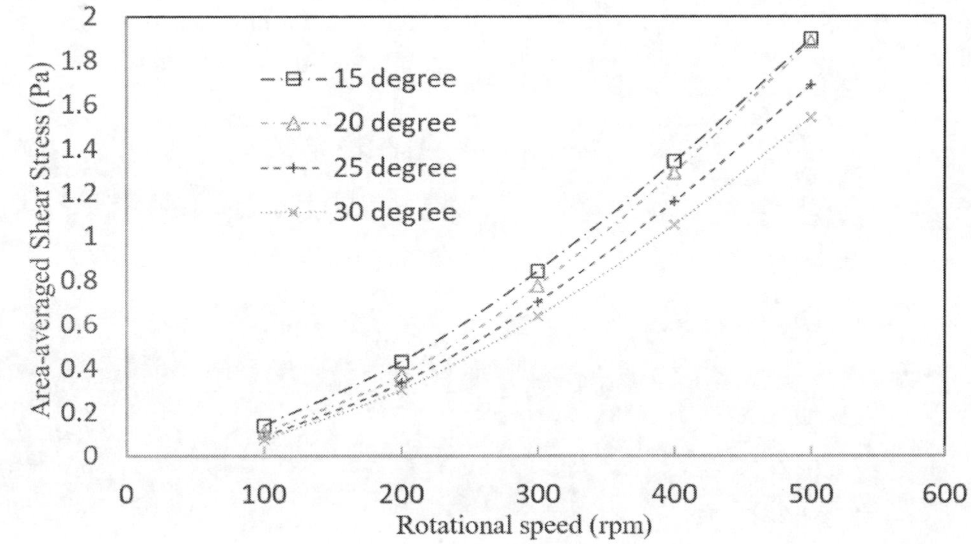

Figure 3 The relationship between the average shear stress and the tilt angle at various rotational speeds

For the uniformity, the flathead cone-shape rotating disks could produce the uniformity in the range of 0.73-0.82, which was quite good, considering the maximum value of 1.00. In contrast to the results of the average shear stress, the uniformity was lower when the tilt angles became narrower. This was due to the closer surface of the rotating disk to the plate bottom that contributed to the higher angular velocity. As a result, the difference in local shear stress around the center and the edge of the rotating disk became greater, leading to lower uniformity. Therefore, to select an optimum design of the flathead cone-shaped rotating disk, the balance in average shear stress and uniformity must be taken into consideration.

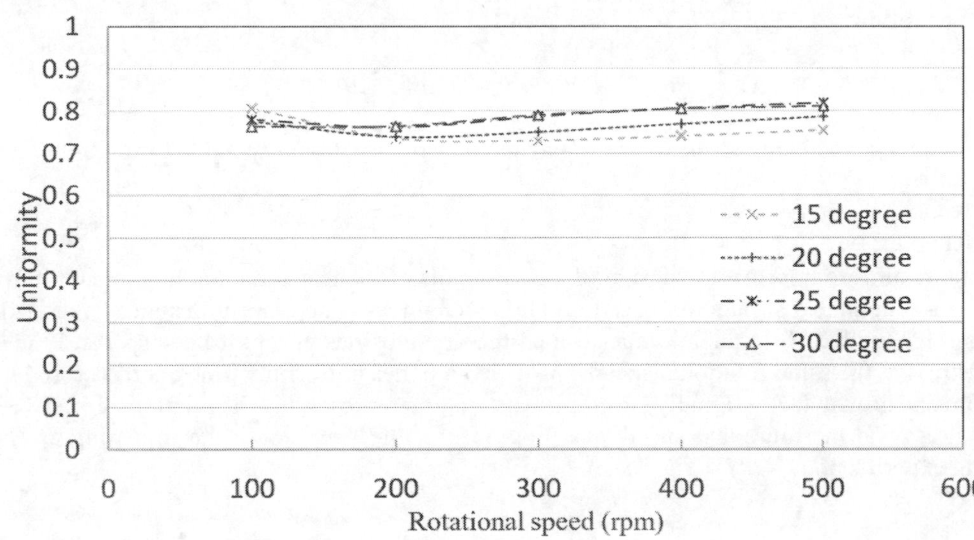

Figure 4 The relationship between the uniformity and the tilt angle at various rotational speeds

From figure 4, the uniformity of 25 and 30 degrees were almost identical at any rotational speeds. The maximum uniformity of 0.820 was reached with the 25 degree at 500 rpm. Therefore, the optimum condition suggested in this study was of $\theta = 25°$, as it had a merit of using lesser medium volume vis-à-vis the 30° (see table 1), while maintaining the same high uniformity at any rotational speeds. With this condition, the new design (flat head) was successfully improved in all viewpoints as compared with the previous design (normal cone head [5]) as at the same rotational speed (500 rpm), it generated greater average shear stress (1.69 Pa vs. 1.62 Pa), provided higher uniformity (0.820 vs. 0.816), and used less medium (4.8 cm^3 vs. 5.0 cm).

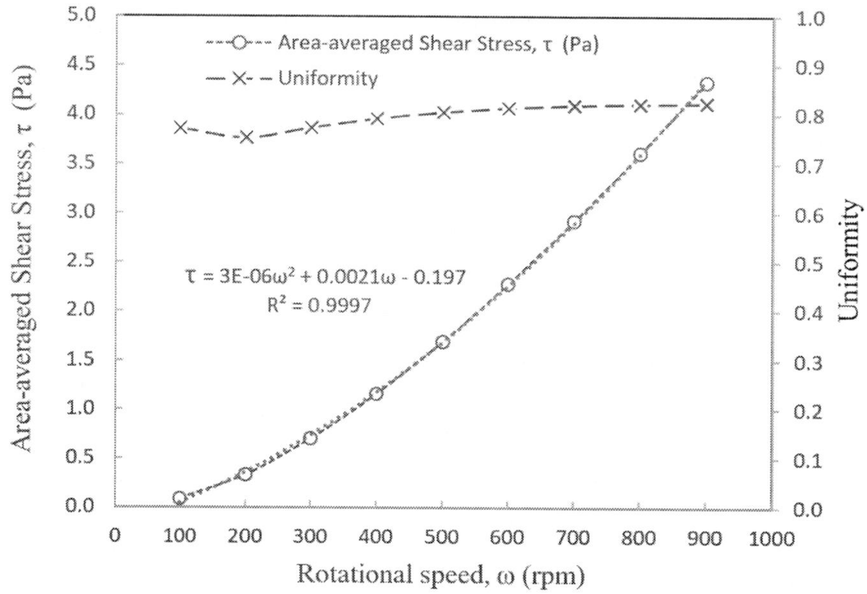

Figure 5 Relationship between the average shear stress and the rotational speed of the suggested design

Consequently, we further investigated the average shear stress generated by the suggested flathead cone-shaped ($\theta = 25°$) at higher rotational speeds up to 900 rpm. This was to determine an empirical equation that facilitates a user to set up their experiments easily. From figure 5, the average shear stress can be expressed as a function of rotational speed using a second degree polynomial function with a high R^2 of 0.9997 as follows:

$$\tau = 3E\text{-}06\omega^2 + 0.0021\omega - 0.197 \quad (4)$$

where τ is the area-averaged shear stress (Pa) and ω is the rotational speed between 0-900 rpm.

4 Conclusion

This study aims to improve the design of a rotating disk for shear force loading on cell in the cell culture plate using numerical simulation based on CFD techniques. The current design of flathead cone-shaped rotating disk with 25° tilt angle outperformed the previous design in all key aspects. It achieved greater shear stress at the same rotational speeds, provided a better uniformity index and required less amount of medium and growth factor. Finally, an empirical equation representing the relationship between the average shear stress and the rotational speed was proposed for the simplicity sake in setting up the shear force loading experiment.

References

[1] Kim S G, Kim S G, Viechnicki B, Kim S and Nah, H. D 2011 *Journal of clinical* **38** pp 1130–1136
[2] Zheng L, Huang Y, Song W, Gong X, Liu M, Jia X, Zhou G, Chen L, Li A and Fan Y 2012 *Journal of Biomechanics* **45** pp 2368–2375
[3] Oest M E, Miller M A, Howard K I and Mann K A, 2014 *Journal of biomechanics* **47**(2) pp 518-525

[4] Kamioka H, Sugawara Y, Murshid S A, Ishihara Y, Honjo T and Takano-Yamamoto T 2006 *Journal of bone and mineral research* **21**(7) pp 1012-1021
[5] Boonrit Keawprachum, Pornsak Srisungsitthisunti, and Nuttapol Limjeerajarus 2016 *Proc. Int. Conf. TSME 2016* (Chaingmai, Thailand) 7
[6] ANSYS® Fluent, Release 16.2 *ANSYS Workbench Help* (ANSYS, Inc., USA)
[7] Kuru L, Parkar M H, Griffiths G S, Newman H N and Olsen I 1998 *J. Dent. Res.* **77** pp 555–564

Comparison of two methods to determine fan performance curves using computational fluid dynamics

Patinya Onma* and Tonkid Chantrasmi*

Department of Mechanical and Aerospace Engineering, King Mongkut's University of Technology North Bangkok, Bangkok 10800, Thailand

*Corresponding Author: Patinya@artith.com, tonkid.c@eng.kmutnb.ac.th

Abstract. This work investigates a systematic numerical approach that employs Computational Fluid Dynamics (CFD) to obtain performance curves of a backward-curved centrifugal fan. Generating the performance curves requires a number of three-dimensional simulations with varying system loads at a fixed rotational speed. Two methods were used and their results compared to experimental data. The first method incrementally changes the mass flow late through the inlet boundary condition while the second method utilizes a series of meshes representing the physical damper blade at various angles. The generated performance curves from both methods are compared with an experiment setup in accordance with the AMCA fan performance testing standard.

1. Introduction

Centrifugal fans or blowers are a type of turbo machinery crucial to many applications in a wide variety of industries as well as residential areas. All of them are used to move air or gas from one place to another, but it is important to design and/or select the fans that are appropriate to the intended applications. For instances, a majority of fans in chemical industries must be made of special materials to withstand high-temperature, corrosive or hazardous gases [1]. Some fans will be used to transport air with particles such as grain in rice mills, kernels in corn factory farms, and sawdust in wood factories. Besides the above specifications, one has to consider the capacity and efficiency of the centrifugal fans according to the system load in the desired applications. As a selection guideline, one can consider what type of blade shapes will be likely appropriate first. In general there are six blade types: airfoil (AF), backward-curved (BC), backward-inclined (BI), radial-tip (RT), forward-curved (FC) and radial-blade (RB) [2].

The first three categories generally have high efficiency (around 80 percent) and their performance curves are stable over a relatively wide range; however, they are primarily suitable only for clean air or clean gas or perhaps air with small non-sticky dust particles. Centrifugal fans with FC blades are well suited for handling high flow rates and low heads. Their performance curves are typically stable from free-delivery point to peak-pressure point, but there is usually a pressure dip slightly to the left of the peak. As for centrifugal fans with RB and RT blades, they have moderate to low efficiency (around 60-75 percent) but can generate a high head than other types. They can be made to handle a wide range of contaminants and can be easily modified or designed to accommodate high temperature.

In this work, a centrifugal fan with BC blades is considered with the main focus on the two primary performance variables, namely volume flow rate and head (or fan pressure). Other variables such as power output and mechanical efficiency can, of course, be calculated afterward for comparison. These two variables at a series of operating points will be obtained both by simulations using Computational Fluid Dynamics (CFD) as well as by experiments according to industrial fan testing standards [3-5]. In this case, AMCA fan testing standard was chosen for the experiment but other standards are quite similar and comparable results can be expected. The goal is to validate some simulation techniques, establish a trustworthy range of operating conditions for each of them, and recommend an appropriate and efficient simulation technique for generating a fan performance curve.

Recently, many research works employed CFD to simulate centrifugal fans usually with a goal to optimize certain design aspects of the fan such as the number of the blades and their physical dimensions [6-10]. Some researches focused on varying the rotational speed of the fan [7-8, 11] while other may study the effects of fan blade types in some specific applications [11-12]. Many of these works performed the simulations only at the design points [9, 13-15], while a few generated an entire performance curve [6, 10, 16-17] with or without comparing to experiment.

In this work, two simulation techniques were used to compute the same performance curve. The first method was similar to many researches in the past. It was done by enforcing varying inlet air velocities, or effectively the volume flow rate, and extracting the fan pressure values at different inlet velocities to generate the performance curve. The second method used a series of computational domains resembling the throttling device in the experiment (a butterfly damper at different opening angles) and imposed the pressure at the boundaries to extract the corresponding volume flow rate. The second method was not involved as it required a different mesh at each condition, but it was more intuitive and could be extended to study the transient behaviors of the flow.

In the next section, the centrifugal fan in this study was described along with the experimental setup that was done according to the AMCA fan testing standard to be used as a reference for comparing the simulation results from the two techniques. Next, in the following section, the two computational techniques were described in details. Then, the simulations from both methods as well as from the experiment were compared and discussed. Finally, in the last section, the conclusion and recommendation were made.

2. Problem Description and Experimental Setup

Figure 1 shows an actual photograph and cross-sectional drawings of the centrifugal fan used in this work. The fan consists of twelve BC blades. The radius of the blade assembly is 762 mm with the diameter at the inlet of 450 mm. The height of blade assembly at the outlet is 165 mm giving the fan outlet of 0.746 m^2. The casing is 1,340 mm by 1,265 mm with geometrical shape as shown in the figure 1. The outlet from the casing to the system is 785 mm tall and 590 mm wide.

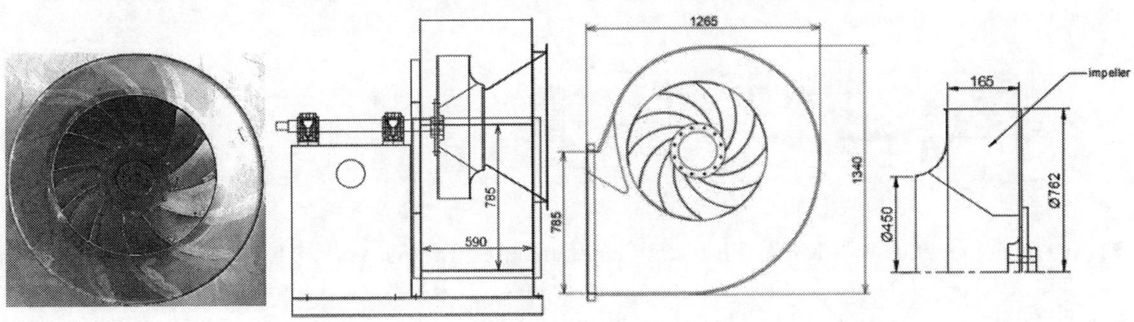

Figure 1. A frontal photograph and cross-sectional drawings of the fan used in this study.

Figure 2 shows the fan performance curves at 1,080-1,755 RPM as provided by the designers (not from an actual testing). In this work, the fan will be run at fan speed of 1,220 RPM in all cases and in the experiment. Note that once a performance curve at one rotational speed is known, one can use Fan Affinity Laws to create performance curves at other speeds over a relatively wide range. According to the provided performance curves in the below figure, the maximum volume flow rates is approximately 325 m^3/min and the peak (static) pressure is about 1,130 Pa (shown as 115 mmAq in the figure).

Note that the performance curve given by the designers does not include the points near the shut-off nor the free-delivery point. It only provides the data around the design point.

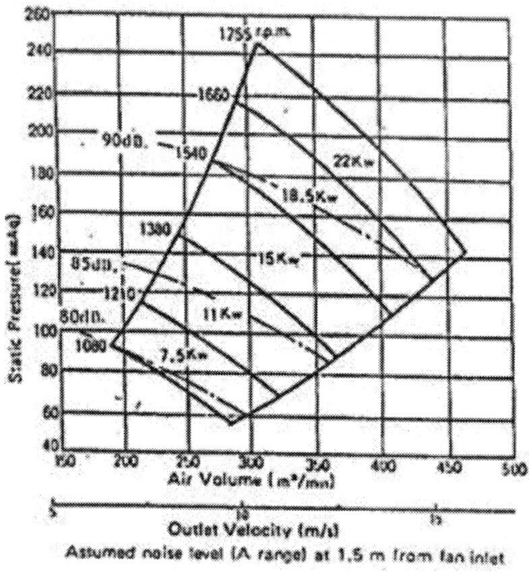

Figure 2. Fan performance curves from design [18]

The outlet of the centrifugal fan was connected to a straight horizontal duct with an adjustable throttling device at the end to simulate various system loads when the fan was installed in the intended application. The dimensions of the duct as well as various components such as the transformation connecting piece, the flow straightener, and the measuring point location are in accordance with the AMCA 210-99 international fan testing standard [3]. Figure 3 shows the schematics of this experimental setup. The length of the duct is 10 times the duct diameter (D), which is the minimum length required by the testing standard.

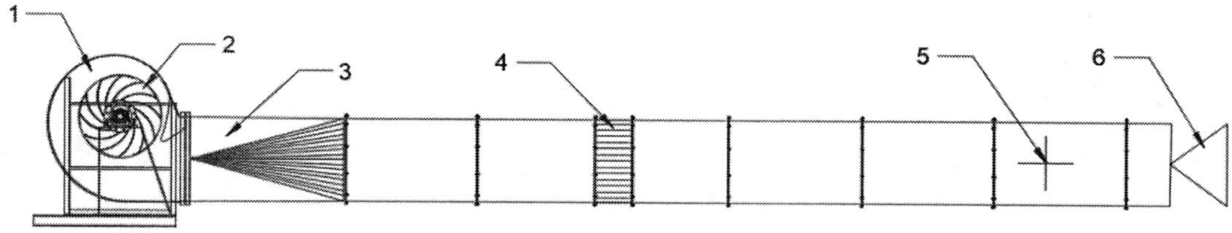

Figure 3. Schematics of AMCA fan testing standard used in this work. The numbers in the diagram indicate each component (1) fan casing (2) fan blades (3) transformation connecting piece (4) flow straightener (5) measuring point, and (6) throttling device.

IOP Conf. Series: Materials Science and Engineering **297** (2018) 012026 doi:10.1088/1757-899X/297/1/012026

The data measurement is done using 24 pitot-static tubes at the positions given in the standard (6 radii and 4 positions per radius) with tolerance of no more than ± 0.0025D. From the (averaged) pressure readings of the pitot tubes, the volume flow rate as well as pressure at the measurement can be accurately calculated. The results from the test are shown in table 1.

Table 1. Experimental data showing volume flow rate (Q), fan total pressure (P_t) and fan overall efficient (η) at various operating points at 1,220 RPM

Q [m³/s]	P_t [Pa]	η [%]
6.71	149	14
6.61	194	18
6.36	232	21
5.81	553	45
4.99	828	59
4.11	1,155	79
2.53	1,395	68
1.86	1,463	65
1.15	1,545	42
0.00	1,521	0

According to the experimental data above, the free-delivery point should be slightly more than 6.71 m^3/s (it is not exact because it is impossible to have zero pressure loss), the shut-off pressure is 1,521 Pa, and the peak efficiency is approximately 79% at volume flow rate of 4.11 m^3/s and total fan pressure of 1,155 Pa.

3. Computational Techniques

Three-dimensional CFD simulations were setup to compute the desired quantities at various operating points along the performance curve at the given rotational speed of 1,220 RPM. ANSYS Fluent commercial software was used to solve for the solution based on finite volume formulation [19 -21]. The simulation settings were steady, pressure-based, and incompressible. The turbulence model was the realizable k-epsilon with standard parameters.

Two different approaches were employed to generate the same performance curve (volume flow rate vs. fan total pressure). They will be discussed in details in the subsections below. For each method, eight simulations were performed to generate eight different points on the performance curve.

All the simulation was done on a dedicated workstation with Intel Xeon Processor E5 1660 v3 (8-core, 20MB cache, 3.0GHz) with 32GB of DDR RAM.

3.1. Method 1: Varying Inlet Velocity

In the first method, the main idea is to use the same geometry and mesh files for all eight simulations. In each of the eight simulations, the volume flow rate was fixed by prescribing a uniform inlet velocity profile at the inlet. The magnitude of the velocity will be varied across each individual simulation. Once the solution had converged, (area-averaged) total pressure difference across the fan was extracted to be plotted on the performance curve.

To avoid the effect of the prescribed uniform velocity profile, a straight section upstream of the inlet was added to give some additional length for the flow to develop. Figure 4 shows the overall geometry used in this method. Since the desired pressure output was measured across the fan only, this straight section should not introduce additional pressure loss to the output. Figure 5 shows a contour of velocity magnitude and that of static pressure near the blades from the simulation with the inlet velocity of 25.4 m/s.

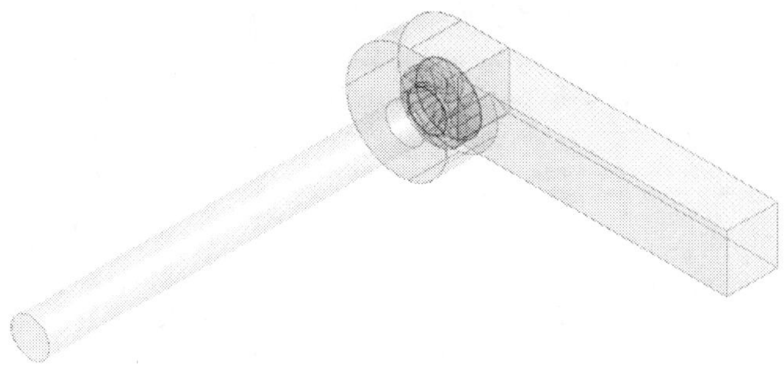

Figure 4. Overall geometry used in Method 1

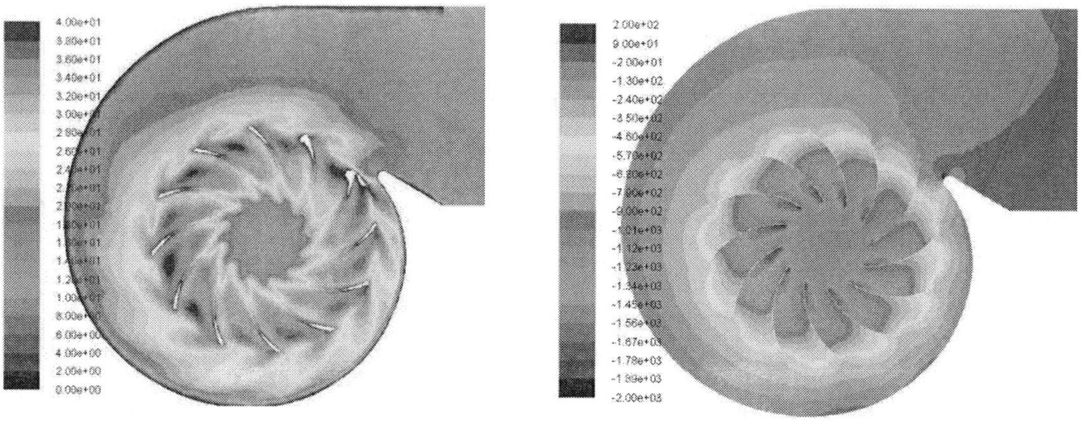

Figure 5. Example of contours of velocity magnitude (left) and static pressure (right) near the blades

The total number of cells in the mesh used was around 500,000 cells. Figure 6 shows 4-point coarser performance curves calculated with these different meshes. The current mesh was deemed to be sufficiently accurate for the purpose of generating the performance curve.

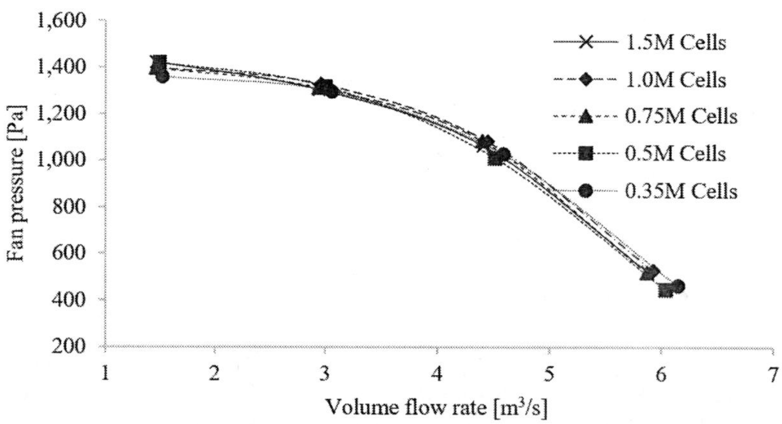

Figure 6. Performance curves generated with different mesh resolutions

3.2 Method 2: Modeling the Throttling Device

In the second method, the geometries of the throttling device at different angle were modeled as an angled flat plate. Since the geometry changes from one simulation to another, one needs to create a new mesh for each simulation at a new point on the performance curve; however, since it more closely resembles the real world, it has a potential to match the experimental data better. Figure 7 shows an example of the overall geometry for this method as well as a side view of the geometry of the throttling device at 50° angle.

The boundary conditions were set to be atmospheric at both inlet and outlet boundaries without prescribing any velocity or mass flow rate. This is also realistic and more intuitive than the first method. Also, there was no additional straight section added upstream of the inlet. Once the solution had converged, both the (area-averaged) total pressure difference across the fan *and* the (integral) volume flow rate were extracted to be plotted on the performance curve.

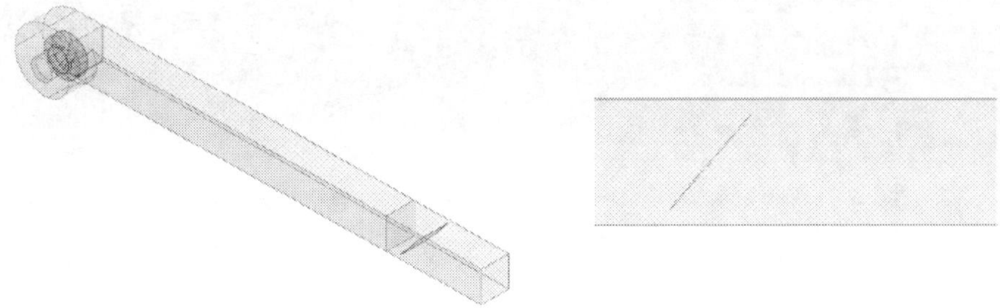

Figure 7. Overall geometry used in Method 2 (left) and geometry of the throttling device at 50° angle

4. Results and Discussions

Table 2 shows the computational from the two methods mentioned above. Note that even though the results from both methods contain eight data points, they are not at the same volume flow rates (nor pressure). In the first method, the volume flow rate can be effectively chosen precisely. In the second method, the input is the angle of the throttling device, and both the flow rate and the fan pressure are not known beforehand.

Table 2. Computational results from Methods 1 and 2

Method 1		*Method 2*	
Q [m^3/s]	P_t [Pa]	Q [m^3/s]	P_t [Pa]
6.0	446	6.4	212
5.3	762	6.3	261
4.5	1,011	6.1	387
3.8	1,187	5.5	689
3.0	1,319	4.5	1,073
2.3	1,415	3.1	1,426
1.5	1,421	1.7	1,580
0.8	1,429	0.8	1,589

Figure 8 compares the computational results from the two methods with the experimental data. For the high volume flow rates (more than 4 m^3/s or 240 m^3/min), the data from both computational methods and the experiment are in good agreement. The experimental results fluctuate from point to point was likely due to some experimental errors, but the magnitude of this fluctuation is small enough to be negligible (relative to the larger differences elsewhere). It is worth noting that this region with high volume flow rates is unsurprisingly the region shown in the fan performance curve from designers (figure 2). Both computational methods can predict the performance curve in this region quite accurately.

At a volume flow rate lower than 4 m^3/s, a clear discrepancy between the two computational methods can be clearly observed. At the same flow rate, the first method predicts lower fan pressure than the second method and the difference increases closer to the shut-off point. The fan pressure from the experiment is between those from the two methods, but given the fluctuation in the experimental data it is difficult to decisively judge which method is more accurate in this case. The authors suspect that the difference is due to the different velocity profiles near the front of the centrifugal fan. This could potentially create a different performance as the flow interacts with the rotating fan blades; however, further investigation is needed to definitively explain this behavior.

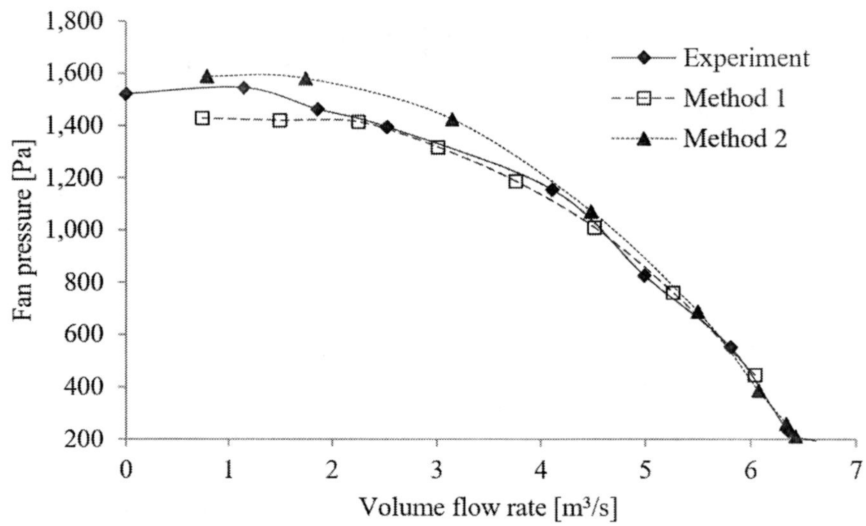

Figure 8. Comparison of the performance curves from the two methods and the experiment

5. Conclusions

In conclusions, this work compares two computational techniques using CFD to construct a fan performance curve. A centrifugal fan with twelve backward-curved blades was used as a model and the results were compared with the experiment following the AMCA fan testing standard. For each technique, eight simulations were performed at different conditions, each of which resulted in a data point along the performance curve. The first method fixed the geometry, added the extra straight section upstream of the fan, varied the flow rate, and extracted the fan pressure. The second method included the throttling device at different angles into the computational domains, which needed changing for each simulation. For this second method, both flow rate and fan pressure were extracted at each simulation. At a flow rate higher than 4 m^3/s, all results were in quite good agreement; however at a flow rate lower than 4 m^3/s, there was a clear difference in fan pressure at the same flow rate. The first method consistently gave a lower fan pressure than the second method. Also, it was inconclusive from the experimental data which of the two methods was more accurate.

Acknowledgment

The authors would like to thank Artith Machinery Co.,Ltd. and Artith Ventilators Ltd., Part. for providing the centrifugal fan and supporting the fan testing experiments in accordance with the AMCA testing standard.

IOP Conf. Series: Materials Science and Engineering **297** (2018) 012026 doi:10.1088/1757-899X/297/1/012026

References

[1] Fran P B 1997 *Fan Handbook* (New York: McGraw-Hill)
[2] CEATI International Inc. 2008 *Fans & Blowers: Energy efficiency reference guide Online*
[3] Laboratory Methods of Testing Fans for Rating *ASHRAE Standard (51-1985), AMCA Standard 210-85*
[4] ISO 5801:2007 *Industrial Fans - Performance Testing using Standardized Airways*
[5] Japanese Industrial Standard *Testing Methods for Turbo-Fans and Blowers (JIS B 8330)*
[6] Chunxi L, Song Ling W and J Yakui J 2011 *J. Energy Conversion and Management* **52** p 2902
[7] Keyur K and Prajesh M 2013 *Int. J. of Advanced Engineering Research and Studies* **2** p 1
[8] Jayapragasan C N, Sumedh J and Reddy K, 2016 *APPN Journal of Engineering and Applied Science* **9** p 1637
[9] Lin S and Huang C 2002 *Experimental Thermal and Fluid Science* **26** 421
[10] Amjadimanesh A, Ajam H and Hossein Nezhad A 2015 *Int. J. of Mechatronics, Electrical and Computer Technology (IJMEC)* **5** p 2109
[11] Singh O P, Khilwani R, Sreenivasulu T and Kannan M 2011. *Int. J. of Advances in Engineering & Technology* **1** p 33
[12] Jayapragasan C N and Reddy K J 2016 *Scientific and Industrial Research* **75** p 638
[13] Chaudhari D R and Patal HN 2015 *Int. J. of Advanced Technology in Engineering and Science* **3** p 134
[14] Lin S and Huang C 2002 *Experimental Thermal and Fluid Science* **26** p 421
[15] Atre Pranav C and Thundil Karuppa Raj R 2012 *Research Journal of Recent Sciences* **1** p 7
[16] Gholamian M, Rao G.K.M and Panitapu B 2013*Case Studies in Thermal Engineering* **1** p 26
[17] Siwek T, Gorski J and Fortuna S 2014 *Environ. Stud* **23** p 2359
[18] ARTITH Ventilator 2016 *SD-EN-01.* **1** p 26
[19] Moukalled F, Mangani L and Darwish M 2016 *The Finite Volume Method in Computational Fluid*
[20] ANSYS Fluent Theory Guide 2013 (ANSYS, Inc)
[21] Stephen B P, Turbulent Flows, (Cambridge: Cambridge University Press)

CFD modelling of a membrane reactor for hydrogen production from ammonia

San Shwe Hla* **and Michael D. Dolan**

CSIRO Energy, PO Box 883, Pullenvale, Brisbane, QLD 4069, Australia

* Corresponding Author: san.hla@csiro.au

Abstract. Despite the growing use of hydrogen (H_2) as a transport fuel, one of the major barriers still remaining is efficient and inexpensive fuel distribution and storage. Current approaches, such as compression, liquefaction or metal hydride formation, incur a significant energy penalty. Ammonia (NH_3) has long been considered a prospective H_2 medium, exhibiting a higher volumetric H_2 density than liquid H_2, through liquid-phase storage at mild pressure. Decomposition of NH_3 into H_2 and N_2 can be achieved via use of catalytic reactors and fuel-cell-grade H_2 can be produced using metal membranes at H_2 distribution sites.In this study, a 3-Dimensional (3D) Computational Fluid Dynamics (CFD) model has been developed to understand the performance of the H_2 separation process in gas mixtures derived from an NH_3-cracking reaction. The reactor consists of 19 tubular membrane tubes, each 470 mm long, inside a tubular shell with an inner diameter of 130 mm. Standard transport and energy equations governing a 3D, pressure-based, steady-state model were derived from the laws of conservation of mass, momentum and energy. The governing equations were solved using commercial CFD software ANSYS Fluent 18.0. Gas flow and mixing were modelled by the two-equation standard k-epsilon model for closure. Coupled solver was used for pressure-velocity coupling, enabling a pseudo-transient option with pseudo time steps of 0.01 s. To estimate H_2 permeation through the metal membrane, a constant H_2 permeability of 3.0E–07 mol.m^{-1} s^{-1} Pa$^{-0.5}$ derived from series of experiments tested under a range of industrial conditions, was used. Model simulations were conducted for an adiabatic temperature of 300 °C, a feed-side pressure of 7.8 bar$_a$ and a permeate side pressure of 0.1 bar$_a$. A parametric analysis was carried out to explore the effects of variation in total feed-gas flow and effects of changes in NH_3-cracking efficiency on H_2 production rates and H_2 yields. The model estimated that 4.6–11.6 kg H_2/day can be produced from a 30–70 L min^{-1} NH_3 inlet flow with 80–90% NH_3-cracking efficiency. At lower NH_3 inlet flow rates, higher H_2 yields can be obtained within a shorter distance of the membrane tubes due to relatively slower velocities and longer residence times. At high inlet flow rates, H_2 yields were significantly lower due to their faster velocities and shorter resident times, but high yields (>95%) were still observed at the membrane reactor outlet. A sensitivity analysis of the model showed that even if metal membranes functioned at only 50% of the maximum permeability, a high H_2 yield similar to that estimated using 100% permeability can still be achieved at the H_2 outlets.

1. Introduction

Hydrogen (H_2) is always considered to be a clean fuel, because it is free of carbon and does not release any CO_2 when utilised in a fuel cell or in combustion systems [1–3]. Today, the transition to a hydrogen society is led by Japan, where the number of stationary polymer electrolyte membrane fuel cells (PEMFC) deployed is around quarter of a million, and increasing dramatically [4]. The application and interest of use of H_2 as a transport fuel is also growing. However, one of the main barriers still remains: how to store and distribute H_2 in an efficient and economical way [5]. All current approaches for H_2

storage and distribution, such as compression, liquefaction or metal hydride formation require a significant amount of energy input [6]. Alternatively, H_2 can be stored in chemicals such as ammonia (NH_3), methanol and, ethanol. Of these options, NH_3 is a very attractive option from a CO_2 aspect as it contains no carbon.

Around 200 Mt/year of NH_3 are produced globally, primarily as an agricultural fertilizer, but also for chemical production and refrigeration [5]. NH_3 also has several promising advantages for use as H_2 storage medium. NH_3 has a higher volumetric H_2 density than even liquid H_2 (120 kg/m^3 *c.f.* 70 kg/m^3), it can be easily liquefied at a mild pressure, and the volumetric energy density of liquid NH_3 is 11.5MJL^{-1} which is higher than the 8.491MJL^{-1} for liquid H_2 and 4.5 MJL^{-1} for compressed H_2 at 690 bar and 15 °C [7]; and it can be stored in solid form for convenient transportation and to limit possible toxic exposure [6, 8, 9].

The limitation of using NH_3 as an H_2 carrier is that it must be decomposed back to nitrogen and H_2 at or near to the point of use, followed by separation of the resulting H_2/N_2 mixture, as illustrated in figure 1.

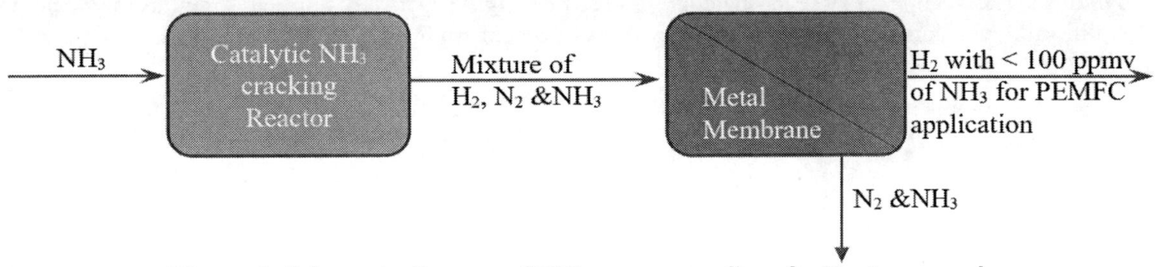

Figure 1. Schematic diagram of NH_3 use as a medium for H_2 storage and on-site H_2 distribution

NH_3 decomposition can be done via a catalytic NH_3-cracking reaction, as described in equation (1).

$$NH_3 \leftrightarrow \frac{3}{2}H_2 + \frac{1}{2}N_2 \quad [\Delta H^\circ_{300K} = 45.9 \text{ kJ/mol}] \tag{1}$$

Because NH_3 decomposition is slightly endothermic, additional heat is required for the reaction. However, the associated energy penalty is comparable with compressed or liquefied H_2 [10]. The resulting gas mixture can be purified to meet ISO14687-2 and 14687-3 standards, for mobile and stationary PEM fuel cells respectively, through the use of metal membranes at the H_2 distribution sites. H_2-selective alloy membranes are an emerging technology which allows the continuous separation of H_2 mixed gas streams at elevated temperatures and pressures with potential application to a number of processes, such as natural gas reforming and coal-derived syngases processing [11, 12].

Palladium (Pd)-based alloy membranes are the clear benchmark technology, having already entered commercial production and undergone numerous long-term industrial trials. However, due to the high price of Pd, these membranes can only be commercially viable by minimising Pd consumption through developing thinner membranes and alloying with less expensive metals [12]. Vanadium (V)-based alloys are an emerging alternative to Pd membranes, and potentially offer high H_2 flux at much lower cost [12, 13]. Details of development of those alloy membranes are outside the scope of this study; here, it is sufficient to mention that V-based alloy membrane tubes were used in this study for H_2 separation from gas mixtures derived from NH_3 decomposition. Our study focused on development and simulation of a 3-dimensional (3D) computational fluid dynamics (CFD) model to understand the performance of the H_2 separation process in $H_2/N_2/NH_3$ gas mixtures, representing the second stage of the process described in figure 1.

IOP Conf. Series: Materials Science and Engineering **297** (2018) 012027 doi:10.1088/1757-899X/297/1/012027

2. Selecting the Reaction Level for the NH_3 Decomposition Process

Pioneering studies of the NH_3 synthesis equilibrium (the reverse reaction of equation (1)), over a large range of temperatures and pressures, were completed by Haber [15] and others. They proposed a formula to calculate a temperature (T)-dependent equilibrium constant (K_p), as shown in equation (2):

$$\log_{10} K_p = \frac{\alpha}{T} + \beta log_{10}T + \gamma T + \delta T^2 + I \quad (2)$$

The values of α, β, γ, δ and I can be determined from experimental data. Gillespie and Beattie [16] used a large amount of experimental data to develop these empirical parameters, thereby enabling K_p to be determined for wide temperature (325–952°C) and pressure (1–1000 atm) range, as described in equation (3).

$$\log_{10} K_p = \frac{2001.6}{T} - 2.69112 log_{10}T - 5.519265 \times 10^{-5}T + \delta 1.848863 \times 10^{-7}T^2 + 2.6899 \quad (3)$$

Since then, this correlation has been widely applied to estimate the equilibrium constants for NH_3 synthesis (i.e. Gaine [17]; Elnashaie *et al.* [18]; Twigg [19]) and applied to indirectly calculate equilibrium constants for NH_3 cracking reactions in recent studies [20–22].

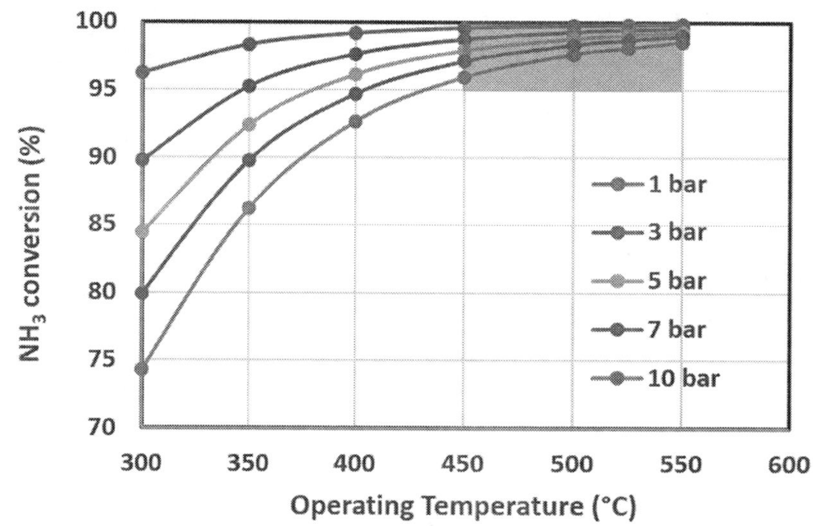

Figure 2. Conversion levels of NH_3 under different operating conditions at chemical equilibrium

The equilibrium constant (K_p) in equation (3), can be used to calculate levels of NH_3 conversion at chemical equilibrium under different temperatures and pressures. The results are plotted in figure 2. As shown in the figure, decomposition of NH_3 increases as the temperature rises, but decreases as pressure increases. The shaded area in figure 2 represents the operating conditions under which we intend to run our NH_3 decomposition reactor, which is the first stage of the H_2 generation process described in figure 1. According to chemical equilibrium calculation, NH_3 cracking of >95% can be achieved at 450–550 °C and 1–10 bar operating pressure. Therefore, it is reasonable to assume that NH_3 cracking conversions (80–90%) used in the following calculations are feasible.

3. Modelling of Membrane Reactor

3.1. Reactor configuration

An isometric diagram of the H_2 membrane reactor used in this study is presented in figure 3(a). The reactor consists of 19 tubular membrane tubes (each 470 mm long and 9.5 mm in diameter) inside a tubular shell with an inner diameter of 130 mm. A side view of the reactor presenting the arrangement of the membrane tubes inside the reactor is given in figure 3(b). The reactor geometry was constructed for a 3D model using the 3D-modelling software SpaceClaim. The mesh of the fluid volume inside the reactor was generated using ANSYS Meshing. A cross-section view and an isometric view of the resulting meshes consisting of 600,000 cells are shown in figure 4. As shown in figure 4, finer meshes were created in the near-wall region to achieve accurate local wall shear predictions.

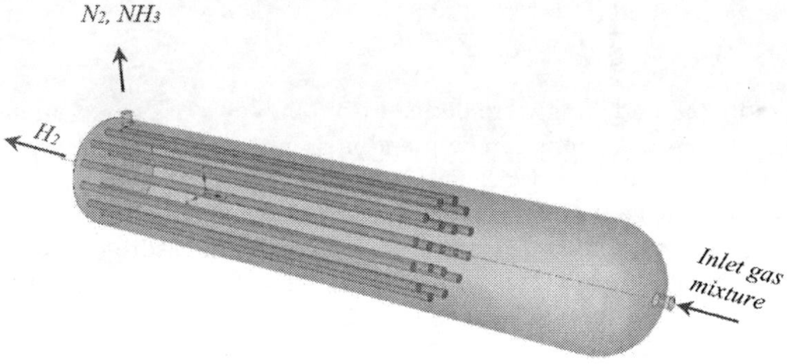

Figure 3(a). Isometric diagram of H_2 membrane reactor

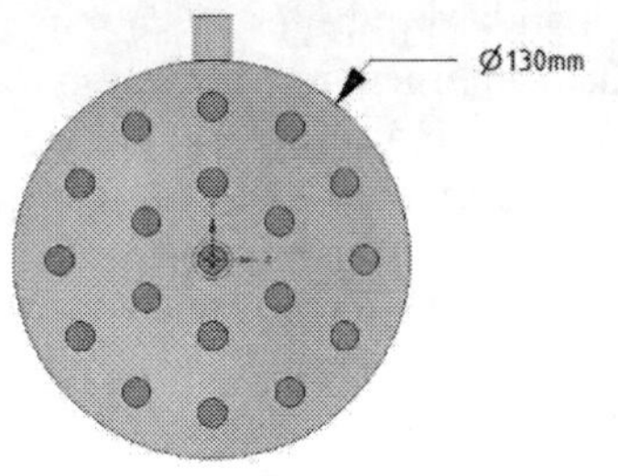

Figure 3(b). Side view of the membrane reactor

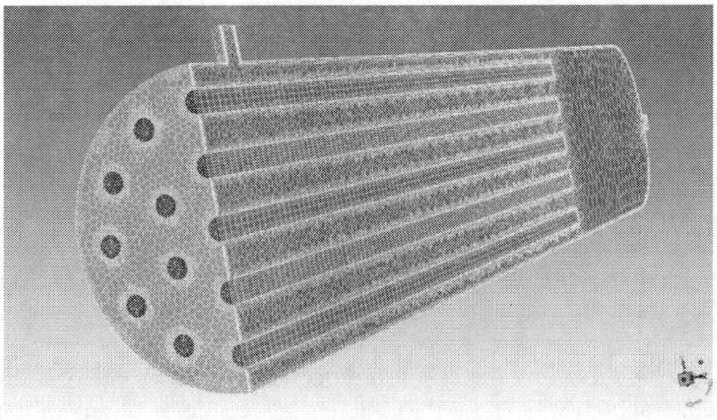

Figure 4. Mesh at symmetry surface and back side of fluid volume inside the membrane reactor

3.2. H_2 permeation submodel

H_2 permeation flux through the metal membrane can be expressed by:

$$J_{H_2} = \frac{P_e}{\delta}\left(P_{H_2,ret}^n - P_{H_2,perm}^n\right) \tag{4}$$

where

J_{H2} = the H_2 diffusion flux through the metal membrane (mol m^{-2} s^{-1})

IOP Conf. Series: Materials Science and Engineering **297** (2018) 012027 doi:10.1088/1757-899X/297/1/012027

δ = thickness of membrane (m)
P_{H2} = H_2 partial pressure (Pa) in the retentate (ret) and permeate (perm) streams
P_e = the H_2 permeability (mol.m m^{-2} s^{-1} $Pa^{-0.5}$)

The H_2 permeability (P_e) is the temperature-dependent parameter, which is usually defined by the Arrhenius equation. However, in this study, our focus was to investigate the performance of the membrane reactor under isothermal conditions at a constant operating temperature of 300 °C. Therefore, a constant H_2 permeability of 3.0E–07 mol.m m^{-2} s^{-1} $Pa^{-0.5}$ measured from our previous experimental work has been used for model simulations. The thickness of membrane used was 0.25 mm. In equation (4), n is the power dependency of the H_2 concentration on its partial pressure. The 'n' factor can have a value of between 0.5 and 1, and can be calculated from the correlation between H_2 pressure and permeation flux [23, 24]. In this study, the mass transport of H_2 is assumed to be controlled by diffusion of atomic hydrogen, so that the process can be described by the Sieverts-Fick law, and n factor is taken as 0.5.

3.3. Model formulation

The standard transport equations governing 3D modelling were derived from the laws of conservation of mass, momentum and energy. Details of these equations can be found in any heat and mass transfer texts and are also outlined in the ANSYS Fluent manual [25]. The governing equations were solved using commercial CFD software ANSYS Fluent 18.0. Gas flow and mixing were modelled using the two-equation standard k-epsilon model for closure. Coupled solver was used for pressure–velocity coupling, enabling a pseudo transient option with pseudo time steps of 0.01 s. The thermophysical and transport properties of gas species and gas mixtures are defined using submodels readily available in Fluent [25].

The calculations for H_2 permeation rates across the V-supported membrane tubes were separately developed in C++ programming language and exported to Fluent as a user-defined function. All of the governing equations considered in the pressure-based 3D model were then solved simultaneously using ANSYS Fluent version 18.0.

4. Model simulation results and discussion

The main objective of this study is to investigate the performance of membrane reactors focusing on H_2 yield and total H_2 generation under different NH_3 inlet flow rates and different NH_3 decomposition levels in the first stage of the system. The baseline input parameters to the 3D membrane reactor model are listed in table 1. As discussed Section 3, the assumption of NH_3 decomposition was justified by the chemical equilibrium calculation. The operation of the membrane reactor was assumed to be isothermal, which means inlet temperature of the gas mixture, membrane reactor temperature and outlet gas temperature are constant at the value listed in table 1. The inlet gas pressure was assumed to be 7.8 bar_a, which is 1 bar below the vapour pressure of NH_3 at 21 °C. As described in equation (4), H_2 separation through the membrane is driven by the H_2 partial pressure difference across the membrane. In order to maximise H_2 yield, a vacuum was applied to the permeate side of the membrane and the pressure was taken as 0.1 bar_a.

Table 1. Baseline operating parameter values used for model simulations

Parameter	Value
NH_3 inlet to the system	35 L min^{-1}
NH_3 decomposition	90 %
H_2 inlet to membrane reactor	66.7 mol%
N_2 inlet to membrane reactor	22.2 mol%
NH_3 inlet to membrane reactor	11.1 mol%
Operating temperature (isothermal)	300°C
Total pressure of the gas mixture	7.8 bar_a
Pressure on permeate side of membrane	0.1 bar_a

Gas velocity magnitudes with vectors are shown in figure 5(a), while figure 5(b) highlights the cross-sectional profiles of gas velocity contours at the outlet of the metal membranes. As more than half of the inlet gas flowed through much smaller spaces inside the membrane tubes, significantly higher velocities of H_2 were found inside the membrane tubes, especially in those located closer to the reactor's centre. It is expected that the flow profile of the inlet gas mixture reached a maximum along the centreline of the reactor, which yielded higher H_2 separation rates and consequently higher H_2 flows inside the membrane tubes around the centreline.

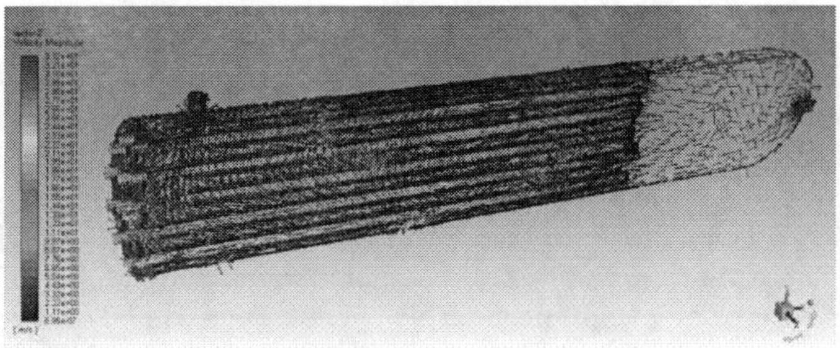

Figure 5(a). Velocity magnitude with vectors

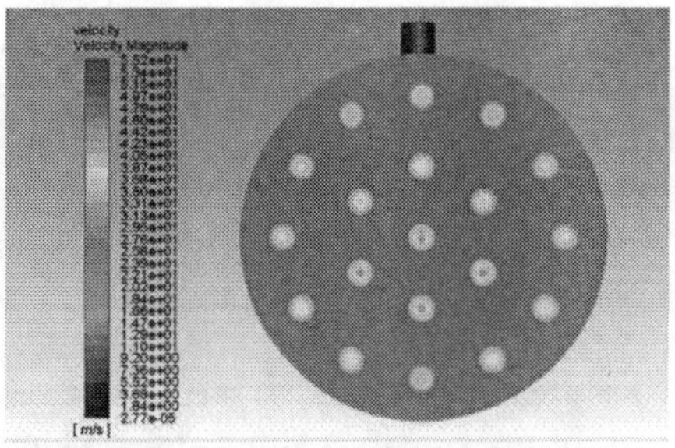

Figure 5(b). Velocity contour at H_2 outlets

The contours of the molar concentration of H_2 are shown in figure 6. Pure H_2 flows inside the membrane tubes, and a very low concentration of H_2 is present in the retentate stream at the outlet.

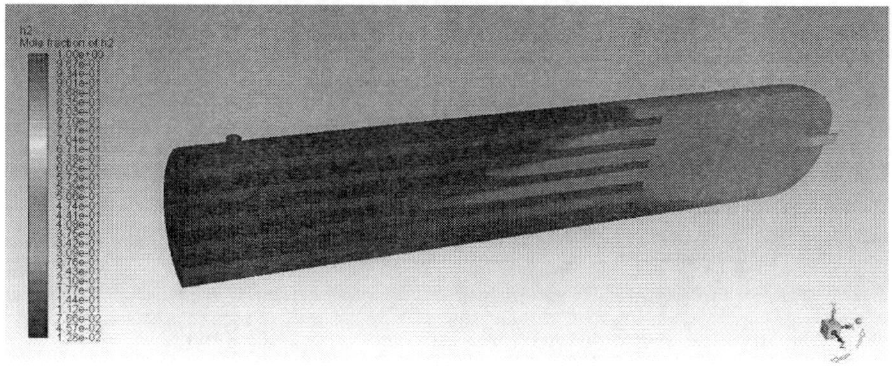

Figure 6. Contours of H_2 molar fraction

4.1. Effect of feed-gas flow rates

To enhance our understanding on the performance of the membrane reactor, a parametric analysis was carried out to explore the effects of variation in total feed-gas flow rate and level of NH_3 cracking efficiency on total H_2 production rates and H_2 yield. The results are shown in figure 7 and figure 8.

In this study, H_2 yield for the membrane rector is defined as:

$$H_2\ yield(\%) = \frac{H_2(mol)_{out\ (ermeate)}}{H_2(mol)_{feed\ gas}} \times 100 \tag{5}$$

The shaded area in figure 7 represents the range of expected H_2 production (4.6–11.6 kg/day) for a range of NH_3 consumption rates (43−108 Nm^3/day) and NH_3 cracking efficiencies (80−90%), while the dotted lines represent the theoretical maximum possible H_2 production rate under the same operating conditions. The theoretical maximum H_2 production rate is defined as a condition where the H_2 pressure on feed side equalises with H_2 pressure on permeate side. As shown in figure 7, H_2 production rates at lower NH_3 inlet flow rates were found to be much closer to theoretical maximum values. For higher NH_3 inlet flow rates, the gaps between maximum possible H_2 production rates and estimated H_2 production rates of the membrane reactor became more significant. This phenomenon can be explained by figure 8, which shows the accumulated H_2 yield along the membrane tubes for different NH_3 flow rates. As expected, at lower NH_3 inlet flow rates, gas mixture velocities were slower and residence times were consequently longer. As a result, higher H_2 yields could be obtained within a shorter distance. Conversely, at high flow rates, H_2 yield is significantly lower due to shorter residence times. Even though relatively high H_2 yields were possible at the outlet of the membrane reactor under higher NH_3 inlet flow rates, the calculated H_2 yields were found to be lower than the theoretical maximum values.

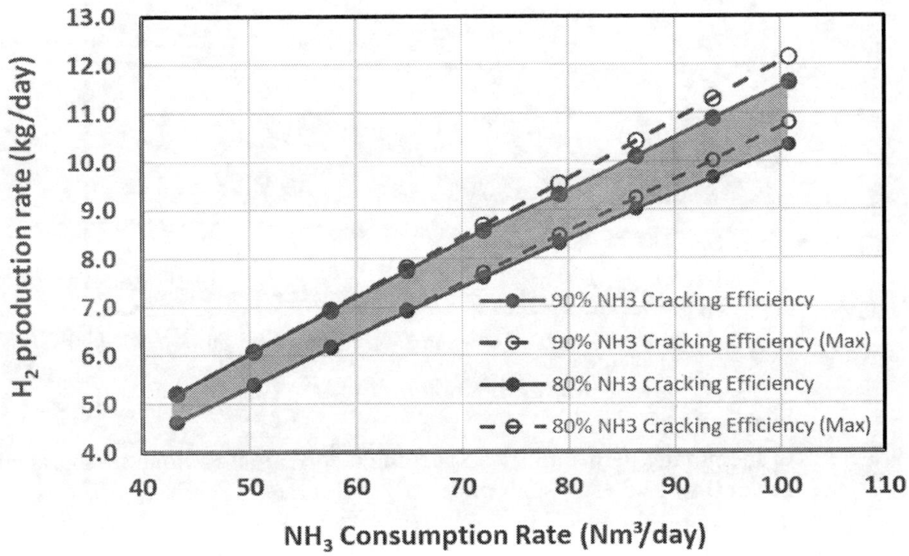

Figure 7. H_2 production rates calculated for different NH_3 inlet flow rates

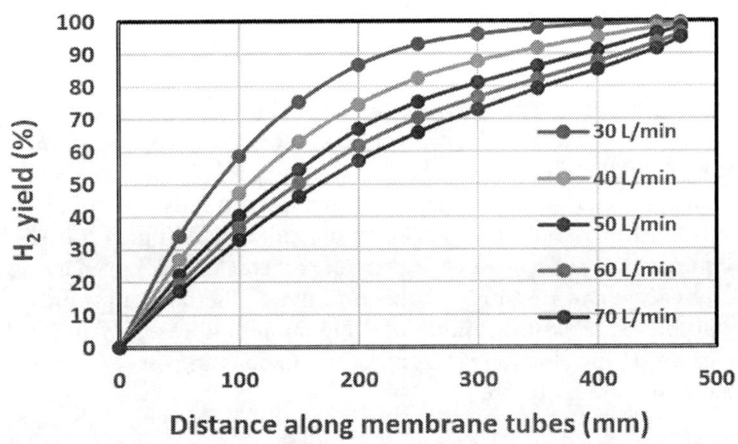

Figure 8. Accumulated H_2 yield along membrane tubes for different NH_3 inlet flow rates (90% NH_3 decomposition efficiency)

4.2. Effects of permeability

In this study, a constant H_2 permeability value of 3.0E–07 mol.m m^{-2} s^{-1} $Pa^{-0.5}$ was used for model simulations. This value was derived from experimental testing of the metal membranes at 300°C. A parametric analysis was carried out under selected operating conditions (NH_3 inlet flow of 50.4 Nm^3h^{-1} and 90% NH_3 cracking efficiency) to explore the effects of changes in permeability of membrane tubes on membrane reactor performance. In the simulations, we varied the permeability to 50, 20 and 10% of the original set value (figure 9). The results show that similar level of high H_2 yields could still be obtained at the H_2 outlets even when the metal membranes functioned at only 50% of their empirical permeability. H_2 yields were significantly reduced at lower levels of membrane permeabilities; for example, with metal membranes functioning only at 10% of base-case permeability, H_2 yield was reduced to 52%.

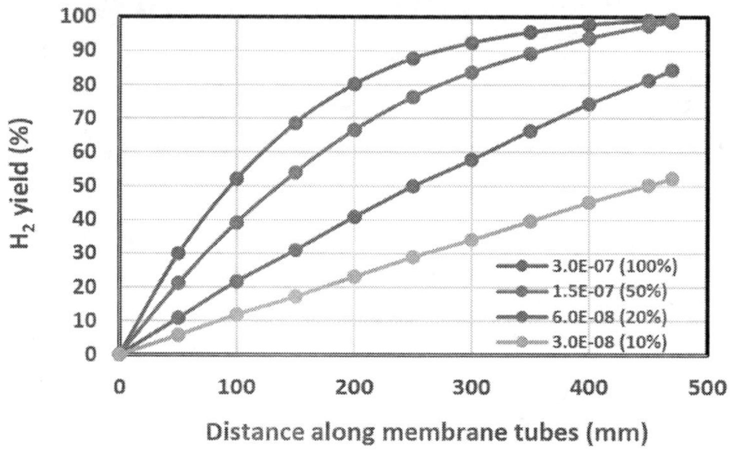

Figure 9. Effects of membrane permeability on accumulated H_2 yields along membrane tubes (NH_3 inlet flow rate of 50.4 Nm^3h^{-1}; 90% NH_3 cracking efficiency)

5. Summary

A 3D CFD model has been developed to investigate the performance of a metal membrane reactor for H_2 production via NH_3 decomposition. Model simulation estimated that a range of H_2 production rates of 4.6−11.6 kg/day are possible for the range of operating conditions investigated. Model results showed that close to theoretical maximum H_2 yields could be achieved when the reactor was operated under lower inlet NH_3 flows. A parametric study was conducted to investigate the effects of membrane permeability on the H_2 yield. The results showed that a similar level of high H_2 yield could be still obtained at the H_2 outlets even if metal membranes functioned at only 50% of their original permeability.

The main objective of this study is to enhance our understanding of the H_2 separation process in $H_2/N_2/NH_3$ gas mixtures, and to design a H_2 membrane reactor with H_2 production rates of 5−10 kg/day from ammonia. Fabrication and assembly of this prototype plant are in progress. The next step of this study, therefore, will be to check the validity of CFD modelling results from this study by comparing with performance of the H_2 production reactor via ammonia using an experimental system which will be available for operation in near future.

Acknowledgment

This research is supported by the Science and Industry Endowment Fund.

References

[1] Dutta S 2014 *J. Ind. Eng. Chem.* **20**(4) pp 1148−56
[2] Jain IP 2009 *Int. J. Hydrogen Energy* **34**(17) pp 7368−78
[3] Mayer H 1999 *Atmos Environ* **33**(24) pp 4029−37
[4] Nathan S. How the hydrogen economy could make a come back 2017 *The Engineer* 12th June 2017 https://www.theengineer.co.uk/how-the-hydrogen-economy-could-make-a-comeback/
[5] Afif A, Radenahmad N, Cheok Q, Shams S, Kim J and Azad A2016 *Renew Sustain Energy Rev* **60** pp 822−835.
[6] Gurz M, Baltacioglu E, Hames Y and Kaya K 2017 *Int J Hydrogen Energy* **Article in Press** pp 1−13
[7] Lan R and Tao S 2014 *Fuel Cells* **2** pp 35−38
[8] MacKenzie J and Avery W 1996 *Proc. of the 31st intersociety energy conversion engineering conf.* pp 1761−66

[9] Chakraborty D, Petersen H, Elkjær C, Cagulada A and Johannessen T *2009 Fuel Cells Bulletin* **2009**(10) pp 12–15.
[10] Appl M 2006 Amonia: *in Ullmann's Encyclopedia of Industrial Chemistry*, ed. B. Elvers (Wiley-VCH).
[11] Hwang K, Lee C, Ryi S, Lee S and Park J 2012 *Int. J. Hydrogen Energy* **37**(8) pp 6601–6607
[12] Dolan, M, Donelson R and Dave N 2010 *Int. J. Hydrogen Energy* **20** pp 10994–11003
[13] Viano D, Dolan, M, Weiss F and Adibhatla A 2015 *J. Membr. Sci.* **487** pp 83–89
[14] Dolan, M, McLennan K, Song G, Liang D and Kellam M 2013 *J. Membr. Sci.* **446** pp 405–409
[15] Haber V Untersuchungen Dber NH_3k. Sieben Mitteilungen. 1914 *Zeitschrift für Elektrochemie* **20**(32) pp 597–603
[16] Gillespie L, Beattie J.1930*Physical Review* **36** 743–753
[17] Gaines L. 1979 *Chem. Eng. Sci.* **34** 37–50
[18] Elnashaie S, Abashar M and Al-Ubaid A 1988 *Ind. Eng. Chem. Res.* **27**(11) 2015–2022
[19] Twigg M 1989 *Catalyst Handbook* (London: Wolfe Scientific)
[20] Hacker V and Kordesch K 2003 NH_3 crackers *Handbook of Fuel Cells – Fundamentals, Technology and Applications* Edited by eds. W Vielstich, A Lamm, H Gasteiger (John Wiley & Sons, Ltd, Chichester)
[21] Dashti A, Khorsand K, Marvast M and Kakavand M 2006 *Petroleum & Coal* **48**(2) 15–23
[22] Abashar M 2016 *Journal of King Saud University – Engineering Sciences* 1–10.
[23] Mendes D, Chibante V, Zheng J, Tosti S, Borgognoni F, Mendes A and Madeira L 2010 *Int. J. Hydrogen Energy* **35** 12596–12608
[24] Basile A, Campanari S, Manzolini G, Iulianelli A, Longo T, Liguori S, De Falco M and Piemonte V 2011 *Int. J. Hydrogen Energy* **36** 1531–1539
[25] ANSYS Inc 2017 *ANSYS Fluent Theory Guide* (ANSYS Inc, Canonsburg, PA, USA)

IOP Conf. Series: Materials Science and Engineering **297** (2018) 012028 doi:10.1088/1757-899X/297/1/012028

Large Eddy Simulation of turbulence induced secondary flows in stationary and rotating straight square ducts

W. Sudjai[1], V. Juntasaro[1,*] and V. Juttijudata[2]

[1]Department of Mechanical Engineering, Faculty of Engineering, Kasetsart University, Bangkhen, Bangkok 10900, Thailand
[2]Department of Aerospace Engineering, Faculty of Engineering, Kasetsart University, Bangkhen, Bangkok 10900, Thailand

* Corresponding Author: varangrat.j@ku.th

Abstract. The accuracy of predicting turbulence induced secondary flows is crucially important in many industrial applications such as turbine blade internal cooling passages in a gas turbine and fuel rod bundles in a nuclear reactor. A straight square duct is popularly used to reveal the characteristic of turbulence induced secondary flows which consists of two counter rotating vortices distributed in each corner of the duct. For a rotating duct, the flow can be divided into the pressure side and the suction side. The turbulence induced secondary flows are converted to the Coriolis force driven two large circulations with a pair of additional vortices on the pressure wall due to the rotational effect. In this paper, the Large Eddy Simulation (LES) of turbulence induced secondary flows in a straight square duct is performed using the ANSYS FLUENT CFD software. A dynamic kinetic energy subgrid-scale model is used to describe the three-dimensional incompressible turbulent flows in the stationary and the rotating straight square ducts. The Reynolds number based on the friction velocity and the hydraulic diameter is 300 with the various rotation numbers for the rotating cases. The flow is assumed fully developed by imposing the constant pressure gradient in the streamwise direction. For the rotating cases, the rotational axis is placed perpendicular to the streamwise direction. The simulation results on the secondary flows and the turbulent statistics are found to be in good agreement with the available Direct Numerical Simulation (DNS) data. Finally, the details of the Coriolis effects are discussed.

1. Introduction

The understanding of the characteristic of turbulence induced secondary flows and its change when involved with the Coriolis effects is crucially important since the flows appear in many industrial applications such as turbine blade internal cooling passages in a gas turbine and fuel rod bundles in a nuclear reactor. In order to reveal the mechanism of the flows, the simple straight square duct is popularly used as the compatible observation case of these phenomena.

In the literature, the turbulent flows through a stationary straight square duct exposed the characteristic of turbulence induced secondary flows which consists of two counter rotating vortices distributed in each corner of the duct. The flows is relatively weak as about 2 per cent of the bulk velocity but it has significant effects on the momentum transfer and the wall shear stress distribution [1-5]. Gavrilakis [2] provided the useful database of the mean velocity and the turbulence statistics of the fully developed flows in a stationary straight square duct at the friction Reynolds number of 300 by using Direct Numerical Simulation (DNS). Thereafter, Huser [3] performed DNS on the stationary duct at the friction Reynolds number of 600, and reported the mechanism of the turbulence induced secondary flows. Huser [3] concluded that the reduction of mean shear stress in the corner bisector allows the high momentum fluid flows from the center region of the duct towards the corners. Consequently, the secondary flows penetrate more toward the corners as the Reynolds number increases [5, 6]. Recently, the highest Reynolds number of the turbulent flows simulation in the stationary duct of DNS has been performed up to 1200 of the friction Reynolds number [6], by the way, the widest range of Large Eddy Simulation (LES) of the flows has been carried out from 300 to 10550 of the friction Reynolds number [5].

Coriolis effects on the turbulence induced secondary flows was presented in the turbulent flows through a rotating straight square duct. In the rotating duct, the flows can be divided into a pressure side and a suction side. As the rotation number increases, the turbulence is enhanced on the pressure side whereas the flows is stabilized on the suction side [7-11]. The rotational effects convert the turbulence induced secondary flows of the stationary duct into the Coriolis force driven two large circulations with a pair of additional vortices on the pressure side. In different rotation numbers, these additional vortices were appeared and disappeared with fixed reattachment points and various positions of separation points on the pressure side [10]. Taylor-Proudman zone, where the streamwise velocity along the spanwise direction is constant, is located at the central region of the duct.

The Taylor-Proudman zone is expanded in the spanwise direction as the rotation number increases [10, 11]. The present of Ekman layers on the lateral walls create suction regions on the pressure side close to the corners. In addition, the positions of the suction regions behave as fixed boundary of the additional vortices since it does slightly change due to the suppression of turbulence near the corners [10]. The Coriolis effects causes the additional resistance on the streamwise mean momentum [11]. Therefore, bulk velocity and flow rate in the duct decrease as the rotation number increases [7-11]. Recently, at the low friction Reynolds number of 300, Dai et al. [10] performed DNS of the turbulent flows in the rotating duct with the range of the friction rotation number based on the mean friction velocity and the full duct width from 0 to 40. Moreover, Fang et al. [11] carried out DNS of the turbulent flow in the rotating duct at the friction Reynolds number based on the mean friction velocity and the half duct width of 150 with the wide range of the friction rotation number based on the mean friction velocity and the half duct width from 0 to 27. Pallares and co-workers [7-9] performed LES with the localized one-equation dynamic subgrid scale model [12] to study the rotation effects on the secondary flows and heat transfer in the square duct at different rotation numbers.

In this paper, ANSYS FLUENT 18.1 commercial software was used to reveal the characteristic of turbulence induced secondary flows in stationary and rotating straight square ducts. LES with a dynamic turbulent kinetic energy subgird scale model [12] was carried out to describe the three-dimensional incompressible fully developed turbulent flows in the ducts. The Reynolds number based on the friction velocity and the hydraulic diameter is 300. For the rotating case, the two different friction rotation numbers of 1.3 and 5 were considered. The former is where the turbulence induced secondary flows is suppressed and the latter is where the additional vortices are located on the pressure side [8]. Moreover, the accuracy of the software has been reported.

2. Numerical simulation

2.1. Physical model and boundary conditions

The coordinate system and the computational domain of the straight square duct are shown on figure 1. The coordinate system (x, y, z) represent streamwise, normal, and spanwise direction, respectively. The velocity components (u, v, w) denote velocity component in x, y and z direction, respectively. The rotation axis is placed at the center of the computational domain with the spanwise rotation direction. The size of a cross sectional area of the duct is D X D where D is the hydraulic diameter. The streamwise length of 6.4D is used in this work. In literature, the simulation of force convection heat transfer in a rotating square duct at high rotation numbers using a dynamical localized subgrid scale model [12] with the shortest of the streamwise length computational domain of 6D was carried out by Pallares et al. [9].

Therefore, the streamwise length computational domain in this work is sufficiently long enough to capture the effects of the secondary flows. No-slip conditions is imposed at each walls. The friction Reynolds number is defined as $Re_\tau = u_\tau D/\nu$ and the friction rotation number is defined as $Ro_\tau = 2\Omega D/u_\tau$ where u_τ denotes the friction velocity, ν denotes the kinematic viscosity and Ω denotes the rotational angular velocity. In order to observe the effects of Coriolis force in rotating square ducts, the friction Reynolds number, Re_τ, is fixed at 300, and the two different friction rotation numbers, Ro_τ, of 1.3 and 5 are considered. The fully developed flow is assumed by imposing periodic boundary condition with the constant pressure gradient, $dP/dx = -4$, along the x-axis.

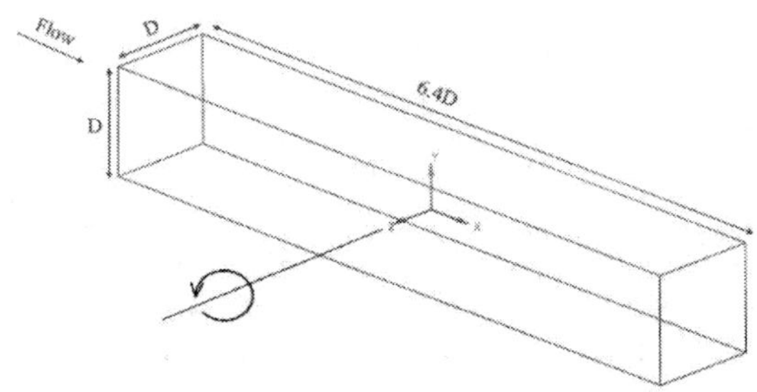

Figure 1. Computational domain

2.2. Numerical methods

In Large Eddy Simulation (LES), turbulent flows are decomposed into large eddies, which contains the most of energy, and universal isotropic small eddies by applying the filtering operation. The large eddies are directly resolved while the small eddies are modeled.

The filtering incompressible Navier-Stoke equations are obtained for the resolved large eddies. Thus, the governing equations are

$$\frac{\partial \rho}{\partial t} + \frac{\partial}{\partial x_i}(\rho \bar{u}_i) = 0 \tag{1}$$

and

$$\frac{\partial}{\partial t}(\rho \bar{u}_i) + \frac{\partial}{\partial x_j}(\rho \bar{u}_i \bar{u}_j) = \frac{\partial}{\partial x_j}(\sigma_{ij}) - \frac{\partial \bar{p}}{\partial x_i} - \frac{\partial \tau_{ij}}{\partial x_j} \tag{2}$$

where $\bar{u}_i$ is the resolved velocity field, $\bar{p}$ is the resolved pressure, ρ is the density and t is time, and σ_{ij} is the stress tensor due to molecular viscosity defined by

$$\sigma_{ij} \equiv \left[\mu\left(\frac{\partial \bar{u}_i}{\partial x_j} + \frac{\partial \bar{u}_j}{\partial x_i}\right)\right] - \frac{2}{3}\mu \frac{\partial \bar{u}_l}{\partial x_l}\delta_{ij} \tag{3}$$

and τ_{ij} is the subgrid scale stress defined by

$$\tau_{ij} \equiv \rho\overline{u_i u_j} - \rho \bar{u}_i \bar{u}_j \tag{4}$$

The unresolved small eddies in the subgrid scale stress must be modeled. In this paper, the LES with a dynamic kinetic energy subgrid scale model proposed by Kim and Menon [12] is implemented using the ANSYS FLUENT 18.1 CFD software.

The transport equation of the subgrid scale kinetic energy, k_{sgs}, is obtained as

$$\rho\frac{\partial k_{sgs}}{\partial t}+\rho\frac{\partial \bar{u}_j k_{sgs}}{\partial x_j}=-\tau_{ij}\frac{\partial \bar{u}_i}{\partial x_j}-C_\varepsilon\rho\frac{k_{sgs}{}^{3/2}}{\Delta_f}+\frac{\partial}{\partial x_j}\left(\frac{\mu_t}{\sigma_k}\frac{\partial k_{sgs}}{\partial x_j}\right) \quad (5)$$

where C_ε is a dynamically adjustable coefficient.

The subgrid scale kinetic energy is obtained by solving the transport equation of Eq. (5)

$$k_{sgs}=\frac{1}{2}\left(\overline{u^2{}_k}-\overline{u_k}^2\right) \quad (6)$$

The subgrid scale eddy viscosity, μ_t, is obtained by computing the k_{sgs} as

$$\mu_t=C_k\rho k_{sgs}{}^{1/2}\Delta_f \quad (7)$$

where C_k is a dynamically adjustable coefficient and Δ_f is the filter size computed from $\Delta_f \equiv V^{1/3}$. Therefore, the subgrid scale stress, τ_{ij}, is modeled as follows,

$$\tau_{ij}-\frac{2}{3}\rho k_{sgs}\delta_{ij}=-2C_k\rho k_{sgs}{}^{1/2}\Delta_f\bar{S}_{ij} \quad (8)$$

Pallares and co-workers [7-9] coded the localized dynamic subgrid scale model or also called the dynamic kinetic energy subgrid scale model by themselves. Their codes were done successfully in the simulations of the turbulent flow and heat transfer in a square duct with and without rotation. On the other hand, this paper used the commercial software ANSYS FLUENT 18.1 to describe the physics of the turbulent induced secondary flows in stationary and rotating straight square ducts and also to ensure that the software is capable to characterize the secondary flows and the effects of Coriolis force due to the rotation in the ducts.

ANSYS Meshing was used to build up the proper computational mesh in order to capture the complexity of the secondary flows in the ducts. Multizone method was used to generate the hexahedral mesh. Uniform mesh was created, in order to capture the streamwise streaks of turbulent structure, by using the edge sizing with selected, no bias, number of divisions. In the cross sectional area of the ducts, the edge sizing with the selected number of divisions with the smooth transition bias growth rate was used along the perimeter of the cross sectional area to generate the proper mesh which can capture the secondary flows. Two types of computational mesh were considered. A coarse mesh type has 66 X 40 X 40 number of divisions along the edge in x, y and z direction, respectively, with the bias growth rate of 1.2 along the cross sectional area. This corresponds to a uniform grid spacing of $\Delta x^+ \approx 29.09$ along the streamwise direction and $(\Delta y^+)_{min} = (\Delta z^+)_{min} = 0.8$ and $(\Delta y^+)_{max} = (\Delta z^+)_{max} \approx 25.55$ on the cross sectional area. For a fine mesh type, the computational domain was divided into 80 X 80 X 80 number of divisions with the bias growth rate of 1.0678 along the cross sectional area. The corresponding grid spacing are a uniform grid spacing of $\Delta x^+ \approx 24$ along the streamwise direction and $(\Delta y^+)_{min} = (\Delta z^+)_{min} = 0.8$ and $(\Delta y^+)_{max} = (\Delta z^+)_{max} \approx 10.33$ on the cross sectional area.

Large Eddy Simulations of both the stationary and rotating straight square ducts were started from precursor solutions of RANS (Reynolds Averaged Navier-Stoke) with a superimposed synthetic turbulence on the mean flow at the individual rotation rates ($Ro_\tau = 0, 1.3$ and 5). The precursor RANS solutions with the perturbations were used as initial condition in each coarse mesh LES cases. Only the stationary duct case was computed up to the fine mesh due to the computational resources. The initial condition of the fine mesh LES stationary duct case was obtained from the coarse mesh LES case at $Ro_\tau = 0$. The sampling procedure was started after the solution reached the statistically steady state to obtain the averaged velocity field and the turbulent intensities. The sampling rate does not exceed $0.1\,D/u_\tau$. For the fine mesh stationary duct case, the mean flow field was obtained over $280\,D/u_\tau$ period of time. Likewise, the mean flow field of the coarse mesh rotating duct cases were obtained over $270\,D/u_\tau$ and $280\,D/u_\tau$ for the friction rotating number of 1.3 and 5, respectively. Time steps were chosen to keep the maximum CFL number below 0.6.

ANSYS FLUENT 18.1's pressure based solver was used to obtain the solutions. For pressure-velocity coupling, the Fractional Step Method was used. For spatial discretization scheme, least squares cell based was used for gradient scheme, second order for pressure scheme, central differencing for momentum and subgrid kinetic energy scheme. Second order implicit was used for transient formulation scheme. Non-iterative time advancement was enabled.

3. Results and discussions

The results of the present ANSYS FLUENT 18.1's LES with a dynamic kinetic energy subgrid scale model proposed by Kim and Menon [12] of the turbulent flows in the stationary and rotating straight square ducts are presented comparing with the available LES and DNS databases. As shown on table 1, the simulated Reynolds number and rotation number based on the bulk velocity, U_b, and the hydraulic diameter, D, at different rotation rates are agree reasonable well with the available databases with the different about 1.5 percent.

3.1. Stationary straight square duct

For the stationary case, the present ANSYS FLUENT 18.1's LES with a dynamic kinetic energy subgrid scale model proposed by Kim and Menon [12] has been compared with the available database of DNS of Gavrilakis [2] and LES of Yao et al. [5]. In addition, Yao et al. [5] performed LES of turbulent flows through a stationary straight square duct from 300 to 10550 of the friction Reynolds numbers with a dynamic Smagorinsky subgrid scale model proposed by Germano et al. [13]. Thus, the LES with another dynamic subgrid scale model [13] at the friction Reynolds number of 300 was chosen for the comparison. The comparisons of the stationary straight square duct case were presented in figure 2-8. The y/D denotes the dimensionless distance along normal direction from -0.5 to 0.5 and the z/D denotes the dimensionless distance along the spanwise direction from -0.5 to 0.5.

3.1.1. Mean flow field and turbulent intensities. The mean flow field was obtained over a time period of $280\ \mathrm{D/u_\tau}$ with the fine mesh. The mean velocity profiles were normalized by the mean streamwise velocity at the centerline of the duct, u_0 since the mean centerline streamwise velocity scales were available from the reference databases. Figure 2 shows the normalized mean streamwise velocity profiles at five locations of the straight square duct, i.e. z/D = -0.45, -0.35, -0.25, -0.15, and 0. The present LES mean streamwise velocity profiles have better agreement with DNS of Gavrilakis [2] than the other LES [5] at all five locations of the duct. Since, solving the flow field with the one transport equation of the dynamic kinetic energy subgrid scale model [12] would give more realistic characteristic of the flows than just only solving the flow field with the algebraic equation of the dynamic Smagorinsky subgrid scale model [13].

On the other hand, the secondary flows were obtained with the good agreement between the present LES and both simulation databases of normalized mean spanwise velocity at five locations of the duct, i.e. z/D = -0.42, -0.35, -0.25, -0.15, and -0.1, as shown on figure 3. Furthermore, the maximum mean secondary velocity is about 2 per cent of the bulk streamwise velocity.

IOP Conf. Series: Materials Science and Engineering **297** (2018) 012028 doi:10.1088/1757-899X/297/1/012028

Table 1. Comparison between the simulated bulk Reynolds number and bulk rotation number and the available databases with the three different friction rotation number of 0, 1.3 and 5 at the fixed friction Reynolds number of 300.

Authors	$Ro_\tau = 0$		$Ro_\tau = 1.3$		$Ro_\tau = 5$	
	Re_b	Ro_b	Re_b	Ro_b	Re_b	Ro_b
Present LES	4480.48	0	3974.833	0.098117	3066.302	0.489189
Gavrilakis [2]	4410	0	-	-	-	-
Yao et al. [5]	4410	0	-	-	-	-
Pallares et al. [9]	-	-	-	-	3080	0.49
Dai et al. [10]	4418	0	3924	0.099	3124	0.48

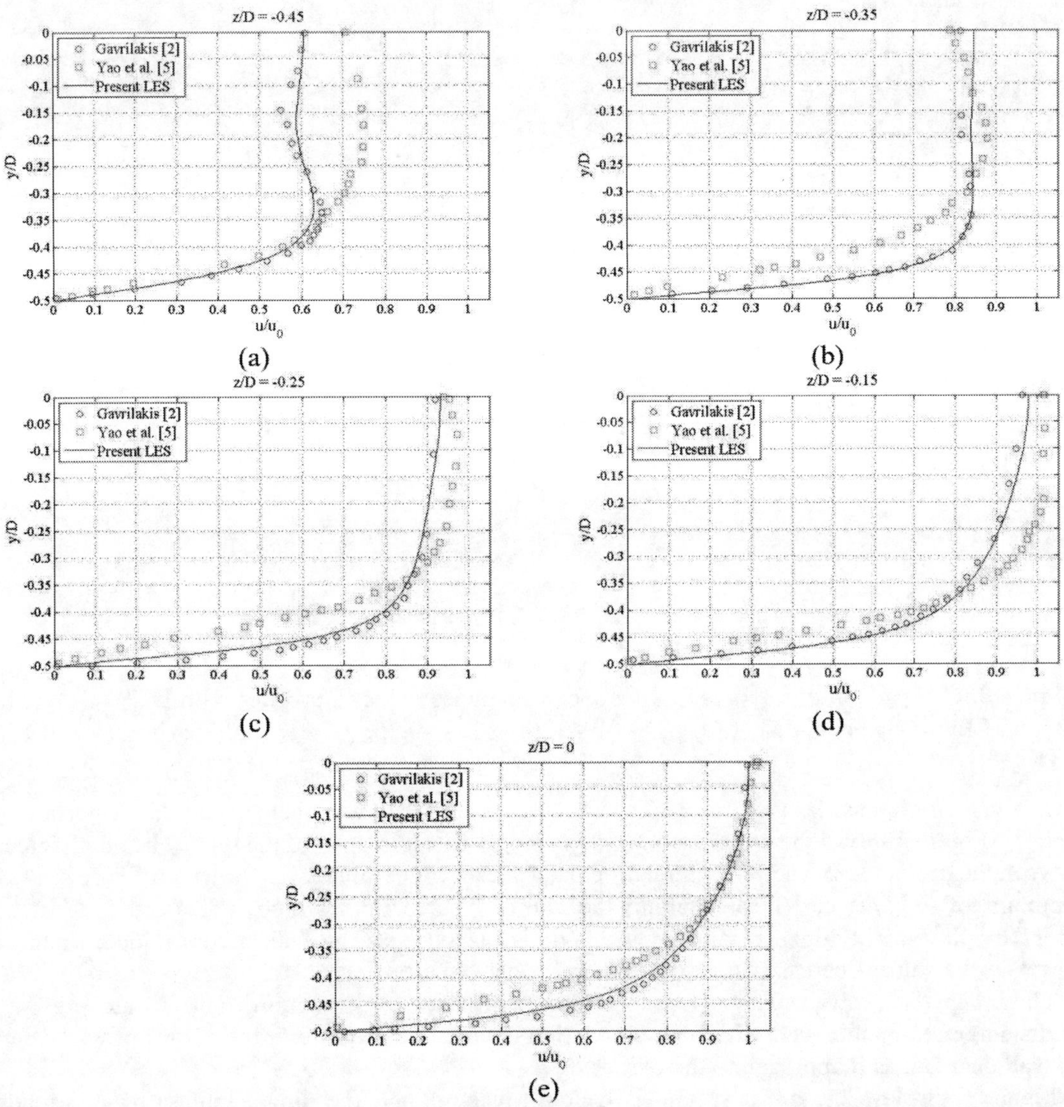

Figure 2. Comparison of the present LES mean streamwise velocity profiles with DNS of Gavrilakis [2] and LES of Yao et al. [5] at (a) z/D = -0.45; (b) z/D = -0.35; (c) z/D = -0.25; (d) z/D = -0.15; (e) z/D = 0.

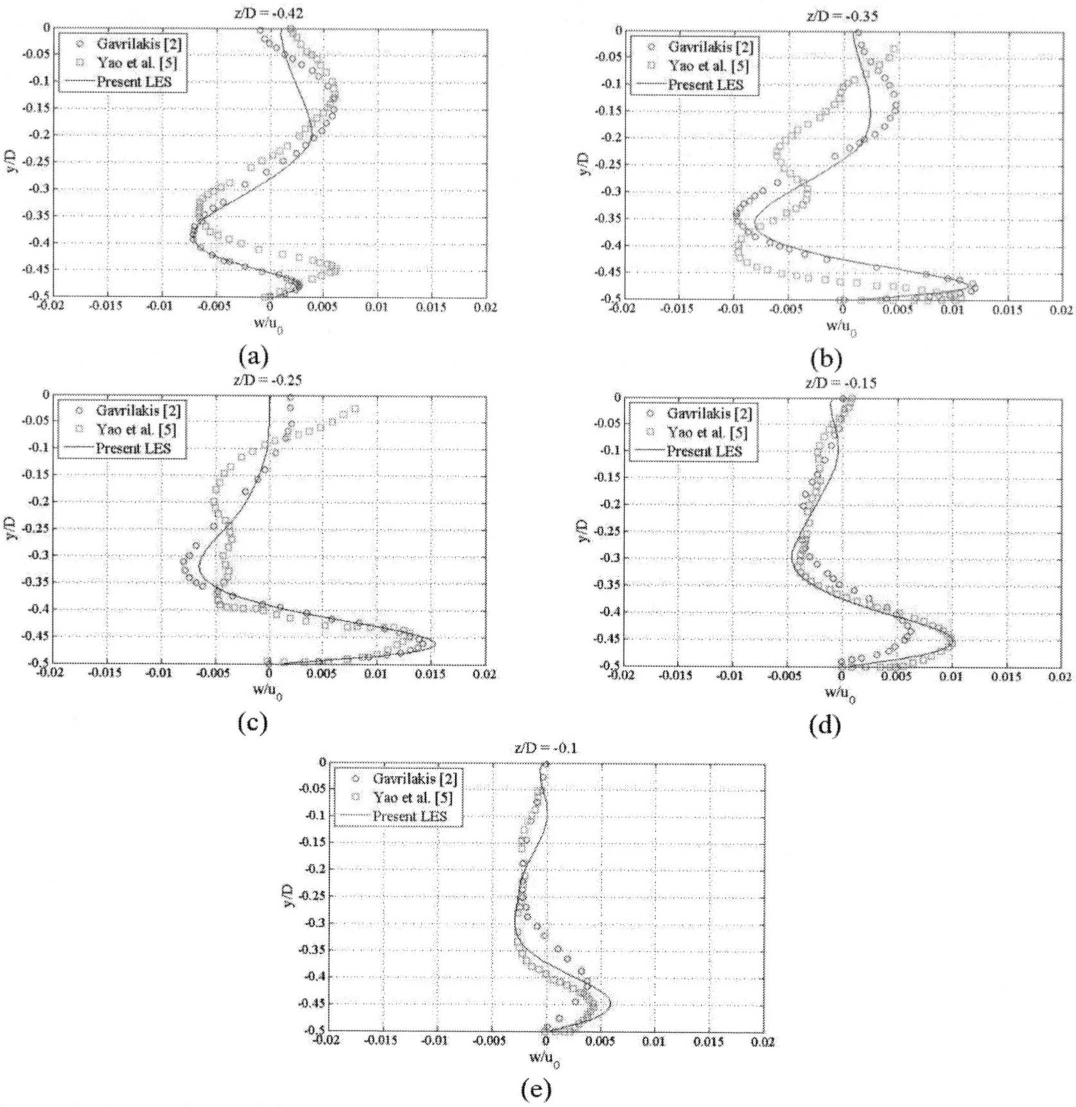

Figure 3. Comparison of the present LES mean spanwise velocity profiles with DNS of Gavrilakis [2] and LES of Yao et al. [5] at (a) z/D = -0.42; (b) z/D = -0.35; (c) z/D = -0.25; (d) z/D = -0.15; (e) z/D = -0.1.

The velocity intensities were normalized by the mean streamwise velocity at the centerline of the duct, u_0. Figure 4 shows the comparison of u_{rms}/u_0 at five locations of the duct. The good agreement between the present LES and both simulation databases were obtained. As shown on figure 4, the local minimum of u_{rms}/u_0 are located along the corner bisector and which indicate the effect of the secondary flows. The highest value of u_{rms}/u_0 is placed near the wall at the middle of the duct. However, the values decrease as its location approach to the corner.

The comparison of w_{rms}/u_0 at five locations of the duct are shown on figure 5. The present LES results agree reasonable well with both simulation databases. In figure 5, the values of w_{rms}/u_0 near the wall decrease as it approach to the corner.

Figure 6 displays the mean secondary velocity vectors and the dimensionless mean streamwise velocity contours and the mean secondary velocity streamline. As shown on figure 6(a) and figure 6(b), the turbulence induced secondary flows consists of two counter rotating vortices distributed in each corner of the duct. The high dimensionless mean streamwise velocity,$u^+ = u/u_\tau$, zone placed at each

IOP Conf. Series: Materials Science and Engineering **297** (2018) 012028 doi:10.1088/1757-899X/297/1/012028

corner of the duct represent the effect of the turbulence induced secondary flows. The secondary flows transfer the high momentum fluid from the center region of the duct towards the corners along the corner bisectors where the local minimum of u_{rms}/u_0 are located, as mentioned in figure 4.

Figure 7 represents the distribution of wall shear stress, τ_w, along the duct wall. The wall shear stress distribution of the present LES was compared with only DNS of Gavrilakis [2] since there were no available wall shear stress distribution results of any LES. Both results shown two local maximum of wall shear stress with some different of local maximum wall shear stress at the middle of the duct due to the sampling procedure. Near the corner, the present LES wall shear stress rapidly rises up to the local maximum at the dimensionless distance along the duct wall of about -0.35 due to the presence of the secondary flows which transfer the high momentum fluid from the center region of the duct towards the corners. Furthermore, the occurrence of the local maximum wall shear stress at the middle of the duct wall represents the Reynolds number effect [2].

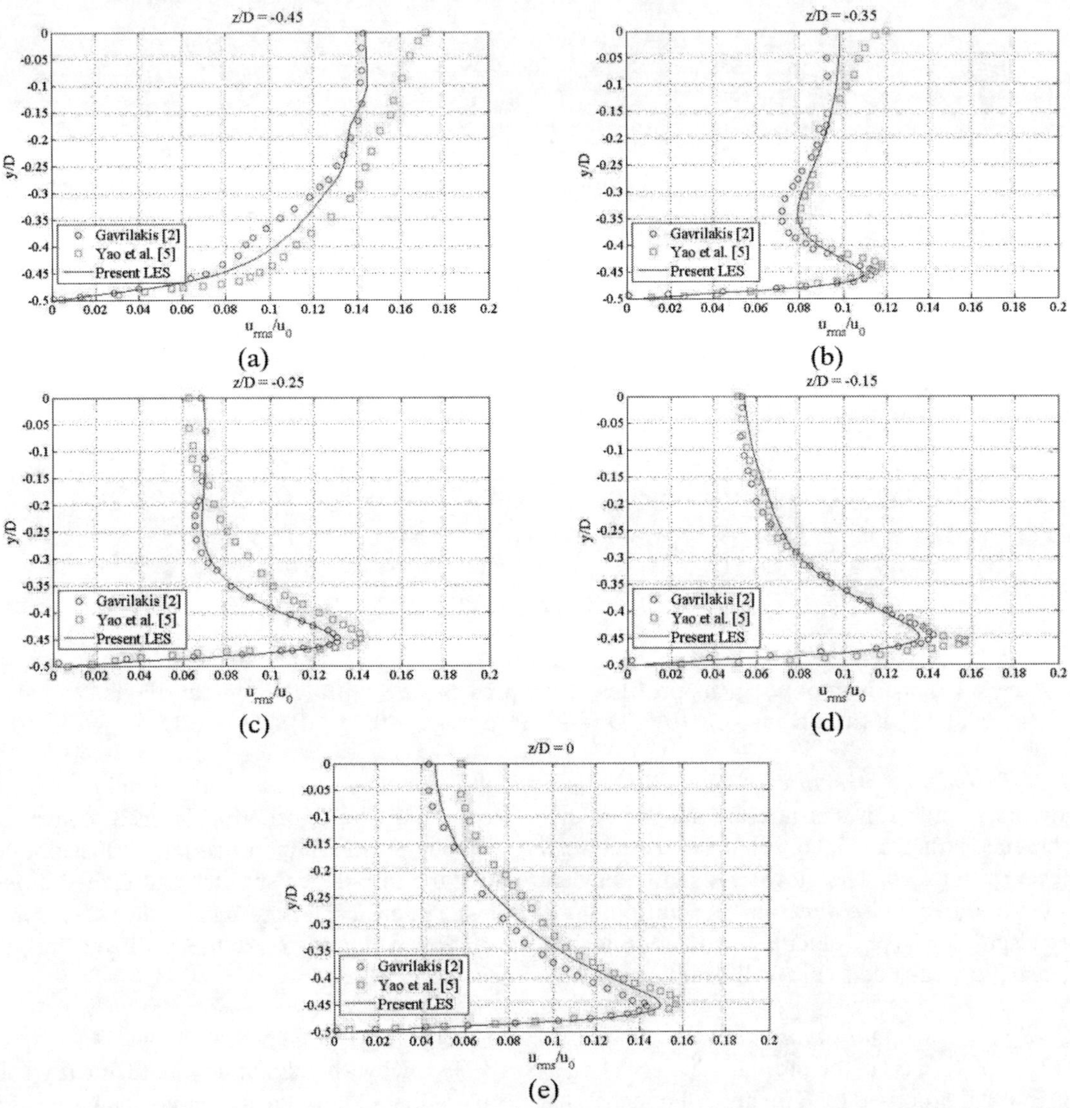

Figure 4. Comparison of u_{rms}/u_0 profiles of the present LES with DNS of Gavrilakis [2] and LES of Yao et al. [5] at (a) z/D = -0.45; (b) z/D = -0.35; (c) z/D = -0.25; (d) z/D = -0.15; (e) z/D = 0.

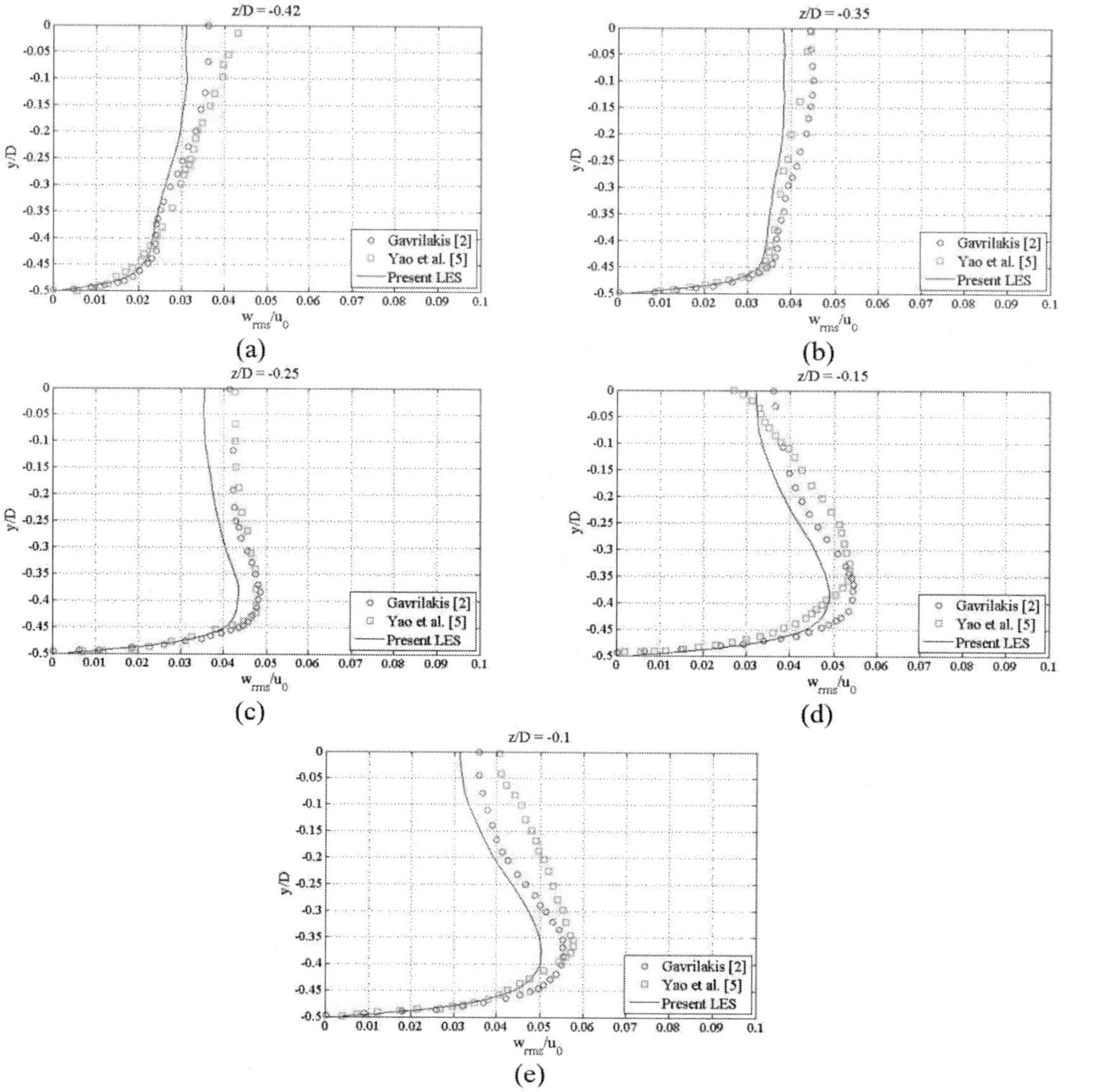

Figure 5. Comparison of w_{rms}/u_0 profiles of the present LES with DNS of Gavrilakis [2] and LES of Yao et al. [5] at (a) z/D = -0.42; (b) z/D = -0.35; (c) z/D = -0.25; (d) z/D = -0.15; (e) z/D = -0.1.

3.1.2. Turbulence structures. Figure 8(a) displays turbulence structures with turbulence intensity contours for the flows inside the stationary duct ($Ro_\tau = 0$). The Q-criterion is used to expose the turbulence structures. The turbulence structures are thoroughly distributed inside the duct due to the effects of the secondary flows. As shown on figure 8(a), the turbulence structures stretched along the duct in the streamwise direction. In addition, the turbulence intensity, I, is defined as the ratio of the root mean square of the velocity fluctuations to the mean flow velocity. Near the wall, the turbulence intensity is weak due to the wall damping effects.

3.2. Rotating straight square duct

For the rotating case, the present ANSYS FLUENT 18.1's LES with a dynamic kinetic energy subgrid scale model proposed by Kim and Menon [12] has been compared with the available databases of DNS of Dai et al. [10] and DNS of Fang et al. [11]. The comparison of the rotating straight square ducts at two different friction rotation numbers of 1.3 and 5 with the constant friction Reynolds number of 300 were presented in figures 6, 8, 9, 10, and 11. The y/D denotes the dimensionless distance along normal direction from -0.5 to 0.5 and the z/d denotes the dimensionless distance along the spanwise direction from -0.5 to 0.5. The DNS of Fang et al. [11] at the friction rotation number of 1 was chosen for comparison due to the lack of data of DNS of Fang et al. [11] at the friction rotation number of 1.3.

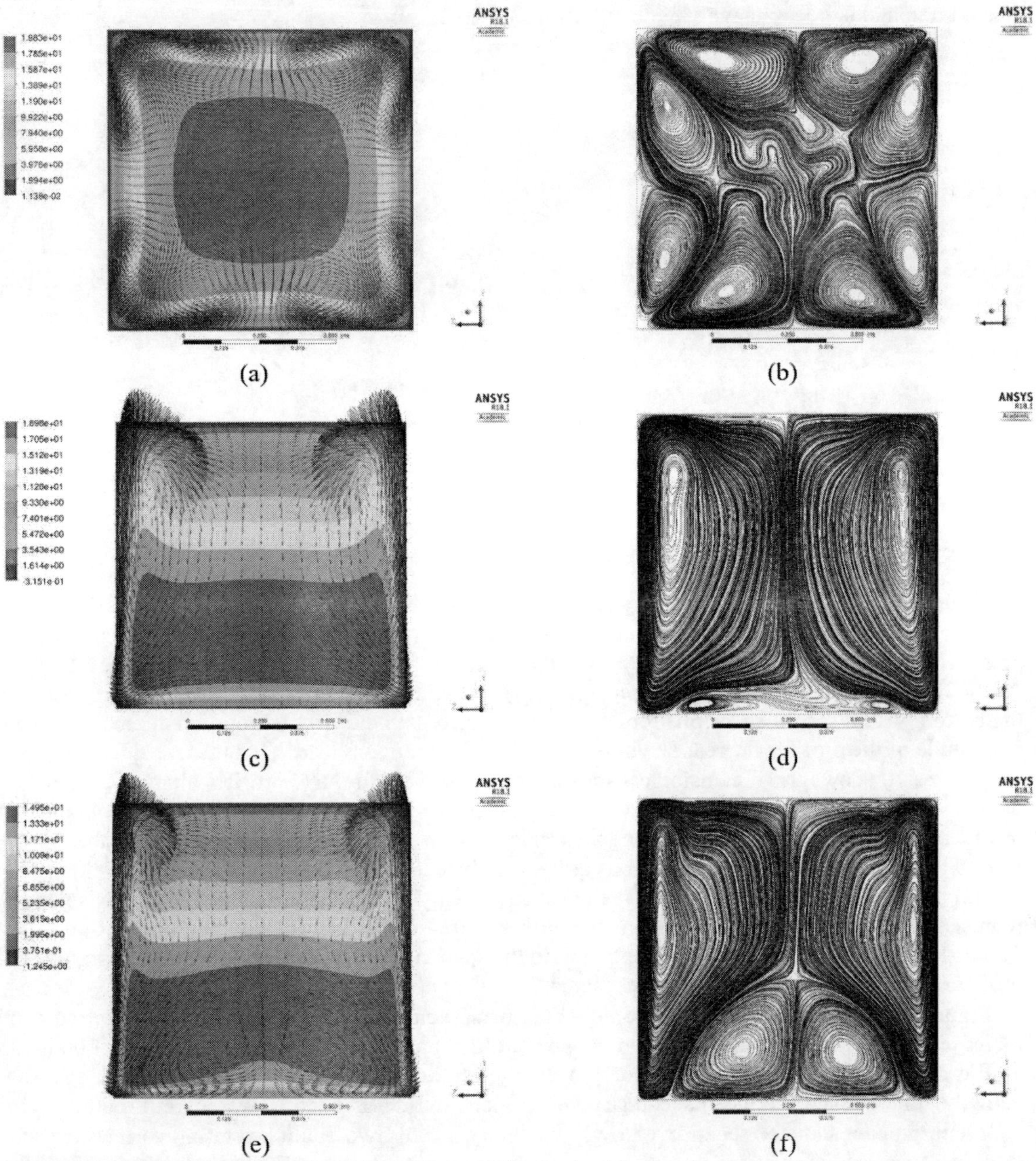

Figure 6. Mean secondary velocity vectors and dimensionless mean streamwise velocity contours at (a) Ro_τ =0; (c) $Ro_\tau = 1.3$; (e) $Ro_\tau = 5$, and Mean secondary streamline at (b) Ro_τ =0; (d) $Ro_\tau = 1.3$; (f) $Ro_\tau = 5$.

3.2.1. *Mean flow field*

The mean flow field were obtained over a time period of $270\ \mathrm{D}/\mathrm{u}_\tau$ and $280\ \mathrm{D}/\mathrm{u}_\tau$ with the coarse mesh for the friction rotation numbers of 1.3 and 5, respectively. The mean velocity profiles along the wall bisectors in normal and spanwise directions of the present LES at $Ro_\tau = 0, 1.3$ and 5 were compared with DNS of Dai et al. [10] at $Ro_\tau = 1.3$ and 5 and DNS of Fang et al. [11] at $Ro_\tau = 1$ and 5. In addition, the friction velocity scales were used since the friction velocity can directly indicate the shear stress distribution and there were available DNS results of the flows.

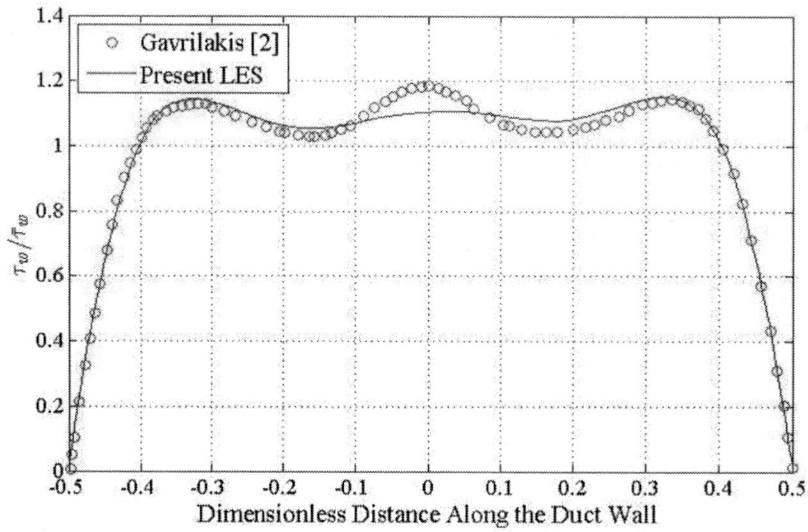

Figure 7. Comparison of wall shear stress of the present LES with DNS of Gavrilakis [2].

Figure 9 shows the dimensionless mean streamwise velocity, u^+, profiles along the y axis at the wall bisector. The good agreement between the present LES and both DNS databases was obtained. For the present LES stationary duct case ($Ro_\tau = 0$), the profile is symmetrically distributed along the y axis. As the friction rotation number increases up to 1.3, the profile tend to move toward the pressure side on the bottom wall due to the Coriolis force. By further increases the friction rotation number up to 5, the magnitude of the profiles decreases while the profiles still move toward the pressure side.

Figures 10 shows the dimensionless mean streamwise velocity, u^+, profiles along the z axis at the wall bisector. The good agreement between the present LES and both DNS databases was obtained. The profile is symmetrically distributed along the z axis at $Ro_\tau = 0$. As the friction rotation number increases from 0 to 1.3, the profiles along the z axis at the wall bisector become flat with constant velocity in the central region of the duct. In addition, the constant velocity profile region along the spanwise direction represents the Taylor-Proudman region. By further increases the friction rotation number up to 5, the Taylor-Proudman region increasingly expands in the spanwise direction. However, the magnitude of the profiles decrease as the rotation number increases.

Figure 11 clarifies that the dimensionless bulk mean velocity, $U_b{}^+$ decreases as the rotation number increases. The good agreement between the present LES and DNS of Fang et al. [11] was obtained.

Figure 6 displays the mean secondary velocity vectors and the dimensionless mean streamwise velocity contours and the mean secondary velocity streamline. For the flows without rotation ($Ro_\tau = 0$), the turbulence induced secondary flows which consists of two counter rotating vortices distributed in each corner of the duct are presented. As the friction rotation number increases from 0 to 1.3, the vortices above the suction side are completely suppressed while the turbulence induced two counter rotating vortices on bottom wall of the stationary duct are reduced in size. Hence, the Coriolis force driven two large circulations with a pair of additional vortices on the pressure wall are presented. The Ekman layers are formed on the lateral walls. The descending current transfer the low momentum fluid from the suction side towards the pressure side while the ascending current appear on the lateral wall.

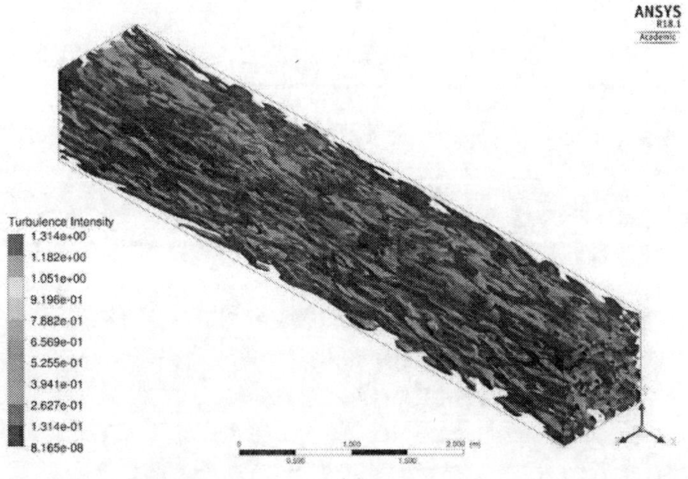

(a)

(b)

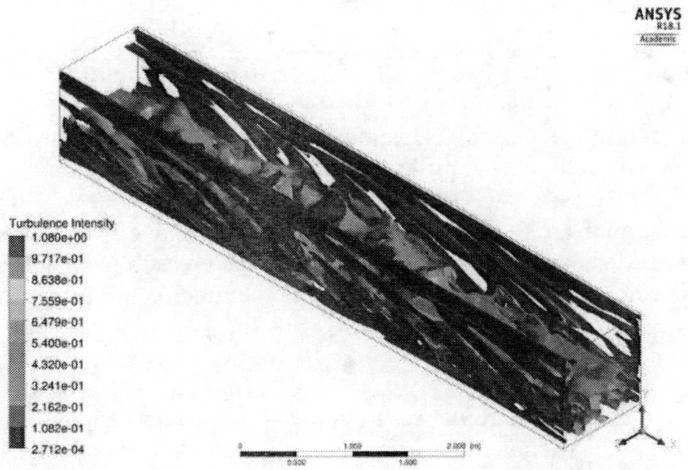

(c)

Figure 8. Turbulence structures with turbulence intensity contours (Q-criterion) at (a) $Ro_\tau = 0$; (b) $Ro_\tau = 1.3$; (c) $Ro_\tau = 5$.

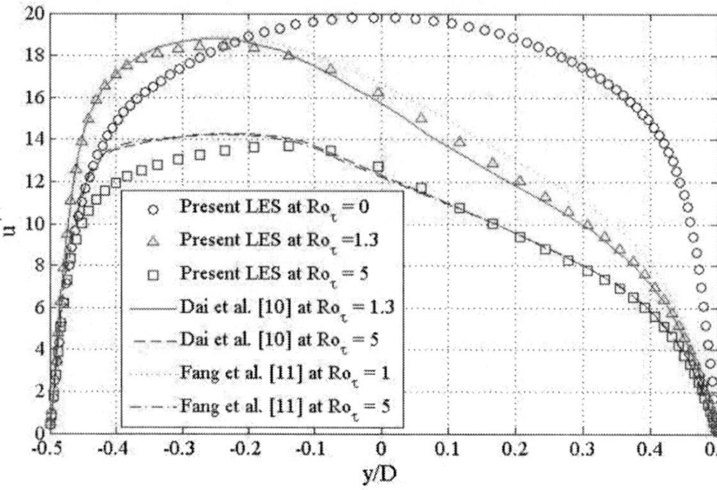

Figure 9. Comparison of mean vertical velocity profiles of the present LES with DNS of Dai et al. [10] and DNS of Fang et al. [11] at different friction rotation numbers.

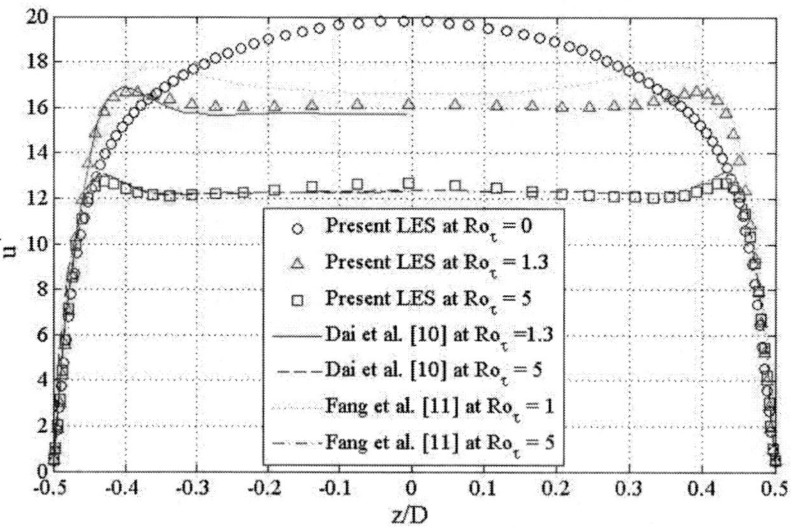

Figure 10. Comparison of mean spanwise velocity profiles of the present LES with DNS of Dai et al. [10] and DNS of Fang et al. [11] at different friction rotation numbers.

The Taylor-Proudman region are found in the center region of the duct. By further increases the friction rotation number up to 5, the pair of additional vortices on the pressure wall are grown again. However, the mean streamwise velocity continuously decreases as the rotation number increases.

3.2.2. Turbulence structures

Figure 8(b) and figure 8(c) display turbulence structures with turbulence intensity contours for the flows inside the rotating duct at the friction rotation number of 1.3 and 5, respectively. The Q-criterion is used to expose the turbulence structures. As shown on both figure 8(b) and figure 8(c), the turbulence structures are totally suppressed on the top wall due to the effects of Coriolis force. Likewise, the turbulence intensity is weak on the top wall where the turbulent is suppressed whereas the high turbulence intensity occurred on the bottom wall where the turbulent is enhanced. As the rotation number increases, the turbulence structures are much more stretched out on the top wall. By the way, the overall turbulence intensity is reduced as the rotation number increases due to the effects of Coriolis force. Therefore, the flows are laminarized at high rotation number.

4. Conclusion

In this paper, the incompressible fully developed turbulent flows in the stationary and rotating straight square ducts with the two different friction rotation number of 1.3 and 5 at the fixed friction Reynolds number of 300 was presented using ANSYS FLUENT 18.1's LES with a dynamic kinetic energy subgrid scale model proposed by Kim and Menon [12]. The present LES results were compared with the available simulation databases of DNS and LES.

The good agreement between the present LES results and the reference data were obtained. However, the different result of the local maximum wall shear stress at the wall bisector of the stationary duct was obtained due to the sampling procedure.

For non-rotating duct ($Ro_\tau = 0$) case, the turbulence induced secondary flows are presented which consists of two counter rotating vortices distributed in each corner of the duct.

As the rotation number increases, the turbulence induced secondary vortices are suppressed on the suction side while reduced in size, at $Ro_\tau = 1.3$, and grown again, at $Ro_\tau = 5$, on the pressure side. Thereby, the Coriolis force driven two large circulations with the pair of additional vortices are presented. By the way, the Taylor-Proudman region is found in the center region of the duct and the Ekman layers are formed on the lateral wall. Furthermore, the turbulent flows are enhanced on the pressure side whereas the flows are laminarized on the suction side. Thus, the turbulence structures are stretched out on the suction side due to the effects of rotation. Moreover, the overall turbulence intensity is reduced as the rotation number increases. Hence, the flows are laminarized at high rotation number. However, the magnitudes of bulk streamwise velocity are found to be decreased as the rotation number increases due to the Coriolis effects.

Finally, the physics of the secondary flows in the stationary and rotating straight square ducts can be represented by using ANSYS FLUENT CFD software with the realistic accuracy of LES with one transport equation dynamic kinetic energy subgrid scale model.

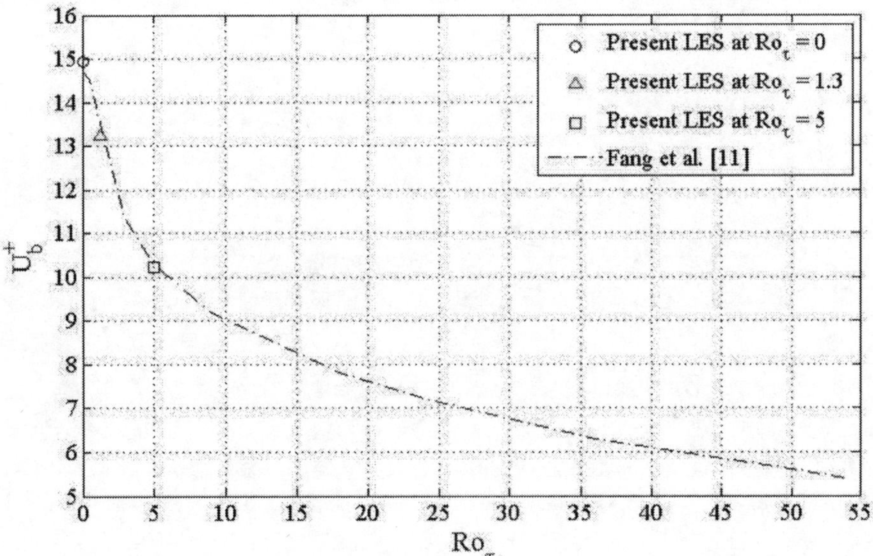

Figure 11. Comparison of dimensionless bulk mean velocity of the present LES with DNS of Fang et al. [11] at different friction rotation numbers.

Acknowledgement
This work was supported by the graduate research assistant scholarship granted by the faculty of engineering, Kasetsart University. The simulations were performed on Wata Cluster supercomputer at the faculty of engineering, Kasetsart University.

References
[1] Madabhushi R K and Vanka S P 1991 *Physics of Fluids A* **3** pp 2734-2745
[2] Gavrilakis S 1992 *J. of Fluid Mechanics* **244** pp 101-129
[3] Huser A 1993 *J. of Fluid Mechanics* **257** pp 65-95
[4] Joung Y, Choi S U and Choi J I 2007 *J. of Engineering Mechanics* **133** pp 213-221
[5] Yao J, Zhao Y and Fairweather M 2015 *Applied Thermal Engineering* **91** pp 800-811
[6] Zhang H, Trias F X, Gorobets A, Tan Y and Oliva A 2015 *Int. J. of Heat and Fluid Flow* **54** pp 258-267
[7] Pallares J and Davidson L 2000 *Physics of Fluids* **12** pp 2878-2894
[8] Pallares J and Davidson L 2002 *Physics of Fluids* **14** pp 2804-2816
[9] Pallares J, Grau F X and Davidson L 2005 *Physics of Fluids* **17** pp 1-11
[10] Dai Y J, Huang W X, Xu C X and Cui G X 2015 *Physics of Fluids* **27**
[11] Fang X, Yang Z, Wang B C and Bergstrom D J 2017 *Int. J. of Heat and Fluid Flow* **63** pp 88-98
[12] Kim W W and Menon S 1997 *the 35th Aerospace Sciences Meeting and Exhibit* (Reno)
[13] Germano M, Piomelli U, Moin P and Cabot W 1991 *Physics of Fluids A* **3** pp 1760-1765

IOP Conf. Series: Materials Science and Engineering **297** (2018) 012029 doi:10.1088/1757-899X/297/1/012029

Prediction of thermal characteristics of turbulent spot using Large Eddy Simulation

Warakorn Pichitpreecha, Jirachai Mingbunjerdsuk and Weerachai Chaiworapuek*

Department of Mechanical Engineering, Faculty of Engineering, Kasetsart University,50 Ngamwongwan Road, Ladyao, Chatuchak, Bangkok 10900, Thailand

* Corresponding Author: E-mail: fengwcc@ku.ac.th, Telephone Number: 0-2797-0999#1868, Fax Number: 0-2579-4576

Abstract. This research presents the prediction of thermal behavior of a turbulent spot on a flat plate having a constant heat flux of 1893 W/m2 using Large Eddy Simulation (LES) with 7.1 million grid points. In this study, the water is used as a working fluid. The mainstream flow on the heating plate having the width of 0.2 m x the length of 0.4 m, corresponding to the local Reynolds number between 42,200 and 98,400. A laminar boundary layer on the test plate was transformed into Bypass transition by injecting the water to initiate the turbulent spot in upward direction and perpendicular to the mainstream flow with the velocity of 26.31 m/s and the period of 0.01 seconds through a 1 mm diameter hole. The results are presented as the contours of Nusselt number and temperature of the spot in the top and elevation views, respectively. They show that the heated near wall water is accumulated by the spot and cause the increase of the temperature inside the spot body in the higher layer flow. The near wall water is replaced by the water from the upper layer and makes the decrease of the surface temperature underneath the spot. Consequently, this leads to the increase of the Nusselt number within the spot bound above the laminar state. The yielded convective coefficient, spot celerities, and half spreading angle from the LES agree well with experimental results reported by other researchers. Thus, this obtained information is a strong evidence to confirm that the LES can provide an accurate prediction of the characteristics of the artificially initiated turbulent spot.

1. Introduction

A turbulent patch or turbulent spot under the boundary layer transition over flat surface can be directly initiated by injecting the water jet in upward direction, perpendicular to the mainstream flow. The spot has been found to have a potential to enhance the heat transfer because it can boost the heat transfer up to 25% above the laminar state with a small input power [1]. The heat transfer is still increasing when the spot propagates further downstream. This turbulent patch or turbulent spot was discovered in 1951 [2]. The turbulent spot plays a key role on boundary layer transition over flat plate and the turbulent flow is performed when a single spot merges with the neighbours. The turbulent spot is a combination of the horseshoe or hairpin vortices [3]. Its structure is arrowhead like shape and this shape alters under the streamwise pressure gradient [4, 5]. The leading and trailing edges of the turbulent spot propagates downstream with the celerity of 74% and 53% of the freestream velocity, respectively [6]. Because the turbulent spot is under the unsteady flow and depends on the instability of the flow, Direct Numerical Simulation – DNS has been used for the prediction of its behavior [7, 8]. However, since the average characteristics of the turbulent spot at a specific moment was reported by Rakpakdee and Chaiworapuek

[9]. Thus, the Large Eddy Simulation-LES is found to have a big potential to simulate the presence of the turbulent spot. It also consumes the computing resource less than the DNS. The Large Eddy Simulation is a mathematical model, used in computational fluid dynamics - CFD. It was first proposed by Smagorinsky [10] in 1963 to simulate atmospheric air current and first utilized by Deardoff [11] in 1970. It is currently applied in many engineering applications such as combustion [12], acoustics [13] and boundary-layer flows [14] etc.

The LES employs the Navier-Stokes equation to resolve a very wide range of time and length scales, directly affect the fluid flows. The detailed flow behaviour such as the velocity fluctuation can be calculated by the LES instead of the Reynolds Averaged Navier Stokes – RANS while the computational cost can be reduced by ignoring the small length scales comparing with the DNS. In LES, these small length scales are neglected by low-pass filtering. They can be viewed as time and spatial averaging from the numerical simulation. Thus, the average behaviour of the detailed turbulent flow can be captured by the LES. This consequently leads to the simulation capability to predict the average behaviour of a turbulent patch, surrounding by the laminar flow. Hence, the current study focuses on the numerical simulation of an artificially turbulent spot over a constant heat flux surface in 3D flow domain. The simulation will be set up as the experiment of Rakpakdee and Chaiworapuek [9] and the results are mutually compared and discussed. The obtained information will unveil the performance of LES in order to predict the characteristics of turbulent spot under the boundary layer transition.

2. Numerical methods and validation

The computational domain in the present study models the transitional flow from the experiment of Rakpakdee and Chaiworapuek [9]. It is rectangular domain, having 0.514 m length in x direction, 0.08 m height in y direction, and 0.2 m width in z direction. The x, y, and z axes refer to the streamwise, heightwise, and spanwise directions, respectively as shown in figure 1. From the setup of Rakpakdee and Chaiworapuek [9], this numerical domain is modeled during $x = 0.186 - 0.7$ m from the plate leading edge to reduce the computing resource. This streamwise distance corresponds to the local Reynolds number, Rex between 26,200 and 98,400. A 1 mm diameter hole locates on the bottom wall at $x = 0.3$ m or the Rex of 42,200. It is used to inject the water in perpendicular direction of the mainstream with the speed and duration of 26.31 m/s and 0.01 s, respectively. Beyond the location of the injection hole, the heat flux of 1,893 W/m2 is applied on the bottom wall to establish a constant heat flux condition of the flat plate. The outlet is set as pressure outlet of 0 Pa. The left and right boundaries are symmetry while the upper wall and the inlet are the velocity inlet, having the magnitude of 0.179 m/s in the direction of the mainstream flow.

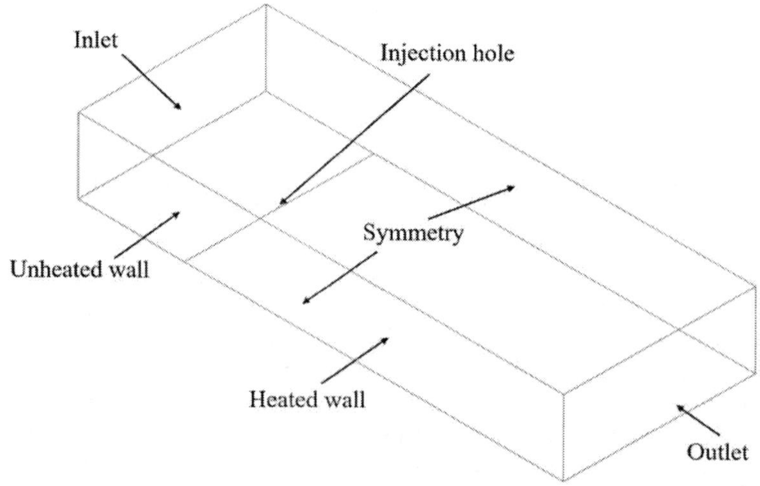

Figure 1. A water flow domain.

The cubic grid, having the size of $l_x = 0.001$ m, $l_y = 0.001$ m, and $l_z = 0.001$ m was meshed through the flow domain. Thus, these caused the 434x81x201 elements or 7.1 million grid points as depicted as figure 2. This study resolved 80% of the turbulent kinetic energy, causing $l / l_0 = 0.42$. The l_0 is the integral length scale that the turbulent kinetic energy is maximum. It can be given as:

$$l_0 = k^{0.5} / (C_\mu \omega) \tag{1}$$

where k = 0.000071, resolved from RANS
$C_\mu = 0.09$
ω = 13.022408, resolved from RANS

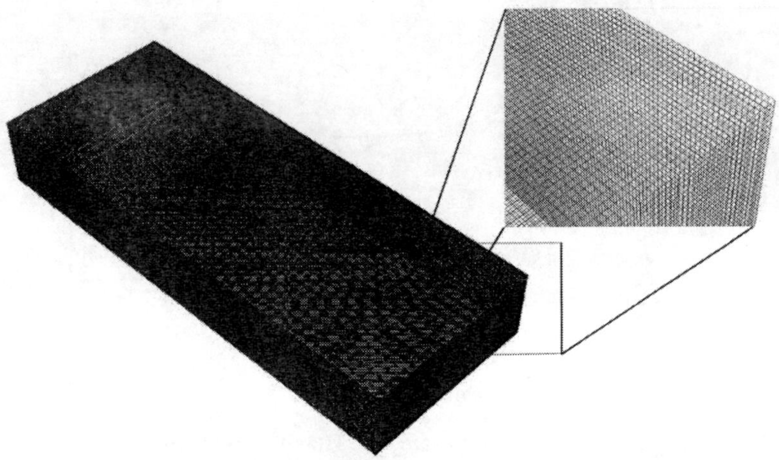

Figure 2. The grid in the flow domaian.

In the current study, the local Reynolds number on the flat plate can be determined as:

$$Re_x = \rho u_\alpha x / \mu \tag{2}$$

where Re_x = local Reynolds number
ρ = water density (kg/m^3)
U_α = freestream velocity (m/s)
x = streamwise distance from the leading edge (m)
μ = dynamic viscosity of water (kg/m•s)

As the boundary condition of the flow domain, the local Nusselt number can be predicted following Kays et al. [15] as:

$$Nu_x = (0.453 Re_x^{1/2} Pr^{1/3}) / [1-(\xi/x)^{3/4}]^{1/3} \tag{3}$$

where Nu_x = local Nusselt number
Pr = Prandtl number
ξ = unheated starting length (m)

In this simulation, the convective coefficient and the local Nusselt number are determined as:

$$h_x = q / (T_s - T_\alpha) \tag{4}$$

and

$$Nu_x = h_x x / k \tag{5}$$

where h_x = local heat transfer coefficient (W/m^2•°C)

q = heat flux (W/m^2)
T_s = surface temperature (°C)
T_α = freestream temperature (°C)
k = thermal conductivity (W/m•°C)

The local Nusselt number from this simulation is validated by comparing with those determined from equation (3) as depicted in figure 3. The figure shows that they agree well with each other.

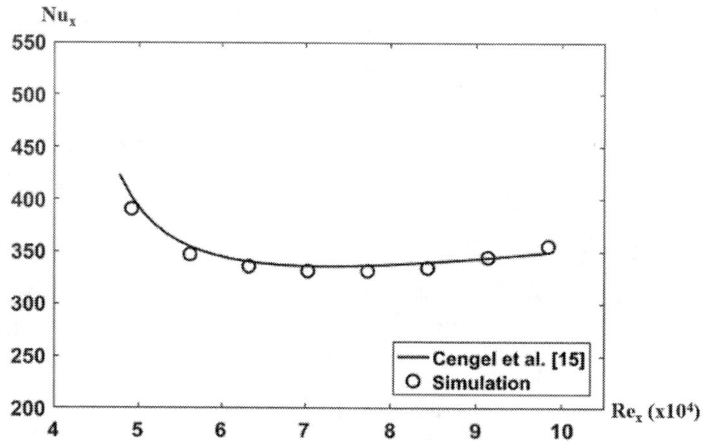

Figure 3. The comparison between the local Nusselt number from the simulation and the correlation of Kays et al. [15].

For laminar boundary layers over a flat plate, the Blasius solution gives:

$$\delta(x) = 5x / Re_x^{1/2} \quad (6)$$

where δ = the overall thickness of the boundary layer (m)

The thickness and velocity can be normalized as:

$$\eta = y / \delta(x) \quad (7)$$

and

$$df/d\eta = u/u_\alpha \quad (8)$$

where η = normalized thickness
y = the distance from the wall to a point (m)
$df/d\eta$ = normalized velocity

The velocity profiles from the simulation are validated with Blasius profile at Rex = 42200, 56200, 70300, 84300 and 98400 as illustrated in figures 4. The comparison shows that the obtained profiles are consistent with the Blasius ones as shown in figure 5. At the location of spot generator, the boundary layer displacement thickness is 0.002 m. It is employed to normalize the length of the flow domain through the simulation.

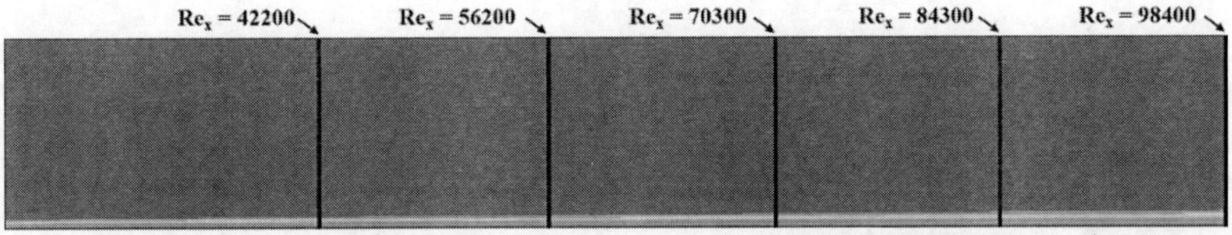

Figure 4. The location of the flow domain at Re_x = 42200, 56200, 70300, 84300 and 98400.

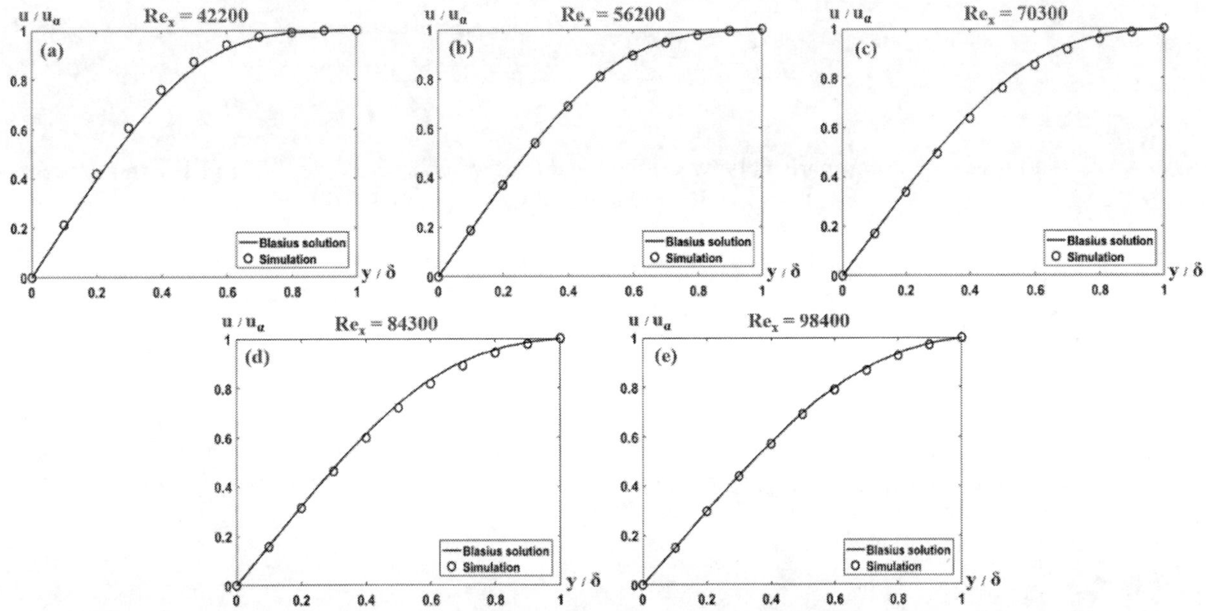

Figure 5. The comparison of the obtained velocity profile and Blasius profile at Re_x = (a) 42200, (b) 56200, (c) 70300, (d) 84300 and (e) 98400.

Moreover, the average convective coefficient calculated inside the bound of the turbulent spot from the LES is validated with the experimental results of Rakpakdee and Chaiworapuek [9] as shown in figure 6. The figure shows that the average heat transfer coefficient from both simulation and experiment is decreasing with time but the numerical results are approximately 10% higher. This is because the heat power, reported by Rakpakdee and Chaiworapuek [9] is the total power of the heater. This power can loss to the other directions even it was well insulated. The actual power, transferring into the water is only partial amount. Thus, the difference, occurring through the comparison can be caused by this reason.

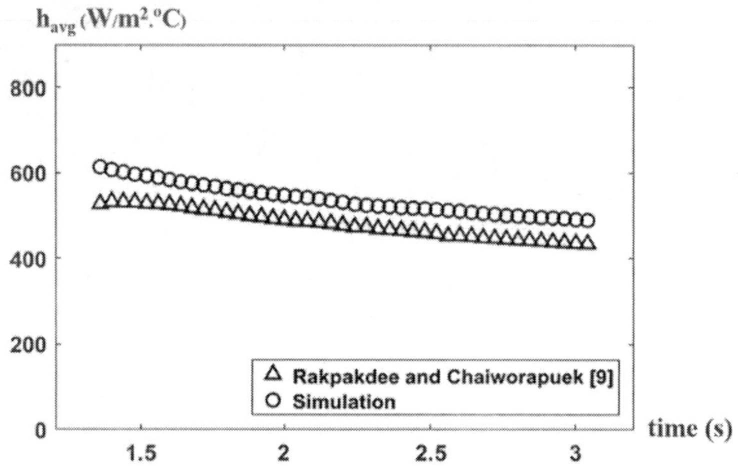

Figure 6. The comparison between the local heat transfer coefficient from LES and experiment of Rakpakdee and Chaiworapuek [9]

3. Numerical results

Figure 7(a) – (b) shows the contour of the instantaneous convective coefficient at the region of the turbulent spot at $t = 3.04$ s. The turbulent spot appears as streaky structure and the maximum magnitude of 650 W/m^2•°C locates at the core of frontal streaks as depicted in figure 7(a). These are in agreement with the contour of the spot footprint, ensemble averaged from 30 turbulent spots by Rakpakdee and Chaiworapuek [9] as illustrated in figure 7(b). The size of both spots is not different at this time. However, the maximum convective coefficient from the ensemble averaging method is less than the instantaneous turbulent spot.

IOP Conf. Series: Materials Science and Engineering **297** (2018) 012029 doi:10.1088/1757-899X/297/1/012029

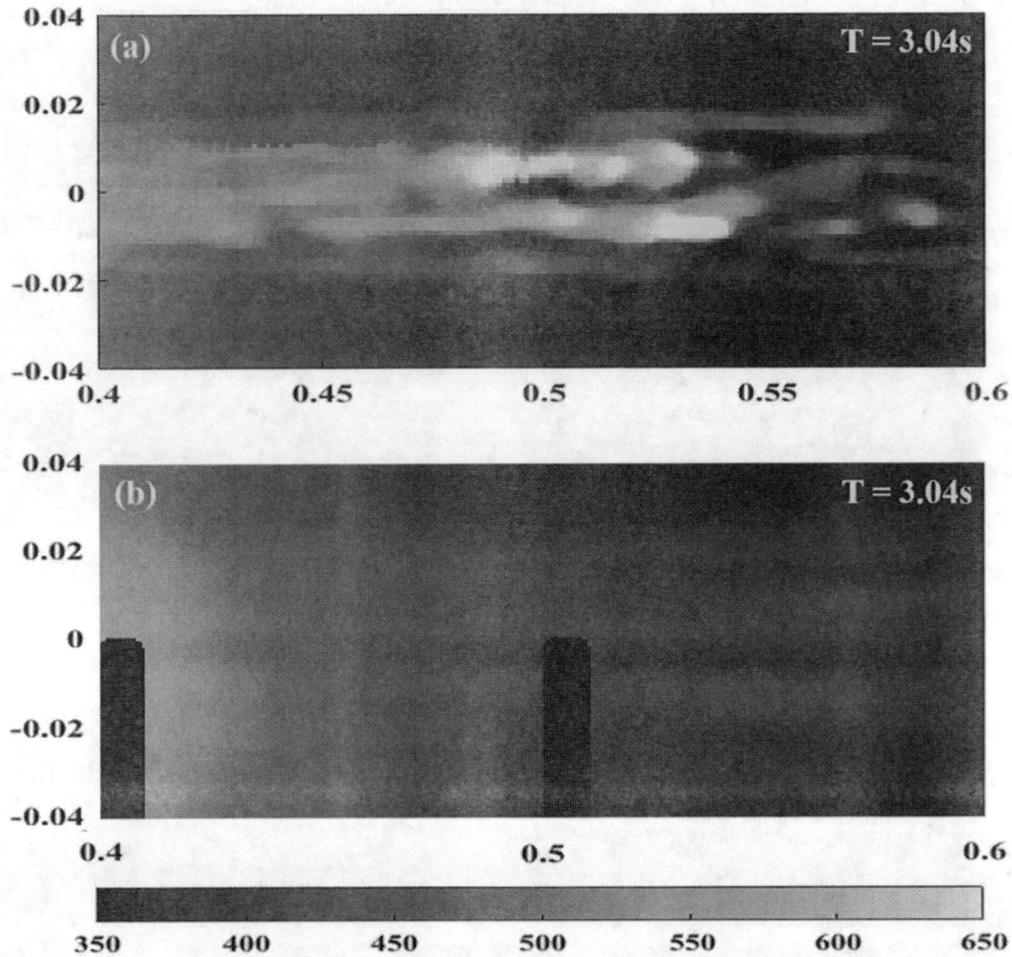

Figure 7. Convective coefficient contours from (a) LES and (b) experiment of Rakpakdee and Chaiworapuek [9] in the region of turbulent spot.

In this simulation, the time is counted after injecting the water through a 1 mm diameter hole and the dimensionless time was defined by

$$\tau = t/(\delta^*/u_\alpha) \tag{9}$$

where τ = the dimensionless time
t = time (s)
δ^* = boundary layer displacement thickness at the water injection (m)

Figure 8(a) - (f) presents the contour of local Nusselt number on the heating surface at τ = 42, 81, 123, 162, 204 and 243, respectively. This can refer to the turbulent spot footprint. The x axis is the local Reynolds number from 42,200 to 98,400 while y axis shows the normalized spanwise distance, $Z = z/\delta^*$ between -50 and 50. The magnitude of the local Nusselt number can be obtained from the below colour bar. At τ = 42 - 123, the relatively high Nusselt number appears inside the spot bound, consistent with the results reported by Chaiworapuek and Kittichaikarn [1]. This occurs from the process of the spot generation. Beyond this time, the maximum Nu_x inside the spot decreases because the effect from the spot initiation fade out and it increases again after τ = 204 by the spot maturity. Furthermore, the spot structure spreads out in spanwise direction with the half spreading angle of 6.45° and stretches in streamwise direction when the spot convects further downstream.

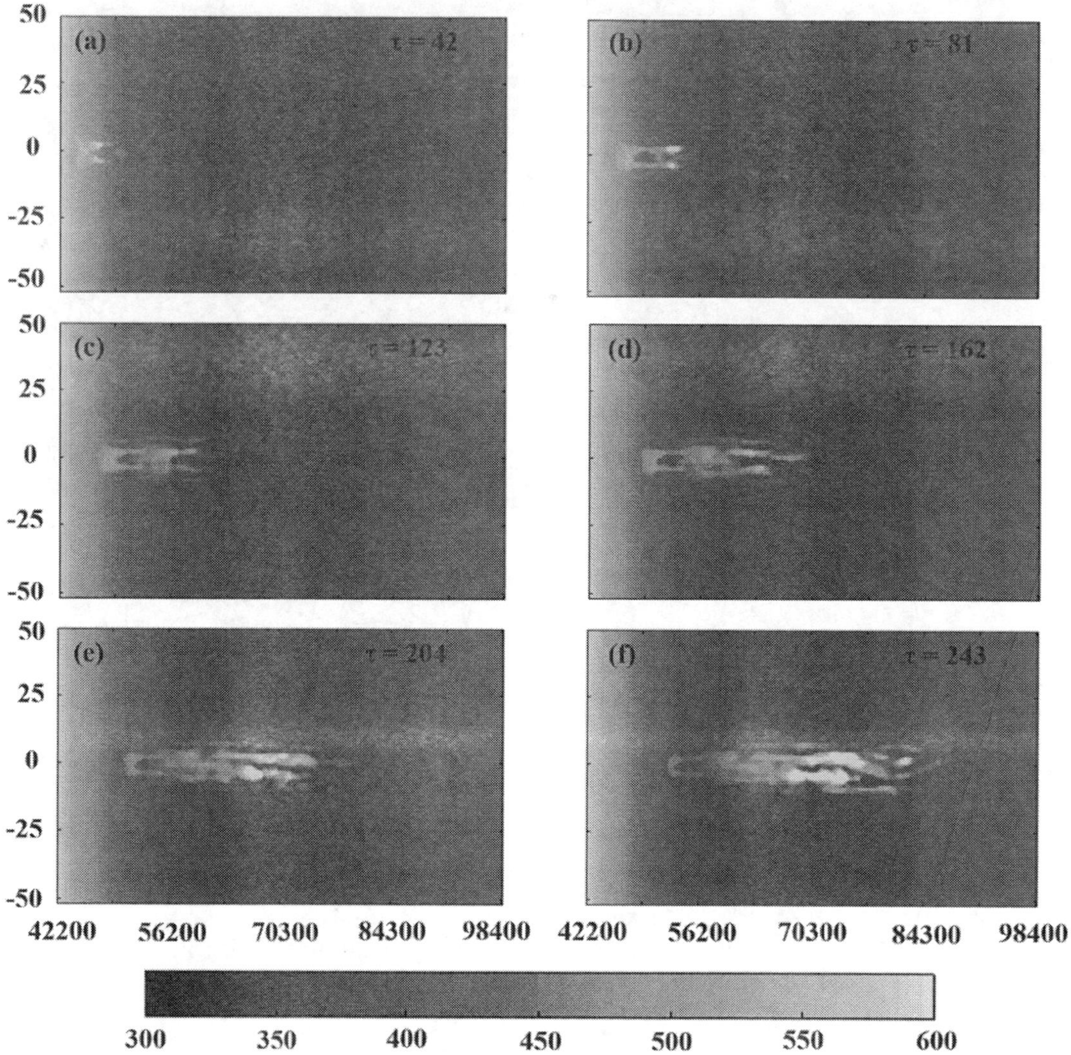

Figure 8. The contour of the Nusselt number at τ = (a) 42, (b) 81, (c) 123, (d) 162, (e) 204, and (f) 243

Figure 9(a) – (d) shows the contour of the instantaneous temperature inside the turbulent spot at τ = 123, 162, 204, and 243. In this figure, x axis is Re_x, in the same range as figure 8. Meanwhile, y axis presents the normalized heightwise distance, $Y = y / \delta^*$ *between* 0 and 30. It is noted that these figures are yielded from the subtraction process between temperature contours with and without the presence of the spot. Then, these distributions can present the turbulent spot structure in heightwise direction and the colour bar presents the temperature difference between 0° and 1°. The results show that the instantaneous turbulent spot accumulates the heated near wall water inside itself when it convects downstream. The spot shape, appearing in the figure is not the single structure because the turbulent spot is a combination of the hairpin vortices. These vortices alternately emerge and are consistent to the substructures, described by Sankaran et al. [6]. If the spot structure is reconstructed using the ensemble averaging method, the shape will appear as those reported by Van Atta and Helland [16]. Besides, the results show that the spot also grows in y direction, meaning that this turbulent patch enlarges in streamwise, spanwise, and heightwise directions simultaneously.

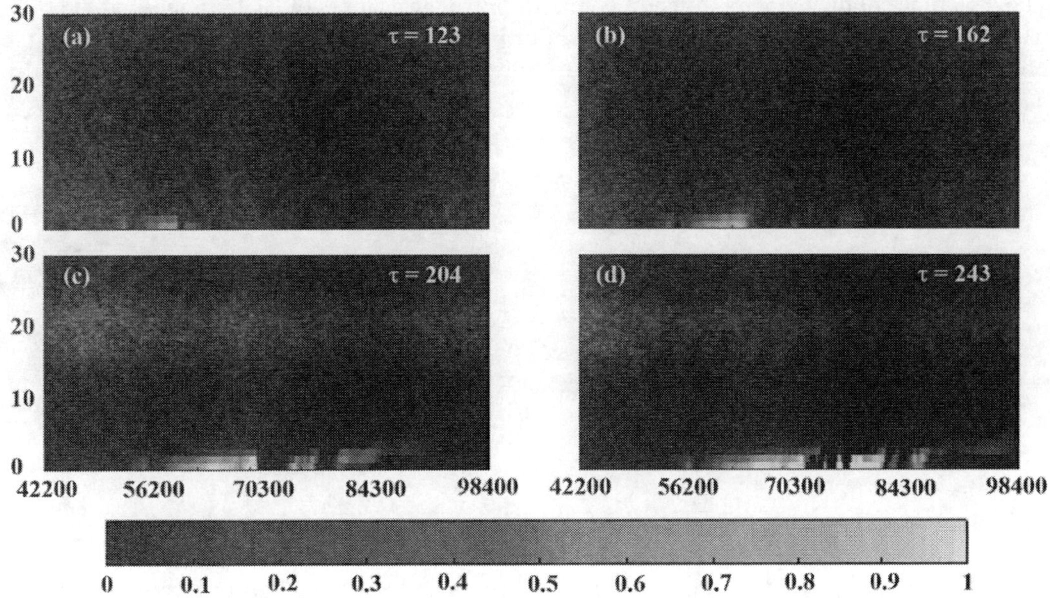

Figure 9. The contour of the temperature difference (side view) at $\tau = 123, 162, 204, 243$

The average Nusselt number, Nu_{avg} is then calculated within the turbulent spot bound and plotted with the Nusselt number of the heating surface, averaged in the same area as shown in figure 10. The figure presents that the Nu_{avg} increases with time and the turbulent spot boosts the heat transfer above the laminar state. At $\tau = 250$, the heat transfer capability of the spot is higher than the level of laminar approximately 21.86% and this difference is increasing with the time. Hence, this increase will eventually lead to the heat transfer level of the turbulent flow.

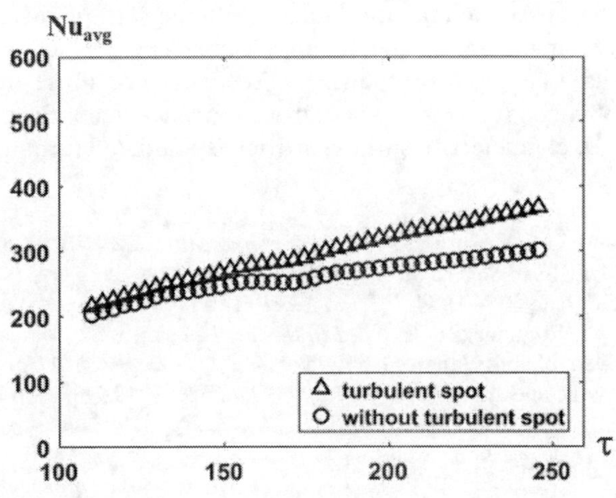

Figure 10. The average Nusselt number within the turbulent spot bound and the Nusselt number of the heating surface, averaged in the same area.

Figure 11 presents the position in streamwise direction of the leading and trailing edges of turbulent spot against the dimensionless time. With this distances, the velocity of the leading and trailing edges of the turbulent spot footprint can be calculated as 0.74 and 0.49 of the freestream velocity, respectively during $\tau = 110 - 250$. These velocities are also in agreement with those obtained by Sankaran et al. [6]. The difference of the velocity of leading and trailing edges causes the enlargement of the turbulent spot in streamwise direction. Thus, the obtained information is a strong evidence to confirm that the LES can provide an accurate prediction in order to determine the characteristics of the artificially initiated turbulent spot.

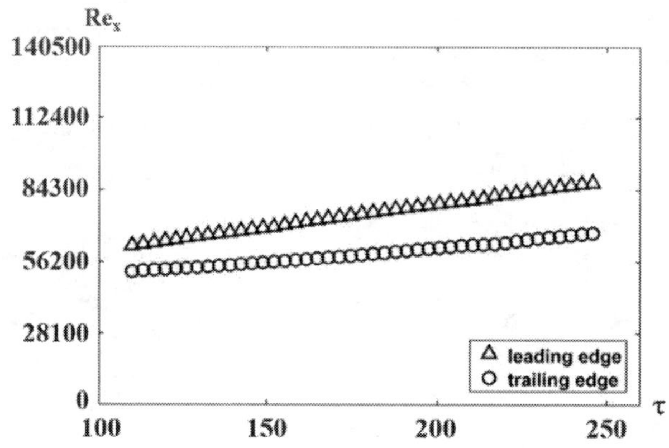

Figure 11. The position of the leading and trailing edges of turbulent spot.

4. Conclusion

This study presents the numerical investigation of the turbulent spot characteristics over a constant heat flux surface using Large Eddy Simulation during the Re_x between 42,200 and 98,400. In this simulation, the spot was initiated by injecting the water in upward direction and perpendicular to the mainstream flow with the velocity of 26.31 m/s and the period of 0.01 s through a 1 mm diameter hole. The results are presented as Nusselt number and temperature distributions of the turbulent spot. It shows that the heated near wall water is accumulated by the spot and causes the increase of the temperature inside the spot body. The near wall water is replaced by the water from the upper layer flow and this mechanism makes the decrease of the surface temperature underneath the spot. Consequently, it gains the Nusselt number within the spot bound above the laminar state. The yielded convective coefficient, spot celerities, and half spreading angle from the LES agree well with those from the experiment of other researchers. Thus, this obtained information is a strong evidence to confirm that the LES can provide an accurate prediction for the characteristics of the artificially initiated turbulent spot.

Reference

[1] Chaiworapuek W and Kittichaikarn C 2016 *J. Heat and Mass Transfer* **92** pp 850-858
[2] Emmons H W 1951 *J. Aero. Sci.* **18** pp 490-498
[3] Itsweire E C and Van Atta C W 1984 *J. Fluid Mech.* **148** pp 319-348
[4] Katz Y, Seifert A and Wygnanski I 1990 *J. Fluid Mech.* **221** pp 1-22
[5] Gostelow J P, Melwani N and Walker G J 1995 *ASME J. Turbomach.* **118** pp 737-743
[6] Sankaran R, Sokolov M and Antonia R A 1988 *J. Fluid Mech.* **197** pp 389-414
[7] Singer B A and Joslin R D 1994 *Physics of Fluids* **6** p 3724
[8] Singer B A 1996 *Physics of Fluids* **8** p 509
[9] Rakpakdee W and Chaiworapuek W 2016 *The 7th TSME Int. Conf. on Mechanical Engineering (Chiang mai: Thailand)*
[10] Smagorinsky J 1963 *Monthly Weather Review* **91**(3) pp 99-164
[11] Deardorff J 1970 *J. Fluid Mech.* **41**(2) pp 453-480
[12] Pitsch H 2006 *Annual Review of Fluid Mechanics* **38** pp 453-482

[13] Wagner C, Hüttl T and Sagaut P 2007 *Large-Eddy Simulation for Acoustics* (Cambridge: Cambridge University Press)
[14] Sullivan P P, McWilliams J C and Moeng C H 1994 *Boundary-Layer Meteorology* **71**(3) pp 247-276
[15] Kays W M, Crawford M E and Weigand B 2005 *Convection Heat and Mass Transfer 4th edition* (New York: McGraw-Hill)
[16] Van Atta C W and Helland K N 1980 *J. Fluid Mech.* **100(2)** pp 243-255

A rotary drum dryer for palm sterilization: preliminary study of flow and heat transfer using CFD

S. Hanifarianty[1], A. Legwiriyakul[2], A. Alimalbari[2], C Nuntadusit[1,2], T. Theppaya[1,2] and M. Wae-Hayee[1,2,*]

[1]Energy Technology, Faculty of Engineering, Prince of Songkhla University, Hatyai, Songkhla 90112, Thailand
[2]Energy Technology Research Center and Department of Mechanical Engineering, Faculty of Engineering, Prince of Songkhla University, Hatyai, Songkhla 90112, Thailand

* Corresponding Author: E-mail: wmakatar@eng.psu.ac.th

Abstract. Preliminary study in this article, the flow and the heat transfer of rotary drum dryer were simulated by using Computational Fluid Dynamics (CFD). A 3D modelling of rotary drum dryer including ambient air was created by considering transient simulation. The temperature distributions on rotary drum dryer surfaces of experimental setup during heating detected by using infrared camera were given to be boundary conditions of modelling. The average temperature at the surface of the drum lids was 80°C, and the average temperature on the heated surface of the drum was 130°C. The results showed that the internal temperature of air in drum modelling was increased relating on time dependent. The final air temperature inside the drum modelling was similar to the measurement results.

1. Introduction

Oil palm is an edible of vegetable oil extracted from mesocarp of oil palm fruit in throughout several processing. This kind of raw material was mostly planted in Southeast Asia, such as Malaysia, Indonesia, and Thailand. After harvesting process, fresh fruit bunch must be processed directly to milling factory not more than 24 hours. Sterilization process plays important role in oil palm factory processing [1]. The purposes of sterilization process are to soften the fruit, enzyme inactivation, reduce water content in fruit and facilitate oil extraction.

The sterilization process uses steam machines with the pressure applied at 40 psi (140°C) for 75 – 90 minutes [2]. From the process, grade A oil palm could be produced for consumption, such as pure vegetable oil. This kind of process consume a lot of water due to oil palm processing in the factory.

The second way to process oil palm is using batch dryer. This process does not require the steam as the first one. A batch dryer system can be placed in small factories. This process will heat oil palm by grilling. Hot air flow up by burning fuel, such as firewood or fuel oil. This kind of process needs much workers to flip oil palm from bottom to upper side. It resulted oil palm not cooked well and it took long time process around 30 hours (figure 1).

Umudee et al [3] studied about oil palm sterilization using microwave irradiation. It was found that sterilization of oil palm using microwave at 50°C is the optimum temperature for heating oil palm. But, for this kind of irradiation heating process is limited to industrial for future used. Noerhidayat et al [4] investigated about oil palm sterilization using high pressure. It proved that at 70 psi (temperature of about 150°C) for 30 minutes showed better in sugar content, release more oil and absorb more water.

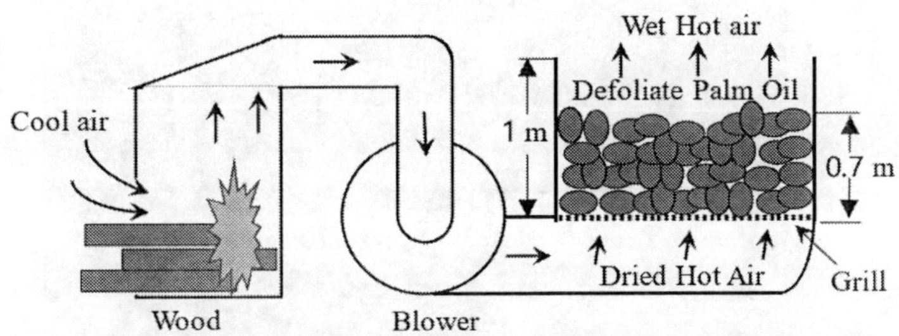

Figure 1. Grilling pattern of oil palm in factory

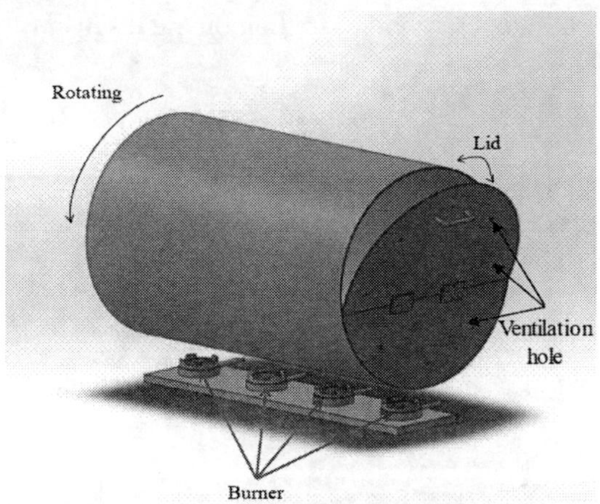

Figure 2. Model of rotary drum dryer.

Rotary drum dryer is being developed as a new tool for drying of medicinal and aromatic plant [5] which is to reduce specific drying energy consumption, keeping good quality and accelerating drying process. Temperature inside in rotating drum is important factor in drying, which depend on the amount of ventilation holes of rotating drum (figure 2). Therefore, the aim of this research was to understand behavior of the temperature inside the rotating drum by using CFD analysis.

2. Method

2.1 Rotary Drum

The rotary drum dryer made of metal sheet with thickness of 3 mm. The dimensions of the drum was 90 cm in length and 57.5 cm in diameter. It is set on the structure which is supported by the bottom roller. The motor with gear that installed on the side of the drum rotates the rotary drum dryer. For the experiments, the fresh fruit palm in the drum is heated at the bottom rotary drum, which use LPG as fuel sources of thermal energy.

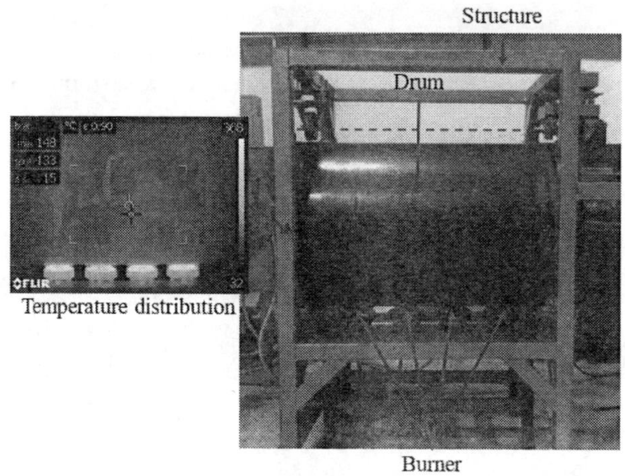

Figure 3. Appearance of temperature distribution on the drum surface using infrared camera.

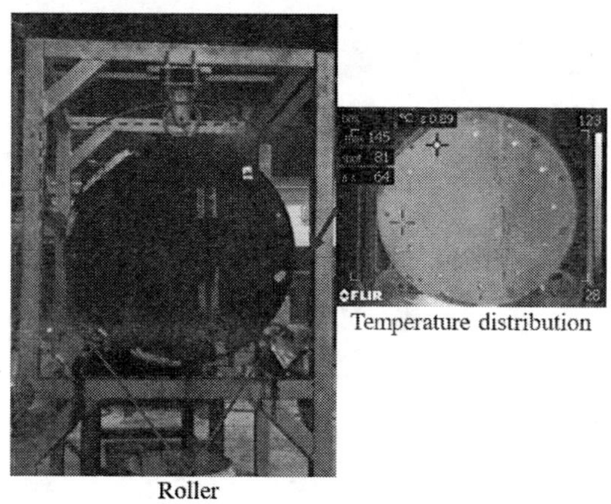

Figure 4. Appearance of temperature distribution on the cap surface of drum using infrared camera.

The figures 3 and 4 above shows the temperature on the surface of the rotary drum dryer and the cap of the drum which in steady conditions. The temperature at the surface of the cap was about 80°C and the temperature of the drum surface was about 130°C. The rotating drum was heated from the combustion of fuel LPG by the direct contact with the drum.

2.2 Simulation model and grid generation

ANSYS Fluent ver. 15.0 was used to simulate the flow characteristics. A 3D numerical model based on the finite-volume method was adopted to solve governing equations with boundary conditions. The geometry of the numerical model was the same as that of the experiment and this model of the rotary drum dryer in CFD will be reduced 10 times from actual dimensions shown in figure 5.

The details of the generated grid are shown in figure 6. The cutting plane along the centerline of the drum is shown to expose the internal grid system. The majority of elements were even rectangular. The simulation was analyzed with four elements cases and 1.2 millions cases and 1.57 millions cases are in the same trend. Therefore, 1.2 millions elements case was considered in this simulation.

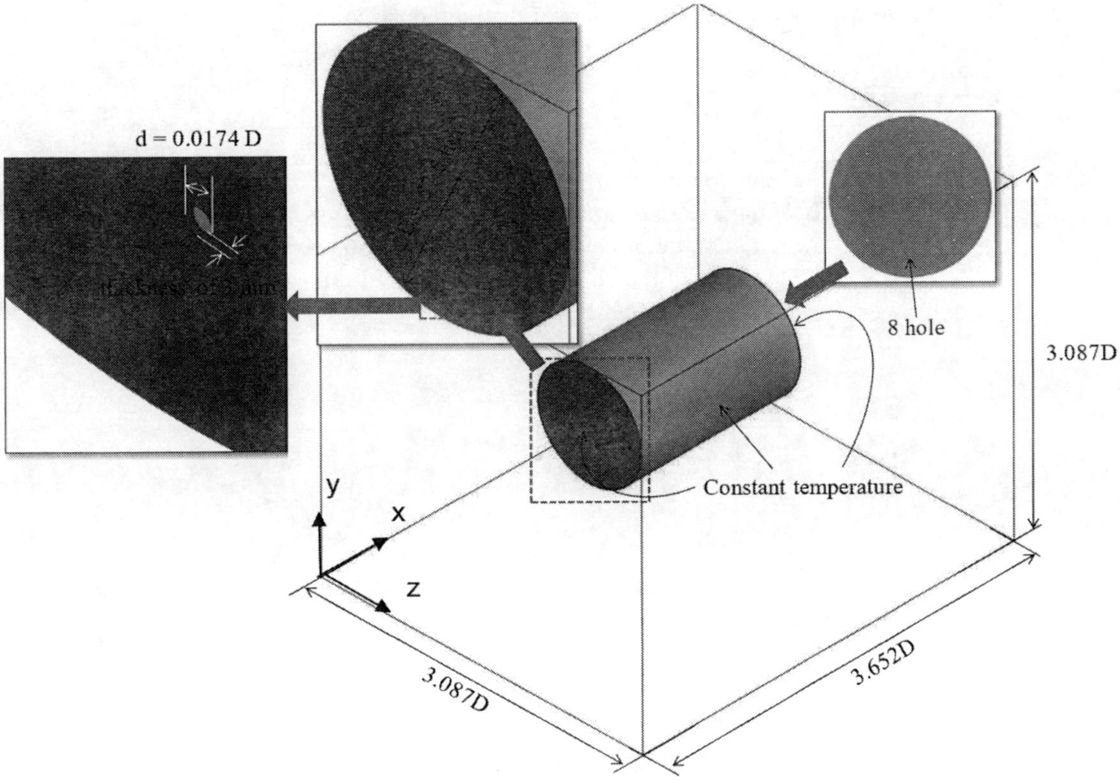

Figure 5. Model of rotary drum including surrounding air

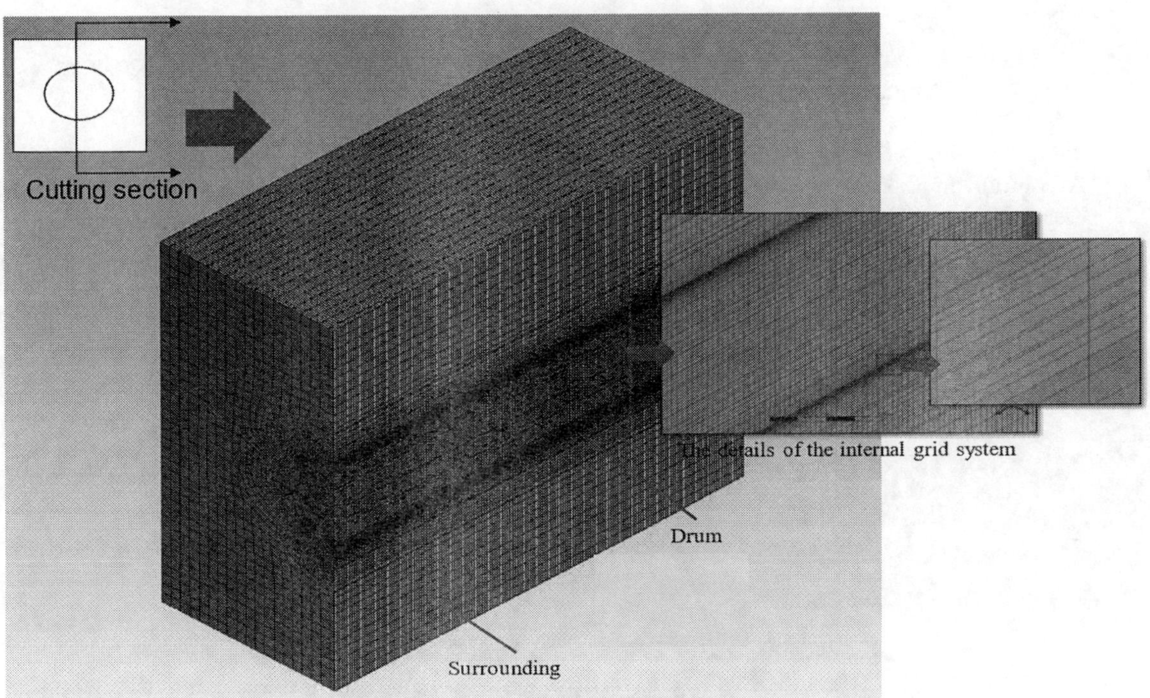

Figure 6. Internal Grid System.

IOP Conf. Series: Materials Science and Engineering **297** (2018) 012030 doi:10.1088/1757-899X/297/1/012030

The boundary conditions were identically specified to the experiment. For the rotary drum dryer, the number of hole was introduced. The details of the boundary condition are specified in table 1.

Table 1. Details of boundary conditions.

Conditions	Setting
The temperature of drum dryer	400 K
The temperature of drum dryer lid	353 K
The temperature of environment	303 K
The pressure of the environment	1 atm

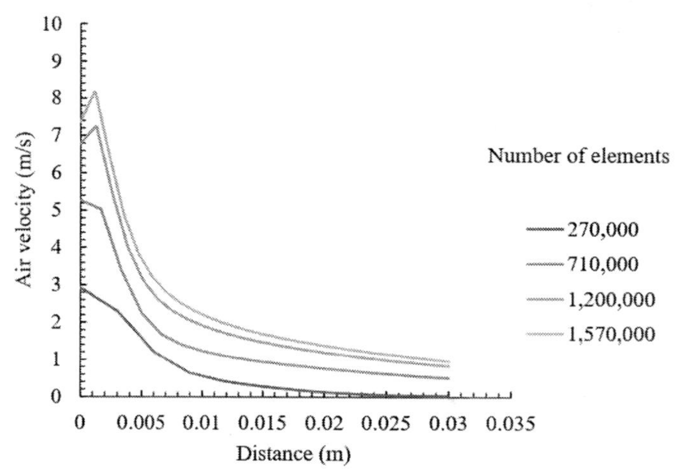

Figure 7. The effect number of elements on air velocity of the centre of ventilation holes to the surrounding

2.3 Calculation Method

A solution method was based on the SIMPLE algorithm with a second-order upwind for all spatial discretizations. The solution was considered to be transient. It was operated until the temperature in the middle of the drum dryer to the steady state. We set the timer to run transient which time period to calculate out as 2 seconds on each occasion, or called time step size was 2 seconds and each period will repeat it for 20 times. It has a number of calculations 1200 times or called numbers of time step is 1200. Therefore, it has the total time in 2400 seconds.

3. Results and Discussions

Temperature distribution and velocity vector profile was analysed with various time in xz plane at centre line cross section of rotary drum model (figure 8). At temperature 400 seconds, the temperature distribution was increase slowly from the edge side of rotary drum. At the same time, the velocity flow was also increase compare with 200 seconds case. According to the theory, volume at air is directly proportional to the temperature, so the airflow inside the rotary drum was pass through the hole. When the time become longer, the flow velocity also higher. Near outside of hole orifice the highest velocity value was found because of the pressure gradient of hole.

At 1000 second, the higher temperature distribution was found, above section of rotary drum while the bottom section was lower. For velocity, flow in above section showed stronger than below section of the rotary drum. Then for 1200 seconds, temperature distribution was the highest for whole surface of cross section of rotary drum. Near the hole of bottom section, it was showed non uniformed of temperature distribution.

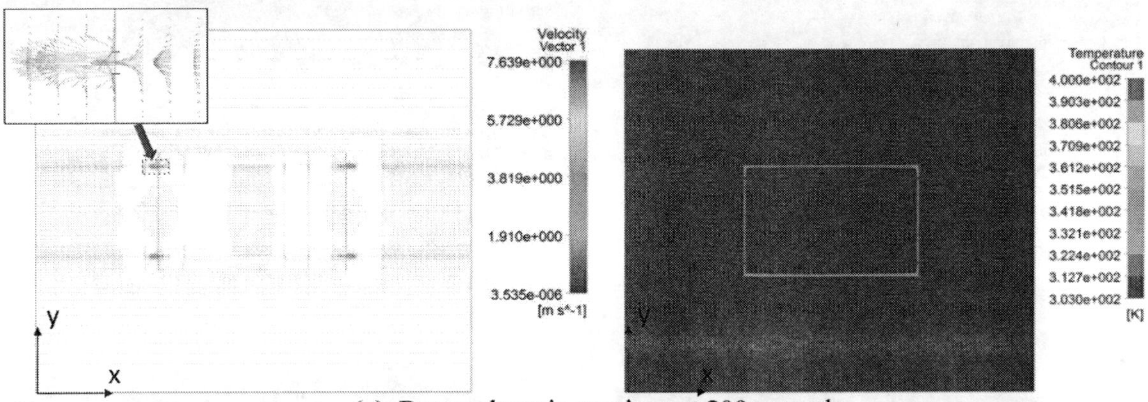

(a) Rotary drum in xy plane at 200 seconds.

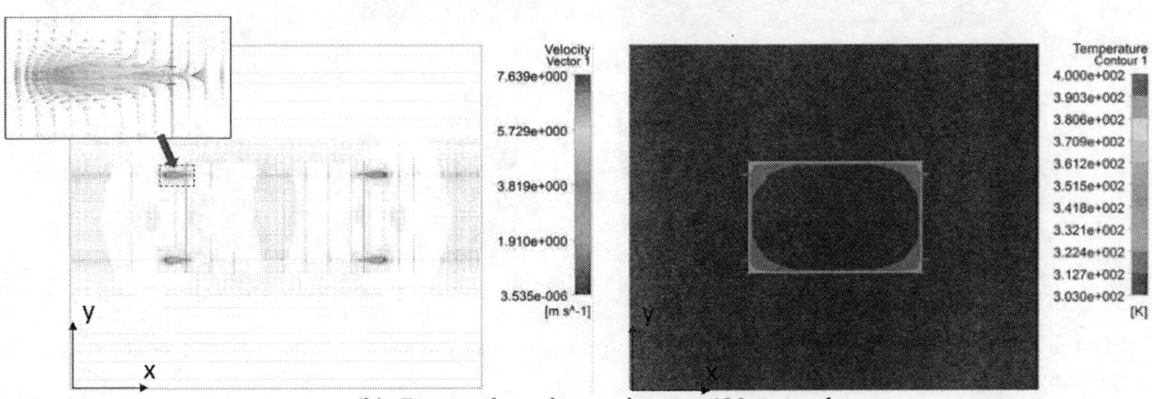

(b). Rotary drum in xy plane at 400 seconds.

Figure 8. Velocity vector profile and temperature distribution of rotary drum in xz plane at various time.

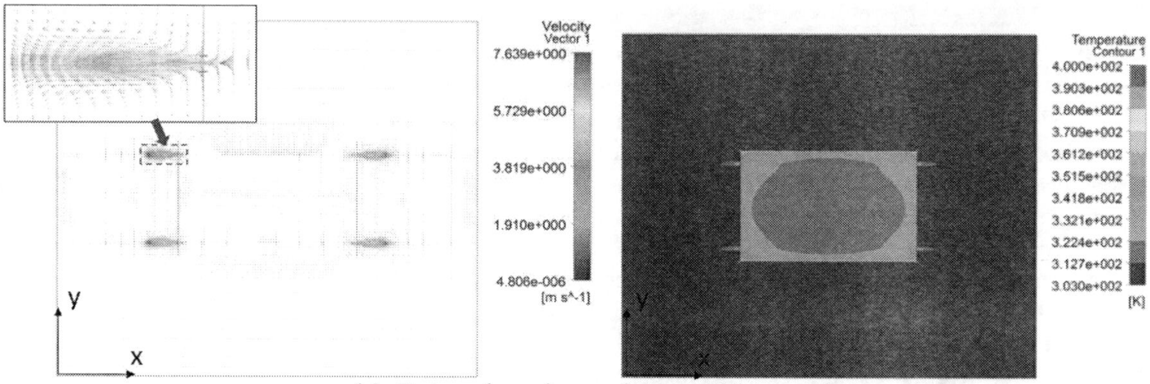

(c). Rotary drum in xy plane at 600 seconds.

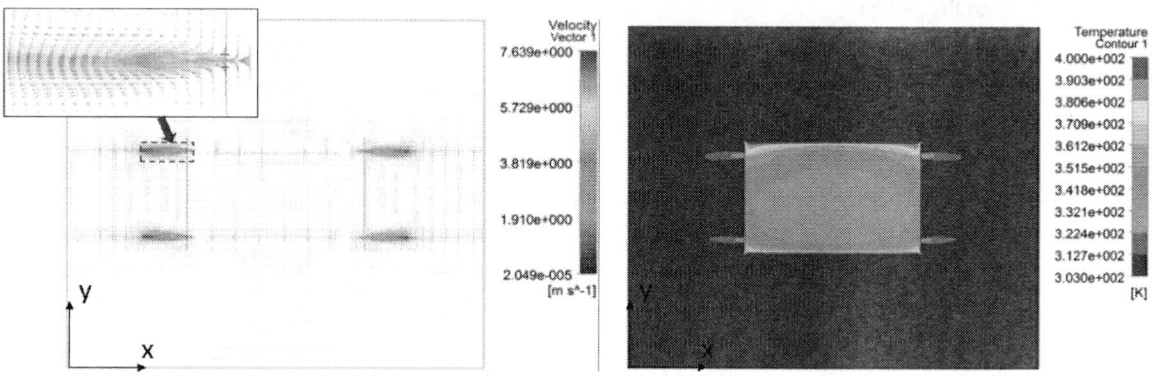

(d). Rotary drum in xy plane at 800 seconds.

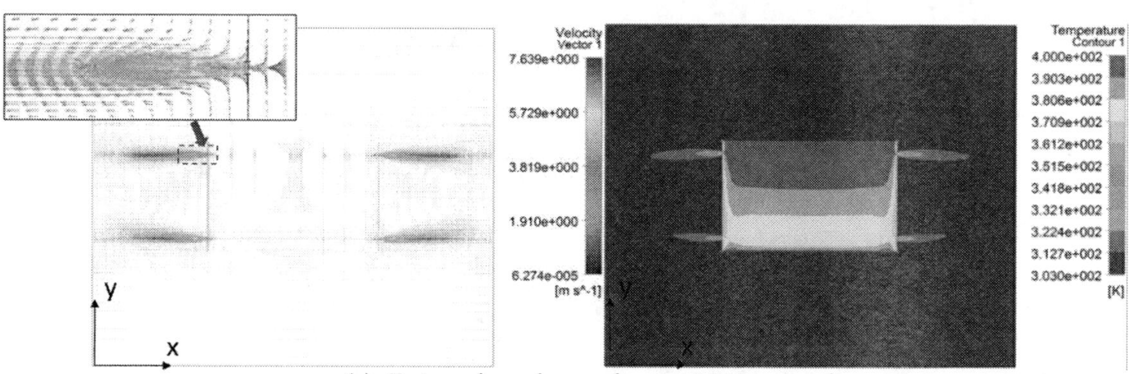

(e). Rotary drum in xy plane at 1000 seconds.

Figure 8. Velocity vector profile and temperature distribution of rotary drum in xz plane at various time (continued)

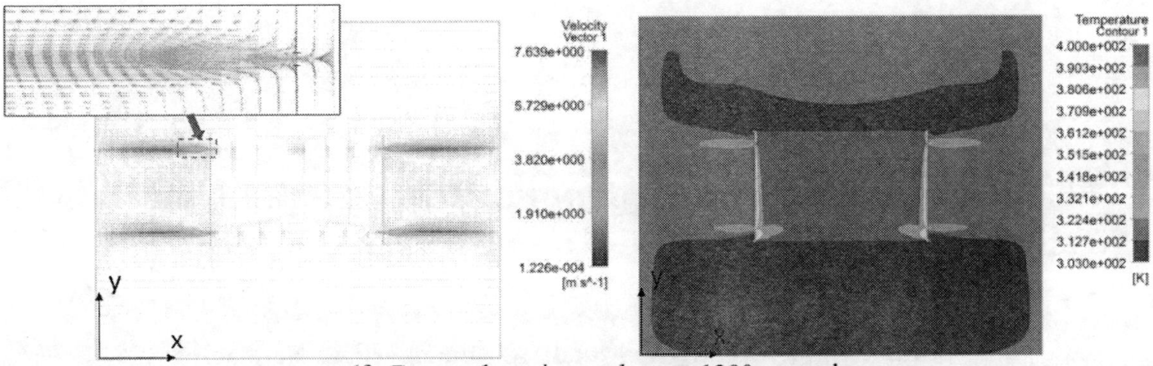

(f). Rotary drum in xy plane at 1200 seconds.

Figure 8. Velocity vector profile and temperature distribution of rotary drum in xz plane at various time (continued)

4. Conclusion

In this work, flow and temperature distribution of rotary drum was analyzed at various times using CFD simulations. The highest and more uniform temperature and flow velocity distribution was found in 1200 seconds for this work. The results was analyzed with fix hole and constant time range (200 to 1200 seconds). For further study about different hole and more temperature range will be analyzed for temperature and velocity flow of rotary drum.

Acknowledgement

The research grant was supported by Energy Policy and Planning Office (EPPO), Ministry of Energy, Thailand, grant No. EE-PSU-59-02, PERIN (PSU Energy System Research Institute), ETRC (Energy Technology Research Centre), and Faculty of Engineering, Prince of Songkla University.

References

[1] Adb. Aziz MK, Morad NA, Mohd. Yunus S and Abd. Razsk AS 2003 Online: http://eprints.utm.my/2759/

[2] Sivasothy K 2000 *Advances in Oil palm Research* **1** pp 745 – 775

[3] Umudee I, Chongcheawchamnan M, Kiatweerasakul M and Tongurai C 2013 *Int. J. of chemical engineering and applications* **4**(3)

[4] Noerhidayat, Yunus R, Zurina ZA, Syafie S, Ramanaidu V and Rashid U 2016 *Int. food research J.* **23**(1) pp 129-134

[5] Tarhan S, Yildirim S, Tuncay MT and Telco I 2008 *J. of Agricultural Machinery Science* **4** (3) pp 295 – 300

The development of rotary drum dryer for palm fruit sterilization

S. Hanifarianty[1], A. Legwiriyakul[2], A. Alimalbari[2], C. Nuntadusit[1,2], T. Theppaya[1,2], and M. Wae-Hayee[1,2,*]

[1] Energy Technology, Faculty of Engineering, Prince of Songkhla University, Hatyai, Songkhla 90112, Thailand
[2] Energy Technology Research Center and Department of Mechanical Engineering, Faculty of Engineering, Prince of Songkhla University, Hatyai, Songkhla 90112, Thailand

* Corresponding Author: E-mail: wmakatar@eng.psu.ac.th

Abstract. The aim of this research was to design and develop a rotary drum dryer for palm fruit sterilization. In this article, the results of the effect of ventilation hole number on the reduction of moisture content in palm fruit were presented. The experimental set up was a drum dryer which has 57.5 cm in a diameter and 90 cm in a length (the size was similar to 200-littre steel drum container). A driving gear and a gear motor rotated the drum dryer. The ventilation hole were drilled on the lateral side of the drum. The diameter of ventilation hole was 10 mm, and the number of ventilation hole were 18, 36 and 72 hole (each side was 9, 18 and 36 hole, respectively). In the experiment, the palm fruit was dried by using LPG to burn and heat the bottom of the drum. The flow rate of LPG was controlled to keep the temperature inside the drum steadily at 120°C.

1. Introduction

Oil palm is economic crop in many countries such as Indonesia, Thailand and Malaysia. It could provide mean income and economic development for living people. Oil palm could use 71 percent for foods (margarine, processed food, chocolate, etc), 24 percent for consumer products (cosmetics, detergents, candles, etc) and 5 percent for energy uses (electricity, heating fuels, etc). Oil palm become raw material for many kind of purposes because of its high pressure oxidation resistance, capability to dissolving chemicals insoluble, high coating power and does not irritation to the body for cosmetic uses.

Oil palm industry undergo two type of processes, the first one is the steam process which uses steam of 140°C for 75 – 90 minutes to stop free fatty acid formation and to soften the palm [1]. In this process it produces crude oil palm (CPO) and crude palm kernel oil (CPKO). Crude oil palm is obtained from mesocarp in orange color of oil palm while for crude palm kernel oil is came from kernel in the nut in white color. Crude oil palm is processed at oil palm milling machine and kernel were removed from nut is kind of byproduct of oil palm milling machine.

The second process is the batch sterilization that is mostly used in small factories. Mostly small factories in Thailand accept falling/defoliate fruit from farmers. That kind of falling fruit/defoliate fruit separated from fresh fruit bunch. During batch sterilization, oil palm was heating. This process will heat by grill. Hot air flow up by burning fuel, namely firewood or fuel oil (figure 1).

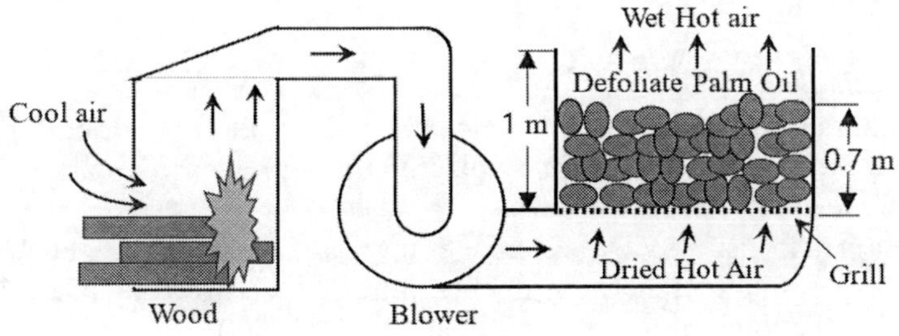

Figure 1. Grilling pattern of oil palm in factory

In the grilling process, several problems have been found in that oil palm was burned due to exposure to direct fire and not heated well. The palm fruit on the ground layer is dried faster than on the upper layer. During grilling process, workers should flip the oil palm from ground layer to the upper layer. In addition, it takes about 30 hours to dry oil palm fruit.

This kind of problem could be solved by designing a rotary drum dryer as shown in figure 2. The falling palm fruit is put into a rotating drum. Then it will be heated from below site. In commercial use, the heating fuel may be firewood. The rotating drum will rotate slowly. The oil palm is mixed together in rotating drum, which will transfer heat to oil palm fruit.

In past, rotary drum principle has been applied to herbal steam [2] however but the baking bin was sieve. This will lose a lot of hot air. In addition, Computational fluid dynamic (CFD) in study of particle dispersion behavior in rotary drum has been used [3]. The purpose of this research was to develop a rotary drum drying of oil palm, which has the advantage: (1) the drying oil palm not expose to smoke, (2) the drying oil palm is mixed in lower layer and upper layer. During drying process, the fuel of rotary drum are LPG. Initially, the effect of number 18, 36 and 72 (two sides) ventilation hole of the rotary drum on the moisture content of the palm fruit was studied.

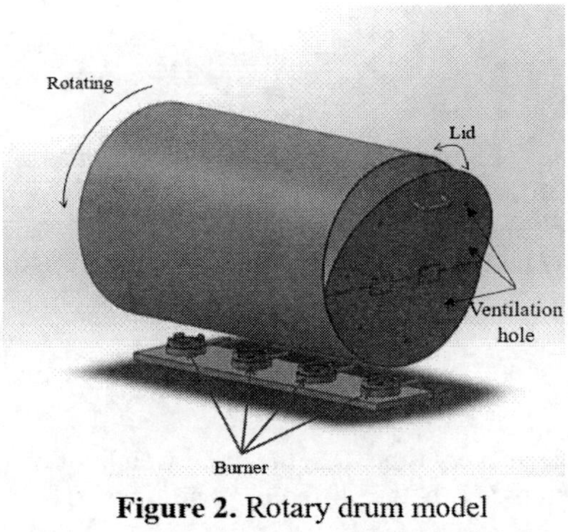

Figure 2. Rotary drum model

2. Method

2.1 Experimental design

Figure 2 show prototype of rotary drum dryer used in the experiment. The rotating drum is made of sheet steel and rolls into a cylindrical shape. With a diameter of 57.5 cm and a length of 90 cm, the rotary drum is mounted on four supported bearings, with one top roller located opposite the drive gear. It serves the drum to rotate in a fixed position.

The roller on top and the gear set pressed with the spring set to adjust the press on the drum. The side of rotating drum has a gear set and a gear motor. It could adjust the rotation speed of the rotating drum

IOP Conf. Series: Materials Science and Engineering **297** (2018) 012031 doi:10.1088/1757-899X/297/1/012031

using the inverter. The rotary drum slowly rotated at 1.68 rev/h and 8.34 rev/h to prevent palm fruit bruise from contusion that result in high FFA.

2.2 Experimental methods

Figure 3 shows a diagram of a rotary drum design. Oil palm fruit 50 kg dried for each batch. LPG was used to burn at the bottom of rotating drum by placing the four gas stoves at the bottom of the rotating drum, which directly heated.

In addition, a rotameters installed to measure the flow rate of LPG fuel for controlling the temperature inside the drum. Every gas pipe in the experiment controlled the flow rate of LPG fuel in order to keep the temperature inside the rotary drum constantly at 120°C (±3°C).

The 50-kg oil palms were load into the drum for drying. The 200-g oil palm sample was put into the small grill. This 200-g oil palm sample was included in those 50-kg oil palms. After finishing the drying, the 200-g oil palm sample was dried again in electric oven to evaluate the dry matter of the oil palm. During drying process, the drum including oil palm was weighed every 1 hour to evaluate moisture content of the oil palm. A data logger which was rotated together with drum was used to detect temperature inside the drum. The parameters in this study were (1) rotating speed of drum at 1.68 rev/h and 8.34 rev/h and (2) ventilation hole number (for releasing moisture) at 9, 18 and 36 for each cap of the drum.

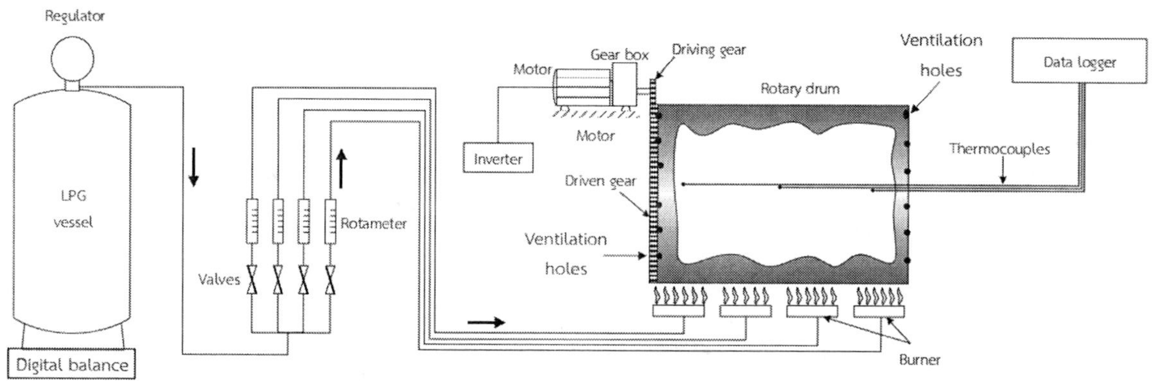

Figure 3. Rotary drum experimental setup

2.3. Calculation of moisture content and moisture ratio

Percentage of moisture content (kg of water/kg of dry matter) was determined using the following equation:

$$\%MC\,(db) = \left(\frac{W-D}{D}\right) \times 100 \qquad (1)$$

where W is weight of sample at every time, D is the dried weight of oil palm. In drying experiment, the initial moisture content compared to the different conditions. The moisture ratio is derived from following equation:

$$MR = \frac{(M_t - M_{eq})}{(M_{in} - M_{eq})} \qquad (2)$$

where M_t is the moisture content at time of drying, M_{eq} is final percentage of moisture content, M_{in} is initial of moisture content.

3. Results and Discussions

Figure 4 shows the effect of ventilation hole and rotating speed on the moisture content at different times. The results showed trend that various drying period of oil palm sample weight decreased. There is small effect of the number of ventilation hole and rotating speed, because of moisture content inside drum. Rotating speed of 1.68 rev/h with ventilation hole of 18 and 9 indicated similar trend but after 6 hours the temperature was increase. The initial moisture content was 12.69 % and reduced to 10.18 % and the ventilation hole 9 the moisture content reduced from 11.52 % to 7.25 % respectively.

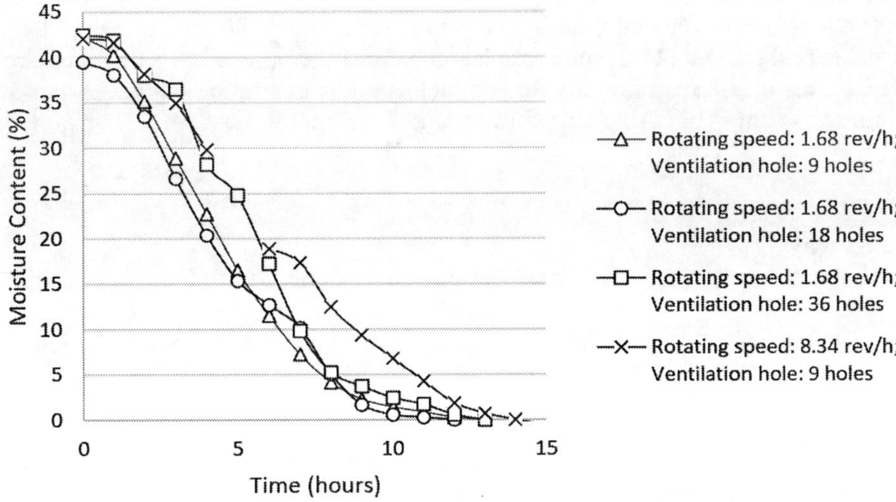

Figure 4. Moisture content of each cases

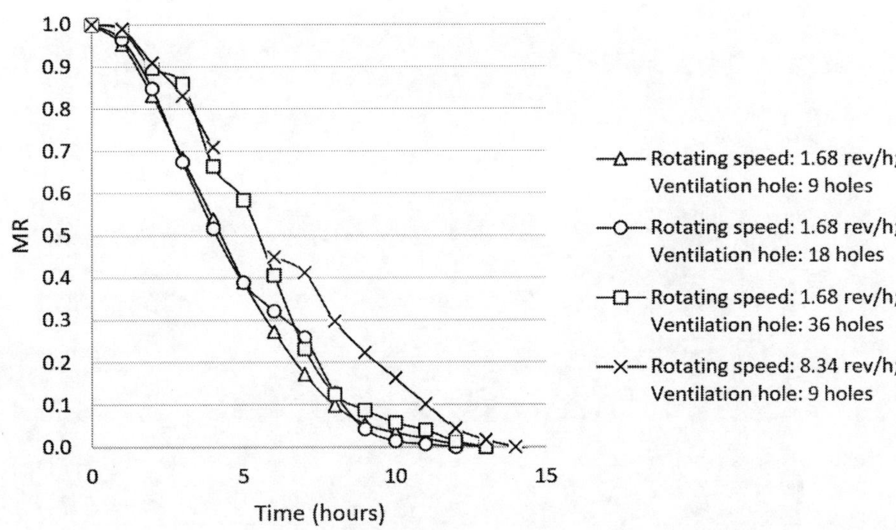

Figure 5. Moisture ratio of each cases

The moisture content of rotation speed 1.68 rev/h with ventilation hole of 36 and rotating speed of 8.34 rev/h and ventilation hole of 9 it also changes due to temperature changes at 6 hours of drying. The initial moisture content was 17.23 % to 9.82 % followed by moisture content of rotary speed of 8.34 rev/h and ventilation hole of 9 also decrease from 18.89 % to 17.33 %. When the drying process throughout passed 12 hours, it was found that, the sample weight dropped almost linearly then almost constant. In addition, LPG fuel consumption per hour in each cases ventilation hole of 36, 18, 9 namely 0.53 kg, 0.61 kg, 0.61 kg, while for rotating speed of 8.34 rev/h with 9 ventilation hole case was 0.56 kg. For drying process, it took about 13-14 hours for changing the color from yellow to brown and palm kernel from white to colorless.

4. Conclusion

In this research, the effect of ventilation hole and rotating speed of rotary drum dryer were studied on moisture content and moisture ratio of oil palm. According to this study, it has been found that for all ventilation hole and rotating speed showed small effect of moisture content and moisture ratio of each cases. For the lowest fuel consumption is case ventilation hole of 36 which is 0.53 kg per hour. This kind of visual appearance oil palm represent that oil palm could proceed to screw pressure machine. For drying in real factory, it takes up to 30 hours to achieve this condition. However, this article presents only preliminary result. This is needed further study for energy cost of drying process compare to factory.

Acknowledgment

The research grant was supported by Energy Policy and Planning Office (EPPO), Ministry of Energy, Thailand, grant No. EE-PSU-59-02, PERIN (PSU Energy System Research Institute), ETRC (Energy Technology Research Centre), and Faculty of Engineering, Prince of Songkla University.

References

[1] Sivasothy K 2000 *Advances in Oil palm Research* **1** pp 745 - 775
[2] Tarhan S, Yildirim S, Tuncay MT and Telci I 2008 *J. of Agricultural Machinery Science* **4(3)** pp 295-300
[3] Santos DA, Duarte CR and Barrozo MAS 2016 *Powder Technology* **294** pp 1-10

The evolution of homogeneous isotropic turbulence in background rotation

Jiratrakul Tunkeaw and Watchapon Rojanaratanangkule*

Faculty of Engineering, Department of Mechanical Engineering, Chiang Mai University, Chiang Mai 50200, Thailand

*Corresponding Author: watchapon.roj@eng.cmu.ac.th

Abstract. Direct numerical simulation is employed to investigate the effect of background rotation on the evolution of homogeneous isotropic turbulence. The frame of reference rotates only in one direction with a rotation numbers varied from 0 to 4, representing both slow and fast rotations. In a stationary frame of reference, the turbulence structures are isotropic and its energy spectrum exhibits a famous $\kappa^{-5/3}$ in the inertial subrange. The background rotation alters the isotropic behaviour of the turbulence structures, leading to the formation of columnar vortices. The slope of the energy spectrum strongly depends on the rotation number and ranges between -5/3 and -3.78.

1. Introduction

Turbulence in background rotation is a fascinating problem because of its appearance in a broad variety of geophysical, astrophysical and engineering applications (e.g. turbomachinery, planetary rotation, etc.). Experiments of grid-generated decaying turbulence in a rotating tank reveals that the slope of energy spectrum $E(\kappa)$ exhibits a rotation number dependent ranging between -5/3 (slow rotation) and about -2.3 (fast rotation) [1]. Early numerical simulations of decaying isotropic turbulence in background rotation observed the reduction in the energy flux and the two-dimensionalisation of the flow [2–4]. However, the conclusive results on the spectral properties of turbulence were still missing because only turbulence at low Reynolds numbers could be considered at that time due to the limitation of the computational resources. Godeferd & Lollini [5] and Morinishi *et al.* [6] performed direct numerical simulation (DNS) at higher Reynolds numbers and observed the similar behaviour of the correlation lengths and the decay properties of the kinetic turbulent energy as those found in the experiments of Hopfinger *et al.* For isotropic turbulence with large scale forcing, the energy spectrum behaves as κ^{-2} in the inertial subrange [8, 9]. In contrast, once the turbulence is forced at intermediate scales, the double cascade was found [9, 10]. The energy spectrum exhibits an inverse cascade with a slope of κ^{-3} scaling in a region of wavenumbers $\kappa < \kappa_{\text{forcing}}$ [9, 10], together with κ^{-2} behaviour when $\kappa > \kappa_{\text{forcing}}$ [10–11]. Baqui & Davidson [12] predicted that the growth in anisotropy increases linearly as $l_{\parallel}/l_{\perp} \sim \Omega t$ (where $l_{\perp}$ and $l_{\parallel}$ respectively being the integral scale in the direction perpendicular and parallel to the rotation axis). Furthermore, they showed that the rate of dissipation of energy changes as $\varepsilon \sim u^3/l_{\perp}$ in the inertial range for turbulence in a rotating background. Large eddy simulations were also used to investigate the effects of background rotation on homogeneous isotropic turbulence (HIT) by Yang & Domaradzki [13]. They found that at high Reynolds numbers, the energy spectrum exhibits a κ^{-2} slope for the quasi-isotropic state, while a κ^{-3} spectrum forms for a strongly anisotropic state. Additionally, the Reynolds stress tensor becomes anisotropic after long enough time for low rotation rates but returns to an isotropic state for very strong rotation rates.

The aim of this work is to investigate the response of isotropic turbulence under uniform background rotation in a vertical direction to explore the evolution of the turbulence structures and the dynamic of the energy spectrum.

2. Numerical Approach

The flow considered in this work is governed by the incompressible continuity and Navier–Stokes equations written in a Cartesian tensor notation as

$$\frac{\partial u_i}{\partial x_i} = 0 \tag{1}$$

$$\frac{\partial u_i}{\partial t} - \varepsilon_{ijk} u_j \omega_k = -\frac{1}{\rho}\frac{\partial P_{eff}}{\partial x_i} + \nu \frac{\partial^2 u_i}{\partial x_j \partial x_j} + +2\varepsilon_{ijk} u_j \Omega_k \tag{2}$$

where $x_i = (x, y, z)$ and $u_i = (u, v, w)$ are respectively Cartesian coordinates and the corresponding velocity vector, $\omega_i = \varepsilon_{ijk} \partial u_k / \partial x_j$ is the vorticity vector and ε_{ijk} is the Levi–Civita symbol. The quantity t denotes time and the effective pressure $P_{eff} = p + \rho u_j u_j / 2$ includes the thermodynamics pressure p and the kinetic energy (the summation over the subscript j being implied). The fluid properties, which are density ρ and kinematic viscosity ν, are assumed constant. The frame of reference is rotated in the vertical direction with a constant angular velocity $\Omega_i = (0, 0, \Omega_z)$.

The simulations are conducted in a periodic cubic box of length $2\pi \times 2\pi \times 2\pi$. It is thus suitable to Fourier transform the governing equations to perform the spatial derivative. The viscous term is integrated analytically using an integrating-factor approach [14]. The Fourier-transformed Navier–Stokes equations reduce to

$$\frac{\partial}{\partial t}\left(I_\nu \hat{u}_i\right) = I_\nu \left(-\hat{i}\kappa_i P_{\text{eff}} + \hat{H}_i\right) \tag{3}$$

where $\hat{u}_i = (\hat{u}, \hat{v}, \hat{w})$ is the velocity in the Fourier space, $I_\nu = \exp\left[\nu \kappa_j \kappa_j t\right]$ is the integrating factor, $\kappa_i = (\kappa_x, \kappa_y, \kappa_z)$ is the wavenumber in each direction of the Cartesian coordinates, $\hat{i} = \sqrt{-1}$ and $\hat{H}_m$ is the nonlinear term. Equation (3) is advanced in time with a low-storage third-order Runge–Kutta scheme [15]. The divergence-free constrain is enforced via a standard pressure-projection method. The nonlinear and Coriolis terms $\varepsilon_{ijk} u_j (\omega_k + 2\Omega_k)$ are computed in a real space and then dealiased using the 3/2-rule in the Fourier space.

The initial condition is obtained from the preliminary simulation, whose initial spectrum is given as

$$E(\kappa, 0) = \frac{q^2 \kappa^\sigma}{2A\kappa_p^{\sigma+1}} \exp\left[-\frac{\sigma}{2}\left(\frac{\kappa}{\kappa_p}\right)^2\right] \tag{4}$$

where $\sigma = 2$ is a free parameter, $\kappa_p = 13$ is the peak location of the spectrum, $q^2 = 3$ is twice the initial kinetic energy, $\kappa = (\kappa_j \kappa_j)^{1/2}$ is the wavenumber magnitude and

$$A = \int_0^\infty \frac{\kappa^\sigma}{\kappa_p^{\sigma+1}} \exp\left[-\frac{\sigma}{2}\left(\frac{\kappa}{\kappa_p}\right)^2\right] d\kappa \,. \tag{5}$$

The preliminary simulation was performed with the resolution of $160\times160\times160$ and was advanced until $t=6$, at which the realistic HIT develops. It should be noted that the grid used throughout this work is uniform. It can be seen from figure 1 that the energy spectrum at this time is in very good agreement with that of Orlandi [16] and Mansour & Wray [17] at almost the same Taylor-scale Reynolds number $R_\lambda = (u'u')^{1/2}\lambda/\nu$ (where λ and u' respectively being the Taylor's microscale and the fluctuation velocity).

To investigate the effect of the background rotation, the rotation number $N = 2\Omega_z L_z / 2\pi u_0'$ (where L_z being the length of the computational domain in the vertical direction and u_0' being the initial fluctuating velocity) is varied from 0 to 4.

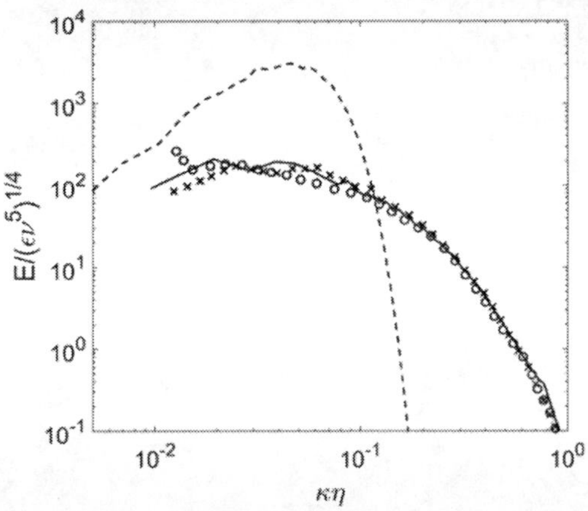

Figure 1. Energy spectra normalized by Kolmogorov units at; - - - - initial spectrum, —— present simulation ($Re_\lambda = 55.8$), × Orlandi ($Re_\lambda = 54.3$) [16] and ○ Mansour & Wray ($Re_\lambda = 54.1$) [17].

3. Results

3.1. *Flow Visualizations*

The structure of the three-dimensional turbulence with and without the effect of external rotation is visualised by means of the isosurfaces of the second invariant of the velocity gradient tensor $Q = -0.5u_{i,j}u_{j,i}$, normalised by $\omega_0^2/4$ (where ω_0 being the magnitude of the initial vorticity). The non-dimensional time is defined as $t^* = t\varepsilon_0 / ke_0$ (where the initial dissipation rate and kinetic energy are respectively denoted by ε_0 and ke_0).

Figures 2(a), 2(c) and 2(e) respectively show the evolution of the turbulence structures at $N = 0$, 0.25 and 0.5 at $t^* = 1.7$. At this time, the turbulence structures for these three cases develop in the same manner and look isotropic. The Taylor's length scale at this time is almost equal at about $\lambda \approx 0.15$. In contrast, at $t^* = 8.5$, the Taylor's length scale increases with an increase in the rotation

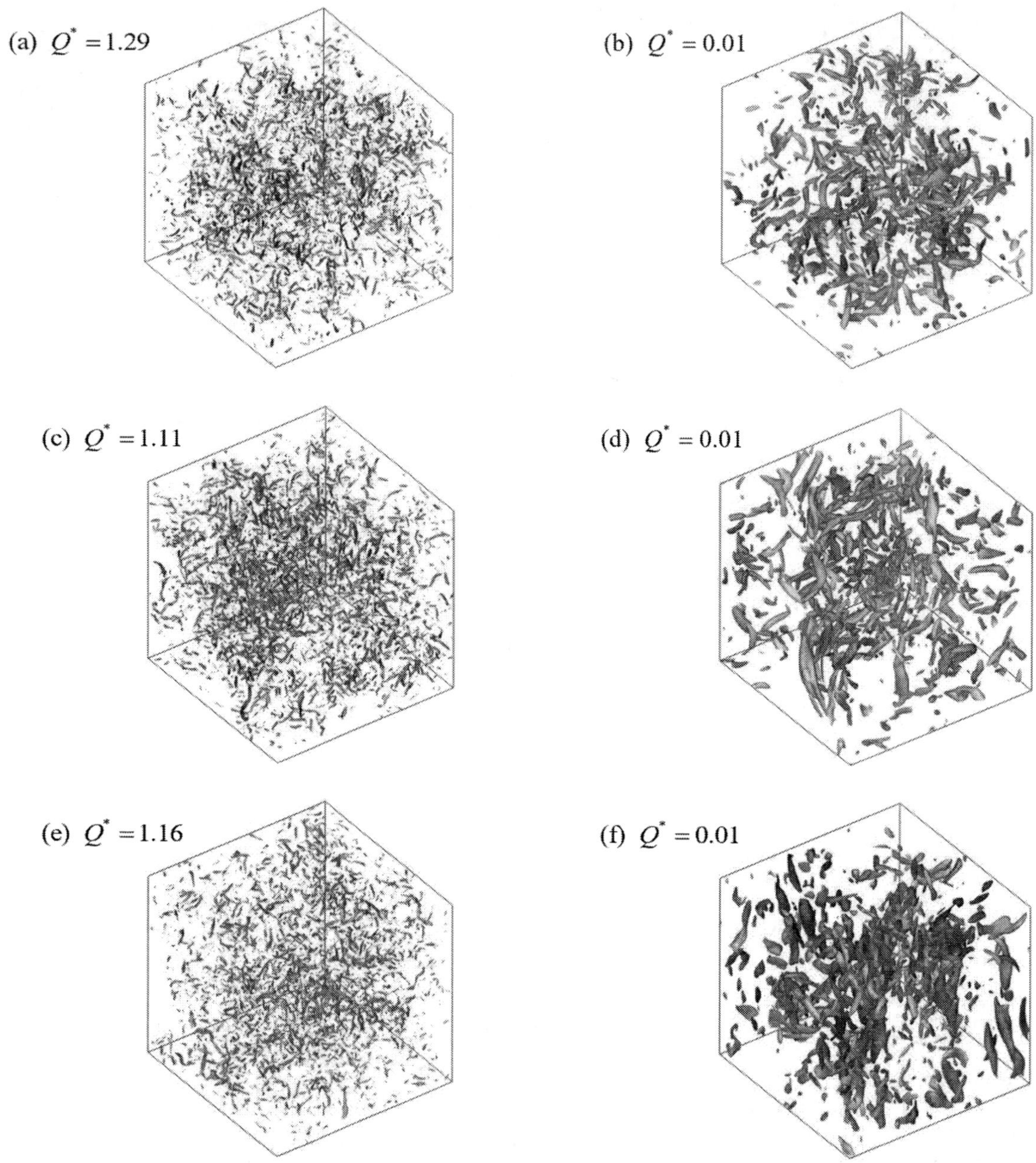

Figure 2. Isosurface of the second invariant of the velocity gradient tensor (Q^*), coloured by the vertical vorticity, at (left) $t^* = 1.7$ and (right) $t^* = 8.5$ for (a, b) $N = 0$, (c, d) $N = 0.25$ and (e, f) $N = 0.5$.

number. For the case of $N = 0$, the Taylor's microscale is about 0.33 and increases to 0.4 and 0.46 once the rotation number increases to 0.25 and 0.5, respectively. For a very low angular velocity ($N = 0.25$), the turbulence structures are still isotropic. The anisotropic of the turbulence structures seems to appear when the rotation number is about 0.5. This leads to the appearance of the columnar vortices, as illustrated in figure 2(f). Thus, it can be concluded that at the early time ($t^* = 0 - 8.5$) the

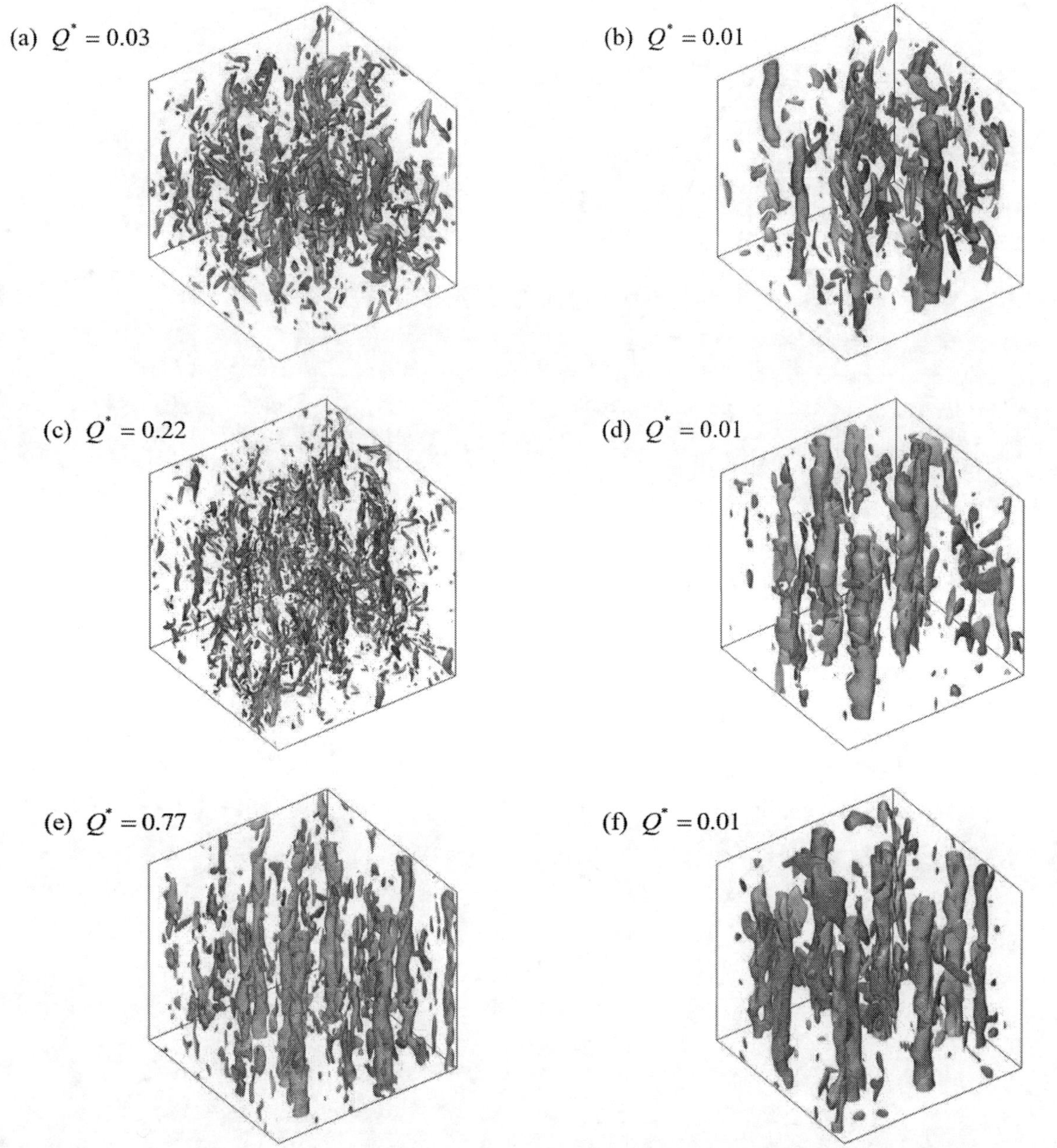

Figure 3. Isosurface of the second invariant of the velocity gradient tensor (Q^*), coloured by the vertical vorticity, at (a, b) $N = 1; t^* = 4.7, 8.5$, (c, d) $N = 2; t^* = 2.7, 8.5$ and (e, f) $N = 4; t^* = 1.2, 8.5$ respectively.

turbulence structures are not influenced much when a frame of reference rotates slowly ($N < 0.5$). For strong background rotation cases ($N = 1$, 2 and 4), the turbulence structures develop in a different way and look more anisotropic, as shown in figures 3(a), 3(c) and 3(e). Additionally, the formation of the columnar vortices seems to form very quickly. It should be noted that the time that the columnar vortices appear depends strongly on the rotation number. They form earlier with an increase in the

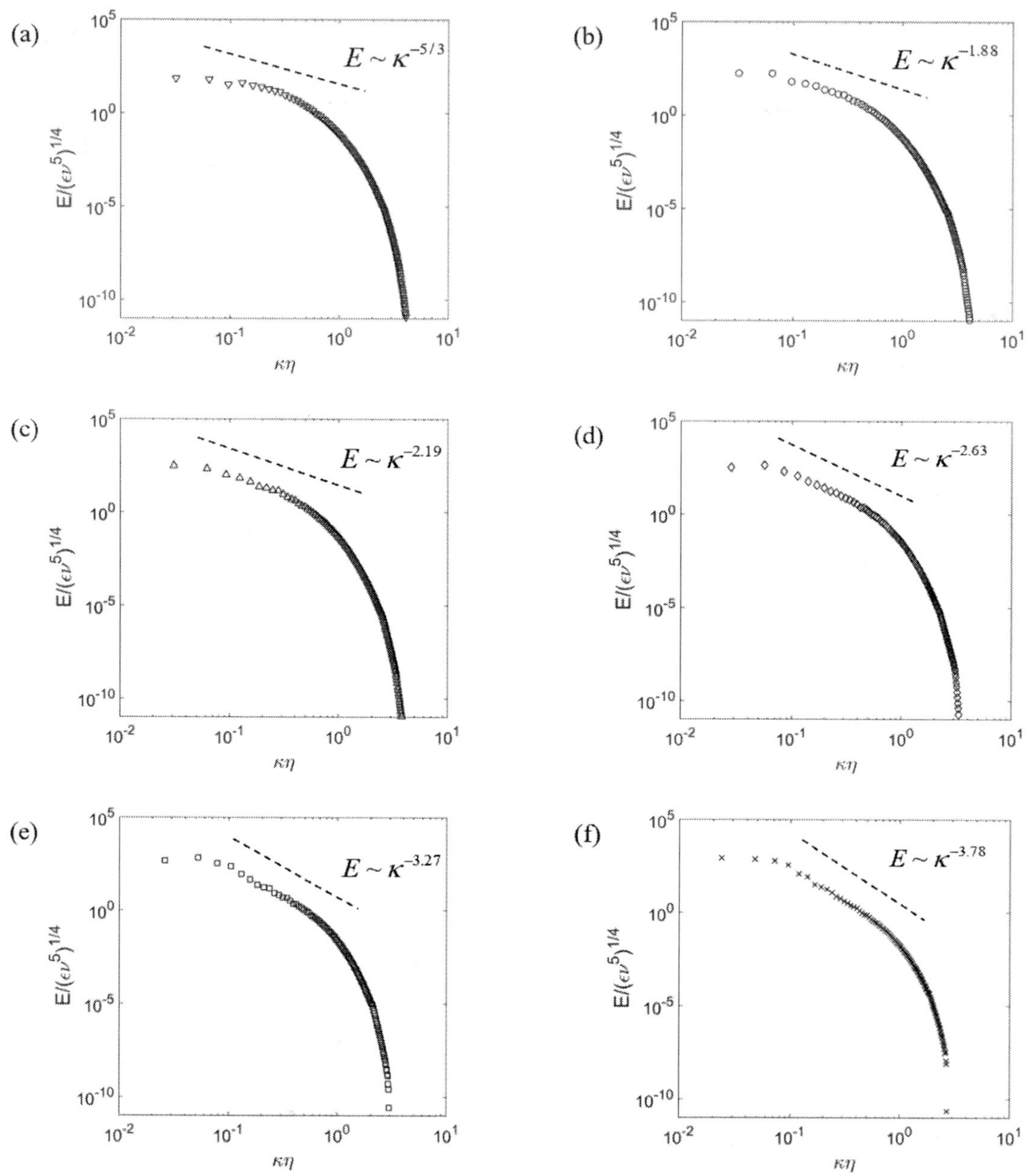

Figure 4. The energy spectrum normalised in Kolmogorov variables at $t^* = 8.5$ of; (a) $N = 0$, (b) $N = 0.25$, (c) $N = 0.5$, (d) $N = 1$, (e) $N = 2$ and (f) $N = 4$.

rotation rate. With time, the small-scale structures are merged with each other, resulting in the formation of quantitatively large vertical vortex. These large-scale coherent structures are stretched in the vertical direction, leading to the appearance of the columnar vortices.

At $t^* = 8.5$, the length of the columnar vortices increases with an increase in the external rotation rate ($\lambda = 0.45$, 0.47 and 0.51 for $N = 1$, 2 and 4), while their number reduces. This suggests that the fast background rotation rate quickly alters the isotropic behaviour of the turbulence structures, yielding the occurrence of the columnar vortices.

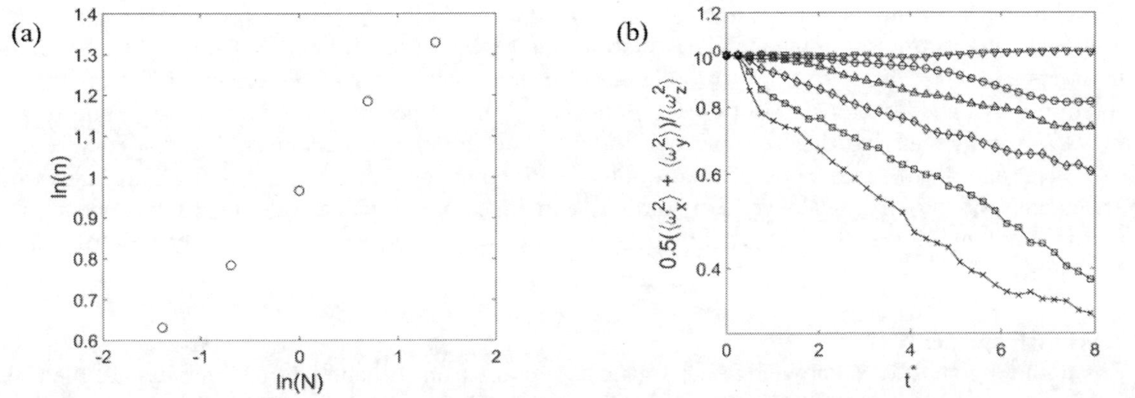

Figure 5. (a) The Slopes of each case on figure 6 and (b) The history of integral of $0.5\left(\langle\omega_x^2\rangle+\langle\omega_y^2\rangle\right)/\langle\omega_z^2\rangle$ for ∇ $N=0$, ○ $N=0.25$, Δ $N=0.5$, ◊ $N=1$, □ $N=2$ and × $N=4$.

3.2. *Spectral Analysis*

The changes from isotropic to anisotropic behaviour of the turbulent structures can be observed by the plot of the three-dimensional energy spectrum, normalised by Kolmogorov variables, written as

$$E(\kappa,t)=\int_0^\infty\int_0^\infty\int_0^\infty \frac{1}{2}\left|\hat{u}_i(\kappa,t)\right|^2 d\kappa_x d\kappa_y d\kappa_z \tag{6}$$

where ε is dissipation rate and η is Kolmogorov length scale. Figure 4 illustrates the normalised energy spectrum for all cases at $t^*=8.5$. For $N=0$, the energy spectrum in the inertial subrange exhibits a slope $E\sim\kappa^{-5/3}$ [18], as shown in figure 4(a). When the frame of reference rotates in the vertical direction (figures 4b – 4f), the energy spectrum decays very quickly. The slopes are dropped to $E\sim\kappa^{-3.78}$ for the case of $N=4$. The range of inertial subrange seems to increase with an increase in the rotation number. This leads to the occurrence of an inverse energy cascade (recalled that the energy is usually transferred from low-wave numbers to high-wave numbers for normal turbulence, in contrast when an inverse energy cascade occurs, the energy is transferred from high-wave numbers to low-wave numbers).

Figure 5(a) shows the slope (n) of the energy spectrum in the inertial subrange of each case. The slope increases with the rotation number as $n\sim N^{0.26}$. The time history of the ratio of the lateral component of the volume-integrated vorticity to its vertical component $0.5\left(\langle\omega_x^2\rangle+\langle\omega_y^2\rangle\right)/\langle\omega_z^2\rangle$ is illustrated in figure 5(b). For zero and low rotation numbers ($N=0$, 0.25 and 0.5) the turbulence structures are also isotropic at the early time ($t^*=1.7$) due to the fact that $\langle\omega_x\rangle\approx\langle\omega_y\rangle\approx\langle\omega_z\rangle$. On the other hands, for moderate and strong rotation numbers ($N=1$, 2 and 4), the vertical vorticity ω_z is greater than the lateral components, suggesting that the turbulence structures are anisotropic. This is confirmed by the flow visualisation displayed in figure 3. With time, the effect of the external rotation is more influenced, resulting in $\langle\omega_z\rangle >> \langle\omega_x\rangle, \langle\omega_y\rangle$. The anisotropy of the turbulence structures becomes stronger with increasing rotation number.

4. Conclusion

Direct numerical simulation is used to investigate the effect of background rotation on the evolution of homogeneous isotropic turbulence. The rotation number is varied from $N = 0$ to 4. At low rotation numbers ($N = 0.25$ and 0.5), the turbulence structures does not feel an effect of the background rotation much and develops as isotropic turbulence at early time. With time, the background rotation plays a role in changing the isotropic behaviour of the turbulence structures. For fast rotation rates ($N = 1$, 2 and 4), the anisotropy of the turbulence structures develops very quickly. This leads to the appearance of the large-scale vertical vortices. An inverse energy cascade seems to appear in the inertial subrange once the background rotation is active. The slope of the energy spectrum in the inertial subrange increases with the rotation number as $n \sim N^{0.26}$.

Acknowledgement

The authors gratefully acknowledge the financial support from Chiang Mai University and the use of HPC services of the Faculty of Engineering, Chiang Mai University.

References

[1] Morize C, Moisy F and Rabaud M 2005 *Physics of Fluids* **17**
[2] Bardina J, Ferziger J H and Rogallo R S 1985 *J. of Fluid Mechanics* **154**
[3] Mansour N N, Cambon C and Speziale C G 1992 *Theoretical and computational study of rotating isotropic turbulence Studies in Turbulence* (New York: Springer)
[4] Hossain M 1994 *Physics of Fluids* **6**
[5] Godeferd F S and Lollini L 1999 *Journal of Fluid Mechanics* **393**
[6] Morinishi Y, Nakabayashi K and Ren S Q 2001 *Physics of Fluids* **13**
[7] Hopfinger E J, Browand F K and Gagne Y 1982 *J. of Fluid Mechanics* **125**
[8] Yeung P K and Zhou Y 1998 *Physics of Fluids* **10**
[9] Thiele M and Müller W C 2009 Structure and decay of rotating homogeneous turbulence *J. of Fluid Mechanics* **637**
[10] Smith L M and Waleffe F 1999 *Physics of Fluids* **11**
[11] Mininni P D, Alexakis A and Pouquet A 2009 *Physics of Fluids* **21**
[12] Baqui Y B and Davidson P A 2015 *Physics of Fluids* **27**
[13] Yang X and Domaradzki J A 2004 Large eddy simulations of decaying rotating turbulence *Physics of Fluids* **16**
[14] Rogallo R S 1981 *NASA Technical Memorandum*
[15] Spalart P E, Moser R D and Rogers M M 1991 *J. of Computational Physics* **96**
[16] Orlandi P 2000 *Fluid Flow Phenomena a Numerical Toolkit* (Rome: Springer)
[17] Mansour N N and Wray A A 1994 *Physics of Fluids* **6**
[18] Smith L M and Reynolds W C 1991 *Phys. Fluids A* **3**

Parametric system identification of catamaran for improving controller design

Surasak Timpitak[1,*], Pradya Prempraneerach[2], and Eakkachai Pengwang[1]

[1]Institute of FIeld roBOtics (FIBO), King Mongkut's University of Technology Thonburi, 126 Pracha Uthit Rd., Bang Mod, Thung Khru, Bangkok 10140, Thailand
[2]Faculty of Engineering, Department of Mechanical Engineering, Rajamangala University of Technology Thanyaburi, 39 Moo 1, Rangsit-Nakhonayok Rd., Klong 6, Thanyaburi, Pathumthani 12110, Thailand

* Corresponding Author: E-mail : surasakk16@gmail.com, Telephone : 02-549-4415

Abstract. This paper presents an estimation of simplified dynamic model for only surge- and yaw- motions of catamaran by using system identification (SI) techniques to determine associated unknown parameters. These methods will enhance the performance of designing processes for the motion control system of Unmanned Surface Vehicle (USV). The simulation results demonstrate an effective way to solve for damping forces and to determine added masses by applying least-square and AutoRegressive Exogenous (ARX) methods. Both methods are then evaluated according to estimated parametric errors from the vehicle's dynamic model. The ARX method, which yields better estimated accuracy, can then be applied to identify unknown parameters as well as to help improving a controller design of a real unmanned catamaran.

1. Introduction

Periodical river/canal surveys are one of the most important operations for agricultural and water-resource management in Thailand. However, to survey vast area of river/canal using small number of official personels, enormous budget as well as lengthy survey period are needed. An Unmanned Surface Vehicle (USV) could be an instrument to assist in collecting survey data. So, autonomous control system design as well as intelligent system of USV are important for vast field operation in different environments. However, effectiveness of autonomous control system is depended on either accuracy of USV dynamic parameters or robustness of controller. In this research work, estimation of parameters in USV dynamic equations is examined to increase the accuracy of the controller design.

Examples of the implementation of Unmanned Surface Vehicle (USV) are in several research. Zhi Li and Ralf Bachmayer [1] developed autonomous way-point control system of catamaran motion for testing unknown variables of a catamaran model using Least-Square method. Consequently, a comparisons of the measured and estimated values are shown for the technical parameters. Moreover, researcher Jeong-Hong Park, Hyung-Won Shim and team [2] demonstrated the technique for predicting parameters in dynamic model using Linear-Quadratic system-identification method. By applying specified force to motor propellers, the parameters of vehicle could be estimated with good accuracy.

Moreover, the parameter estimations are implemented in several application for enhancing accuracy of the systems. For example, Huaiyu Wu, Dong Sun and Zhaoying Zhou used a small control board for performing flight control in Micro Air Vehicle (MAV) [3]. By using a multi-channel remote control, the correlation of the vertical and horizontal flight of the MAV could be recorded. In this study, three inputs types are: 1) four inputs (throttle, rudder, aileron, elevator) 2) two inputs (aileron, elevator) and 3) one input (aileron) that are used for identifying variables of MAV dynamic equation with the Autoregressive exogenous (ARX) method in MATLAB Toolbox. Characteristic equation is used in modeling and analyzing of Pitch-Control with MATLAB Toolbox. Moreover, from China Chongqing University by Yupeng Yuan, Shan Liang and Min Gao [4], also applied ARX method to a microwave heating process in vast industries. This microwave heating process, a closed system, requires a precision control and high safety. Parameters of microwave heating can be calculated using main input variable is the time delay. Experimental data was analyzed for unknown parameters base on the ARX model technique. In addition, Bito Irie from University of Texas at Arlington [5], studied the input/output characteristics relationship between measured dynamic tire loads and road profiles of the tractor (Tractor-semitrailer) in the vehicle model, which approach to the system identification by polynomial. These state-space black-box modeling are formulated and compared with the measured values. The parameters are estimated with ARX, ARMAX based on estimation to obtain data for simulations and compared for the best techniques. It is observed that the ARMAX technique as well as the most accurate method for the tractor model.

This paper presents an importance of predicting and identify unknown parameters of the motion equations unmanned surface vehicles (USV). This research are organized as follows. Section 2 describes kinematic and dynamics equations of USV. The numerical parameter estimation methods both LSM and ARX are presented in Section 3. Section 4 analyzes dynamic system based on magnitude as function of frequency in order to select proper frequency input range. Last, USV parameter estimation using both LSM and ARX methods are demonstrated and compared in Section 5.

2. Kinematic and dynamic equations of USV

For analyzing USV motion in simulation, kinematics and dynamics models, particularly for catamaran, are described [6-9]. Generally, USV motion consists of 6 Degrees of Freedoms (DOFs): 3 positions and 3 rotations. Corresponding parameters of vehicle motion and Earth- and Body-fixed coordinate frames are shown in figure 1.

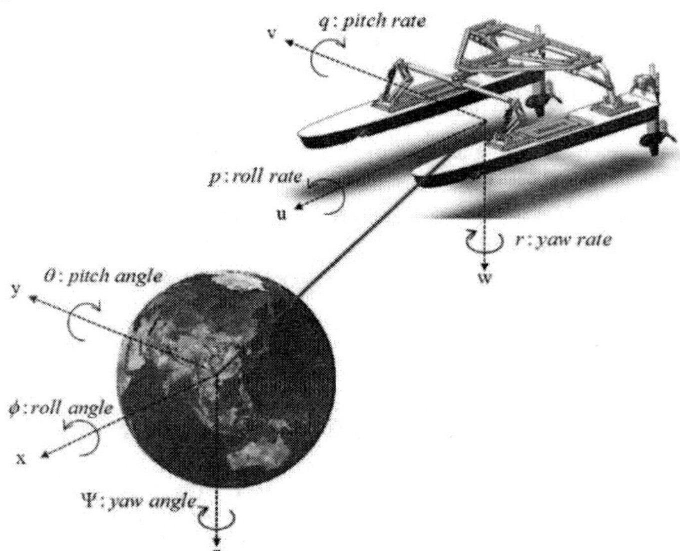

Figure 1. Vehicle dynamic models based on earth-fixed and body-fixed coordinate frames.

2.1. USV kinematic model

The variables can be described [6] by the body-fixed frame and the earth-fixed frame in terms of distance, velocity for linear motions, and angular motions of the vehicle, which can be summarized as shown on Table 1.

IOP Conf. Series: Materials Science and Engineering **297** (2018) 012033 doi:10.1088/1757-899X/297/1/012033

Table 1. Notations used for kinematic models of unmanned surface vehicles. [6]

DOF	Motion Description	Positions & Euler angles		Linear & Angular Velocities	
1	Motion in x-direction (surge)	x		u	
2	Motion in y-direction (sway)	y	$\boldsymbol{\eta}_1$	v	$\mathbf{v}_1$
3	Motion in z-direction (heave)	z		w	
4	Rotation about x-axis (roll)	ϕ		p	
5	Rotation about y-axis (pitch)	θ	$\boldsymbol{\eta}_2$	q	$\mathbf{v}_2$
6	Rotation about z-axis (yaw)	ψ		r	

Combining these kinematic variables in Earth- and Body-fixed frames into two sets of column vectors, the previous parameters can be expressed in equation (1).

$$\boldsymbol{\eta} = \left[\boldsymbol{\eta}_1^T, \boldsymbol{\eta}_2^T\right]^T; \quad \mathbf{v} = \left[\mathbf{v}_1^T, \mathbf{v}_2^T\right]^T; \tag{1}$$

Linear-velocity vector (v_1) with respect to vehicle's center of mass (CM) can be transformed to linear-velocity vector in the Earth-fixed frame ($\dot{\eta}_1$) using a Jacobian matrix ($J_1(\eta_2)$) in equation (2)-(3), which is a function of Euler angles.

$$\dot{\boldsymbol{\eta}}_1 = \mathbf{J}_1(\boldsymbol{\eta}_2)\,\mathbf{v}_1, \tag{2}$$

where

$$\mathbf{J}_1(\boldsymbol{\eta}_2) = \begin{bmatrix} c\psi c\theta & -s\psi c\phi + c\psi s\theta s\phi & s\psi s\phi + c\psi c\phi s\theta \\ s\psi c\theta & c\psi c\phi + s\phi s\theta s\psi & -c\psi s\phi + s\theta s\psi c\phi \\ -s\theta & c\theta s\phi & c\theta c\phi \end{bmatrix} \tag{3}$$

Similarly, a Jacobian matrix ($J_2(\eta_2)$) can transform the angular-velocity vector (v_2) at vehicles' center of mass (CM) to the angular-velocity vector ($\dot{\eta}_2$) in the Earth-fixed frame, as shown in equation (4).

$$\dot{\boldsymbol{\eta}}_2 = \mathbf{J}_2(\boldsymbol{\eta}_2)\,\mathbf{v}_2 \tag{4}$$

where the Jacobian matrix ($J_2(\eta_2)$) is

$$\mathbf{J}_2(\boldsymbol{\eta}_2) = \begin{bmatrix} 1 & s\phi t\theta & c\phi t\theta \\ 0 & c\phi & -s\phi \\ 0 & s\phi / c\theta & c\phi / c\theta \end{bmatrix} \tag{5}$$

Integrating equation (3) and (5), the 6-DOF kinematics equation of vehicle can be expressed as the following:

$$\dot{\boldsymbol{\eta}} = \mathbf{J}(\boldsymbol{\eta})\mathbf{v} \tag{6}$$

$$\begin{bmatrix} \dot{\boldsymbol{\eta}}_1 \\ \dot{\boldsymbol{\eta}}_2 \end{bmatrix} = \begin{bmatrix} \mathbf{J}_1(\boldsymbol{\eta}_2) & 0_{3\times3} \\ 0_{3\times3} & \mathbf{J}_2(\boldsymbol{\eta}_2) \end{bmatrix} \begin{bmatrix} \mathbf{v}_1 \\ \mathbf{v}_2 \end{bmatrix} \tag{7}$$

2.2. USV dynamics model [6]

To identify USV parameters using dynamics simulation, the motions of the surface vessel are assumed to be restricted only in the horizontal plane without roll motion (or $p = 0$) and pitch motion (or $q = 0$). Thus, there are translation motions in surge (along x-axis) and sway (along y-axis), while heave motion along z-axis (or $w = 0$) is negligible small. As a result, 3-DOF dynamic equations of USV in the body-fixed frame, which are derived from the Newton second law, are written in equation (8).

$$\mathbf{M}\dot{\mathbf{v}} + \mathbf{C}(\mathbf{v})\mathbf{v} + \mathbf{D}(\mathbf{v})\mathbf{v} = \boldsymbol{\tau} \tag{8}$$

where **M** matrix is a combination of the rigid-body mass and inertia ($\mathbf{M}_{RB}$) and hydrodynamics forces and moments ($\mathbf{M}_A$) and m is mass the vehicle, x_G is center of gravity in x-direction, I_z is moment of inertia in about z-axis Or, $\mathbf{M} = \mathbf{M}_{RB} + \mathbf{M}_A$ are expressed in equation (9).

$$\mathbf{M}_{RB} = \begin{bmatrix} m & 0 & 0 \\ 0 & m & mx_G \\ 0 & mx_G & I_z \end{bmatrix}; \mathbf{M}_A = \begin{bmatrix} -X_{\dot{u}} & 0 & 0 \\ 0 & -Y_{\dot{v}} & -Y_{\dot{r}} \\ 0 & -Y_{\dot{r}} & -N_{\dot{r}} \end{bmatrix} \tag{9}$$

Similarly, coriolis and centripetal matrix ($\mathbf{C} = \mathbf{C}_{RB} + \mathbf{C}_A$) denote ($X_{\dot{u}}, Y_{\dot{v}}, Y_{\dot{r}}, N_{\dot{r}}$) by reference [6] including from rigid-body motion ($\mathbf{C}_{RB}$) and hydrodynamic force and moments, created by surrounding fluid, ($\mathbf{C}_A$) can be written in equation (10).

$$\mathbf{C}_{RB}(v) = \begin{bmatrix} 0 & 0 & -m(x_G r + v) \\ 0 & 0 & mu \\ m(x_G r + v) & -mu & 0 \end{bmatrix}; \mathbf{C}_A(v) = \begin{bmatrix} 0 & 0 & Y_{\dot{v}} v + Y_{\dot{r}} r \\ 0 & 0 & -X_{\dot{u}} u \\ -Y_{\dot{v}} v - Y_{\dot{r}} r & X_{\dot{u}} u & 0 \end{bmatrix} \tag{10}$$

The hydrodynamic damping matrix (**D**) exhibits uncoupled between surge- motion and sway- and yaw-motion, damping forces and moment (neglect heave, roll, pitch) [6] formulated in equation (11).

$$\mathbf{D}(v) = \begin{bmatrix} -X_u & 0 & 0 \\ 0 & -Y_v & -Y_r \\ 0 & -N_v & -N_r \end{bmatrix} \tag{11}$$

Lastly, external forces and moments acting on USV, denoted by $\boldsymbol{\tau} = [X\ Y\ N]^T$, can be described by a multiplication of thruster-configuration matrix (**B**) and controlled input-force vector ($\mathbf{u} = [T_L\ T_R]^T$), as shown in equation (12).

$$\boldsymbol{\tau} = \mathbf{B}\mathbf{u} \tag{12}$$

According to Figure 2, USV has two thrusters, fixed at the stern, therefore a forward force ($X = T_L + T_R$) is a sum of two controlled input-forces and a sway force (Y) is zero and yaw moment $N = T_L X_L - T_R X_R$ around the center of mass (CM) or the body-fixed reference frame. X_L and X_R are distances along y-axis from the CM of USV to left- and right-thrusters, respectively. Thus, the thruster-configuration matrix can be written as in equation (13). [10].

$$\mathbf{B} = \begin{bmatrix} 1 & 1 \\ 0 & 0 \\ X_L & X_R \end{bmatrix}, \mathbf{u} = \begin{bmatrix} T_L \\ T_R \end{bmatrix} \tag{13}$$

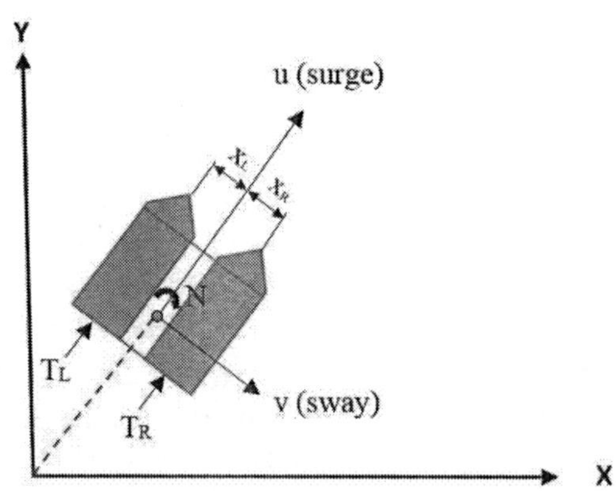

Figure 2. Earth-fixed and body-fixed coordinate frames of USV with two fixed thrusters at stern.

Nevertheless, the sway motion is comparable smaller than linear surge-motion and angular yaw-motion, thus USV's dynamic equation can be further simplified into 2-DOF dynamic equation in surge- and yaw-directions, as written in Equation (14).

$$\begin{aligned}\dot{u} &= \left(\frac{X_u}{m - X_{\dot{u}}}\right)u + \left(\frac{1}{m - X_{\dot{u}}}\right)T_L + \left(\frac{1}{m - X_{\dot{u}}}\right)T_R \\ \dot{r} &= \left(\frac{N_r}{I_{zz} - N_{\dot{r}}}\right)r + \left(\frac{X_L}{I_{zz} - N_{\dot{r}}}\right)T_L - \left(\frac{X_R}{I_{zz} - N_{\dot{r}}}\right)T_R\end{aligned} \tag{14}$$

3. Method Identification

This section briefly explains two system-identification techniques for estimating unknown parameters of USV dynamic system. The Least-Square Method (LSM) and AutoRegressive eXogenous (ARX) methods described [11] are based on an input-output description of the dynamic system. The simplified USV model in equation (14) can be reformatted into a state-space form for surge- and yaw-motions in equation (15) and (16), respectively.

$$\mathbf{Y}_1 = \mathbf{\theta}_1\mathbf{X}_1 = \left[\left(\frac{X_u}{m - X_{\dot{u}}}\right) \quad \left(\frac{1}{m - X_{\dot{u}}}\right) \quad \left(\frac{1}{m - X_{\dot{u}}}\right)\right]\begin{bmatrix} u \\ T_L \\ T_R \end{bmatrix} + e_1 \tag{15}$$

$$\mathbf{Y}_2 = \mathbf{\theta}_2\mathbf{X}_2 = \left[\left(\frac{N_r}{I_{zz} - N_{\dot{r}}}\right) \quad \left(\frac{X_L}{I_{zz} - N_{\dot{r}}}\right) \quad \left(\frac{X_R}{I_{zz} - N_{\dot{r}}}\right)\right]\begin{bmatrix} u \\ T_L \\ T_R \end{bmatrix} + e_2 \tag{16}$$

where e_1 and e_2 are random external disturbances.

3.1. Least-Square Method (LSM) [11]

The Least-Squares Method (LSM) could be used to estimate unknown parameters in the dynamic system by minimizing a weighted least-square objective function (J), defined as in equation (17).

$$J = \mathbf{e}^T W\mathbf{e} = \left(\mathbf{Y} - \hat{\mathbf{Y}}\right)^T \mathbf{W}\left(\mathbf{Y} - \hat{\mathbf{Y}}\right) \tag{17}$$

where a prediction error (e) is defined as: $\mathbf{e} = \mathbf{Y} - \hat{\mathbf{\theta}}\mathbf{X}$ .By minimizing the cost function (J), optimal values of unknown parameters can be computed from equation (18).

$$\hat{\mathbf{\theta}} = \left(\mathbf{X}^T\mathbf{W}\mathbf{X}\right)^{-1}\mathbf{X}^T\mathbf{W}\mathbf{Y} \tag{18}$$

Moreover, these estimated parameters ($\hat{\mathbf{\theta}}$) could be employed to evaluate the prediction error as well as to adjust the weight so that **e** can be further reduced.

3.2. AutoRegressive Exogenous (ARX) Method [11]

The AutoRegressive Exogenous (ARX) method is a one of the effective system-identification technique. The ARX method can be applied with discrete-time model, expressed in a linear difference equation with external disturbance ($e(t)$), as show in equation (19).

$$y(t) + a_1 y(t-1) + \ldots + a_{n_a} y(t-n_a) = b_1 u(t-1) + \ldots + b_{n_b} u(t-n_b) + e(t) \tag{19}$$

where $y(t-i)$ and $u(t-i)$ are delayed output and input by i-time step, respectively. Reformulating this difference equation in a form of a backward shift operator (q^{-1}) or $q^{-1}u(t) = u(t-1)$, the discrete-time model can be described as a ratio of polynomial functions, $A(q)$ and $B(q)$, in equation (20).

$$y(t) = \frac{B(q)}{A(q)}u(t) = G(q)u(t) \tag{20}$$

Polynomial functions, $A(q)$ and $B(q)$, are correspondingly composed of finite constant coefficients of output and input, which are truncated upto n_a and n_b terms. In ARX method, n_k indicates an input-output delay, n_a are order of $A(q)$ polynomial, related to number of poles and n_b are order of $B(q)$ polynomial, related to number of zeroes plus one. These coefficients, a_i and b_i, are unknown parameters of the specified model structure in a form of linear difference equation.

$$A(q)=\sum_{k=0}^{n_a} a_k q^{-k}=1+a_1q^{-1}+a_2q^{-2}+...+a_{n_a}q^{-n_a} \tag{21}$$

$$B(q)=\sum_{k=1}^{n_b} b_k q^{-k}=b_1q^{-1}+b_2q^{-2}+...+b_{n_b}q^{-n_b} \tag{22}$$

Therefore, the model structure can be rewritten as the linear regression below.

$$\mathbf{y}(t,v)=\boldsymbol{\theta}(t)^T\mathbf{v} \tag{23}$$

The unknown discrete parameters, which are lumped as a vector: $\mathbf{v}=[a_1,a_2,...,a_{n_a},b_1,b_2,...,b_{n_b}]^T$, can be estimated form input-output relation using the LSM. The output vector is denoted as $\mathbf{y}(t)=[y(\max(n_a,n_b)),...,y(N)]^T$, where $N >> \max(n_a,n_b)$ and a matrix $\phi(t)^T$ is a regression matrix for the case with $n_a \geq n_b$, as shown in equation (24).

$$\phi=\begin{bmatrix} -y(n_a-1) & \cdots & -y(0) & u(n_a-1) & \cdots & u(n_a-n_b) \\ -y(n_a) & & -y(1) & u(n_a) & \cdots & u(n_a-n_b+1) \\ -y(n_a+1) & & \vdots & & \vdots & \\ \vdots & & & & & \\ -y(N-1) & \cdots & -y(N-n_a) & u(N-1) & \cdots & u(N-n_b) \end{bmatrix} \tag{24}$$

For the case with $n_b \geq n_a$, the n_a can be replaced n_b and vice versa. As a result, the ARX method can estimate these unknown parameters, as expressed in Equation (25).

$$\hat{\mathbf{v}}=(\phi^T\phi)^{-1}\phi^T\hat{\mathbf{y}} \tag{25}$$

4. Analytical Frequency Range

In order to perform system-identification technique, two sinusoidal inputs or thrust forces with different and varying frequencies are imposed on the USV dynamics models in equation (14), thus system resonance or natural frequency must be recognized beforehand. Using Bode diagram, natural frequencies can be determined from the dynamic equation in state-space form, shown in Equation (26).

$$\begin{aligned}\dot{X}&=AX+BU\\ Y&=CX\end{aligned} \tag{26}$$

where $\mathbf{X}=[u\ \mathrm{r}]^T$ is a state vector and $\mathbf{U}=[T_L\ T_R]^T$ is an input force vector, consisting of input1 (T_L) and input2 (T_R). The magnitude and phase responses for states (u, r) are plotted as function of varying frequency, as demonstrated in Figure 3 and Figure 4. There is no natural frequency for these two first-order systems, there are only corner frequencies at $\omega_{n,1}$ = 0.0115 rad/sec and $\omega_{n,2}$ = 1.1049 rad/sec. The frequency range of the forcing sinusoidal functions from Bode diagram can be guideline for selecting forcing frequency of the system-identification analysis in next section.

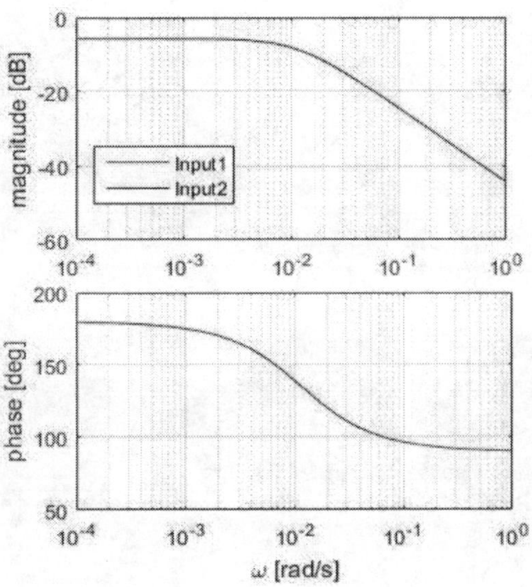

Figure 3. Bode diagram for surge motion (u).

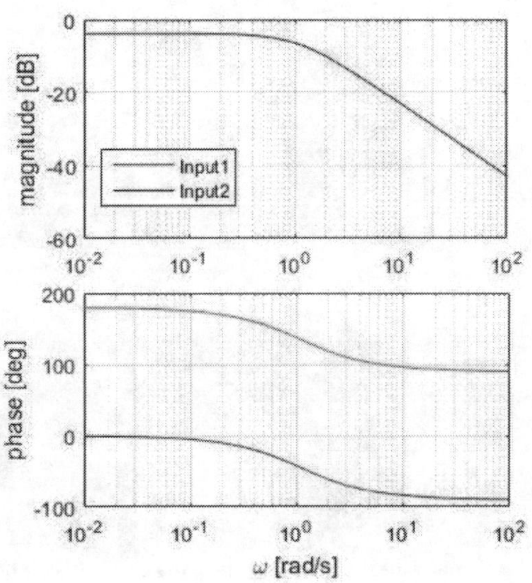

Figure 4. Bode diagram for angular motion (r).

5. Performance Analysis of System identification

The dynamic model of the USV, defined in equation (14), is employed as a performance evaluation between LSM and ARX system-identification techniques. Rewriting the USV system in Equation (27) and (28) can lumped USV parameters together in a form of α's and β's.

$$\dot{u} = \alpha_1 u + \alpha_2 T_L + \alpha_3 T_R \tag{27}$$

$$\dot{r} = \beta_1 r + \beta_2 T_L + \beta_3 T_R \tag{28}$$

The estimated USV parameters using the LSM and ARX methods are validated against known USV parameters such that parametric estimation errors can be considered as function of input frequency and response time. The coupled sinusoidal thrust forces with varying frequency ($\omega = 2\pi f$) are specified as inputs in MATLAB/Simulink. Estimated parameters from dynamic response in time-domain are then compared with known parameters.

5.1. Parametric estimation error by LSM

The Least-Square Method (LSM) is a numerical analysis technique, commonly used in the system identification. According to figure 5, estimated surge parameter (α_1) and yaw parameter (β_1) improve over increasing response time, but these two parameters are insensitive to varying input frequencies. DC gains of the USV transfer function (α_2, α_3) and (β_2, β_3) are depended on both response time and input frequency, moreover estimated parameters from LSM can be improved by using low input-frequency. With appropriate input-frequency, parametric estimation errors decrease or converge to zero with time.

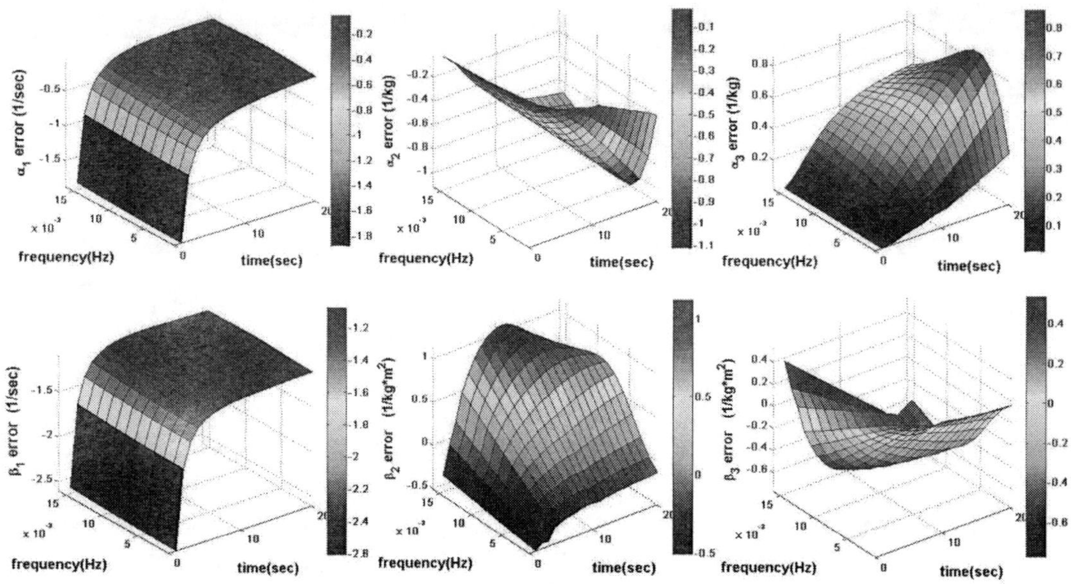

Figure 5. Parametric estimation errors using LSM as function of response time and varying input frequency.

5.2. Parametric estimation error by ARX

To apply ARX method, dynamic continuous-time model must be converted to discrete-time model in the z-domain, described by a_i and b_i coefficients in equation (21) to (22). The AutoRegressive eXogenous (ARX) is simple and the most efficient numerical technique, which can be utilized with high-order polynomial model. Similar to section 5.1, results of estimated parameters in this section demonstrate as function of input-frequency and time, as shown in Figure 6. It can be observed that estimation errors of poles and DC gains of surge-and yaw-motions using the ARX technique are small and unchanged. Accuracy of the ARX approach is much better than that of the LSM technique.

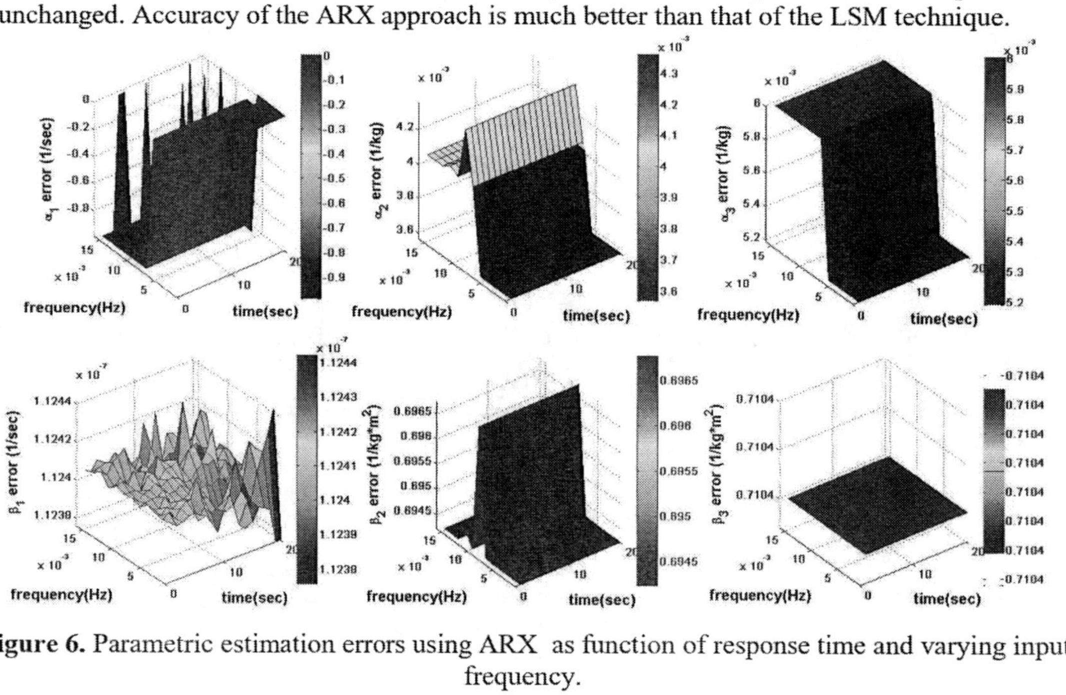

Figure 6. Parametric estimation errors using ARX as function of response time and varying input frequency.

5.3. Comparison of parametric estimation error between LSM and ARX

Sources of error from both LSM and ARX methods come from numerical time-integration of dynamic differential system, Computing pseudo inverse numerically in equation [18] is another source of error for LSM. In the ARX method, delayed time (nk) of discrete input structure and transforming discrete-time to continuous-time model contribute to estimation error of system lumped parameters of transfer function. Using LSM, estimation error decreases as integration time increases, because large input-output responses with appropriate frequency excitation can discover more accurate system parameters.

This section presents estimation of surge- and yaw-parameter error (using both LSM and ARX techniques). The accuracy comparison of parametric estimation error are summarized as minimum and maximum errors, as shown in table 2. The results show that the ARX approach yields much better accuracy of parameter estimation that of LSM for most lumped parameters.

Table 2. Maximum and minimum parametric estimation errors between reference parameters and estimated parameters from both system identification techniques

Axis	Parameter (unit)	LSM		ARX	
		Min error	Max error	Min error	Max error
Surge Motion	$\alpha_1\left(1/\sec\right)$	-1.879711	-0.058874	-0.988450	0.95789×10^{-11}
	$\alpha_2\left(1/kg\right)$	-1.114721	-0.0211549	0.003565	0.0043563
	$\alpha_3\left(1/kg\right)$	0.007522	0.859333	0.0051886	0.006000
Yaw Motion	$\beta_1\left(1/\sec\right)$	-2.604681	-1.083689	0.11237×10^{-6}	0.11244×10^{-6}
	$\beta_2\left(1/kg\cdot m^2\right)$	-0.508471	1.116051	0.694235	0.696756
	$\beta_3\left(1/kg\cdot m^2\right)$	-0.773391	0.528940	0.710399	0.710399

6. Conclusion

This paper presents system identification (SI) methods: Least-Square Method (LSM) and AutoRegressive eXogenous (ARX), based on simplified dynamic USV model with two fixed input-thrusters input-thrusters ($\mathbf{u} = [T_L\ T_R]^T$). The appropriated frequency range is also investigated by using Bode diagrams in order to choose input frequency. Then, the numerical methods: 1) Least-square method (LSM) and 2)AutoRegressive eXogenous (ARX) techniques are implemented to estimate lumped parameters of the USV model. Results reveal that parametric estimation error from LSM can be improved with large response time data with proper input frequency excitation. On the other hand, the ARX approach provide faster parameter convergence to true parameter values using only short interval of input-output response time for this simplified dynamic system, since the accuracy of parameter estimation in discrete-time domain of ARX method is better than that of LSM.

Acknowledgment

This research work is financially supported by Office of the Higher Education Commission for fiscal year of 2016 under "Remotely operated surveillance catamaran for suspicious object detection from images" research project.

References

[1] Li Z and Bachmayer R 2013 *Proc.. Oceans 2013 MTS/IEEE* (San Diego, United States)
[2] Park J H, Shim H W, Jun B H, Kim S M, Lee P M and Lim Y K 2010 *Oceans 2010 IEEE-Sydney* pp 1-5
[3] Wu H, Sun D and Zhou Z 2004. Model *IEEE Trans. on Robotics* **20**(4) pp 702-712

[4] Yuan Y, Liang S, Zhong J, Xiong Q and Gao M 2015 *Proc. 27th Chinese IEEE Control and Decision Conf. (CCDC)* pp 4116-4120
[5] Irie B 2009 *MS Thesis and Dissertations at The University of Texas at Arlington*.
[6] Fossen T I 1994 *Guidance and Control of Ocean Vehicles* (New York: John Wiley & Sons)
[7] Yang B, Liu Y, Liang F and Dong Z 2016 *Proc. Int. Archives Photogrammetry, Remote Sensing and Spatial Information Sciences (XXIII ISPRS2016 Congress)* Vol. XLI-B3 Prague (Czech Republic)
[8] Mertz C, Navarro-Serment L E, MacLachlan R, Duggins D and Gowdy J 2013 *J. of Field Robotics* **30** pp 17-43
[9] Pandey J and Hasegawa K 2015 *Proc. Int. Assc. Institutes of Navigation World Congress* (Prague, Czech Republic)
[10] Prempraneerach P 2016 *Proc. 2016 IEEE Int. Computer Science and Engineering Conf. (ICSEC2016)* pp 1-6
[11] Keesman K J 2011 *System Identification* (New York: Springer-Verlag)

Experimental and numerical investigations of heat transfer and thermal efficiency of an infrared gas stove

A. Charoenlerdchanya[1], P. Rattanadecho[2], and P. Keangin[1,*]

[1] Department of Mechanical Engineering, Faculty of Engineering, Mahidol University, 25/25 Phutthamonthon 4 Road, Salaya, Nakhon Pathom 73170, Thailand.
[2] Department of Mechanical Engineering, Faculty of Engineering, Thammasat University (Rangsit Campus), 99 mu 18, Paholyothin Road, Klong Nueng, Klong, Pathumthani 12120, Thailand.

* Corresponding Author: pornthip.kea@mahidol.ac.th

Abstract. An infrared gas stove is a low-pressure gas stove type and it has higher thermal efficiency than the other domestic cooking stoves. This study considers the computationally determine water and air temperature distributions, water and air velocity distributions and thermal efficiency of the infrared gas stove. The goal of this work is to investigate the effect of various pot diameters i.e. 220 mm, 240 mm and 260 mm on the water and air temperature distributions, water and air velocity distributions and thermal efficiency of the infrared gas stove. The time-dependent heat transfer equation involving diffusion and convection coupled with the time-dependent fluid dynamic equation is implemented and is solved by using the finite element method (FEM). The computer simulation study is validated with an experimental study, which is use standard experiment by LPG test for low-pressure gas stove in households (TIS No. 2312-2549). The findings revealed that the water and air temperature distributions increase with greater heating time, which varies with the three different pot diameters (220 mm, 240 mm and 260 mm). Similarly, the greater heating time, the water and air velocity distributions increase that vary by pot diameters (220, 240 and 260 mm). The maximum water temperature in the case of pot diameter of 220 mm is higher than the maximum water velocity in the case of pot diameters of 240 mm and 260 mm, respectively. However, the maximum air temperature in the case of pot diameter of 260 mm is higher than the maximum water velocity in the case of pot diameters of 240 mm and 220 mm, respectively. The obtained results may provide a basis for improving the energy efficiency of infrared gas stoves and other equipment, including helping to reduce energy consumption.

1. Introduction

In the present, the domestic gas stoves are separated with three types, which are low-pressure stove, high-pressure stove and infrared gas stove. Process method of infrared gas stove is infrared radiation (IR). Infrared radiation is classified with wavelength in 3 types, namely near-infrared radiation: NIR (0.75 - 1.5 μm), middle-infrared radiation: MIR (1.5 - 5.6 μm) and far-infrared radiation: FIR (5.6 - 1000 μm) [1].

Most of burner type of infrared gas stove that use in Thailand is ceramic burner. Because of burner type ceramic match with domestic cooking more than the other types. From table 1, wavelength and maximum burner temperature of ceramic burner is appropriate for using when compared with the other types. Consider magnetic wavelength reverse variation with temperature, the burner in far-infrared radiation will make food absorb the radiation as well. In the other side, the burner in far-infrared radiation will use more time to make food ripe. In addition, the burner in near-infrared radiation will use less time to make food ripe but maybe inside of food is not ripe. Therefore, the ceramic burner is gas stove with infrared radiation medium that is suitable.

Table 1. Comparison of the properties of burner [1].

Parameters	Types of burner			
	Catalytic burner	Ceramic burner	Metal fiber burner	Porous burner
Wavelength category	Long-wave	Medium-wave	Medium-wave	Short-wave
Peak wavelength	3.5 μm	2.4 μm	2.2 μm	1.7 μm
Maximum burner temperature attained	850 K	1100 K	1200 K	1400 K
Maximum thermal load	30 kWm^{-2}	120 kWm^{-2}	200 kWm^{-2}	1000 kWm^{-2}

Infrared gas stove is compared with the other domestic gas stove and the result is infrared gas stove saves energy and decreases gas consumption more than the other types. From data of entrepreneur of infrared gas stove found that commercial quantity is 430,000 stoves per year. Comparing to the commercial quantity of infrared gas stove 2 years before, the quantity increases by 14 % [1]. This report shows that the people are increasingly interested in infrared gas stove and realize the way to save energy and decrease gas consumption.

According to previous researches, there is little research about infrared gas stove. Many previous researches studied about the thermal efficiency and gas emission of burner by experiment such as comparison between the types of burner and percentage of thermal efficiency that porous radiant recirculated burner (PRRB) has thermal efficiency more than standard burner (SD) about 10 % [2]. Porous radiant recirculated burner (PRRB) was compared with conventional domestic burner (CB) in air temperature 300 °C condition; it was found that the thermal efficiency of PRRB increases about 12 % [3]. In swirling central flame ring condition, the maximum thermal efficiency of PRRB is about 60 % [3]. Comparing porous radiant burner (PRB) with LPG conventional domestic stove found that PRB has higher thermal efficiency than LPG conventional domestic stove and CO and NO_x emissions of PRB were less than LPG conventional domestic stove [4, 5].

Several researchers studied the parameters that effect on the thermal efficiency. Thermal efficiency of conventional open flame gas cooker with swirling central flame burner was higher than conventional radial flow burner about 15 % [6]. Different design of burner heads were compared and found that flat face burner (Brass) has highest thermal efficiency [7]. From three-dimensional computational fluid dynamic (CFD), thermal efficiency of conventional domestic burner with modified design increased about 2.5 % [8]. There were some numerical studies of the domestic burner model. One-dimensional numerical model of reticulated ceramic radiant burner can predict burner flame structure by compared temperature with thermocouples measurements [9].

In this paper aim to study about the effect of the diameters of pot on the water and air temperature distributions, water and air velocity distributions and thermal efficiency of the infrared gas stove during heating by using infrared gas stove. In order to verify the accuracy of the present model, the computer simulation study is validated with an experimental study. The test method is referenced from the low-pressure gas stove in households with LPG test (TIS No. 2312-2549) and the Development of energy efficiency standards infrared burner project. An axially symmetric model of pot and infrared gas stove are considered in this study, which minimized the computation time while maintaining good resolution and represent the full three-dimensional results. The governing equations as well as initial and boundary conditions are solved by using the axisymmetric finite element method (FEM). The range of temperature and velocity are separated by the different colours that are convenient to analyse the results. The results will be the basis for the study and analysis about the effects of various parameters on temperature distribution, velocity distribution and thermal efficiency during heating by using infrared gas stove. The complete mathematical model is useful for the development of heating process by infrared gas stove technologies and guide the development of effective infrared gas stove to lead the way in energy saving way.

2. Experimental Study

For experimental, the test method is referenced from the low-pressure gas stove in households with LPG test (TIS No. 2312-2549) and the Development of energy efficiency standards infrared burner project. Figure 1 shows an infrared gas stove used in the experiment. The infrared gas stove is composed of ceramic plate that it emits heat to the pot.

Experimental methods to determine the thermal efficiency of the infrared gas stove in this study are in the following steps. Starting with check and prepare experimental equipment, set up stove and connect the stove with gas gauge, pressure equipment, gas temperature gauge and valve gas. Then adjust the gas valve to the maximum gas inject position at gas pressure of 280 millimeter water (mm H_2O). Warm the burner for 5 minutes to remove stains dirt and oil. Turn off the stove and place the pot on the center of stove then put the thermometer to measure temperature. The initial water temperature should be set up at 25 ± 2 °C. In addition, the testing room is controlled room temperature at 25 ± 2 °C. Before the test, the temperature of gas and water temperature are recorded. The position for measuring the temperature of the center of the water at the height of 5 mm from the bottom of the pot is considered. Follow this procedure until the temperature of water increase from initial for 50 °C. Finally, turn off the stove; record the last temperature of water and quantity of gas. An experiment is repeated 5 times. The data of the test will be used to calculate the thermal efficiency of infrared gas stoves. Before do the test again, pot cleaner clearly to minimize heat loss. Figure 2 shows the experimental setup. The temperature measured in experimental is recorded by using digital thermometer during heating. The dimension of pot and infrared gas stove and conditions in this experiment setup is used for the simulation.

Figure 1. Infrared gas stove used in the experiment.

Figure 2. The experimental setup.

3. Simulation Study

Analysis of heat transfer in water during heating by using infrared gas stove is presented. The system of governing equations as well as initial and boundary conditions are solved numerically using the FEM via COMSOL^{TM} Multiphysics.

3.1. Physical model

In order to verify the accuracy of the present model, the simulation results are validated against the experimental results with the same conditions. The mathematic model used in computer simulation is divided into two parts, namely a container (pot) and infrared gas stove. The pot made of aluminum material. Because of the pot and burner are cylinder shape and the pot is located on the center of burner,

IOP Conf. Series: Materials Science and Engineering **297** (2018) 012034 doi:10.1088/1757-899X/297/1/012034

an axially symmetric model are considered in this study for save the computation time. As shown in figure 3, the axially symmetric model of pot and infrared gas stove for heat transfer analysis of water heating by using infrared gas stove is proposed. Heat absorbed by the water and converted into internal heat generation, which causes its temperature of water to rise. The convergence test is carried out to identify the suitable numbers of element required. The number of elements where solution is independent of mesh density is found to be 431,110 elements. It is reasonable to confirm that, at this number of element, the accuracy of the simulation results is independent from the number of elements through the calculation process.

3.2. Equations for heat transfer analysis

The heat transfer model is developed to predict the temperature distribution in the water and air. The mathematical model in this study is shown in figure 3. The assumptions have been offered for computer analysis is as follows: lack of the combustion of gas fuel, unconsidered a container base and no phase change and no chemical reaction in the water. Also, thermal properties of water, aluminum pot and air are constant. The initial condition of water is defined as T_0 = 25 °C at t = 0 s in order to comply with the experiment. The governing equation describing the heat transfer phenomenon is given in equation (1):

$$\rho c \frac{\partial T}{\partial t} + \rho c \vec{u} \cdot \nabla T = \nabla \cdot (k \nabla T) = 0 \tag{1}$$

where ρ = 1000 kg/m^3 is the density, c = 0.004186 MJ/kg.K is the specific heat capacity, T is the temperature (°C), t is the time (s), $\vec{u}$ is the velocity (m/s) and k = 0.6065 W/m·K is the thermal conductivity.

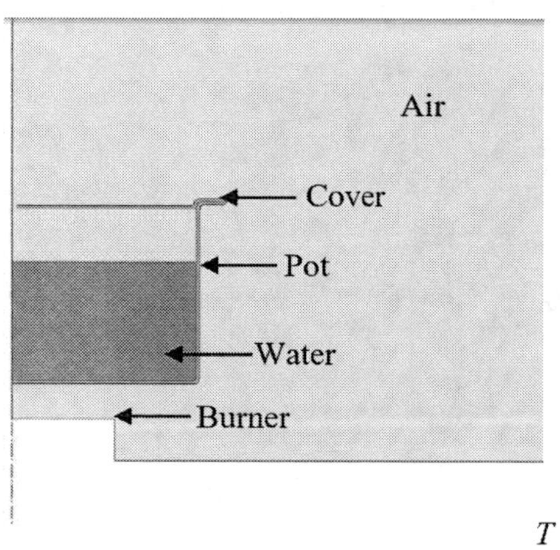

Figure 3. Physical model for this study.

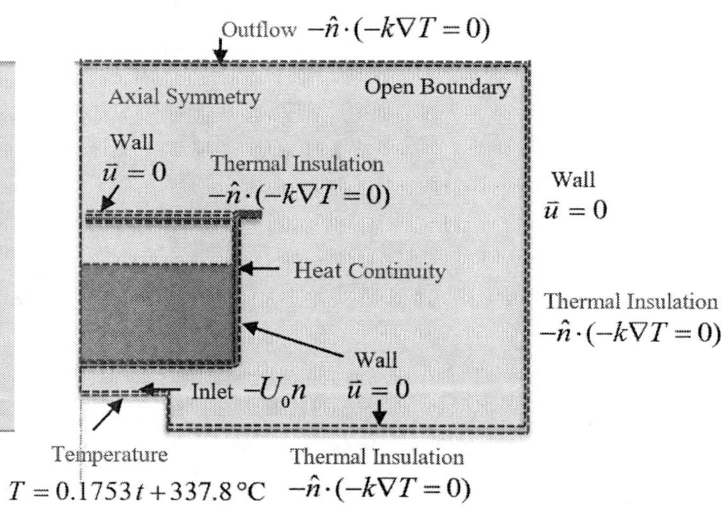

Figure 4. Boundary condition for numerical analysis.

3.3. Equations for fluid flow analysis

Corresponding to heat transfer analysis, fluid flow analysis inside the water and air is assumed in axially symmetric model. Fluid flow analysis is formulated to describe the water and air velocity distributions. Using standard symbols, the governing equations describing the fluid flow are given as follows:

Continuity equation:

$$\frac{\partial \rho}{\partial t}+\nabla\cdot\left(\rho\vec{u}\right)=0 \tag{2}$$

Navier-Stokes equation:

$$\rho\frac{\partial u}{\partial t}+\rho(\vec{u}\cdot\nabla)\vec{u}=\nabla\cdot\left[-pI+\mu\left(\vec{u}+\nabla\vec{u}\right)^{T}-\frac{2}{3}\mu\left(\nabla\cdot\vec{u}\right)I\right] \tag{3}$$

where ρ is the density (kg/m^3), t is the time (s), $\vec{u}$ is the velocity (m/s), p is the pressure (Pa), μ is the viscosity (Ns/m^2) and I is the identity matrix tensor.

The equation for calculate the thermal efficiency is given in equation (4) [1]:

$$\eta=\frac{m\times c\times(T_2-T_1)}{V\times Q}\times\frac{273+T_g}{298}\times\frac{101.3}{P_s+P_m-P_{sat}}\times 100\% \tag{4}$$

where η is the thermal efficiency (-), m is the mass (kg), c = 0.004186 MJ/kg.K is the specific heat capacity, T_1= 25 ± 2 °C is the initial temperature of water, T_2 = 75 ± 2 °C is the final temperature of water, V is the amount of gas used test (m^3), Q = 107.86 MJ/m^3 is the lower heating value, T_g is the temperature of the gas during the test (°C), P_s is the atmospheric pressure during test (kPa), P_m is the pressure of the gas during the test (kPa), P_{sat} is the pressure of saturated stream at a temperature of gas during test (kPa). The boundary conditions for heat transfer analysis are shown in figure 4.

4. Results

In this section, the experiment of water temperature distribution results and the results of the numerical are focused on the water and air temperature distributions, water and air velocity distributions and thermal efficiency during heating.

4.1. The comparison of water temperature of simulation results with the experiment results

The accuracy of numerical model is verified by the validation against the experimental under the same geometric model and same conditions. The comparison of water temperature of simulation results with the experiment results, which varies with the three different pot diameters at 220 mm, 240 mm and 260 mm during heating by using infrared gas stove as presented in figure 5. It can be seen that the water temperature of simulation results are in good agreement with the water temperature of experiment results. There are 10.167, 5.454 and 6.411 average percentage errors in case of using diameters of pot at 220 mm, 240 mm and 260 mm, respectively. Certain amount of mismatch between the simulation results and the experiment results is caused by the numerical scheme. Figure 6 also indicates that an increase in the heating times results in an increase temperature. This favorable comparison lends confidence in the accuracy of the present numerical model and ensures that the numerical model can accurately represent the phenomenon of heat transfer between heating the water using infrared gas stove.

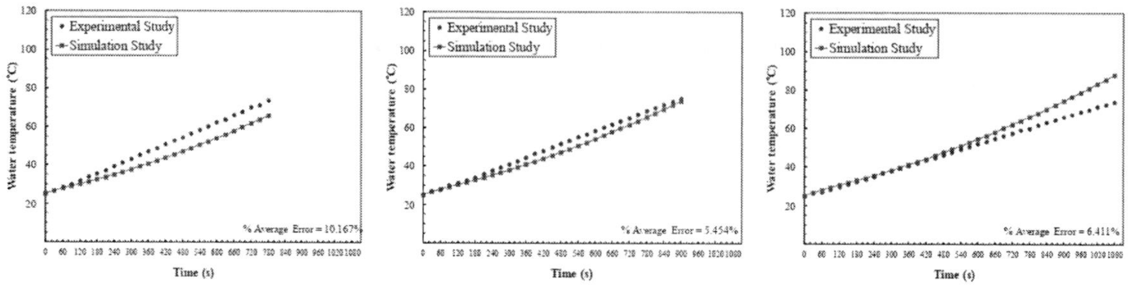

(a) (b) (c)

Figure 5. The comparison of water temperature of simulation results with the experiment results that vary by diameters of pot using (a) Pot diameter 220 mm, (b) Pot diameter 240 mm and (c) Pot diameter 260 mm.

4.2. The comparison of air temperature of simulation results with the experiment results

As same as the conditions from the comparison of water temperature of simulation results with the experiment results, the comparison of air temperature of simulation results with the experiment results are considered. Because of the surrounding temperature will show the heat loss to the air. The results of the validation test cases are illustrated in figure 6 and clearly show good agreement of air temperature between the simulation results and that of experimental results. For all pot diameters, the results of these simulations are consistent with experimental results. There are 1.894, 2.483 and 2.021 average percentage errors in case of using diameters of pot at 220 mm, 240 mm and 260 mm, respectively. It is found that an increase in the heating times results in an increase air temperature. In addition, it can be seen that the air temperature increase corresponds to higher diameters of pot in the same time.

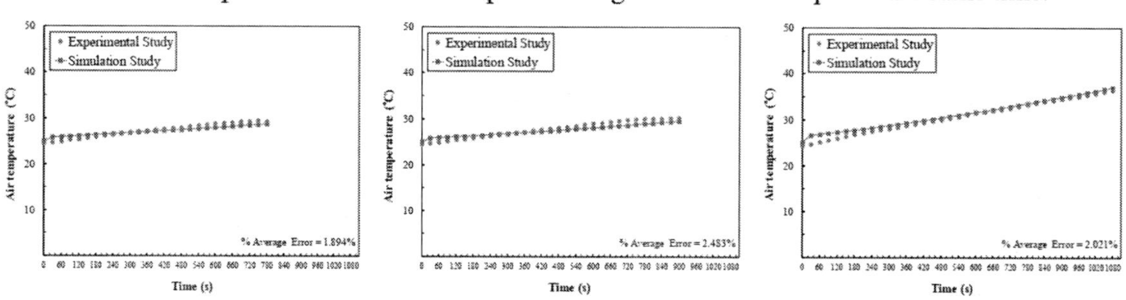

(a) (b) (c)

Figure 6. The comparison of air temperature of simulation results with the experiment results that vary by diameters of pot using (a) Pot diameter 220 mm, (b) Pot diameter 240 mm and (c) Pot diameter 260 mm.

4.3. The effects of diameter of pot

Figures 7(a)–(c) show the effects of pot diameter on the patterns of the water temperature simulated results at heating times of 780 s during water heating by using infrared gas stove. The diameter of pots of 220 mm, 240 mm and 260 mm are considered. It is evident that the water temperature in the case of diameters of pot at 260 mm has the highest temperature because the bottom of the pot is larger which heat is received from the burner than other cases. Heat energy is passed to a container and is delivered to water and thereafter the absorbed energy is converted to thermal energy, which increases the water temperature. Figures 8 display the simulation results of water velocity distribution at different pot diameters (220 mm, 240 mm and 260 mm). The distributions of the water velocity simulated results at heating times of 780 s during water heating by using infrared gas stove are investigated. It can be found that the water velocity in the case of diameters of pot at 220 mm is higher than the water velocity in the case of diameters of pot at 240 mm and 260 mm, respectively. Because of the velocity of the big area will less than the velocity of the small area. The patterns of the air velocity as shown in figure 9(a)–(c). The flowing air with higher velocity acts as a heat sink and results in dissipates the thermal heat to the surrounding air. From the simulation results are presented in figure 8 and figure 9, the water velocity is very small as compared to the conventional problems. However, the main idea behind this research is to propose the completed model in order to completely explain the actual heat transfer during water heating.

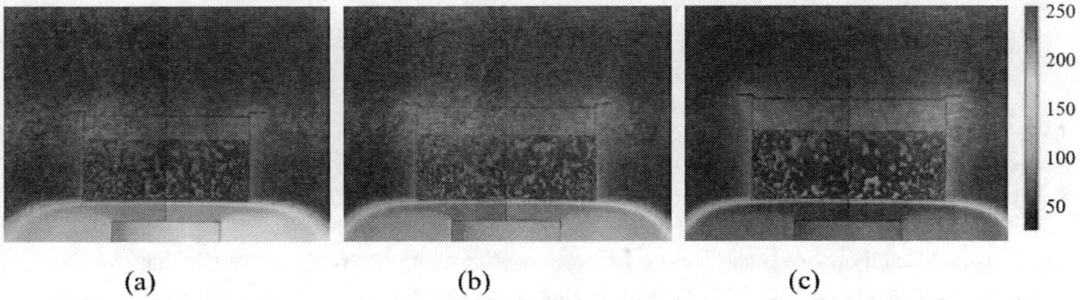

Figure 7. The water temperature patterns of simulation results that vary by diameters of pot using (a) Pot diameter 220 mm, (b) Pot diameter 240 mm and (c) Pot diameter 260 mm.

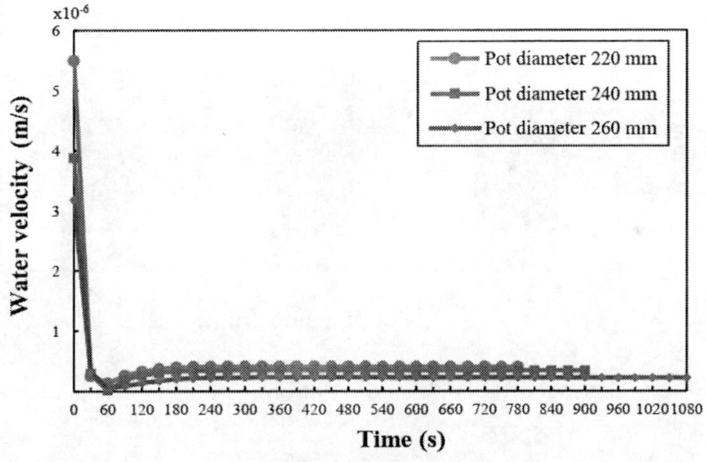

Figure 8. Comparison of the water velocity of simulation results at various diameters of pot at heating times of 780 s.

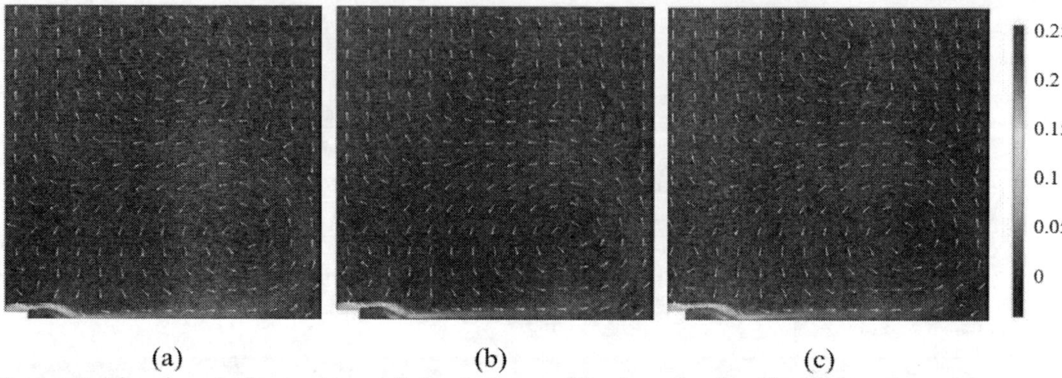

Figure 9. The air velocity patterns of simulation results that vary by diameters of pot using (a) Pot diameter 220 mm, (b) Pot diameter 240 mm and (c) Pot diameter 260 mm.

4.4. The thermal efficiency

The thermal efficiency of simulation results that use the quantity of gas consumed by experiment as showed in figure 10. It is found that the thermal efficiency of pot diameter 260 mm is higher than thermal efficiency of pot diameter 240 mm and 220 mm, respectively because the lager pot size has a large temperature gradient produced by the heat energy, causing a strong impact of natural convection that gives a buffer characteristic to the water temperature during water heating process.

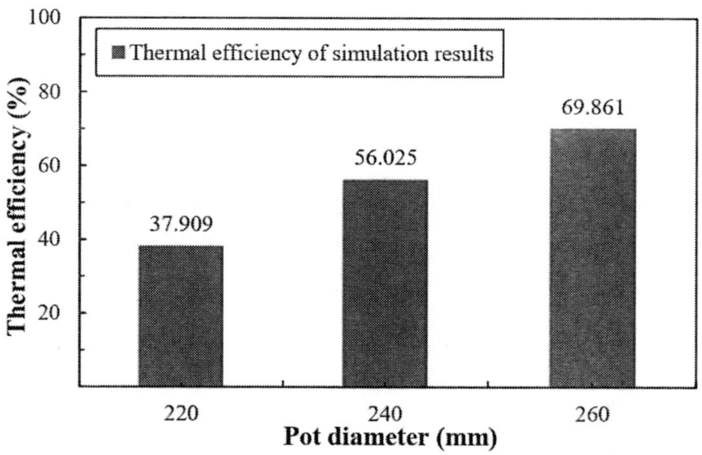

Figure 10. The thermal efficiency of simulation results at various diameters of pot.

5. Conclusion

This research presents the experimental and numerical analysis of water heating using infrared gas stove. The water and air temperature distributions, water and air velocity distributions that vary by three different pot diameters (220 mm, 240 mm and 260 mm) are considered. The conclusions of this study can be summarized as follows:

1) The results from the simulation are in agreement with the experimental data.

2) The water temperature distribution increase with greater heating time that vary by three different pot diameters (220 mm, 240 mm and 260 mm). The pot with larger area has higher quantity of water, so it makes more time to increase temperature to the final temperature (75 ± 2 °C).

3) The air temperature distribution increase with greater heating time that vary by three different pot diameters (220 mm, 240 mm and 260 mm). The air temperature measured point of every pot diameters is at the same position, so the larger pot diameter was increase the air temperature more than the smaller

pot diameter. Because of the smaller pot has long destination between the air temperatures measured point and pot.
4) The water velocity distribution decreases with greater pot diameters that vary by three different pot diameters (220 mm, 240 mm and 260 mm). The reason of velocity trend is about the size of pot diameter and quantity of water, so the water velocity of the pot diameter 220 mm which is less area is more than the water velocity of 240 mm and 260 mm in the same time.
5) The air velocity distribution decrease with greater pot diameters that vary by three different pot diameters (220 mm, 240 mm and 260 mm) corresponds to water velocity distribution. The convective heat transfer characteristic has a strong effect on the air velocity distribution.
6) The average percentages thermal efficiency of three different pot diameters of 220 mm, 240 mm and 260 mm are 37.909, 56.025 and 69.861, respectively. From the equation (4), the different water temperature between max and min values is effect to thermal efficiency, so the pot diameter 260 mm which has highest temperature at the last time has highest thermal efficiency.

Acknowledgements
The authors gratefully acknowledge Mahidol University for supporting this research.

References
[1] Department of Mechanical Engineering, Faculty of Engineering, Mahidol University, Final Report, Development of energy efficiency standards infrared burner project
[2] Jugjai S, et al. 1996 *RERIC Int. Energy J.* **18**(2) pp 97-101
[3] Jugjai S, et al. 2002 *Experimental Thermal and Fluid Science* **25** pp 581-592
[4] Pantangi V K, et al. 2011 *Energy* **36** pp 6074-6080
[5] Muthukumar P and Shyamkumar P I 2011 *Fuel* **112** pp 562-566
[6] Jugjai S, et al. 2001 *Int. J. of Energy Research* pp 657-674
[7] Khan M Y and Saxena A 2013 *Int. J. of Engineering Research & Technology* **2** pp 656-659
[8] Boggavarapu P, et al. 2013 *Fuel* pp 709-715
[9] Rumminger M and Dibble R "1996 *Combustion/The Combustion Institute* pp 1755-1762

IOP Conf. Series: Materials Science and Engineering **297** (2018) 012035 doi:10.1088/1757-899X/297/1/012035

Hot air impingement on a flat plate using Large Eddy Simulation (LES) technique

C. Plengsa-ard* and M. Kaewbumrung

Mechanical Engineering Department Kasetsart University, Chatuchak, Bangkok 10900, Thailand

* E-mail: fengcpp@ku.ac.th, Tel: 02-797-0999 Ext 1833

Abstract. Impinging hot gas jets to a flat plate generate very high heat transfer coefficients in the impingement zone. The magnitude of heat transfer prediction near the stagnation point is important and accurate heat flux distribution are needed. This research studies on heat transfer and flow field resulting from a single hot air impinging wall. The simulation is carried out using computational fluid dynamics (CFD) commercial code FLUENT. Large Eddy Simulation (LES) approach with a subgrid-scale Smagorinsky-Lilly model is present. The classical Werner-Wengle wall model is used to compute the predicted results of velocity and temperature near walls. The Smagorinsky constant in the turbulence model is set to 0.1 and is kept constant throughout the investigation. The hot gas jet impingement on the flat plate with a constant surface temperature is chosen to validate the predicted heat flux results with experimental data. The jet Reynolds number is equal to 20,000 and a fixed jet-to-plate spacing of $H/D = 2.0$. Nusselt number on the impingement surface is calculated. As predicted by the wall model, the instantaneous computed Nusselt number agree fairly well with experimental data. The largest values of calculated Nusselt number are near the stagnation point and decrease monotonically in the wall jet region. Also, the contour plots of instantaneous values of wall heat flux on a flat plate are captured by LES simulation.

1. Introduction

In recent years, heat transfer of a hot jet impinging on a solid wall is widely applied in a cooling and heating process. Turbulent jets are initiated and discharged from the nozzle exit. High heat transfer rate at a surface is promoted. These rates are affected by many parameters for example Reynolds number, the distance from a stagnation point in radial direction, and a nozzle-plate spacing. There are many previous experimental studies on jet impingement reported in the literature. Wang et al. [1] investigated heat transfer characteristics during jet impingement on a flat surface while Zhou et al. [2] showed the numerical analysis of turbulent round jet impingement. Heat transfer by liquid jets impinging on a hot flat surface was proposed by Hosain et al. [3]. Moreover, Lyttle et al. [4] studied air jet impingement for small values of nozzle-plate spacing, while the circular air jet impingement on a flat plate with a uniform heat flux on the surface was reported by Baughn et al. [5], Singh et al. [6], and Mohanty et al. [7]. Physics of the jet impingement was explained by Gardon et al. [8]. Moreover, Zuckerman et al. [9] proposed the average heat transfer correlation from a plate and heat transfer rate was investigated in more details. The experiment of Cooper et al. [10] showed the mean velocity and turbulence statistics for a single jet impingement from a long pipe. Only a few papers for three dimensional numerical investigation of hot air jet impingement on a flat plat were published in the literature, thus the investigation of hot air jet impingement is needed. In the present study, CFD code is used to calculate

IOP Conf. Series: Materials Science and Engineering **297** (2018) 012035 doi:10.1088/1757-899X/297/1/012035

the flow of jet over the flat plate and to analyze heat transfer from the surface. Modeling and meshing are created in FLUENT.

2. Turbulence model

To simulate flow field, the Large Eddy Simulation (LES) is used for the research. The flow field is separated into large-scale (resolved or filtered) and small scale (sub-grid) motion by filtering (spatially averaging) local flow variables in LES. A subgrid-scale scale Smagorinsky-Lilly model is present and the eddy viscosity is modeled. The Smagorinsky constant is set to 0.1 and is kept constant throughout the investigation. More detail information for a particular turbulence model can be found in the literature. ANSYS-Fluent usually has included the standard law of the wall in the code (LES-SL). However, the classical Werner-Wengle wall model (LES-WW) is focused in this research. Since the wall shear stress is strongly dependent on wall heat transfer, the effect of heat transfer is investigated. The air impingement jet is chosen to be a test case.

3. Werner-Wengle wall shear stress model formulation [11]

$$\tau_w = \frac{2\rho \nu U}{\Delta y}\ ,\ U \le \frac{\nu}{2\Delta y} A^{\frac{2}{1-B}} \tag{1}$$

$$\tau_w = \rho \left\{ \frac{1-B}{2} A^{\frac{1+B}{1-B}} \left(\frac{\nu}{\Delta y} \right)^{1+B} + \frac{1+B}{A} \left(\frac{\nu}{\Delta y} \right)^{B} U \right\}^{\frac{2}{1+B}}\ ,\ U > \frac{\nu}{2\Delta y} A^{\frac{2}{1-B}} \tag{2}$$

where τ_w is wall shear stress, U is the magnitude of gas velocity at the first grid point parallel to the wall, ρ is fluid density, Δy is the height of the first cell near walls , ν is kinematic viscosity, A and B are equal to 8.3 and 1/7 respectively.

4. Test case description

The computational domain for impinging jets is divided into the two regions. First, in the top pipe region, the initial velocity profiles at the inlet are assumed by a fully developed flow. To generate turbulent motions, the initial condition for sub-grid scale turbulent kinetic energy is scaled with 5% of the jet bulk velocity in the pipe. The length of the pipe is 30D, which can provide a fully developed flow at the exit, and the adiabatic wall is applied at the pipe surfaces.

In the base region, the distance between the pipe exit and the wall surfaces is H=2D and the turbulent jet impinges normally on the surface with size of 25D. The inlet and initial condition of air in the pipe and the base region is set to 700 K. The numerical study of impinging jets is tested with a computational grid, as shown in figures 1 and 2. In addition, figure 3 shows the CFD mesh at inlet of the computational domain. In order to be consistent with available experimental data, constant wall temperature boundary conditions are used on the top wall and the bottom, or impingement, wall, while the outlet pressure boundary conditions are applied at the ambient exits. The constant wall temperature is chosen to be 450 K, which matched with experimental data. The boundary conditions of the sub-grid scale turbulent kinetic energy are the Neumann boundary conditions.

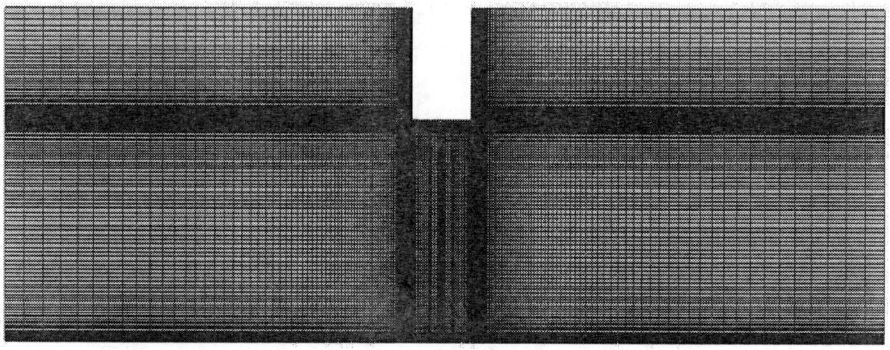

Figure 1. CFD mesh of the impingement jet on the flat plate

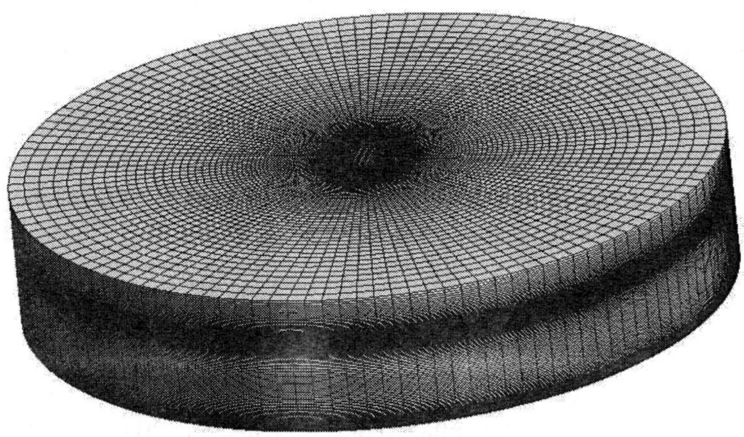

Figure 2. CFD mesh of the computational domain

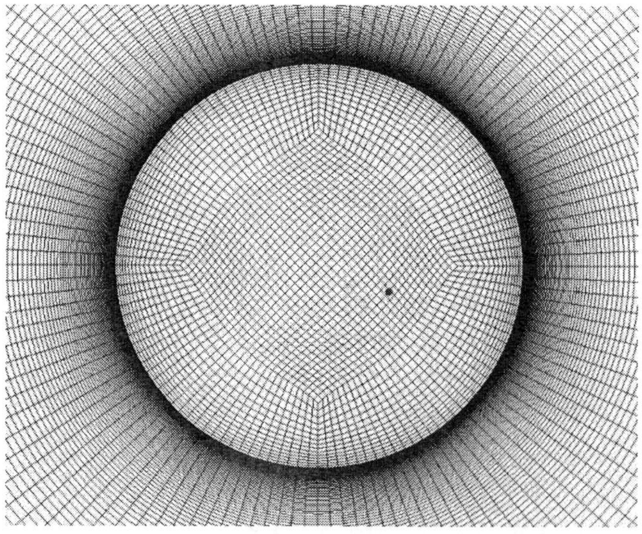

Figure 3. CFD mesh at inlet of the computational domain

IOP Conf. Series: Materials Science and Engineering **297** (2018) 012035 doi:10.1088/1757-899X/297/1/012035

Cylindrical coordinate is used for the computational domain of the impinging jet. The original point (r=0) is located at the center line. The velocity components in the axial (z), radial (r) and azimuthal (θ) directions are denoted by W, U and V. The research also focuses heat transfer on the impingement wall, when Werner-Wengle wall model is applied in the code. The non-dimensional heat transfer coefficient, i.e., Nusselt number on the impingement surface are calculated and compared with existing experimental data.

5. Results and discussion

The comparisons of computed and measured time averaged Nusselt number with three different mesh sizes can be seen in figure 4. The experiment work from Baughn et al. [5] are used for comparison. First, the numerical study of impinging jets is tested with three different computational grids i.e., LES-grid I, LES- grid II and LES-grid III. There is no significant difference between LES-grid II and LES- grid III. To obtain a better result, grid II is chosen for the simulation in this research. The axial and radial mean velocity components at the stagnation point (r/D =0) and r/D=0.5 as predicted by LES-SL and LES-WW models are shown in figures 5 and 6, respectively. Also, the root mean square (rms) of axial and radial velocities as predicted by LES-SL and LES-WW models are presented in figures 7 and 8. These axial mean velocity and rms values are normalized by the jet bulk velocity, W_b. The experiment data are selected from works of Geers et al. [12] to compare with calculated values. The mean axial velocity, W, and the mean radial velocity, U, at r/D=0 (centerline) are shown in figure 5. On the centerline, the axial mean velocity remains nearly constant from the pipe exit (Z/D=2.0) up to Z/D=0.4, while, below the position Z/D=0.4, the flow decelerates while the jet flows approach to the wall. As can be seen in figure 6, at the position r/D =0.5, the strong flow deflection due to the impingement wall makes the mean radial velocity (U/W_b) start to increase and reaches the maximum value near the wall whereas the mean axial velocity (W/W_b) approaches to zero.

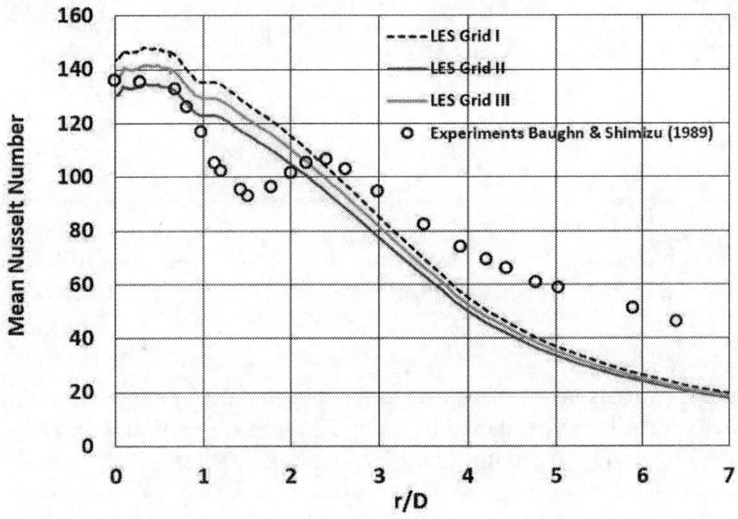

Figure 4. Comparisons of computed and measured time average Nusselt number with three different mesh sizes.

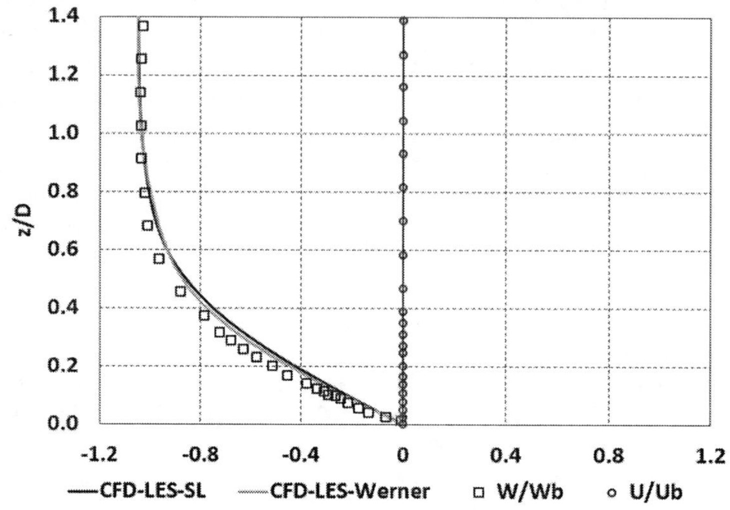

Figure 5. Comparisons of measured and computed mean axial and radial velocity profiles as predicted by LES-SL and LES-WW models at a stagnation point (r/D = 0) with Re =20,000 and H/D =2.0

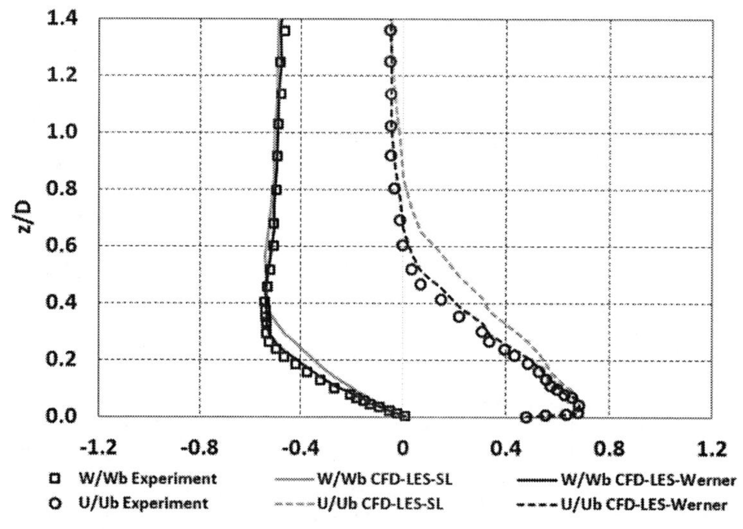

Figure 6. Comparisons of measured and computed mean axial and radial velocity profiles as predicted by LES-SL and LES-WW models at r/D = 0.5 with Re =20,000 and H/D =2.0

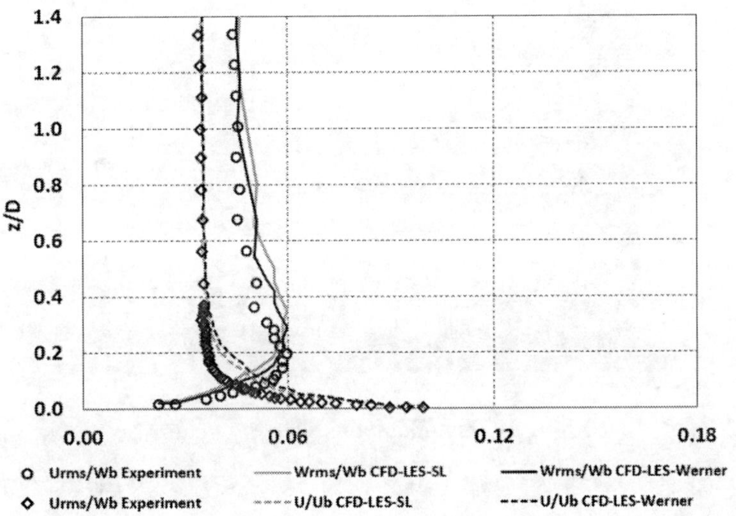

Figure 7. Comparisons of measured and computed root mean square (rms) of axial and radial velocities as predicted by LES-SL and LES-WW models at a stagnation point (r/D = 0) with Re =20,000 and H/D =2.0

For the rms values as shown in figure 7 and 8, the mean axial and radial velocity fluctuations decrease down to zero due to wall effect. At the centerline (r/D=0) in figure 7, the rms values of the axial and radial velocity fluctuation are over-predicted in the region Z/D < 0.3 while the results agree fairly well when Z/D > 0.3. At r/D = 0.5, the velocity fluctuations are higher than those of values at the centerline. The rms values at the point r/D = 0.5 agree fairly well with the selected experiment data until Z/D < 0.4. The discrepancies between the rms values of the velocity fluctuations predicted by LES-SL, LES-WW and experiment data can be observed in figure 8 in the region Z/D > 0.4.

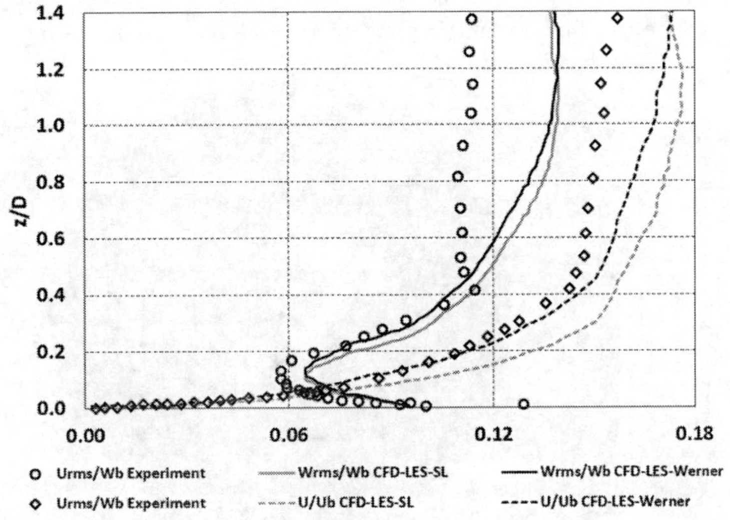

Figure 8. Comparisons of measured and computed root mean square (rms) of axial and radial velocities as predicted by LES-SL and LES-WW models at r/D = 0.5 with Re =20,000 and H/D =2.0

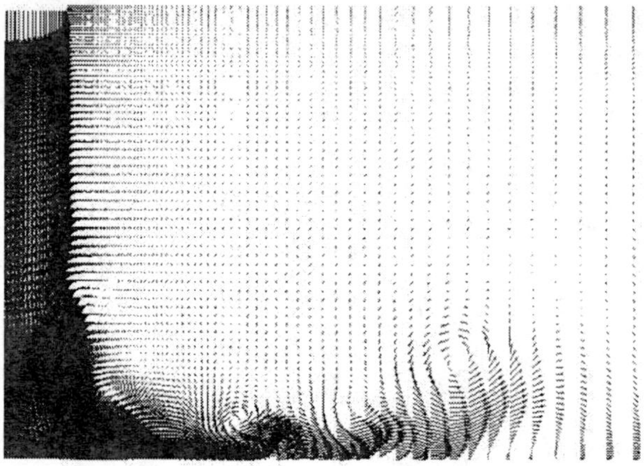

Figure 9. Instantaneous velocity vector field

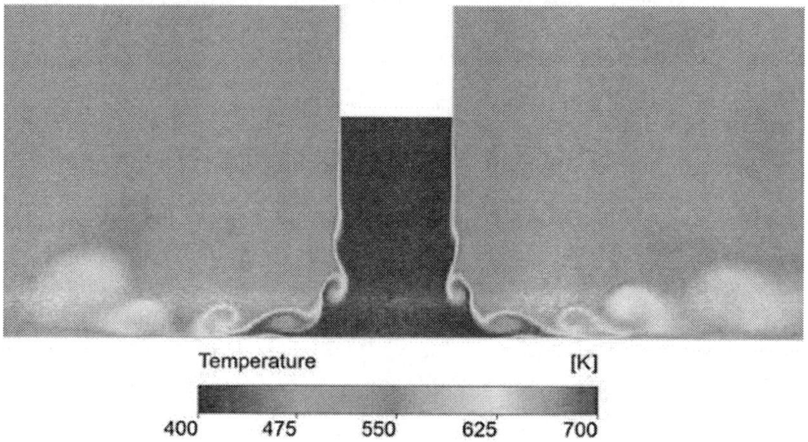

Figure 10. Contours of the instantaneous jet temperature as predicted by LES-WW model

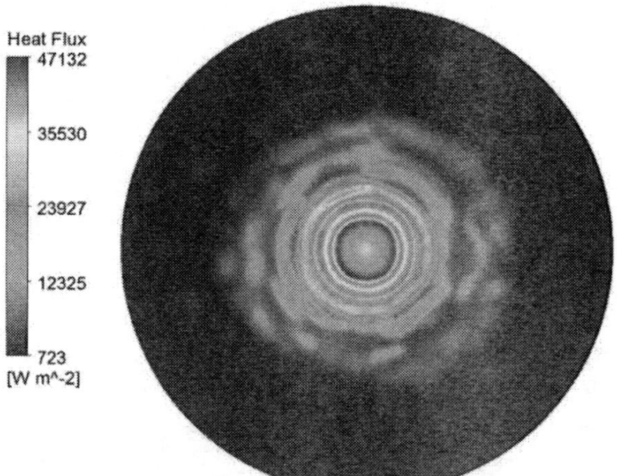

Figure 11. Contours of the instantaneous heat flux as predicted by LES-WW model

The instantaneous velocity vector field of wall jet are illustraed in figure 9. The development of wall jet can be seen. The instantaneous vectors field reveals the absence of the small eddy motion within the stagnation zone. The thin boundary layer near walls are formed in the radial direction after jet deflection, while the flow patterns of acceration wall jet are observed. Figures 10 and 11 demonstrate the contour plots of the instantaneous jet temperature and heat flux as predicted by LES-WW model. The plate surface temperature is chosen to be 400 K in the research. The transfer of heat is from the higher wall jet temperature to the lower temperature of the plate. Figure 11 shows the contour plots of the instantaneous heat flux as predicted by LES-WW model. The magnitude of plotted heat fluxes are varied from 723 to 47,132 W/m^2, while the peak heat transfer rate is near the stagnation zone. The contour of heat fluxes is non-uniform and the heat flux results from LES-WW model can provide more flow structure and more local information on the impingement wall.

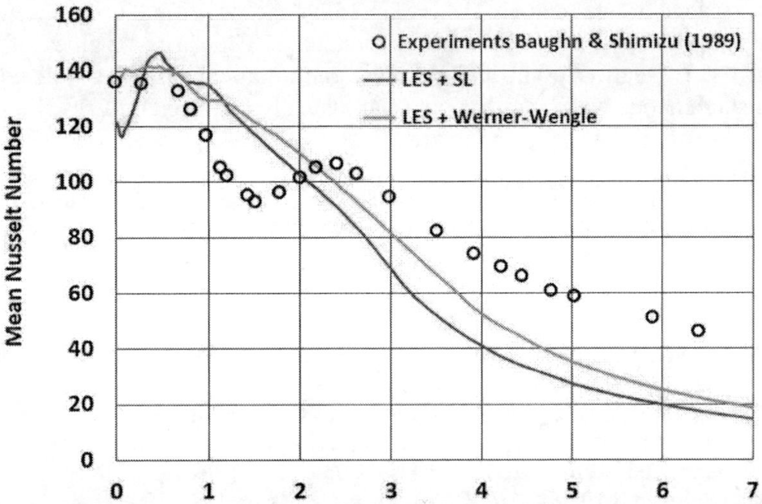

Figure 12. Time averaged Nusselt number as predicted by LES-SL and LES-WW models with Re =20,000 and H/D =2.0

During the post-processing step, the predicted Nusselt number is stored in each wall surface of the impingement wall. The plots of the time-averaged values of these surface properties with the centroid of those surfaces are shown in Figure 12. In this research, Nusselt number defined as

$$Nu = \frac{hD_h}{k} = \frac{q_w D_h}{(T_{jet} - T_w)k} \tag{3}$$

where q_w is the convective heat transfer from hot gas to wall, T_{jet} is gas temperature of wall cells and T_w is wall temperature. The physical properties are evaluated at the film temperature which is defined as

$$T_f = \frac{1}{2}(T_{jet} + T_w) \tag{4}$$

Figure 12 presents the time averaged Nusselt number as predicted by LES-SL and LES-WW models with Reynolds number is 20,000 and H/D is about 2.0. It is clear that the values of Nusselt number from the measurement are largest at the stagnation point and decrease monotonically in the wall jet region.

The Nusselt number predictions from the LES-SL and LES-WW models agree fairly well with the experimental data within the stagnation zone, while both models show under-prediction when r/D > 2.0.

6. Conclusions

Heat transfer and flow field resulting from a single hot air impinging wall is investigated in this study. The simulation is carried out using computational fluid dynamics (CFD) commercial code FLUENT. To simulate fluid flows, Large Eddy Simulation (LES) approach with a subgrid-scale Smagorinsky-Lilly model is employed. The classical Werner-Wengle wall model is used to compute the predicted results of velocity near walls. The Smagorinsky constant in the turbulence model is set to 0.1 and is kept constant throughout the investigation. The hot gas jet impingement on the flat plate with a constant surface temperature is chosen to validate the predicted heat flux results with experimental data. The jet Reynolds number is equal to 20,000 and a fixed jet-to-plate spacing of H/D = 2.0. Nusselt number on the impingement surface are calculated. As predicted by the wall model, the instantaneous computed Nusselt number agree fairly well with experimental data. The largest values of calculated Nusselt number are near the stagnation point and decrease monotonically in the wall jet region. The local information of instantaneous wall heat flux on a flat plate are captured by LES simulation.

Acknowledgements

This material is based upon work supported by Faculty of Engineering Kasetsart University. In addition, the authors are grateful for the intellectual support from PhD. candidate Mongkol Kaewbumrung from Applied Combustion Lab, Mechanical Engineering Kasetsart University.

References

[1] Wang B, Lin D, et al. 2016 *Applied Thermal Engineering* **100** pp 902-910
[2] Zhou T, Xu D, et al. 2016 *Applied Thermal Engineering* **100** pp 55-61
[3] Hosan M L, Fdhila R B, et al. 2015 *Applied Energy* **164** pp 934-943
[4] Lytle D and Webb B W 1991 *In Experimental Heat Transfer, Fluid Mechanics and Thermodynamics* ed. Keffer J F, Shah R K and Ganic E N (Elsevier)
[5] Baughn, J and Shimizu S 1989 *ASME J. Heat Transfer* **111** pp 1096-1098
[6] Singh D, Premachandran B, et al. 2013 *Numer. Heat Transfer* Part A **64**(2) pp 153-185
[7] Mohanty A K and Tawfek A 1992 *Int. J. Heat Mass Transfer* **36**(6) pp 1639-1647
[8] Gardon R and Cobonpue, J 1962 *Int. Developments in Heat Transfer Int. Heat Transfer Conf. ASME* pp 454-460
[9] Zuckerman N and Lior N 2005 *ASME J. Heat Transfer* **127**(5) pp 544-552
[10] Cooper D, Jackson D C, et al. 1993 *Intl J. Heat Fluid Flow* **36** pp 2675-2684
[11] Werner H and Wengle H 1989 *In Proc. of 7th Symposium on Turbulent Shear Flows* (Stanford University, California) pp 10.21-10.26
[12] Geers L, Hanjalic K, et al. 2004 *Exps. Fluids* **36** pp 946-958

IOP Conf. Series: Materials Science and Engineering **297** (2018) 012036 doi:10.1088/1757-899X/297/1/012036

CFD modelling wall heat transfer inside a combustion chamber using ANSYS forte

C. Plengsa-ard* and M. Kaewbumrung

Mechanical Engineering Department Kasetsart University,
Chatuchak,Bangkok 10900,Thailand

* E-mail: fengcpp@ku.ac.th, Tel: 02-797-0999 Ext 1833

Abstract. A computational model has been performed to analyze a wall heat transfer in a single cylinder, direct injection and four-stroke diesel engine. A direct integration using detailed chemistry CHEMKIN is employed in a combustion model and the Reynolds Averaged Navier Stokes (RANS) turbulence model is used to simulate the flow in the cylinder. To obtain heat flux results, a modified classical variable-density wall heat transfer model is also performed. The model is validated using experimental data from a CUMMINs engine operated with a conventional diesel combustion. One operating engine condition is simulated. Comparisons of simulated in-cylinder pressure and heat release rates with experimental data shows that the model predicts the cylinder pressure and heat release rates reasonably well. The contour plot of instantaneous temperature are presented. Also, the contours of predicted heat flux results are shown. The magnitude of peak heat fluxes as predicted by the wall heat transfer model is in the range of the typical measure values in diesel combustion.

1. Introduction

It is known that heat transfer in diesel engines affects engine efficiency and emission. An increase of heat transfer to the combustion chamber walls is lower the in-cylinder pressure and the average gas temperature and this reduces the work per cycle transferred to the piston [1]. Thus, the magnitude of engine heat transfer is strongly dependent on engine efficiency. The changes in the gas temperature due to heat loss through the combustion chambers also have an impact on the pollutant emission formation. A higher in-cylinder temperature promotes the NOx emissions and the lower in-cylinder temperature causes the particulates during the exhaust. Since the engine heat transfer is important, the accurate predicted engine heat transfer results are needed.

There are many previous experimental and numerical studies on engine heat transfer reported in the literature.With experimental study, Hendricks [2] and Gingrich [3] showed experimental data of piston heat transfer under conventional diesel engine. Factors affecting heat transfer in a diesel engine also were investigated by Das [4]. The numerical modelling of engine wall heat transfer was researched and improved over the last few decades. Woshini [5] and Annand [6] proposed the empirical correlation to calculate the overall heat transfer coefficient. One-dimensional wall heat transfer model was introduced by Isshiki et al. [7] and Yang et al. [8]. The core regions and near wall layers were separated within these one-dimensional models. Heat flux results was predicted by solving an energy equation without using a heat transfer coefficient, while the core region was considered a global region with uniform properties without spatial distinction. However, in-cylinder core regions for internal combustion engine cannot be considered globally since there are local changes in the core regions due to turbulence, swirling and tumbling. Also, in order to obtain an accurate values of heat flux, multi-dimensional model for core regions and one-dimensional heat transfer near walls are necessary. In these multi-dimensional models, conservation of mass, momentum and energy were directly solved in the core region, whereas the wall models were generally used within the wall layers. Since the mechanism of wall layers are similar to the physics in the boundary layers, the predicted heat fluxes were solved by using thermal boundary layer assumptions. Han et al. [9] derived the wall heat transfer model from the one-dimensional energy equation which included the variation of gas density and turbulent Prandtl number across the boundary layer. These models provided a better predicted heat flux results when compared with those of previous works.

Therefore, the purpose of this present work is to demonstrate how engine wall heat transfer model in the commercial code i.e., ANSYS forte predicts the heat flux results. The test case operates in diesel engine condition. The data from experiment is compared with those of heat flux prediction. The heat transfer in this study is based on the model work from Han et al. [9]. Also, the heat flux results through the combustion chamber walls due to gas phase heat convection is the major concern.

2. Wall heat transfer formulation

For ANSYS- Forte, the wall heat transfer model from Han et al. [9] is used to calculate the gas-phase heat transfer near wall regions. The analysis of the formulations of this model is initiated from the one-dimensional energy equation across the boundary layer which is described below in equation (1).

$$c_p\rho\frac{\partial T}{\partial t} + c_p\rho v\frac{\partial T}{\partial y} = \frac{\partial}{\partial y}\left[(k+k_t)\frac{\partial T}{\partial y}\right] + \frac{dP}{dt} + \rho Q \quad (1)$$

where ρ and T are the fluid density and temperature near walls respectively, c_p is the specific heat at constant pressure, p is pressure, v is fluid velocity, k is laminar thermal conductivity, k_t is turbulent thermal conductivity and Q is volumetric heat release rate. Several assumptions are introduced in the derivation of the model, including the following: (1) fluid velocity near walls is parallel to the wall, and only the temperature and density gradients normal to the wall are considered, (2) the gas is assumed to be ideal, (3) the pressure is uniform in space, (4) viscous dissipation is neglected, and (5) radiation heat transfer is not included. Based on these previous assumptions, the wall heat flux in the near wall regions is computed by equation (2).

$$q_W = \frac{\rho u_* c_p T \ln(\frac{T}{T_W})}{\{2.1 * \ln(y^+) + 2.5\}} \quad (2)$$

$$y^+ = \frac{u^* y}{\nu} \quad (3)$$

$$u^* = (c_\mu^{0.25} tke^{0.5}) \quad (4)$$

where q_w is heat flux on the wall, u^* is friction velocity, y^+ is the dimensionless distance, T_w is the wall temperature, ν is the kinematic viscosity, tke is turbulent kinetic energy and C_μ is equal to 0.09.

3. Turbulence and Combustion models

In order to simulate in-cylinder flows, turbulence is modeled using the RNG k-ε turbulence model in ANSYS forte commercial code. These consider velocity dilation in the ε-equation and spray induced source terms for both k and ε equations. For the combustion model, ANSYS forte uses a detailed chemistry CHEMKIN PRO solver. A skeletal reaction mechanism for n-heptane fuel, which has similar ignition characteristics with diesel fuel, is used to simulate diesel fuel chemistry. The mechanism includes 36 species and 76 reactions. However, the physical properties of the fuel in this work are represented with those of tetra-decane in the calculations.

4. Validation setup

4.1 Engine geometry and operating condition

The measurement data from a single cylinder direct injection 4 stroke diesel engine by Cummins N-series with 2.34L displacement is employed to validate computational results in this study.[10] A summary of the main geometrical data is listed in table 1, while a schematic diagram of the engine is presented in figure 1. Also, the engine condition for the test case is summarized in table 2.

Table 1. Cummins N-series specification.[10]

Geometry data	
Engine type	DI diesel
Displacement (L)	2.34
Bore (cm)	13.97
Stroke (cm)	15.24
Swirl ratio	0.5
Connecting rod length (cm)	30.48
Number of cylinders	1
Fuel injector type	Common rail
Number of nozzles	8, equally space

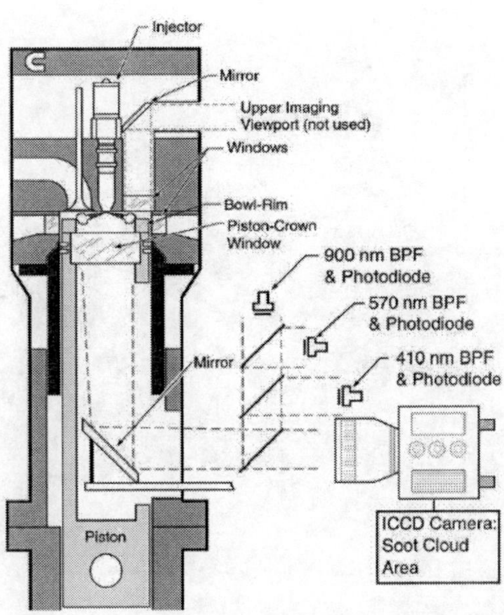

Figure 1. A schematic diagram of Cummins N-series engine [10]

Table 2. The operating engine conditions [10]

Engine speed (rpm)	1200
IMEP (bar)	4.4
Intake pressure (kPa)	233
Intake temperature (K)	384
SOI (°ATDC)	-7
DI duration (CA°)	10

4.2 *Numerical approach*

In this paper, sector symmetry is assumed and periodic boundary conditions were applied for each injector hole. A 45 degree sector mesh was used to represent one-eighth of the engine combustion chamber. There were 60 cells in the radial direction, 42 cells in the azimuthal direction and 62 cells in the axial direction, with 6 cells in the squish region at the top dead center. This mesh resolution, which can be seen in figure 2, has been found to give adequately grid-independent results. Figure 3 illustrates the variation of simulated in-cylinder pressure on the test engine condition for the three different mesh sizes. It can be seen that there is no significant difference between grid II and grid III. To save computational time, grid II is chosen for the simulation in this research. Moreover, the computational domain consisted of one section of the modeled engine cylinder which is used for simulations between inlet valve close (IVC) and exhaust valve open (EVO). Therefore, the model computes only the closed volume part of the engine cycle. The simulations are started at the intake valve closure (IVC) with uniform in-cylinder mixture distribution assumptions. The swirl is initialized based on the wheel flow velocity profile and the turbulent kinetic energy is initialized at IVC by scaling with the mean piston speed. The pressure and temperature at IVC are given based on the experimental data. The boundary conditions of turbulent kinetic energy are the Neumann boundary conditions. The averaged wall temperature was 460 K for the test case.

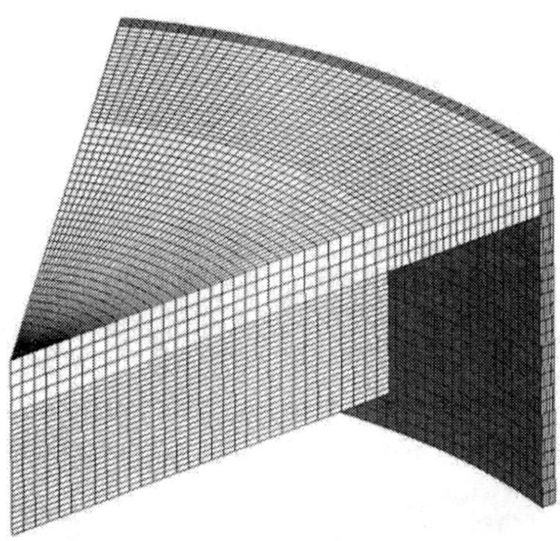

Figure 2. CFD 45 degree sector mesh of a simulated engine.

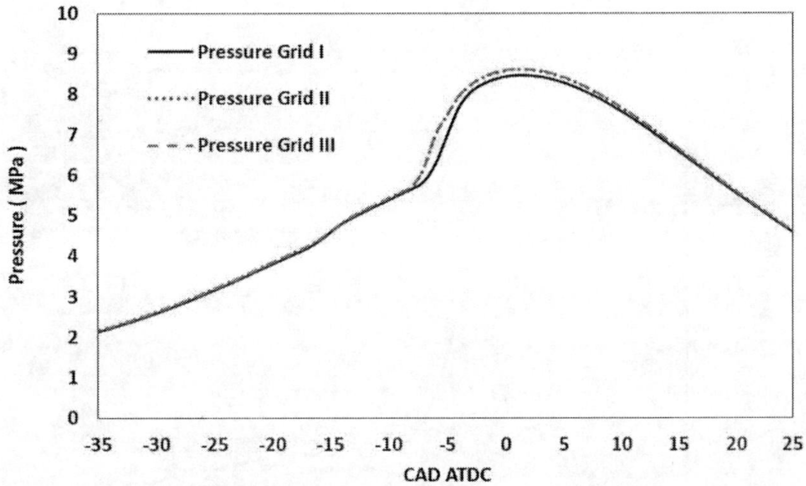

Figure 3. Variation of simulated in-cylinder pressure with crank angles for the test case with three different mesh sizes

5. Results and discussion

For model validation, comparison between computed and measured in-cylinder pressure and heat release rate as predicted by ANSYS forte are shown in figure 4. The start of injection is 7 degrees before the top dead center. Good agreement with the magnitude of peak in-cylinder pressure can be seen in the figure, while the over prediction of the cool flame location occurs. The trend of heat release curve matches with the conventional diesel combustion. The predicted peak heat release rate during the premixed burn occurs slightly before the top dead center and the overestimation of calculated heat release rate can be seen during the mixing control combustion regime.

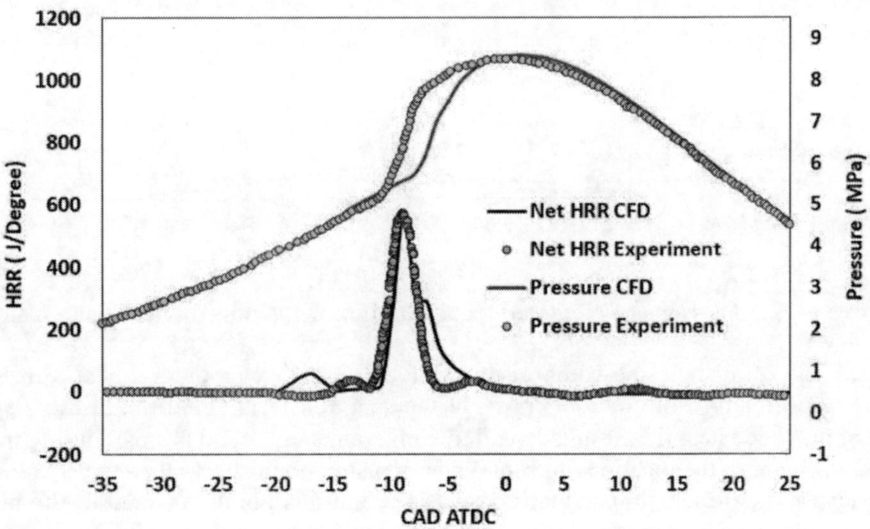

Figure 4. Comparisons of computed and measured in-cylinder pressure and heat release rate with crank angles. The start of injection is 7 degrees before TDC

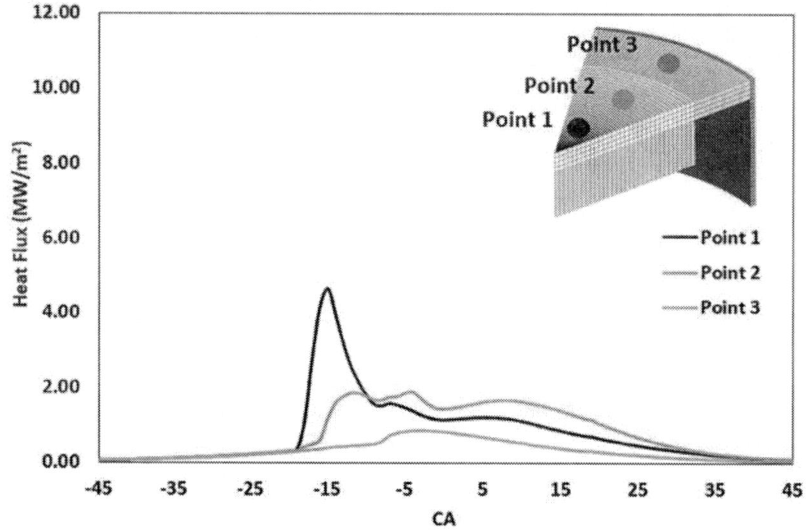

Figure 5. Comparisons of predicted local heat flux with crank angles near a piston head

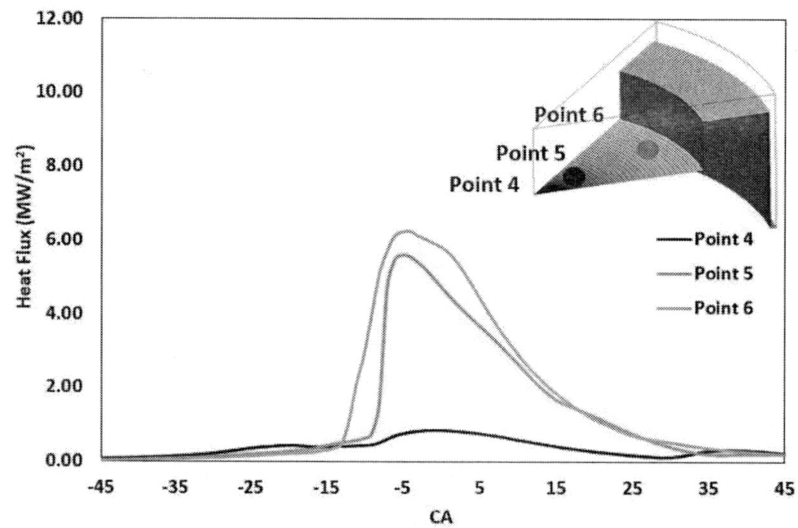

Figure 6. Comparisons of predicted local heat flux with crank angles near a piston bowl

Figures 5 and 6 illustrate the comparisons of local heat flux prediction near a piston head and a piston bowl respectively. Both pictures provide the spatial effects along the primary spray axis. The magnitude of predicted heat flux results is varied with crank angles and is small during the compression and strokes. In figure 6, the location of point 4 near a piston bowl shows the lowest values of predicted heat flux because the diesel spray actually targets beyond this point. At point 1, the highet values of predicted heat flux occurs near a piston head close to an injector due to the high temparature from the flame during the combustion. At the locations of ponit 5 and 6, the magnitude of predicted heat flux is higher when compared with other locations. The distributions of heat transfer rate for three area of the combustion chamnber are presented in figure 7. As can be seen, most of heat transfer loss to combustion chamber walls are from the piston bowl. This is true because the flame fronts will reach the area of

IOP Conf. Series: Materials Science and Engineering **297** (2018) 012036 doi:10.1088/1757-899X/297/1/012036

piston during the combustion events. Figures 8 and 9 show the contour plots of calculated in-cylinder temperature and heat flux results at 5 degrees after the top dead center.

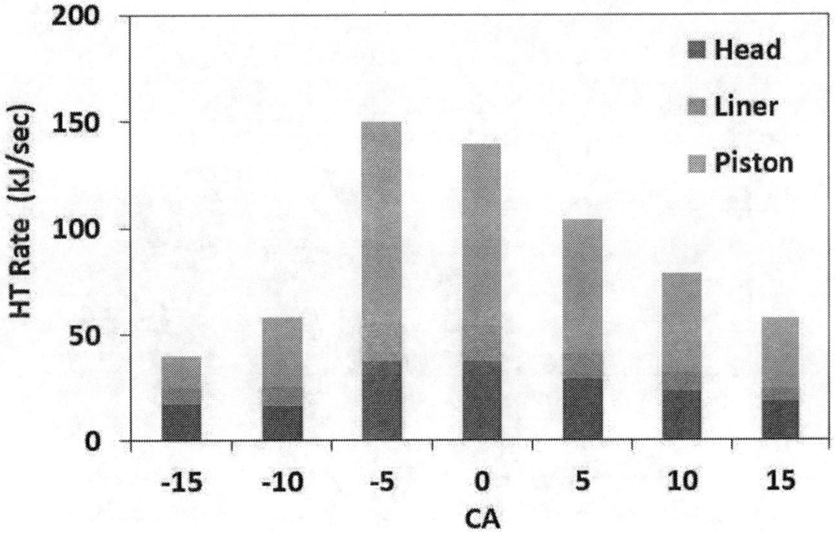

Figure 7. Distributions of heat transfer rate for the three area of the combustion chamber

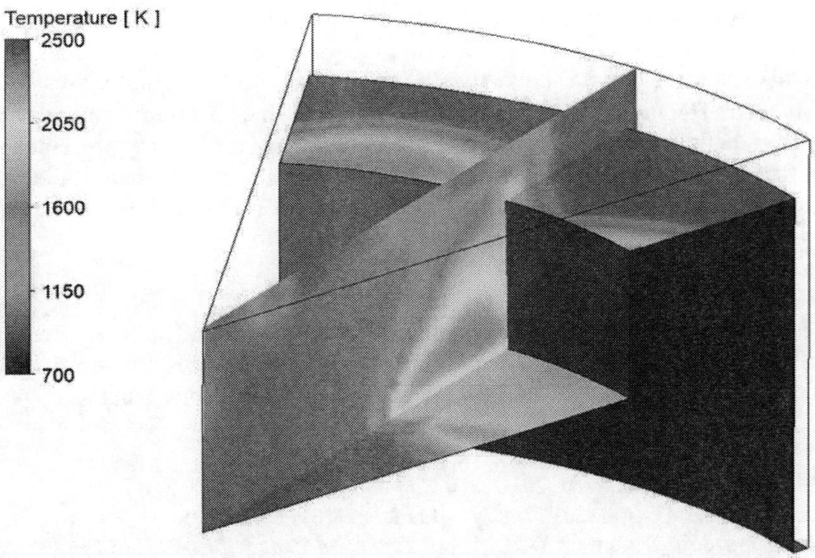

Figure 8. Contours of predicted peak in-cylinder temperature at 5 degrees after TDC

The maximum temperature is about 2300 K which is in a range of typical conventional diesel engine. Moreover, the non-uniform of temparature and surface heat flux is presented. The location of locally maximum and minimum values of temperature and heat flux results on the piston bowl can be captured by the wall model. The contour plots of wall heat flux information are consistent with the local heat flux results in figure 6.

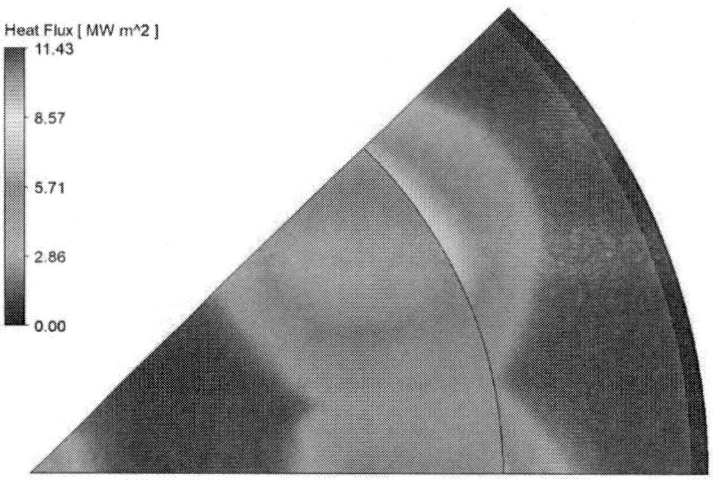

Figure 9. Contours of calculated heat fluxes at 5 degrees after TDC (top view)

6. Conclusions

The wall heat transfer models from Han et al.[10] is tested. The model is validated using experimental data from a CUMMINs engine operated with a conventional diesel combustion. The operating engine condition is tested at 5 degrees before TDC. Comparisons of simulated in-cylinder pressure and heat release rates with experimental data shows that the model predicts the cylinder pressure and heat release rates reasonably well. The model can capture the location of the cool flame and main heat release. The contour plot of predicted temperature and computed heat flux results are presented. Comparisons of predicted heat flux near a piston head and a piston bowl are shown. The magnitude of peak heat fluxes as predicted by the wall heat transfer model is in the range of the typical measure values in diesel combustion.

Acknowledgements

This material is based upon work supported by Faculty of Engineering Kasetsart University. In addition, the authors are grateful for the intellectual support from Mr. Mongkol Kaewbumrung, a PhD. candidate from Applied Combustion Lab, Mechanical Engineering Kasetsart University.

References

[1] Jia M, Gingrich E, Wang H, et al. 2016 *Int J Engine Res* **17**(3) pp 331-346
[2] Hendricks T, Splitter D, et al. 2014 *Int J Engine Res* **15**(6) pp 684-705
[3] Gingrich E, Ghandhi J B, Reitz R D 2014 *SAE Technical Paper 2014-01-1182*
[4] Das S and Roberts CE 2013 *SAE Technical Paper 2013-01-0875*
[5] Woschni G 1967*SAE Technical Paper* 670931
[6] Annand W J D 1963 *Proc. of the Institution of Mechanical Engineers* vol. 177 pp 973-996
[7] Isshiki, N and Nishiwaki N 1970 *Proc. of the 4th Int. Heat Transfer Conference*, FC3.5 pp 1-10
[8] Yang J, Pierce P, Martin, J, et al. 1988*SAE Technical Paper* 881314
[9] Han Z and Reitz R D 1997 *Int. J. Heat and Mass transfer* **40**(3) pp 613-625
[10] Singh S, Reitz R D and Musculus M P B 2006 *SAE Technical Paper* 2006-01-0055

IOP Conf. Series: Materials Science and Engineering **297** (2018) 012037 doi:10.1088/1757-899X/297/1/012037

Analysis of temperature profile and electric field in natural rubber glove due to microwave heating: effects of waveguide position

P. Keangin[1,*], U. Narumitbowonkul[1] and P. Rattanadecho[2]

[1]Department of Mechanical Engineering, Faculty of Engineering, Mahidol University, 25/25 Phutthamonthon 4 Road, Salaya, Nakhon Pathom, 73170, Thailand.
[2]Department of Mechanical Engineering, Faculty of Engineering, Thammasat University (Rangsit Campus), 99 mu 18, Paholyothin Road, Klong Nueng, Klong, Pathumthani, 12120, Thailand.

* Corresponding Author: E-mail: pornthip.kea@mahidol.ac.th

Abstract. Natural rubber (NR) is the key raw material used in the manufacture of other products such as rubber band, tire and shoes. Recently, the NR is used in natural rubber glove (NRG) manufacturing in the industrial and medical fields. This research aims to investigate the electromagnetic wave propagation and heat transfer in NRG due to heating with microwave energy within the microwave oven at a microwave frequency of 2.45 GHz. Three-dimensional model of NRG and microwave oven are considered in this work. The comparative effects of waveguide position on the electric field and temperature profile in NRG when subjected to microwave energy are discussed. The finite element method (FEM) is used to solve the transient Maxwell's equation coupled with the transient heat transfer equation. The simulation results with computer programs are validated with experimental results. The placement of waveguides in three cases are left hand side of microwave oven, right hand side of microwave oven and left and right hand sides of microwave oven are investigated. The findings revealed that the placing the waveguide on the right side of the microwave oven gives the highest electric field and temperature profile. The values obtained provide an indication toward understanding the study of heat transfer in NRG during microwave heating in the industry.

1. Introduction

Today, rubber is as widely used due to its beneficial proprieties like strength, long lasting, water resistance and heat resistance [1]. Commercially, the most important rubbers are natural rubber (NR); it is an elastomer that has greater flexibility compared to synthetic rubbers. NR is used as the main material in many everyday applications, for example, rubber band, hoses, vehicle tires, automotive industries and medical equipment, especially the main materials of natural rubber gloves (NRG) [2, 3]. The NRG industry becomes bigger and wider. The basic knowledge on rubber technology leads to manufacturing process improvement and development of new products, especially the knowledge on heating of NRG.

In the past, rubber glove heating processes used conventional heating method using hot air. Many previous studies on hot-air vulcanized of rubber glove [4-6]. Due to the lower conductivity of rubber material, frequently, inner temperature of rubber products is less than surface temperature in conventional vulcanization process lead to the nonuniformity in temperature distribution of rubber material [7, 8]. Several NRG heating processes were incorporated into combined infrared and hot-air vulcanization of rubber glove [9]. Khamdaeng et al. [9] analyzed the rubber deformation to describe the deformation behavior of the combined infrared and hot-air vulcanized of rubber glove. However, the infrared can potentially penetrate into the rubber material with their shorter wavelengths, the heat and mass transfer inside material is subjected to the emission. Infrared combined with hot-air heating will provide the complex phenomena than infrared or hot-air heating only [10-12]. In addition, the infrared heating is complicated setting because of the less space using and expensive equipment but also rapidly responses to temperature controlling [9]. Finding ways to develop the efficiency of rubber material heating processes is an essential for research and particular industry.

A new advancing technique for rubber material heating such as NRG heating is microwave heating. The major advantages of microwave heating are rapidly emerging, precise control, and fast start-up due to the microwave energy that penetrates into the material will be absorbed and converted into internal energy in the material, so-called volumetric heating. In recent years, the utilization of microwave energy on applications of NR has been increasing rapidly in daily life, including drying, vulcanization and pre-heating. Many previous studied on vulcanization of the NR with the microwave energy can be found in the research [13-15]. Martin et al. [13] vulcanized rubber mixtures by using electron beam and microwave irradiation. Sombatsompop and Kumnuantip [14] compared physical and mechanical properties of NR, carbon black and reclaimed rubber blends vulcanized by conventional thermal and microwave irradiation methods. Chen et al. [15] analyzed temperature distribution in rubber material during microwave heating using experimental. There are some experimental studied of the NR heating process using rectangular wave guide [16, 17]. Doo-ngam et al. [16] developed microwave pre-heating process of NR-compounding with various sulphur contents using a rectangular wave guide. Makul and Rattanadecho [17] presented a new method to pre-cure NR-compounding (NRc) with a rectangular wave guide, significant results showed that microwave energy can produce partial cross-linking at temperatures below the actual vulcanizing process.

Although many previous researches have studied the simulation or experimental of microwave heating in NR, there are few research studies simulation or experimental of NRG microwave heating, especially a detailed study of the parametric effects, i.e. placement of waveguides effects on electric field and temperature profile. This is because complicated of the shape of gloves and complicated of problem that microwave heating is a multiphysics phenomenon involving both electromagnetic waves and heat transfer. A comprehensive understanding of the fundamentals of electromagnetic waves and heat transfer interactions will help to develop the efficiency of NRG microwave heating.

The aim of the proposed study is to investigate the influences of waveguide position on the electric field and temperature profile in a NRG during microwave heating within the microwave oven. The variations in three cases e.g. left hand side of microwave oven, right hand side of microwave oven and left and right hand sides of microwave oven are investigated. The microwave power input of 100 W, microwave frequency of 2.45 GHz, heating times of 300 s and NRG size M are used for this problem. The transient Maxwell's equation coupled with the transient heat transfer equation is solved by using the finite element method using COMSOL^{TM} Multiphysics. The NRG and microwave oven models are built and simulated via a three-dimensional finite element computational model. The results obtained from the simulation are examine and compared with the experimental results to verify the accuracy of the present numerical model. The obtained results of this work will lead to understanding electric field and temperature profile within NRG during microwave heating and may be of assistance in determining practical guideline to improve heating process of NRG in the process industry.

2. Experimental Study

Figure 1 illustrates the schematic diagram of the experimental apparatus for the measurement of the temperature in the NRG during microwave heating. Microwave cavity with dimensions 310 mm x 280 mm x 190 mm (axial: x, y and z, respectively) is used in experimental. The magnetron with microwave power 100 W radiates to the interior of the microwave oven. The working frequency of microwave is 2.45 GHz, a widely used frequency inside the microwave oven. To set up an experiment, start from the dipping ceramic hand shaped former into natural rubber latex (NRL) with thickness 0.5 mm to mold

NRG. A thin layer of the NRL covering the hand shaped mold is vertically placed in the center of the microwave oven. Microwave energy which is generated by microwave generator propagates through the waveguide (cross sectional area 79 mm x 43 mm) to the NRG that is located at the center of microwave oven. The temperature of NRG measured in experimental using thermocouple K-type at various heating times during microwave heating [18, 19] and recorded temperature data by data logger (Wisco DL2200 Data Logger). Figure 2 shows the NRG temperature measurement inside the microwave oven at different positions (position A1-A5). NRG is formed using the medium (M) size of hand mold, it is standard size commonly used. The dimensions of NRG size M are shown in figure 3. The NRG has a width of 110 mm and a length of 200 mm. Temperature measurement position (position A1-A5) inside NRG from experimental is located at the 45 mm, as displayed in figure 3. The initial NRG temperature is set at 24 °C for all cases. During microwave heating, the heating time of 300 are performed. Experimental results of temperature distribution various temperature measuring positions (A1-A5) at microwave power input of 100 W and NRG size M at heating times of 300 s are systematically investigated.

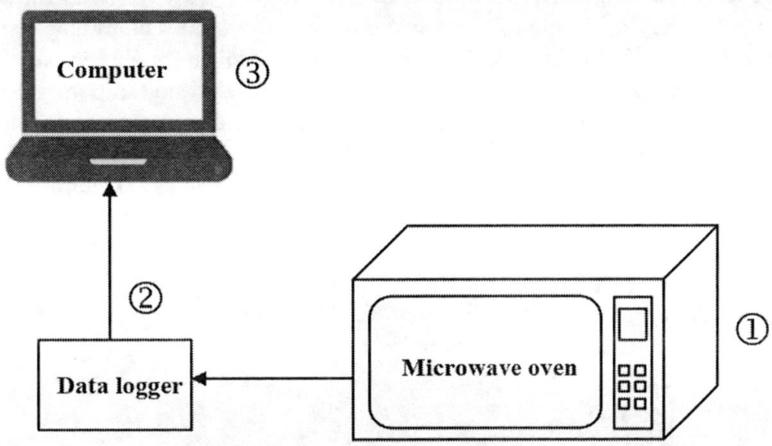

Figure 1. Equipment setup; ① Microwave oven ② Data logger (Wisco DL2200) and ③ Thermocouple

Figure 2. Measuring temperature of NRG inside the microwave oven

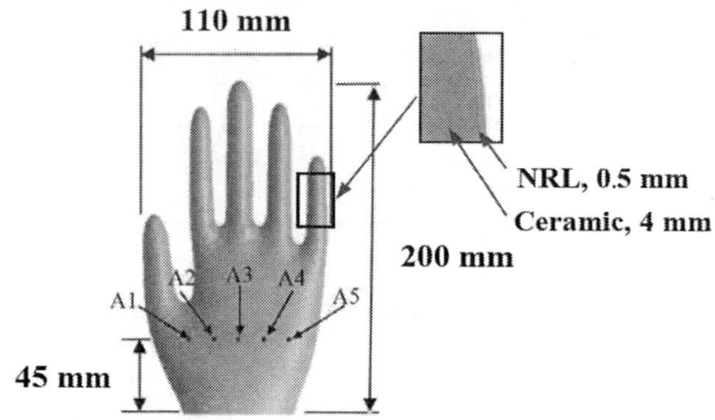

Figure 3. Dimensions of NRG and temperature measurement position (position A1-A5) inside NRG

3. Simulation Study

3.1 Physics Model

The schematic diagram of the physics models of NRG and microwave oven for numerical analysis are shown in figure 4. The dimension of microwave oven cavity and rectangular wave guide in numerical study is similar to the dimension in experimental study. Full three-dimensional models of NRG and microwave oven are considered to demonstrate the electric field and temperature profile in NRG due to microwave heating. The thickness of NRL is 0.5 mm and thickness of ceramic is 4 mm in experiment analysis used in numerical simulation analysis with the same conditions. The NRG is subjected to a uniform microwave through a wave guide. Microwave irradiation can penetrate the NRG surface and is converted into thermal energy within the NRG.

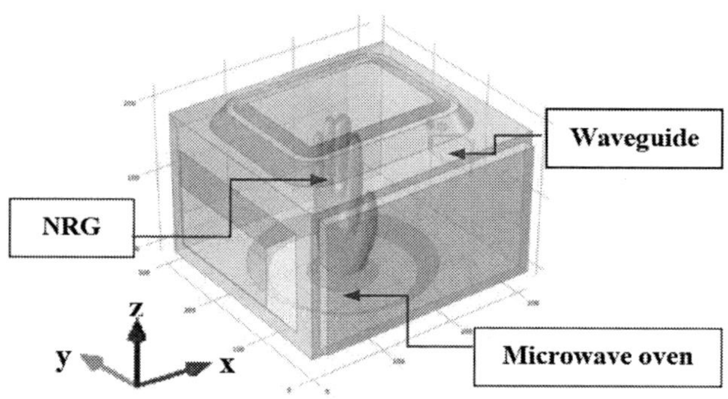

Figure 4. Physical models of NRG and microwave oven for numerical analysis

3.2 Analysis of Electromagnetic Wave Propagation

A mathematical model is formulated to predict the temperature profiles and electric field within the NRG during the microwave heating process. This research is carried on the following assumptions:

1) The mathematical model of heating of NRG by microwave in the multi-mode is considered.
2) The absorbed energy by air in a microwave oven is negligible.
3) The walls of the microwave oven are impedance and waveguide is perfect conductors.
4) The model assumes that dielectric properties of NRG and ceramic properties are constant.

The electromagnetic wave propagation in NRG is calculated using Maxwell's equations which mathematically described the interdependence of the electromagnetic waves. The general form of Maxwell's equations is simplified to demonstrate the electric fields penetrated in NRG which can be written as follows [20]:

$$\nabla\times\left(\frac{1}{\mu_r}\nabla\times\bar{E}\right)-k_0^2\left(\varepsilon_r'-\frac{j\sigma}{\omega\varepsilon_0}\right)\bar{E}=0 \tag{1}$$

where $\bar{E}$ is the electric field intensity (V/m), μ_r is the relative magnetic permeability (H/m), ε_r' is the relative dielectric constant, ε_0 = 8.8542x10^{-12} (F/m) is the permittivity of free space, and σ is the electric conductivity (S/m), $\varepsilon_r = n^2$ and n is the refractive index.

The boundary condition for analyzing the electric field, as presented in figure 5, is considered as follows:

In numerical simulation analysis, the NRG subjected to a uniform microwave power input via waveguide, so as to NRG is heated by microwave. The combination of microwave power input and heating time can be set, which is the advantage of this microwave heating device. The geometric models and condition in numerical simulation analysis corresponded to the experimental. The port power level and port phase of wave excitation at this port of microwave oven. The power flux associated with a propagating electromagnetic wave is represented by the poynting vector. Therefore, at the left boundary of the considered domain, an electromagnetic wave propagation simulator employs are the port boundary condition as follows [21]:

$$S=\frac{\int\left(\bar{E}-\bar{E}_1\right)\cdot\bar{E}_1}{\int\bar{E}_1\cdot\bar{E}_1} \tag{2}$$

Furthermore, the propagation constant is defined as:

$$\beta=\frac{2\pi}{c}\times\sqrt{f^2-\frac{c^2}{4\times a^2}} \tag{3}$$

where β is the propagation constant, a = 3.4 inch is the depth dimension of waveguide, c = 3x10^8 (m/s) is the speed of light and f = 2.45 GHz is the operating frequency of the microwave generator.

The walls surfaces of the microwave oven are impedance boundaries are applied:

$$\sqrt{\frac{\mu_0\mu_r}{\varepsilon_0\varepsilon_r-j\dfrac{\sigma}{\omega}}}\hat{n}\times\bar{H}+\bar{E}-\left(\hat{n}\cdot\bar{E}\right)\hat{n}=\left(\hat{n}\cdot\bar{E}_s\right)\hat{n}-\bar{E}_s \tag{4}$$

The waveguide of the microwave oven are good conductors. Therefore, the wall of waveguide as perfect conductor boundaries:

$$\hat{n}\times\bar{E}=0 \tag{5}$$

The internal boundary along the interfaces between NRL and the ceramic is considered as continuity boundary conditions (no contact resistant occurs) as shown in figure 6 and as follows:

$$\hat{n}\times\left(E_{NRL}-E_{air}\right)=0 \tag{6}$$

$$\hat{n}\times\left(H_{NRL}-H_{air}\right)=0 \tag{7}$$

3.3 Analysis of Heat Transfer

To solve the thermal problem, the coupled model of the electromagnetic wave propagation and unsteady heat transfer as well as the boundary conditions is investigated. However, the heat transfer analysis of microwave heating is considered only in a NRG.

The temperature profile is corresponded to the electric field. This is because when the electric field propagates in the NRG, it was absorbed by the medium and converted into internal heat generation, which causes its temperature to rise. This research is based upon the following assumptions:

1) The NRG is assumed to be homogeneous and thermally isotropic.
2) There is no phase change of substance and no chemical reaction in the NRG.
3) The contact surface of NRG is assumed to be smooth.
4) The effect of shrinkage of NRG is negligible.
5) The model assumes that thermal properties of NRG and ceramic are constant.

The heat transfer equation describing the heat transport in a NRG is identical as the following equation [22]:

$$\rho C_p \frac{\partial T}{\partial t} - k\nabla^2 T = Q \tag{8}$$

where ρ is the density (kg/m^3), C_p is the heat capacity (J/kg·K), T is the temperature (°C), k is the thermal conductivity (W/m·K) and Q is the external heat source term (W).

In equation (8), the first and second on the left-hand side of equation denoted the transient term and heat conduction term. While, the term Q appeared on the right-hand side of equation denote the external heat source term, respectively. The external heat source term is equal to the resistive heat generated by the electric field (microwave power absorbed), which is a function of the electric field, relative dielectric loss factor, loss tangent coefficient and frequency are defined as equation (9):

$$Q = 2\pi f \varepsilon_0 \varepsilon_r' (\tan\delta) \vec{E}^2 \tag{9}$$

where $\tan\delta$ is the loss tangent coefficient. The electric field in our external heat source term is calculated through the electromagnetic wave propagation analysis.

Consistent with the assumption, the boundary condition for analyzing heat transfer, as shown in figure 5, is considered as follows:

The surface of NRL and ceramic are assumed to be a thermal insulation boundary condition:

$$-\hat{n}\cdot(-k\nabla T) = 0 \tag{10}$$

Boundary conditions along the interfaces between NRL and ceramic glove mold former are considered as continuity boundary condition:

$$T_{NRL}(x_0,t) = T_{Ceramic}(x_0,t) \tag{11}$$

$$-k_{NRL}\frac{\partial T(x_0,t)}{\partial x} = -k_{Ceramic}\frac{\partial T(x_0,t)}{\partial x} \tag{12}$$

where k_{NRL} and $k_{Ceramic}$ are the thermal conductivities of the NRL layer and ceramic layer, respectively.

The dielectric properties can be written in the general form of complex permittivity (ε). The complex permittivity is a function of dielectric constant and dielectric loss factor as defined by [23]:

$$\varepsilon = \varepsilon' - j\varepsilon'' = \varepsilon_0\left(\varepsilon_r' - j\varepsilon_r''\right) \tag{13}$$

where ε'' is the dielectric constant, ε' is the dielectric loss factor, ε_r' is the relative dielectric constant and ε_r'' is the relative dielectric loss factor.

The NRG is assumed to be homogeneous and electrically as well as thermally isotropic. The dielectric and thermal properties of NRG and ceramic are constant. Table 1 tabulates the dielectric and thermal properties of NRG and ceramic used in the computations.

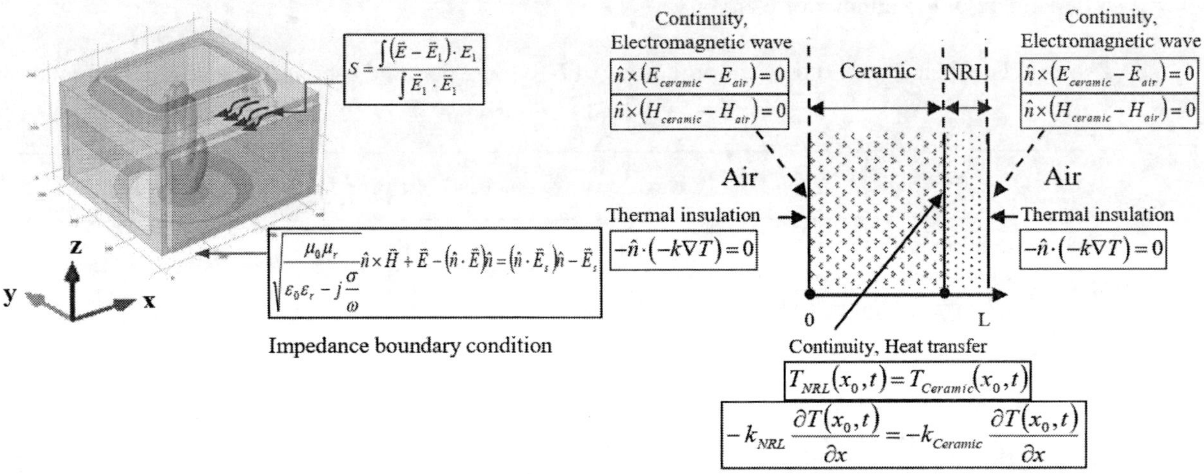

(a) (b)

Figure 5. The boundary condition for numerical analysis; (a) inside of microwave oven and (b) inside of NRG

4. Formulation of the Problem

In this study, the effects of waveguide position on the electric field and temperature profile in NRG are studied. The placement of waveguides in three case are left hand side of microwave oven, right hand side of microwave oven and left and right hand sides of microwave oven are examined. The conditions of waveguide positions are shown in figure 6. The simulation results of electric field vector and temperature profile in NRG during microwave heating based on microwave power input of 100 W and heating time of 300 s are presented. The frequency of microwave of 2.45 GHz and NRG size M are used in this study.

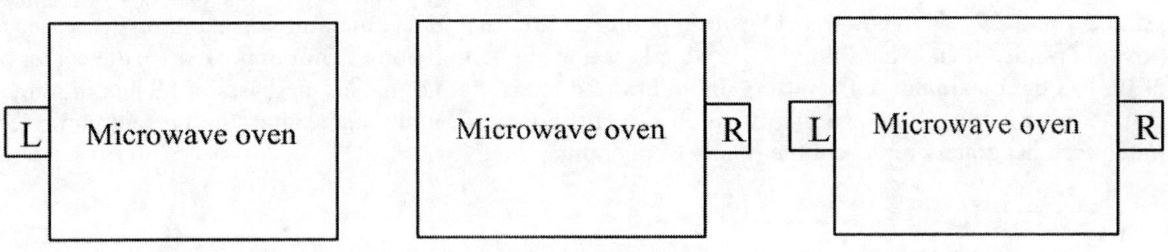

(a) Case 1 (b) Case 2 (c) Case 3

Figure 6. Physical models equipped with various waveguide positions

IOP Conf. Series: Materials Science and Engineering **297** (2018) 012037 doi:10.1088/1757-899X/297/1/012037

5. Calculation Procedure

In this study, the finite element formulations of the coupled electromagnetic wave propagation-heat transfer model in the NRG are carried out over the entire domain. The transient Maxwell's equation coupled with the transient heat transfer equation and related boundary conditions are implemented using finite element method with COMSOL™ Multiphysics software. To obtain a good approximation, a fine mesh is specified in the sensitive areas. The three-dimensional model is discretized using triangular elements and the Lagrange quadratic elements are then used to approximate the temperature and electric field variations across each element. A grid independence test is carried out to identify the appropriate number of elements required. This grid independence test leads to a mesh with approximately 431,110 elements. It is reasonable to assume that, at this element number, the accuracy of the simulation results is independent of the number of elements.

Table 1. The dielectric and thermal properties of NRG and ceramic used in the numerical simulation at microwave frequency of 2.45 GHz [24-28]

Property	Emblem	Natural Rubber Latex	Ceramic
Electric conductivity (S/m)	σ	0	0
Relative dielectric constant (-)	ε_r'	20	3.4
Relative dielectric loss factor (-)	ε_r''	0.3	0
Thermal conductivity (W/m·K)	k	0.13	1.1
Density (kg/m^3)	ρ	975	2,200
Heat capacity (J/kg·K)	C_p	1894	480

6. Results and Discussion

The accuracy of the present numerical model is verified by the validation against the experiment results with the same geometry and same conditions. The comparison of NRG temperature distribution of simulation results with the experiment results based on microwave power input of 100 W, microwave frequency of 2.45 GHz, heating times of 300 s and NRG size M at the position A1, A2, A3, A4 and A5 are shown in figure 7. It is found that the temperature distributions in NRG of simulation results are in excellent agreement with the temperature distributions in NRG of experiment data. Certain amounts of mismatch between the simulation results and the experiment results are caused by the numerical scheme. The temperature distribution from the simulation and experiment results gradually increase at position point A1 and A2 to a maximum at position point A3 after that the temperature distribution decreases at position point A4 and A5. Figure 7 also displays that at position point A3 or approximate the center of NRG has the maximum temperature. In addition, it is also found that an increase in the heating times results in an increase temperature distribution. This comparison guarantees that the present numerical model can accurately represent the transport phenomena in NRG.

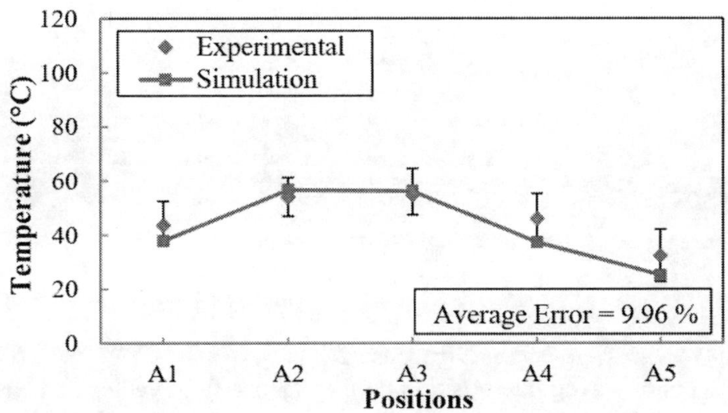

Figure 7. The comparison of NRG temperature distribution of simulation results with the experiment results at heating times of 300 s.

The effects of waveguide position on the electric field vector and temperature profile of NRG based on microwave power input of 100 W, microwave frequency of 2.45 GHz, heating times of 300 s and NRG size M are shown in figure 8. Figures 8(a)–(c) show the electric field vector and temperature profile of NRG in Case 1, Case 2 and Case 3 as described in section 4, respectively. These figures illustrate the volumetric heating effect expected from microwave heating. From figures 8(a), it is found that the electric field intensity in the palm of the NRG and intensity toward the area on the middle finger, it resulted in very high temperatures in this zone. However, the hot spot zone occurs at the position near the middle finger at low temperature in Case 1. The hot spot zone occurs at the position near the middle of palm, little finger and middle finger in Case 2, as shown in figure 8(b). While, the hot spot zone occurs at the position near the middle of palm and middle finger in Case 3, as shown in figure 8(c) due to in case of the waveguide is in place on both sides of the microwave oven, the overlap between their combined waves. Considering the influences of the waveguide position on the hot spot zone in each case, it can be seen that in Case 2 has more hot spot zone than Case 3 and Case 1, respectively.

Figure 9 show the effects of waveguide position on the electric field vector and temperature profile of NRG in top view under same condition of figures 8. Figures 9(a)–(c) show the electric field vector and temperature profile of NRG in Case 1, Case 2 and Case 3, respectively. From the figures, electric field is propagated through the waveguide toward NRG, which causes its temperature to rise. It is found that the electric field vector and temperature profile for all cases have the same pattern with figure 8.

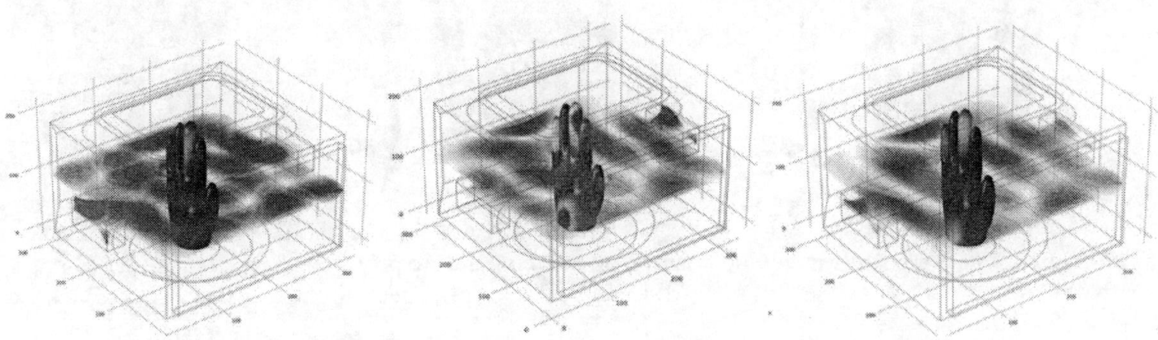

(a) Case 1 (b) Case 2 (c) Case 3

Figure 8. The effects of waveguide position on the electric field vector and temperature profile of NRG; (a) Case 1(b) Case 2 and (c) Case 3

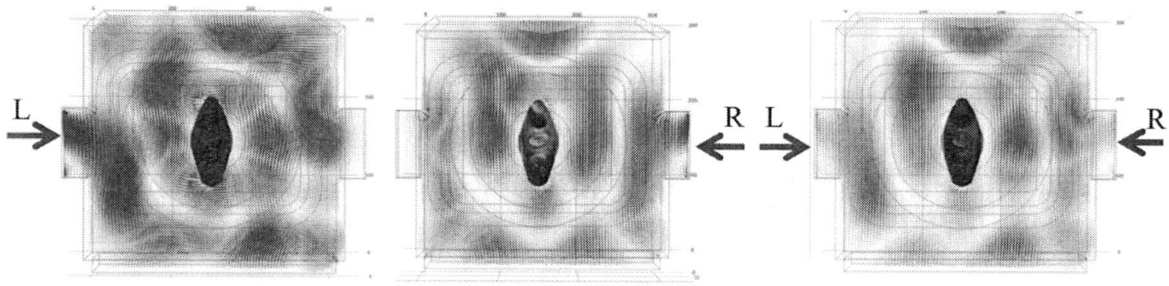

(a) Case 1 (b) Case 2 (c) Case 3

Figure 9. The effects of waveguide position on the electric field vector and temperature profile of NRG in top view; (a) Case 1(b) Case 2 and (c) Case 3

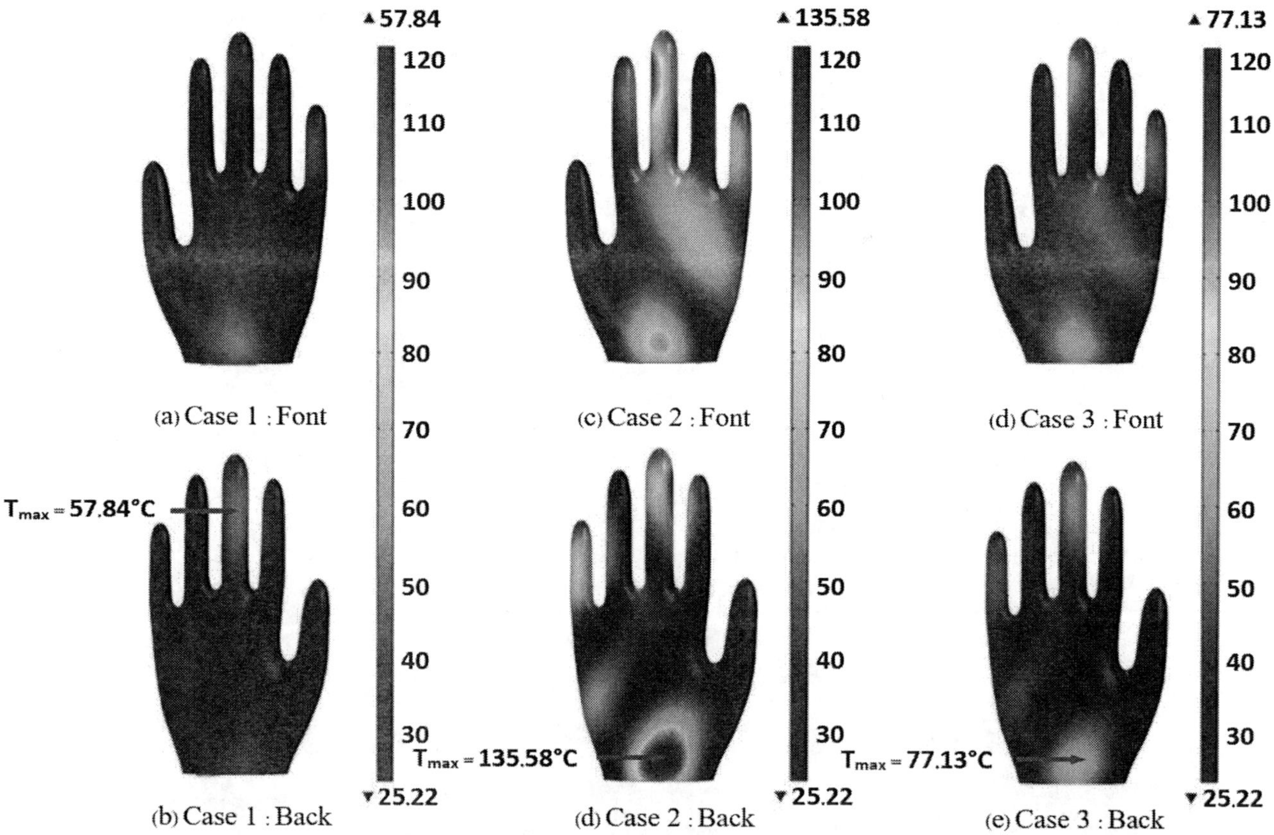

Figure 10. The effects of waveguide position on the temperature profile of NRG; (a) Case 1 : Font , (b) Case 1 : Back, (c) Case 2 : Font , (d) Case 2 : Back, (e) Case 3 : Font and (f) Case 3 : Back

Figure 10 displays the effects of waveguide position on the temperature profile of NRG based on microwave power input of 100 W, microwave frequency of 2.45 GHz, heating times of 300 s and NRG size M. It is found that the temperature profile follows the electric field profile. The hot spot zone occurs at the position near the middle of palm, little finger and middle finger in Case 2, while, the hot spot zone occurs at the position near the middle of palm and middle finger in Case 3. However, the temperature profile varies little at the position near the middle of palm in font of NRG and at the position on the middle finger in back of NRG in Case 1. It is seen that the maximum temperatures in Case 1 is 57.84 °C, in Case 2 is 135.58 °C and in Case 3 is 77.13 °C, respectively. The temperature distribution of NRG at various positions of waveguide at heating times of 300 s is shown in figure 11. It can be observed that the Case 2 has highest temperature distribution more than the temperature distribution in Case 3 and Case 1, respectively corresponds to the results in figure 10 because the waveguide is placed on the left side of the microwave oven with a concave surface in the case 1 and it is found that the wave overlap occurred in the case 3.

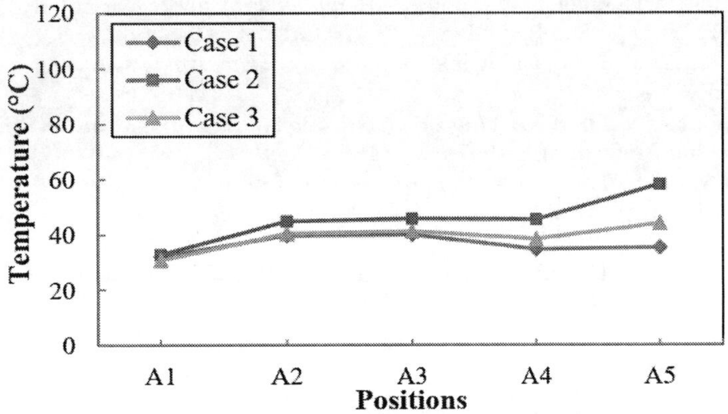

Figure 11. The temperature distribution of NRG at various positions of waveguide at heating times of 300 s

7. Conclusions

This research is carried out to observe the effects of waveguide position on the electric field and temperature profile in NRG. The results obtained represent the phenomena occurring in NRG during microwave heating. The governing equations in the study are analyzed by the finite element method. The simulation results with computer programs are validated with experimental results. The placement of waveguides in three cases, namely left hand side of microwave oven, right hand side of microwave oven and left and right hand sides of microwave oven are investigated. The results show that the hot spot zone occurs at the position of the middle finger in Case 1, however, hot spot zone occurs at the position near the middle of palm, little finger and middle finger in Case 2. In addition, the hot spot zone occurs at the position of the middle of palm and middle finger in Case 3. It is found that the placing the waveguide on the right side of the microwave oven (Case 2) gives the highest electric field and temperature profile. The advanced results in this research can be used in applications such as it provides guidance for the study of microwave heating of NRG manufacturing process.

References

[1] Lee S Y, You J H, Lee Y S, Kim H S, Cheong S K and Shin K H 2016 *Journal of the Korean Society of Safety* **31**(1) pp 19–24
[2] Chankrachang M, Limphirat W, Yongyingsakthavorn P, Nontakaew U and Tohsan A 2017 *The 3rd Electronic and Green Materials International Conference 2017* (EGM 2017)
[3] Bittner C, Garrido M V, Krach L H and Harth V 2016 *Allergy and Respiration* pp 37–44
[4] Zhang B L, Chen M, Ao N J, Deng W Y and Liu H L 2004 *Journal of Applied Polymer Science* **92**(5) pp 3196–3200.
[5] Chen M., Ao N J, Zhang B L, Den C M, Qian H L and Zhou H L 2005 *Journal of Applied Polymer Science* **98**(2) pp 591–597

[6] Ehntholt D J, Cerundolo D L, Bodek I, Schwope A D, Royer M D and Nielsen A P 1990 *American Industrial Hygiene Association Journal* **51**(9) pp 462–468
[7] M and Mirzaee O 2010 *Journal of Alloys and Compounds* **494**(1–2) pp 175–189
[8] Chen H L, Li T, Liang Y, Sun B and Li Q L 2017 *Heat and Mass Transfer* **53**(3) pp 1051–1060
[9] Khamdaeng T, Panyoyai N and Wongsiriamnuay T 2014 *American J.of Applied Sciences* **11**(4) pp 648–655
[10] Salagnac P, Glouannec P and Lecharpentier D 2004 *International J. of Heat and Mass Transfer* **47**(19–20) pp 4479–4489
[11] Lakhal W, Trabelsi S, Sediki E and Moussa M 2009 *American J. of Engineering and Applied Sciences* **2**(4) pp 590–602
[12] Dongbang W and Matthujak A 2013 *American J. of Applied Sciences* **10**(4) pp 353–360
[13] Martin D, Ighigeanu D, Mateescu, Craciun G and Ighigeanu A 2002 *Radiation Physics and Chemistry* **65**(1) pp 63–65
[14] Sombatsompop N and Kumnuantip C 2006 *J. of Applied Polymer Science* **100**(6) pp 5039–5048
[15] Chen H L, Li T, Liang Y, Sun B and Li Q L 2017 *Heat and Mass Transfer* **53(3)** pp 1051–1060
[16] Doo-ngam N, Rattanadecho P, and Klinklai W 2007 *Songklanakarin J. Science and Technology* **29**(6) pp 1599–1608
[17] Makul N and Rattanadecho P 2010 *Int. Communication in Heat and Mass Transfer* **37**(7) pp 914–923
[18] Olmstead W E and Brodwin M E 1997 *Int. J. of Heat and Mass Transfer* **40**(7) pp 1559–1565
[19] Potter D 1996 *National Instruments Corporation* pp 1–15
[20] Wessapan T, Srisawatdhisukul S and Rattanadecho P 2011 *Int. Communications in Heat and Mass Transfer* **38** pp 255–262
[21] COMSOL Multiphysics 1998-2012 Reference Guide COMSOL Protected by U.S. Patents 7,519,518; 7,596,474; and 7,623,991. Patents pending.
[22] Narumitbowonkul U, Keangin P and Rattanadecho P 2014 *The 5th TSME Int. Conf. on Mechanical Engineering 2014*
[23] Rattanadecho P and Makul N 2016 *Technology An Int. J.* **34**(1) pp 1–38
[24] Rands R D, Ferguson W J and Prather J L 1944 *J. of Research of the National Bureau of Standards* **33** pp 63–70
[25] Khalid K B 1992 *The Microwave Conf., APMC 92. Asia-Pacific* 2 pp 611–616
[26] Hassan J, Khalid K, Mohammad W and Yusoff D W 1997 *Pertanika J. of Science and Technology* **5**(2) pp 179–190
[27] Sridee J 2006 *Rheological Properties of Natural Rubber Latex* (A Thesis Submitted in Partial Fulfillment of the Requirements for the Degree of Master of Engineering in Polymer Engineering Suranaree University of Technology Academic Year 2006, ISBN 974 – 533 – 588 – 6)
[28] Ortiz-Serna P, Díaz-Calleja R, Sanchis M J, Floudas G, Nunes R C, Martins A F and Visconte L L 2010 *Macromolecules* **43**(11) pp 5094–5102

IOP Conf. Series: Materials Science and Engineering **297** (2018) 012038 doi:10.1088/1757-899X/297/1/012038

Finite-time stabilization of chaotic gyros based on a homogeneous supertwisting-like algorithm

Pitcha Khamsuwan*, Teerawat Sangpet, Suwat Kuntanapreeda
Department of Mechanical and Aerospace Engineering,
King Mongkut's University of Technology North Bangkok, Bangkok 10800, Thailand

* Corresponding Author: pitcha007@gmail.com

Abstract. This paper presents a finite-time stabilization scheme for nonlinear chaotic gyros. The scheme utilizes a supertwisting-like continuous control algorithm for the systems of dimension more than one with a Lipschitz disturbance. The algorithm yields finite-time convergence similar to that produces by discontinuous sliding mode control algorithms. To design the controller, the nonlinearities in the gyro are treated as a disturbance in the system. Thanks to the dissipativeness of chaotic systems, the nonlinearities also possess the Lipschitz property. Numerical results are provided to illustrate the effectiveness of the scheme.

1. Introduction

Gyros are important components in navigational, aeronautical and space systems. However, gyros can exhibit chaotic behavior, which will adversely affect the performance of the systems [1]. Recently, many authors have discussed the problem of controlling the chaotic gyros [2–4].

Chaotic control is challenging and has been extensively studied in the past decades. The study can be classified into chaos stabilization and chaos synchronization. Chaos stabilization attempts to eliminate the chaotic behavior of systems while chaos synchronization is to control a chaotic system so that it follows another chaotic system. Since Ott et al. [5] introduced the idea of controlling chaotic systems, a number of approaches have been proposed to control the chaotic systems [6–8]. Finite-time control has recently gained the attention of the chaos control community [9–11]. The main attractions of the finite-time control include faster convergence, robustness, and disturbance rejection.

Sliding mode control (SMC) has attracted interest in nonlinear control because of its numerous attractive properties. The properties include robustness against uncertainties and external disturbances, a fast dynamic response and a simple design. However, chattering phenomenon is one of the major obstacles for the implementation of SMC in a wide range of applications. A number of solutions have been proposed to eliminate or reduce chattering, including the boundary layer technique [12] and high-order SMC (HOSMC) methods [13, 14]. The supertwisting algorithm is one of effective HOSMC methods. HOSMC has recently attracted interest in chaos control [15–17].

In this paper, we examine the problems of stabilization of nonlinear chaotic gyros based on the HOSMC. To achieve this goal, a recently established homogeneous supertwisting-like modification of twisting algorithm [18] is employed. The algorithm is highly effective for the stabilization of both system state and its derivative in a finite time.

The rest of the paper is organized as follows. Preliminaries are presented in Section 2. Control design is described in Section 3. In Section 4, numerical results are given to illustrate the effectiveness of the control scheme. Finally, conclusions are drawn in the last section.

2. Preliminaries

2.1. Chaotic gyro

Consider the symmetric gyro mounted on a vibrating base as shown in figure 1. Here, (X, Y, Z) and (x, y, z) are the reference and body frames, respectively. The dynamics of the gyro is described by z-x-z Euler angles: ϕ (precession), θ (nutation), ψ (spin) and body rates: $\omega_x, \omega_y, \omega_z$. The rotational sequence of the Euler angles, starting by assuming (X, Y, Z) and (x, y, z) coincide, is listed as [19]:

1. Rotation about Z axis through angle ϕ to produce (x', y', z') frame.
2. Rotation about x' axis through angle θ to produce (x'', y'', z'') frame.
3. Rotation about z'' axis through angle ψ to produce (x, y, z) frame.

In the figure, M_g is the gravity force, l is the distance of the centre of gravity (CG) of the gyro from the origin, τ_m is the external torque applied to the gyro and $\bar{l}\sin(\omega t)$ is the motion of the vibrating base.

The kinetic energy of the gyro can be expressed as:

$$T = \frac{1}{2} I_1 \left(\omega_x^2 + \omega_y^2 \right) + \frac{1}{2} I_3 \omega_z^2 \tag{1}$$

where I_1 and I_3 are the polar and equatorial moments of inertia of the gyro, respectively. Based on the z-x-z Euler angle transformation, the body rates can be expressed in terms of Euler rates $\dot{\phi}$, $\dot{\theta}$, $\dot{\psi}$ as [19]:

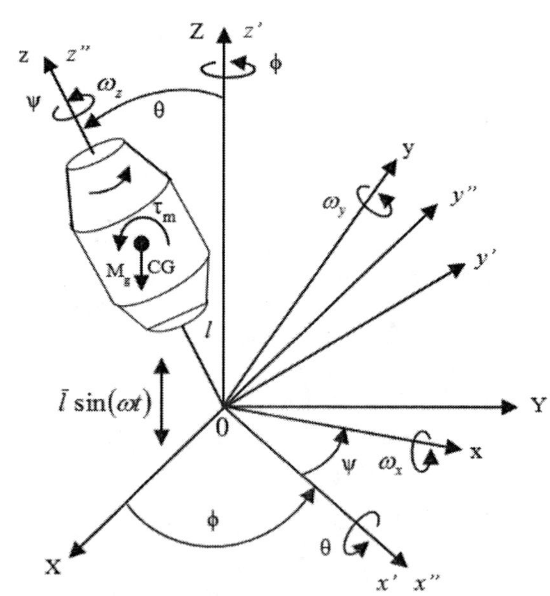

Figure 1. A symmetric gyroscope mounted on a vibrating base.

$$\omega_x = \left(\dot{\phi}\sin\theta\sin\psi + \dot{\theta}\cos\psi\right), \tag{2}$$
$$\omega_y = \left(\dot{\phi}\sin\theta\cos\psi - \dot{\theta}\sin\psi\right), \tag{3}$$
$$\omega_z = \left(\dot{\phi}\cos\theta + \dot{\psi}\right). \tag{4}$$

By substituting equations (2), (3) and (4) in equation (1), the kinetic energy becomes

$$T = \frac{1}{2}I_1\left(\dot{\phi}^2\sin^2\theta + \dot{\theta}^2\right) + \frac{1}{2}I_3\left(\dot{\phi}\cos\theta + \dot{\psi}\right)^2. \tag{5}$$

The potential energy of the gyro can be expresses as:

$$V = M_g\left(l + \bar{l}\sin\omega t\right)\cos\theta, \tag{6}$$

Thus, the Lagrangian of the system can be expressed as [20]:

$$L = T - V = \frac{1}{2}I_1\left(\dot{\theta}^2 + \dot{\phi}^2\sin^2\theta\right) + \frac{1}{2}I_3\left(\dot{\phi}\cos\theta + \dot{\psi}\right)^2 - M_g\left(l + \bar{l}\sin\omega t\right)\cos\theta. \tag{7}$$

Since ϕ and ψ do not appear explicitly in the Lagrangian, they are cyclic coordinates. Thus, the momentum integrals:

$$P_\phi = \frac{\partial L}{\partial \dot{\phi}} = I_1\dot{\phi}\sin^2\theta + I_3\left(\dot{\phi}\cos\theta + \dot{\psi}\right)\cos\theta, \tag{8}$$
$$P_\psi = \frac{\partial L}{\partial \dot{\psi}} = I_3\left(\dot{\phi}\cos\theta + \dot{\psi}\right), \tag{9}$$

are conserved. Then,

$$I_1\dot{\phi}\sin^2\theta + I_3\left(\dot{\phi}\cos\theta + \dot{\psi}\right)\cos\theta = \beta_\phi, \tag{10}$$
$$I_3\left(\dot{\phi}\cos\theta + \dot{\psi}\right) = I_3\omega_z = \beta_\psi, \tag{11}$$

where β_ϕ and β_ψ are constants. From equations (10) and (11), the expressions for $\dot{\phi}$ and $\dot{\psi}$ can be written as:

$$\dot{\phi} = \frac{\beta_\phi - \beta_\psi\cos\theta}{I_1\sin^2\theta}, \tag{12}$$
$$\dot{\psi} = \frac{\beta_\psi}{I_3} - \cos\theta\left(\frac{\beta_\phi - \beta_\psi\cos\theta}{I_1\sin^2\theta}\right). \tag{13}$$

The Routhian of the system is expressed as [20]:

$$R = L - \beta_\phi\dot{\phi} - \beta_\psi\dot{\psi}. \tag{14}$$

Substitute equations (7), (12) and (13) in equation (14) yields

$$R = \frac{1}{2}I_1\dot{\theta}^2 - \left\{\frac{\left(\beta_\phi - \beta_\psi\cos\theta\right)^2}{2I_1\sin^2\theta} + \frac{\beta_\psi^2}{2I_3} + M_g\left(l + \bar{l}\sin\omega t\right)\cos\theta\right\}. \tag{15}$$

By assuming the torque τ_m comprises the linear-plus-cubic dissipative torque $\tau_d = -D_1\dot{\theta} - D_2\dot{\theta}^3$ and the control torque τ_c, the equation of motion can be expressed in the coordinate of θ as:

$$\frac{d}{dt}\left(\frac{\partial R}{\partial \dot{\theta}}\right) - \frac{\partial R}{\partial \theta} = -D_1\dot{\theta} - D_2\dot{\theta}^3 + \tau_c \tag{16}$$

where D_1 and D_2 are constants. Substituting (15) in (16) yields

$$I_1\ddot{\theta}+\frac{(\beta_\phi-\beta_\psi\cos\theta)(\beta_\psi-\beta_\phi\cos\theta)}{I_1\sin^3\theta}+M_g(l+\bar{l}\sin\omega t)(\sin\theta)=-D_1\dot{\theta}-D_2\dot{\theta}^3+\tau_c \quad (17)$$

Note that, from equations (10) and (11), $\beta_\phi=\beta_\psi$ when $\theta=0$. Since β_ϕ and β_ψ are constants, the equality $\beta_\phi=\beta_\psi$ is true for all θ. By using $\beta_\phi=\beta_\psi$, equation (17) becomes

$$\ddot{\theta}+\frac{\beta_\phi^2}{I_1^2}\frac{(1-\cos\theta)^2}{\sin^3\theta}+\frac{D_1}{I_1}\dot{\theta}+\frac{D_2}{I_1}\dot{\theta}^3-\frac{M_g l}{I_1}\sin\theta=\frac{M_g\bar{l}}{I_1}(\sin\omega t)(\sin\theta)+\frac{1}{I_1}\tau_c\,. \quad (18)$$

By expressing equation (18) in a state equation form as in [1], we obtain:

$$\begin{aligned}\dot{x}_1&=x_2\\ \dot{x}_2&=-\alpha^2\frac{(1-\cos x_1)^2}{\sin^3 x_1}-c_1x_2-c_2x_2^3+\beta\sin x_1+(f\sin\omega t)(\sin x_1)+u\end{aligned} \quad (19)$$

where $x_1=\theta$, $x_2=\dot{\theta}$ are the state variables, $u=\tau_c/I_1$ is the control input, $\alpha=\beta_\phi/I_1=I_3\omega_z/I_1$, $c_1=D_1/I_1$, $c_2=D_2/I_1$, $\beta=M_g l/I_1$, and $f=M_g\bar{l}/I_1$ are positive constants. When $32<f<36$, $\alpha=10$, $c_1=0.5$, $\beta=1$, $c_2=0.05$, and $\omega=2$, the gyro with zero input becomes chaotic [1]. An example of the chaotic behaviour for $f=35.5$ and the initial conditions $(x_1(0),x_2(0))=(1,-1)$ is shown in figure 2. Note that all trajectories are bounded since chaotic systems are dissipative.

2.2. *Homogeneous supertwisting-like algorithm*

Consider the dynamical system:

$$\begin{aligned}\dot{x}_1&=x_2\\ \dot{x}_2&=\rho(t)+u\end{aligned} \quad (20)$$

where x_1 and x_2 are the state variables, u is the control input and $\rho(t)$ is the disturbance. It is assumed that $\rho(t)$ satisfies the Lipschitz condition with constant L.

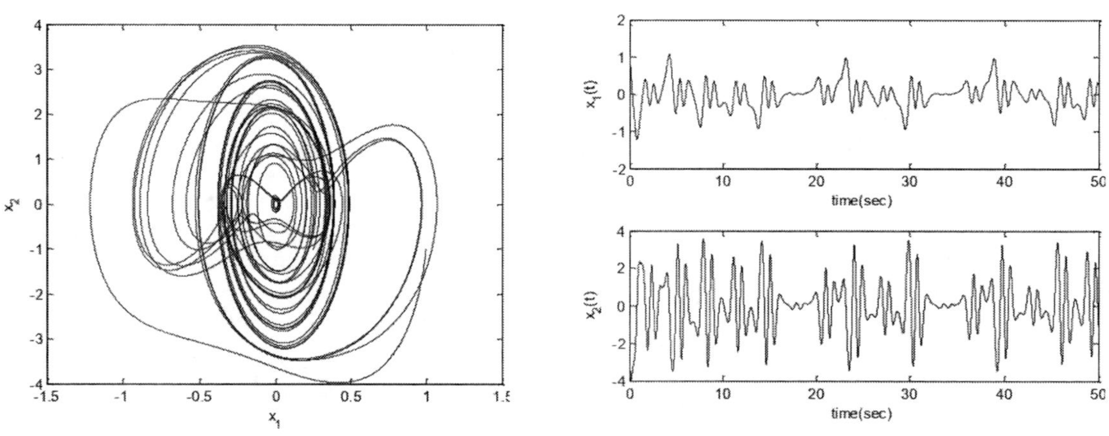

(a) (b)

Figure 2. Chaotic behaviour: (a) Phase trajectory and (b) Time history

A homogeneous continuous supertwisting-like algorithm for systems of dimension more than one has recently been introduced in [18]. The algorithm generalizes the well-known supertwisting algorithm to higher dimensional systems. It yields finite-time convergence similar to that produced by discontinuous sliding mode control algorithms.

Based on the homogeneous continuous supertwisting-like algorithm [18], the control law for systems of dimension two with a Lipschitz disturbance is given as:

$$u=-\lambda_0\left|\int_{t_0}^{t} x_1(s)ds\right|^{1/4}\operatorname{sgn}\left(\int_{t_0}^{t} x_1(s)ds\right)-\lambda_1|x_1(t)|^{1/3}\operatorname{sgn}(x_1(t))-\lambda_2|x_2(t)|^{1/2}\operatorname{sgn}(x_2(t))-\alpha_c\int_{t_0}^{t}\operatorname{sgn}(x_2(s))ds \quad (21)$$

where $\lambda_0,\lambda_1,\lambda_2>0$ and $\alpha_c>0$ are control parameters. The control law (21) yields finite-time convergence of both state variables x_1 and x_2 to the origin if the following conditions hold for the control parameters [18]:

$$\alpha_c>L \text{ and } \lambda_2^2>2(\alpha_c+L)^2/(\alpha_c-L). \quad (22)$$

3. Control design

The control objective is to suppress the chaos governed by equation (19) by driving the state variables x_1 and x_2 to the origin in finite time. To achieve the objective, we define the control law as:

$$u=c_1x_2+c_2x_2^3+u_{st}. \quad (23)$$

where u_{st} adopts the supertwisting-like control law (21).

Substituting the control law (23) into equation (19) yields

$$\begin{aligned}\dot{x}_1&=x_2\\ \dot{x}_2&=-\alpha^2\frac{(1-\cos x_1)^2}{\sin^3 x_1}+\beta\sin x_1+(f\sin\omega t)(\sin x_1)+u_{st}.\end{aligned} \quad (24)$$

The system (24) can be written in the form (20) by defining:

$$\rho(t)=-\alpha^2\frac{(1-\cos x_1)^2}{\sin^3 x_1}+\beta\sin x_1+(f\sin\omega t)(\sin x_1). \quad (25)$$

In order to adopt the control law (21), $\rho(t)$ must possess the Lipschitz property. A sufficient condition for a function to be Lipschitz is its derivative being bounded. Thus, we examine the derivative of $\rho(t)$ instead.

By taking the derivative of $\rho(t)$ with respect to time, it yields

$$\begin{aligned}\dot{\rho}(t)&=\alpha^2\dot{x}_1\left\{\frac{2[\cos x_1-1]}{\sin^2 x_1}-\frac{3\cos x_1[\cos x_1-1]^2}{\sin^4 x_1}\right\}+\beta\dot{x}_1\cos x_1+\omega f(\cos\omega t)(\sin x_1)\\ &\quad+f\dot{x}_1(\sin\omega t)(\cos x_1)\\ &=\alpha^2x_2\left\{\frac{2[\cos x_1-1]}{\sin^2 x_1}-\frac{3\cos x_1[\cos x_1-1]^2}{\sin^4 x_1}\right\}+\beta x_2\cos x_1+\omega f(\cos\omega t)(\sin x_1)\\ &\quad+fx_2(\sin\omega t)(\cos x_1)\end{aligned} \quad (26)$$

Since chaotic systems are dissipative, x_1 and x_2 are bounded. It is obvious that $\dot{\rho}(t)$ can become unbounded only if $\frac{2[\cos x_1 - 1]}{\sin^2 x_1}$ and $\frac{3\cos x_1[\cos x_1 - 1]^2}{\sin^4 x_1}$ have a singularity. In this case, the singularity can occur only at $x_1 = 0$. However, since

$$\lim_{x_1 \to 0}\left\{\frac{2[\cos x_1 - 1]}{\sin^2 x_1}\right\} = -1$$

and

$$\lim_{x_1 \to 0}\left\{\frac{3\cos x_1[\cos x_1 - 1]^2}{\sin^4 x_1}\right\} = 0.75,$$

$\dot{\rho}(t)$ has no singularity. Thus, we can conclude that $\dot{\rho}(t)$ is bounded (i.e., $\max|\dot{\rho}(t)| \le L,\ \ 0 < L < \infty$). Therefore, $\rho(t)$ is Lipschitz with the constant L.

4. Numerical simulations

In this section, numerical simulations are utilized to illustrate the effectiveness of the supertwisting-like algorithm-based control scheme to suppress the chaos of chaotic gyros. The parameters of the gyros are set as follows: $\alpha = 10,\ \beta = 1,\ c_1 = 0.5,\ c_2 = 0.05,\ \omega = 2$ and $f = 35.5$. The fourth-order Runge-Kutta method with a time step of 0.001 second was used in all simulations.

First, the Lipschitz constant L is estimated. From equation (26), we obtain

$$\begin{aligned}
\max|\dot{\rho}(t)| &= \max\left|\alpha^2 x_2\left\{\frac{2[\cos x_1 - 1]}{\sin^2 x_1} - \frac{3\cos x_1[\cos x_1 - 1]^2}{\sin^4 x_1}\right\} + \beta x_2 \cos x_1 + \omega f(\cos \omega t)(\sin x_1) + f x_2 (\sin \omega t)(\cos x_1)\right| \\
&\le \alpha^2 \max\left|x_2\left\{\frac{2[\cos x_1 - 1]}{\sin^2 x_1} - \frac{3\cos x_1[\cos x_1 - 1]^2}{\sin^4 x_1}\right\}\right| + \beta \max|x_2 \cos x_1| + \omega f \max|(\cos \omega t)(\sin x_1)| \\
&\quad + f \max|x_2 (\sin \omega t)(\cos x_1)| \\
&\le \alpha^2 \max|x_2| \max\left|\frac{2[\cos x_1 - 1]}{\sin^2 x_1} - \frac{3\cos x_1[\cos x_1 - 1]^2}{\sin^4 x_1}\right| + (\beta + f)\max|x_2| + f\omega \\
&\le \alpha^2 \max|x_2| \max\left\{\left|\frac{2[\cos x_1 - 1]}{\sin^2 x_1}\right| + \left|\frac{3\cos x_1[\cos x_1 - 1]^2}{\sin^4 x_1}\right|\right\} + (\beta + f)\max|x_2| + f\omega \\
&= \alpha^2 X_2 R_{max} + (\beta + f)X_2 + f\omega
\end{aligned} \tag{27}$$

where X_2 and R_{max} are upper bounds of $|x_2|$ and $R(t) = \left|\frac{2[\cos x_1 - 1]}{\sin^2 x_1}\right| + \left|\frac{3\cos x_1[\cos x_1 - 1]^2}{\sin^4 x_1}\right|$, respectively. From figure 2(b), it can be found that $X_2 \le 4$. By examining the value of $R(t)$ over $x_1 \in (-\frac{\pi}{2}, \frac{\pi}{2})$ (see figure 3), it results in $R_{max} \le 2.1$. From equation (27), it yields $\max|\dot{\rho}(t)| \le 1057$. Here, the Lipschitz constant $L = 1100$ is chosen.

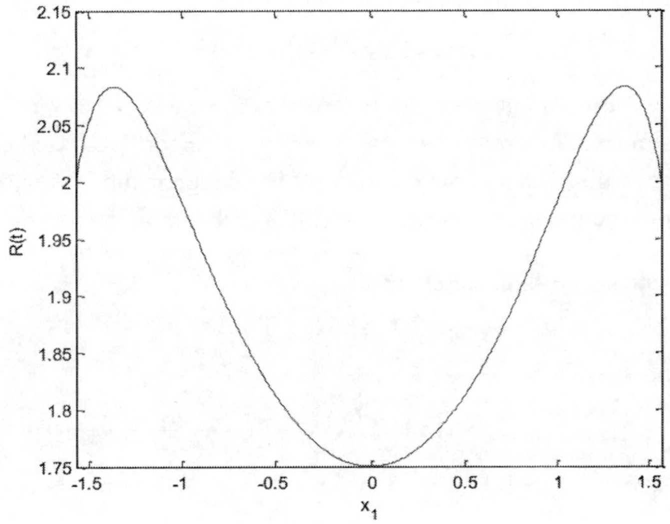

Figure 3. Plot of $R(t)$ versus x_1

Next, the parameters α_c, λ_0, λ_1 and λ_2 of u_{st} are selected. Here, $\alpha_c = 1200$ and $\lambda_2 = 400$ were taken satisfying the condition (22) and $\lambda_0 = 200$ and $\lambda_1 = 100$ were chosen such that the desired transient response of the control system is achieved.

Simulation results are depicted in figure 4. The initial condition of the system was set as $(x_1(0), x_2(0)) = (1, -1)$. The controller was activated at time equal to 5 sec. The result shows that the state variables x_1 and x_2 converged to the origin in less than 0.5 sec after the controller was activated.

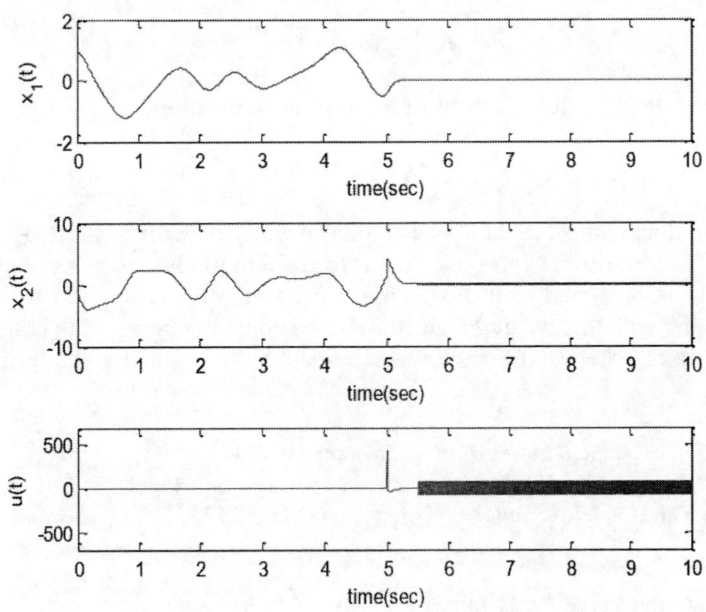

Figure 4. State variables and control input of the controlled chaotic gyro.

In addition, a comparison with a sliding mode controller based on the twisting algorithm (TA) [14, 15] was made. The twisting algorithm control law was given as:

$$u(t) = -k_1 sign(x_1(t)) - k_2 sign(x_2(t)) \tag{28}$$

where $k_1, k_2, > 0$ were control parameters. The parameters $k_1 = 100$, $k_2 = 50$ were chosen such that the control system had a similar response to the supertwisting-like algorithm-based control system (i.e., x_1 and x_2 converged to the origin within about 0.5 sec). A simulation result is depicted in figure. 5.

From figures 4 and 5, both controllers (23) and (28) yield similar responses of the state variables, whereas the controller (23) provides markedly smaller chattering control signal. This illustrates the benefit of the supertwisting-like algorithm.

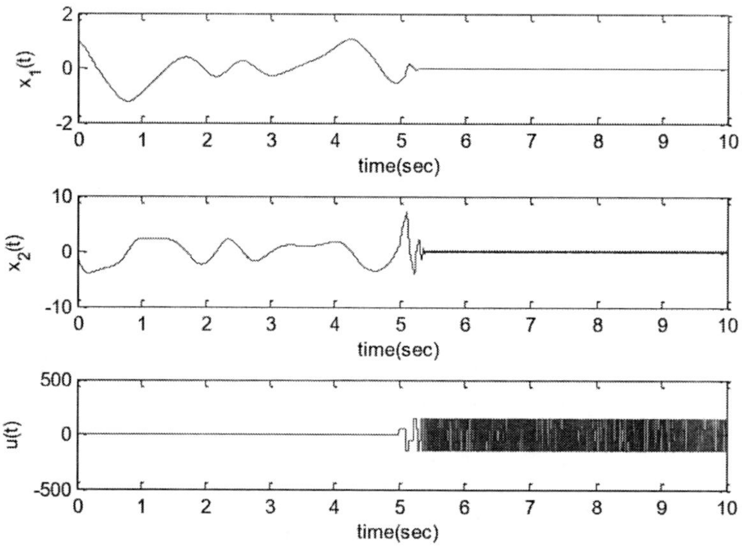

Figure 5. State variables and control input of the controlled chaotic gyro using TA.

5. Conclusions

In this paper, a finite time control scheme based on a recently developed homogeneous supertwisting-like continuous control algorithm is proposed for stabilization of chaotic gyros. Numerical simulations have illustrated that the proposed control scheme can be effectively applied to a chaos stabilization problem. By comparing with the twisting algorithm based control scheme, both control schemes provide similar control responses, whereas the proposed scheme results in smaller chattering control signal.

References

[1] Chen H K 2002 *J. of Sound and Vibration* **255**(4) pp 719–740
[2] Yan J-J, Hung M-L and Liao T-L 2006 *J. of Sound and Vibration* **298**(1-2) p 298–306
[3] Wang C-C and Yau H-T 2011 *Nonlinear Dynamics* **66**(1) pp 53–65
[4] Yang C-C, Ou C-J 2013 *Communications in Nonlinear Science and Numerical Simulation* **18**(3) pp 682–691
[5] Ott E, Grebogi C and Yorke J A 1990 *Physical Review Letters* **64**(11-12) pp 1196–1199
[6] Chen G 2011 *Applied Mathematics and Computation* **217**(17) pp 7258–7264
[7] Chadl M and Zelinka I 2014 *Computers and Mathematics with Applications* **68**(12) pp 2142–2147
[8] Khamsuwan P and Kuntanapreeda S 2016 *J. of Computational and Nonlinear Dynamics* **11**(5) p 051021

[9] Aghababa M P and Aghababa H P 2012 *Nonlinear Dynamics* **69**(4) pp 1881–1888
[10] Wang J, Chen X and Fu J 2014 *Nonlinear Dynamics* **78**(2) pp 1321–1328
[11] TranX-T and Kang H-J 2015 *Nonlinear Dynamics* **80**(1) pp 637–651
[12] Slotine J J E and Li W 1991 *Applied Nonlinear Control* (Engle Wood Cliffs : Pretice-Hall)
[13] Bartolini G, Ferrara A and Usai E 1998 *IEEE Transactions on Automatic Control* **43**(2) pp 241–246
[14] Levant A 2003 *Int. J. of Control* **76**(9-10) pp 924–941
[15] Li H, Liao X and Li C 2011 *Neurocomputing* **74**(17) pp 3212–3222
[16] Xiang W, Yugao Huangpu Y 2010 *Communications in Nonlinear Science and Numerical Simulation* **15**(11) pp 3241–3247
[17] Zhankui S and Sun K 2013 *Communications in Nonlinear Science and Numerical Simulation* **18**(9) pp 2540–2548
[18] Basin M V and Rodríguez-Ramírez P C 2014 *IEEE Trans. on Industrial Electronics* **61**(11) pp 6472–6480
[19] James R W 1978 *Space Attitude Determination and Control* (Dordercht : Kluwer Academic Publishers)
[20] Goldstein H 1980 *Classical Mechanics* (New York : Addison-Wesley)

An analytical comparison of optimal output transition and polynomial trajectories

Arom Boekfah

Department of Mechanical Engineering, Faculty of Engineering, Mahidol University
25/25 Salaya, Phuttamonthon, Nakhon Pathom 73170 Thailand

* Corresponding Author: arom.boe@mahidol.ac.th

Abstract. This paper compares the solution of the optimal output transition approach with polynomial trajectories. Although the optimal output transition approach can guarantee the optimal output trajectory, the complex problem formulation tends to be required, especially for nonlinear systems. The aim of this paper is to investigate differences between the optimal output transition approach and the polynomial-based approach. Based on the standard first and second order linear system examples, the output transition trajectory can be obtained by using the polynomial-based approach without any significant differences when compared to the optimal output transition approach.

1. Introduction

Changing the output of the system y from the initial value y_I to the final value y_F during transition time T_{tt} is a fundamental control problem [1] and is of particular interest. Applications of the output transition problem can be found, for example, in nanopositioning using a piezo-based flexible structure system [2] and in a vertical take-off and landing (VTOL) aircraft [3]. One of the output transition approaches is the optimal output transition approach where the acceleration of the output trajectory is minimized [4]. However, obtaining the optimal output transtion trajectory normally requires a complex problem formulation, especially for nonlinear systems [5, 6]. The objective of this paper is to observe differences, if any, between the solution of the optimal output transition approach [4] and polynomial trajectories [7, 8]. Examples of the standard first and second order linear systems are provided to show the exact solution of the optimal output transition approach. According to the optimal solution of the example systems, the output trajectory can be obtained by using the polynomial-based approach without any significant differences when compared to the optimal output transition approach. This work extends the previous work that studied simulation results by using the flexible structure example in [9] to analytical results.

2. Output Transition Problem

Consider the invertible transfer function G of the linear system

$$G(s)=\frac{y(s)}{u(s)}, \tag{1}$$

where y is the output and u is the input of the system. The output transition problem is to find the output transition input u_{tran} that can be used to transition the output y from the initial value y_I to the final value y_F during transition time T_{tt}, as shown in figure 1.

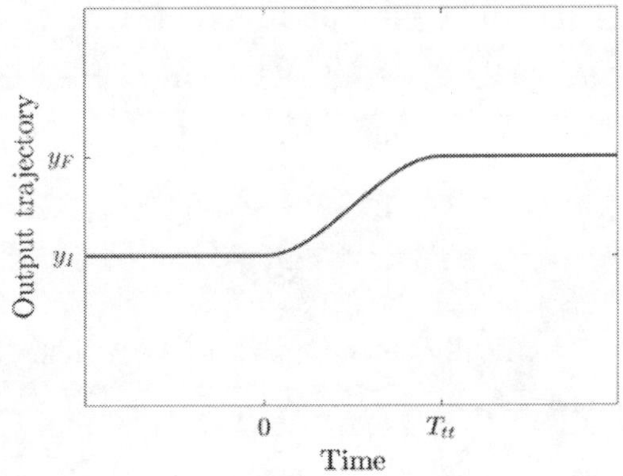

Figure 1. The output of the system y is transitioned from the initial value $y(t)=y_I, \forall t \le 0$ to the final value $y(t)=y_F, \forall t \ge T_{tt}$.

2.1. *Optimal Output Trajectory*

The optimal output transition approach as given in [4] is summarized in this section. Let the associated system dynamics be given by

$$\dot{\tilde{x}}(t)=\frac{d}{dt}\left(\tilde{x}(t)\right)=\tilde{A}\tilde{x}(t)+\tilde{B}\tilde{u}(t) \tag{2}$$

during transition time $t\in\left[0,T_{tt}\right]$. The associated state $\tilde{x}$ is given by

$$\tilde{x}(t)=\begin{bmatrix} y_{tran}(t) & \frac{d}{dt}\left(y_{tran}(t)\right) & \frac{d^2}{dt^2}\left(y_{tran}(t)\right) & \cdots & \frac{d^r}{dt^r}\left(y_{tran}(t)\right) & \frac{d^{(r+1)}}{dt^{(r+1)}}\left(y_{tran}(t)\right)\end{bmatrix}' \tag{3}$$

where $'$ represents the transpose of the matrix, r is the relative degree of the system, y_{tran} is the desired output transition trajectory with the boundary conditions

$$\tilde{x}(0)=\begin{bmatrix} y_{tran}(0) \\ \dot{y}_{tran}(0) \\ \ddot{y}_{tran}(0) \\ \vdots \\ y_{tran}^{(r+1)}(0)\end{bmatrix}=\begin{bmatrix} y_I \\ 0 \\ 0 \\ \vdots \\ 0\end{bmatrix} \text{ and } \tilde{x}(T_{tt})=\begin{bmatrix} y_{tran}(T_{tt}) \\ \dot{y}_{tran}(T_{tt}) \\ \ddot{y}_{tran}(T_{tt}) \\ \vdots \\ y_{tran}^{(r+1)}(T_{tt})\end{bmatrix}=\begin{bmatrix} y_F \\ 0 \\ 0 \\ \vdots \\ 0\end{bmatrix}, \tag{4}$$

and matrices $\tilde{A}$ and $\tilde{B}$ are given by

$$\tilde{A}=\begin{bmatrix}0&1&0&\cdots&0&0\\0&0&1&\cdots&0&0\\\vdots&\vdots&\vdots&\ddots&\vdots&\vdots\\0&0&0&\cdots&1&0\\0&0&0&\cdots&0&1\\0&0&0&\cdots&0&0\end{bmatrix}_{(r+1)\times(r+1)} \text{ and } \tilde{B}=\begin{bmatrix}0\\0\\\vdots\\0\\0\\1\end{bmatrix}_{(r\times1)} . \tag{5}$$

Optimal output trajectory can be obtained by minimizing the cost functional J

$$\min_{\tilde{u}}\left[J_{T_{tt}}=\int_0^{T_{tt}}\left[\tilde{u}(\tau)\right]^2 d\tau\right], \tag{6}$$

where the optimal input $\tilde{u}$ can specifically be chosen as the $r+1$ time derivative of the output $y^{(r+1)}$ and as a function of transition time T_{tt}

$$\tilde{u}(t)=y_{T_{tt}}^{(r+1)}(t)=\frac{d^{(r+1)}}{dt^{(r+1)}}\left(y_{T_{tt}}(t)\right)=\tilde{B}'\left[e^{\tilde{A}'(T_{tt}-t)}\right]\left[G_{tran}^{-1}(T_{tt})\right]\left\{\tilde{x}(T_{tt})-e^{\tilde{A}T_{tt}}\tilde{x}(0)\right\}. \tag{7}$$

The invertible controllability Gramian G_{tran}, as a function of transition time T_{tt}, is given by

$$G_{tran}(T_{tt})=\int_0^{T_{tt}} e^{\tilde{A}(T_{tt}-\tau)}\tilde{B}\tilde{B}'e^{\tilde{A}'(T_{tt}-\tau)}d\tau . \tag{8}$$

Finally, the desired output transition trajectory y_{tran} can be found by integrating the solution of equation (7) $r+1$ times with initial conditions at time $t=0$ as given in equation (4). It is noted that equation (7) is slightly different from the filtered version of the rth time derivative of the output $y_{T_{tt}}^{(r)}(t)=d^r y_{T_{tt}}(t)/dt^r$ that is used in the previous work [4].

2.2. Polynomial Trajectories

The polynomial-based approach was originally developed in [7] and is summarized in this section. Polynomial trajectories can be obtained by assuming the polynomial of the form

$$y_{tran}(t)=a_0+a_1\left(\frac{t}{T_{tt}}\right)+a_1\left(\frac{t}{T_{tt}}\right)^2+\ldots+a_n\left(\frac{t}{T_{tt}}\right)^n, t\in\left[0,T_{tt}\right], \tag{9}$$

where coefficients a 's can be obtained by matching the output y_{tran} and its time derivatives with the boundary conditions given in equation (4) and $n=2r+1$ for the system with relative degree r. Alternatively, polynomial trajectories can be written as

$$y_{tran}(t)=y_I+\left(y_F-y_I\right)\sum_{i=r+1}^{2r+1}a_i\left(\frac{t}{T_{tt}}\right)^i, t\in\left[0,T_{tt}\right] \tag{10}$$

where coefficients a_i 's are given by

$$a_i=\frac{(-1)^{i-r-1}(2r+1)!}{i\times r!(i-r-1)!(2r+1-i)!} \tag{11}$$

as studied in [8]. It is noted that the output trajectory, obtained by using equation (7), is expected to be equivalent to the $r+1$ time differentiable polynomial trajectories, given in equation (10).

2.3. Output Transition Input
Substituting the output transition trajectory y_{tran} obtained by using the optimal output transition approach from equations (2) to (8) or using the polynomial-based approach in equation (10) into the system in equation (1), the output transition input u_{tran} becomes

$$u_{tran}(s) = G^{-1}(s)y_{tran}(s) \tag{12}$$

with the use of the inverse system G^{-1}.

3. First and Second Order System Examples
The standard first and second order linear systems are used to compare the solution of the optimal control approach with polynomial trajectories.

3.1. First Order System Example.
One of the systems that has relative degree $r=1$ is the first order system

$$G_1(s) = \frac{y(s)}{u(s)} = \frac{b}{s+b}, \tag{13}$$

where b is constant. The output transition input u_{tran} for the first order system in equation (13) can be given by

$$u_{tran}(t) = \frac{1}{b}\dot{y}_{tran}(t) + y_{tran}(t), t \in [0, T_{tt}], \tag{14}$$

where the output transition trajectory y_{tran} can be obtained by using the optimal output trajectory or polynomial trajectories.

3.1.1. Optimal Output Trajectory. Since the relative degree $r=1$, the matrices $\tilde{A}$ and $\tilde{B}$ in equation (5) for the first order system becomes

$$\tilde{A} = \begin{bmatrix} 0 & 1 \\ 0 & 0 \end{bmatrix} \text{ and } \tilde{B} = \begin{bmatrix} 0 \\ 1 \end{bmatrix}. \tag{15}$$

By substituting equation (15) into equation (8), the controllability Gramian G_{tran} can be found as

$$G_{tran}(T_{tt}) = \int_0^{T_{tt}} \begin{bmatrix} (T_{tt}-\tau)^2 & T_{tt}-\tau \\ T_{tt}-\tau & 1 \end{bmatrix} d\tau = \begin{bmatrix} T_{tt}^3/3 & T_{tt}^2/2 \\ T_{tt}^2/2 & T_{tt} \end{bmatrix}. \tag{16}$$

For the system with relative degree $r=1$, the optimal input $\tilde{u}$ can be computed by using equation (5)

$$\tilde{u}(t) = \ddot{y}_{T_{tt}}(t) = \begin{bmatrix} 0 & 1 \end{bmatrix} \begin{bmatrix} 1 & 0 \\ T_{tt}-t & 1 \end{bmatrix} \begin{bmatrix} 12/T_{tt}^3 & -6/T_{tt}^2 \\ -6/T_{tt}^2 & 4/T_{tt} \end{bmatrix} \left\{ \begin{bmatrix} y_F \\ 0 \end{bmatrix} - \begin{bmatrix} 1 & T_{tt} \\ 0 & 1 \end{bmatrix} \begin{bmatrix} y_I \\ 0 \end{bmatrix} \right\} = (y_F - y_I)\left(\frac{6}{T_{tt}^2} - \frac{12}{T_{tt}^3}t \right) \tag{17}$$

for $t \in [0, T_{tt}]$. The optimal output trajectory y_{tran} can be found by integrating equation (17) twice with the initial conditions $\dot{y}_{tran}(0) = 0$ and $y_{tran}(0) = y_I$ as

IOP Conf. Series: Materials Science and Engineering **297** (2018) 012039 doi:10.1088/1757-899X/297/1/012039

$$y_{tran}(t) = y_I + (y_F - y_I)\left(3\left(\frac{t}{T_{tt}}\right)^2 - 2\left(\frac{t}{T_{tt}}\right)^3\right), t \in [0, T_{tt}]. \quad (18)$$

3.1.2. Polynomial Trajectories. For the system with relative degree $r = 1$, the coefficients a_i can be computed by using equation (11) as

$$a_2 = 3 \text{ and } a_3 = -2 \quad (19)$$

and the polynomial trajectory given in equation (10) with the coefficients a_i in equation (19) becomes equation (18) exactly.

3.2. Second Order System Example.
The second order system G_2 which has a general form of

$$G_2(s) = \frac{y(s)}{u(s)} = \frac{\omega_n^2}{s^2 + 2\zeta\omega_n s + \omega_n^2}, \quad (20)$$

where ω_n is the natural frequency and ζ is the damping ratio, has relative degree $r = 2$. The output transition input u_{tran} for the second order system in equation (20) can be given by

$$u_{tran}(t) = \frac{1}{\omega_n^2}\ddot{y}_{tran}(t) + \frac{2\zeta}{\omega_n}\dot{y}_{tran}(t) + y_{tran}(t), t \in [0, T_{tt}], \quad (21)$$

where the output transition trajectory y_{tran} can be obtained by using the optimal output trajectory or polynomial trajectory.

3.2.1. Optimal Output Trajectory. The matrices $\tilde{A}$ and $\tilde{B}$ in equation (5) for the second order system with relative degree $r = 2$ becomes

$$\tilde{A} = \begin{bmatrix} 0 & 1 & 0 \\ 0 & 0 & 1 \\ 0 & 0 & 0 \end{bmatrix} \text{ and } \tilde{B} = \begin{bmatrix} 0 \\ 0 \\ 1 \end{bmatrix}. \quad (22)$$

By substituting equation (22) into equation (8), the controllability Gramian G_{tran} can be found as

$$G_{tran}(T_{tt}) = \int_0^{T_{tt}} \begin{bmatrix} (T_{tt}-\tau)^4/4 & (T_{tt}-\tau)^3/2 & (T_{tt}-\tau)^2/2 \\ (T_{tt}-\tau)^3/2 & (T_{tt}-\tau)^2 & T_{tt}-\tau \\ (T_{tt}-\tau)^2/2 & T_{tt}-\tau & 1 \end{bmatrix} d\tau = \begin{bmatrix} T_{tt}^5/20 & T_{tt}^4/8 & T_{tt}^3/6 \\ T_{tt}^4/8 & T_{tt}^3/3 & T_{tt}^2/2 \\ T_{tt}^3/6 & T_{tt}^2/2 & T_{tt} \end{bmatrix}. \quad (23)$$

For the system with relative degree $r = 2$, the optimal input $\tilde{u}$ can be computed by using equation (5)

$$\tilde{u}(t) = \ddot{y}_{T_{tt}}(t) = \begin{bmatrix} 0 & 0 & 1 \end{bmatrix} \begin{bmatrix} 1 & 0 & 0 \\ T_{tt}-t & 1 & 0 \\ (T_{tt}-t)^2/2 & T_{tt}-t & 1 \end{bmatrix} \begin{bmatrix} 720/T_{tt}^5 & -360/T_{tt}^4 & 60/T_{tt}^3 \\ -360/T_{tt}^4 & 192/T_{tt}^3 & -36/T_{tt}^2 \\ 60/T_{tt}^3 & -36/T_{tt}^2 & 9/T_{tt} \end{bmatrix}$$
$$\left\{ \begin{bmatrix} y_F \\ 0 \\ 0 \end{bmatrix} - \begin{bmatrix} 1 & 0 & 0 \\ T_{tt} & 1 & 0 \\ T_{tt}^2/2 & T_{tt} & 1 \end{bmatrix} \begin{bmatrix} y_I \\ 0 \\ 0 \end{bmatrix} \right\} = (y_F - y_I)\left(\frac{60}{T_{tt}^3} - \frac{360}{T_{tt}^4}t + \frac{360}{T_{tt}^5}t^2\right) \quad (24)$$

for $t \in [0, T_{tt}]$. The optimal output trajectory y_{tran} can be found by integrating equation (24) three times with the initial conditions $\ddot{y}_{tran}(0) = \dot{y}_{tran}(0) = 0$ and $y_{tran}(0) = y_I$ as

$$y_{tran}(t) = y_I + (y_F - y_I)\left(10\left(\frac{t}{T_{tt}}\right)^3 - 15\left(\frac{t}{T_{tt}}\right)^4 + 6\left(\frac{t}{T_{tt}}\right)^5\right), t \in [0, T_{tt}]. \quad (25)$$

3.2.2. Polynomial Trajectories. For the system with relative degree $r = 2$, the coefficients a_i can be computed by using equation (11) as

$$a_3 = 10,\ a_4 = -15, \text{ and } a_5 = 6 \quad (26)$$

and the polynomial trajectory given in equation (10) with the coefficients a_i in equation (26) is exactly similar to equation (25).

4. Conclusion and Discussion

Optimal output trajectory was compared with polynomial trajectories. Based on the standard first and second order system examples, the polynomial-based approach can be used to obtain the output transition trajectory without any significant difference when compared to the optimal output trajectory. Since this work seeks to find the exact solution of the optimal output transition approach, only the examples of the first and second order linear systems are considered. However, polynomial trajectories could be used not only for higher order linear systems, as simulated with the flexible structure example in [9], but also for nonlinear systems, as studied in [6, 8], instead of the optimal output transition approach, as proposed in [4].

References

[1] Perez H and Devasia S 2003 *Automatica* **39** p 181
[2] Clayton G, Tien S, Leang K, Zou Q and Devasia S 2009 *ASME J. Dyn. Syst. Meas. Control* **131** p 1
[3] Hauser J, Sastry S and Meyer G 1992 *Automatica* **28** p 665
[4] Boekfah A and Devasia S 2016 *IEEE Trans. Control Syst. Technol.* **24** p 265
[5] Devasia 2011 *Automatica* **47** p 1379
[6] Boekfah A 2017 *Proc. 2nd Int. Conf. on Control and Robotics Engineering* p 97
[7] Piazzi A and Visioli A 2000 *IEEE/ASME Trans. Mechatronics* **5** p 12
[8] Graichen K, Hagenmeyer V and Zeitz M 2005 *Automatica* **41** p 2033
[9] Boekfah A 2017 *Proc. 31st Conf. of Mechanical Engineering Network of Thailand* p 1000

IOP Conf. Series: Materials Science and Engineering **297** (2018) 012040 doi:10.1088/1757-899X/297/1/012040

Two models for identification and predicting behaviour of an induction motor system

Chien-Hsun Kuo*

Department of Mechanical and Automation Engineering, Kao Yuan University, No.1821, Jhongshan Rd., Lujhu Dist., Kaohsiung City 821, Taiwan

* Corresponding Author: t30081@cc.kyu.edu.tw

Abstract. System identification or modelling is the process of building mathematical models of dynamical systems based on the available input and output data from the systems. This paper introduces system identification by using ARX (Auto Regressive with eXogeneous input) and ARMAX (Auto Regressive Moving Average with eXogeneous input) models. Through the identified system model, the predicted output could be compared with the measured one to help prevent the motor faults from developing into a catastrophic machine failure and avoid unnecessary costs and delays caused by the need to carry out unscheduled repairs. The induction motor system is illustrated as an example. Numerical and experimental results are shown for the identified induction motor system.

1. Introduction

System identification deals with the problem of building the mathematical models of dynamical systems according to the input and output data. This technique is important in a diversity of fields such as manufacturing process, communication, economics, and system dynamics and control. For an unknown system, it is mostly required to identify the system model before one can perform the control design. Shah et al. [1] proposed an algorithm which balances a data fidelity term with a norm induced by the set of single pole filters to identify the unknown system from noisy linear measurements. Smith [2] used a nuclear norm minimization based method for frequency domain subspace identification. The induction machines are widely used in industry, but the possibility of faults is unavoidable. Fault identification and diagnosis schemes are intended to provide advanced warnings of incipient faults, so that appropriate maintenance action can be taken at an early stage [3, 4]. In this paper, the induction motor system is identified by both ARX and ARMAX model methods. Through the identified system model, the predicted output could be compared with the measured one to help prevent the motor faults from developing into a catastrophic machine failure and avoid unnecessary costs and delays caused by the need to carry out unscheduled repairs.

2. Model Structures

A finite-dimensional, linear, discrete-time, time-invariant stochastic system can be expressed as:

$$x_{k+1} = Ax_k + Bu_k + w_k,\ y_k = Cx_k + v_k \tag{1}$$

where $x \in R^{n \times 1}$, $u \in R^{ni \times 1}$, $y \in R^{no \times 1}$ are state, input and output vectors, respectively; [A,B,C] are the state-space system matrices. The sequences of process noise $w \in R^{n \times 1}$ and measurement noise $v \in R^{no \times 1}$ are assumed white, Gaussian, and zero mean. The noises w_k and v_k are also assumed uncorrelated with covariance Q and R, respectively.

2.1. ARX Model

If one defines the error between the actual output y_k and the estimated output $C\hat{x}_k$ as residual ε_k, equation (1) can also be expressed as

$$\hat{x}_{k+1} = A\hat{x}_k + Bu_k + AK\varepsilon_k,\ y_k = C\hat{x}_k + \varepsilon_k. \tag{2}$$

where $\hat{x}$ is the estimated state vector and $K \in R^{n \times no}$ is the steady-state Kalman filter gain.
The relation between signal input and output with zero initial condition could be described as

$$y_k = \sum_{i=1}^{\infty} C\bar{A}^{i-1}AKy_{k-i} + \sum_{i=1}^{\infty} C\bar{A}^{i-1}Bu_{k-i} + \varepsilon_k \tag{3}$$

where $\bar{A} = A - AKC$. Since $\bar{A}$ is asymptotically stable, $\bar{A}^{i-1} \approx 0$ if $i \geq q$ for a sufficient large number q [5]. Thus, equation (3) becomes

$$y_k = \sum_{i=1}^{q} a_i y_{k-i} + \sum_{i=1}^{q} b_i u_{k-i} + \varepsilon_k, \tag{4}$$

where

$$a_i = C\bar{A}^{i-1}AK, b_i = C\bar{A}^{i-1}B, i = 1,2,\ldots,q. \tag{5}$$

The model described by equation (4) is the open-loop ARX model. q is the open-loop system ARX model order, and the ARX model parameters a_i and b_i can be estimated through least squares method. Suppose that there are N data points of y_k and u_k, $k = 0,1,\ldots,N-1$, are given. The batch least squares solution for estimating the parameters a_i and b_i is

$$\phi = YV^T(VV^T)^{-1} \tag{6}$$

where $Y = [y_0 \quad y_1 \quad \cdots \quad y_q \quad \cdots \quad y_{N-1}]$, $\phi = [b_1 \quad a_1 \quad \cdots \quad b_q \quad a_q]$

$$V = \begin{bmatrix} 0 & u_0 & \cdots & u_{q-1} & \cdots & u_{N-2} \\ 0 & y_0 & \cdots & y_{q-1} & & y_{N-2} \\ \cdots & \cdots & \cdots & \cdots & \cdots & \cdots \\ 0 & 0 & 0 & u_0 & \cdots & u_{N-q-1} \\ 0 & 0 & 0 & y_0 & \cdots & y_{N-q-1} \end{bmatrix} \tag{7}$$

The open-loop system Markov parameters $Y_s(k) = CA^{k-1}B$, and the open-loop Kalman filter Markov parameters $Y_k(k) = CA^{k-1}AK$, $k = 1,2,\ldots,q$ can be obtained from the coefficients a_i and b_i

$$Y_s(k) = b_k + \sum_{i=1}^{k} a_i Y_s(k-i),\ Y_k(k) = a_k + \sum_{i=1}^{k-1} a_i Y_k(k-i) \tag{8}$$

where $Y_s(0) = 0$ and $Y_k(0) = I$ which is the identity matrix.
Then, system matrices (A,B,C) and Kalman filter gain K can be realized by the Eigensystem Realization Algorithm (ERA) [6].

2.2. ARMAX Model

A modified model structure is formulated to reduce the requirement of model order. To add and subtract the term My_k to the right-hand side of the state equation in equation (2) to yield

$$\begin{aligned}\hat{x}_{k+1} &= A\hat{x}_k + Bu_k + AK\varepsilon_k + My_k - My_k = (A+MC)\hat{x}_k + (AK+M)\varepsilon_k + Bu_k - My_k \\ y_k &= C\hat{x}_k + \varepsilon_k\end{aligned} \tag{9}$$

Now the new relationship of reference input and output becomes

$$y_k = \sum_{i=1}^{\infty} C\tilde{A}^{i-1}(-M)y_{k-i} + \sum_{i=1}^{\infty} C\tilde{A}^{i-1}Bu + \sum_{i=1}^{\infty} C\tilde{A}^{i-1}\overline{M}\varepsilon_{k-i} + \varepsilon_k \tag{10}$$

where $\tilde{A} = A + MC$, $\overline{M} = M + AK$ (11)

Make $\tilde{A}$ asymptotically stable, $\tilde{A}^{i-1} \approx 0$ if $i \geq p$ for a sufficient large number p, equation (10) becomes

$$y_k = \sum_{i=1}^{p} h_i y_{k-i} + \sum_{i=1}^{p} t_i u_{k-i} + \sum_{i=1}^{p} s_i \varepsilon_{k-i} + \varepsilon_k \tag{12}$$

where $h_i = C\tilde{A}^{i-1}(-M)$, $t_i = C\tilde{A}^{i-1}B$, $s_i = C\tilde{A}^{i-1}\overline{M}$, $i = 1,2,\ldots,p$ (13)

Equation (12) is an ARMAX model containing the dynamics of residual which is different from equation (4).

Define $\theta = \begin{bmatrix} t_1 & h_1 & \ldots & t_p & h_p \end{bmatrix}$, $\Psi = \begin{bmatrix} s_1 & s_2 & \ldots & s_p \end{bmatrix}$,

$$W = \begin{bmatrix} 0 & \varepsilon_0 & \varepsilon_1 & \cdots & \varepsilon_{p-1} & \cdots & \varepsilon_{N-2} \\ 0 & 0 & \varepsilon_0 & \cdots & \varepsilon_{p-2} & \cdots & \varepsilon_{N-3} \\ \cdots & \cdots & \cdots & & \cdots & \cdots & \cdots \\ 0 & 0 & 0 & \cdots & \varepsilon_o & \cdots & \varepsilon_{N-p-1} \end{bmatrix}, \quad R = \begin{bmatrix} \varepsilon_o & \varepsilon_1 & \cdots & \varepsilon_{N-1} \end{bmatrix},$$

where Y and V are shown in equation (7). Then equation (12) could be written in a matrix form

$$Y = \theta V + \Psi W + R \tag{14}$$

The least squares solution of the parameters $\hat{\theta}$ and $\hat{\Psi}$ that minimizes the cost function $J = \sum_{k=0}^{N-1} \varepsilon_k^T \varepsilon_k = tr(RR^T)$ is

$$\begin{bmatrix} \hat{\theta} & \hat{\Psi} \end{bmatrix} = Y \begin{bmatrix} V \\ W \end{bmatrix}^{+} = Y \begin{bmatrix} V^T & W^T \end{bmatrix} \cdot \begin{bmatrix} VV^T & VW^T \\ WV^T & WW^T \end{bmatrix}^{-1}$$

The estimated parameters could also be expressed as

$$\hat{\theta} = Y\Lambda - \hat{\Psi}W\Lambda, \quad \hat{\Psi} = Y(I - \Lambda V)W^T\lambda^{-1} \tag{15}$$

where $\Lambda = V^T(VV^T)^{-1}$, $\lambda = W(I - \Lambda V)W^T$.

Examining the first term of $\hat{\theta}$ in equation (15) which is the ordinary least squares solution in open loop system; and in the second term one may consider it as a bias term. One can write, hence,

$$\hat{\theta} = \theta^{LS} - \theta^{bias} \tag{16}$$

where $\theta^{LS} = Y\Lambda = YV^T(VV^T)^{-1}$, $\theta^{bias} = \hat{\Psi}W\Lambda = \hat{\Psi}WV^T(VV^T)^{-1}$.

The parameter matrix, $\hat{\Psi}$, in Equation (15) could also be written as

$$\begin{aligned}\hat{\Psi} &= Y(I - \Lambda V)W^T\lambda^{-1} \\ &= (Y - \theta^{LS}V)W^T[WW^T - WV^T(VV^T)^{-1}VW^T]^{-1} \\ &= \hat{e}^{LS}W^T[WW^T - WV^T(VV^T)^{-1}VW^T]^{-1}\end{aligned} \tag{17}$$

where $\hat{e}^{LS}$ denotes the colored residual between the measurement Y and the least squares estimation $\theta^{LS}V$.

2.3. Iterative Procedure for Identification

Step 1. An initial estimate of the parameter θ, denoted by $\hat{\theta}^{LS}$, is computed from the ordinary least squares solution. It is assumed that an estimate of the parameter matrix, Ψ, denoted by $\hat{\Psi}_0$ is zero.

$$\hat{\theta}^{LS} = YV^T(VV^T)^{-1},\ \hat{\Psi}_0 = 0.$$

Step 2. Calculate the colored residual sequence $\hat{e}^{LS}$ which corresponds to the initial estimate ARMAX model parameter $\hat{\theta}^{LS}$. $\hat{e}^{LS} = Y - \hat{\theta}^{LS}V$. This is also an initial estimate of the white residual sequence denoted by $\hat{e}_0 = \hat{R}_0$.

Step 3. Construct the residual matrix $\hat{W}_0$ and then the parameter $\hat{\Psi}$ could be updated by equation (17)

$$\hat{\Psi}_1 = \hat{e}^{LS}\hat{W}_0^T[\hat{W}_0\hat{W}_0^T - \hat{W}_0V^T(VV^T)^{-1}V\hat{W}_0^T]^{-1}$$

The updated parameter $\hat{\Psi}_1$ is used to correct the initial bias as follows

$$\hat{\theta}_1^{bias} = \hat{\Psi}_1\hat{W}_0\Lambda = \hat{\Psi}_1\hat{W}_0V^T(VV^T)^{-1},\ \hat{\theta}_1 = \hat{\theta}^{LS} - \hat{\theta}_1^{bias}$$

Step 4. Compute the new whitened residual sequence $\hat{R}_1$ by using the estimated parameters $\hat{\theta}_1$ and $\hat{\Psi}_1$ as follows: $\hat{R}_1 = Y - \hat{\theta}_1V - \hat{\Psi}_1\hat{W}_0$

Step 5. Iterate the procedure from step 3 to step 4 by generating the new residual matrix $\hat{W}_1$ and using the updated parameters $\hat{\theta}_1$ and $\hat{\Psi}_1$. The next cycle is to calculate $\hat{\Psi}_2, \hat{\theta}_2^{bias}$, *and* $\hat{\theta}_2$.

After having obtained the ARMAX model estimated parameters θ and Ψ [7], one can use the estimated coefficients to construct the open-loop system Markov parameter $Y_s(k) = CA^{k-1}B$, Kalman filter Markov parameters $Y_k(k) = CA^{k-1}AK$, and $Y_m(k) = CA^{k-1}M$, $k = 1,2,\ldots,p,p+1,\ldots$ as follows:

$$Y_s(k) = t_k + \sum_{i=1}^{k} h_iY_s(k-i),\ Y_m(k) = -h_k + \sum_{i=1}^{k-1} h_iY_m(k-i),$$

$$Y_k(k) = h_k + s_k + \sum_{i=1}^{k-1} h_iY_k(k-i)$$

$\{A \quad B \quad C \quad M \quad K\}$ can also be realized by the Eigensystem Realization Algorithm (ERA) [6].

3. Experimental Setup

The experimental apparatus used in this study consists of a 3-hp, 1800-rpm (i.e. *fr* = 30 Hz), 4-pole three phase induction machine driving a 5-kW DC generator via a flexible coupling, as shown in figure 1. The generator is used to absorb the energy generated by the motor. A piezo-electric accelerometer is mounted on the housing of the induction electrical machine to measure the vibration signal. Three current sensors were used to record the three phase current signals from the inverter. All measured signals were sampled at 6 kHz via a real-time data acquisition device (Type NI 6062E) under the motor no-load condition. The frequency of inverter varies from 1 Hz to 60 Hz.

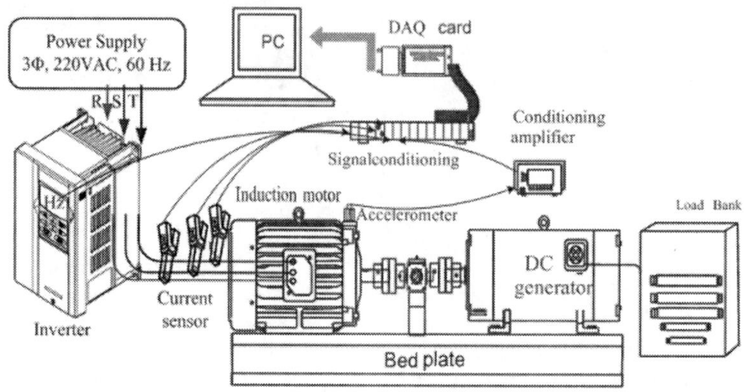

Figure 1. Experimental setup.

4. Numerical Validation

The inputs consist of three currents into the motor and the output is the vibration signal from the accelerometer. Each data set contains 122,880 points. Two sets of input-output data are recorded, one set is implemented for system identification, and the other is used for comparison between the predicted outputs and the true one. Figure 2 shows new test data (red) and predicted outputs of ARX (blue) and ARMAX (green) from 8,000 to 8,500 points. (Here, both identified models use model order=15, realized states=4 and the number of iterations for ARMAX=4.) Figure 3 shows the frequency plot of the real (blue) and predicted outputs (ARX (blue) and ARMAX (green)) from Fourier Transform.

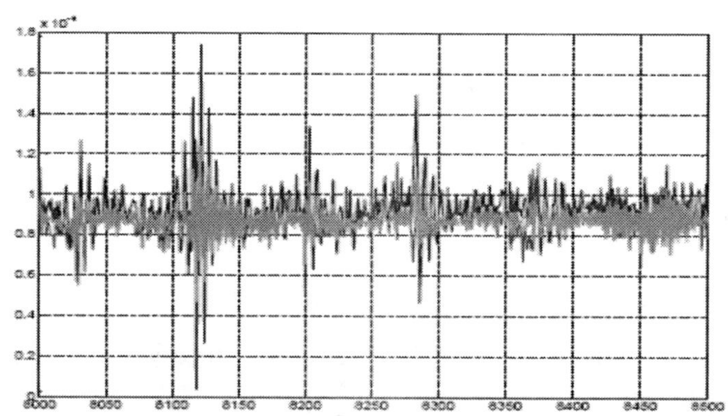

Figure 2. New test data (red), ARX (blue) and ARMAX (green) predicted outputs from 8,000 to 8,500 points.

IOP Conf. Series: Materials Science and Engineering **297** (2018) 012040 doi:10.1088/1757-899X/297/1/012040

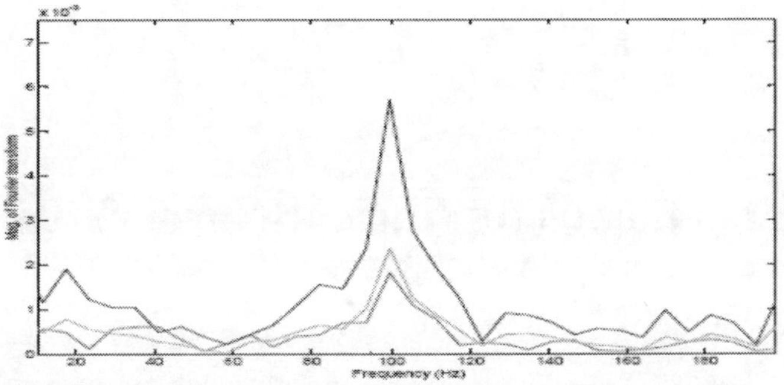

Figure 3. The frequency plot of the real (red), ARX (blue) and ARMAX (green) predicted outputs from Fourier Transform.

5. Conclusion

In this paper, both the ARX and ARMAX system models show an effectiveness approach for identifying the induction motor system. The ARMAX model has a better result compared with the ARX model when the model order is small; however, there is not much difference between these two when the model order is large enough. Actually, for the cases with high level of noise, the most significant error is from the original estimated parameters which have been corrupted by the disturbance. Hence, it is desirable to design a filter that can reduce the influence of disturbances so that the system identification will have a more accurate model.

References

[1] Shan P, Bhaskar B, Tang G, and Recht B 2012 *IEEE 51st Annual Conference on Decision and Control*

[2] Smith R 2012 *American Control Conference(Montreal)* p 2689

[3] Tavner P, Ran L, Penman J, and Sedding H 2008 *The Institution of Engineering and Technology*

[4] Vachtsevanos G, Lewis F L, Roemer M, Hess A and Wu B 2006 *Intelligent Fault Diagnosis and Prognosis for Engineering Systems.* (New Jersey: John Wiley & Sons)

[5] Chen C W, Juang J N, and Huang J K 1993 *Engineering Systems: Design and Optimization Techniques and Their Application* (vol 57) ed C T Leondes (New York: Academic Press) pp 331-368

[6] Juang J N 1994 *Applied System Identification* (New Jersey: PRT Prentice-Hall)

[7] Phan M., Horta L G, Juang J N, and Longman R W 1995 *Journal of Vibrations and Acoustics* **117** p 232

IOP Conf. Series: Materials Science and Engineering **297** (2018) 012041 doi:10.1088/1757-899X/297/1/012041

System identification using Nuclear Norm & Tabu Search optimization

Asif A. Ahmed, Marco P. Schoen*, and Ken W. Bosworth

Idaho State University, Department of Mechanical Engineering, 921 South 8th Avenue, Stop 8060, Pocatello, ID, 83209, USA

* Corresponding Author: schomarc@isu.edu

Abstract. In recent years, subspace System Identification (SI) algorithms have seen increased research, stemming from advanced minimization methods being applied to the Nuclear Norm (NN) approach in system identification. These minimization algorithms are based on hard computing methodologies. To the authors' knowledge, as of now, there has been no work reported that utilizes soft computing algorithms to address the minimization problem within the nuclear norm SI framework. A linear, time-invariant, discrete time system is used in this work as the basic model for characterizing a dynamical system to be identified. The main objective is to extract a mathematical model from collected experimental input-output data. Hankel matrices are constructed from experimental data, and the extended observability matrix is employed to define an estimated output of the system. This estimated output and the actual – measured – output are utilized to construct a minimization problem. An embedded rank measure assures minimum state realization outcomes. Current NN-SI algorithms employ hard computing algorithms for minimization. In this work, we propose a simple Tabu Search (TS) algorithm for minimization. TS algorithm based SI is compared with the iterative Alternating Direction Method of Multipliers (ADMM) line search optimization based NN-SI. For comparison, several different benchmark system identification problems are solved by both approaches. Results show improved performance of the proposed SI-TS algorithm compared to the NN-SI ADMM algorithm.

1. Introduction

In recent years, nuclear norm system identification methods have found a renewed interest in the research community due to advances in convex optimization heuristics. Nuclear norm by definition is the sum of singular values of a matrix. In the context of this paper, it involves estimating the system matrices *A, B, C* and *D* of a state-space realization from given input-output data [1]. The method is aimed at utilizing a discrete-time state-space model, as well as the properties of observability and controllability matrices to develop corresponding Hankel and Toeplitz matrices, respectively [1]. Amongst the benefits of this method is that of minimizing the nuclear norm of a given matrix and thereby minimizing its rank without losing the linear matrix structure in the low-rank approximation [3, 4]. Tabu Search optimization involves an iterative search – similarly to the Local Search algorithm – without its shortcomings. A single agent explores the cost surface and moves from its current to a more minimal solution iteratively [5]. Searching is performed within tabu balls of fixed respective radii, to find a number of solutions, from which the solution of least cost is deemed as a probable globally optimal solution [5].

2. Mathematical System Structure

In order to perform minimization, the following deterministic state-space system is used as the basis model [2]:

$$x(k+1) = Ax(k) + Bu(k) \tag{1}$$

$$y(k) = Cx(k) + Du(k) \tag{2}$$

In the above two equations, $x(k)$ represents state sequence, $u(k)$ represents input, $y(k)$ represents the output, and A, B, C, and D represent system matrices, while k is the time index. Hankel $H_{u,s}$ and Toeplitz $T_{u,s}$ matrices for the system are given in equations (3) and (4), respectively [2]:

$$H_{u,s} = \begin{pmatrix} 0 & \cdots & 0 & D \\ 0 & \cdots & D & CB \\ \vdots & \iddots & \vdots & \vdots \\ D & \cdots & CA^{s-3}B & CA^{s-2}B \end{pmatrix} \tag{3}$$

$$T_{u,s} = \begin{pmatrix} D & 0 & \cdots & 0 \\ CB & D & \cdots & 0 \\ \vdots & \vdots & \ddots & \vdots \\ CA^{s-2}B & CA^{s-3}B & \cdots & D \end{pmatrix} \tag{4}$$

Observability matrix O_s and controllability matrix Ω_s for the system of rank s are defined in equations (5) and (6), respectively [2]:

$$O_s = \begin{pmatrix} C \\ CA \\ CA^2 \\ \vdots \\ CA^{s-1} \end{pmatrix} \tag{5}$$

$$W_s = \begin{pmatrix} B & AB & A^2B & \cdots & A^{s-1}B \end{pmatrix} \tag{6}$$

3. Iterative Alternating Direction Method of Multipliers

Nuclear Norm minimization is applied through Iterative ADMM Method outlined in [2]. In this method, the cost function to be minimized is shown in equation (7):

$$\underset{y}{\text{minimize}} \quad \gamma \,||\, H(y)\, U^{\perp} \,||_{*} + \frac{1}{2} \,||\, y\text{-}b \,||^2 \tag{7}$$

In (7), γ is the regularization parameter, $H(y)$ is the output Hankel matrix, and $U^{\perp}$ is the orthogonal complement of the input Hankel matrix, $||\, H(y)\, U^{\perp} \,||_{*}$ is the nuclear norm of $H(y)\, U^{\perp}$, y is the output data matrix, while b is the measured data matrix, $||\, y\text{-}b \,||$ is the 2-norm of $y\text{-}b$ [2]. The cost function as given in equation (7) is constrained as shown in (8) and (9), with $Z = H(y)U^{\wedge}$ [2]:

$$\underset{y,\, Z}{\text{minimize}} \quad \gamma \,||Z||_{*} + \frac{1}{2} \,||\, y\text{-}b \,||^2 \tag{8}$$

$$\text{subject to } H(y)U^{\wedge} - Z = 0 \tag{9}$$

In the line search algorithm described in [1], augmented Lagrangian is used at each step of the ADMM iteration for obtaining the vector y that yields the minimum solution [2]. The augmented Lagrangian and its gradient are as given in equations (10) and (11) [2]. In these equations, y is the output data vector, L is the Lagrange multiplier, tr means trace of matrix, ^{T} represents the transpose of a matrix, ρ is a positive scalar, $_F$ represents Frobenius norm, H^{*} represents adjoint matrix containing the sum of entries on respective reverse diagonals of the given Hankel matrix, $H\left(y^{k+1,l}\right)$ is the Hankel matrix of $y^{k+1,l}$, k and l represent the outer loop iteration and inner loop iteration, respectively [2].

$$L_{\rho}(y,Z,L)=\gamma\|Z\|_{*}+\frac{1}{2}\|y-b\|^{2}+tr\left(L^{T}\left(H(y)U^{\wedge}-Z\right)\right)+\frac{\rho}{2}\|H(y)U^{\wedge}-Z\|_{F}^{2} \tag{10}$$

$$\nabla_{y}L_{\rho}\left(y^{k+1,l}\right)=\rho H^{*}\left(\left(H\left(y^{k+1,l}\right)U^{\wedge}-\left(Z^{k}-\left(\rho^{-1}\right)L^{k}\right)\right)\left(U^{\wedge}\right)^{T}\right)+\left(y^{k+1,l}-b\right) \tag{11}$$

The step size is computed at each iteration using values from present and past iterations employing the Barzilai Borwein method as given by equation (12) [2]. In this formula, x^{l} represents the current output, x^{l-1} represents the output from the previous iteration, ∇f^{l} represents the current gradient, ∇f^{l-1} represents the gradient from the previous iteration, and $<x^{l}-x^{l-1},\ \nabla f^{l}-\nabla f^{l-1}>$ represents the dot product of $x^{l}-x^{l-1}$ and $\nabla f^{l}-\nabla f^{l-1}$ [2]. Also, define

$$\alpha^{(l)}=\frac{<x^{l}-x^{l-1},\ \nabla f^{l}-\nabla f^{l-1}>}{<\nabla f^{l}-\nabla f^{l-1},\ \nabla f^{l}-\nabla f^{l-1}>}. \tag{12}$$

The use of the gradient descent method for performing a line search involves the use of matrix D^{l}, which in this work is set to an identity matrix. The following are the steps for the Iterative ADMM method, [2]:

- Preliminary Step: Assign initial values for y, Z, Λ, and assign ρ to be a positive scalar.
- Step 1: Iteratively calculate y using inner and outer line search routines as follows:
  Let $y^{(k+1,0)}=y^{k}$ and $y^{(k+1,-1)}=y^{k}$ (if applicable)
  In the case that there is no fixed termination criteria, compute the following:
  $$y^{(k+1,l+1)}=y^{(k+1,\ l)}-\alpha^{(l)}(D^{(l)})^{-1}\nabla_{y}L_{\rho}(y^{k+1,l}) \tag{13}$$
- Step 2: Perform singular-value decomposition and increment Z in the outer line search as shown in (14) and (15). In these equations, W, Σ, V^{T} matrices are the result of performing singular value decomposition on the right-hand side of equation (14). I in equation (15) is the identity matrix.
  $$W\Sigma V^{T}=H\left(y^{k+1}\right)U^{\wedge}+(\rho)^{-1}L^{k} \tag{14}$$
  $$Z^{k+1}=W\,max\{\Sigma-\frac{\gamma}{\rho}I,0\}V^{T} \tag{15}$$
- Step 3: The Lagrange multiplier is incremented in the outer line search as follows:
  $$L^{k+1}=L^{k}+\rho(H\left(y^{k+1}\right)U^{\wedge}-Z^{k+1}) \tag{16}$$

4. Tabu Search Method

Tabu Search Optimization aims at finding the best possible solution through iteratively searching and moving from a current solution to a better solution. Through the search, paths taken are stored in the tabu list. With this list, directions for new moves are determined that would yield a better local optimum solution without cycling, in turn ultimately leading to the probable globally optimal solution. Tabu Search can be broken down into the following steps [5, 6]:

Step1: Choose a random initial solution *So* from a search area having an n-dimensional radius R, where n is the dimension of the problem. Solution *So* will be the current local optimum solution.

Step 2: By random choice, N new solutions are selected through moving the current solution around *So* in the search space. Set $S_{1}(r)$ comprises N solutions.

Step 3: Utilize the objective function to evaluate the cost value of each member in $S_{1}(r)$. From these, choose the best solution which is the one with the minimum cost.

Step 4: Compare the cost of the best solution obtained in step (3) to the cost of the solution obtained in step (1).

Step 5: If the cost of the best solution from step (3) is greater than or equal to the cost of the solution from step (1), then leave *So* unchanged. On the other hand, if the cost of the solution from step (3) is less than the cost of the solution from step (1), store the best solution from step (3) in *So*. The solution with the minimum cost is the best solution.

Step 6: Store the best solution from step (5) in the tabu list.

Step 7: Repeat the process from step (2) through step (6) until the maximum number of iteration is reached, or if the global minimum is obtained.

The TS algorithm also incorporates a diversification and intensification function by incorporating a promising list – similar to the Tabu list – storing a set number of recent solutions. The Tabu list is used to avoid cycling and hence drive the search to remote places in the search field. The promising list is utilized in a second search, where individual items in the list are further explored. Each item on either list comprises an n-dimensional sphere with a constant radius.

5. Eigen System Realization

Eigensystem Realization Algorithm (ERA) is utilized in order to identify the system using estimated output from the two estimation techniques discussed in Sections III and IV [7]. This algorithm aims to use the discrete impulse response h for constructing a square block Hankel matrix $H(0)$ as shown in (17) [7, 8]. A second square block Hankel matrix $H(1)$ is obtained by shifting every column in the original block Hankel matrix $H(0)$ by one unit to the left. Singular-value decomposition is performed on $H(0)$, the result of which are the matrices P_n, D_n, and Q_n^T. From the results of singular-value decomposition, a matrix $D_{nn} = D_n(1:n,1:n)$ is obtained, where n represents the order of the system. Observability and Controllability matrices are calculated by using (18) and (19) with respect to the order of the system n. System matrices *A, B, C,* and *D* are calculated through the use of (20) through (23). Once system matrices are known, it is possible to calculate a discrete-time or a continuous-time transfer function [7].

$$H(0) = \begin{pmatrix} h(1) & h(2) & \cdots & h(n) \\ h(2) & h(3) & \cdots & h(n+1) \\ \vdots & \vdots & \iddots & \vdots \\ h(n) & h(n+1) & \cdots & h(2n-1) \end{pmatrix} \tag{17}$$

$$W_o = P_n(:,1:n)*(D_{nn}^{0.5}) \tag{18}$$

$$C_o = (D_{nn}^{0.5})*(Q_n^T(:,1:n))^T \tag{19}$$

$$A = (D_{nn}^{-0.5})*(P_n(:,1:n))^T * H(1)*(Q_n^T(:,1:n))*(D_{nn}^{-0.5}) \tag{20}$$

$$B = C_0(:,1) \tag{21}$$

$$C = W_0(1,:) \tag{22}$$

$$D = h(1) \tag{23}$$

6. Benchmark Simulated Systems

A system associated with a benchmark research problem that was chosen for applying and comparing the two system identification techniques. They yielded estimation results as shown in (24), (25), (26), table 1, figure 1, and figure 2 [9]. Equation (24) represents discrete-time transfer function of the true system, (25) is the estimated discrete-time transfer function from the Iterative ADMM method, and (26) is the estimated discrete-time transfer function using the Tabu Search method. Table 1 represents key system identification metrics; and simulation plots for the Iterative ADMM method and the Tabu Search method are given in figure 1 and figure 2, respectively.

System 1 (P. Dreesen, M. Ishteva, and J. Schoukens) [9]:

$$G\_true(z)=\frac{8.4853z^2+25.4558z+16.9706}{z^3-1.5z^2+z-0.25} \tag{24}$$

$$G\_ItADMM(z)=\frac{0.0004246z^2+14.16z+13}{z^3-1.682z^2+1.19z-0.3228} \tag{25}$$

$$G\_TabuSearch(z)=\frac{2.203e\text{-}05z^2+38.14z+7.02}{z^3-1.548z^2+1.049z-0.2695} \tag{26}$$

Table 1. Estimation & Simulation Results.

Metric	System Identification Results		
	Iterative ADMM	Tabu Search	True System
Iterations	201	100	-
CPU Time	0.8775	128.7338	-
Gamma Value	427.3409	427.3409	-
Error	Identification Error=42.1145%	Percent Error=0%	-
A Eigenvalue 1	0.5445 + 0.4982i	0.5096 + 0.4997i	0.5000 + 0.5000i
A Eigenvalue 2	0.5445 – 0.4982i	0.5096 - 0.4997i	0.5000 - 0.5000i
A Eigenvalue 3	0.5927 + 0.0000i	0.5289 + 0.0000i	0.5000 + 0.0000i

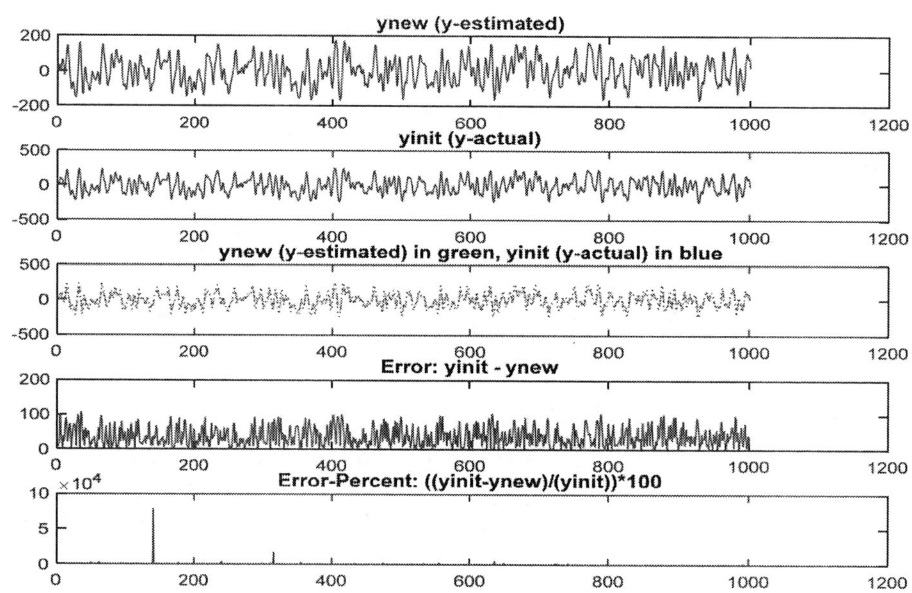

Figure 1. Iterative ADMM Estimation & True System Compared

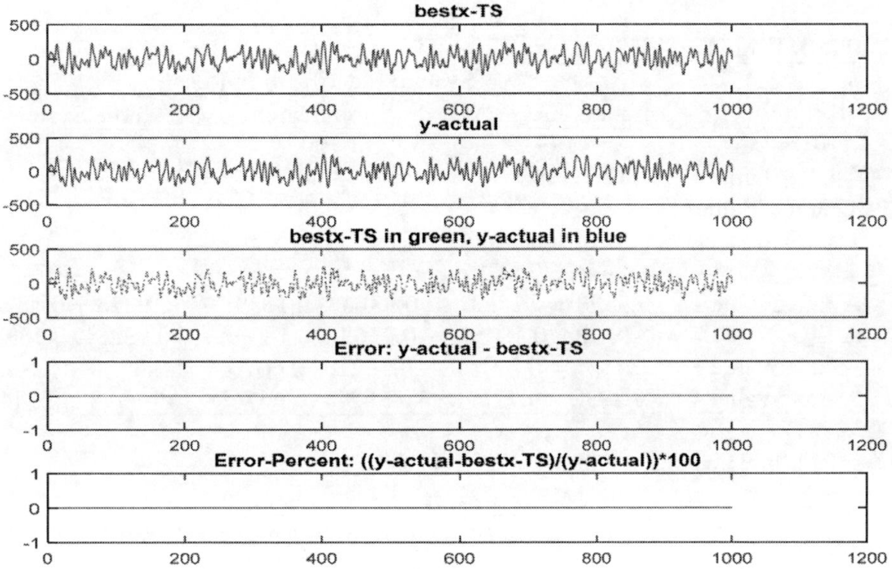

Figure 2. Tabu Search Estimation & True System Compared

Eleven additional benchmark systems are analyzed in a similar manner as System 1 from above [9 – 13]. Respective system identification results are presented as follows in (27) – (59), and tables 2 – 12 [9 – 13]:

System 2 (P. Dreesen, M. Ishteva, and J. Schoukens) [9]:

$$G\_true(z)=\frac{0.7071z+4.2426}{z^2+0.750z+0.125} \tag{27}$$

$$G\_ItADMM(z)=\frac{1.717}{z^2+0.7677z+6.903e\text{-}07} \tag{28}$$

$$G\_TabuSearch(z)=\frac{3.748}{z^2+0.7714z+0.1436} \tag{29}$$

Table 2. Estimation & Simulation Results.

METRIC	System Identification Results		
	Iterative ADMM	Tabu Search	True System
Iterations	151	100	-
CPU Time	0.3828	136.4569	-
Gamma Value	25	25	-
Error	Identification Error=58.0721%	Percent Error=0%	-
A Eigenvalue 1	-0.7677	-0.4576	-0.5
A Eigenvalue 2	0	-0.3138	-0.25

System 3 (A. Padilla, H. Garnier, and M. Gilson) [10]:

$$G\_true(z)=\frac{-0.7757z^3-1.367z^2+1.624z+0.6028}{z^4-1.198z^3+0.3238z^2-0.6383z+0.6015} \tag{30}$$

$$G\_ItADMM(z)=\frac{-0.000235z^3-1.821z^2+2.043z-0.1704}{z^4-1.493z^3+0.7493z^2-0.6952z+0.5122} \tag{31}$$

$$G\_TabuSearch(z)=\frac{-0.0001543z^3-2.381z^2+1.817z+0.7007}{z^4-1.148z^3+0.1312z^2-0.4208z+0.5251} \quad (32)$$

Table 3. Estimation & Simulation Results.

METRIC	System Identification Results		
	Iterative ADMM	Tabu Search	True System
Iterations	179	100	-
CPU Time	1.0068	173.6016	-
Gamma Value	8.9437	8.9437	-
Error	Identification Error=36.7502%	Percent Error=0%	-
A Eigenvalue 1	0.9398 + 0.1843i	0.9363 + 0.1839i	0.9328 + 0.1834i
A Eigenvalue 2	0.9398 – 0.1843i	0.9363 – 0.1839i	0.9328 – 0.1834i
A Eigenvalue 3	-0.1932 + 0.7219i	-0.3621 + 0.6676i	-0.3338 + 0.7444i
A Eigenvalue 4	-0.1932 – 0.7219i	-0.3621 – 0.6676i	-0.3338 – 0.7444i

System 4 (A. Padilla, H. Garnier, and M. Gilson) [10]:

$$G\_true(z)=\frac{-0.7526z^3-1.329z^2+1.58z+0.5876}{z^4-1.194z^3+0.3258z^2-0.6458z+0.6059} \quad (33)$$

$$G\_ItADMM(z)=\frac{-3.801e\text{-}06z^3-1.63z^2+0.6493z-0.9706}{z^4-0.08649z^3+0.3589z^2-0.2013z+0.2105} \quad (34)$$

$$G\_TabuSearch(z)=\frac{-0.000158z^3-2.31z^2+1.77z+0.6786}{z^4-1.145z^3+0.1359z^2-0.4308z+0.5303} \quad (35)$$

Table 4. Estimation & Simulation Results.

METRIC	System Identification Results		
	Iterative ADMM	Tabu Search	True System
Iterations	187	100	-
CPU Time	0.9257	135.1128	-
Gamma Value	17.6022	17.6022	-
Error	Identification Error=69.9599%	Percent Error=0%	-
A Eigenvalue 1	-0.4482 + 0.000i	0.9354 + 0.1865i	0.9319 + 0.1861i
A Eigenvalue 2	-0.0708 – 0.8303i	0.9354 – 0.1865i	0.9319 – 0.1861i
A Eigenvalue 3	-0.0708 - 0.8303i	-0.3626 + 0.6719i	-0.3349 + 0.7476i
A Eigenvalue 4	0.6764 + 0.0000i	-0.3626 - 0.6719i	-0.3349 – 0.7476i

System 5 (A. Padilla, H. Garnier, and M. Gilson) [10]:

$$G\_true(z)=\frac{0.05257z+0.04601}{z^2-1.637z+0.6703} \quad (36)$$

$$G\_ItADMM(z)=\frac{0.04767z^2-4.084e\text{-}05z+5.438e\text{-}05}{z^2-1.732z+0.7571} \quad (37)$$

$$G\_TabuSearch(z)=\frac{0.08385z+2.908e\text{-}05}{z^2-1.69z+0.7183} \quad (38)$$

Table 5. Estimation & Simulation Results.

METRIC	System Identification Results		
	Iterative ADMM	Tabu Search	True System
Iterations	149	100	-
CPU Time	0.7612	135.1274	-
Gamma Value	0.9346	0.9346	-
Error	Identification Error=43.3266%	Percent Error=0%	-
A Eigenvalue 1	0.8662 + 0.0820i	0.8449 + 0.0665i	0.8185 + 0.0189i
A Eigenvalue 2	0.8662 – 0.0820i	0.8449 – 0.0665i	-0.8185 – 0.0189i

System 6 (A. Y. K. Yong, A. H. Tan, and C. L. Cham) [11]:

$$G\_true(z)=\frac{z+0.5}{z^2-1.5z+0.7} \tag{39}$$

$$G\_ItADMM(z)=\frac{-2.751e\text{-}05z+0.1489}{z^2-1.58z+0.6947} \tag{40}$$

$$G\_TabuSearch(z)=\frac{-1.384e\text{-}05z+1.66}{z^2-1.411z+0.6503} \tag{41}$$

Table 6. Estimation & Simulation Results.

METRIC	System Identification Results		
	Iterative ADMM	Tabu Search	True System
Iterations	153	100	-
CPU Time	0.7766	136.7983	-
Gamma Value	75	75	-
Error	Identification Error=89.4514%	Percent Error=0%	-
A Eigenvalue 1	0.7898 + 0.2662i	0.7054 + 0.3907i	0.7500 + 0.3708i
A Eigenvalue 2	0.7898 - 0.2662i	0.7054 - 0.3907i	0.7500 - 0.3708i

System 7 (A. Y. K. Yong, A. H. Tan, and C. L. Cham) [11]:

$$G\_true(z)=\frac{0.0644z^2+0.0249z-0.0045}{z^3-2.451z^2+2.129z-0.6666} \tag{42}$$

$$G\_ItADMM(z)=\frac{-1.498e\text{-}07z^2+0.004713z+0.001619}{z^3-2.517z^2+2.202z-0.6771} \tag{43}$$

$$G\_TabuSearch(z)=\frac{1.727e\text{-}06z^2+0.1604z-0.08228}{z^3-2.43z^2+2.077z-0.6368} \tag{44}$$

Table 7. Estimation & Simulation Results.

METRIC	System Identification Results		
	Iterative ADMM	Tabu Search	True System
Iterations	125	100	-
CPU Time	0.5013	142.2242	-
Gamma Value	25	25	-
Error	Identification Error=90.3023%	Percent Error=0%	-
A Eigenvalue 1	0.9461 + 0.0000i	0.9438 + 0.0000i	0.9427 + 0.0000i
A Eigenvalue 2	0.7854 + 0.3142i	0.7431 + 0.3500i	0.7542 + 0.3720i
A Eigenvalue 3	0.7854 – 0.3142i	0.7431 - 0.3500i	0.7542 – 0.3720i

System 8 (A. Y. K. Yong, A. H. Tan, and C. L. Cham) [11]:

$$G\_true(z)=\frac{0.4479z^2+0.1254z-0.0076}{z^3-2.114z^2+1.622z-0.4319} \quad (45)$$

$$G\_ItADMM(z)=\frac{8.524e\text{-}06z^2+0.04927z+0.009867}{z^3-2.474z^2+2.156z-0.6554} \quad (46)$$

$$G\_TabuSearch(z)=\frac{-2.562e05z^2+0.9902z-0.7127}{z^3-2.3z^2+1.888z-0.5486} \quad (47)$$

Table 8. Estimation & Simulation Results.

METRIC	System Identification Results		
	Iterative ADMM	Tabu Search	True System
Iterations	159	100	-
CPU Time	0.7058	134.9639	-
Gamma Value	46.4	46.4	-
Error	Identification Error=87.1189%	Percent Error=0%	-
A Eigenvalue 1	0.8275 + 0.3400i	0.7574 + 0.3540i	0.7452 + 0.3705i
A Eigenvalue 2	0.8275 - 0.3400i	0.7574 - 0.3540i	0.7452 - 0.3705i
A Eigenvalue 3	0.8189 – 0.0000i	0.7849 + 0.0000i	0.6236 + 0.0000i

System 9 (H. Ase, and T. Katayama) [12]:

$$G\_true(z)=\frac{-0.03z+0.1}{z^2-1.8z+0.85} \quad (48)$$

$$G\_ItADMM(z)=\frac{-3.009e\text{-}06z+0.02164}{z^2-1.791z+0.8437} \quad (49)$$

$$G\_TabuSearch(z)=\frac{-1.278e\text{-}05z+0.06723}{z^2-1.811z+0.8585} \quad (50)$$

Table 9. Estimation & Simulation Results.

METRIC	System Identification Results		
	Iterative ADMM	Tabu Search	True System
Iterations	152	100	-
CPU Time	0.7088	145.6996	-
Gamma Value	7.02	7.02	-
Error	Identification Error=71.2422%	Percent Error=0%	-
A Eigenvalue 1	0.8956 + 0.2038i	0.9056 + 0.1959i	0.9000 + 0.2000i
A Eigenvalue 2	0.8956 – 0.2038i	0.9056 - 0.1959i	0.9000 - 0.2000i

System 10 (H. Ase, and T. Katayama) [12]:

$$G\_true(z)=\frac{0.06875z+0.13775}{z^5-2.34z^4+3.08z^3-2.53z^2+1.24z-0.368} \quad (51)$$

$$G\_ItADMM(z)=\frac{\left[\begin{array}{l}-2.547e\text{-}06z^4-0.01145z^3+0.09295z^2\\-0.1272z+0.1945\end{array}\right]}{\left[\begin{array}{l}z^5-2.112z^4+2.747z^3\\-2.287z^2+1.149z-0.4002\end{array}\right]} \quad (52)$$

$$G\_TabuSearch(z)=\frac{\begin{bmatrix}2.052e-09z^4-1.44e-09z^3\\+4.646e-09z^2+0.06875z+0.1378\end{bmatrix}}{\begin{bmatrix}z^5-2.34z^4+3.08z^3\\-2.53z^2+1.24z-0.368\end{bmatrix}} \tag{53}$$

Table 10. Estimation & Simulation Results.

METRIC	System Identification Results		
	Iterative ADMM	Tabu Search	True System
Iterations	144	100	-
CPU Time	0.7452	133.8792	-
Gamma Value	4.4725	4.4725	-
Error	Identification Error=51.4888%	Percent Error=0%	-
A Eigenvalue 1	0.9131 + 0.0000i	0.9012 + 0.0000i	0.9012 + 0.0000i
A Eigenvalue 2	0.2068 + 0.7576i	0.3037 + 0.7491i	0.3037 + 0.7491i
A Eigenvalue 3	0.2068 - 0.7576i	0.3037 - 0.7491i	0.3037 - 0.7491i
A Eigenvalue 4	0.3926 + 0.7460i	0.4157 + 0.6724i	0.4157 + 0.6724i
A Eigenvalue 5	0.3926 - 0.7460i	0.4157 - 0.6724i	0.4157 - 0.6724i

System 11 (M. Galrinho, C. R. Rojas, and H. Hjalmarsson) [13]:

$$G\_true(z)=\frac{1}{z^2-0.2z+0.5} \tag{54}$$

$$G\_ItADMM(z)=\frac{-1.43e\text{-}07z+0.6673}{z^2-0.5887z+0.4752} \tag{55}$$

$$G\_TabuSearch(z)=\frac{-1.623e\text{-}08z+1}{z^2-0.2z+0.5} \tag{56}$$

Table 11. Estimation & Simulation Results.

METRIC	System Identification Results		
	Iterative ADMM	Tabu Search	True System
Iterations	175	100	-
CPU Time	0.7678	159.8861	-
Gamma Value	1.677	1.677	-
Error	Identification Error=67.9169%	Percent Error=0%	-
A Eigenvalue 1	0.2944 + 0.6233i	0.1000 + 0.7000i	0.1000 + 0.7000i
A Eigenvalue 2	0.2944 – 0.6233i	0.1000 - 0.7000i	0.1000 - 0.7000i

System 12 (A. Padilla, H. Garnier, and M. Gilson) [10]:

$$G\_true(z)=\frac{-0.7691z^3-1.359z^2+1.616z+0.6008}{z^4-1.2z^3+0.3309z^2-0.6484z+0.6065} \tag{57}$$

$$G\_ItADMM(z)=\frac{-3.257e\text{-}06z^3-0.3884z^2+0.3775z+9.63e\text{-}06}{z^4-3\text{-}.022z^3+3.712z^2-2.255z+0.5847} \tag{58}$$

$$G\_TabuSearch(z)=\frac{-0.0001684z^3-2.366z^2+1.813z+0.6935}{z^4-1.151z^3+0.1404z^2-0.4324z+0.5304} \tag{59}$$

Table 12. Estimation & Simulation Results.

METRIC	System Identification Results		
	Iterative ADMM	Tabu Search	True System
Iterations	182	100	-
CPU Time	0.9399	134.3384	-
Gamma Value	23.351	23.351	-
Error	Identification Error=72.2935%	Percent Error=0%	-
A Eigenvalue 1	0.5818 + 0.5584i	-0.3611 + 0.6721i	-0.3333 + 0.7479i
A Eigenvalue 2	0.5818 – 0.5584i	-0.3611 – 0.6721i	-0.3333 – 0.7479i
A Eigenvalue 3	0.9292 + 0.1889i	0.9367 + 0.1838i	0.9333 + 0.1834i
A Eigenvalue 4	0.9292 - 0.1889i	0.9367 - 0.1838i	0.9333 - 0.1834i

7. Optimization Methods Compared

System identification techniques of Iterative ADMM presented in Section III, and Tabu Search presented in Section IV were applied, tested, simulated, analyzed, and compared for twelve systems from benchmark research problems, the results for which are presented in Section VI [9 – 13]. Key findings from the results are summarized as follows:

- Number of Iterations: Iterative ADMM uses more iterations than Tabu Search, due to Tabu Search method arriving at an optimal solution with minimum cost, and Iterative ADMM method not being able to arrive at fully optimal solution, thereby requiring more iterations for its computation.
- Computation Time: Iterative ADMM is faster, using less computation time than Tabu Search.
- Error: Iterative ADMM has certain spikes of very high percent error caused due to a non-optimal value of gamma being used. This could be improved by testing the Iterative ADMM method for a broader range of regularization parameter gamma, to arrive at the value that would yield optimal results [2]. Tabu Search, being fully optimized, is able to give zero percent error for all systems. Identification error for Iterative ADMM gives a measure of how close the estimation is to the actual output, and is calculated using (60) as follows:

$$e_I = 100 \times \frac{\left\| \hat{y} - \bar{y} \right\|_F}{\left\| \bar{y} \right\|_F} \tag{60}$$

- Eigenvalues of A Matrix: The extracted eigenvalues of the system matrix A of Iterative ADMM estimation are close to the true system A matrix eigenvalues for some systems, but not for all systems. That is caused by a non-optimal value of gamma being used, thereby decreasing the accuracy of discrete-time impulse response used for eigensystem realization to obtain system matrices. Eigenvalues of the A matrix of Tabu Search estimation are closer – but not exactly equal for all systems – to those of the true system. Slight inaccuracy in this case is caused by the chosen sample time and the number of samples used for obtaining discrete-time impulse response and eigensystem realization for obtaining system matrices.

8. Conclusions

Applications of two system identification techniques of Nuclear Norm and Tabu Search were investigated in this research. Key metrics show that Tabu Search minimization method yields optimal estimation results, while Iterative ADMM minimization method is able to perform system identification which matches the behavior of the true system to some degree of accuracy but not optimally. A perfect match in a stochastic system is usually a suspect; however, considering the finite data utilized in the algorithm, Tabu Search will find a model that exactly fits the finite set of input/output data with a given model. It does not mean that, for example, if a different realization of the input/output data is supplied, this "perfect" model will have 0% error, but should yield a reasonable result.

Future work in this research could involve further enhancement of solution for Iterative ADMM to perform estimation for a broad range of values for regularization parameter gamma while observing the identification error, in order to search for a value for gamma that could yield closer to optimum solution for a given system.

In this research, the gradient descent method, which uses D^l as an identity matrix is chosen in the development of the Iterative ADMM method involving inner and outer line search [2]. Other methods that could be investigated for development of Iterative ADMM solution include Newton and Quasi-Newton methods [2]. Newton's method involves using D^l as Hessian $\nabla^2 \mathcal{L}_\rho(y)$ [2]. Quasi-Newton methods use D^l which is an approximation to the Hessian, evaluated from information about gradient computed at present and past iterates [2]. For the Quasi-Newton approach, the BFGS formula for approximating inverse of Hessian is used [2]; the naming of which stems from the inventors of the formula: Broyden, Fletcher, Goldfarb, and Shanno [2]. Hessian inverse $\Pi^l = (D^l)^{-1}$ is approximated from (61) - (63) [2]. In these equations, $\Delta y^{k+1,l-1}$ is the change in y computed by evaluating difference between current inner line-search y and the y from previous inner line-search iteration [2]. $\Delta y^{k+1,l-1}$ is the change in g, calculated by subtracting gradient of Lagrangian at current inner line-search iteration from gradient of Lagrangian of previous inner line-search iteration [2]. I is the identity matrix, ^{T} means transpose of matrix, $\Pi^l$is the Hessian inverse at the current inner line-search iteration, and $\Pi^{l-1}$is the Hessian inverse from the previous inner line-search iteration [2].

$$\Delta y^{k+1,l-1} = y^{k+1,l} - y^{k+1,l-1} \tag{61}$$

$$\Delta g^{k+1,l-1} = \nabla_y \mathbf{L}_\rho(y^{k+1,l}) - \nabla_y \mathbf{L}_\rho(y^{k+1,l-1}) \tag{62}$$

$$\Pi^l = \left(I - \frac{\Delta y^{k+1,l-1} (\Delta g^{k+1,l-1})^T}{(\Delta g^{k+1,l-1})^T \Delta y^{k+1,l-1}}\right)' \Pi^{l-1} \left(I - \frac{\Delta g^{k+1,l-1} (\Delta y^{k+1,l-1})^T}{(\Delta g^{k+1,l-1})^T \Delta y^{k+1,l-1}}\right) + \frac{\Delta y^{k+1,l-1} (\Delta y^{k+1,l-1})^T}{(\Delta g^{k+1,l-1})^T \Delta y^{k+1,l-1}} \tag{63}$$

This research utilizes the Barzilai – Borwein method for evaluation of step size as represented in (12) [2]. Two other methods that could be researched that involve different methods for calculation of step-size are Backtracking line search method and Wolfe line search [2]. Backtracking line search is based on the requirement that step-size α^l meets the Armijo condition, also known as the sufficient decrease condition in (64) [2]. In this equation, x^l is the value of the output y at current inner line-search iteration, α^l is the step-size at current inner line-search iteration, d^l is the descent direction, c_1 is a constant as $c_1 \in (0,1)$, ∇f^l is the gradient evaluated at current inner line-search iteration, and ^{T} means transpose of matrix [2].

$$f(x^l + \alpha^l d^l) \le f(x^l) + c_1 \alpha^l (\nabla f^l)^T d^l \tag{64}$$

Wolfe line search is based on the requirement that step-size α^l meets the curvature condition and the Armijo condition [2]. Curvature condition is given in (65) [2].

$$(\nabla f(x^l + \alpha^l d^l))^T d^l \ge c_2 (\nabla f^l)^T d^l \tag{65}$$

Acknowledgment

This work is partially supported by the National Science Foundation under Grant No. NSF 1403688. The content is solely the author's responsibility. The support is greatly appreciated.

References

[1] Sadigh D, Ohlsson H, Sastry S S and Seshia S A 2014 *Proc. 19th IFAC World Congress* (Cape Town, South Africa) pp 9510-9515
[2] Deshmane H 2014 *System identification via nuclear norm regularization* M. S. thesis (Univ. of Minnesota, Minneapolis, MN)
[3] Fazel M 2002 *Matrix rank minimization with applications* Ph.D. dissertation (Stanford Univ., Stanford, CA)
[4] Liu Z, Hansson A and Vandenberghe L 2013 *Systems & Control Lett* **62**(8) pp 605-612
[5] Areerak K N, Srikaew A, Sujitjorn S and Totarong P 2002 *Proc. Intern. Conf. on Indus. Tech., IEEE ICIT,* (Bangkok, Thailand) pp 915-920
[6] Sujitjorn S, Kluabwang J, Puangdownreong D and Sarasiri N 2010 *ECTI Trans. on Elec. Eng., Electronics, and Communications.* **8**(1) pp 1-10
[7] Juang J.N and Suzuki H 1988 *ASME J. of Vibr. Acous, Stress, and Reliab. in Design* **110**(1) pp 24-29
[8] Xu W and Qiao S 2006 *Proc. 2nd Inter. Conf. on Struct. Matr.* (Kowloon Tong, Hong Kong) pp 550-563
[9] Dreesen P, Ishteva M and Schoukens J 2015 *Proc. 17th IFAC Symp. on System Identification* (Beijing, China) pp 951-956
[10] Padilla A, Garnier H and Gilson M 2015 *Proc. 17th IFAC Symp. on System Identification* (Beijing, China) pp 757-762
[11] Yong A Y K, Tan A H and Cham C L 2015 *Proc. 17th IFAC Symp. on System Identification* (Beijing, China) pp 939-944
[12] Ase H and Katayama T 2015 *Proc. 17th IFAC Symp. on System Identification* (Beijing, China) pp 638-643
[13] Galrinho M, Rojas C R and Hjalmarsson H 2015 *Proc. 17th IFAC Symp. on System Identification* (Beijing, China) pp 98-103
[14] Liu Z and Vandenberghe L 2010 *SIAM J. on Matrix Ana. and Applic.* **31**(3) pp 1235 – 1256
[15] Hansson A, Liu Z and Vandenberghe L 2012 *Proc. 2012 IEEE 51st IEEE Conf. on Decision and Control* (Maui, Hawaii) pp 3439-3444

IOP Conf. Series: Materials Science and Engineering **297** (2018) 012042 doi:10.1088/1757-899X/297/1/012042

A neural network based implementation of an MPC algorithm applied in the control systems of electromechanical plants

Piotr M. Marusak[1] and Suwat Kuntanapreeda[2,*]

[1] Institute of Control and Computation Engineering, Warsaw University of Technology, Nowowiejska 15/19, 00-665 Warsaw, Poland
[2] Department of Mechanical and Aerospace Engineering, King Mongkut's University of Technology North Bangkok, Bangkok 10800, Thailand

* Corresponding Author: suwat@kmutnb.ac.th

Abstract. The paper considers application of a neural network based implementation of a model predictive control (MPC) control algorithm to electromechanical plants. Properties of such control plants implicate that a relatively short sampling time should be used. However, in such a case, finding the control value numerically may be too time-consuming. Therefore, the current paper tests the solution based on transforming the MPC optimization problem into a set of differential equations whose solution is the same as that of the original optimization problem. This set of differential equations can be interpreted as a dynamic neural network. In such an approach, the constraints can be introduced into the optimization problem with relative ease. Moreover, the solution of the optimization problem can be obtained faster than when the standard numerical quadratic programming routine is used. However, a very careful tuning of the algorithm is needed to achieve this. A DC motor and an electrohydraulic actuator are taken as illustrative examples. The feasibility and effectiveness of the proposed approach are demonstrated through numerical simulations.

1. Introduction

Model predictive control (MPC) is widely used in many practical applications due to many advantages they offer [1–6]. MPC algorithms generate the control signal in such a way that future behavior of a control system, predicted many time instants ahead, fulfills assumed criteria. Usually, the algorithms solve an optimization problem with constraints to minimize the performance function at each iteration numerically in real time. However, the real-time optimization is time-consuming. Thus, it might not be suited for control systems of electromechanical plants in which the control sampling time is very short. Analytical MPC algorithms in which the control law is obtained offline might be considered as alternative solutions [7]. In such an approach, however, the algorithms should be appropriately formulated so the output constraints could be imposed on the whole prediction horizon.

Constraints in input and output variables are common in most physical control systems. The existence of constraints should be taken into account during design of the controller, otherwise the control performance may deteriorate significantly. One of the advantages of the MPC is that it allows taking the constraints into consideration quite easily.

Dynamic neural networks for solving constrained optimization have been developed in the past decades [8–10]. The neural network approach converts the optimization problem into an ordinary differential equation whose equilibrium point is the solution of the optimization problem. Based on the convergence property of the neural network, the solution will be obtained. It has been shown that the neural network approach has potential advantages over traditional numerical algorithms for real-time applications.

MPC using neural networks-based optimization has been recently investigated [11–15]. In [11], a neural network-based MPC scheme was presented for tracking of underactuated vessels. The scheme iteratively solves an optimization problem using a projection neural network. In [12], the MPC was investigated for the problem of missile interception. The constrained optimization problem was solved using a linear variational inequality-based primal-dual neural network. A discrete-time recurrent neural network for solving four-tank benchmark problem based on a MPC technique was investigated in [13]. Applications of the neural network-based optimization MPC to mobile robots were presented in [14, 15].

This paper considers applications of neural network-based optimization MPC to electromechanical plants. The rest of the paper is organized as follows. A review for the MPC is provided in the next section. The neural network based implementation of the MPC algorithm adopted in this paper is described in Section 3. Simulation results are presented in Section 4. The descriptions of the electromechanical plants that are used as illustrative examples are also provided in this section. The last section concludes the paper.

2. Model Predictive Control

The main concept of the MPC is using a model of the control plant to predict some future outputs and then calculating the control signal to optimize a predefined performance function.

Consider the following optimization problem:

$$\underset{\Delta \boldsymbol{u}}{\arg\min} \left\{ J_{MPC} = \sum_{i=1}^{p} \left(\bar{y}_k - y_{k+i|k}\right)^2 + \sum_{i=0}^{s-1} \lambda \cdot \left(\Delta u_{k+i|k}\right)^2 \right\} \tag{1}$$

$$\text{subject to} \quad \Delta \boldsymbol{u}_{min} \le \Delta \boldsymbol{u} \le \Delta \boldsymbol{u}_{max}, \; \boldsymbol{u}_{min} \le \boldsymbol{u} \le \boldsymbol{u}_{max}, \; \boldsymbol{f}_{min} \le \boldsymbol{F} \cdot \boldsymbol{y} \le \boldsymbol{f}_{max}$$

where $\bar{y}_k$ is the set-point value, $y_{k+i/k}$ is the output of the control plant at the $(k+i)^{\text{th}}$ sampling instant predicted at the kth sampling instant, $\Delta u_{k+i/k}$ is the future control signal increment, p and s denote prediction and control horizons, respectively, $\lambda \ge 0$ is the tuning parameter, $\Delta \boldsymbol{u} = \left[\Delta u_{k|k}, \ldots, \Delta u_{k+s-1|k}\right]^T$, $\boldsymbol{u} = \left[u_{k|k}, \ldots, u_{k+s-1|k}\right]^T$, $\boldsymbol{y} = \left[y_{k+1|k}, \ldots, y_{k+p|k}\right]^T$, $\boldsymbol{F}$ is the output constraint matrix, $\Delta \boldsymbol{u}_{\min}$, $\Delta \boldsymbol{u}_{\max}$, $\boldsymbol{u}_{\min}$, $\boldsymbol{u}_{\max}$, $\boldsymbol{f}_{\min}$, $\boldsymbol{f}_{\max}$ are vectors of lower and upper limits. As a solution to the optimization problem (1), the optimal vector of changes of the control signal is obtained. From this vector, the $\Delta u_{k|k}$ element is applied to the control plant and the optimization is repeated at the next sampling instant.

There are various MPC algorithms [1–3]. The way the predicted output values $y_{k+1|k}$ are derived depends on the dynamic control plant model that the predictive algorithm is based on. If the linear model is used, the superposition principle can be applied and the vector of predicted output values is given by:

$$\boldsymbol{y} = \tilde{\boldsymbol{y}} + \boldsymbol{A} \cdot \Delta \boldsymbol{u}, \tag{2}$$

where $\tilde{\boldsymbol{y}} = \left[\tilde{y}_{k+1|k}, \ldots, \tilde{y}_{k+p|k}\right]^{\mathrm{T}}$ is called the free response; $\boldsymbol{A}$ is called the dynamic matrix and is composed of the coefficients of the step responses of the control plant:

$$\boldsymbol{A} = \begin{bmatrix} a_1 & 0 & \cdots & 0 & 0 \\ a_2 & a_1 & \cdots & 0 & 0 \\ \vdots & \vdots & \ddots & \vdots & \vdots \\ a_p & a_{p-1} & \cdots & a_{p-s+2} & a_{p-s+1} \end{bmatrix}, \tag{3}$$

where a_i $(i = 1,\ldots,p)$ are step response coefficients of the control plant, p is the dynamics horizon equal to the number of sampling instants after which the coefficients of the step responses can be assumed as settled. The control plant model in the form of step responses is used in the Dynamic Matrix Control (DMC) algorithm–a popular MPC algorithm widely used in industrial applications.

The performance function from (1) can be expressed in the matrix-vector form as:

$$J_{MPC} = (\bar{y} - y)^T \cdot (\bar{y} - y) + \lambda \cdot \Delta u^T \cdot \Delta u . \tag{4}$$

After substituting (2) into (4), one obtains:

$$J_{MPC} = (\bar{y} - \tilde{y} - A \cdot \Delta u)^T \cdot (\bar{y} - \tilde{y} - A \cdot \Delta u) + \lambda \cdot \Delta u^T \cdot \Delta u , \tag{5}$$

where $\bar{y} = [\bar{y}_k,\ldots,\bar{y}_k]^{\mathrm{T}}$ is the vector of p elements being the current set-point value. After assuming that $\Lambda = \lambda \cdot I$, where I is the identity matrix, the performance function (5) can be expressed as:

$$J_{MPC} = \Delta u^T \cdot (A^T \cdot A + \Lambda) \cdot \Delta u - 2 \cdot (\bar{y} - \tilde{y})^T A \cdot \Delta u + (\bar{y} - \tilde{y})^T \cdot (\bar{y} - \tilde{y}) . \tag{6}$$

Since the free response depends only on the past, the performance function (6) is the quadratic function of the decision variables Δu. After transformation of the constraints, the problem (1) becomes:

$$\underset{\Delta u}{\arg\min}\ \Delta u^T \cdot (A^T \cdot A + \Lambda) \cdot \Delta u - 2 \cdot (\bar{y} - \tilde{y})^T A \cdot \Delta u + (\bar{y} - \tilde{y})^T \cdot (\bar{y} - \tilde{y}) \tag{7}$$

$$\text{subject to}\quad \Delta u_{min} \le \Delta u \le \Delta u_{max} , \tag{8}$$

$$u_{min} \le u_{k-1} + J \cdot \Delta u \le u_{max} , \tag{9}$$

$$f_{min} \le F \cdot (\tilde{y} + A \cdot \Delta u) \le f_{max} , \tag{10}$$

where $u_{k-1} = [u_{k-1},\ldots,u_{k-1}]^T$ is the vector of s elements equal to the control value from the last iteration,

$$J = \begin{bmatrix} 1 & 0 & \cdots & 0 \\ 1 & 1 & \cdots & 0 \\ \vdots & \vdots & \ddots & \vdots \\ 1 & 1 & \cdots & 1 \end{bmatrix}^T .$$

Note that all the constraints (8)–(10) depend linearly on the decision variables Δu. Thus, the problem (7)–(10) is the linear-quadratic optimization problem. Such a problem can be solved by using standard numerical routines. Note that, in the case when the sampling time is very short, finding the solution numerically may be too time-consuming.

In the case when output constraints (10) are taken into consideration in the algorithm, the problem of empty set of admissible solutions may occur. In such a case a mechanism of soft constraints may be applied. It consists of adding a variable $\varepsilon \ge 0$ into the problem (7)–(10) by means of which the output constraints are relaxed. Thus, the optimization problem will have the following form:

$$\underset{\Delta u, \varepsilon}{\arg\min}\ \Delta u^T \cdot (A^T \cdot A + \Lambda) \cdot \Delta u - 2 \cdot (\bar{y} - \tilde{y})^T A \cdot \Delta u + (\bar{y} - \tilde{y})^T \cdot (\bar{y} - \tilde{y}) + \gamma \cdot \varepsilon^2 \tag{11}$$

$$\text{subject to}\quad \Delta u_{min} \le \Delta u \le \Delta u_{max} , \tag{12}$$

$$u_{min} \le u_{k-1} + J \cdot \Delta u \le u_{max} , \tag{13}$$

$$f_{min} \le F \cdot (\tilde{y} + A \cdot \Delta u) + \varepsilon , \qquad \varepsilon = [\varepsilon,\ldots,\varepsilon]^T \tag{14}$$

$$F \cdot (\tilde{y} + A \cdot \Delta u) - \varepsilon \le f_{max} . \tag{15}$$

3. Neural Network Based Implementation of MPC Algorithm

The MPC algorithm used in the paper is implemented using a neural network approach to quadratic programming (QP) problem solving, proposed in [8]. The approach makes use of a piecewise formulation of the QP problem solved by the numerical MPC algorithm and the solution of which is generated by a recurrent neural network.

3.1. A piecewise formulation of the basic QP problem solved in numerical MPC algorithm

Let us rewrite the optimization problem (7)–(10) as:

$$\arg\min_{\Delta u} \tfrac{1}{2}\Delta u^T \cdot Q \cdot \Delta u + c \cdot \Delta u + (\bar{y}-\tilde{y})^T \cdot (\bar{y}-\tilde{y}) \qquad (16)$$

$$\text{subject to} \quad d \le E \cdot \Delta u \le h, \qquad (17)$$

where

$$Q = 2 \cdot (A^T \cdot A + \Lambda), \qquad (18a)$$

$$c = -2 \cdot (\bar{y}-\tilde{y})^T A, \qquad (19)$$

$$E = \begin{bmatrix} I \\ J \\ F \cdot A \end{bmatrix}, \quad d = \begin{bmatrix} \Delta u_{min} \\ u_{min} - u_{k-1} \\ f_{min} - F \cdot \tilde{y} \end{bmatrix}, \quad h = \begin{bmatrix} \Delta u_{max} \\ u_{max} - u_{k-1} \\ f_{max} - F \cdot \tilde{y} \end{bmatrix}. \qquad (20a)$$

Note that ***d*** and ***h*** define the lower and upper bounds, respectively.

As it is shown in [8], by using the Lagrange multipliers, the solution of the problem (16)–(20) is given by:

$$\Delta u^* = R \cdot x^* + a, \qquad (21a)$$

where $R = Q^{-1} \cdot E^T$, $a = -Q^{-1} \cdot c$, and x^* is the solution of the following piecewise equation:

$$W \cdot x + q = P_X(W \cdot x + q - x), \qquad (22)$$

where

$$W = E \cdot Q^{-1} \cdot E^T, \quad q = -E \cdot Q^{-1} \cdot c, \qquad (23)$$

$$P_X(x_i) = \begin{cases} d_i & \text{if } x_i < d_i \\ x_i & \text{if } d_i \le x_i \le h_i \\ h_i & \text{if } x_i > h_i \end{cases}, \qquad (24)$$

where x_i are the elements of the vector ***x***, d_i are elements of the vector ***d***, and h_i are elements of the vector ***h***.

3.2. A piecewise formulation of the QP problem with soft output constraints

In the case when the mechanism of soft output constraints is used, the following matrices and vectors change as:

$$Q = \begin{bmatrix} 2 \cdot (A^T \cdot A + \Lambda) & 0 \\ 0 & \varepsilon \end{bmatrix}, \qquad (18b)$$

$$E = \begin{bmatrix} I & 0 \\ J & 0 \\ F \cdot A & 1 \\ F \cdot A & -1 \\ 0 & 1 \end{bmatrix}, \quad d = \begin{bmatrix} \Delta u_{min} \\ u_{min} - u_{k-1} \\ f_{min} - F \cdot \tilde{y} \\ -\infty \\ 0 \end{bmatrix}, \quad h = \begin{bmatrix} \Delta u_{max} \\ u_{max} - u_{k-1} \\ \infty \\ f_{max} - F \cdot \tilde{y} \\ \infty \end{bmatrix}, \tag{20b}$$

where $0 = [0,\ldots,0]^T$, $1 = [1,\ldots,1]^T$, $-1 = [-1,\ldots,-1]^T$, $\infty = [\infty,\ldots,\infty]^T$, $-\infty = [-\infty,\ldots,-\infty]^T$ are the vectors of appropriate dimensions.

3.3. Obtaining the solution by means of a recurrent neural network

To find the solution x^* of the piecewise equation (22), a recurrent neural network (RNN) proposed in [8] will be utilized. The equation describing this RNN is as follows:

$$\dot{x} = \mu \cdot (P_X(W \cdot x + q - x) - W \cdot x - q), \tag{25}$$

where μ is a scaling constant, which must be set appropriately in order to obtain convergence. It is shown in [8] that the state trajectory of the proposed RNN is globally convergent to the solution x^* of the piecewise equation (22).

In the current work instead of the continuous-time RNN, a discrete-time RNN was used. Therefore, the state equation was transformed, using the discrete integrator, into:

$$x_{v+1} = x_v + \bar{\mu} \cdot (P_X(W \cdot x_v + q - x_v) - W \cdot x_v - q), \tag{26}$$

where v is the discrete iteration index and $\bar{\mu}$ is the parameter which must be tuned appropriately to obtain convergence of the RNN. Like in other iterative optimization routines, the iteration stops when some predefined accuracy criterion is met or the maximum iteration number is reached. After x^* is obtained, the optimum control signal increment Δu^* of the optimization problem (16)–(20) can be directly determined from (21a).

In the case when the mechanism of soft output constraints is used, the solution (21a) transforms to:

$$\begin{bmatrix} \Delta u^* \\ \varepsilon \end{bmatrix} = R \cdot x^* + a \cdot \tag{21b}$$

4. Numerical Results

Numerical simulations are conducted in this section to illustrate feasibility and effectiveness of the proposed approach. Two electromechanical plants are used as illustrative examples. They are a DC motor and an electrohydraulic linear actuator as shown in figures 1 and 2, respectively. DC motors are commonly used as illustrative control plants, allowing a reader to easily follow the proposed new ideas used in control systems of these plants [16–18]. Electrohydraulic actuators are extensively used in the applications that need large action forces. They serve as very efficient actuators because they possess a high force-to-weight ratio, fast response, high stiffness, and high load capability. Control of electrohydraulic actuators is challenging since their dynamics are highly nonlinear, thus limiting the performance of traditional linear controllers [18–20].

4.1. DC Motor

As shown in figure 1, the control plant is a permanent magnet DC motor driving an inertial load through a gear train. A speed control problem is considered here. The speed is directly measured by the tachometer-generator (TG). The speed signal (ω) is fed back to the computer through the interface board. The control signal (V_{in}) generated by the control algorithm implemented on the computer is converted to an analog signal by the interface board, and is transmitted to the power amplifier to drive the motor.

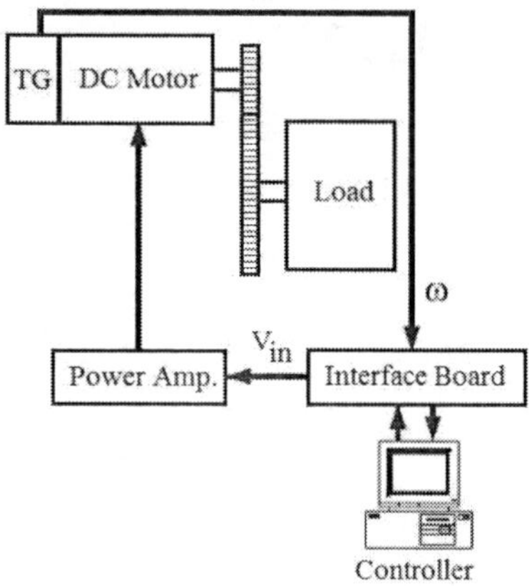

Figure 1. DC motor control plant.

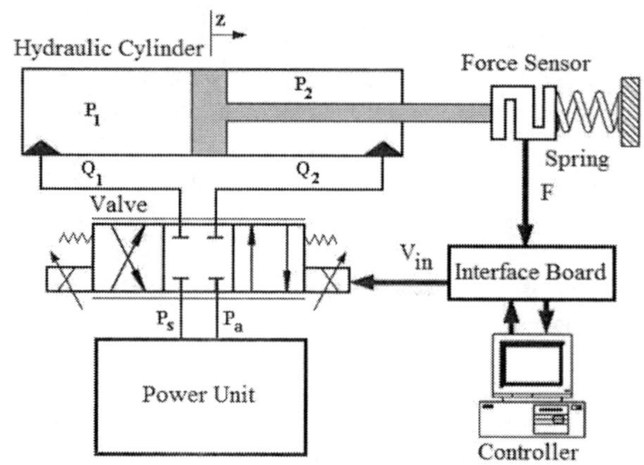

Figure 2. Electrohydraulic control plant.

4.1.1. Mathematical Model. The mathematical model of the DC motor used to simulate the control plant consists of the model of the electrical part and that of a mechanical moving part [16]. The model of the electrical part is given by:

$$L\cdot\frac{dI}{dt}+K_e\cdot\omega+R\cdot I=V_{in} \tag{27}$$

where I the armature current, R is the rotor resistance, L is the rotor inductance, and K_e is the back-emf constant. The mechanical moving part is described by:

$$J\cdot\frac{d\omega}{dt}+c\cdot\omega=K_T\cdot I-T_{dist} \tag{28}$$

where K_T is the torque constant, $J=J_m+N^2\cdot J_L$ is the inertia moment of the moving part, J_m is the rotor inertia moment, J_L is the load inertia moment, N is gear ratio, $c=c_m+N^2\cdot c_L$ is the damping constant of the moving part, c_m is the motor damping constant, c_L is the load damping constant and T_{dist} is the external torque disturbance. The values of the parameters of the control plant used in this study are given in table 1.

4.1.2. Controller Design. The design objective is to drive the motor such that its speed (ω) tracks a set-point signal (r). The sampling time (T_s) is 5 msec. Here, we selected the prediction horizon p = 37, the control horizon s = 2 and $\bar{\mu}$ = 1. The input u_k and output y_k of the control plant are V_{in} and ω, respectively.

Table 1. Parameters of the DC motor plant [18].

Parameter	Value	Parameter	Value
R	3.77 Ω	J_L	1.356e-3 kg-m^2
L	8.04 mH	c_m	1.41e-5 Nm/rad/s
K_e	0.1203 V/rad/s	c_L	7.06e-6 Nm/rad/s
K_T	0.1203 Nm/Amp	N	1/4
J_m	3.81e-5 kg-m^2		

4.1.3. Results. The set-point signal $r = 10$ rad/sec was used. The disturbance $T_{dist} = 0.01$ Nm was also applied at t $\geq$ 0.3 sec. Figure 3 shows the results when there was no constraint imposed. It is seen that the speed converges to the set-point very fast. Indeed, the design objective has been achieved. The controller was also able to successfully eliminate the effect of the disturbance. Note that the acceleration of the motor has very large amplitude that is close to 1000 at its maximum.

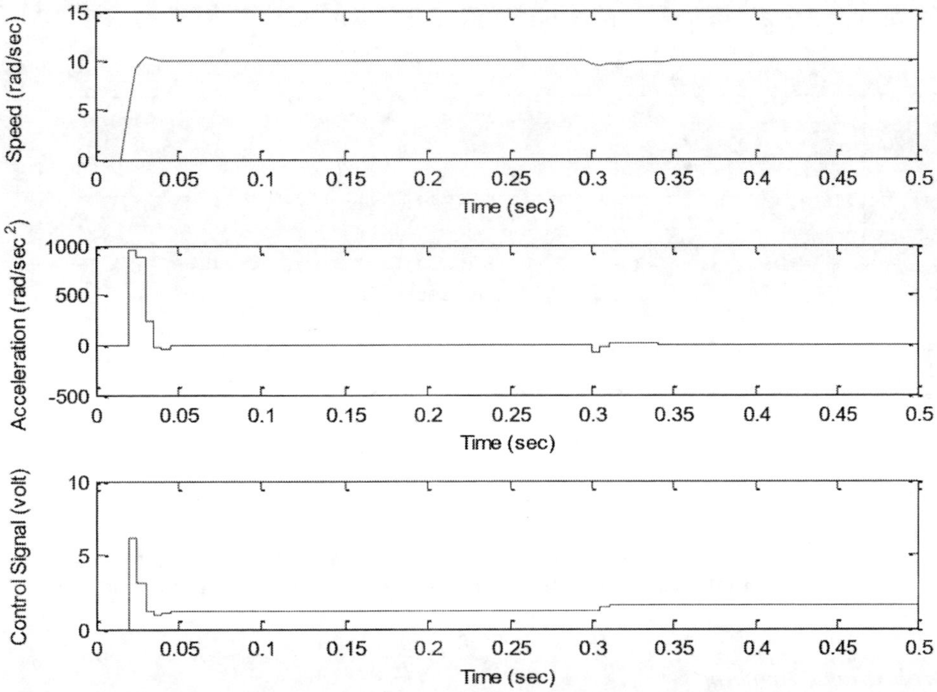

Figure 3. DC motor control results: without constraint.

Next, assume that the acceleration amplitude has limits at ±200 rad/sec^2 due to safety constraints. So, we set the output constraint as:

$$-200 \leq \Delta y_k = (y_k - y_{k-1}) / T_s \leq 200 \tag{29}$$

with $\gamma = 1e9$. Moreover, in order to simplify the task and consequently to reduce the computational load, the only one sample at k is imposed as the constraint in each iteration, instead of all the future samples. The control results with the acceleration constraints are shown in figure 4. The speed satisfactorily converges to the set-point, and the acceleration is approximately bounded by the limits as desired. Note that the convergence of the speed is slower than in the previous case as expected since the controller maintains the amplitude of the acceleration under the assumed limits.

Note that we also compared the results with those obtained with the MPC using a standard quadratic programming routine. It was found that the results are precisely the same provided that the appropriate accuracy of the solution is assumed. However, if one prefers faster generation of the control signal over accuracy (optimality of the solution), there is a possibility to obtain the demanded tradeoff. The MPC using the neural network based optimization may work faster (even much faster) than its standard counterpart. It depends, however, on parameters of the MPC algorithm (control horizon and number of constraints in the optimization problem) and on the parameters of the algorithm solving the optimization problem, like discrete iteration accuracy criterion and μ (or $\bar{\mu}$) parameter. In summary, a careful tuning of the algorithm is needed to obtain the algorithm which works fast. Otherwise, it may be slower than the algorithm employing the standard optimization routine.

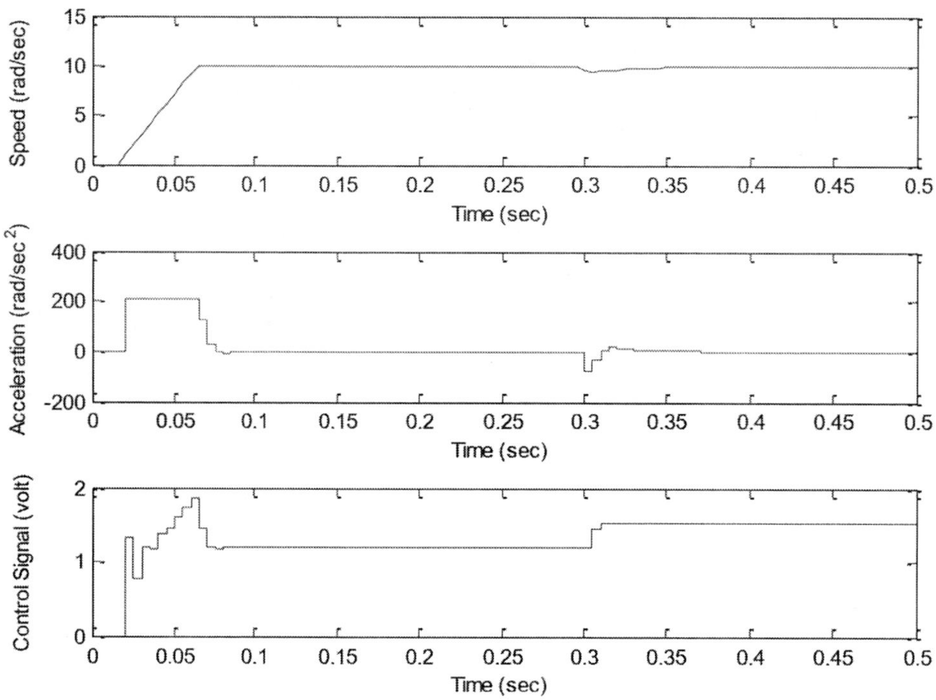

Figure 4. DC motor control results: with acceleration constraint consideration

4.2. Electrohydraulic Actuator

As shown in figure 2, the control plant consists of a hydraulic power unit, a servo valve, a single-rod hydraulic cylinder, a force sensor, an interface board, a PC computer, and a spring (representing the environment or load). A force control problem is considered here. The force (F) is fed back to the computer through the interface board. The control signal (V_{in}) generated by the control algorithm implemented on the computer is converted to an analog signal by the interface board, and is transmitted to the servo valve.

4.2.1. Mathematical Model. The dynamics of the actuator's moving part is given by

$$m_a \cdot \frac{d^2 z}{dt^2} = P_1 \cdot A_1 - P_2 \cdot A_2 - c_a \cdot \frac{dz}{dt} - F - F_{dist} \tag{30}$$

where z is the actuator's displacement, P_1 and P_2 are the pressures inside the extend and retract chambers of the cylinder, A_1 and A_2 are the piston areas facing the extend and retract chambers, and m_a and c_a are the mass of the moving part and equivalent viscous damping coefficient, respectively; F is the environment or load's force and F_{dist} is the external force disturbance. After neglecting the mass and stiffness of the force sensor, the force can be modeled as

$$F = k_s \cdot z \tag{31}$$

where k_s is the stiffness of the spring. The cylinder dynamics can be written as [19]

$$\frac{V_1}{\beta}\cdot\frac{dP_1}{dt}=-A_1\cdot\frac{dz}{dt}+Q_1 \text{ and } \frac{V_2}{\beta}\cdot\frac{dP_2}{dt}=A_2\cdot\frac{dz}{dt}-Q_2, \quad (32)$$

where $V_1 = V_{01} + A_1z$ and $V_2 = V_{02} - A_2z$ are the total volumes of the chambers, V_{01} and V_{02} are the extend and retract chamber volumes when $z = 0$, β is the effective bulk modulus of the hydraulic oil, and Q_1 and Q_2 are the flows. It is assumed that the servo valve has a linear flow gain characteristic. Thus, by neglecting the valve dynamics, the flows Q_1 and Q_2 can be related to the voltage input V_{in} by a linear mapping:

$$Q_1=\begin{cases}K\cdot V_{in}, & dz/dt\geq 0\\ \gamma\cdot K\cdot V_{in}, & dz/dt<0\end{cases} \text{ and } Q_2=\begin{cases}\frac{K}{\gamma}\cdot V_{in}, & dz/dt\geq 0\\ K\cdot V_{in}, & dz/dt<0\end{cases}, \quad (33)$$

where K is the flow/signal gain of the valve and $\gamma = A_1/A_2$ is the flow factor.

In this study, equations (30)–(33) were used as the mathematical model to simulate the control plant. The values of the parameters of the model are given in table 2.

4.2.2. Controller Design. The design objective is to drive the actuator so that its force (F) tracks a set-point signal (r). The sampling time (T_s) is 0.1 msec. Here, we selected the prediction horizon p = 400, the control horizon s =1 and $\bar{\mu}$ = 1e6. The input u_k and output y_k of the control plant are V_{in} and F, respectively.

4.2.3. Results. The set-point signal r = 800 N was chosen. The disturbance F_{dist} = 200 N was also applied at t ≥ 0.6 sec. Figure 5 shows the results when there is no constraint imposed. It shows that the design objective has been achieved. However, in real systems, there are limits on the control signal due to, for example, the saturation of the actuator. Here, we assume that the control signal has limits at ±10 V. By using the controller from the previous case, in which the limits are not considered in the design process, the controlled force has a significant overshoot as shown in figure 6.

Next, we re-designed the controller with consideration of the following constraints on the control signal:

$$-10\leq u_k\leq 10. \quad (34)$$

After taking the constraints into consideration in the controller, satisfactory results are obtained as shown in figure 7. The control signal is properly limited, and there is no significant overshoot in the response.

Table 2. Parameters of the electrohydraulic plant [18].

Parameter	Value	Parameter	Value
m_a	6 kg	k_s	1.206e5 N/m
c_a	500 N/ms^{-1}	K	1.026e–5 m^3/V
A_1	2.463e–3 m^2	V_{01}	3.079e–4 m^3
A_2	1.455e–3 m^2	V_{02}	1.82e–4 m^3
β	1.5e9 N/m^2		

IOP Conf. Series: Materials Science and Engineering **297** (2018) 012042 doi:10.1088/1757-899X/297/1/012042

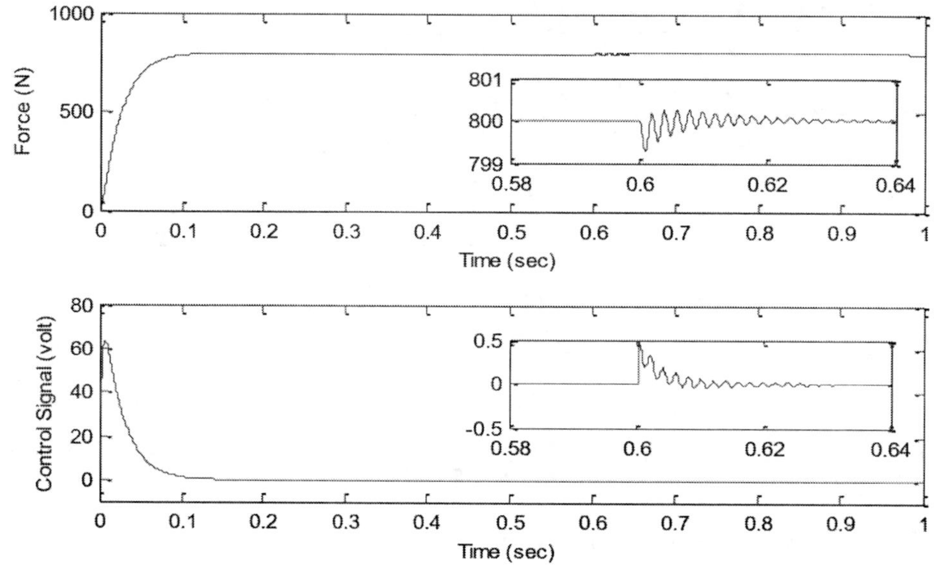

Figure 5. Electrohydraulic actuator control results: without constraint.

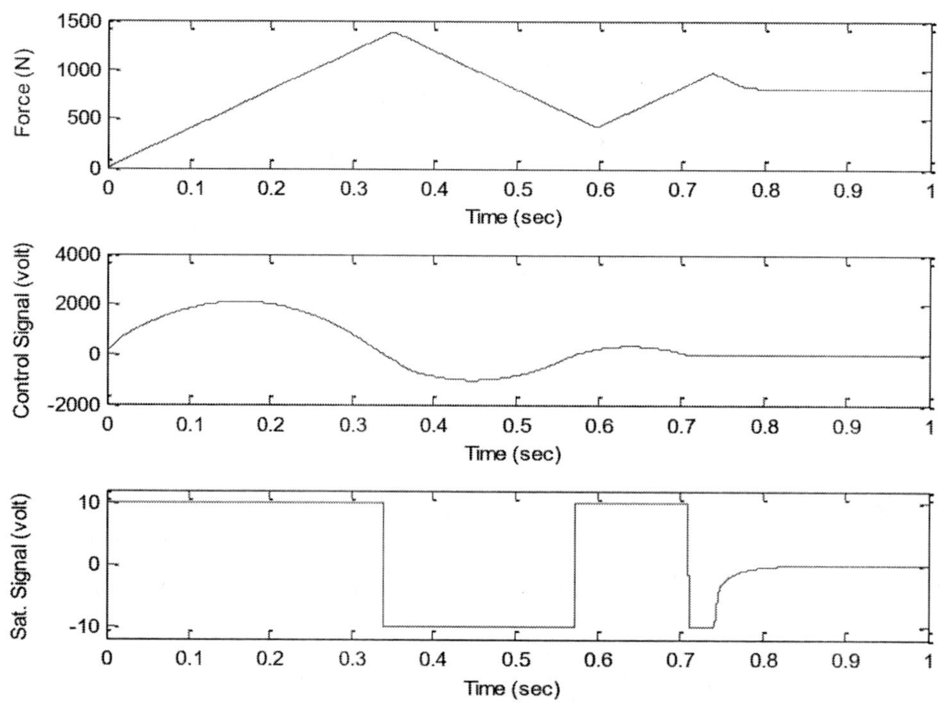

Figure 6. Electrohydraulic actuator control results: without constraint consideration.

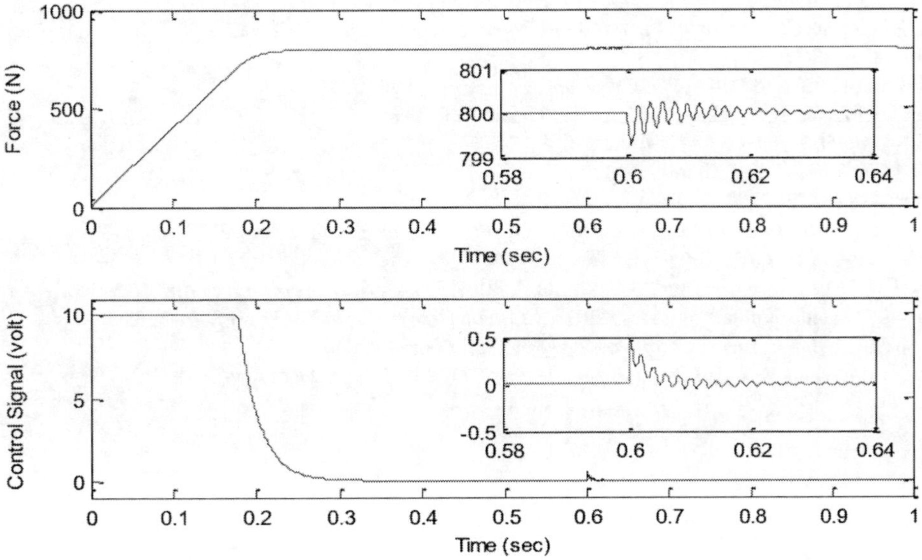

Figure 7. Electrohydraulic actuator control results: with constraint consideration.

5. Conclusions

This paper has presented results of tests of a neural network-based optimization MPC scheme for applications in control systems of electromechanical plants. The scheme is based on transforming the MPC optimization problem into a set of differential equations, which can be interpreted as a recurrent neural network. Based on its properties, the neural network, if it is convergent, is globally convergent to the solution of the original optimization problem. Speed control of a DC motor and force control of an electrohydraulic linear actuator were used as illustrative examples. For the DC motor, the MPC controller with imposed acceleration constraints successfully kept the acceleration within the limits. For the electrohydraulic actuator, the constraints were imposed on the control signal. Satisfactory results were obtained.

During the conducted experiments the following observations were made:

- the MPC using the neural network based optimization may generate the same results as the algorithm employing the standard quadratic programming routine (if the sufficiently high discrete iteration accuracy criterion is assumed), moreover
- the MPC using the neural network based optimization may work faster than its standard counterpart (offering smaller sampling times); however,
- it is very important to properly choose the parameters of the algorithm, μ (or $\overline{\mu}$) in particular. Otherwise, the algorithm may work much too slow or it can even diverge.

References

[1] Camacho E F and Bordons C 1999 *Model Predictive Control* (New York: Springer)
[2] Maciejowski J M 2002 *Predictive Control With Constraints* (Harlow: Prentice Hall)
[3] Tatjewski P 2007 *Advanced Control of Industrial Processes:Structures and Algorithms* (London: Springer)
[4] Yuan L, Zhao H, Chen H and Ren B 2016 *Mechatronics* **38** pp 1–15
[5] Chatrattanawet N, Hakhen T, Kheawhom S and Arpornwichanop A 2017 *J. Clean. Prod.* **148** pp 934–947
[6] Mirshams M and Khosrojerdi M 2017 *Acta Astronaut.* **132** pp 143–149
[7] Marusak P M and Kuntanapreeda S 2011 *Control Eng Pract.* **19** pp 62–73
[8] Xia Y, Feng G and Wang J 2004 *Neural Netw.* **17** pp 1003–1015

[9] Xia Y, Feng G and Wang J 2008 *IEEE Trans. Neural Netw.* **19**(8) pp 1340–1353
[10] Liu Q, Huang T and Wang J 2014 *IEEE Trans. Neural Netw. Learn. Syst.* **25**(7) pp 1308–1318
[11] Yan Z and Wang J 2012 *IEEE J. Ocean. Eng.* **37**(4) pp 717–726
[12] Li Z, Xia Y, Su C-Y, Deng J, Fu J and He W 2015 *IEEE Trans. Neural Netw. Learn. Syst.* **26**(8) pp 1803–1809
[13] Xu J, Li C, He X and Huang T 2016 *Neurocomputing* **190** pp 172–178
[14] Xiao H, Li Z, Yang C, Zhang L, Yuan P, Ding L and Wang T 2017 *IEEE Trans. Ind. Electron.* **64**(1) pp 505–516
[15] Ke F, Li Z, Xiao H and Zhang X 2017 *IEEE Trans. Syst. Man Cybern. A* Syst. Hum. **47**(7) pp 1428–1438
[16] Friedland B 1987 *Control System Design: An Introduction to State-Space Methods* (New York: McGraw–Hill)
[17] Yousef H A, Elkhatib M E and Sebakhy O A 2009 *Expert Syst. Appl.* **37**(2) pp 1522–1527
[18] Marusak P M and Kuntanapreeda S 2015 *Control Theory Tech.* **13**(3) pp 245–255
[19] Merritt H E 1967 *Hydraulic Control Systems* (NewYork: Wiley)
[20] Tang Y, Zhu Z and Shen G 2016 *J. Mech. Eng. Sci.* **230**(12) pp 1951–1967

IOP Conf. Series: Materials Science and Engineering **297** (2018) 012043 doi:10.1088/1757-899X/297/1/012043

Spherical beamforming for spherical array with impedance surface

Khemapat Tontiwattanakul*
Department of Mechanical and Aerospace Engineering,
King Mongkut's University of Technology North Bangkok,
1518 Pracharat 1 Rd, Wongsawang, Bangsue, Bangkok 10800, Thailand

* Corresponding Author: khemapat.t@eng.kmutnb.ac.th

Abstract. Spherical microphone array beamforming has been a popular research topic for recent years. Due to their isotropic beam in three dimensional spaces as well as a certain frequency range, the arrays are widely used in many applications such as sound field recording, acoustic beamforming, and noise source localisation. The body of a spherical array is usually considered perfectly rigid. A sound field captured by the sensors on spherical array can be decomposed into a series of spherical harmonics. In noise source localisation, the amplitude density of sound sources is estimated and illustrated by mean of colour maps. In this work, a rigid spherical array covered by fibrous materials is studied via numerical simulation and the performance of the spherical beamforming is discussed.

1. Introduction

A spherical microphone array is a versatile device that is mainly used in noise source localisation and beamforming. The entirely analysis of spherical array with rigid surface is shown in following most-cited articles authored by Rafaely [1, 2]. The working bandwidth of the array is limited by the number of sensors of the array and the size of the spherical body. This leads the array to a narrow working bandwidth that is regarded as a major disadvantage of this technology. There are some attempts to extend the bandwidth of the array. Meyer and Elko suggested that the bandwidth of the array can be extended if the sensors are more directive at high frequency and the best performance can be obtained when the beam width of the sensor is close to the beam width of eigenbeam [3]. Epain and Daniel attempt to push the frequency limit of a spherical array by making a conical waveguide then placing an omnidirectional sensor at the end [4]. That makes the sensors become more directive and the working bandwidth of the array was extended as a consequence. The extension of the spherical beamforming was studied further. Tontiwattanakul suggests that a beamforming technique based on singular value decomposition namely SVD beamforming has capability to extend the bandwidth of the array. It was verified by both numerical study and experiment that arrays with waveguides have slightly better performance [5].

The scattering sound field on the surface of a sphere changes depending on the boundary condition. In microphone array application, sound hard boundary condition is usually assumed. However, there are some studies about concerns on the scattering sound field on a sphere with impedance surface. Treeby *et al.* studied sound field scattered by a sphere with the impedance surface [7]. The total sound field is derived based on spherical harmonics decomposition. Treeby also considered a sphere with an impedance surface as a human head with hair. The effect of the impedance surface to interaural time difference (ITD) and interaural level difference (ILD) in comparison to one with rigid surface was discussed in [8].

In this work, a spherical array with impedance surface is studied. Spherical beamforming is employed to a rigid spherical array with or without impedance surface. The performance of the two arrays is evaluated and discussed.

2. Sound field on the surface of a rigid sphere with or without impedance surface

A sound field on the surface of a spherical object can be expressed in term of spherical harmonics. When a given plane wave with single frequency ω impinges on the surface of a sphere of radius r_0 located at the centre of a coordinate system, the total sound field at any position in three-dimensional space denoted by $\mathbf{x}$ can be computed by a summation of the incident plane wave $p^{inc}(\mathbf{x},\omega)$ and the corresponding scattered field $p^{sc}(\mathbf{x},\omega)$

$$p^{tot}(\mathbf{x},\omega) = p^{sc}(\mathbf{x},\omega) + p^{inc}(\mathbf{x},\omega). \tag{1}$$

The incident sound field can be represented by a series of spherical harmonics. It is also known as *Jacobi-Anger expansion* [9] given by

$$p^{tot}(\mathbf{x},\omega) = 4\pi p_0 \sum_{\nu=0}^{\infty}\sum_{\mu=-\nu}^{\nu} (-i)^{\nu} j_{\nu}(kx) Y_{\nu}^{\mu}(\hat{\mathbf{x}}) Y_{\nu}^{\mu}(\hat{\mathbf{y}})^{*}, \tag{2}$$

where $k = \omega / c_0$ denotes the wave number, c_0 denotes the speed of sound in air, i denotes the imaginary unit, $\hat{\mathbf{x}} = \mathbf{x}/x$ denotes a unitary vector indicating the direction such that the pressure on the sphere is evaluated at position $\mathbf{x}$ with magnitude $|\mathbf{x}| = x$, $\hat{\mathbf{y}}$ denotes a unitary vectors indicating the *direction of arrival* of the incident wave, $j_{\nu}(\cdot)$ denotes a spherical Bessel function of the first kind of degree ν, and $Y_{\nu}^{\mu}(\hat{\mathbf{x}})$ and $Y_{\nu}^{\mu}(\hat{\mathbf{y}})$ denotes a spherical harmonic of order ν and degree μ and is evaluated in the directions $\hat{\mathbf{x}}$ and $\hat{\mathbf{y}}$, respectively.

Note that, the time convention $e^{-i\omega t}$ is adopted throughout this paper. Therefore, the scattered field $p^{sc}(\mathbf{x},\omega)$ can be represented by means of the spherical harmonic expansion, given by

$$p^{sc}(\mathbf{x},\omega) = \sum_{\nu=0}^{\infty}\sum_{\mu=-\nu}^{\nu} C_{\nu\mu}(kx) h_{\nu}^{(1)}(kx) Y_{\nu}^{\mu}(\hat{\mathbf{x}}), \tag{3}$$

where $h_{\nu}^{(1)}(\cdot)$ denotes Hankel function of the first kind. When the total pressure on the sphere is considered, Fourier coefficient $C_{\nu\mu}$ can be determined depending on the boundary condition. In case of the sphere with rigid surface, $C_{\nu\mu}$ can be obtained by applying Neumann boundary condition given as the following

$$\frac{\partial}{\partial x} p^{tot}(\mathbf{x},\omega) = 0. \tag{4}$$

For the sphere with uniform impedance surface with the specific impedance $Z_c(\omega)$, the boundary condition is

$$\frac{\partial}{\partial x} p^{tot}(\mathbf{x},\omega) + \frac{i\rho_0 c_0}{Z_c(\omega)} p^{tot}(\mathbf{x},\omega) = 0. \tag{5}$$

By combining $C_{\nu\mu}$ with equations (2) and (1), the total sound field on the surface of the sphere is given by

$$p^{tot}(r_0,\omega) = \sum_{\nu=0}^{\infty} r_\nu \sum_{\mu=-\nu}^{\nu} Y_\nu^\mu(\hat{\mathbf{x}}) Y_\nu^\mu(\hat{\mathbf{y}})^*, \tag{6}$$

where r_ν is called *radial function* and it is given by

$$r_\nu = 4\pi(-i)^\nu \left(j_\nu(kr_0) - \frac{j'_\nu(kr_0)}{h'^{(1)}_\nu(kr_0)} h_\nu^{(1)}(kr_0) \right), \tag{7}$$

for the rigid sphere and

$$r_\nu = 4\pi(-i)^\nu \left(j_\nu(kr_0) - \frac{j'_\nu(kr_0) + i\rho_0 c_0 j_\nu(kr_0)/Z_c(\omega)}{h'^{(1)}_\nu(kr_0) + i\rho_0 c_0 h_\nu^{(1)}(kr_0)/Z_c(\omega)} h_\nu^{(1)}(kr_0) \right), \tag{8}$$

for the impedance surface.

Equations (7) and (8) suggest that sound field at any point on the spheres can be computed. The derivation of the expression in equations (7) and (8) is shown in many works by previous authors, for examples, in [2, 7, 9].

Note that when a hard surface is covered by fibrous material, some of the acoustic power convert into active and reactive power. The surface is now considered as an impedance surface associated with the specific acoustic impedance (or complex impedance) as a function of frequency. A model that is widely used to predict the specific impedance $Z_c(\omega)$ of fibrous materials is the one proposed by Delany and Bazley [10] and given by

$$Z_c(\omega) = \rho_0 c_0 \left(1 + 0.571 \frac{\rho_0 f^{-0.754}}{\sigma} + i0.087 \frac{\rho_0 f^{-0.732}}{\sigma} \right), \tag{9}$$

where f is the frequency of the incident wave and σ is the flow resistivity of the fibrous material [10]. For a given apparent density ρ_{av} and the mean fibre diameter d_{av} then the flow resistance can be computed by the following formula [11]

$$\sigma = \frac{3.18 \times 10^{-9} \times \rho_{ap}^{1.53}}{d_{ap}^2}. \tag{10}$$

The relations in equations (9) and (10) are empirical formula obtained by measurement of a large number of samples. The measurement data was then curve-fitted. Sometimes the specific impedance is referred to in the normalised version denoted by ζ such that $Z_c(\omega) = \rho_0 c_0 \zeta$ and is given by

$$\zeta = \gamma + i\xi, \tag{11}$$

where γ and ξ denote resistance and reactive components of the specific impedance, respectively.

3. Directivity of a sensor mounted on the sphere with or without covering materials

In this section, the directivity of the sensor as it is mounted on the surface of the rigid sphere is studied. The sphere of interest has its radius of 0.070 m. There is an omnidirectional microphone, as a sensor, flush mounted on the surface of the sphere in the direction $\theta = \pi / 2$ and $\phi = 0$, where θ and ϕ denote elevation angle and azimuth angle, respectively. The elevation angle starts from the north pole of the sphere and the azimuth angle starts from the x-axis of the Cartesian coordinate. The numerical study is performed in three cases as follows: firstly, the rigid sphere without covering of material; secondly, the rigid sphere with covering of lightweight glass wool with ρ_{av} = 12 kg/m^3 and d_{av} = 10 μm; and thirdly, the rigid sphere with covering of felt with density of 190 kg/m^3 and thickness of 10.3 mm.

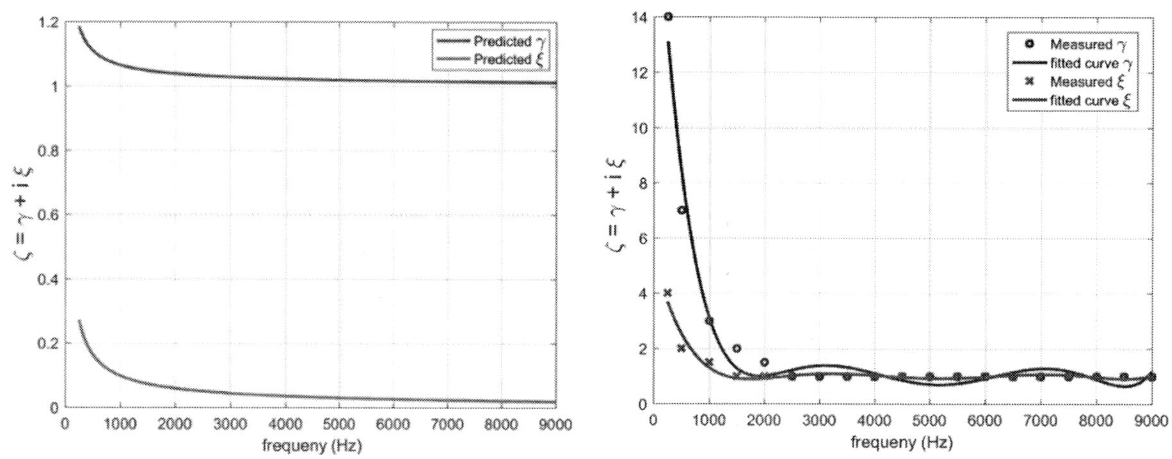

Figure 1. The specific impedance of lightweight glass wool predicted by Delany and Bazley's formula (left) and curve-fitted values of the specific impedance of felt with density of 190 kg/m^3 and 10.3 mm thick obtained from a measurement (right)

In the first case, the specific acoustic impedance is calculated using Delany and Bazley's formula as given by equation (9). It should be noted that Delany and Bazley's formula provide accurate prediction when 0.01 $< \rho_0 f / \sigma <$ 1.0 or 10 $< f <$ 1,200 Hz, approximately. Therefore, a big assumption is then made at this state such that Delany and Bazley's formula is used to predict the specific impedance of the glass wool up to 9,000 Hz. By considering the plot of the specific impedance of the glass wool as shown in figure 3, both resistance and reactive parts of the specific impedance converge at high frequencies. This suggests that the specific impedance at high frequency can be considered as a constant complex number.

In case of the sphere covered by felt, the tested data as reported in [8] is used in this study. The specific impedance is measured using impedance tube method according to ISO 10534-2. The data was curve-fitted by 6-th order polynomial in the range between 250 Hz to 9,000 Hz with 95% confidence level. Figure 3 shows the tested data overlaid with the interpolated data. The curve-fitted functions are given by

$$\gamma = 7.473 \cdot 10^{-22} f^6 - 2.31 \cdot 10^{-17} f^5 + 2.806 \cdot 10^{-13} f^4 - 1.693 \cdot 10^{-9} f^3 + 5.272 \cdot 10^{-6} f^2 - 0.0079 f + 5.361,$$

and

$$\xi = 2.882 \cdot 10^{-21} f^6 - 8.915 \cdot 10^{-17} f^5 + 1.085 \cdot 10^{-12} f^4 - 6.578 \cdot 10^{-9} f^3 + 2.074 \cdot 10^{-5} f^2 - 0.0319 f + 19.9.$$

The total pressure picked up by the sensor is then calculated. In order to study the directivity of the sensor, it is assumed that there are plane waves with unitary amplitude impinging on the spheres from all directions on the azimuth plane ($\theta = 0$). The corresponding total pressure field can be computed by using equations (7) and (8). It should be noted that the order of the spherical harmonics must be properly high in order to guarantee the convergence of the series. Treeby *et al.* discussed in [7] that 30 orders of spherical harmonics are enough to obtain converging result. Nevertheless, 50 orders of spherical harmonics are used in the calculation of this study.

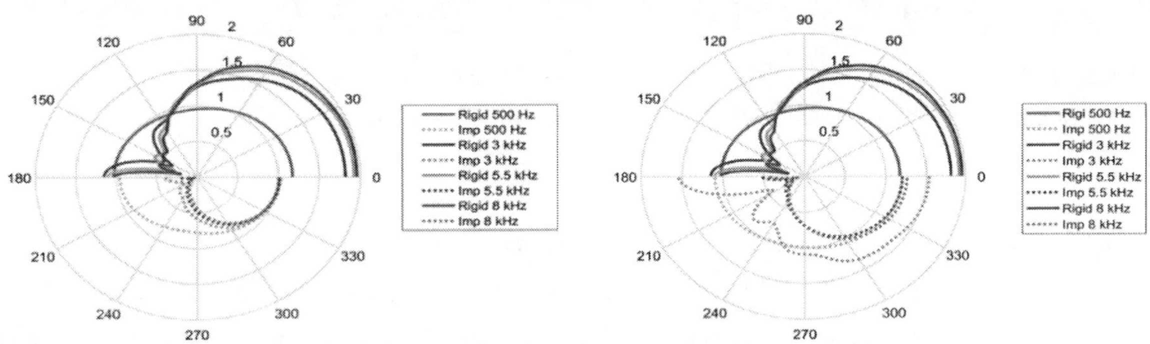

Figure 2. Directivity pattern of a sensor flush mounted on a rigid sphere (top half of the left and the right plot), on a rigid sphere with glass wool (bottom half of the left plot), on a rigid sphere with felt (bottom half of the right plot)

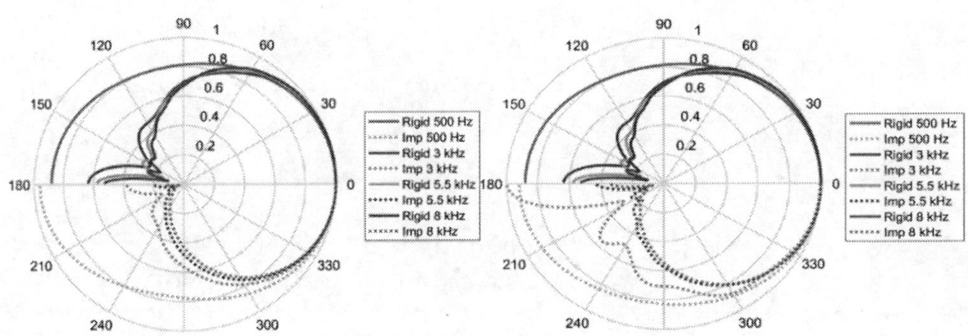

Figure 3. Normalised directivity pattern of a sensor flush mounted on a rigid sphere (top half of the left and the right plot), on a rigid sphere with glass wool (bottom half of the left plot), on a rigid sphere with felt (bottom half of the right plot)

IOP Conf. Series: Materials Science and Engineering **297** (2018) 012043 doi:10.1088/1757-899X/297/1/012043

A polar plot of the magnitude of total pressure at all azimuth angles is now referred to as directivity pattern. Figure 2 shows comparisons of directivity when the sensor mounted on the rigid sphere with or without covering material. It can be seen that the magnitude of total field of the sphere with impedance surface is smaller in comparison to that of the rigid sphere. The total pressure on the circumference is then normalised by the maximum value and is shown in figure 3 in order to provide better illustration for the directiveness. The sensor seems to be slightly more directive when there is covering material.

4. Discrete spherical array beamforming

Let a spherical array of radius r_0 have M sensors flush mounted on the surface and the sensors are populated in a nearly uniform scheme. The coordinates of the nearly uniform points on the sphere can be found in [12]. Let plane waves arriving to the array from S directions and $S > M$. Such sound sources distribute uniformly as well as the sensors. The sound pressure picked up by m-th sensor can be regarded as a contribution of the sound sources impinging to the sphere from all S directions. This approach is sometimes referred to as *plane waves superposition* [5, 13] given by

$$p(\mathbf{x},\omega) = \int_{\Omega} H(\mathbf{x},\hat{\mathbf{y}},\omega) a(\hat{\mathbf{y}},\omega) d\Omega(\hat{\mathbf{y}}), \tag{12}$$

where $a(\hat{\mathbf{y}},\omega)$ is the amplitude density of plane wave, and $\Omega(\hat{\mathbf{y}})$ is a solid angle corresponding to the direction $\hat{\mathbf{y}}$. It should be noted that when the sound sources $a(\hat{\mathbf{y}},\omega)$ are distributed uniformly then the solid angle is equal. $H(\mathbf{x},\hat{\mathbf{y}},\omega)$ denotes acoustic transfer function. In case of rigid sphere, it is given by

$$H(\mathbf{x},\hat{\mathbf{y}},\omega) = \sum_{\nu=0}^{\infty} b_\nu(kx) \sum_{\mu=\nu}^{\nu} Y_\nu^\mu(\hat{\mathbf{x}}) Y_\nu^\mu(\hat{\mathbf{y}})^*, \quad |\mathbf{x}| = x = r_0. \tag{13}$$

When the sound sources are distributed in nearly uniform configuration, equation (12) can be rewritten in a matrix equation as

$$\mathbf{p} = \mathbf{Ha}, \tag{14}$$

where

$$\mathbf{p} = \left[p(\mathbf{x}_1,\omega) \;\; p(\mathbf{x}_2,\omega) \;\; \cdots \;\; p(\mathbf{x}_M,\omega) \right]^T,$$
$$\mathbf{a} = \left[a(\hat{\mathbf{y}}_1,\omega)\Delta\Omega(\hat{\mathbf{y}}_1) \;\; a(\hat{\mathbf{y}}_2,\omega)\Delta\Omega(\hat{\mathbf{y}}_2) \;\; \cdots \;\; a(\hat{\mathbf{y}}_L,\omega)\Delta\Omega(\hat{\mathbf{y}}_L) \right]^T,$$

and

$$\mathbf{H} = \begin{bmatrix} H(\mathbf{x}_1,\hat{\mathbf{y}}_1,\omega) & H(\mathbf{x}_1,\hat{\mathbf{y}}_2,\omega) & \cdots & H(\mathbf{x}_1,\hat{\mathbf{y}}_L,\omega) \\ H(\mathbf{x}_2,\hat{\mathbf{y}}_1,\omega) & H(\mathbf{x}_2,\hat{\mathbf{y}}_2,\omega) & \cdots & H(\mathbf{x}_2,\hat{\mathbf{y}}_L,\omega) \\ \vdots & \vdots & \ddots & \vdots \\ H(\mathbf{x}_M,\hat{\mathbf{y}}_l,\omega) & H(\mathbf{x}_M,\hat{\mathbf{y}}_2,\omega) & \cdots & H(\mathbf{x}_M,\hat{\mathbf{y}}_L,\omega) \end{bmatrix},$$

where m and l denote the index number of the sensor and that of the direction of arrival of the impinging plane wave, respectively, and $\Delta\Omega(\hat{\mathbf{y}}_l)$ denotes a solid angle corresponding to l-th sound source. Position $\mathbf{x}_m$ denotes the position of the m-th sensor.

The elements of matrix $\mathbf{H}$ can be computed using equation (13) and can be rewritten by

$$H(\mathbf{x}_m, \hat{\mathbf{y}}_l, \omega) = \sum_{\nu=0}^{\infty} b_\nu(kr_0, r_0) \sum_{\mu=\nu}^{\nu} Y_\nu^\mu(\hat{\mathbf{x}}_m) Y_\nu^\mu(\hat{\mathbf{y}}_l)^*. \tag{15}$$

Note that matrix $\mathbf{H}$ can be rewritten in terms of spherical harmonic matrices $\mathbf{Y}_x$ and $\mathbf{Y}_y$, and the corresponding radial weighting matrix $\mathbf{B}$ as

$$\mathbf{H} = \mathbf{Y}_x \mathbf{B} \mathbf{Y}_y^H. \tag{16}$$

The working bandwidth of a spherical array is limited by size of the array and number of sensors. The array tends to suffer the so-called *spatial aliasing* when the size of the array is comparable to the wavelength of an impinging wave, in other words, when $kr_0 > N$ where $k = \omega / c_0$ is the wave number of the sound field impinging on the array, ω is the angular velocity, c_0 is the speed of sound, r_0 is the radius of the array, and N is the order of spherical harmonic. N is given by the relation $N = \sqrt{(M-1)}$, where M is the number of sensors distributed uniformly around the sphere [2]. In order to avoid spatial aliasing, it is important to note that matrix $\mathbf{H}$ must be *order limited* according to the criterion $N = \sqrt{(M-1)}$ [2].

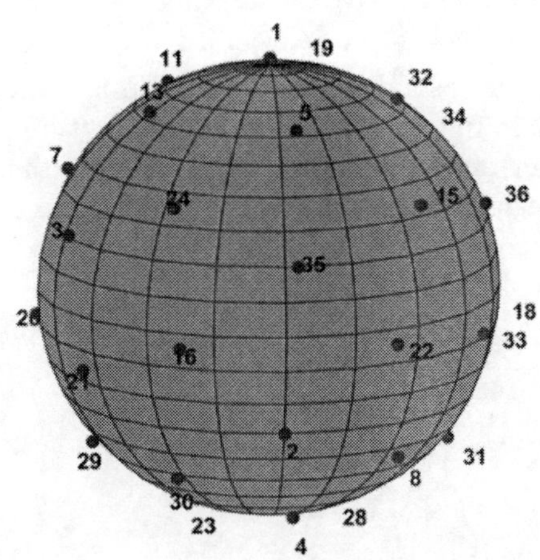

Figure 4. 36 sensors distributed on the surface of the sphere in nearly uniform scheme, where the numbers denote microphone indices.

By substituting equation (16) into (14), thus

$$\mathbf{p} = \mathbf{Y}_x \mathbf{B} \mathbf{Y}_y^H \mathbf{a}, \tag{17}$$

and the amplitude density of the plane waves can be computed by the following relation

$$\mathbf{a}_{sp} = \mathbf{Y}_y \mathbf{B}^\dagger \mathbf{Y}_x^H \mathbf{p}, \tag{18}$$

where $(\cdot)^\dagger$ and $(\cdot)^H$ denote Moore-Penrose pseudoinverse and Hermitian transpose of matrix, respectively. Equation (18) is now referred to as *spherical beamforming* and $\mathbf{a}_{sp}$ is the output of the beamformer.

By substituting equation (17) into (18), an analysis can be taken at this state

$$\mathbf{a}_{sp} = \mathbf{Y}_y \mathbf{B}^\dagger \mathbf{Y}_x^H \mathbf{Y}_x \mathbf{B}^\dagger \mathbf{Y}_y^H \mathbf{a}. \tag{19}$$

It should be noted that the multiplication $\mathbf{Y}_x^H \mathbf{Y}_x$ does not result an identity matrix due to the lag of the completeness of an order limited system. Thus, some modification can be taken by replacing Hermitian transpose by pseudoinverse. Pseudoinverse beamforming is given as below

$$\mathbf{a}_{inv} = (\mathbf{Y}_y^H)^\dagger \mathbf{B}^\dagger \mathbf{Y}_x^\dagger \mathbf{p}, \tag{20}$$

where $\mathbf{a}_{inv}$ denotes beamformer output of the pseudoinverse beamformer.

5. Performance of the spherical beamformer

In this work, there are 36 sensors distributed nearly uniformly all around the sphere. Figure 4 depicts the spherical body with sensors arranged in nearly uniform scheme. By using the criterion discussed in the previous section, the rigid spherical array does not suffer from spatial aliasing for the frequency up to 3,900 Hz. The performance of the array above the aliasing limit is discussed in this section.

The total pressure picked up by the sensors of the array in three following cases is computed using equation (6). The output from the pseudoinverse beamformer, which is given by equation (20), is calculated and plotted as shown in figures 5 through 8 in the dynamic range of 20 dB in reduction. Note that the black dots overlaid on the colour map plots indicate the positions of the 36 sensors. By considering the directivity pattern of the beamformer output, it can be seen that large side lobes appear at frequencies above aliasing frequency. However, this phenomena is acceptable in some applications, for example, in noise source localisation.

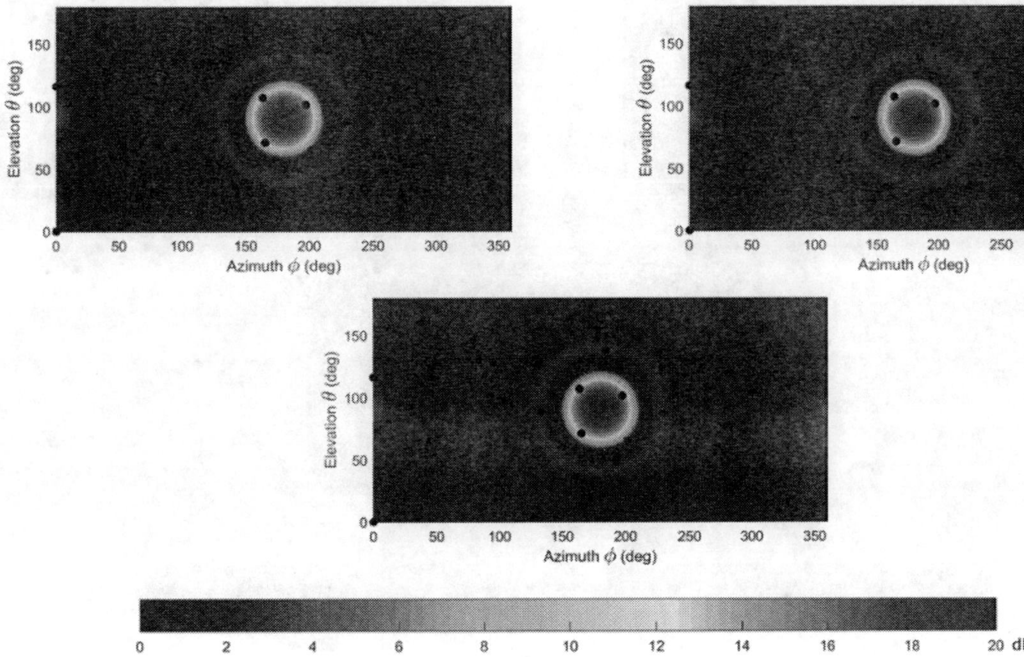

Figure 5. The directivity pattern beamformer output at 500 Hz of the rigid sphere (top left plot), the rigid sphere with glass wool (top right plot), and the rigid sphere with felt (bottom)

The performance of the array can be evaluated based on a quantitative index. The directivity index (DI) is widely used as an indicator and is given by

$$DI(\omega) = 10\log_{10}\left\{\frac{4\pi\,|\,p_{ax}\,|^2}{\int_0^{2\pi}\int_0^{\pi}|p(\theta,\phi)|^2\, sin\theta d\theta d\phi}\right\}, \quad (21)$$

where p_{ax} denotes magnitude of the beamformer output at direction of the main lobe, $p(\theta,\phi)$ denotes magnitude of the beamformer output at direction (θ,ϕ). The directivity index of the spherical array with or without impedance surface is computed from 250 to 9,000 Hz and is plotted in figure 9. The decaying of the DI in figure 9 suggests that the array loses its performance at around 3,500 Hz.

In practice, the low frequency limit of the array is usually constrained by the measurement error of the sound field, i.e. numerical error when a sound field is computed for the numerical simulation and the noise of sensor. However, this point is beyond the discussion of this paper.

IOP Conf. Series: Materials Science and Engineering **297** (2018) 012043 doi:10.1088/1757-899X/297/1/012043

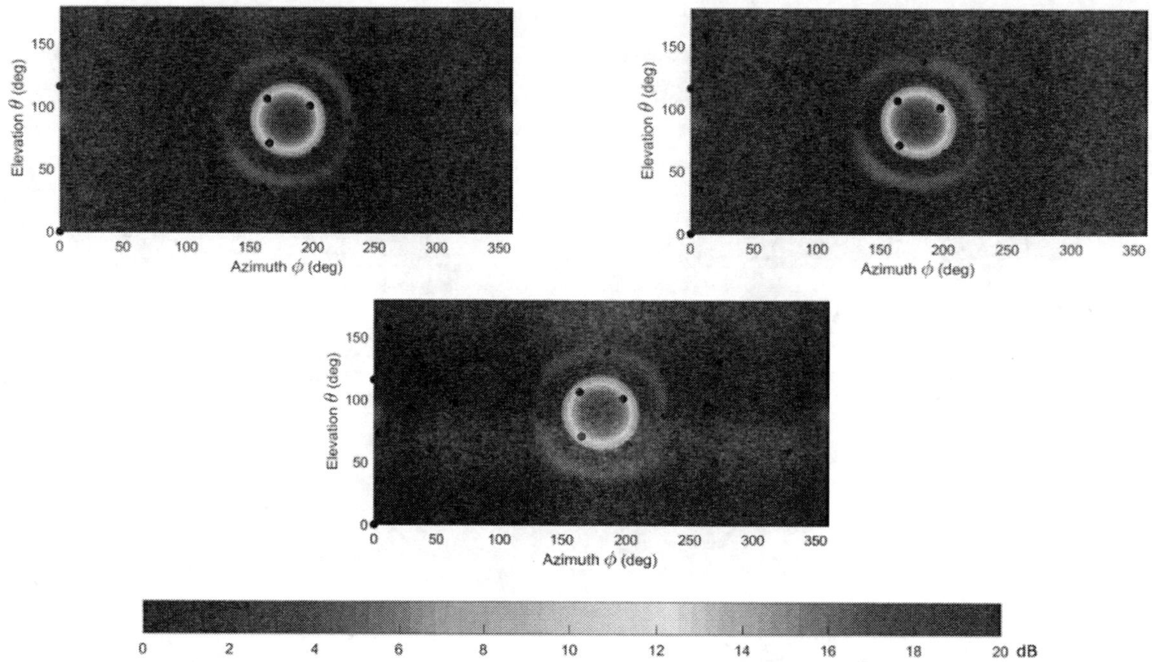

Figure 6. The directivity pattern beamformer output at 3,000 Hz of the rigid sphere (top left), the rigid sphere with glass wool (top right), and the rigid sphere with felt (bottom)

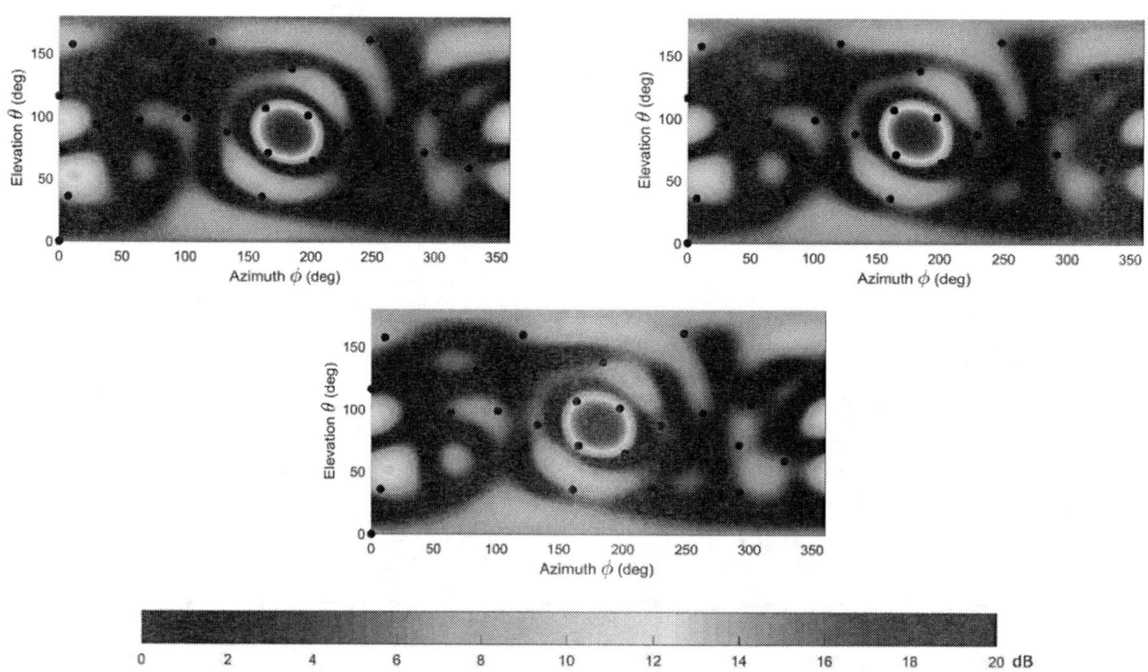

Figure 7. The directivity pattern beamformer output at 5,500 Hz of the rigid sphere (top left), the rigid sphere with glass wool (top right), and the rigid sphere with felt (bottom)

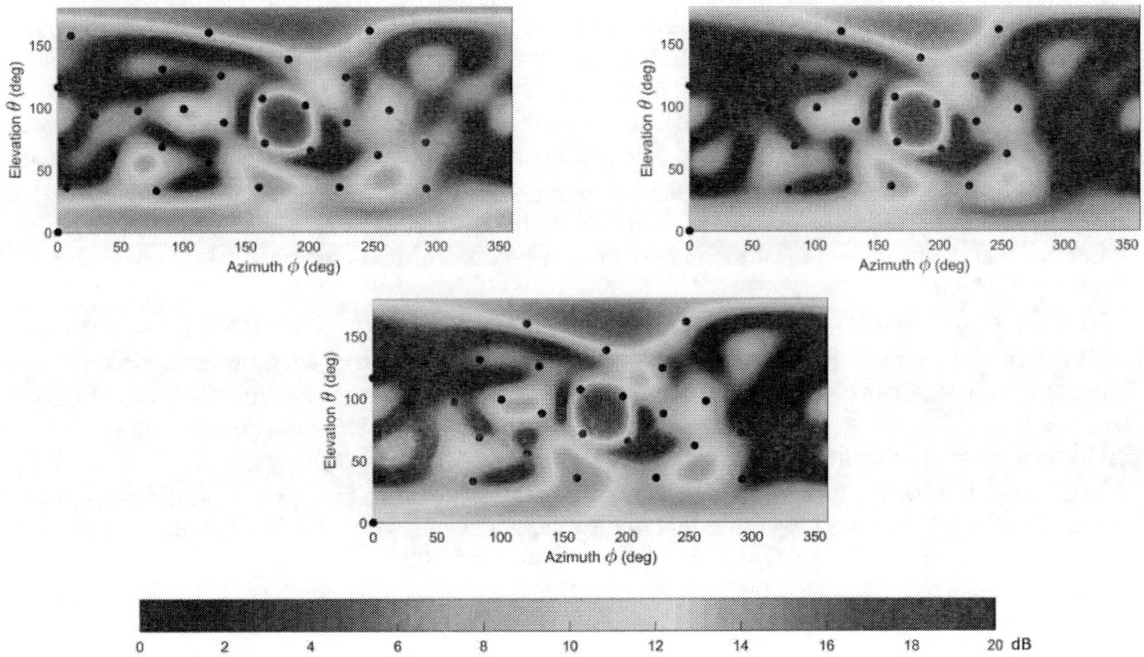

Figure 8. The directivity pattern beamformer output at 8,000 Hz of the rigid sphere (top left), the rigid sphere with glass wool (top right), and the rigid sphere with felt (bottom)

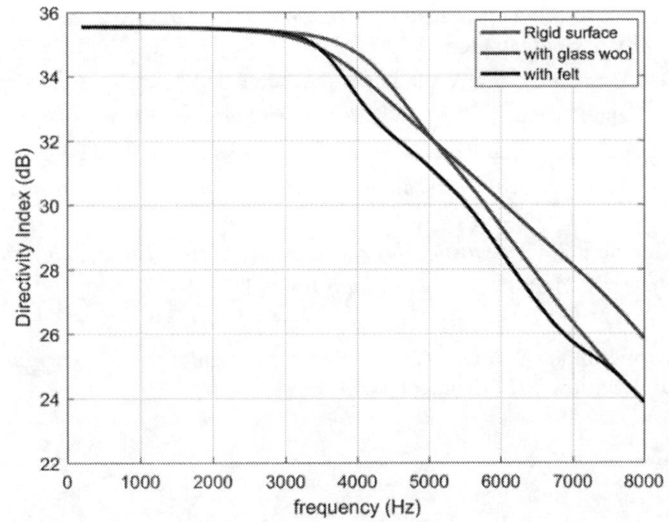

Figure 9. The directivity index (DI) of the beamformer outputs as a function of frequency

6. Conclusion

This paper discusses the effect of the spherical array with or without impedance surface. Some numerical studies were carried out. It can be concluded that the impedance surface play an important role to the scattering of sound field on the surface of the sphere. The numerical studies carried out in this work suggest that the impedance surface also affects the performance of the spherical array. The flush mounted sensor on the impedance surface tends to be more directive than one without the impedance surface.

By considering figure 1, it suggests that magnitude of the resistance part of the two materials is not much different as it seems to level off at around 1 but the difference in the reactive part is considerable. It implies that the reactive part of the material might be a factor that obstructs the improvement of the sensor's directivity.

Figures 5 and 6 illustrate the directivity pattern of the beamformer below the aliasing frequency. It can be seen that the main lobe distinguishably indicates the direction of sound source. Figures 7 and 8 illustrate the performance of the three beamformer outputs. Some large sidelobes appear and become larger at higher frequency. Although, sidelobes seem to be smaller for the case that the array is covered by fibrous material, it is difficult to assess the performance by visualising.

The DI plot in figure 9 suggests that the array with felt seem to have lower performance than the others since its DI is always lower than that of the others. The one with glass wool seems to have slightly better performance (2 dB higher) at high frequencies above 5,000 Hz.

References

[1] Rafaely B 2004 *The Journal of the Acoustical Society of America* **116** 2149

[2] Rafaely B 2005 *IEEE Transactions on speech and audio processing* **13** 135–143

[3] Meyer J and Elko G W 2008 Handling Spatial Aliasing in Spherical Array Applications *Hands-Free Speech Communication and Microphone Arrays, 2008. HSCMA 2008* 2 pp 8 – 11

[4] Epain N and Daniel J 2008 Improving Spherical Microphone Arrays *AES 124th Convention* (Amsterdam, The Netherlands) pp 1 – 9

[5] Tontiwattanakul K 2016 *Signal processing for microphone arrays with novel geometrical design* Ph.D. thesis University of Southampton

[6] Beranek L L 1996 *Acoustics* (New York, USA: the Acoustic Society of America)

[7] Treeby B E, Pan J and Paurobally R M 2007 *J. Acoust. Soc. Am.* **122** 46 – 57

[8] Treeby B E, Paurobally R M and Pan J 2007 *J. Acoust. Soc. Am.* **121** 2217–2226

[9] Williams E G 1999 *Fourier acoustics: sound radiation and nearfield acoustical holography* (academic press)

[10] Cox T and D'Antonio P 2009 *Acoustic Absorbers and Diffusers: Theory, Design and Application* vol 25

[11] Bies D A and Hansen C H 1980 *Topics in Catalysis* **13** 357–391

[12] Fliege J and Maier U 1999 *IMA Journal of Numerical Analysis* **19** 317–334

[13] Fazi F M, Noisternig M and Warusfel O 2012 Representation of sound fields for audio recording and reproduction *Acoustics 2012* (Nantes) p 859

Software-in-the-loop simulation of a quadcopter portion for hybrid aircraft control

Shoaib Mansoor[1], **Mana Saedan**[2,*]
[1] Graduate School, Chiang Mai University, Chiang Mai 50200, Thailand
[2] Department of Mechanical Engineering, Chiang Mai University, Chiang Mai 50200, Thailand

[*]Corresponding Author: mana.saedan@cmu.ac.th, Tel: +66-(0)5394-414,
Fax: +66-(0)5394-4145

Abstract. In this paper, we present the design of the software-in-the-loop simulation framework for a quadcopter that is incorporated in our hybrid aircraft. The hybrid aircraft comprises a quad-copter and a fixed wing with one forward thrust rotor. We need to develop a split control system that utilizes a typical quadcopter controller to control four motors/propellers and a supervisor controller to control a forward thrust rotor. The supervisor controller shall take feedback signals from the quadcopter and will command the fifth rotor for stabilizing the hybrid aircraft and resolves problems like thrust saturation. The simulation simulates the control algorithm and verifies the quadcopter's behavior using MATLAB and Simulink together. Achieving these results, we come to know how our hybrid controller will be implemented, what results to expect once the forward thrust rotor is attached to the quadcopter. The software-in-the-loop simulation of a quadcopter is one of the most effective methods for verifying overall control performance and safety of the hybrid aircraft before actual hardware implementation and flight test.

1. Introduction

Hybrid aircrafts are unique type of aircrafts that combine the hover capability of helicopters or quadcopters along with the speed and range of airplanes. Computational tools presently are a prime source of saving efforts, money and time. Artificial stability can be given to aircraft via such control systems. Stability and maneuverability have an inverse effect to each other. In this article we present the modular control of a quadcopter portion of a hybrid aircraft. The aircraft comprises a standard quadcopter, a fixed wing and one forward thrust rotor. A mathematical model is used for modeling the simulations of quadcopter control part of the whole system. Figure 1 shows a prototype model for the hybrid aircraft. The purpose for designing the quadcopter controller is to obtain Thrust Percentage values in relation to the motor speed of each motor of the quadcopter. The basic idea is to have a supervisor controller, which takes the motion command, and the necessary data from a standard quadcopter controller and in result controls the forward thrust rotor. The reason for having a control for a forward thrust motor is to deal with problems like thrust saturation, and a better stability for the hybrid aircraft. Initially the mathematical model for the whole system is developed.

Figure 1: Hybrid Aircraft comprising a quad-copter, fixed wing and a froward thrust rotor

For the successful progress of our research, we carried out a detailed literature review of the previous works related to our project. A detailed summary of the literature review is discussed in section 1.1.

1.1. Literature Review

One type of very basic UAVs (Unmanned Air Vehicles) is quadcopters. Starting with dynamics of UAVs, the research study conducted on Robotics and Automation by Marcus, Heinrich and Paolo [1] mainly explains the concept of a quadrotor UAV with tilting propellers. Another similar research article by Pounds and Paul from Australian National University [2] has defined the progress towards realizing practical quadrotor robot helicopters and, in particular, the conceptual design of Australian National University's 'X-4 Flyer' platform. Two challenges faced by the 'X-4' were generating sufficient thrust and managing unstable dynamic behavior. They have addressed these issues with a rotor design technique for maximizing thrust and the application of a novel rotor mast configuration. Relating to kinematics of UAVs, the research article presented by Dongming, Cai, Jorge and Lakmal from Institute Khalifa University UAE [3] considered the work on attitude control of quadrotor UAVs. They applied an intuitive kinematic representation, called the rotation vector. Based on the property that the rotation vector rate is equivalent to the body angle velocity when the rotation is small, a simple and intuitive attitude reference is proposed. A proportional-derivative (PD) law was used by integrating the new attitude reference for the attitude control of quad-rotor UAVs. Along with considerable knowledge of kinematics and dynamics of quadrotor vehicles, we proceeded further with the work on control method and control theory. The work by Shakev, Andon, Kaynak and Shiev on stabilization of quadrotor using feedback controllers [4] showed that five different types of nonlinear feedback laws with saturation elements, previously proposed for global control of systems with multiple integrators, are applied and tested to control the quad-rotor rotorcraft's roll and pitch angles. Moving forward with the control theory, the research by Adriano on SIL (Software In Loop) simulation using MATLAB (Simulink) and X-plane [5] introduced a new method to simulate a guidance algorithm running on Simulink that controls a fixed wing unmanned aircraft model running on the flight simulator X-Plane, which simulates the vehicle dynamics, sensors, and actuators.

2. Mathematical Model

A mathematical model of the aircraft was identified and constructed. It is then implemented in the MATLAB Simulink environment, to be tested as a simulation. After verifying the correctness of the model, the control algorithms are crafted to control a simulated aircraft. The mathematical model of the hybrid aircraft has been derived from the quadcopter dynamics and control [6-7], and the thrust vector in forward direction (x-axis) has been added from the fifth propeller. Figure 2 shows the inertial and body axes of the hybrid aircraft along with the propeller numbering.

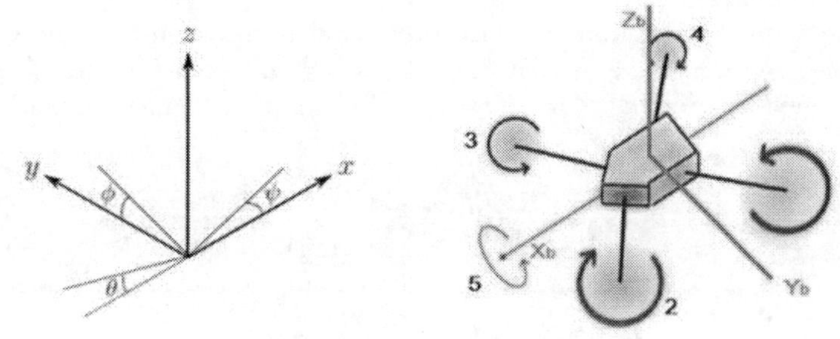

Figure 2. Inertial and body axes

Before delving into the physics of quadcopter motion, the kinematics in the body and inertial frames are formalized. We define the position and velocity of the quadcopter in the inertial frame as $\boldsymbol{x} = (x, y, z)^T$ and $\dot{\boldsymbol{x}} = (\dot{x}, \dot{y}, \dot{z})^T$, respectively. Similarly, we define the roll, pitch, and yaw angles in the body frame as $\theta = (\phi, \theta, \psi)^T$ with corresponding angular velocities equal to $\dot{\boldsymbol{\theta}} = (\dot{\phi}, \dot{\theta}, \dot{\psi})^T$. However, note that the angular velocity vector is $\boldsymbol{\omega} \neq \dot{\theta}$. The angular velocity is a vector pointing along the axis of rotation, while $\dot{\boldsymbol{\theta}}$ is just the time derivative of yaw, pitch, and roll. In order to convert these angular velocities into the angular velocity vector, we can use the following relation:

$$\boldsymbol{\omega} = \begin{Bmatrix} 1 & 0 & -sin\theta \\ 0 & cos\phi & cos\theta sin\phi \\ 0 & -sin\phi & cos\theta cos\phi \end{Bmatrix} \dot{\theta} \tag{1}$$

where $\boldsymbol{\omega}$ is the angular velocity vector in the body frame.

We can relate the body and inertial frame by a rotation matrix **R** which goes from the body frame to the inertial frame. This matrix is derived by using the Euler angle conventions and successively "undoing" the yaw, pitch, and roll.

$$\boldsymbol{R}_B^I = \begin{Bmatrix} cos\theta cos\psi & sin\phi sin\theta cos\psi - cos\psi & cos\phi sin\theta cos\psi + sin\phi sin\psi \\ cos\theta sin\psi & sin\phi sin\psi sin\theta + cos\phi cos\psi & cos\phi sin\psi sin\theta - cos\psi sin\phi \\ -sin\theta & sin\phi cos\theta & cos\phi cos\theta \end{Bmatrix} \tag{2}$$

In order to properly model the dynamics of the system, we need an understanding of the physical properties that govern it. We will begin with a description of the motors being used for our hybrid aircraft, and then use energy considerations to derive the forces and thrusts that these motors produce on the entire quadcopter. All motors on the hybrid aircraft are identical, so we can analyze a single one without loss of generality. Note that adjacent propellers, however, are oriented opposite to each other; if a propeller is spinning "clockwise," then the two adjacent ones will be spinning "counter-clockwise," so that torques are balanced if all propellers are spinning at the same rate.

Considering that our major component of solution is thrust, we already know that brushless motors are used for all hybrid copter applications. Hence, solving for the thrust magnitude T, we obtain that thrust is proportional to the square of angular velocity of the motor:

$$\mathbf{T} = \left(\frac{K_V K_T \sqrt{2\rho A}}{K_T}\omega\right)^2 = k\omega^2 \tag{3}$$

where k is some appropriately dimensioned constant, K_T is the torque proportionality constant, K_v is a proportionality constant (indicating back-EMF generated per RPM), ρ is the density of the surrounding air and A is the area swept out by the rotor. Summing over all the motors, we find that the total thrust on the hybrid aircraft (in the body frame) is given by

$$\boldsymbol{T_B} = \textstyle\sum_{i=1}^{4} T_i = k \begin{bmatrix} T_X \\ 0 \\ \sum \omega_i{}^2 \end{bmatrix} \quad T_X = k\omega_x{}^2 \tag{4}$$

Since the system of the hybrid aircraft consists of nonlinear coupled dynamics, we need to define it in the form of $\dot{x} = f(x, u)$.

IOP Conf. Series: Materials Science and Engineering **297** (2018) 012044 doi:10.1088/1757-899X/297/1/012044

2.1. Equations of Motion

In the inertial frame, the acceleration of the hybrid aircraft is due to thrust, gravity, and linear friction. We can obtain the thrust vector in the inertial frame by using our rotation matrix $\boldsymbol{R}$ to map the thrust vector from the body frame to the inertial frame. Thus, the linear motion can be summarized as

$$[\mathrm{m}\ddot{\mathrm{x}}]^I = \begin{bmatrix} 0 \\ 0 \\ -\mathrm{mg} \end{bmatrix} + \boldsymbol{R}_B^I \mathrm{T_B} + \mathbf{F_D} \tag{5}$$

where x is the position of the aircraft, g is the acceleration due to gravity, $\mathbf{F_D}$ is the drag force, and $\mathbf{T_B}$ is the thrust vector in the body frame. We derive the rotational equations of motion from Euler's equations for rigid body dynamics. Expressed in vector form, Euler's equations are written as

$$\boldsymbol{I}\dot{\boldsymbol{\omega}} + \boldsymbol{\omega} \times (\boldsymbol{I}\boldsymbol{\omega}) = \boldsymbol{\tau} \tag{6}$$

where $\boldsymbol{\omega}$ is the angular velocity vector, $\boldsymbol{I}$ is the inertia matrix, and $\boldsymbol{\tau}$ is a vector of external torques. We can rewrite this as

$$\dot{\boldsymbol{\omega}} = \begin{bmatrix} \dot{\omega}_x \\ \dot{\omega}_y \\ \dot{\omega}_z \end{bmatrix} = \mathbf{I}^{-1}(\boldsymbol{\tau} - \boldsymbol{\omega} \times (\mathbf{I}\boldsymbol{\omega})) \tag{7}$$

We can model our hybrid aircraft as two thin uniform rods crossed at the origin with a point mass (motor) at the end of each. With this in mind, it is clear that the symmetries result in a diagonal inertia matrix of the form

$$\boldsymbol{I} = \begin{bmatrix} \mathrm{I_{xx}} & 0 & 0 \\ 0 & \mathrm{I_{yy}} & 0 \\ 0 & 0 & \mathrm{I_{zz}} \end{bmatrix} \tag{8}$$

Therefore, we obtain our final result for the body frame rotational equations of motion

$$\dot{\boldsymbol{\omega}} = \begin{bmatrix} \tau_\varphi \mathrm{I_{xx}}^{-1} \\ \tau_\theta \mathrm{I_{yy}}^{-1} \\ \tau_\psi \mathrm{I_{zz}}^{-1} \end{bmatrix} - \begin{bmatrix} \frac{\mathrm{I_{yy}} - \mathrm{I_{zz}}}{\mathrm{I_{xx}}} \omega_y \omega_z \\ \frac{\mathrm{I_{zz}} - \mathrm{I_{xx}}}{\mathrm{I_{yy}}} \omega_x \omega_z \\ \frac{\mathrm{I_{xx}} - \mathrm{I_{yy}}}{\mathrm{I_{zz}}} \omega_x \omega_y \end{bmatrix} \tag{9}$$

Considering the Newton-Euler equations, the hybrid aircraft is considered as a rigid body. In the body frame, the force required for the acceleration of mass and the centrifugal force are described as follows:

$$m\boldsymbol{v}_B + \boldsymbol{v} \times (m\boldsymbol{v}_B) = \boldsymbol{R}^T G + \boldsymbol{T}_B \tag{10}$$

where $\boldsymbol{R}^T = \boldsymbol{R}^{-1}$ and $\boldsymbol{T}_B$ = Thrust of the body.

In the inertial frame the centrifugal force is nullified. Thus, only the gravitational force and direction of thrust are contributing in the acceleration of the hybrid aircraft.

$$\begin{bmatrix} \ddot{x} \\ \ddot{y} \\ \ddot{z} \end{bmatrix} = -g \begin{bmatrix} 0 \\ 0 \\ 1 \end{bmatrix} + [\boldsymbol{R}]\frac{T}{m} \tag{11}$$

This becomes:

$$\begin{bmatrix} \ddot{x} \\ \ddot{y} \\ \ddot{z} \end{bmatrix} = -g \begin{bmatrix} 0 \\ 0 \\ 1 \end{bmatrix} + \frac{1}{m} \begin{bmatrix} T_x(\cos\theta\cos\psi) + T_z(\cos\phi\sin\theta\cos\psi + \sin\phi\sin\psi) \\ T_x(\cos\theta\sin\psi) + T_z(\cos\phi\sin\psi\sin\theta - \cos\psi\sin\phi) \\ T_x(-\sin\theta) + T_z(\cos\phi\cos\theta) \end{bmatrix} \tag{12}$$

In the body frame, the angular acceleration of inertia $\boldsymbol{I}\dot{\boldsymbol{v}}$, the centripetal forces $v \times Iv$ and the gyroscopic forces Γ are equal to the external torque $\boldsymbol{\tau}$

$$\boldsymbol{I}\dot{\boldsymbol{v}} + \boldsymbol{v} \times \boldsymbol{I}\boldsymbol{v} + \boldsymbol{\Gamma} = \boldsymbol{\tau} \tag{13}$$

Therefore,

$$\dot{v} = I^{-1}\left(-\begin{bmatrix} p \\ q \\ r \end{bmatrix} \times \begin{bmatrix} I_{xx}p \\ I_{yy}q \\ I_{zz}r \end{bmatrix} - I_r \begin{bmatrix} p \\ q \\ r \end{bmatrix} \times \begin{bmatrix} 0 \\ 0 \\ 1 \end{bmatrix} w_\Gamma + \tau \right) \tag{14}$$

$$\begin{bmatrix} \dot{p} \\ \dot{q} \\ \dot{r} \end{bmatrix} = \begin{bmatrix} \frac{(I_{yy}-I_{zz})}{I_{xx}} qr \\ \frac{(I_{zz}-I_{xx})}{I_{yy}} pr \\ \frac{(I_{xx}-I_{yy})}{I_{zz}} pq \end{bmatrix} - I_r \begin{bmatrix} \frac{q}{I_{xx}} \\ \frac{-p}{I_{yy}} \\ 0 \end{bmatrix} w_\Gamma + \begin{bmatrix} \frac{\tau_\varphi}{I_{xx}} \\ \frac{\tau_\theta}{I_{yy}} \\ \frac{\tau_\psi}{I_{zz}} \end{bmatrix} \tag{15}$$

where $w_\Gamma = \omega_1 - \omega_2 + \omega_3 - \omega_4$.

With the dynamics and kinematics of the system solved, we now need to create a control scheme which will give the solution to our system, in terms of position and angular velocity. The next section contains the control schematic which is the solution to our system, and the simulation is required to give us output in terms of thrust of the rotor, motor velocity, position and attitude of the UAV.

3. Control Structure

The hybrid aircraft works on the schematic shown in figure 3, a control structure for extracting the position, velocity and attitude of hybrid aircraft such that the thrust is fed into the dynamics block along with the feedback of angular velocity and velocity in x, y, z axis of the hybrid aircraft. The dynamics are solved, integrated and fed into feedback system. The integrated data is fed into kinematics block for the kinematics to be solved so that we can get the positional velocity (p, q and r). The data from the dynamics block is doubly integrated so that we can have the position of the hybrid aircraft.

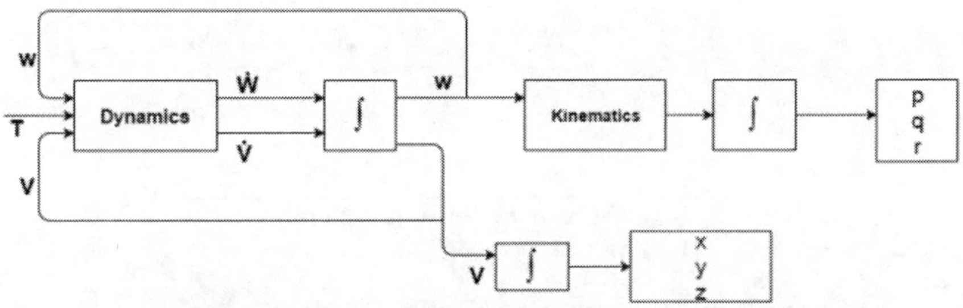

Figure 3. Control schematic of Hybrid Aircraft

Furthermore, figure 4 explains the control structure in more detail. Our idea is to use the previously designed quadcopter control and simulations for our split control. We shall check in the results if the quadcopter control system works well for our control system. We will begin with the simulation of quadcopter to maintain its height and cover some distance in x-axis. Regarding the split control structure, our input for the controller will be the velocity command (thrust for motors), this command will be processed using the dynamics and state equations. The output will be the angular velocities, roll pitch and yaw angles and the position of the quadcopter. The results will let us know what changes to expect when we implement our forward thrust rotor.

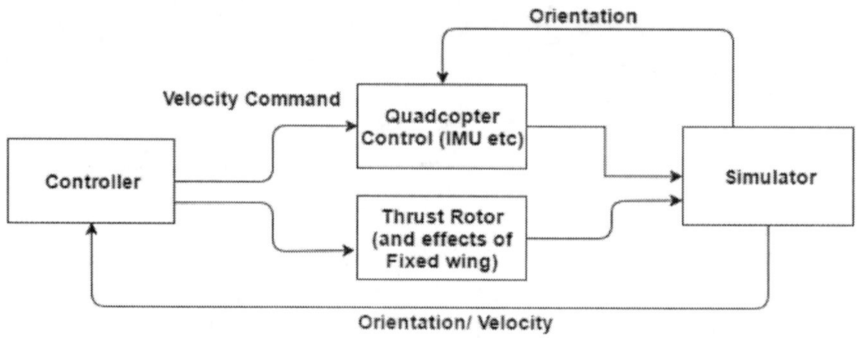

Figure 4. Split control structure

3.1. Parameters

Initial conditions are established for the programmed simulations. Table 1 shows the parameters for the initial conditions for which the simulation is accomplished.

Table 1. Parameters for the control algorithm and simulations

Parameters	Value	Units
g	9.81	m/s^2
m	4.468	kg
l	0.225	m
k	$2.980 \cdot 10^{-6}$	
b	$1.140 \cdot 10^{-7}$	
I_M	$3.357 \cdot 10^{-5}$	kg m^2
I_{xx}	$4.856 \cdot 10^{-3}$	kg m^2
I_{yy}	$4.856 \cdot 10^{-3}$	kg m^2
I_{zz}	$8.801 \cdot 10^{-3}$	kg m^2
A_x	0.35	kg/s
A_y	0.25	kg/s
A_z	0.25	kg/s
T_{max}	20	N
H	10	m

T_{max} = Maximum thrust of each rotor
H = Altitude

The control algorithm was designed in MATLAB. The Simulink-based supervisory controller runs the kinematic and dynamic models of quad-copter based on the initial conditions of the model. The initial conditions are aligned with the parameters in table 1. The quadcopter will start from the height of 10 meters and maintain its height while following a path of 275 meters in the direction of the x-axis. We expect a constant positive pitch and stable behavior of the quadcopter in the conditions applied.

3.2 Control Algorithms

The designed control algorithms are based on the equations explained in Section 2 (Mathematical Modeling) and the reference was established in figure 2.

An algorithm is established for altitude commands in which step commands are given to the quadcopter system. These commands are processed via the quadcopter dynamic controller. The quadcopter control includes the command and control of altitude along with the dynamics. The simulation is run and the results are obtained. The altitude controller takes the altitude commands, processes the data and applies correction using the designed PID controller. The roll, pitch and yaw corrections are applied in real time to the quadcopter for constant altitude and attitude improvement. Figure 6 shows the basic control structure, it is the input-output summary of the designed controller. We command the thrust, which is converted into

RPM. The quadcopter dynamic equations account for the commanded input and the disturbances. Hence the output results are given in the form of figure 7.

4. Simulation Results and Analysis

The results shown are for the quadcopter maintaining its height of 10 meters, the run time of simulation is 10 seconds. The quadcopter initially starts from 10 meters, maintains the altitude of 10 meters throughout and covers the length in x-direction of 275 meters. Initially there is a constant throttle command for every motor, to hover at the height of 10 meters, therefore each motor has a considerable motor speed.

The results help us understand the behavior of the quadcopter as shown in figure 7. The quadcopter undergoes the simulation conditions displaying stable behavior. The angular velocities and rotational angles change minutely, as the quadcopter maneuvers itself with stability at the height of 10 meters.

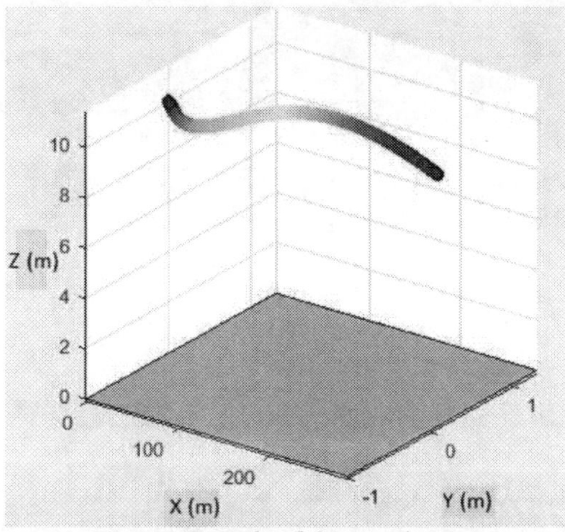

Figure 5. Path of the quadcopter

Figure 5 shows the path of the quadcopter in the simulation.

While maintaining the altitude of 10 meters, the quadcopter continues to move with a constant acceleration of 5.17 m/s^2. It completes the path of 275 meters in x-direction while maintaining the height of 10 meters in z-axis. There is hardly any change in position in y-axis as expected, since it is not commanded to maneuver itself in y-axis. Between 4-6 seconds it experiences a disturbance in y-direction but stabilizes itself well enough. The information retrieved from the results is very beneficial for us. We get to know the behavior of the quadcopter under a specific set of conditions. We get to know the position, velocity and attitude of our aircraft as shown in figure 7. The attitude and velocity of our aircraft along with the rotational angles of roll, pitch and yaw motions can now be determined for any maneuver. Also, as per our discussion in section 3, we now know that, for forward motion at the height of 10 meters, there is a positive pitch angle of 10 degrees, which shows that our simulation is in line with the command given as well as theory. There is a forward pitch, since the quadcopter is moving forward while maintaining the altitude. When we implement our fifth rotor, we know that there should be no positive pitch, in fact the pitch angle should be zero, since the forward thrust rotor will cater for the thrust required in the direction of x-axis.

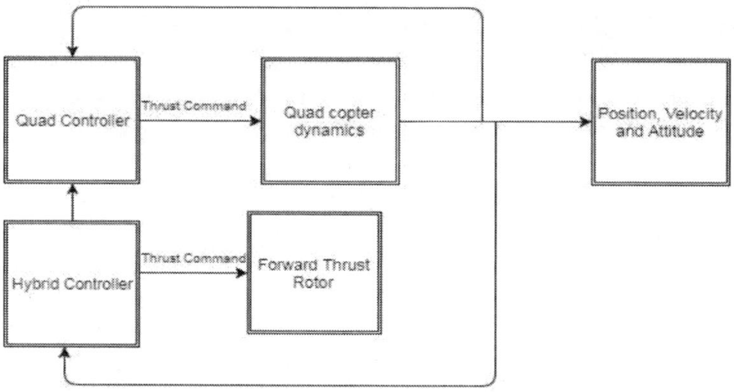

Figure 6. Input-output structure for the complete design of simulation model

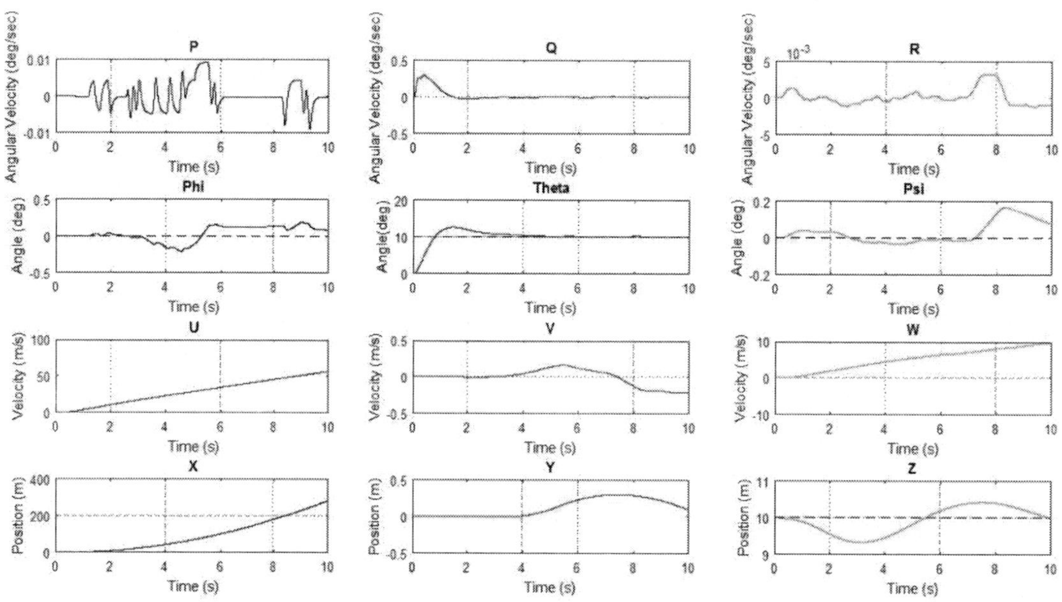

Figure 7. Position, velocity and attitude of the quadcopter

5. Conclusion

The results obtained are in line with the theory. Also, the results are satisfactory as expected according to the literature review.

Considering all the collected data, we are now able to set up the input and feed back to our controller from the simulation results for our forward thrust rotor and thereby our developed control structure can now be further expanded to incorporate a fixed wing and a forward thrust rotor. As a further expansion of the project, a fully developed split control can then be made by incorporating this designed control structure to introduce a totally unique form of hybrid aircraft and its control in the industry. This fully developed hybrid control structure shall be able to provide solutions for problems such as thrust saturation.

Acknowledgment

The support of Mechanical Engineering Department, Faculty of Engineering, Chiangmai University is gratefully acknowledged.

References

[1] Ryll M, Bülthoff H H and Giordano P R 2013 *Proc. IEEE Int. Conf. on Robotics & Automation* pp 295-302

[2] Pounds P, Mahony R and Gresham J 2004 *Proc. of the 2004 Australasian Conf. 2004*

[3] Gan D, Cai G, Dias J, Seneviratne L, Dias J and Seneviratne L 2013 *Proc. IEEE 20th Int. Conf on Electronics Circuits, and Systems (ICECS)* pp 597-600.

[4] Shakev N G, Topalov A V, Kaynak O, Shiev K B 2012 *J. of Intelligent & Robotic Systems* **65**(1-4) pp 389-408

[5] Bittar A, Figuereido H V, Guimaraes P A and Mendes A C 2014 *Proc. Int. Conf. on Unmanned Aircraft Systems (ICUAS)* pp 993-1002

[6] Andrew Gibiansky, *Quadcopter Dynamics, Simulation and Control*

[7] Luukonnen T, "Modeling and control of quadcopter", Independent Research project in Applied Mathematics, Espoo, August 22, 2011.

[8] Cai Guowei et al. 2009 *Mechatronics* **19** 1057-1066

[9] Partovi Ali Reza, et al. 2011 *IFAC Proc.* vol 44 pp 10385-10390

[10] Notarstefano Giuseppe and John Hauser 2010 *IFAC Proc.* vol 43.14 pp 119-124.

[11] Escareno J Salazar S and Lozano R 2006 *Proc. 45th IEEE Conf. on Decision and Control* (San Diego, California)

[12] Hugo Romero et al. 2007 *Proc. 2007 IEEE/RSJ International Conference on Intelligent Robots and Systems* pp 147-152

[13] Sérgio Ronaldo Barros dos Santos et al. 2011 *Proc. 21st Brazilian Conference for Mechanical Engineers 2011, Proceeding COBEM 2011.*

[14] Dongwon Jung and Panagiotis Tsiotras, "Modeling and hardware-in-the-loop simulation for a small unmanned air vehicle", Georgia Institute of Technology Atlanta, GA, 30332-0150.

Railway bogie vibration analysis by mathematical simulation model and a scaled four-wheel railway bogie set

Noppharat Visayataksin*, and Manon Sooklamai

Mechanical Technology Education, Faculty of Industrial Education and Technology King Mongkut's University of Technology Thonburi , Bangkok 10140, Thailand

* Corresponding Author: GuNopvisayataksin@gmail.com, 662-470-8526

Abstract. The bogie is the part that connects and transfers all the load from the vehicle body onto the railway track; interestingly the interaction between wheels and rails is the critical point for derailment of the rail vehicles. However, observing or experimenting with real bogies on rail vehicles is impossible due to the operational rules and safety concerns. Therefore, this research aimed to develop a vibration analysis set for a four-wheel railway bogie by constructing a four-wheel bogie with scale of 1:4.5. The bogie structures, including wheels and axles, were made from an aluminium alloy, equipped with springs and dampers. The bogie was driven by an electric motor using 4 round wheels instead of 2 straight rails, with linear velocity between 0 to 11.22 *m/s*. The data collected from the vibration analysis set was compared to the mathematical simulation model to investigate the vibration behavior of the bogie, especially the hunting motion. The results showed that vibration behavior from a scaled four-wheel railway bogie set significantly agreed with the mathematical simulation model in terms of displacement and hunting frequency. The critical speed of the wheelset was found by executing the mathematical simulation model at 13 *m/s*.

1. Introduction

Hunting motion is a self-oscillation running off the centreline or a swaying motion of the railway vehicles while running on the railway track. This hunting motion is one of the general vibration behaviors of railway vehicles that always occurs; it affects riding comfort of passengers; it is a cause of wheel and rail damage; and, most seriously, it could cause the derailment of railway vehicles [1]. In order to minimize or eliminate the problem of hunting motion, the design of wheel and rail geometry needs to be corrected [2, 3], using suitable primary and lateral suspension stiffness and damping parameters, such as a special camping design and characteristics, nonlinear spring and damper coefficients [4-7], and take into consideration the relevant factors that affect the rail vehicles' stability, and increase the critical speed [8-10] which results in riding comfort and safety.

This work was to investigate the relevant factors that affect the vibration behavior of railway vehicles, especially the hunting motion behavior. Two types of investigation tools were therefore developed, one being a mathematical simulation model, and the other was a scaled four-wheel bogie set. Then the vibration behavior and influence of the bogie's parameters was investigated by varying the variables in a mathematical simulation model, and the experimental result from the scaled four-wheel bogie set were compared with each other to authenticate the simulation results.

2. Methodology

In this research, as mentioned above, two types of investigation tools were developed. The mathematical simulation model was developed based on an existing theory for the dynamics of railway vehicle systems [1]. A scaled four-wheel railway bogie set was constructed in small scale, based on real railway bogie frame and dimensions.

2.1. Mathematical Simulation Model

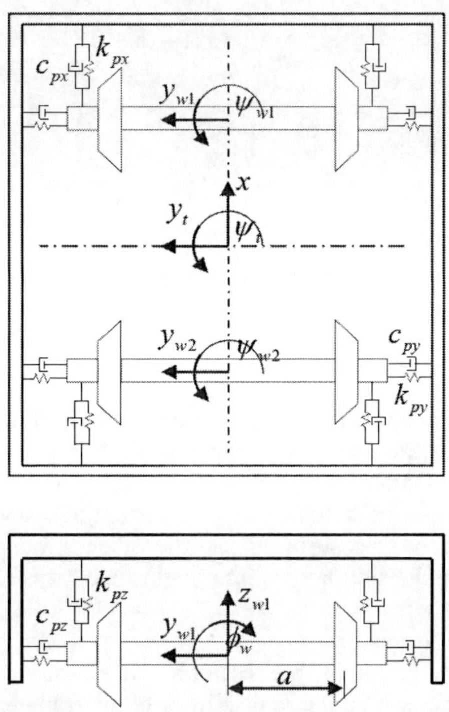

Figure1. Four-wheel, two-axle railway bogie model.

Figure 1 shows the physical model of the bogie with independently rotating wheelsets. In this investigation model, the focusing point is the hunting motion stability. The governing differential equations of the motion of the bogie are given as [7]

$$m_w\ddot{y}+\left(\frac{2f_{11}}{v}+\frac{2f_{11}}{v}\frac{r_0\lambda}{a}\right)\dot{y}-2f_{11}\dot{\psi}+2\frac{f_{12}}{v}\psi+W_a\frac{\lambda}{a}y+F_{sywi}=0 \qquad (1)$$

$$I_{wx}\ddot{\psi}+I_{wy}\frac{v\lambda}{r_0a}\dot{y}+\frac{2af_{33}\lambda}{r_0}y-\frac{2f_{12}}{v}\dot{y}-\frac{2f_{12}r_0}{v}\frac{\lambda}{a}\dot{y}+2f_{12}\psi+\frac{2a^2f_{33}}{v}\dot{\psi}-aW_a\lambda\psi+\frac{2f_{22}}{v}\psi=0 \qquad (2)$$

where F_{sywi} is the suspension force, given as below [2]

$$F_{sywi} = k_{py}y + c_{py}\dot{y} \tag{3}$$

where f_{33}, f_{11}, f_{12}, f_{22} are the nonlinear creep forces according to Kalker's linear creep theory, the creep coefficients are given by [9]

$$f_{33} = Gmn\left[\frac{3\pi\left(1-\sigma^2\right)}{2E\left(A+B\right)}N\right]^{\frac{2}{3}} C_{11} \tag{4}$$

$$f_{11} = Gmn\left[\frac{3\pi\left(1-\sigma^2\right)}{2E\left(A+B\right)}N\right]^{\frac{2}{3}} C_{22} \tag{5}$$

$$f_{12} = Gmn\left[\frac{3\pi\left(1-\sigma^2\right)}{2E\left(A+B\right)}N\right] C_{23} \tag{6}$$

$$f_{22} = Gmn\left[\frac{3\pi\left(1-\sigma^2\right)}{2E\left(A+B\right)}N\right]^{\frac{4}{3}} C_{33} \tag{7}$$

$$\left(A+B\right) = \frac{1}{2}\left(\frac{1}{R_w} + \frac{1}{r_r} + \frac{1}{r_w}\right) \tag{8}$$

where R_w, r_w, r_r, G, C_{ij}, σ, m, n, are the wheelset rolling radius, transverse radius of wheel profile, transverse radius of rail profile, the shear modulus, non-dimensional Kalker coefficient, Poisson ratio and coefficients related to *A* and *B*, respectively. And *N* is the wheel/rail normal force that is affected by the aerodynamic loads and motion of the vehicle. For this work, the important variables used in the mathematical simulation model such as creep force and several other parameters are shown in table 1. These were taken by measuring from the scaled four-wheel railway bogie set that was constructed for this work. In the simulations, the bogie speed was at 6.70 *m/s*, 8.96 *m/s*, and 11.22 *m/s*.

2.2. The Scaled Four-Wheel Bogie Set

One of the original tasks performed in this research was a scaled four-wheel railway bogie set. It was constructed in a small scale based on the real railway bogie frame configuration and dimensions with a scale of about 1:4.5 (see figure 2). The overall size of the scaled bogie is approximately 70 cm long, 55 cm wide, and 25 cm high; the driven wheels are not included in these dimensions. The bogie does not have power to run itself, it is therefore driven by the drive wheels that also support the bogie, instead of a straight rail track. Each drive wheel is supported directly to a driven wheel (train's wheel). Therefore, one set of a scaled four-wheel railway bogie is composed of four drive wheels, and all wheels run at the same rotation speed controlled by an electric motor with a variable speed drive control system. In the experiment, the bogie operated from 0 – 11.22 *m/s* in terms of linear velocity by controlling the VSD control panel. See figure 3.

Table 1. Parameters used in the mathematical simulation model

Parameters	Variable	value
Wheelset mass	m_w	$11.6\,kg$
Roll moment of the inertia of the wheelset	I_{wx}	$60.12\,kg\cdot m^2$
Spin moment of the inertia of the wheelset	I_{wy}	$0.92\,kg\cdot m^2$
Yaw moment of the inertia of the wheelset	I_{wz}	$60.12\,kg\cdot m^2$
Wheel radius	r_0	$0.10\,m$
Half of the distance between contact points on two rails	a	$0.17\,m$
Lateral creep force coefficient	f_{11}	$2{,}589{,}438.65\,N$
Lateral/spin creep force coefficient	f_{12}	$1{,}006.63\,N\cdot m$
Spin creep force coefficient	f_{22}	$1.10\,N\cdot m^2$
Longitudinal creep force coefficient	f_{33}	$2{,}750{,}559.28\,N$
Spring stiffness of lateral primary suspension	k_p	$400{,}000\,N/m$
Viscous damping constant of lateral primary suspension	c_p	$1{,}800\,Ns/m$
Wheel conicity	λ	0.05
Weight of bogie	W_a	$245.25\,N$

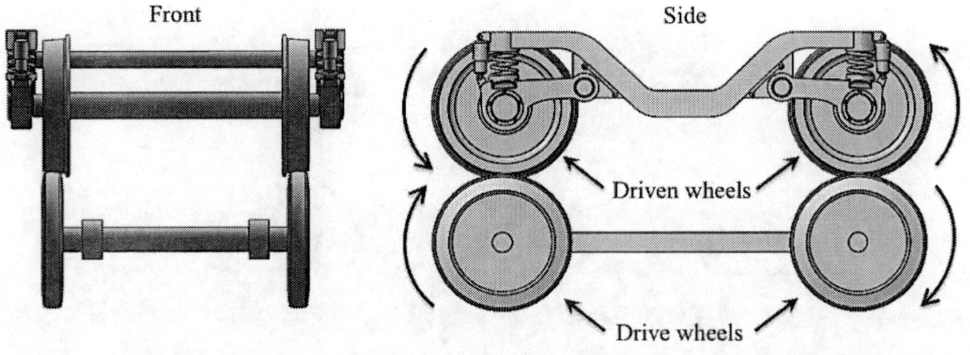

Figure 2. Main components and configuration of the scaled four-wheel railway bogie set.

Figure 3. A scaled four-wheel railway bogie set on a stand and control panel.

In order to observe the bogie motion, for this first phase of the research, we focused on the hunting motion, so the important required data was the oscillation of the wheelset in terms of time-distance domain. Therefore, Arduino UNO board and HC-SR04 distance sensor were installed to collect the required vibration data, as shown in figure 4.

Figure 4. Displacement sensor installation position.

3. Results and Discussions

3.1. Validation of Mathematical Simulation Model

In order to investigate the vibration behavior of the railway bogie by using a mathematical simulation model, it was very important to validate the numerical simulation model before analyzing further steps later. In this work we compared the numerical simulation result with the reference model developed by others [2] for ensuring that the developed numerical simulation model was correct. It was shown that both results were in agreement with each other, as shown in figure 5.

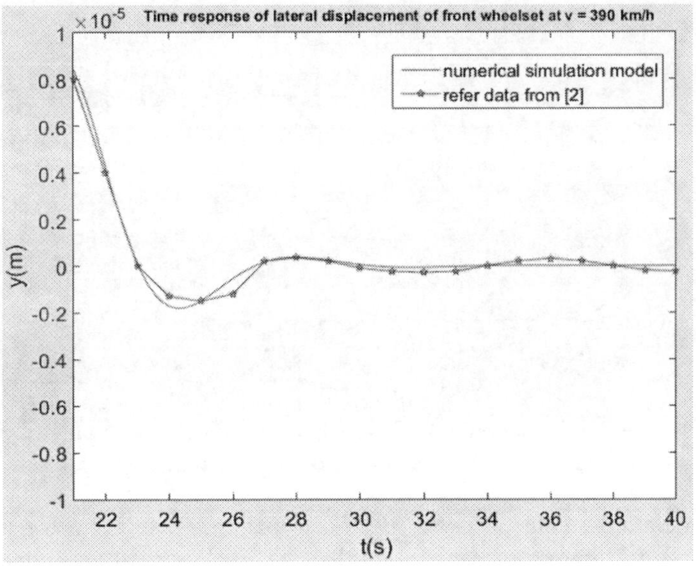

Figure 5. Transversal displacement of the wheelset at the same parameters, between the developed simulation model and the reference model.

3.2. Simulation Results

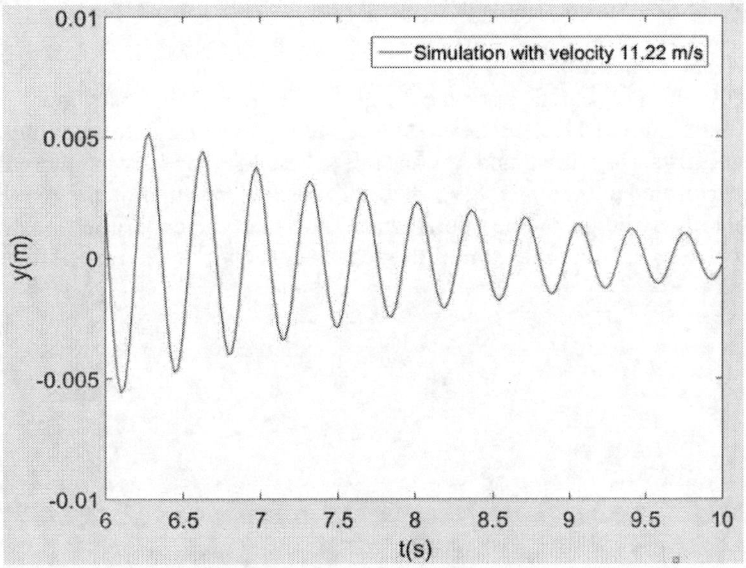

Figure 6. Transversal displacement of the wheelset while running at 11.22 *m/s* which is the maximum speed of a scaled four-wheel railway bogie set.

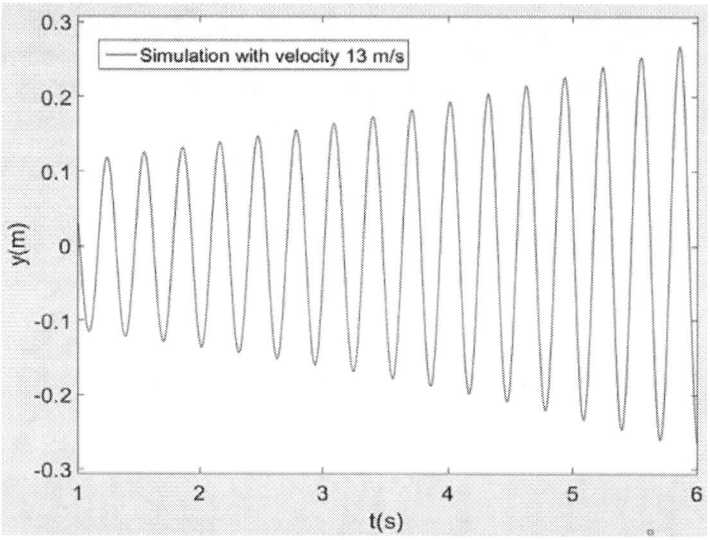

Figure 7. Transversal displacement of the wheelset while running at 13 *m/s*, which is where the critical speed appeared.

From the simulation results at lower speed, for example at 11.22 *m/s* (with the displacement-time relationship shown in figure 6), it was found that the displacement was continuously decreasing. This meant that the wheelset was still stable. However, in the simulation at 13 *m/s* it was found that the displacement was increasing. Therefore, it can be debated that 13 *m/s* is the critical speed where the wheelset loses stability to run on the track. Please note that, these simulation results were based on the parameter values shown in table 1, which are smaller than real railway bogie parameters, e.g. the wheel diameter. Therefore, the linear velocity of the wheelset found in this simulation is too low when compared to a real situation. In the real situation, this critical speed will cause the rail vehicle to be derailed. However, we also compared the results of the experiment that was performed with the scaled four-wheel railway bogie set, the results are shown in the following sections.

3.3. Experimental and Simulation Results

For the experiment with the scaled four-wheel railway bogie set, the instrument could vary the speed from 0 *m/s* to the maximum of 11.22 *m/s*. But, at low speed or velocity, the oscillation or hunting motion of the wheelset is also very low, and increasing when the speed was increased. Therefore, this experiment was performed at 6.70 *m/s*, 8.96 *m/s*, and 11.22 *m/s*, which is the maximum speed that this scaled four-wheel railway bogie can be operated at. After that, the experiment results were put together with the simulation results. A comparison of the results is shown in figures 8-10:

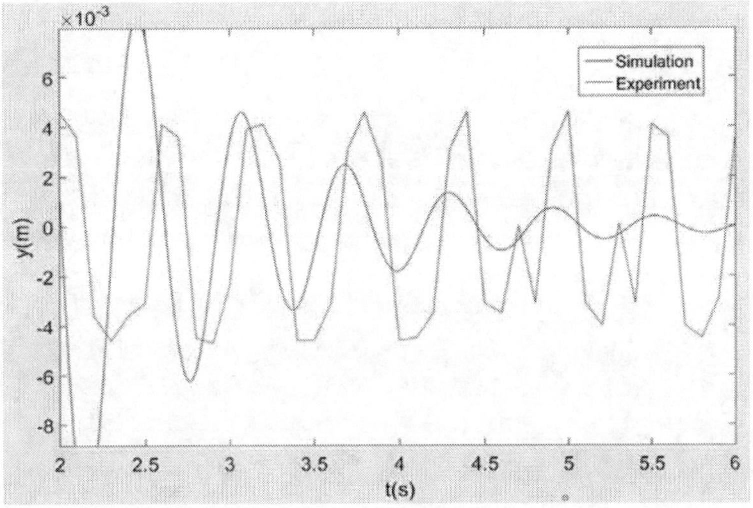

Figure 8. Transversal displacement of the wheelset while running at 6.70 *m/s*, comparison between experiment and simulation.

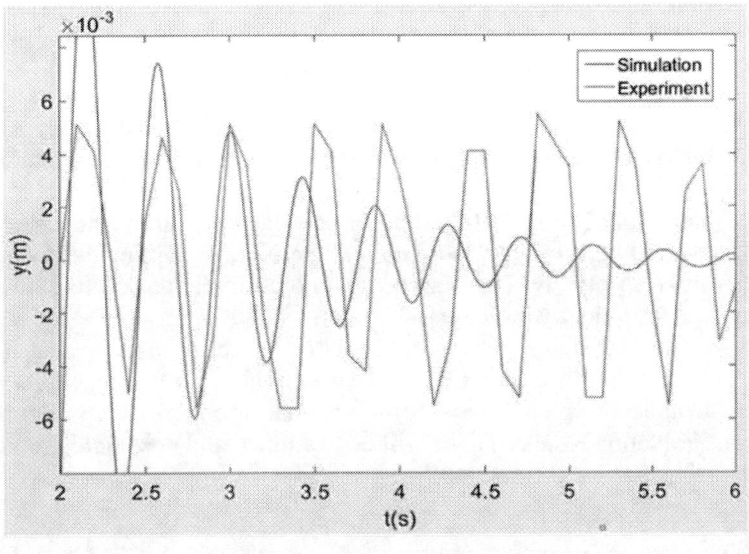

Figure 9. Transversal displacement of the wheelset while running at 8.96 *m/s*, comparison between experiment and simulation.

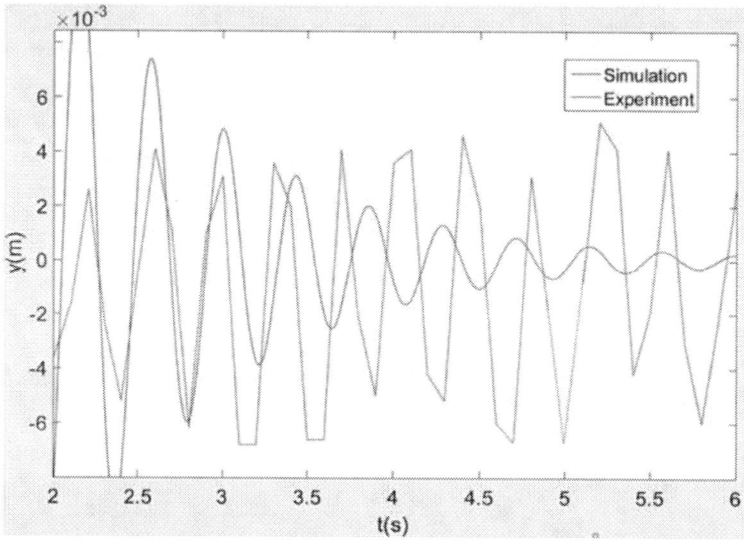

Figure 10. Transversal displacement of the wheelset while running at 11.22 *m/s*, comparison between experiment and simulation.

From the comparison results between the experiment and mathematical simulation model, when compared to each other at the same velocity, of course, the experiment and the simulation results are not perfectly aligned. However, the tendency of the curves is consistent and there are points that are comparable, see figures 8-10.

Firstly, we consider the frequency of the oscillation. At a lower one (figure 8), the frequency was lower when compared to the higher velocity (figure 9 or 10), in which the oscillation frequency was also higher.

Secondly, consider the magnitude of the hunting displacement from the experiment. At velocities of 6.70 *m/s*, 8.96 *m/s*, and 11.22 *m/s*, it was found that the average hunting displacements were 4.6 mm, 5.6 mm, and 6.8 mm, respectively. These agreed with the simulation results showing that the hunting displacement increases the velocity increases.

Finally, we consider the stable convergence of the wheelset. The results showed a difference between the two methods. By this instance of the research, the main factors will be the spring stiffness, damping coefficient, and the creep force coefficient that strongly affects the stability of both the experiment and the simulation results. These will be examined and considered to get the correct values at a later date.

4. Conclusions

In this work, a mathematical simulation model was developed for analyzing the influence of bogie component parameters that affect the stability of the railway vehicles, This work focused on hunting motion behavior. A scaled four-wheel railway bogie set was also constructed. This scaled railway bogie was able to demonstrate the bogie movement more easily and more safely than observing a real railway bogie operation, which is also prohibited due to safety rules of the company.

From the mathematical simulation model, we can ascertain the influence of each parameter that affects the hunting motion, by varying the parameter values. Also, when comparing the simulation results with the experimental results, we found both agreeing and conflicting results. However, the disagreeing results could be because of the still unknown exact parameter values. Therefore, the next step is to get the correct parameter values so the mathematical simulation model can predict and give reliable results.

Acknowledgment
This research work is partially supported by the National Science Technology and Innovation Policy Office (STI) under the pilot project for curriculum and manpower development to support railway transportation growth in Thailand, year 2015.

References
[1] Vijay K G and Rao V D 1984 *Dynamics of Railway Vehicle Systems* (Canada: ACADEMIC PRESS)
[2] Lee S Y and Cheng Y C 2005 *J. Sound. & Vibration* **282** pp 881-898
[3] Karim H, ALI A, and Khan A 2011 *J. Eng. Sci. & Tech.* **6** (2) pp 146 – 160
[4] Hiroshi Y, Hirokazu T, and Hirokazu O 2008 *J. Vibration & Control* pp 209-230
[5] Daniel B and Simona B 2012 *Romanian J. Acoustics & Vibration* **9**(1) p 33
[6] Tudor S, Ioan S, and Daniel B 2010 *Proc. The Romanian Academy, Series A.* **11**(4) pp 355–362
[7] Yabuno H, Okamoto T, and Oshima N 2002 *Meccanica* **37** pp 555-568
[8] Pilkee K, Jeehyun J and Jongwon S 2011 *Int J. Precision Eng. & Manufacturing* **12**(3) pp 505-519
[9] Ranjbar M and Ghazavi M R 2012 *Proc. Int. Conf. on Advances in Robotic, Mechanical Eng. & Design* pp 11-14
[10] Xiao-Hui Z, Han W, Jiang L and Hong-Zhi S 2014 *Proc. Int. Conf, on Acta Mechanica Sinica* pp 889–900

Enhanced control of a flexure-jointed micromanipulation system using a vision-based servoing approach

T. Chuthai[1,2], M. O. T. Cole[1,2,*], T. Wongratanaphisan[1,2] and P. Puangmali [1,2]

[1] Department of Mechanical Engineering, Faculty of Engineering, Chiang Mai University, Chiang Mai 50200, Thailand
[2] Center for Mechatronic Systems and Innovation, Chiang Mai University, Chiang Mai 50200, Thailand

[*] Corresponding Author: motcole@hotmail.com, +66 53 944146

Abstract. This paper describes a high-precision motion control implementation for a flexure-jointed micromanipulator. A desktop experimental motion platform has been created based on a 3RUU parallel kinematic mechanism, driven by rotary voice coil actuators. The three arms supporting the platform have rigid links with compact flexure joints as integrated parts and are made by single-process 3D printing. The mechanism overall size is approximately 250x250x100 mm. The workspace is relatively large for a flexure-jointed mechanism, being approximately 20x20x6 mm. A servo-control implementation based on pseudo-rigid-body models (PRBM) of kinematic behavior combined with nonlinear-PID control has been developed. This is shown to achieve fast response with good noise-rejection and platform stability. However, large errors in absolute positioning occur due to deficiencies in the PRBM kinematics, which cannot accurately capture flexure compliance behavior. To overcome this problem, visual servoing is employed, where a digital microscopy system is used to directly measure the platform position by image processing. By adopting nonlinear PID feedback of measured angles for the actuated joints as inner control loops, combined with auxiliary feedback of vision-based measurements, the absolute positioning error can be eliminated. With controller gain tuning, fast dynamic response and low residual vibration of the end platform can be achieved with absolute positioning accuracy within ±1 micron.

1. Introduction

High-precision micromanipulation systems have been increasingly applied for various automated tasks over the last two decades, such as biological micromanipulation [1, 2, 3], micro-assembly tasks for MEMS [4] and optical components [5], specimen handling [6, 7], micro positioning stages [8], scanning microscopy and atomic force microscopy (AFM) [9]. Most of these applications involve the use of compliant mechanisms (composed of flexure joints instead of bearing joints) due to their advantages in achieving high-fidelity, friction-free motion. Issues of backlash, wear and lubrication are avoided and they can also be easier to fabricate.

Due to assembly error, kinematic model error, thermal drift and elastic effects, a micromanipulation system may still not be sufficiently accurate when using local/decentralized control of actuators. To improve accuracy requires additional sensors to measure the position of the end-effector/platform, ideally in real-time. However, standard sensor systems having micrometric resolution and centimeter ranges have restricted application. For instance, interferometers have large range and very high resolution but can be bulky and are able to measure rectilinear motion only. Multi-direction measurement requires a combination of several sensors, which may not be possible due to restrictions on space or geometry [10]. An alternative for noncontact sensing with several measurement directions is to use a digital microscope vision system. However, this then requires developing a suitable approach for closed-loop visual servoing, which is the focus of the present work.

There have been several works on visual servoing with digital microscopy based on image feature tracking where the working area is limited to the field of view of the microscope [2, 11, 12]. A method to achieve increased measurement range and working area for an X-Y stage was shown in [10] and [13], where a pseudo-periodic pattern (PPP) imprinted on the end platform allowed absolute position information to be obtained by image processing. However, this approach required a specially designed and fabricated patterned plate. Also, the processing algorithm is computationally intensive, which means update rates may not be sufficient for real-time feedback in motion servoing. An alternative considered in this paper is to use a uniform grid pattern on the end platform and implement an incremental tracking algorithm via image processing. This approach enables vision-based measurements over an extended workspace and allows update rates suitable for real-time servoing.

This paper describes a vision-based control approach for a flexure-jointed micromanipulation system with a Delta-like parallel kinematic architecture and rotary voice coil actuation. An initial comparison of linear and nonlinear PID feedback control in joint space is made. Nonlinear PID control is then augmented with task-space visual servoing to achieve enhanced speed, accuracy and stability of motion. The remainder of the paper is organized as follows: section 2 gives a system description. Basic dynamic characteristics and results from step response testing under PID control are presented in section 3. Section 4 describes the vision-based control method. Section 5 introduces an enhanced control combining nonlinear PID control with vision-based servoing. The final section draws conclusions.

2. System Description

2.1. A Flexure-jointed Micromanipulator

A desktop experimental motion platform has been created based on a Delta-type (3RUU) parallel kinematic architecture (figure 1). The system has been designed to achieve high fidelity translational motion of the end platform. The supporting arms for the platform have rigid links with compact flexure joints as integrated parts made by single-process 3D printing (stereolithographic process with photoreactive resin). This allows simple and low-cost manufacture of the system without sacrificing attainable motion precision. The three arms are driven by rotary voice coil actuators having angular range of ±17°. The mechanism overall size is approximately 250x250x100 mm. A selection of flexure-hinge designs are shown in figure 2. A simple notch design with a round cutout from both sides (figure 2a) is commonly used in micro-positioning manipulators, e.g. [8, 14]. Longer beam-like flexures, as shown in figures 2b-2d can achieve higher range of bending motion. To achieve high angular range with a more compact joint, a series connection of flexures may be adopted [5]. The design used for the current system is a compact S-beam flexure, as shown in figure 2e. The joint is sufficiently compliant to give an angular range of motion of approximately ±16°.

2.2. Workspace Simulation

The kinematic equations for the mechanism can be derived using a pseudo-rigid-body modeling (PRBM) approach where a revolute joint is considered to be located at the center of each flexure [15]. Hence, we consider 3 parallel RUU chains (as in [16]) but with some additional joint offset parameters that account for non-collocation of the revolute joints. This leads to an inverse kinematic solution in the form:

$$\theta_i = 2\tan^{-1}\left(\frac{-F_i \pm (E_i^2 + F_i^2 + G_i^2)^{1/2}}{G_i - E_i}\right), \qquad i = 1,2,3 \tag{1}$$

where θ_i are the angles for each actuator and E_i, F_i and G_i are length parameters that are functions of the link parameters and the Cartesian position of the end-platform.

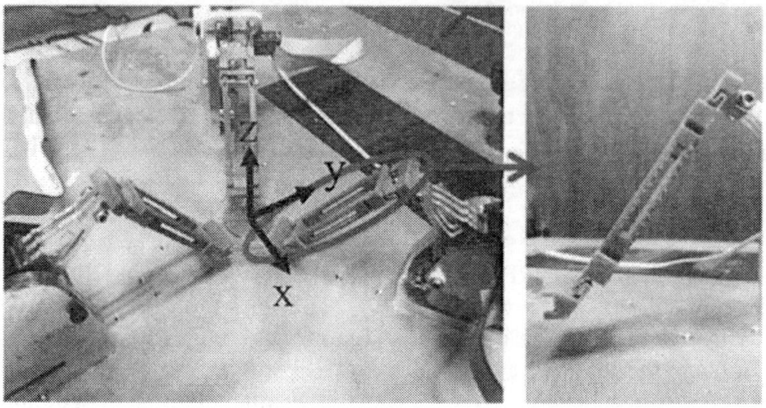

Figure 1. Flexure-jointed micromanipulator with moving end-platform based on PKM architecture

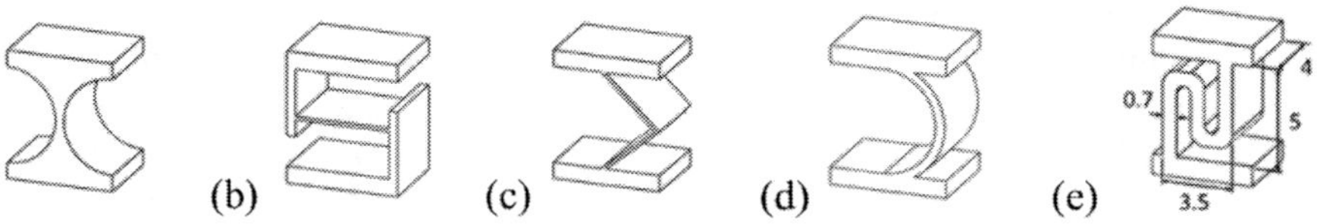

Figure 2. Different types of flexure joints: (a)-(d) are from [5], (e) is current design

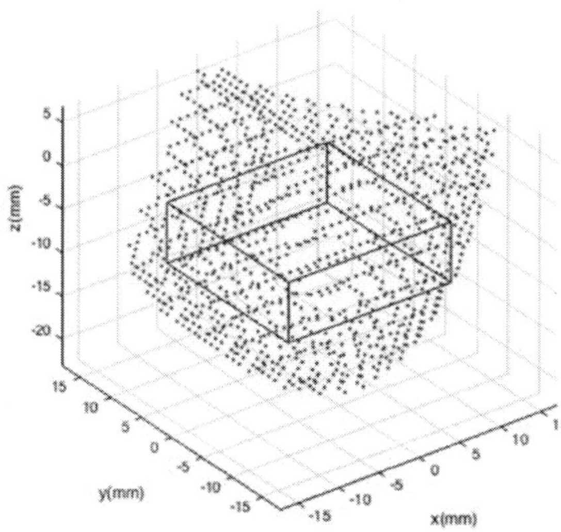

Figure 3. Simulated mechanism workspace: red dots define the reachable workspace and the blue box shows the operating workspace.

By using equation (1), an approximate workspace analysis for the flexure-jointed micromanipulator can be undertaken. Figure 3 shows the reachable workspace with boundary defined by red dots and the target operating workspace with boundary shown by the blue box having dimensions of 20x20x6 mm.

2.3. *Experimental Setup*

The experimental setup shown in figure 4 comprises the Delta mechanism, with three rotary voice coil actuators, placed on a vibration isolation table. A digital microscope (Point Grey FL3-U3-13Y3M-C camera with 10X DIN plan objective lens) is used for end platform observation. Rotation of the actuators is measured by 14-bit rotary magnetic absolute encoders (AS5048A). Joint control servo-loops are implemented on a microcontroller (STM32F4 Discovery) with a sampling frequency of 1000 Hz. The control actuation signal is output through DACs (MCP4922) via SPI communication, and these signals are input to H-bridge driver circuits which power the actuator coils.

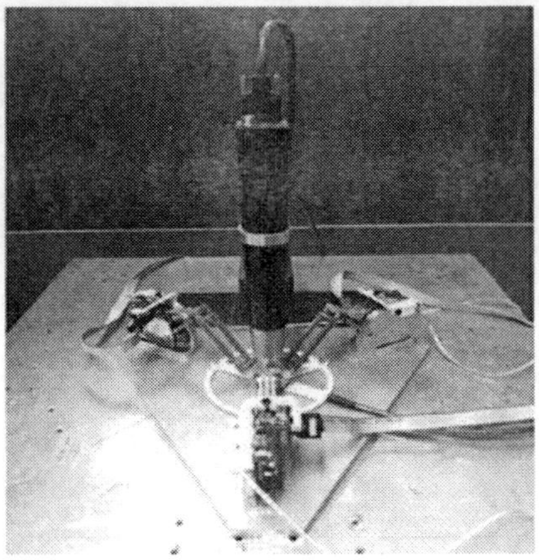

Figure 4. Experimental set-up for visual servoing.

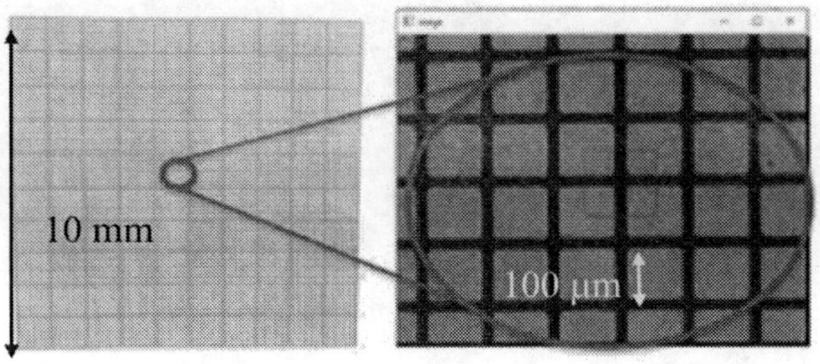

Figure 5. Grid pattern on platform top surface.

Image processing is performed on a desktop PC using Open CV libraries. The camera image is 1280x1048 pixels on a 1/2-inch CMOS detector. Consequently, with 10X objective lens, the image scale factor is 0.48 μm per pixel. The image data can be used

1. to verify the accuracy of kinematic models
2. to assess performance of the joint control loops
3. for feedback control of measured platform position

The platform position measurement is achieved by observing the motion of a uniform grid pattern fixed on the top surface of the platform, as shown in figure 5. This grid pattern is a standard microscope reticle with grid spacing of 100 μm. The image processing technique is to be described in detail in section 4.

3. Behavior under PID Joint Angle Feedback

3.1. *Frequency Response Characteristics*

The dynamic properties of the mechanism can be assessed by frequency response testing. This requires measurement of how the system reacts to a sinusoidal input signal. The frequency of the input signal driving a single actuator was varied between 0.5 Hz and 200 Hz. Response information (amplitude/phase) was extracted from the output signal, which is the rotation angle for the actuated joint. The frequency response is acquired as shown in figure 6. Note that these results are for the open-loop (uncontrolled) mechanism.

Four resonant peaks are seen in the data at 5, 14.5, 16.5 and 75.5 Hz. The three low frequency modes relate to 3-DOF translational motion of the end platform. These natural frequencies are non-zero due to the intrinsic stiffness of the flexure joints: the joints produce elastic forces under rotation that tend to resist motion. The high frequency mode is associated with a resonance of the compliant structure and involves a more complex vibration pattern for the mechanism links.

These results show that the flexure-based mechanism exhibits a complex multi-mode response behavior due to the vibratory dynamics associated with the joint compliance. Although we can anticipate that established approaches to the control of parallel kinematic mechanisms (such as PID control or active disturbance rejection) might be applicable, the selection and tuning of controllers for good performance is more challenging.

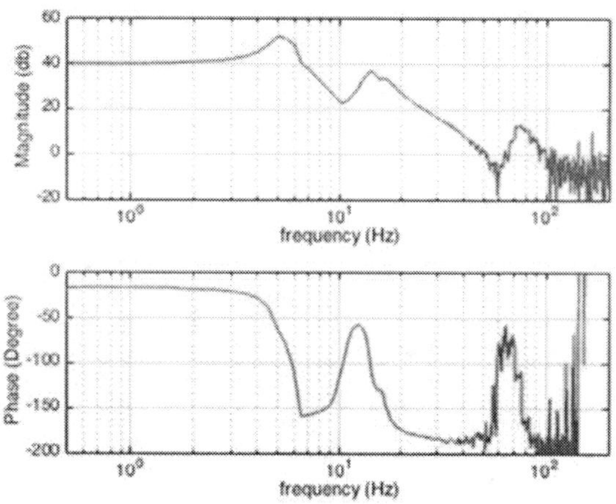

Figure 6. Measured frequency response data for open loop system (excited by single actuator).

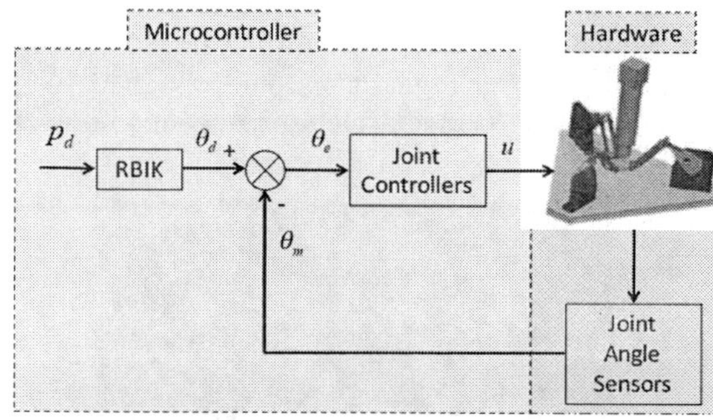

Figure 7. Block diagram for joint feedback control.

3.2. Task Space Transient Response

To control motion of the mechanism, the rigid body inverse kinematic model (RBIK) from equation (1) may be applied in combination with localized feedback control loops for each actuated joint, as shown in figure 7. In this way, the task-space (x-y-z) position demand variables may be specified to achieve the desired motion of the end platform. There are three main considerations for the control performance, which are

1. speed of response (settling time and rise time)
2. steady-state positioning accuracy
3. effect of disturbances, including noise

These aspects should be assessed with respect to the actual motion of the end platform. Without visual servoing, however, the steady-state positioning accuracy is highly dependent on the accuracy of the kinematic model.

An initial investigation of step response behavior under PID control of the actuators was performed focusing on planar (x-y) motion of the end platform. For each step response test, the x-position demand was changed repeatedly from 0 to 100 μm and then back to 0. Each step occurs immediately after the position of the platform has been maintained within δ pixels for 0.5 seconds. This allowed a comparison of basic performance measures for different controller gain settings. These were assessed using vision-based measurements. Rise time is assessed as the time taken to move between 10% and 90% of the final position change. Note that the steady-state response is subject to noise disturbance. Therefore, the settling time is defined as the time to reach and stay within δ pixels of the quasi-steady position (determined by a moving window average). Note that the value of δ for the settling criterion must be chosen to exceed the noise-induced oscillation of the end-platform. Therefore, a suitable value for δ must match the noise attenuation performance of the controller.

Experimental results are presented in figure 8 where the upper graph shows joint angle response at actuator 1 and the lower graph shows the end-platform motion in x direction. The measured angular position (red line) from the magnetic sensor has significant noise and is subject to a fairly coarse quantization. These sensors have 14-bit resolution and so the angle is quantized in steps of 0.022°. The green line shows the same signal but with low-pass filtering. This is shown for visualization only. A red line on the lower graph shows the measured x-position of the end-platform obtained from the digital microscope via image processing (as detailed in the next section). The blue line on both graphs represents the reference (demand) signal in joint-space and task-space respectively.

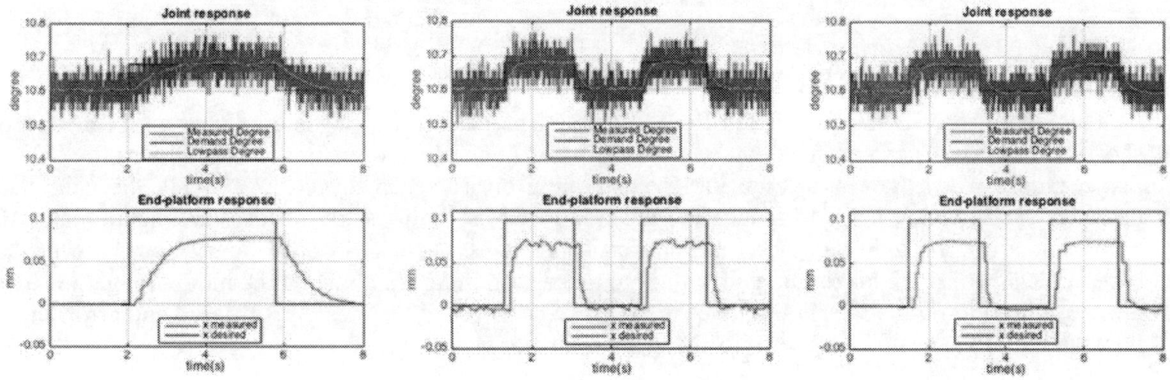

(a) linear, low gain: PID1 (b) linear, high gain: PID2 (c) nonlinear: NPID

Figure 8. Step response measurements for different PID joint controllers.

Figures 8a and 8b show the responses obtained with different linear gain settings: low gain (PID1) and high gain (PID2). In figure 8a, gains are tuned to minimize noise-induced oscillation of the end-platform in steady state. This requires a low-gain setting which then gives a slow rise time but can achieve a small bound on the variation (oscillations) in steady-state positioning (to within 2 pixels or ±0.48 μm). On the other hand, in figure 8b gains were tuned as high as possible before the system became unstable. This gain setting gives fast rise time but there are large noise-induced oscillations under steady-state positioning (within 10 pixels or ±2.4 μm).

In order to exploit the advantage of both gain settings, nonlinear feedback gains are introduced. The nonlinear feedback gains are defined as

$$K(\tilde{e},\alpha,\beta,\gamma,C)=\begin{cases} C\cdot\beta^{\alpha-1} & ; \quad |\tilde{e}|<\beta \\ C\cdot|\tilde{e}|^{\alpha}\, sign(\tilde{e}) & ; \quad \beta\le|\tilde{e}|<\gamma \\ C\cdot\gamma^{\alpha-1} & ; \quad |\tilde{e}|\ge\gamma \end{cases} \tag{2}$$

with

$$\alpha=\ln\left(\frac{k_h\beta}{k_l\gamma}\right)\Big/\ln\left(\frac{\beta}{\gamma}\right), \qquad C=k_h\gamma^{1-\alpha} \tag{3}$$

where k_h is maximum gain value, k_l is minimum gain value. Defining the joint angle error as e and low-pass filtered error $\tilde{e}$ and derivative $\dot{\tilde{e}}$, the control law is given by

$$u = K_p(\cdot)e + \int_0^t K_I(\cdot)ed\tau + K_d(\cdot)\dot{e} \tag{4}$$

$$K_p(\cdot) = K(\tilde{e}, \alpha_p, \beta_p, \gamma_p, C_p)$$

$$K_I(\cdot) = K(\tilde{e}, \alpha_i, \beta_i, \gamma_i, C_i)$$

$$K_d(\cdot) = K(\tilde{e}, \alpha_d, \beta_d, \gamma_d, C_d)$$

This nonlinear PID law is adapted from [17, 18]. The basic form of the nonlinear gain function is shown in figure 9. It can be seen that when the error is large, the gain is large, which helps to achieve fast response when there is a large initial positioning error. When the error is small, the gain also becomes small, which helps to reduce the noise-induced oscillations for steady-state positioning. An upper bound on the gain value (k_h) is imposed to avoid issues with control input saturation. In the experiments, we set the maximum gain values to those used in figure 8b while the minimum gain values are those used in figure 8a. The limits of the transition interval (β, γ) were set by trial and error under observation of the response under step changes in position demand.

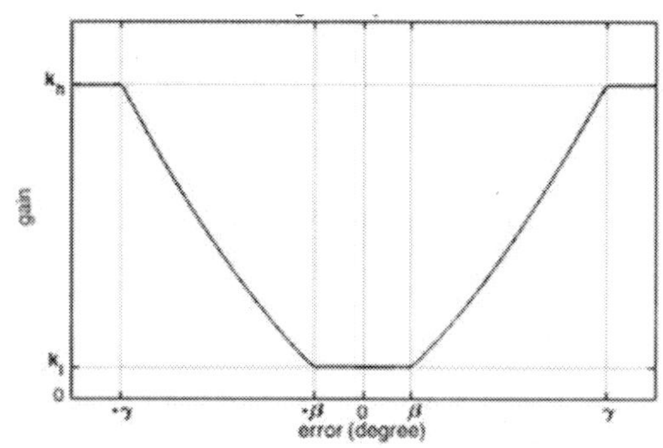

Figure 9. The nonlinear gain as function of error.

The step response under nonlinear PID control is shown in figure 8c. Fast rise time and small platform oscillation in steady-state are achieved. The details of the controller gain settings and step response performances are summarized in table 1.

It is clear from the results in figure 8 that the kinematic model error is significant. This is due to the limitations of the pseudo-rigid-body modeling approach, which does not capture the flexure deformation very accurately. It is seen that although the position demand in x direction is 100 μm, the end platform moves by only 80 μm, approximately. Compensation of errors in the inverse kinematic model can be achieved by the introduction of visual servoing, which will be discussed in the next section.

Table 1. Controller gains and performance.

Parameter		Controller PID1	PID2	NPID
K_p		25	200	25-200
K_i		500	5000	500-5000
K_d		0.7	5	0.7-5
Settling band	δ (pixel) (μm)	2 (0.96)	10 (4.8)	3 (1.44)
Response times	T_{rise} (s)	1.2	0.3	0.3
	T_{settle} (s)	3.5	1	1

IOP Conf. Series: Materials Science and Engineering **297** (2018) 012046 doi:10.1088/1757-899X/297/1/012046

4. Visual Servoing Implementation

4.1. Image Processing

Real-time image processing is used to extract the end-platform position information from the digital microscope image. Previous work on the image processing problem and recognition algorithms for automated cell detection and injection has been reported in [1] and [12]. However, the processing problem here is quite different as motion tracking is achieved using a periodic pattern rather than a unique feature of the image.

The image processing procedure is shown in figure 10. First, a grey-level image is received from the microscope camera (figure 10a). This is converted to a binary image by adjusting the threshold intensity and then using a blurring function to isolate the grid crossing points. The blurring transforms

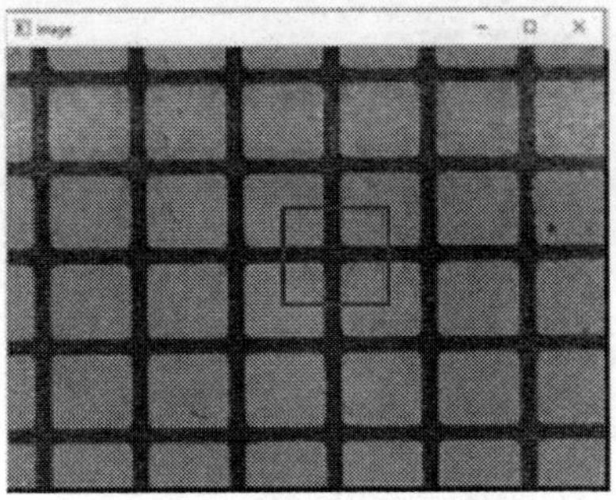

(a) Preprocessing

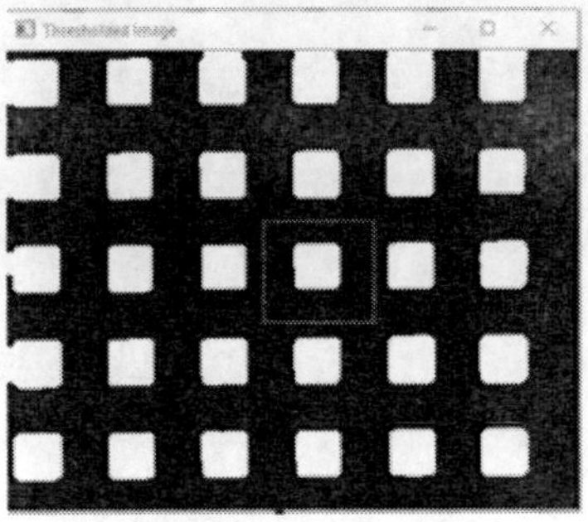

(b) Post processing

Figure 10. Image processing.

the binary image back to grey-scale but with high intensity at the grid crossing points and low intensity elsewhere. Converting back to a binary image gives an array of white squares where each square matches a crossing point on the original grid pattern (figure 10b). By using standard functions from the OpenCV library (i.e. findcontour() and moments()), the center position of each square can be determined. The position measurement algorithm operates by tracking the grid intersection point that falls within a central region of the image, as shown by the red square in figure 10a. When the currently tracked intersection point moves outside the central region, a new intersection point moves inside, and this point is then tracked instead. In this way the position tracking operates similarly to an incremental encoder where the known spacing of the grid can be used to determine the absolute position of the end platform.

The image processing takes about 35 milliseconds. Following communication with the microcontroller to send the position data, the total elapsed time is approximately 45 milliseconds. Therefore, the update/sampling frequency for the vision-based control can be set to 20 Hz.

4.2. Coordinate Relations

To execute the vision-based position measurement, it is necessary to know the relation between the discretized image space and the task space. Here, we consider motion of the end-link platform in the horizontal plane only, i.e. for fixed z position.

The coordinate frames for the image space and task space are shown on figure 11 where the following elements are indicated:

- $\{O\}$ Image coordinate frame (fixed in image)
- $\{S\}$ Fixed frame for task space (fixed in image)
- B Current tracked point on end-link
- A End-link datum (fixed on end-link)
- D Reference point (target for end-link)

The image coordinate frame $\{O\}$ is at the corner of the image space and is fixed with respect to the global reference frame (as the camera is fixed to the base frame). The global datum for the task space is the frame $\{S\}$ and the end-link datum (point A) is initially coincident with its origin. The position of the currently tracked point B in the image space is ${}^{o}B = [u \;\; v]^T$. This may be offset from the end-link datum (point A) as the tracked point changes during motion. The position of the currently tracked point in the task space ${}^{o}B = [x_m \;\; y_m]^T$ is obtained by a transformation ${}^{s}s_o$ involving a shift and rotation operation, followed by a scaling by the image resolution λ:

$$ {}^{s}B = {}^{s}s_o({}^{o}B) = \lambda\, {}^{s}R_o({}^{o}B - {}^{o}t_s) \tag{5} $$

where ${}^{s}t_o$ denotes the location of the origin of frame $\{S\}$ with respect to frame $\{O\}$ and

$$ {}^{s}R_o = \begin{bmatrix} 1 & 0 \\ 0 & -1 \end{bmatrix}, \tag{6} $$

For the experiments, the image resolution is 0.48 μm/pixel.

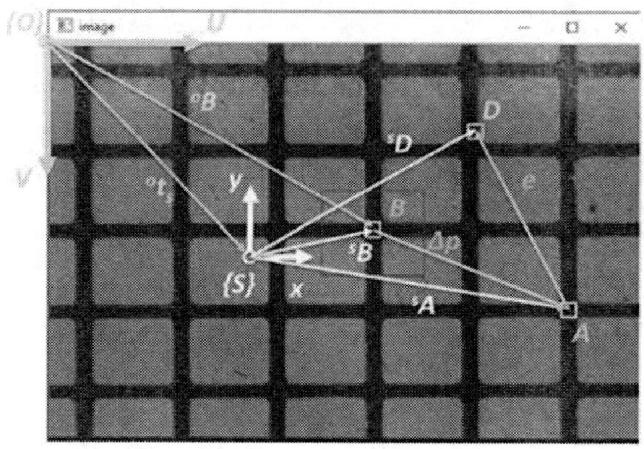

Figure 11. Frame of image and task space on the end-platform.

The current tracked point will be translated from the end-link datum by an integer number of grid spacings. This offset is known if the changes in tracking point are counted during motion (similar to an incremental encoder but in 2D). This offset is given by

$$ \Delta p = [m\Delta x \;\; n\Delta y \;\; -]^T \tag{7} $$

where

m is the grid increments in x direction
n is the grid increments in y direction
Δx is the grid spacing of in x direction
Δy is the grid spacing of in y direction

Combining the current tracked point position with the offset correction gives the absolute position of the platform as

$$ \begin{aligned} p_m = {}^{s}A &= {}^{s}B + \Delta p \\ &= [x_m + m\Delta x \quad y_m + n\Delta y \quad -]^T \end{aligned} \tag{8} $$

Note that position information for the z direction cannot be obtained from the image data and, hence, no image-based correction is applied to the z axis control.
The positioning error, based on the current demand position (point D), is

$$ e = p_d - p_m = {}^{s}D - {}^{s}A \tag{9} $$

This information is used to modify the input to the inverse kinematics, as described in the next subsection.

IOP Conf. Series: Materials Science and Engineering **297** (2018) 012046 doi:10.1088/1757-899X/297/1/012046

4.3. Visual Servoing Strategy

Visual servoing will correct for the residual errors after using the inverse kinematic model for the initial joint commands. The image-based position error is operated on by a PI controller, as shown in figure 12. Definitions of the variables in figure 12 are given in table 2. In this way, the input to the kinematic model is compensated so that zero error can be achieved in steady-state. This type of control structure may be termed direct visual servoing as set-point adjustments for the joint control loops occur continuously (at the image update rate) [19]. There is still a challenge to achieve satisfactory dynamic performance as there is a dynamic coupling between the inner joint control loops and the outer vision-based feedback loop. It can be expected, based on quasi-static conditions, that using low gain settings for the vision feedback will achieve stable operation, but elimination of positioning error will then be slow. An empirically tuned visual servoing control, implemented with the optimized nonlinear PID joint control, is considered as a means to achieve enhanced performance, as detailed in the following section.

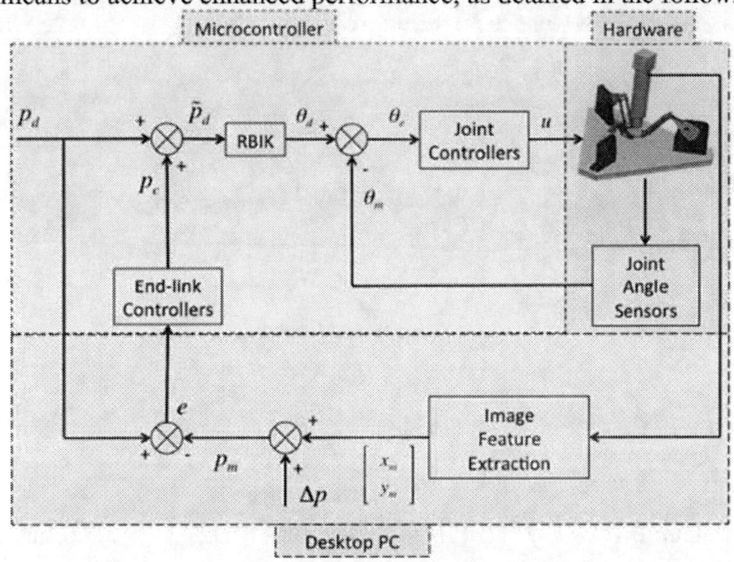

Figure 12. Control block diagram for visual servoing.

Table 2. Definition of variables in figure 12.

Variable	Definition
$\theta_d = \begin{bmatrix} \theta_{m1} & \theta_{m2} & \theta_{m3} \end{bmatrix}^T$	demand angles for actuators
$\theta_e = \begin{bmatrix} \theta_{e1} & \theta_{e2} & \theta_{e3} \end{bmatrix}^T$	angle errors for actuators
$\theta_m = \begin{bmatrix} \theta_{m1} & \theta_{m2} & \theta_{m3} \end{bmatrix}^T$	actuator angles (measured)

5. Enhanced Control

In this section, experimental results are reported for both step response and trajectory tracking tasks where the combination of joint feedback control and visual servoing is employed.

The step response tests involved the same operations as in figure 8, where step changes in demand of 100 microns were introduced. As before, three different cases are considered involving different PID joint controllers. Figures 13a and 13b show results with the linear PID gain settings (PID1 and PID2).

Figure 13c shows results with the nonlinear PID feedback (NPID). The vision-based PI feedback controller was applied for all three cases but was tuned empirically with joint controller PID1.

Note that the vision-based feedback is paused immediately following the step change in demand. This is because the initial motion (driven by the joint feedback loops) is so fast that reliable position information cannot be ensured. The vision feedback is enabled again when the speed of motion has dropped below a threshold level so that accurate vision-based tracking can be resumed. The remaining position error is then compensated through the vision feedback, which is evident as an additional change in the joint angle reference demand (that compensates for the kinematic model error).

Comparing figure 13 with figure 8, it is clear that the error related to kinematic modelling is compensated in steady-state but different rise times and settling times are achieved due to the different gain settings. The nonlinear PID controller again shows the best performance with rise time of approximately 1 second and settling time of approximately 1.5 seconds. Also, the end-platform position can be maintained within a range of 3 pixels in steady state. Note that, due to the vision-based compensation, this figure now relates to the absolute positioning accuracy, which is within ±0.72 microns.

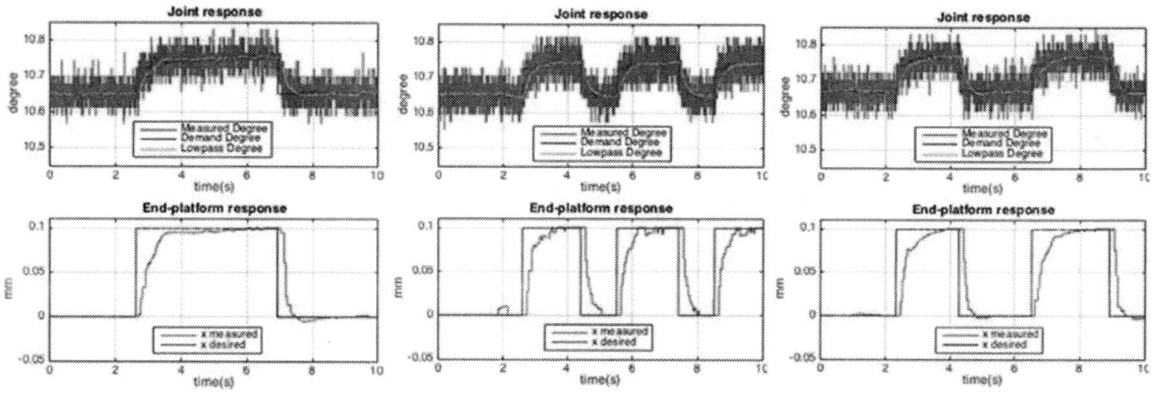

(a) linear, low gain: PID1 (b) linear, high gain: PID2 (c) nonlinear: NPID

Figure 13. Step response measurements for different PID joint controllers with visual servoing.

In further experiments, tracking of a circular trajectory with a diameter of 1 mm was considered, as shown in figure 14 where the upper graph shows the task space trajectory and the lower graph shows the measured errors in x and y coordinates during motion. Test results are shown for vision-based servoing combined with linear (PID1) and with nonlinear (NPID) joint control. Figures 14a and 14b are for path velocities of 52 and 104 microns per second respectively.

It can be seen that tracking errors are similar for both controllers for the slower speed task, and stay within approximately 6 microns (neglecting transients at start/end of motion). However, for the faster velocity task, tracking error is significantly increased for the linear PID controller (within approximately 20 microns), whereas the nonlinear PID controller maintains accuracy within approximately 8 microns.

6. Conclusions

A 3D-printed Delta-like micromanipulator with flexure-joints has been created as an experimental motion platform with nominal workspace of 20x20x6 mm. Using linear PID control of the actuated joints was found to have significant limitations as low gain settings gave good platform stability (noise rejection) but speed of response and tracking accuracy were poor. High gain settings gave improved speed of response but resulted in unacceptable platform vibration. By using a nonlinear PID controller, advantages from both low gain (less oscillation) and high gain settings (fast response time) could be realized in a methodical and effective manner.

With joint feedback control only, significant errors in absolute positioning occurred due to deficiencies in the inverse kinematic model, which was based on pseudo-rigid-body modelling methods. To eliminate steady-state positioning error, a vision-based feedback control was implemented via a digital microscope system. This allowed high accuracy to be achieved in both step response and trajectory tracking tasks. The best performance was obtained in combination with nonlinear PID joint control, for which absolute positioning accuracy within 3 pixels, (1.44 μm) could be achieved and tracking errors were maintained within approximately 8 microns for contouring tests with circular trajectories.

There is scope for improvement for this system, as response times were still affected by error in the kinematic model. Although errors are compensated by the vision-based feedback, the speed of compensation was limited due to the relatively slow sample rates. Further work will aim to improve control performance by refining the kinematic model, either through on-line calibration/adaptation or by improving the theoretical model to better account for compliance behavior of the flexures.

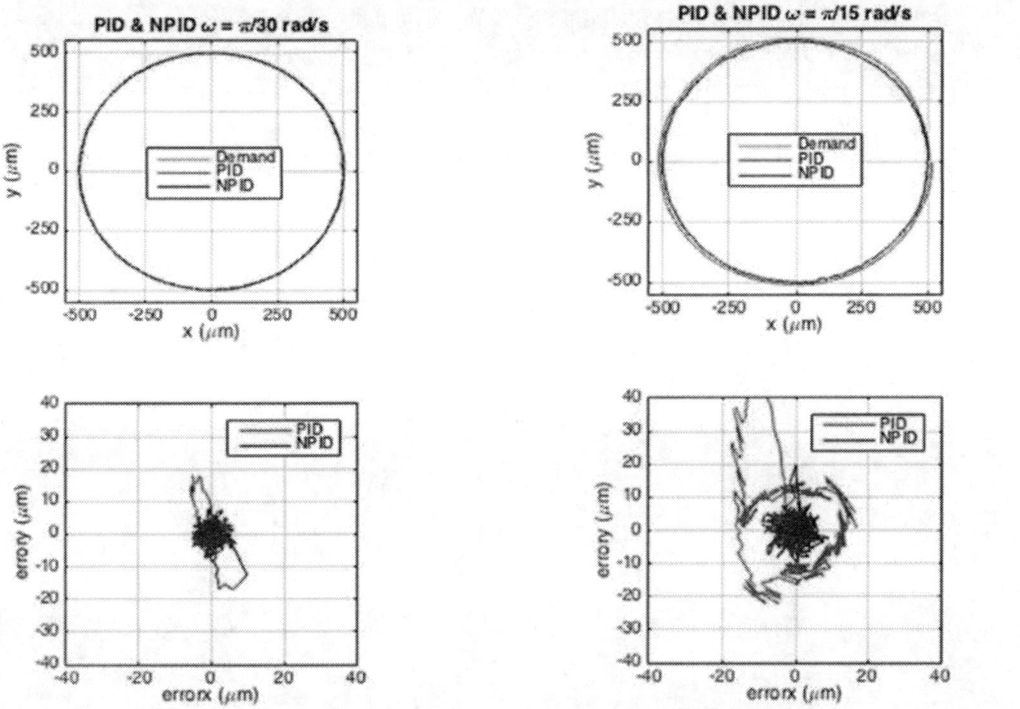

(a) path velocity of 52 microns per second (b) path velocity of 104 microns per second

Figure 14. Tracking performance for circular trajectory.

References

[1] Zhao W, Zhang WJ, Gupta MM, Zong GH and Ouyang PR *Seventh Int. Conf. on Control, Automation, Robotics And Vision (ICARCV)* pp 429-434

[2] Mattos LS and Caldwell DG 2009 *5th Annual IEEE Conf. on Automation Science and Engineering* pp 609-614.

[3] Sakaki K, Dechev N, Park EJ and Burke RD 2007 *Conf. on Intelligent Robots and Systems* pp 3137-3143.

[4] Liu X, Kim K and Sun Y 2007 *Micromechanisms and Microengineering* **17** pp 1796-1802

[5] Kozuka H, Arata J, Okuda K, Onaga A, Ohno M, Sano A and Fujimoto H 2012 *IEEE/RSJ Int. Conf. on Intelligent Robots and Systems* pp 2519-2524

[6] Chang D, Gu GM, and Kim J 2013 *IEEE/RSJ Int. Conf. on Intelligent Robots and Systems (IROS)* pp 413-418.

[7] Wang DH, Yang Q and Dong HM 2013 *IEEE/ASME Trans. on Mechatronics* **18(1)** pp 138-147

[8] Acer M and Sabanovic A 2013 *IEEE Int. Conf.* on *Industrial Technology (ICIT)* pp 118-123

[9] Requicha AAG, Arbuckle DJ, Mokaberi B and Yun J 2009 *Int. J. of Robotics Research* **28**(4) pp 512-522

[10] Tan N, Clevy C, Laurent GJ, Sandoz P and Chaillet N 2015 *IEEE Trans. Robot.* **31**(6) pp 1497-1507

[11] Tan N, Clevy C, Laurent GJ and Chaillet N 2012 *The 2012 IEEE/ASME Int. Conf. on Advanced Intelligent Mechatronics* pp 256-261

[12] Xu Q, Li Y and Xi N 2009 *IEEE Trans. on Automation Science and Engineering* **6**(4) pp 710-719

[13] Tan N, Clevy C, Laurent GJ and Chaillet N 2016 *IEEE Robotics and Automation Letters July* **1(2)** pp 638-645

[14] Xiao S and Li Y 2013 *IEEE Int. Conf. on Robotics and Automation (ICRA)* pp 4506-4511

[15] Howell LL 2001 *Compliant mechanisms* (New York: John Wiley & Sons)

[16] William II RL 2016 *The Delta Parallel Robot: Kinematics Solutions.* [Internet] Available from: www.ohio.edu/people/williar4/html/pdf/DeltaKin.pdf

[17] Han J 2009 *IEEE Trans. on Industrial Electronics* **56**(3) pp 900-906

[18] Bennehar M, Chemori A and Pierrot F 2014 *IEEE/RSJ Int.Conf. on Intelligent Robots and Systems (IROS 2014)* pp 1670-1675

[19] Hutchinson S, Hager G and Corke P 1996 *IEEE Trans. on Robotics and Automation* **12**(5) pp 651-670

IOP Conf. Series: Materials Science and Engineering **297** (2018) 012047 doi:10.1088/1757-899X/297/1/012047

Backstepping boundary control: an application to the suppression of flexible beam vibration

Nipon Boonkumkrong[1,2,*], Pichai Asadamongkon[2], Sinchai Chinvorarat[3]

[1] Department of Mechanical Engineering, Faculty of Engineering, Srinakharinwirot University,63 Moo7 Rangsit-Nakornnayok Rd., Ongkharak, Nakornnayok 26120, Thailand

[2] Department of Mechanical Engineering, Faculty of Engineering, Kasem Bundit University, 1761 Pattanakarn Rd, Bangkok 10250, Thailand

[3] Department of Mechanical and Aerospace Engineering, Faculty of Engineering, King Mongkut's University of Technology North Bangkok,1518 Pracharat 1 Rd., Wongsawang, Bangsue, Bangkok 10800, Thailand

[*] Corresponding Author: nipon413@hotmail.com

Abstract. This paper presents a backstepping boundary control for vibration suppression of flexible beam. The applications are such as industrial robotic arms, space structures, etc. Most slender beams can be modelled using a shear beam. The shear beam is more complex than the conventional Euler-Bernoulli beam in that a shear deformation is additionally taken into account. At present, the application of this method in industry is rather limited, because the application of controllers to the beam is difficult. In this research, we use the shear beam with moving base as a model. The beam is cantilever type. This design method allows us to deal directly with the beam's partial differential equations (PDEs) without resorting to approximations. An observer is used to estimate the deflections along the beam. Gain kernel of the system is calculated and then used in the control law design. The control setup is anti-collocation, i.e. a sensor is placed at the beam tip and an actuator is placed at the beam moving base. Finite difference equations are used to solve the PDEs and the partial integro-differential equations (PIDEs). Control parameters are varied to see their influences that affect the control performance. The results of the control are presented via computer simulation to verify that the control scheme is effective.

1. Introduction

Flexible beams constitute an important problem in many applications such as space structures, industrial robotic arms, cantilever cranes, helicopter rotor, astronomical telescopes, etc. In this paper, we consider a model of the undamped shear beam. Most of the slender beams can be represented by this model. The beam system is distributed parameter in nature, so it is governed by partial differential equations (PDEs). The model consists of a wave equation coupled with a second-order-in-space ODE or can be alternatively represented as a fourth-order-in-space/second-order-in-time PDE. The shear beam is more complex than the Euler-Bernoulli model and slightly simpler than the Timoshenko model. There are many models for flexible beams such as Euler-Bernoulli, Rayleigh, shear and Timoshenko beam equations. Derivations and comparisons of these beam models can be found in [1].

Boundary control is preferable for controlling PDE systems since actuation and sensing are only through the boundary conditions [2]. For its historical development, the reader is referred to [3]. Morgul presented results for boundary control of infinite dimensional systems of Euler-Bernoulli [4]. The passivity property of nonlinear Euler-Bernoulli beams has been studied in Fard [5]. The method guarantees finite gain L_2 stability and passivity of closed-loop systems. Sasaki [6] used a Lyapunov functional of the system to derive the control law by minimizing the time rate of change of the functional at every point in time and using neural networks in tuning a control gain.

Krstic et al. presented the backstepping boundary control for undamped shear beams with an anti-collocated setup in which the sensing and the actuation are at different locations [7, 8, 9, 10]. Gain kernels for both controller and observer are found and used for control law design. The applications include atomic force microscopy (AFM) where the piezo actuation is applied at the beam base. The industrial application is rather limited. Ali and Padhi presented active control of Euler-Bernoulli beams [11]. They proposed two state feedback controllers based on optimal dynamic inversion techniques. Boonkumkrong N. and Kuntanapreeda S. applied the method of backstepping boundary control to the thermal system experiment [12].

This work is motivated by the interests in stabilizing vibrating slender bodies. The structure is modeled by a completely undamped shear beam model. The design is a combination of the classical damping boundary feedback idea with backstepping boundary control. The works of Fard, Sasaki and Krstic are the inspiration of this paper [5, 6, 7].

At present, the application of the backstepping boundary control is rather limited, because the application of controllers to the beam is difficult. An objective of this paper is to apply the method to the problem of the free vibration suppression of the undamped shear beam with controllers applied to the beam by using a moving base. This is new method to apply the backstepping boundary controllers to the system.

The rest of this article is organized as follows. The next section presents a mathematical model of the system. Brief explanations of the beam models are also provided. Backstepping boundary controller design is given in Section 3. In Section 4, an observer design is presented. Numerical studies are presented in Section 5. The results of simulation are given in Section 6. This article is concluded in the last section.

2. Mathematical models

The undamped shear beam model can be expressed by a second-order-in-time, fourth-order-in-space PDE, as follows [1]

$$a\,w_{tt}(x,t) - \varepsilon\, w_{xxtt}(x,t) + w_{xxxx}(x,t) = 0 \tag{1}$$

where $w(x,t)$ is the beam deflection, ε is a constant inversely proportional to the shear modulus and a is a positive constant. The subscripts t and x denote the partial differentiation with respect to time and space, respectively. The deflection due to the beam mass is neglected.

The beam model can also be represented by a wave equation PDE coupled with a second-order-in-space ODE as follows [1, 14]

$$\varepsilon\, w_{tt}(x,t) = w_{xx}(x,t) - \varphi_x(x,t), \quad x \in (0,1) \tag{2}$$

$$\varphi_{xx}(x,t) - b^2\varphi(x,t) + b^2 w_x(x,t) = 0 \tag{3}$$

where φ is the rotating angle as a result of the bending moment of the beam and $b = \sqrt{a/\varepsilon}$. The derivations of beam models equation (1) and equations (2) - (3), can be found in [14]. In [14], one can verify that both models are equivalent.

In figure 1, the beam is free at one end $(x = 0)$ and has the following boundary conditions [1]:

$$w_x(0,t) - \varphi(0,t) = 0 \tag{4}$$

$$\varphi_x(0,t) = 0. \tag{5}$$

Equations (4) and (5) mean that the shear force and the moment are both zero at the free end, respectively. From the Newton's 2nd law of motion, the boundary condition of the other end $(x = 1)$ with a controller $U(t)$ applied at the moving base is as follows [6]

$$m_b \frac{d^2 w_b(t)}{dt^2} = -\left(w_x(1,t) - \varphi(1,t)\right) + U(t), \tag{6}$$

Where m_b is the mass of the moving base, $w_b(t)$ is the displacement of the base, the terms in the parentheses on the right-hand side of equation (6) represent a shear force exerted by the beam on the base [4]. The beam model, equations (2) - (3), is used to design the control law. Note that $w_b(t) = w(1,t)$, i.e. the displacements of the beam right end and the moving base are the same.

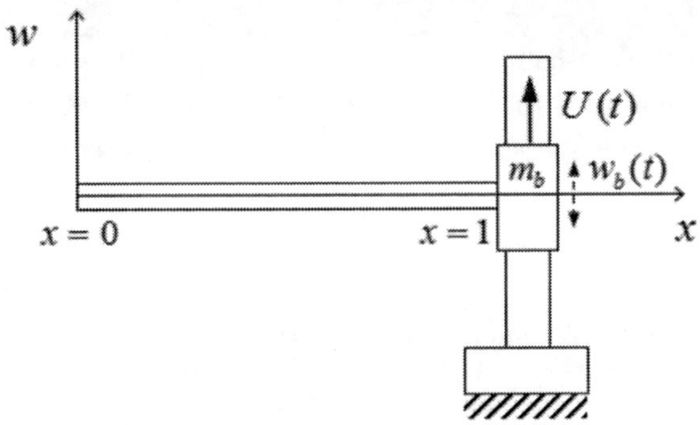

Figure 1. Shear beam with a controller.

3. Backstepping Boundary Control

In this section, backstepping boundary control method is presented and the control law is then formulated. The system, equations (2) – (3), will be first re-written in a hyperbolic partial integro-differential equation (PIDE), as the following,

$$\varepsilon w_{tt}(x,t) = w_{xx}(x,t) + b^2 w(x,t) - b^2 \cosh(bx) w(0,t) + b^3 \int_0^x \sinh\,[b(x-y)]\, w(y,t)\, dy$$
$$-b\frac{\cosh(bx)}{\cosh(b)}\left(\varphi(1,t) - b\sinh(b)\, w(0,t) + b^2 \int_0^1 \cosh\,[b(1-y)]\, w(y,t)\, dy\right) \tag{7}$$

$$w_x(0,t) = \frac{1}{\cosh(b)}\{\varphi(1,t) - b\sinh(b)\, w(0,t) + b^2 \int_0^1 \cosh\,[b(1-y)]\, w(y,t)\, dy\}. \tag{8}$$

Equations (7) and (8) are obtained by solving the ODE equation (3) as a two-point boundary value problem using Laplace transform in the spatial variable x as follows [7]

$$\varphi(x,t) = \cosh(bx)\, \varphi(0,t) - b\int_0^x \sinh\,(b(x-y))\, w_y(y,t)\, dy. \tag{9}$$

The term $\varphi(0,t)$ in equation (9) can be expressed in term of $\varphi(1,t)$ by evaluating equation (9) at $x=1$ to get

$$\varphi(1,t)=\cosh(b)\,\varphi(0,t)-b\int_0^1 \sinh(b(1-y))\,w_y(y,t)\,dy. \tag{10}$$

Solve equation (10) for $\varphi(0,t)$,

$$\varphi(0,t)=\frac{1}{\cosh(b)}\{\varphi(1,t)+b\int_0^1 \sinh(b(1-y))\,w_y(y,t)\,dy\}. \tag{11}$$

Integrate by parts the integral term on the right side of equation (11) to get

$$\varphi(0,t)=\frac{1}{\cosh(b)}\{\varphi(1,t)-b\sinh(b)\,w(0,t)+b^2\int_0^1 \cosh(b(1-y))\,w(y,t)\,dy\}. \tag{12}$$

The integral term on the right-hand side of equation (12) is not spatially casual because the upper limit of integration is 1. To put the system into a strictly feedback form, we eliminate this integral by choosing the first controller as follows

$$\varphi(1,t)=b\sinh(b)\,w(0,t)-b^2\int_0^1 \cosh(b(1-y))\,w(y,t)\,dy. \tag{13}$$

So that $\varphi(0,t)=0$ in equation (12). Then, equation (9) becomes

$$\varphi(x,t)=b\sinh(bx)\,w(0,t)-b^2\int_0^x \cosh(b(x-y))\,w(y,t)\,dy. \tag{14}$$

Note that the upper limit of integration is now x.

Differentiating $\varphi(x,t)$ with respect to x and substituting the results into the wave equation (2), we get the system in the strictly feedback form for control design,

$$\varepsilon w_{tt}(x,t)=w_{xx}(x,t)-b^2\cosh(bx)\,w(0,t)+b^3\int_0^x \sinh(b(x-y))\,w(y,t)\,dy. \tag{15}$$

$$w_x(0,t)=0. \tag{16}$$

Next, use the following transformation [7],

$$v(x,y)=w(x,y)-\int_0^x k(x,y)w(y,t)\,dy, \tag{17}$$

where is $k(x,y)$ is the gain kernel of the system. Equation (17) is used to map the system equation (2) into the following exponentially stable target system

$$\varepsilon v_{tt}(x,t)=v_{xx}(x,t), \tag{18}$$

$$v_x(0,t) = c_0 v(0,t), \tag{19}$$

$$v_x(1,t) = -c_1 v_t(1,t) \tag{20}$$

where c_0 and c_1 are design parameters. Stability proof of equations (18)-(20) can be found in [9].

Substituting the transformation equation (17) into the target system, equations (18)-(20), we can derive the following PDE for gain kernel $k(x,y)$ [13]:

$$k_{xx}(x,y) = k_{yy}(x,y) + b^2 \sinh(b(x-y)) + b^3 \int_0^x k(x,\xi)\, w(\xi,t)\, d\xi \tag{21}$$

$$k(x,x) = -\frac{\lambda}{2} x - c_0 \tag{22}$$

$$k_y(x,0) = b^2 \int_0^x k(x,\xi)\cosh(b\xi)\, d\xi - b^2 \cosh(bx). \tag{23}$$

The second boundary controller is obtained by differentiating equation (17) with respect to x and setting $x = 1$ [13]:

$$w_x(1,t) = k(1,1)\, w(1,t) + \int_0^1 k_x(1,y)\, w(y,t)\, dy - c_1 w_t(1,t) + c_1 \int_0^1 k(1,y)\, w_t(y,t)\, dy \tag{24}$$

Stability proof of the feedback control can be found in [10]. Gain kernels $k(1,y)$ and $k_x(1,y)$ of control law equation (24) are shown in figures 2 and 3, respectively.

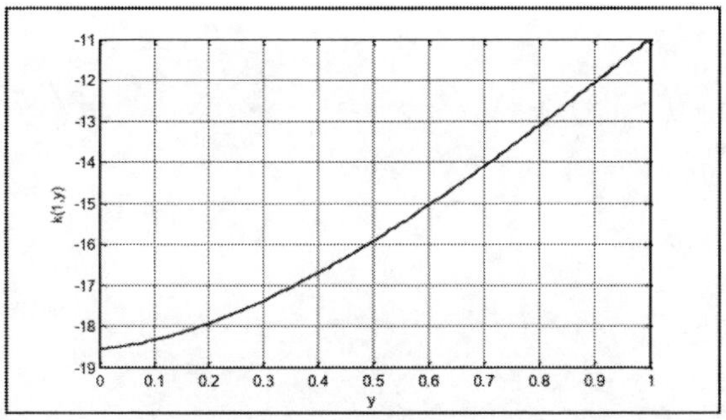

Figure 2 The gain kernel $k(1,y)$.

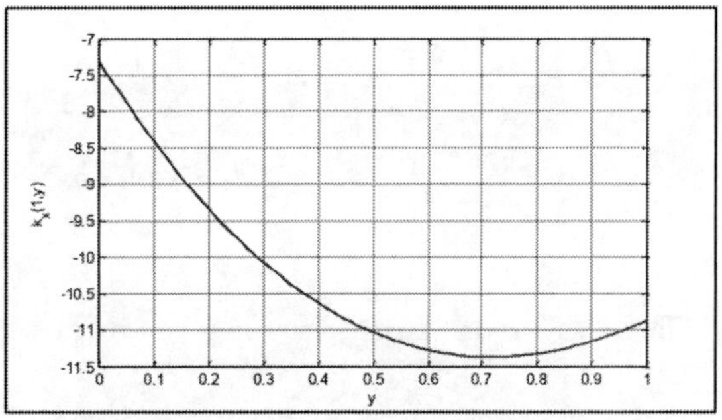

Figure 3 The gain kernel $k_x(1,y)$.

In backstepping boundary control method, we use the controller, which consists of two equations, i.e. equation (13) and equation (24), to control the beam. Figure 4 shows the transformation from cantilever beam to target system. The first part of controller, $\varphi(1,t)$ converts the free end into a moving end, i.e. $\varphi(0,t) = 0$, and then the second part of the controller, $w_x(1,t)$, converts the beam into taut string with a stiff spring at the tip, equation (19) and a tuned damper at the base, equation (20).

The first controller, equation (13) is combined to the beam model, equations (7)-(8), so the controller, equation (24) is applied at the moving base as follows

$$U(t) = k(1,1)\,w(1,t) + \int_0^1 k_x(x,y)\,w(y,t)\,dy - c_1 w_t(1,t) + c_1 \int_0^1 k(1,y)\,w_t(y,t)\,dy + m_b \frac{d^2 w(1,t)}{dt^2} \tag{25}$$

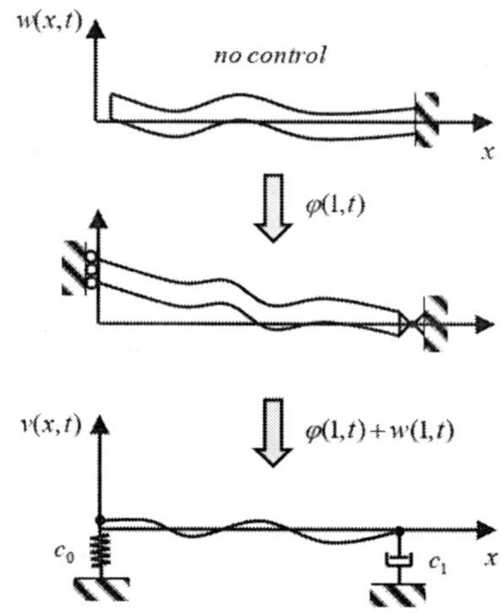

Figure 4. Feedback transformation for beam.

4. Observer Design

For the backstepping boundary control method, the states along the beam are needed for control law calculation, but the only possible measurement of the system is at the boundary $x = 0$. The Luenberger-like observer is used to estimate beam deflections along the beam [10].

Consider the following beam model:

$$\varepsilon\, w_{tt}(x,t) = w_{xx}(x,t) + b^3 \int_0^x \sinh(b(x-y))\, w(y,t)\,dx - b^2 \cosh(bx)\, w(0) - b \sinh(bx)\,\alpha(0) \tag{26}$$

$$w_x(0,t) = \varphi(0,t) \tag{27}$$

The Luenberger-like observer is given by [10]

$$\varepsilon\, \hat{w}_{tt}(x,t) = \hat{w}_{xx}(x,t) + b^2 \hat{w}(x,t) + b^3 \int_0^x \sinh(b(x-y))\, w(y,t)\,dx - b^2 \cosh(bx)\, w(0) - b\sinh(bx)\,\varphi(0) + p_y(x,0)\,[w(0,t) - \hat{w}(0,t)] \tag{28}$$

$$\hat{w}_x(0,t) = \varphi(0) + p(0,0)\,[w(0,t) - \hat{w}(0,t)] - \tilde{c}_0\,[w_t(0,t) - w_t(0,t)] \tag{29}$$

$$\hat{w}(1,t) = w(1,t)\,. \tag{30}$$

The observer gains $p_y(x,0)$ in equation (18) and $p(0,0)$ in equation (29) are determined by solving the PDE [10],

$$p_{yy}(x,y) = p_{xx}(x,y) + b^2 p(x,y) + b^3 \sinh(b(x-y)) + b^3 \int_y^x p(\xi,y) \sinh(b(x-\xi))\,dx \tag{31}$$

$$p(x,x)=\frac{b^2}{2}(x-1) \tag{32}$$

$$p(1,y)=0. \tag{33}$$

Defining the observer error as $\widetilde{w}=w-\hat{w}$ and substracting equations (28)-(29) from equations (26)-(27), we obtain the observer error:

$$\varepsilon\,\widetilde{w}_{tt}(x,t)=\widetilde{w}_{xx}(x,t)+b^2\widetilde{w}(x,t)+b^3\int_0^x \sinh(b(x-y))\,\widetilde{w}(y,t)\,dx+p_y(x,0)\,\widetilde{w}(0,t) \tag{34}$$

$$\widetilde{w}_x(0,t)=-p(0,0)\,\widetilde{w}(0,t)+\widetilde{c}_0\,\widetilde{w}_t(0,t) \tag{35}$$

$$\widetilde{w}(1,t)=0. \tag{36}$$

Using the following transformation,

$$\widetilde{w}(x,t)=\widetilde{v}(x,t)-\int_0^x p(x,y)\widetilde{v}(y,t)dy \tag{37}$$

to convert the error system equations (34)-(36) into

$$\varepsilon\widetilde{v}_{tt}(x,t)=\widetilde{v}_{xx}(x,t) \tag{38}$$

$$\widetilde{v}_x(0,t)=\widetilde{c}_0\,\widetilde{v}_t(0,t) \tag{39}$$

$$\widetilde{v}(1,t)=0, \tag{40}$$

which is known to be exponentially stable (see [10]).

The gain kernel PDEs can be solved numerically, but in this paper, we solve them numerically. The plot of observer gain kernel $p(x,1)$ and $p_y\,(x,1)$ are shown in figures 5 and 6, respectively.

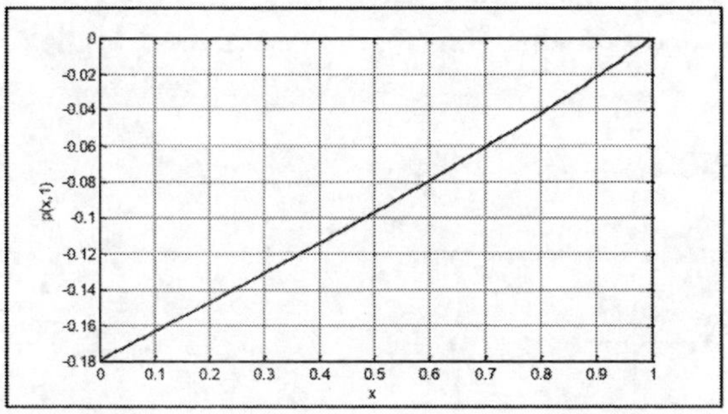

Figure 5. The observer gain kernel $p(x,1)$.

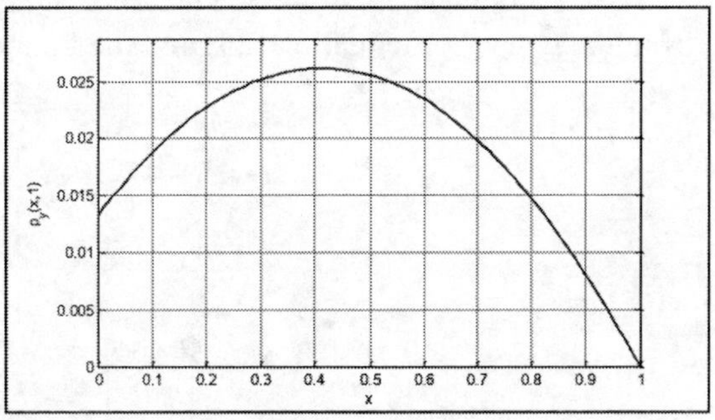

Figure 6. The observer gain kernel $p_y(x,1)$.

5. Numerical Calculations

The beam model with strictly feedback form, equation (15) is a second-order-in-time, second-order-in-space partial integro-differential equation (PIDE) and is used in a numerical calculation. It is a simple second-order partial differential equation with the integration terms, and can be easily solved with finite difference.

The highest order of the PIDE that we study in this paper is second order, so we have the following finite-difference approximations [15]:

$$\left(\frac{\partial^2 w}{\partial x^2}\right)_m^n = \frac{w_{m+1}^n - 2w_m^n + w_{m-1}^n}{(\Delta x)^2} + O[(\Delta x)^2] \tag{41}$$

$$\left(\frac{\partial^2 w}{\partial t^2}\right)_m^n = \frac{w_m^{n+1} - 2w_m^n + w_m^{n-1}}{(\Delta t)^2} + O[(\Delta t)^2] \tag{42}$$

For the integration approximation, we use the trapezoidal integration rule as follows,

$$\int_a^b f(x)\,dx = \frac{\Delta x}{2} f(a) + \sum_{i=a+1}^{b-1} f(i) + \frac{\Delta x}{2} f(b) \tag{43}$$

Substituting equations (41)-(43) into equation (15), we get the following finite difference equation,

$$\begin{aligned} w_m^{n+1} &= \frac{r^2}{\varepsilon}(w_{m+1}^n - 2w_m^n + w_{m-1}^n) + 2w_m^n - w_m^{n-1} + \frac{(\Delta t)^2}{\varepsilon} b^2 w_m^n - \frac{(\Delta t)^2}{\varepsilon} b^2 \cosh(bx(m))\, w_1^n \\ &+ \frac{(\Delta t)^2}{\varepsilon} b^3 \left\{ \frac{\Delta x}{2} \sinh[b(x(m) - y(1))] w(1) + \Delta x \sum_{j=2}^{m-1} \sinh[b(x(m) - y(j))] w(j) \right\} \end{aligned} \tag{44}$$

where Δx and Δt are the spatial and the temporal increments indexed by m and n, which start from 1 to M and N, respectively, and $r = \Delta t / \Delta x$. Figure 7 shows the calculation grid with spacing Δx in the row and Δt in the column. The dark circles (the first and second rows from the bottom) are the values obtained by the initial conditions and the grey circle (w_m^{n+1}) is the point to be calculated.

With the boundary condition, equation (16), one can obtain the first element (w_1^{n+1}) of each row and from the controller, equation (25) and the boundary-condition equation (6), the last element of the row (w_M^{n+1}) is obtained.

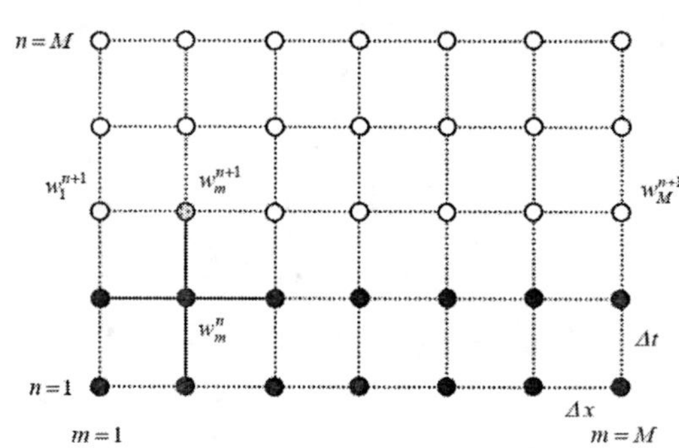

Figure 7. Calculation grid

6. Simulations

In this section, the simulation results are presented. The shear beam is simulated using the finite difference equations mentioned in the last section. The beam started to vibrate under the following initial conditions (see figure 8) [8]:

$$w(x,0) = 0.1(1-x)^2 \sin(1.6\pi(1-x)) \tag{45}$$

$$w_t(x,0) = -0.1(1-x)^2 \sin(1.6\pi(1-x)), \tag{46}$$

where equation (45) is the initial displacement, and equation (46) is the initial velocity at $t = 0$.

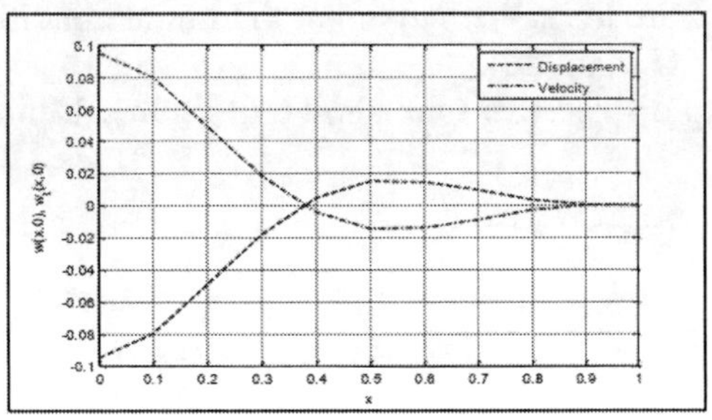

Figure 8 Initial conditions

The beam was simulated with the parameter of beam material $\varepsilon = 1$ and $b = 0.6$. The mass of the moving base was fixed at $m_b = 0.01$, and the spring and damping constants were set at $c_0 = 1$ and $c_1 = 1$, respectively. The parameter for observer equation (29) is set at $\tilde{c}_0 = 15$ [8]. The following values are used: the grid size $M = 20$ (in space), the time step $\Delta t = 0.01$ and the final time $T = 30$. Since there is no damping term in the beam model and no external force exerted, the beam with zero Dirichlet boundary condition at $x = 1$, i.e. with a fixed moving base, will vibrate perpetually as shown in figure 9. The tip displacement at $x = 0$ is shown in figure 10. This is the uncontrolled case.

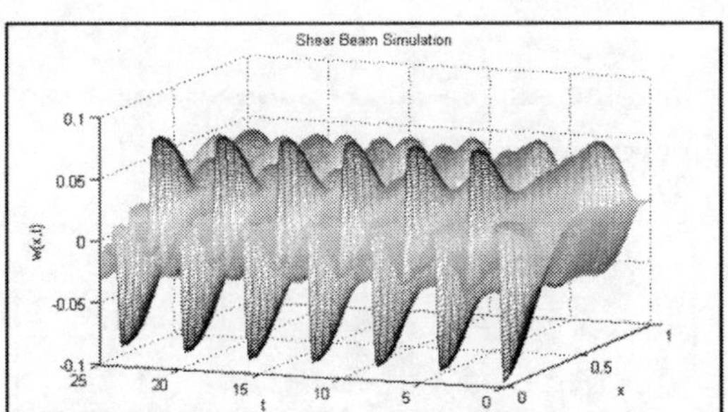

Figure 9. Beam vibration without control in 3-D

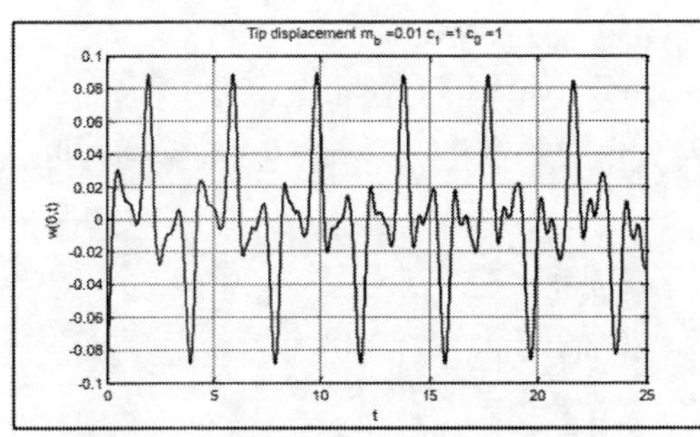

Figure 10. Tip displacement of the Beam

The simulation of the shear beam with the control law, equation (25) is shown. The simulations of the shear beam with full-state control and with observer are shown in figures 11 and 12, respectively.

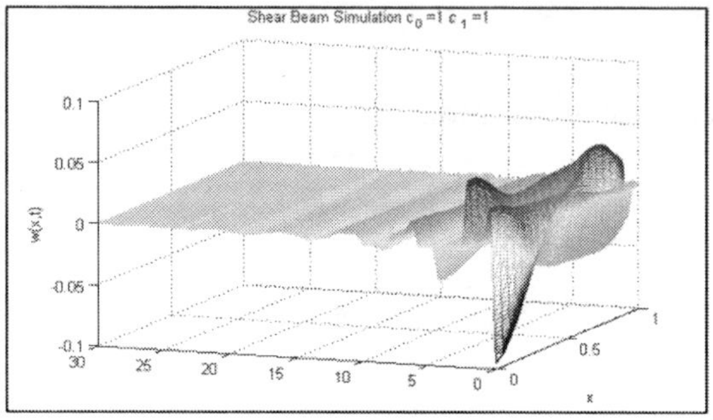

Figure 11. Shear beam simulation with control

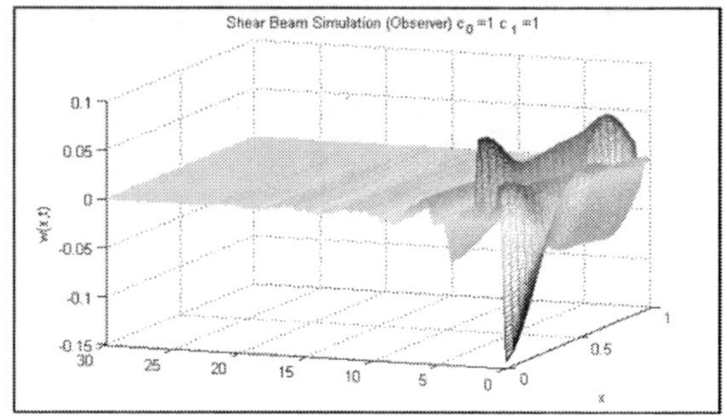

Figure 12. Shear beam simulation with observer

We see that the responses are the same. Note that the initial conditions for observer are 50% more than the values of equations (45) and (46).

The control action and tip displacement are shown in figures 13 and 14, respectively. The settling time is about 20.

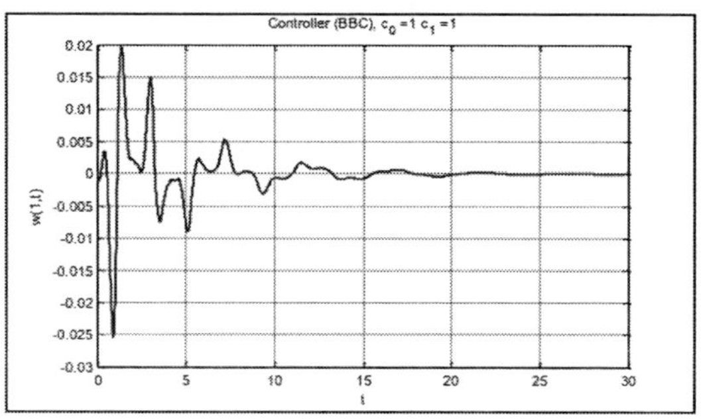

Figure 13. Control action, $w(1,t)$.

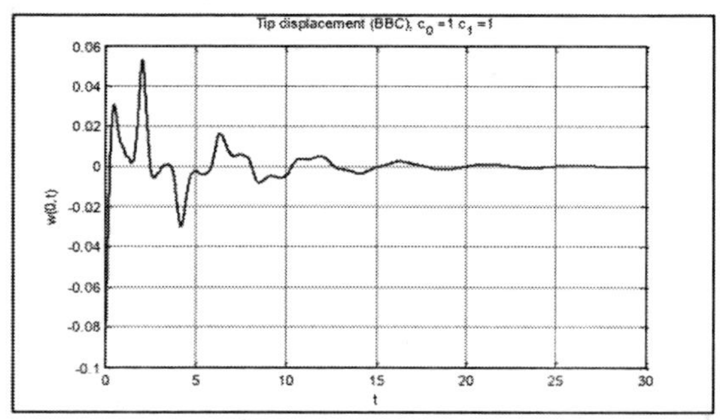

Figure 14. Tip displacement, $w(0,t)$.

6.1 Parameter Tunings

In this subsection, we study parameter changes that affect the control performance. First, we fix damping constant at $c_1 = 1.0$ and vary spring constants at various values, c_0 at 0.5, 1.0 and 2.0. Figure 15 shows tip displacements.

In the case of a weaker value of damping, i.e. $c_0 = 0.5$, the response takes a longer time to settle at $t = 30$ and for the strong values of damping constant, it takes about $t = 20$.

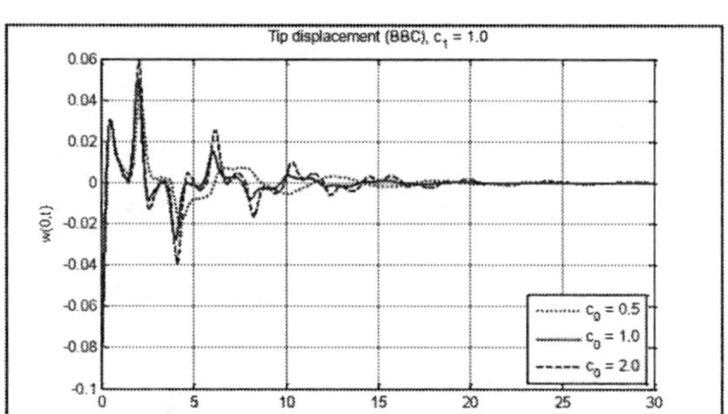

Figure 15. Tip displacement with various spring constants (c_0).

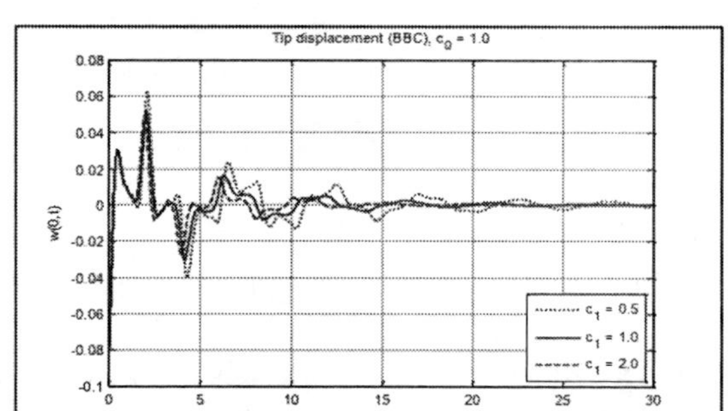

Figure 16. Tip displacements with various damping constants (c_1).

In the second case, we fix spring constant at $c_0 = 1.0$ and vary damping constants at various values, c_1 at 0.5, 1.0 and 2.0. Figure 16 shows tip displacements of the beam. In the case of a weak value of damping, i.e. $c_1 = 0.5$, the response takes a longer time to settle at $t = 30$ and for the strong values of damping constant, it takes about $t = 20$.

7. Conclusion

Backstepping boundary control is applied to the problem of vibration suppression of the undamped shear beam. The beam model consists of a wave equation, coupled with a second-order-in-space ODE. An observer is used to estimate the deflections along the beam. Gain kernel of the system is calculated and then used in the control law design. Finite difference equations are used to solve the PDEs and the partial integro-differential equations (PIDEs). Numerical results for the control of a shear beam are presented to verify that the control scheme is effective in suppressing the vibration of the beam, and the parameters, i.e. c_0 and c_1, can be tuned to obtain the control with the desired behaviours.

For real implementation or experiment, an electromagnetic linear actuator/motor can be used to move or slide the base. The speed of the actuator depends on the frequency of the vibration.

For further research, the other boundary control methods such as passivity-based control method will be applied to the shear beam and their performance will be compared.

References

[1] Han S M, Benaroya H and Wei T 1999 *J. of Sound and Vibration*, **225** pp 935-988
[2] Smyshlyaev A and Miroslav M. 2004 *IEEE Trans. on Automatic Control* **49** pp 2185-2202
[3] Padhi R and Ali Sk F 2009 *Annual reviews in Control 33* pp 59-68
[4] Morgul O 1992 *Automatica* **28** pp 1255-1260
[5] Fard M P 2002 *Modeling, Identification and Control* **23** pp 239-258
[6] Sasaki M, Asai H, Kawafuku M and Hori Y 2000 *Proc. of the IEEE Int. Conf. on Systems, Man and Cybernetics* pp 3259-3264
[7] Krstic M, Balogh A and Smyshlyaev A 2006a *Proc. of the 17th Int. Symp. Math. Theory of Networks and Systems*
[8] Krstic M, Balogh A and Smyshlyaev A 2006b *Proc. of the 45th IEEE Conf. on Decision & Control.*
[9] Krstic M, Siranosian A, Balogh A and Guo BZ 2007 *Proc. of the 2007 American Control Conf.*
[10] Krstic M, Guo BZ 2008 *J. of Control Optimization* **47** pp 553–574
[11] Ali SF and Padhi R 2009 *J. of Systems and Control Engineering* **223** pp 657-672
[12] Boonkumkrong N and Kuntanapreeda S 2014 *J. of Systems and Control Engineering* **228** pp 295-302
[13] Krstic M, Smyshlyaev A 2008 *Boundary Control of PDEs: A Course on Backstepping Designs* (Philadelphia: SIAM)
[14] Rao S S 2011 *Mechanical Vibrations* (New Jersey: Prentice Hall)
[15] Mathews J H and Fink K D 1999 *Numerical Methods Using MATLAB* (New Jersey: Prentice Hall)

Study on design of light-weight super-abrasive wheel

K. Nohara, K. Yanagihara* and M. Ogawa

National Institute of Technology Ariake, Ariake College, Higashi Hagio-Cho 150, Omuta-Shi, 836-8585, Japan

* Corresponding Author: Kiyoshi@ariake-nct.ac.jp

Abstract. Fixed-abrasive tool, also called a grinding wheel, is produced by furnacing abrasive compound which contains abrasive grains and binding powder such as vitrified materials or resins. Fixed-abrasive tool is installed on spindle of grinding machine. And it is given 1,800-2,000 min^{-1} of spindle rotation for the usage. The centrifugal fracture of the compound of fixed-abrasive tool is one of the careful respects in designing. In recent years, however, super-abrasive wheel as a fixed-abrasive tool has been developed and applied widely. One of the most characteristic respects is that metal is applied for the body of grinding-wheel. The strength to hold abrasive grain and the rigidity of wheel become stronger than those of general grinding wheel, also the lifespan of fixed-abrasive tool becomes longer. The weight of fixed-abrasive tool, however, becomes heavier. Therefore, when the super-abrasive wheel is used, the power consumption of spindle motor becomes larger. It also becomes difficult for the grinding-wheel to respond to sudden acceleration or deceleration. Thus, in order to reduce power consumption in grinding and to obtain quicker frequency response of super-abrasive wheel, the new wheel design is proposed. The design accomplishes 46% weight reduction. Acceleration that is one second quicker than that of conventional grinding wheel is obtained.

1. Introduction

Fixed-abrasive tool, what we call grinding wheel, is produced by furnacing abrasive-compound which contains abrasive grains and binding powder such as vitrified materials or resins [4, 5]. Fixed-abrasive tool is installed on spindle of grinding machine. And it is given 1800~2000 min-1 of spindle rotation for the usage. The centrifugal fracture of the compound of fixed-abrasive tool is one of the careful respects in designing. In recent years, however, super-abrasive wheel as a fixed-abrasive tool has been developed and applied widely. One of the most characteristic respects is that metal is applied for the body of grinding-wheel. The strength to hold abrasive grain and the rigidity of wheel become stronger than those of general grinding wheel, also the lifespan of fixed-abrasive tool becomes longer. The weight of fixed-abrasive tool, however, becomes heavier. Therefore, when the super-abrasive wheel is used, the power consumption of spindle motor becomes larger. It also becomes difficult for the grinding-wheel to respond to sudden acceleration or deceleration.

Thus, the purpose of this research is to design lighter grinding wheel while maintaining the rigidity, and to evaluate the effects on grinding performance, especially energy consumption.

2. Wheel design

2.1. Considered aspects in designing

In designing light-weight wheel, the following aspects are considered:

(1) Lower weight, same rigidity
(2) Wheel balance
(3) Avoiding centrifugal destruction
(4) Economical cost for manufacturing

Inasaki et al. [1-4] proposed a sensor-integrated grinding wheel where various sensors such as an acceleration sensor are installed on its carbon-composite body. Although it is an ambitious experimental proposal, from practical view point, utilizing carbon-composite is too expensive and possibility of delamination is one of the concerning aspects for durability.

Therefore, utilization of conventional super-abrasive wheel and weight reduction of the wheel are considered. Figure 1 and table 1 show a picture of utilized super-abrasive wheel and its specifications, respectively. Cubic Boron Nitride (CBN) as abrasive is selected for steel grinding, and resin as binder is selected to prevent centrifugal breakage of wheel.

Figure 2 shows the design of lightened super-abrasive wheel. 6-spoke design is selected according to the balance within production cost budget, strength and the result of FEM analysis. Round shape is also applied to connection between spokes, rim, and hub to avoid stress concentration.

Table 1. Specifications of experimented super-abrasive wheel

Abrasive material	Granularity	Coupling	Concentration	Binder
CBC	120	N	50	BW4
	Wheel diameter [mm]	Wheel thickness [mm]	Hole diameter [mm]	Abrasive layer thickness [mm]
	200.00	20.00	50.80	5.00

Figure 1. Utilized wheel

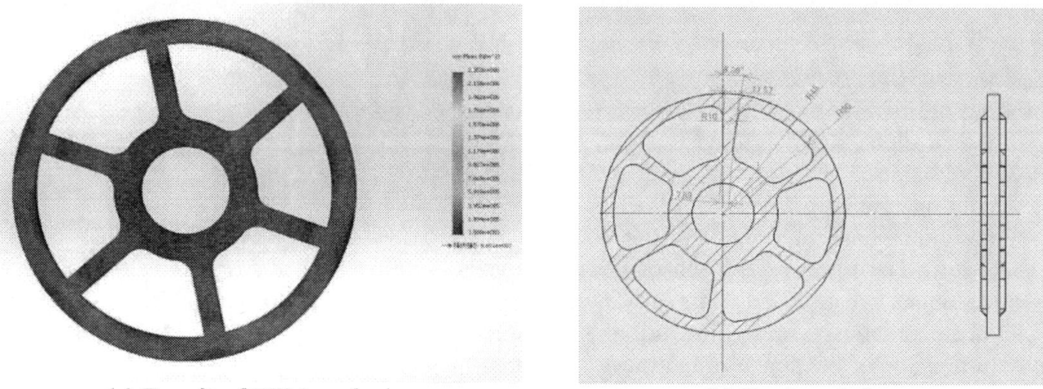

(a) Result of FEM analysis (b) Designed shape

Figure 2. Design of light-weight wheel

2.2. Manufacturing strategy

There are three methods to fabricate the designed shape in figure 2 from the raw shape in figure 1. They are milling, electro-discharge machining, and wire electro-discharge machining. In case of milling, the cost of the fixing tool of the work piece (jig) is a problem. In case of applying Electro- Discharge Machining (EDM), the cost of making carbon pole is a problem. On the other hand, in case of Wire Electro-Discharge Machining (WEDM), fixing the work piece is not needed, and wire pole has sufficient flexibility to trace complicated tool trajectory. Thus, wire electro discharge is selected to fabricate the light-weight wheel.

Figure 3 shows fabricated light-weight super-abrasive wheel. The wheel weight before processed is 1,183.7 g. Meanwhile, the weight after processed is 547.6 g that is 46% reduction, as shown in figure 4.

Figure 3. Light-weight concept

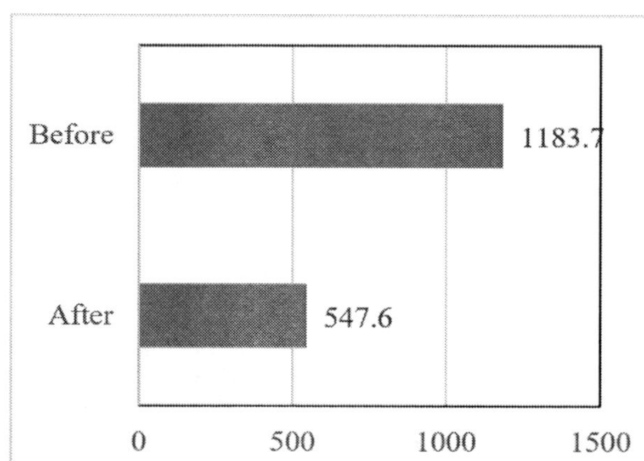

Figure 4. Comparison of wheel weight

3. Grinding Performance

3.1. experimental setup

Figure 5 shows the installation of the developed wheel. The evaluation of the fabricated light-weight wheel is executed with the experimental setup as shown in figure 6.

The specification of the grinder is shown in table 2. In order to monitor power, an ammeter logger (Fluke 120B) is installed on the spindle motor of the grinder.

Figure 5. Installed wheel

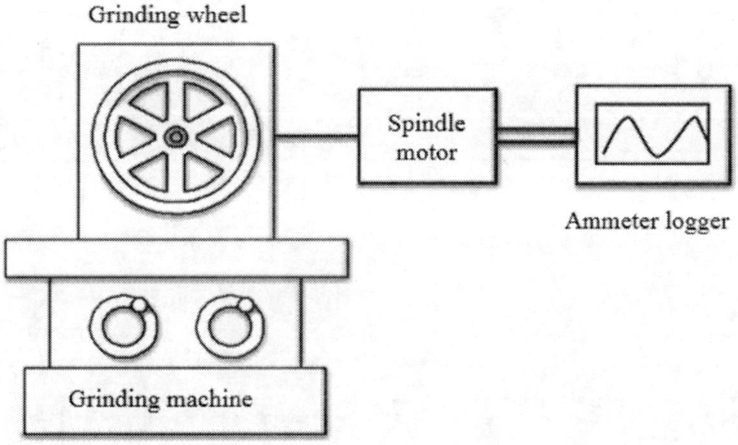

Figure 6. Experimental setup

Table 2. Specifications of utilized Grinder

Grinder	OKAMOTO PSG52DX
Table size	550×200
Wheel size	Φ205×25×Φ50.8
Grinding wheel rotation speed	1800 min^{-1}
Feed stroke	500
Minimum preset-able depth of grinding	0.1μm

3.2. Comparison of spindle-motor current at the start-up

In order to evaluate how the reduction of wheel weight effect on grinding, the current of the spindle motor is measured.

Figure 7 shows the measured spindle motor currents when the motor is started from 0 min^{-1} and accelerated to 2,000 min^{-1}. Figures 7(a) and 7(b) are cases of utilizing normal wheel and light-weight wheel, respectively.

The spindle motor is vector-controlled AC-servo-motor. The controllable parameters are current and frequency. The figures demonstrate almost the same maximum amplitude. However, the durations that the frequency becomes constant in utilization of each wheel are different. The duration of figure 7(b), which is 2.16 sec, is shorter than that of figure 7(a), which is 3.18 sec. The amplitudes of both the current plots do not decrease rapidly but slightly as shown in the figures. This means that the power consumption is large in proportion to the duration. Therefore the result shows that the light-weight wheel is effective in not only reduction of power consumption but also improvement of frequency response of wheel rotation.

IOP Conf. Series: Materials Science and Engineering **297** (2018) 012048 doi:10.1088/1757-899X/297/1/012048

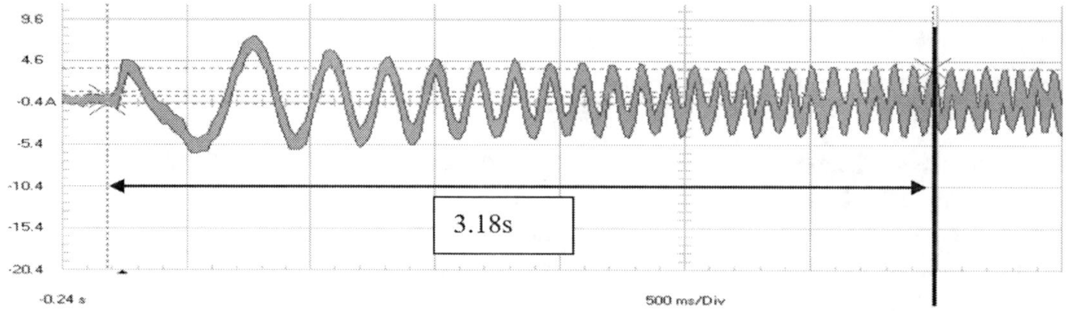

(a) Normal

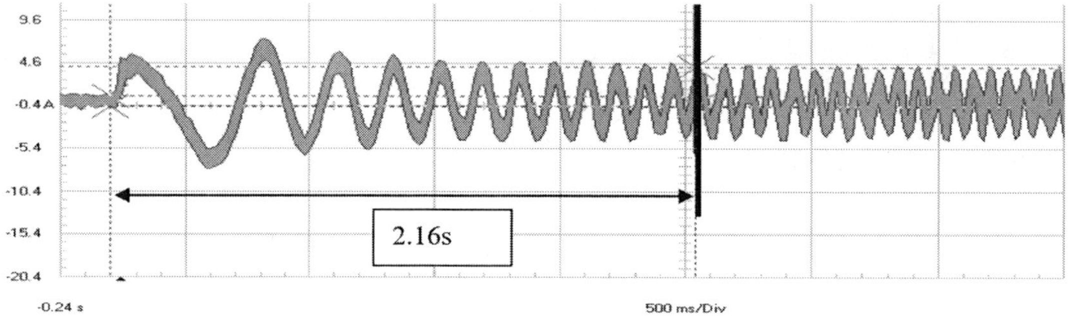

(b) Light-weight wheel

Figure 7. Comparison of waveform of spindle-motor current

4. Effects on processing

Reducing weight of grinding wheel can involve reduction of rigidity as fixed-abrasive tool. In this section, it is investigated how the reduction of the fixed-abrasive tool rigidity affects the ground surface.

Figure 8 shows the ground specimen of S45C with light-weight wheel. Figure 9 shows the roughness curves of work-pieces which are ground with normal and light-weight wheel.

The mean roughness of normal and light-weight wheel is 0.74 μm Ra and 0.23 μm Ra, respectively. It shows that reduction of wheel-weight in our design does not affect for wheel rigidity and grinding performance.

Figure 8. ground specimen

IOP Conf. Series: Materials Science and Engineering **297** (2018) 012048 doi:10.1088/1757-899X/297/1/012048

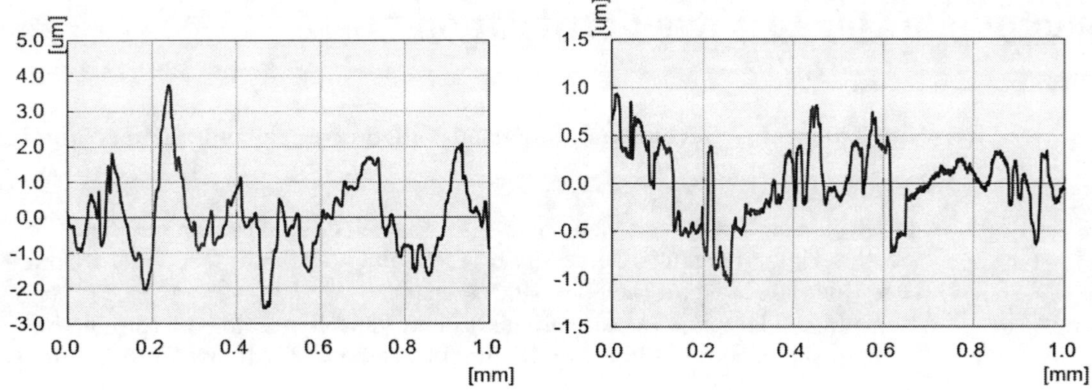

Ra 0.74µm
(a) Normal

Ra 0.23µm
(b) light-weight wheel

Figure 9. comparison of ground surface

5. Conclusion

In order to reduce power consumption and improve response of rotational acceleration or deceleration of spindle, reducing of weight of grinding wheel while maintaining the rigidity is investigated. Obtained conclusions from this study are as follows:

1) To realize a light-weight grinding wheel while maintaining high rigidity, six-spoke shape is designed with CAD. 46% weight-reduction has been accomplished.

2) To confirm the effect of this light-weight grinding wheel, verification experiment is conducted to accelerate the spindle from stop to 2,000 min^{-1} which is the operation speed without any wheel failure.

3) In the experiment, it is confirmed that the time until when the frequency becomes constant by the inverter control is shortened. Improvement of acceleration or deceleration and reduction of power consumption is confirmed.

4) The reduction of wheel weight does not affect wheel rigidity, surface integrity of ground surface, or grinding performance in the case of our design.

References

[1] Inasaki I 1999 *Int. J. Advanced Manufacturing Technology* **15** pp 730-736
[2] Shida A and Inasaki I 1998 *J. of JSME* **65**(629) pp 173-177
[3] Yanaighara K and Tani Y 2000 *Kikai Gijyutsu* **48**(2) pp 59-63
[4] Marinescu I, Hitchiner M, Uhlmann E and Brian Rowe W 2006, *I. Inasaki Handbook of Machining with Grinding Wheels* (CRC Press) p 226
[5] Kawamura S, Yano S, Higuchi N, Sugita T 1984 *Griding and Abrasive process* (in Japanese, Kyoritsu publication)

IOP Conf. Series: Materials Science and Engineering **297** (2018) 012049 doi:10.1088/1757-899X/297/1/012049

Development master arm of 2-DOF planar parallel manipulator for In-Vitro Fertilization

Kitti Thamrongaphichartkul[1], Supachai Vongbunyong[1,*] and Lalana Nuntakarn[2]

[1] Innovation and Advanced Manufacturing Research Group, Institute of Field Robotics, King Mongkut's University of Technology Thonburi, 126 Pracha Uthit Rd., Bang Mod, Thung Khru, Bangkok 10140, Thailand
[2] Ramathibodi Hospital, Deparment of Obstetrics and Gynecology, Reproductive Endocrinology and Infertility Unit, Mahidol University, 272 Rama VI Rd., Thung Phaya Thai, Ratchathewi, Bangkok 10140, Thailand

* Corresponding Author: supachai.von@kmutt.ac.th, 081-431-1968

Abstract. Micromanipulator is a mechanical device used for manipulating miniature objects in the order of micron. It is widely used in In-Vitro Fertilization (IVF) in which sperms will be held in a micro-needle and penetrate to an oocyte for fertilization. IVF needs to be performed by high skill embryologists to control the movement of the needle accurately due to the lack of tactile perception of the user. Haptic device is a device that can transmit and simulate position, velocity and force in order to enhance interaction between the user and system. However, commercially available haptic devices have unnecessary degrees of freedom and limited workspace which are inappropriate for IVF process. This paper focuses on development of a haptic device for using in IVF process. It will be used as a master arm for the master-slave system for IVF process in order to enhance the ability of users to control the micromanipulator. As a result, the embryologist is able to carry out the IVF process more effectively with having tactile perception.

1. Introduction

Currently infertility can be found in 8-10% of couples globally [1]. A number of couples require Assisted Reproductive Technology (ART) treatment for conception. In-Vitro Fertilization (IVF) is a conventional method that allows sperms and an oocyte to fertilize in a petri dish. With this method, proper sperms and oocyte can be carefully selected, for instance sperm with high mobility. This is currently considered one of the most effective treatments for infertility and genetic abnormality. Intracytoplasmic Sperm Injection (ICSI) is a procedure in which the sperm is injected into an oocyte in order to be fertilized. ICSI process is performed by using manual micromanipulator (see figure 1) where a micro needle and a holding pipette are equipped and controlled. ICSI consists of 3 steps: (1) selecting a single sperm by using micro needle; (2) holding the oocyte by using holding pipet, and (3) controlling the position micro needle to penetrate into the oocyte and injection sperm into oocyte.

In general, embryologists with the ICSI expertise manually use a micromanipulator for accurately manipulating sperm and oocyte. The movement need to be scaled down due to the small size of the subjects which are in the order of microns (sperms are 3-5 microns and oocytes are 100 microns see figure 2), the success rate of the current and manual ICSI process are between 50-80% [2]. This is because important physical variables – i.e. position, velocity and tactile perception – are uncontrollable.

From this difficulty, haptic devices can be considered one of the tools that potentially improve this procedure.

Figure 1. Micromanipulator (RI Intragal-3)

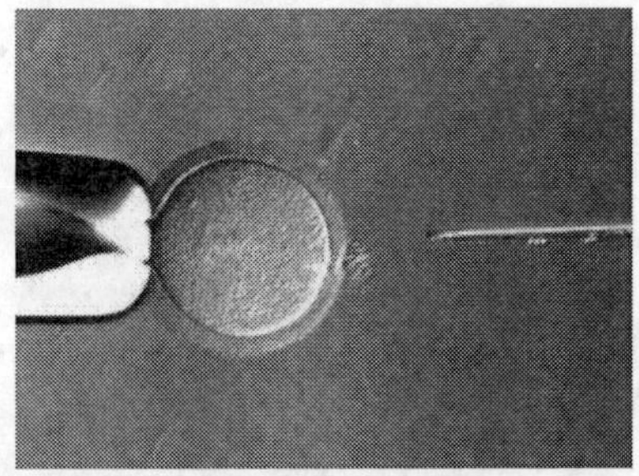

Figure 2. Sperm is deposited into the egg in research

Haptic devices are the devices that are able to simulate an augmented force by providing some feedback to the users. Therefore, the users can perceive the virtual environment and manipulate the virtual objects in more intuitive ways. Research in field of haptic interface, e.g. [5-10] and [12], works on how to design force feedback devices. One of the most notable and commercially available haptic devices is Sensable PHANTOM Omni [13] (see figure 3). It is a motorized device that generates tactile feedback that allows the users to perceive virtual forces and virtual objects. [10] presents a haptic device that is designed based on 5-bar linkage mechanism. This mechanism has 2-DOF which is necessary for controlling a micromanipulator in the direction along x- and y-axes. A number of research works show that haptic devices used with robotic system in ICSI process is able to improve the success rate of 90% and survival rate of 90.7% (n = 120) [4]. In addition, the authors of [6] design a new end-effector to sense the contact force with strain gauge and developed a haptic device for transmit force to the embryologists in ICSI process. As a result, the embryologists can feel injection force and contact of membrane through the device.

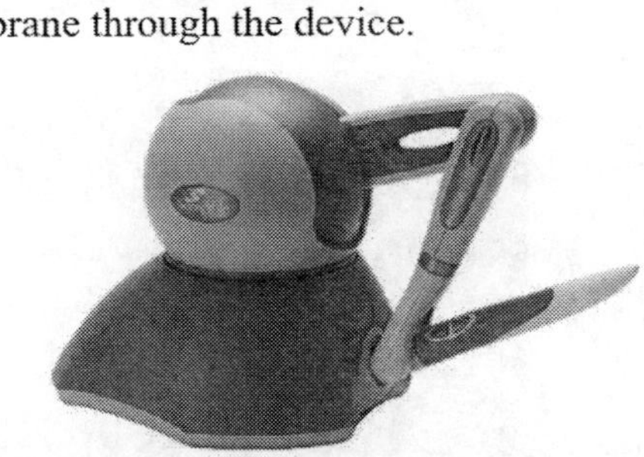

Figure 3. PHANTOM 6-DOF Master Device

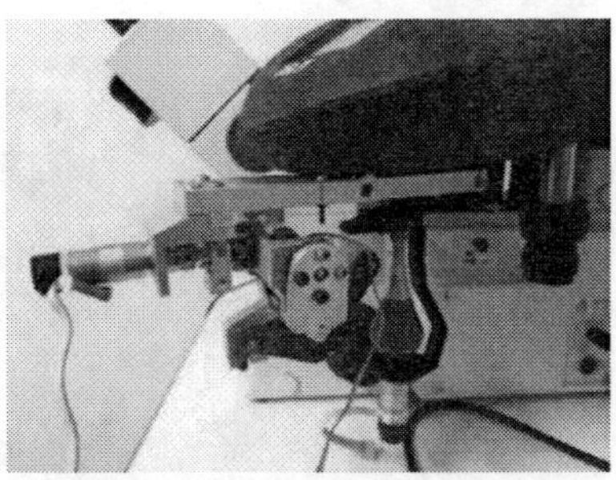

Figure 4. Slave arm is added on micromanipulator

In this research, the manual micromanipulator is enhanced by equipping an added-on system that allows semi-automatic control of the existing manual manipulator. The system primarily consists of (1) a master arm which is a haptic device and (2) a slave arm which is connected to the micromanipulator's control joystick. The slave arm is previously presented in [12] (see figure. 4).

In this article, the master arm with haptic interface is presented. The master arm is haptic device connecting between the embryologist and micro environment. It can help the embryologist to manipulate the subjects by enabling tactile perception. Virtual fixture is enabled in order to guide the user to work easily. In addition, the workspace is designed and customized to be more user-friendly than commercially available haptic devices.

This article is organized as follows. System overview which is an architecture of our system is in Section 2. Design method of the master arm is in Section 3. Final design and implementation is described in Section 4. The section 5 describes an experiment. The discussion and conclusion are described in Section 5 and Section 6, respectively

2. System Overview

In this section the system used for operating haptic application is explained. The system overview is shown on the connection diagram on figure 5. The main components are users, a master arm, a controller, and a semi-auto micromanipulator.

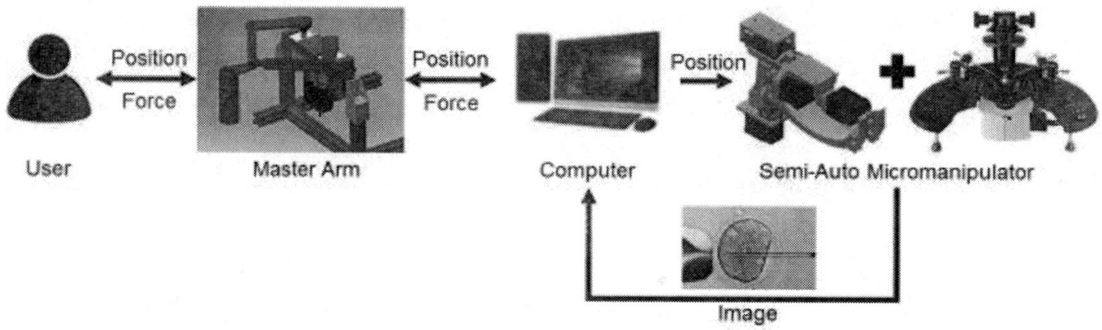

Figure 5 System components in research

2.1. User

User is an embryologist with micromanipulator expertise that can control position of the micro needle and micro pipette via the master arm. User can sense the force between the micro needle and the oocyte.

2.2. Master Arm

The master arm is a haptic device used for implementing force feedback while manipulating the micro needle and the micro pipette. The user can control them without using conventional joystick directly (see figure 6).

Figure 6. Joystick used for controlling a micro needle.

2.3. Controller

The controller consists of 2 control layers, which are implemented on different architectures. The high level system is operated on the computer (C#) for creating haptic interface with user. The low level system based on C++ on a microcontroller for controlling the hardware and mechanisms.

2.3.1. High Level System. At the High level control layer, EMGU Library is used for processing images in order to detect the micro needle and the micro pipette shown in figure 7. The result is used to simulate the impact force that will be emulated on the master arm.

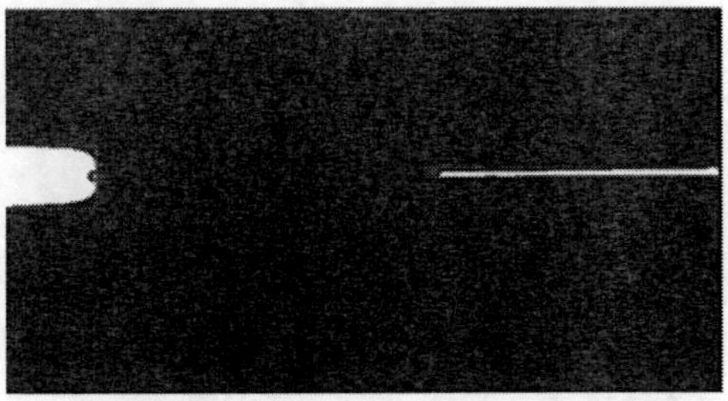

Figure 7. Image obtained after image processing

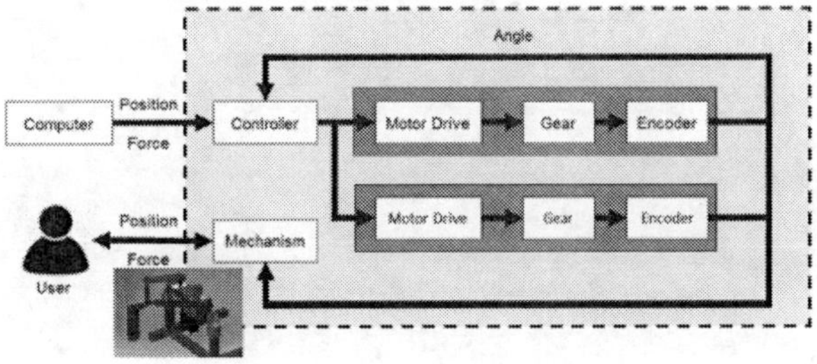

Figure 8 Low level system

2.3.2. Low Level System. Use microcontroller STM32F746 to implement a hardware interface because it has QEI module that is capable of reading encoders and running the control loop at 1 kHz. Figure 8 shows the low level system in this research.

2.4. Semi-Auto Micromanipulator.

Manual micromanipulator is a device in which the user can use to manipulate small objects by controlling the joystick. The hand movement of the user is scaled down to the movement in the order of microns. The slave arm is an added-on mechanism with 3 DOF that is used to control the joystick (see figure 4). It is designed by the authors and presented in [12]. It is equipped on the micromanipulator by clamping between the base of the slave arm and the base of micromanipulator (see figure 9).

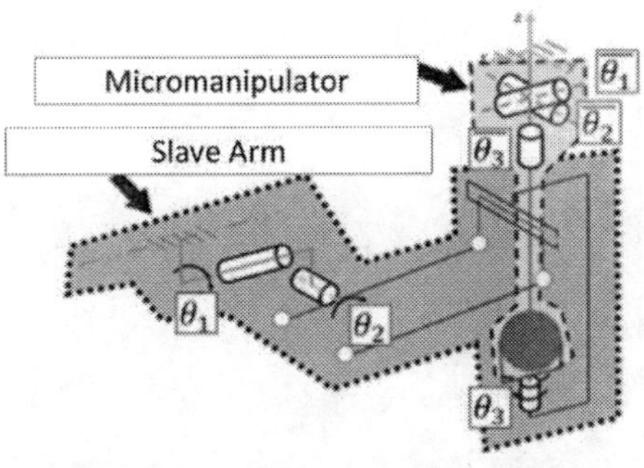

Figure 9. Mechanism of slave arm

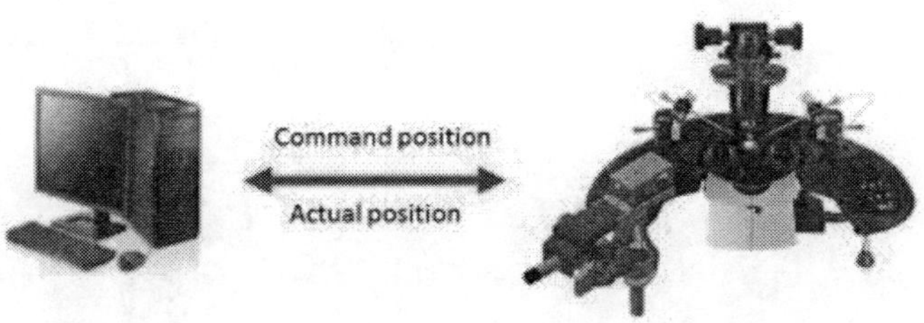

Figure 10. Communication between computer and slave arm

In this research, our system is capable of upgrading a manual micromanipulator to a semi-automatic micromanipulator by adding on the slave arm. It can change the traditional control method to computer control method. The command of position can be sent from the computer to the micromanipulator (see figure 10).

3. Design

As shown in figure 11, a five-bar linkage planar parallel mechanism [10] presents the conceptual design of the master arm. This is a simple yet effective mechanism that allows the movement of 2 DOF which is sufficient for the intended function in this research. The end-effector is located at point P_3 and moves on the x-y plane. At point P_1 and P_2 , motor and encoder for driving θ_1 and θ_2 are located. The force at the end-effector is related to torque applied by motor at joint 1 and joint 2.

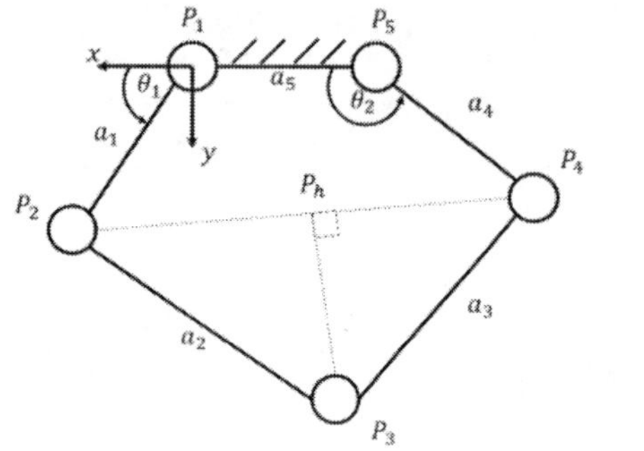

Figure 11. 5-bars linkage mechanism

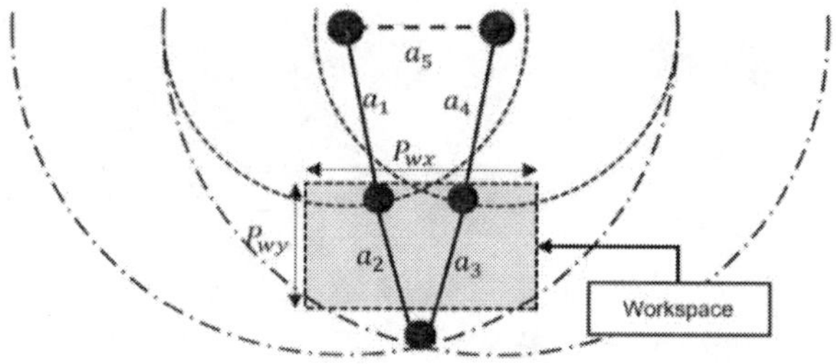

Figure 12. 5-bars linkage mechanism design

3.1. Linkage Design

The mechanism's length is designed by specifying the operating workspace. A graphical method is used to find the length of each link (see figure 12). The determined workspace is $P_{wx} = 100$ and $P_{wy} = 100$.

Therefore, the lengths of linkages are $a_1 = 90$, $a_2 = 130$, $a_3 = 130$, $a_4 = 90$ and $a_5 = 60$. Simulation of the end-effector was done by using a commercial mathematic simulation software. The positions in the designed workspace are shown in figure 13.

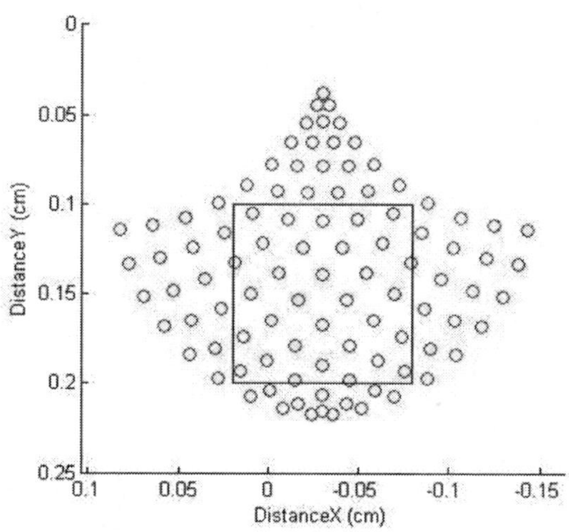

Figure 13 5-bars linkage mechanism simulation

3.2. Forward Kinematics

Forward kinematics is derived in order to describe both the position of end-effector P_3 and the position of the each linkage [10]. In figure 13, P_1 is used as the origin. The end-effector's position $P_3(x_3, y_3)$ is given by equations (1) and (2)

$$x_3 = x_h \pm \frac{\|P_3 - P_h\|}{\|P_2 - P_4\|}(y_4 - y_2) \tag{1}$$

$$y_3 = y_h \mp \frac{\|P_3 - P_h\|}{\|P_2 - P_4\|}(x_4 - x_2) \tag{2}$$

3.3. Jacobian

Jacobian is a method to transform force at end of effector (P_3) to torque of motor at joints 1 and 2 [10]. See equation (3).

$$J = \begin{bmatrix} \frac{\partial x_3}{\partial \theta_1} & \frac{\partial x_3}{\partial \theta_5} \\ \frac{\partial y_3}{\partial \theta_1} & \frac{\partial y_3}{\partial \theta_5} \end{bmatrix} \tag{3}$$

3.4. Torque

The master arm can apply the force to the user at the end-effector. The force is a result of the torque generated by the motors at joint 1 and joint 2. In addition, the force at P_3 is used to create the virtual wall which helps the user by guiding the movement in particular lines. The torque for each joint can be obtained by using equations (4) and (5).

$$\tau = J^T F \tag{4}$$

$$\tau = K_t i \tag{5}$$

The torque can be controlled by the motor driver circuit as it is proportional to the input current. The relationship between voltage and current was obtained by fitting the data shown in figure 14. The relation is in equation (6). (Here, i is current and v is voltage.)

$$i = 0.0444v \tag{6}$$

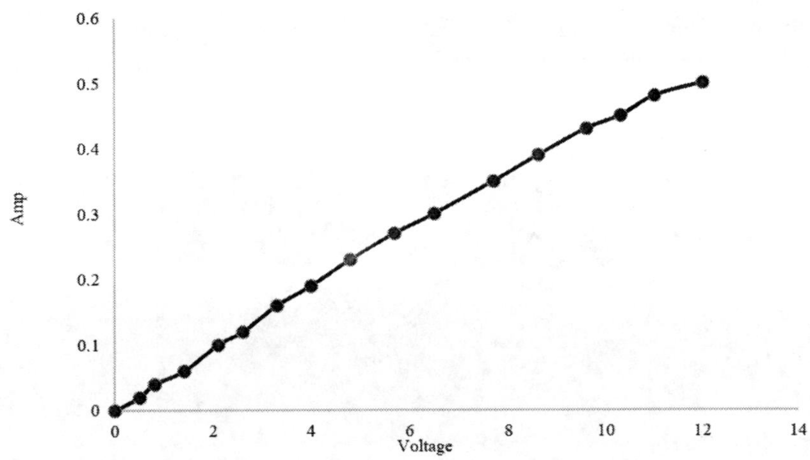

Figure 14. Current and voltage in motor

3.5. Virtual Fixture

According to the movement in micro environment, the user cannot sense the reaction force between an oocyte and the microneedle during the ICSI operation. Visual fixture [3] is one way to model virtual force by using the virtual wall concept that can be modelled the system with spring and damper in equation (7).

$$F = k\delta x + b\delta \dot{x} \tag{7}$$

Figure 15 shows the model of the force feedback. The set point is a target of the end of the microneedle. If the microneedle's end is at the distance from the target, the virtual force will be created according to equation (7) and sent to the master arm. This can help the user to move a microneedle's end along the desired path to the target. The variable δx is distance between position of microneedle and position of the set point.

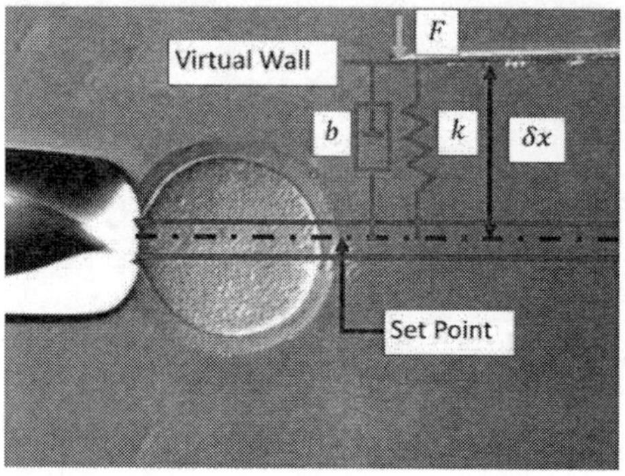

Figure 15. Model of force feedback in virtual fixture

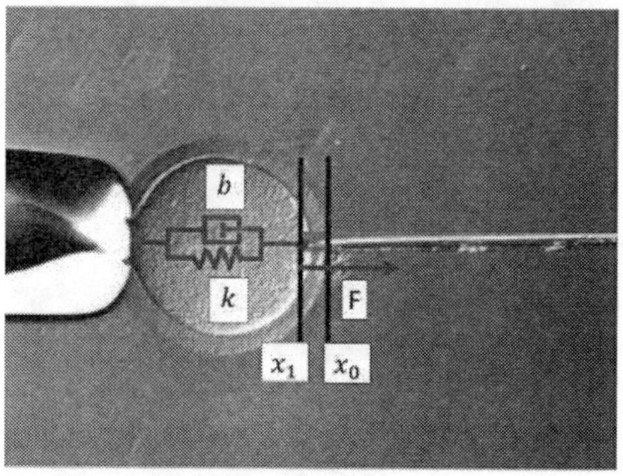

Figure 16 Model of force between microneedle and oocyte

Virtual fixture is used to create the virtual force to enable the tactile perception between the oocyte and the microneedle. In figure 16, the model of virtual force is equation (7) where δx is distance between x_0 and x_1.

4. Implementation

The final prototype is shown in figure 17. The system consists of a controller STM32F746, a motor drive VNH5019, and motor Faulhaber series 3257 to create a master arm prototype. The flowchart is shown in figure 18.

Figure 17. The prototype of master arm

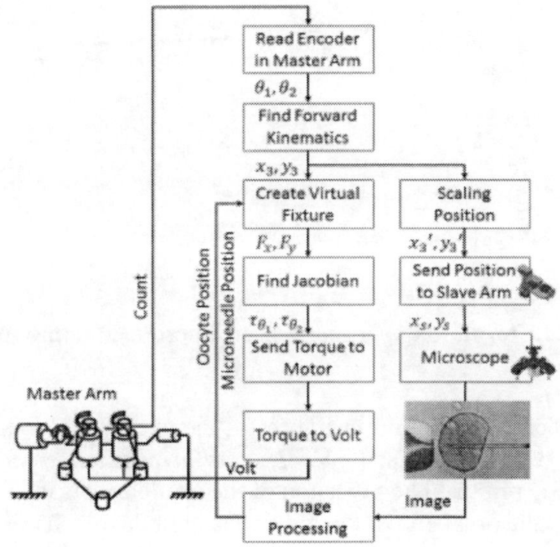

Figure 18. Flow chart diagram of master arm

5. Experiment

The experiment is designed for testing the concept of tactile perception of the user while using the master arm. The experiment was done on the master-slave system shown in figure 19.

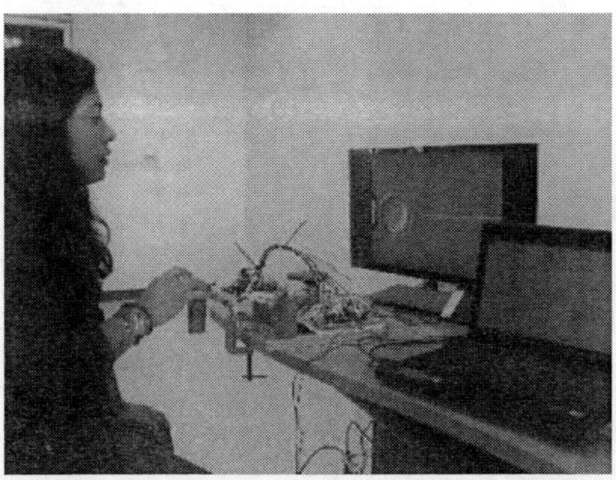

Figure 19. System components in research

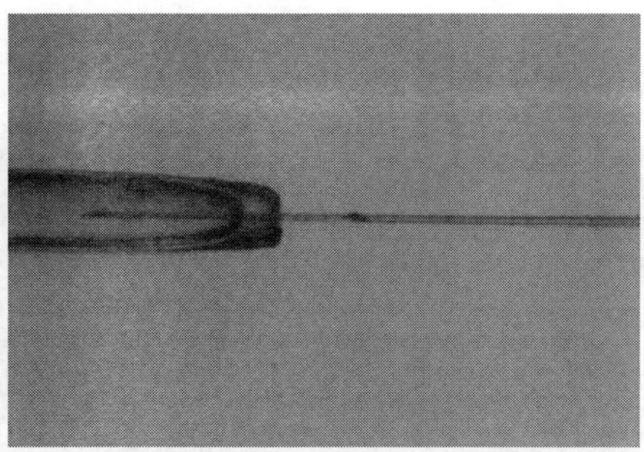

Figure 20. Microneedle is insert into micro pipette.

The goal of the experiment is to test the accuracy and precision of the system while it is controlled by the user. In the experiment, the user needed to control the semi-micromanipulator by using the master arm to move the microneedle in a straight line and insert it into the micropipette. The virtual fixture is used to create a virtual wall with spring and damper to constrain the movement along y-axis. The parameter in experiments are virtual spring constant 0.01 N/m and damping constant 0.01 Ns/m. The procedures in the experiment are:

1.) User control master arm in straight line.
2.) Create virtual fixture to create virtual force with spring and damper to constrain x axis.
3.) Send force generated from virtual fixture to user through master arm.
4.) Use image processing to track the micro needle.

The result of the experiment is presented in figure 20. Where the microneedle was in the micro-pipette. Figure 21 shows that real micro needle position. For the result, the microneedle can be moved along the straight line with the error within ± 10 microns in y-axis.

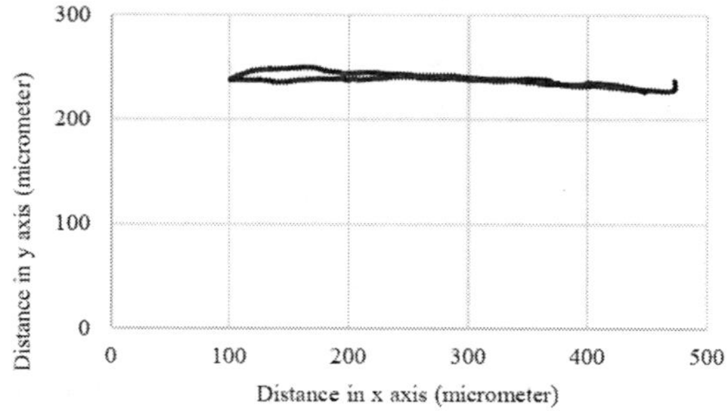

Figure 21. Result in experiment insert microneedle into micropipette.

6. Discussion

From the experiment of tactile perception in figure 19, the user was able to use the master arm control microneedle to move in a straight line; however, certain position error was occurred caused by a mechanism created with 3D printer. The precision of the printed parts is quite low (0.1 m), which leads to the backlash that affects the precision of the joint control in the mechanism. When the master arm has the backlash, it affects the slave arm so that the control of microneedle movement will be less precise. However, this backlash can be eliminated by using the parts manufactured with higher precision. As a result, the accuracy and precision of the system will be improved.

7. Conclusion

In this research, the master arm for using in In-Vitro Fertilization process has been developed. The master arm was able to help embryologists to work more intuitively by creating the virtual force that enables tactile perception. The concept has been proved from the experiment result that the user can control the microneedle in the way that precision and accuracy are needed. However, they can be improved if the precision of the mechanism is improved. For the future work, we focus on bilateral control for IVF and design a new micromanipulator.

Acknowledgment

This work was supported by Asahi Glass Company, Envimed Company Limited and Prof. Teraporn Vutyavanic from Chiang Mai IVF Polyclinic. The authors gratefully acknowledge the contribution of Institute of Field Robotics, King Mongkut's University of Technology Thonburi and Department of Obstetrics and Gynaecology, Reproductive Endocrinology and Infertility Unit, Mahidol University.

References

[1] Palermo G, Joris H, Devroey P and Steirteghem A C V 1992 *Lancet* **340**(8810) pp 17–18

[2] Lu Z, Zhang X, Leung C, Esfandiari N, Casper RF and Sun Y 2011 *IEEE Int.Conf. on Robotics and Automation* (Shanghai International Conference Center)

[3] Viboon Sangveraphunsiri and Supachai Vongbunyong 2009 *The 23rd Conf. of the Mechanical Engineering Network of Thailand* (Chiang Mai)

[4] Zhe Lu, Member, IEEE, Xuping Zhang, Member, IEEE, Clement Leung, Navid Esfandiari, Robert F. Casper, and Yu Sun, 2011 *IEEE TRANS. ON BIOMEDICAL ENGINEERING* **58**(7)

[5] Vu Minh Hung and Uhn Joo Na 2011 *11th Int.Conf. on Control, Automation and Systems* (KINTEX, Gyeonggi-do, Korea)

[6] Sung-Yong Cho and Jae-Hong Shim 2004 *Int. Conf. on Intelligent Robots and Systems* (SBndaI, Japan)

[7] Aude Bolopion and Stéphane Régnie 2013*IEEE TRANS. ON AUTOMATION SCIENCE AND ENGINEERING* **10**(3)

[8] Sun Yu and Bradley J Nelson 2001 *Proceedings of the 2001 IEEE Int. Conf. on Robotics & Automation* (Seoul, Korea)
[9] Lingxuan Zhang, Hongxia Xiong and Liangliang Han 2015 *Proc. of the 2015 IEEE Int. Conf. on Information and Automation* (Lijiang, China)
[10] Gianni Campion, Qi Wang, and Vincent Hayward 2005 *IEEE/RSJ Int. Conf. on Intelligent Robots and Systems*
[11] Gourishetti Ravali and Muniyandi Manivannan 2017 *IEEE Reviews in Biomedical Engineering*
[12] Thamrongaphichartkul K, Vongbunyong S and Nuntakarn L 2017 *TRS Conf. on Robotics and Industrial Technology*
[13] Geomagic Touch Haptic Device, URL : www.geomagic.com

IOP Conf. Series: Materials Science and Engineering **297** (2018) 012050 doi:10.1088/1757-899X/297/1/012050

Lane detection using Randomized Hough Transform

Peerawat Mongkonyong[1,*], Chaiwat Nuthong [1], Supakorn Siddhichai[2] and Masaki Yamakita[3]

[1] International College, King Mongkuts Institute of Technology Ladkrabang (KMITL), Bangkok 10520, Thailand
[2] National Electronics and Computer Technology Center (NECTEC)
[3]Tokyo Institute of Technology, Tokyo (Tokyo Tech)

* Corresponding Author: 56610013@kmitl.ac.th, 0822087697

Abstract. According to the report of the Royal Thai Police between 2006 and 2015, lane changing without consciousness is one of the most accident causes. To solve this problem, many methods are considered. Lane Departure Warning System (LDWS) is considered to be one of the potential solutions. LDWS is a mechanism designed to warn the driver when the vehicle begins to move out of its current lane. LDWS contains many parts including lane boundary detection, driver warning and lane marker tracking. This article focuses on the lane boundary detection part. The proposed lane boundary detection detects the lines of the image from the input video and selects the lane marker of the road surface from those lines. Standard Hough Transform (SHT) and Randomized Hough Transform (RHT) are considered in this article. They are used to extract lines of an image. SHT extracts the lines from all of the edge pixels. RHT extracts only the lines randomly picked by the point pairs from edge pixels. RHT algorithm reduces the time and memory usage when compared with SHT. The increase of the threshold value in RHT will increase the voted limit of the line that has a high possibility to be the lane marker, but it also consumes the time and memory. By comparison between SHT and RHT with the different threshold values, 500 frames of input video from the front car camera will be processed. The accuracy and the computational time of RHT are similar to those of SHT in the result of the comparison.

1. Introduction

Nowadays, demands on vehicles are increasing because of the rapid population growth. This leads to the major causes of higher road accidents [1]. Nearly 3,500 people die on the road every day. It is found that the lane changing is the cause of most accidents. The advantage of LDWS should be an optional tool to relieve such disaster. LDWS is one of the methods that are proposed to solve the problem. LDWS uses the image processing mechanism to detect the lines of the image, selects the current lane markers, warns the driver when the vehicle leaves the current lane and tracks the lane marker for the next frame. In this article, overviews of SHT and RHT are presented and the algorithms are briefly introduced and compared.

Kalviainen et al. [2] presented the probabilistic and non-probabilistic Hough transforms overview and comparisons. They compared a lot of algorithms such as SHT, RHT, Dynamic RHT and Window RHT. They tested line detection with synthetic and real-world images demonstrating the high speed and low memory usage. The result of the comparison was a good model for the road image.

Lim et al. [3] performed lane-vehicle detection and tracking. Vertical mean distribution was used to remove the sky region by averaging the gray values of each row on the image. A big jump of this method indicated the line that divided the image into sky region and road region. But the result would be erroneous if some parts of the car appeared in the image such as a console. Because the averaged value would have changed. To solve this problem, the vanishing point detection was used for this section.

Schreiber et al. [4] presented the lane detection that the vanishing points were discussed. Vanishing points were detected using Hough Transform (HT). This method was used the vanishing point to detect straight line segments. But the vanishing point also can remove the unwanted area to reduce the computational time. In edge detection, Canny edge detection was used in image preprocessing section. In term of computational time, some type of edge detection was better than Canny edge detection [5].

Kultanen et al. [6] introduced the RHT. The mechanisms of a random sampling of point pairs from the x-y plane, converted to the ρ-θ plane and voted the accumulator. In the lane marker detection, the accuracy of this method can increase by assuming one point of the pair to be the vanishing point. Because the vanishing point was the part of the straight line in the image.

Wang et al. [7] proposed a B-Snake based lane detection and tracking algorithm without any parameters of the camera. This article used Canny/Hough estimation of vanishing points (CHEVP) to provide a good initial position for the B-Snake. The snake got drawbacks. It often gets stuck in local minima states which can be counteracted by simulated annealing techniques. Minute features were often ignored during energy minimization over the entire contour. And their accuracy was governed by the convergence criteria used in the energy minimization technique.

To solve the problems, a lane detection is proposed in this article as outlined by the flow diagram in figure 1. Lane boundary detection is divided into 3 parts. The image that obtained from the camera has many ingredients. The image preprocessing is applied to make it appropriate for processing. Image preprocessing is performed on the road image to locate the left-right lane edges by separating the sky region and analyzing road region known as the region of the interest (ROI) in this article to extract the prominent road features such as lane markers. The information of the vanishing point is used to limit the area in line detection and refer the point in RHT. Lane marker detection part attempts to identify the lane marker from the detected lines.

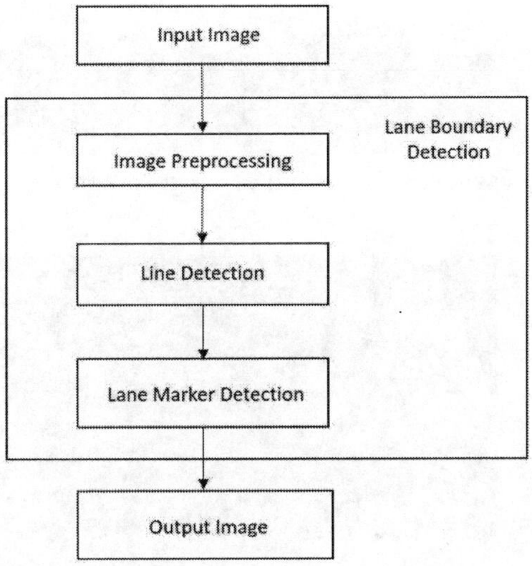

Figure 1. System overview.

This article is organized as follows. Section 2 discusses about the lane boundary detection and explains the detection technique. Some simulation results are shown in section 3 followed by the conclusion and future works.

2. Lane Boundary Detection

Lane boundary detection consists of 3 parts, i.e. image preprocessing prepares a proper image for further processing from the original image as shown in figure 2, line detection to extract the lines contained in the images and lane marker detection to detect the current lane markers.

Figure 2. Original image.

2.1. Image preprocessing

The proposed image preprocessing includes the grayscale image conversion, image filtering, image morphology as shown in figure 3, and sky region removing which applies some technique to remove some part of the image outside the ROI. This can reduce the computational time in the vanishing point detection method. In this article, the sky region removing removes the upper half of the image as shown in figure 4. Consuming some part of the road surface is unnecessary because this method focuses on the lane markers.

Figure 3. The result of image preprocessing.

Figure 4. The result of sky region removing.

2.2. *Line detection*

The line detection part consists of the edge detection that includes a variety of mathematical methods that aim at identifying points in a digital image at which the image brightness changes sharply as shown in figure 5. Many edge enhancement algorithms have been proposed. The Sobel edge detection is used for this article because it is better than Canny edge detection in term of being less time-consuming [5]. After edge detection method, the image contains only white and black pixels.

Figure 5. The result of edge detection.

A vanishing point is a point intersection of the projections of parallel lines in the image as shown in figure 6. HT can evaluate the vanishing point by detecting the lines and the intersection point of those lines in the image.

The objective of HT is to find lines in a given image. They are 2 types of HT to be considered in this article, i.e. SHT and RHT. They are different mainly in the resource usage, which is one of the concerned issues for the researchers.

2.2.1. Standard Hough Transform. The SHT computation consists of calculating the parameter and accumulating the data in the accumulator, finding the local maxima which represent line segments of the image and extracting the line segments from the maxima positions. The disadvantages of SHT are its computational complexity and huge storage memory consumption.

It assumes all of the white pixels of the image to be the points and converts them into ρ-θ plane. ρ is the line connecting the polar coordinate to the origin where the x-axis intersects the y-axis. θ is the angle between ρ and x-axis as shown in figure 7. Each point in the x-y plane will have the infinity lines that pass the point. Those lines are converted by equation (1).

$$\rho = x_0 \cos\theta + y_0 \sin\theta, \tag{1}$$

where x_0 and y_0 are the coordinates corresponding to the angles θ (x_0, y_0). The line plot of those lines will be the sets of point in the ρ-θ plane. So, the point connection will be the sinusoidal wave for each point in the x-y plane as shown in figure 8. The intersection of lines in ρ-θ plane means they have the same value of ρ-θ. By voting θ in the accumulator will get the set of points in the x-y plane that has the high possibility to be on the same line.

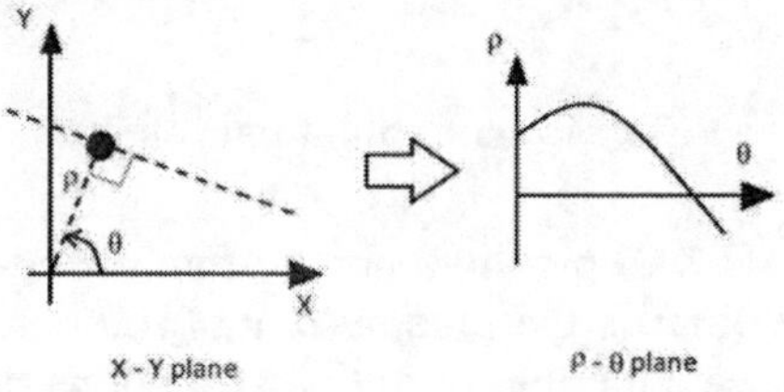

Figure 6. Principle of Hough transform. A point in x-y plane is converted into a curve in ρ-θ plane.

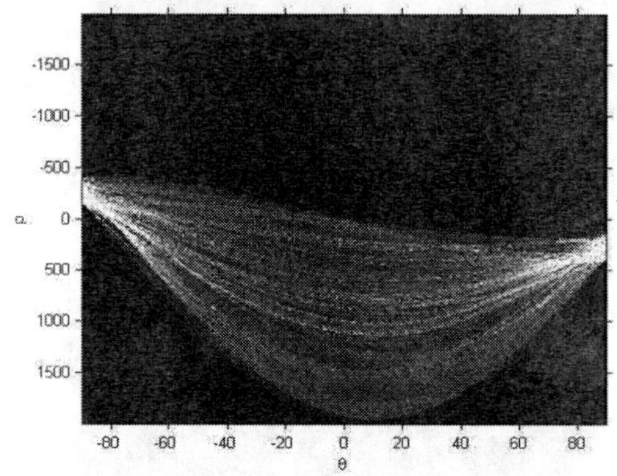

Figure 7. Data voting in ρ-θ plane.

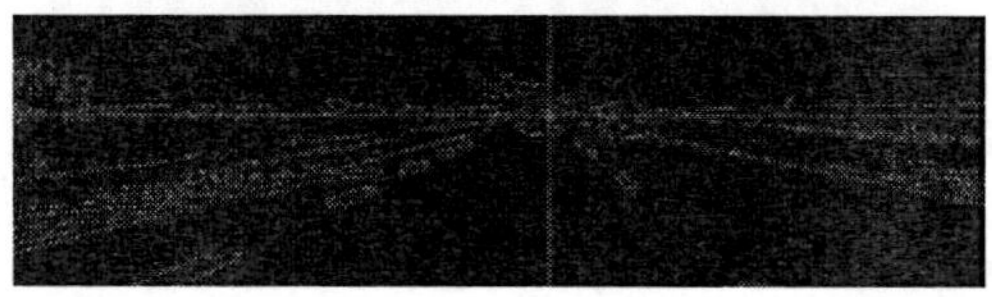

Figure 8. The result of vanishing point detection.

2.2.2 Randomized Hough Transform. On the contrary, The RHT computation consists of selecting 2 random pixels from edge image and calculating the parameter from the line of the point connection as shown in figure 9, accumulating and voting the data in the accumulator, extracting the line segments from that data. The RHT algorithm lies in the fact that each point in ρ-θ plane can be expressed with 2 points or 1 line from the original binary edge image. The advantage of RHT are high parameter resolution, the infinite scope of the parameter space, small storage requirements and high speed.

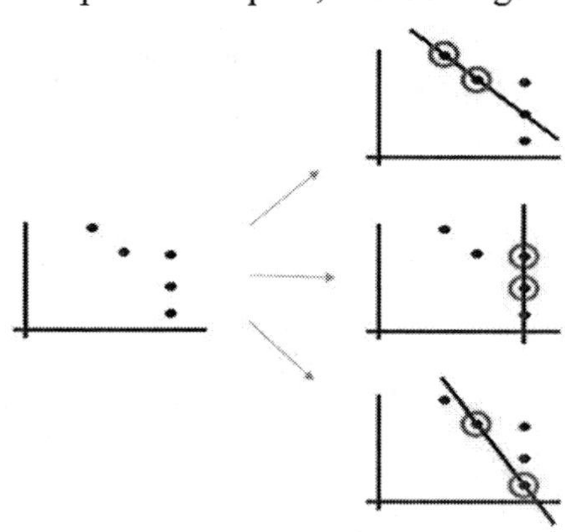

Figure 9. The random point pair method in RHT.

A line detection using the RHT will pick two points from the edge detection. The first point is the vanishing point and the second point is the random point. The random type is uniformly distributed pseudorandom integers. Then, create the line from that two points using linear equation. This method also uses equation (1) for the conversion. After that, the data are accumulated in the accumulator. The vote has 2 types. Firstly limiting the number of input θ and picking the highest one after voting. Secondly, limiting the vote and picking the first θ that is equal to the vote. This article uses the second type because the user can adjust the computation time. On the other hand, SHT uses the first type to vote the accumulator but the limit of the input is all of the data. So, the computation time of RHT should be lower than SHT because it uses only some points of the image.

The main difference between SHT and RHT is that while in SHT a single pixel in the original image is mapped to a curve in ρ-θ plane, in RHT a pair of pixels is mapped to a single cell in ρ-θ plane. SHT generates all parameter combinations, but RHT generates only a small subset of all parameter combinations.

After that, the area removing uses the information of the vanishing point to further remove the unwanted area. Changing the value of that pixel to 0 (black pixel), it can reduce some computational time and some error for this method as shown in figure 10. After the vanishing point and other 2 points at the corner are specified. The corresponding linear equation can be found, i.e.

$$y = mx + c, \tag{2}$$

where m is the slope and c is the y-axis intersection. The vanishing point also can be used to divide the image into the left-hand side and the right-hand side. This method detects the lane marker of two sides as shown in figure 11 and figure 12.

Figure 10. The result of area removing. The image shows only the ROI.

Figure 11. The right-hand side image that divided by the vanishing point.

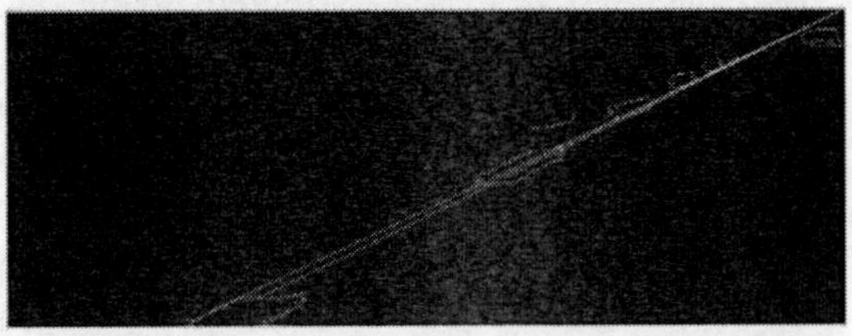

Figure 12. The left-hand side image that divided by the vanishing point.

In case of curve and junction, the difference between straight and curved road is some area under the vanishing line as shown in figure 13.

Figure 13. (2) is the area affected by curved road and (1) is the area not effected by curved road.

However, the high possibility lines would be the lane markers because of the number of the lane marker pixels.

2.3. Lane marker detection

After line extraction, the lane marker detection will select the lowest θ value of each side. They are assumed to be the lane marker as shown in figure 14.

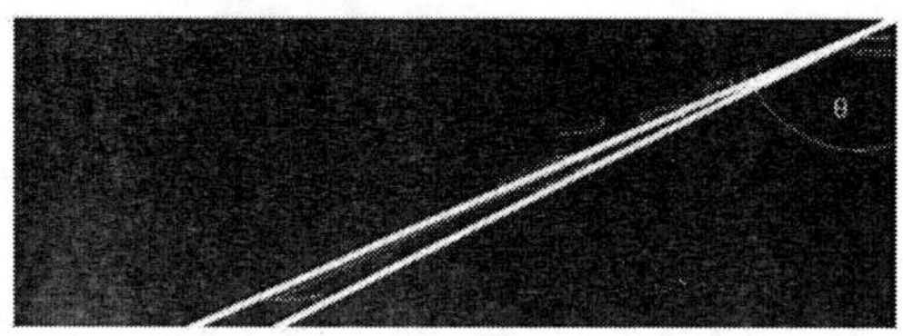

Figure 14. The lowest θ will get the high possibility to be the lane marker.

Generally, the lane marker should be located at the middle of the road surface and bottom of the image. For each side, the system needs to find at least two different θ's from the vote and compares them. The lowest θ will have the high possibility to be the lane marker of that side.

3. Result

The system runs on MATLAB R2013a on ASUS K550J CPU Intel Core i7-4710HQ, up to 3.5 GHz RAM 8 GB. A video input was captured under clear condition on a sunny day about 4 p.m. in a highway environment using iPhone 6 camera.

The accuracy in table 1 is calculated from the corrected image of 500 frames input video. The corrected image is the image that has the edge of current lane region on the lane marker as shown in figure 15. The result of computational time is from MATLAB function that calculates the processing time of the project.

Table. 1 The comparison table.

Type	Accuracy (%)	Computational time (s)
1. SHT	93.0	150.146
2. RHT (threshold = 5)	93.6	142.488
3. RHT (threshold = 10)	93.8	142.639
4. RHT (threshold = 15)	95	146.587
5. RHT (threshold = 20)	95.2	150.047

Figure 16 shows that the pixels of the shadow are longer than the right lane marker. This means that the system has the higher possibility to randomize the point on the shadow and assume it to be a lane marker. To solve this problem, some additional methods for image preprocessing such as image filtering, another type of edge detection and the threshold of RHT can be applied. However, this would increase computation time of the system.

Figure 15. The result of the corrected image.

Figure 16. The mistake from the detection because of the heavy shadow.

The threshold increasing in RHT will increase the accuracy and computational time. At the higher threshold, the value of accuracy tends to be stable as shown in figure 17. On the other hand, the system tends to consume a lot of computational time in the higher threshold state as shown in figure 18.

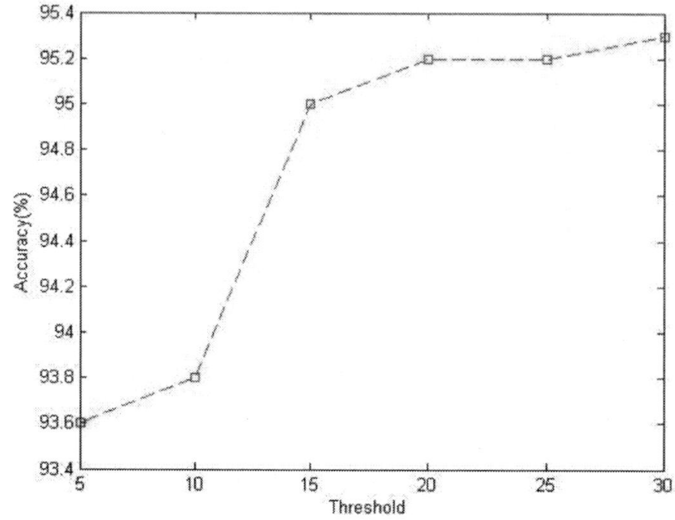

Figure 17. The accuracy of each threshold in RHT.

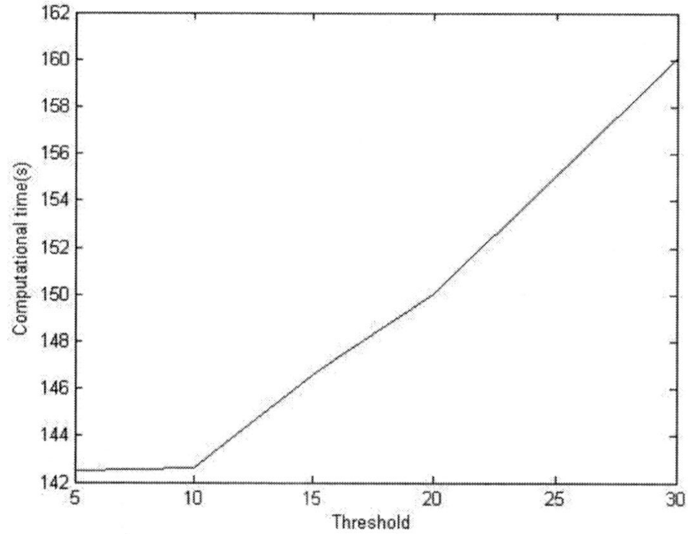

Figure 18. The computational time of each threshold in RHT.

4. Conclusion

Lane detection has been presented in this article image preprocessing, line detection and lane marker detection. It has been shown that the main advantage of the RHT method over the SHT is computational time. The difference in computational time between SHT and RHT with 5 threshold is about 5.1 percent, RHT with 10 threshold is about 4.99 percent, RHT with 15 threshold is about 2.73 percent and RHT with 20 threshold is about 0.06 percent, according to table 1.

The system still has two further parts to complete, i.e. the warning part that finds the condition of lane changing and the tracking part that reduces the computational time of the current frame using the information of the previous frame.

In the future work, the developer can develop the system into the application on a smartphone, which can be attached to the car front.

IOP Conf. Series: Materials Science and Engineering **297** (2018) 012050 doi:10.1088/1757-899X/297/1/012050

References

[1] Royal Thai Police, Thailand (2006). *The situation of traffic accident cause of the accident by a person and environment causes of the equipment used in driving, whole kingdom*, URL: service.nso.go.th/nso/web/stat-series/tables/00000_Whole_Kingdom/14.18.2\\A1.xls, accessed on 23/02/2017

[2] Heikki Kalviainen, Petri Hirvonen, Lei Xu and Erkki Oja 1995 *Image and Vision Computing* **13** pp 239 – 252

[3] King Hann Kim, Li-Minn Ang, Kah Phooi Seng and Siew Wen Chin 2009 *IMECS* **2** pp 1124 – 1129

[4] David Schreiber, Bram Alefs and Markus Clabain 2005 *IEEE Int. Conf. on Intelligent Transportation Systems* pp 1114 – 1119

[5] Raman Maini and Dr. Himanshu Aggarwal 2009 *IJIP* **3** pp 1 – 12

[6] Pekka Kultanen, Lei Xu and Erkki Oja 1990 *IEEE* **1** pp 631 – 635

[7] Yue Wang, Eam Khwang Teoh and Dinggang Shen 2004 *Image and Vision Computing* **22** pp 269 – 280

External force/velocity control for an autonomous rehabilitation robot

Peerayuth Saekow *, Paramin Neranon and Pruittikorn Smithmaitrie

Prince of Songkla University, Faculty of Engineering, Hat Yai, Songkla 90110, Thailand

* Corresponding Author: E-mail: Peerayuth.psu@gmail.com
Telephone Number: 083-6551504

Abstract. Stroke is a primary cause of death and the leading cause of permanent disability in adults. There are many stroke survivors, who live with a variety of levels of disability and always need rehabilitation activities on daily basis. Several studies have reported that usage of rehabilitation robotic devices shows the better improvement outcomes in upper-limb stroke patients than the conventional therapy–nurses or therapists actively help patients with exercise-based rehabilitation. This research focuses on the development of an autonomous robotic trainer designed to guide a stroke patient through an upper-limb rehabilitation task. The robotic device was designed and developed to automate the reaching exercise as mentioned. The designed robotic system is made up of a four-wheel omni-directional mobile robot, an ATI Gamma multi-axis force/torque sensor used to measure contact force and a microcontroller real-time operating system. Proportional plus Integral control was adapted to control the overall performance and stability of the autonomous assistive robot. External force control was successfully implemented to establish the behavioral control strategy for the robot force and velocity control scheme. In summary, the experimental results indicated satisfactorily stable performance of the robot force and velocity control can be considered acceptable. The gain tuning for proportional integral (PI) velocity control algorithms was suitably estimated using the Ziegler-Nichols method in which the optimized proportional and integral gains are 0.45 and 0.11, respectively. Additionally, the PI external force control gains were experimentally tuned using the trial and error method based on a set of experiments which allow a human participant moves the robot along the constrained circular path whilst attempting to minimize the radial force. The performance was analyzed based on the root mean square error (E_RMS) of the radial forces, in which the lower the variation in radial forces, the better the performance of the system. The outstanding performance of the tests as specified by the E_RMS of the radial force was observed with proportional and integral gains of $K_p = 0.7$ and $K_i = 0.75$, respectively.

1. Introduction

Stroke is a state describing the lack of blood in the brain which occurs form having ischemic stroke, arterial occlusive diseases, or Intracranial haemorrhage. This causes tissue of the brain to die, and its function to instantly stop. The body subsequently will be paralyzed, weak, and immobilized which results in the disability. Stroke and its effects are considered the major problem and cause of people's death throughout the world. According to the World Safety Organization (WSO)'s record, it is postulated that at least one person dies in every 6 seconds, and it is the third major reason of death and disability of the world [14]. Therefore, many research papers have studied a new approach to the

IOP Conf. Series: Materials Science and Engineering **297** (2018) 012051 doi:10.1088/1757-899X/297/1/012051

restoration of the original condition of patients with stroke based on the repeated movement to stimulate the muscles with an assistive robot. The robot is able to perform the task repeatedly and efficiently with consistent control. According to studies by Hermano and Krebs [1, 2], they have developed upper-limb robotic rehabilitation devices, namely MIT-MANUS and InMotion2, respectively. Each robot has two degrees of freedom (DOF). The user's paretic forearm is placed in a support provided, which is attached to a robot end-effector. The user can move the handle in the transverse plane and perform goal-directed tasks which focus on human's shoulder and elbow movement. In addition, a monitor was served to display user targets and provide visual feedback.

An upper-limb robotic rehabilitation device which is the Mirror-Image Motion Enabler (MIME) was developed by Peter and Van [3, 11]. The MIME robot is made up of a six-DOF paretic arm robot with the end effector, which has actuators to apply forces in goal-directed movements. The arm is attached to the end effector, which restricts wrist and hand movement. Loureiro and Harwin [4] constructed an upper-limb robotic rehabilitation device called the GENTLE/s with a virtual reality (VR) display provided to motivate users to engage in the rehabilitation therapy. The GENTLE/s includes a chair, shoulder supports, a wrist connector, an elbow orthosis, two computers, a large monitor with speakers, an exercise table, and a keypad. The arm is suspended to overcome the negative effects of gravity on the paretic arm and to help with shoulder subluxation; in the meantime, the VR can simulate three different environments: an empty room, a real room with some general shapes, and a detailed room with several objects. The Guide (ARM) robot is one of the upper-limb robotic rehabilitation devices and it was designed by Reinkensmeyer, Dewald and Rymer [5, 6]. The Guide (ARM) was utilized to assess multi-joint coordination and the workspace deficits of a paretic arm in a quantitative manner. The ARM Guide was attached to the user's forearm with a custom splint and guides the arm on a linear path, and range of motion and constraint force components were simultaneously measured. Furthermore, there are some research papers currently focusing on high-DOF robot in rehabilitation activities. For examples, Therapy Wilmington Robotic Exoskeleton (T-WREX) robot which has five degrees of freedom and it was proposed by Housman and Reinknsmeyer [12]. The robot was initially designed to support a weak human arm in intensive movement training by reducing the effects of gravity. The T-WREX uses rubber bands to provide various levels of arm support and has position sensors at each joint, which can measure arm movement. It is able to be used in conjunction with a computer to allow the user to interact in a VR environment and play games. A six-DOF arm robot, named REHAROB, was used in upper-limb rehabilitation. This robot can perform passive physiotherapy at a constant velocity, and it was trained by therapists [7].

However, there is currently no research having studied on and developed an omni-directional mobile robot used in upper-limb stroke rehabilitation which actively help patients through exercise-based activities. Consequently, this paper highlights on the development of a behavioral control strategy using force/velocity control based on PI algorithm for the autonomous robotic trainer designed to guide a stroke patient through an upper-limb rehabilitation task. By first understanding the principle dynamics and control behavior, two sets of experiments have been conducted. PI velocity control algorithm was firstly designed and optimized to ensure the effective robot movements; secondly, the PI external force control was also experimentally tuned to provide successful and safe stroke rehabilitation tasks. The challenge is further complicated by the dynamic nature of the human-robot environment, which by its nature necessitates very careful design of the control strategy and its implementation in order to protect the human operator from the risk of harm or injury by the robot.

2. The designed robotic

This section introduces the conceptual design of the assistive robotic system as schematically illustrated in Figure 1. The system involves a four-wheel omni-directional mobile robot which coordinates with a set of physical sensors and their feedback signals have been processed by an Arduino controller in real-time. An ATI Gamma multi-axis force/torque sensor was used to measure the interactive force between the human and robot, whereas optical linear encoders were mounted to the motor actuators. The force sensor system consists of an ATI F/T Gamma sensor, an electrically shielded and twisted transducer cable and a stand-alone ATI controller in which optional analogue, parallel and serial outputs have been already provided. The ranges of force/torque measurements are up to ±130 N with 0.1 N resolution and ±10 Nm with 0.0025 Nm resolution, respectively. A faulhaber 12V-DC motor with an encoder which is any of a class of rotary electrical machines that converts direct current electrical energy into mechanical energy was utilized. The motors with 64:1 planetary gearbox and 120RPM with 12CPR Encoder were used for an autonomous robotic platform. A detailed description of Arduino Mega 2560 microcontroller is based on the ATmega2560, which has 54 digital input/output pins (14 of which can be used as PWM outputs), 16 analog inputs, 4 UARTs (hardware serial ports), a 16-MHz crystal oscillator. According to the mobile robotic platform design, anodized aluminium was used for the body structures to ensure a durable and robust, yet lightweight construction, whilst plastic linkages and circuit board carriers ensure lightness yet durability. The omni-wheels adopted are wheels with small discs around the circumference which are perpendicular to the turning direction. The effect is that the wheel can be driven with full force, but it can also slide laterally with great ease.

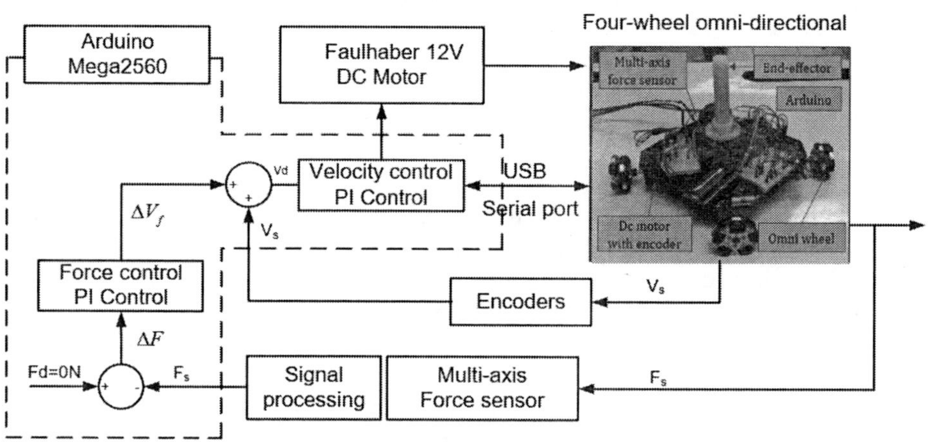

Figure 1. Block diagram and four-wheel omni-directional mobile robot.

3. Robot dynamic model

The robotic dynamic model is essential for describing a mobile robot's position, orientation and motions of the wheels and for analyzing and synthesizing the overall dynamic behavior of the robot. The omni-wheels robot has three degrees of freedom on the x-y plane and can perform forward-reverse movement and left-right movements. The movement in all three directions is due to the sum of and the four wheels. The kinematic equations and relations between robot velocity and various geometry specifications are expressed as the following. Oliveira and Hashemi [8, 13] have suggested that there are two coordinate frames used in the model consisting of the global frame (x, y) which represents the environment of the robot as depicted in Figure 2. The robot's location and orientation in this global frame can be represented as (x, y, θ). The global velocity of the robot can be written as $(\dot{x}, \dot{y}, \dot{\theta})$ and define a local frame $[x_r, y_r]$ that is attached to the robot itself.

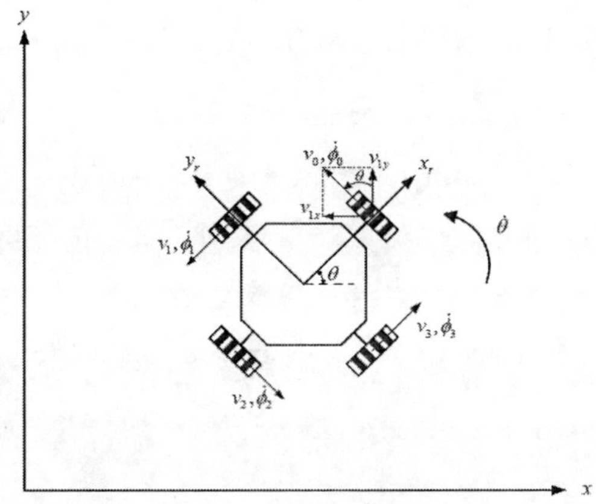

Figure 2. Global and Robot coordinates.

Tranformation matrices from linear velocity on the global frame to linear on the robot frame can be expressed as the following equation:

$$\begin{bmatrix} \dot{x} \\ \dot{y} \\ \dot{\theta} \end{bmatrix} = \begin{bmatrix} cos\theta & sin\theta & 0 \\ -sin\theta & cos\theta & 0 \\ 0 & 0 & 1 \end{bmatrix} \begin{bmatrix} \dot{x}_r \\ \dot{y}_r \\ \dot{\theta} \end{bmatrix} \tag{1}$$

The relationship between the translational velocity of wheel v_i and the global velocity of the robot in the environment $(\dot{x}, \dot{y}, \dot{\theta})$ is presented. The translation of velocity of wheel can be divided into a two parts which are translation and rotation.

$$V_i = V_{trans,i} + V_{rot} \tag{2}$$

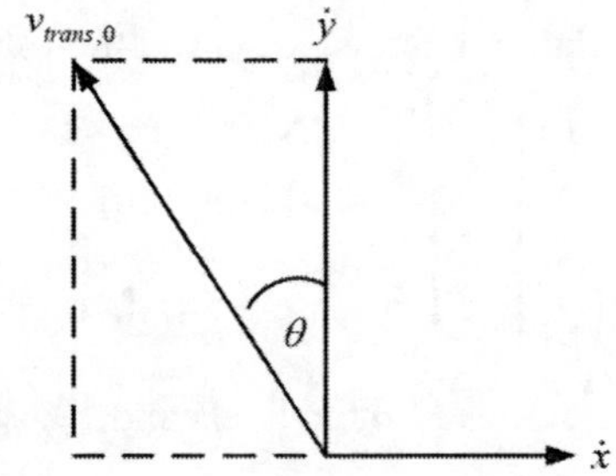

Figure 3. Translation velocity at wheel 0.

Translation as shown in Figure 3 can map the vector $V_{trans,0}$ onto the vectors $\dot{x}$ and $\dot{y}$ as follows:

$$V_{trans,0} = -\sin(\theta)\,\dot{x} + \cos(\theta)\dot{y}. \tag{3}$$

The vector for all wheels takes into consideration that the vector v_i are position and offset by $\theta + \alpha_i$ as follows:

$$V_{trans,i} = -\sin(\theta + \alpha_i)\,\dot{x} + \cos(\theta + \alpha_i)\dot{y}, \tag{4}$$

where α_i = starting point and count degrees in the clockwise direction as positive, $\alpha_0 = 0°$, $\alpha_1 = 90°$, $\alpha_2 = 180°$ and $\alpha_3 = 270°$.
Once the mobile robot has a pure rotation, it gives:

$$V_{rot} = R\dot{\theta}, \tag{5}$$

where R is distance between wheel and center robot (R = 0.155 m).

Substituting the equations (4) and (5) into the equation (2) gives:

$$V_i = -\sin(\theta + \alpha_i)\,\dot{x} + \cos(\theta + \alpha_i)\dot{y} + R\dot{\theta}. \tag{6}$$

The translation velocity of the hub is related to the angular velocity $\dot{\phi}_i$ of the wheel as:

$$v_i = r\dot{\phi}_i, \tag{7}$$

where r is wheel radius (r = 0.029 m) and i = 0,1,2,3 represents indices of the wheels.

Substituting the equation (7) into the equation (6) gives:

$$\dot{\phi}_i = \frac{1}{r}(-\sin(\theta + \alpha_i)\,\dot{x} + \cos(\theta + \alpha_i)\dot{y} + R\dot{\theta}). \tag{8}$$

Then equation (8) can be transformed to matrix representation (9) as folows:

$$\begin{bmatrix}\dot{\phi}_1\\ \dot{\phi}_2\\ \dot{\phi}_3\\ \dot{\phi}_4\end{bmatrix} = \frac{1}{r}\begin{bmatrix}-\sin(\theta) & \cos(\theta) & R\\ -\sin(\theta+\alpha_1) & \cos(\theta+\alpha_1) & R\\ -\sin(\theta+\alpha_2) & \cos(\theta+\alpha_2) & R\\ -\sin(\theta+\alpha_3) & \cos(\theta+\alpha_3) & R\end{bmatrix}\begin{bmatrix}\dot{x}\\ \dot{y}\\ \dot{\theta}\end{bmatrix} \tag{9}$$

Simply converting the global to robot coordinates and substituting equation (1) into the equation (9) lead to:

$$\begin{bmatrix}\dot{\phi}_1\\ \dot{\phi}_2\\ \dot{\phi}_3\\ \dot{\phi}_4\end{bmatrix} = \frac{1}{r}\begin{bmatrix}-\sin(\theta) & \cos(\theta) & R\\ -\sin(\theta+\alpha_1) & \cos(\theta+\alpha_1) & R\\ -\sin(\theta+\alpha_2) & \cos(\theta+\alpha_2) & R\\ -\sin(\theta+\alpha_3) & \cos(\theta+\alpha_3) & R\end{bmatrix}\begin{bmatrix}cos\theta & sin\theta & 0\\ -sin\theta & cos\theta & 0\\ 0 & 0 & 1\end{bmatrix}\begin{bmatrix}\dot{x_r}\\ \dot{y_r}\\ \dot{\theta}\end{bmatrix}. \tag{10}$$

Therefore, the local frame matrix relation can be expressed the following equation:

$$\begin{bmatrix}\dot{\phi}_1\\ \dot{\phi}_2\\ \dot{\phi}_3\\ \dot{\phi}_4\end{bmatrix} = \begin{bmatrix}0.00 & 34.48 & 5.35\\ -34.48 & 0.00 & 5.35\\ 0.00 & -34.48 & 5.35\\ 34.48 & 0.00 & 5.35\end{bmatrix}\cdot\begin{bmatrix}\dot{x_r}\\ \dot{y_r}\\ \dot{\theta}\end{bmatrix}, \tag{11}$$

where $\dot{\phi}_1, \dot{\phi}_2, \dot{\phi}_3, \dot{\phi}_4$ are angular velocities of the wheels and $\dot{x_r}, \dot{y_r}, \dot{\theta}$ are velocities of the robot in local frame.

4. Fundamental of Force/Velocity Control

This section explains robot force/velocity control algorithms, which are a fundamental requirement in the achievement of the robot's real-time control. External force control involves any contact force between the mobile robot and its environment. In the meantime the robot velocity control algorithm has perform satisfactorily stable performance for every robot movement. The key performance issue in design of a robotic force/velocity control system is that of stability and reliability that means the robot has to perform in a stable and reliable manner while operating and contacting with various unknown stiffness environments and moving along the constrained path. Using external force/velocity control (see Figure 4.) developed by De Schutter and Van Brussel [9] is considered to offer a better solution regarding the safety constraints, simplicity and implementation efficiency. The most crucial aspect in this control method is to achieve a suitable compromise between the system response and stability, where the response time was required to be as short as possible.

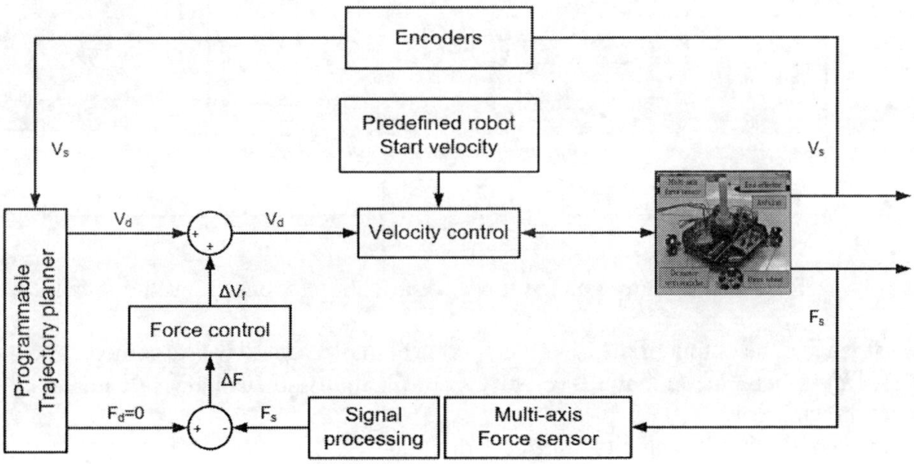

Figure 4. External force/velocity control [9].

Proportional plus integral (PI) control is most widely utilized in industrial process control applications because it is relatively easy to implement and no other type of control can match its simplicity and clear functionality, such as motor speed or position control as suggested by De Schutter and Van Brussel [9]. The PI (proportional plus integral) force control algorithms were successfully implemented with discrete sampling periods. The discrete form of the proportional integral equation is needed to approximate the integral of the error and because this technique facilitates an increase in the accuracy and stability of control system. Therefore, it was decided to apply simple proportional plus integral robot force control in this project.

An incremental discrete-time PI control algorithm [10] with sampling time period τ and the discrete time interval k can be calculated by applying the equations (12) and (13) and it gives:

$$U(k) = k_P e(k) + k_I \sum_{j=0}^{k} e(j). \tag{12}$$

The incremental PI control value represented by $\Delta u(k)$ can be calculated as:

$$\Delta u(k) = k_P e(k) + k_I \sum_{j=0}^{k} e(j) - k_P e(k-1) + k_I \sum_{j=0}^{k-1} e(j) \tag{13}$$

$$\Delta u(k) = k_P [e(k) - e(k-1)] + k_I e(k) \tag{14}$$

$$e(k) = f_d - f_s \tag{15}$$

where $u(k)$ is the PI control output; k_p is proportional gain; k_i is integral gain; f_d is desired force, which is initially defined as 0, and f_s is the actual force (measured by the ATI force control)

A block diagram of external force/velocity control based on the PI control is subsequently shown in Figure 5 where e is the error defined as the difference in magnitude between the desired (f_d) and actual (f_s) forces, while d_e is the change in error (e). The PI control output can be determined as the incremental displacement (ΔU_k) modified by the previous computed value of (ΔU_{k-1}), which is scaled before being transferred to the robot.

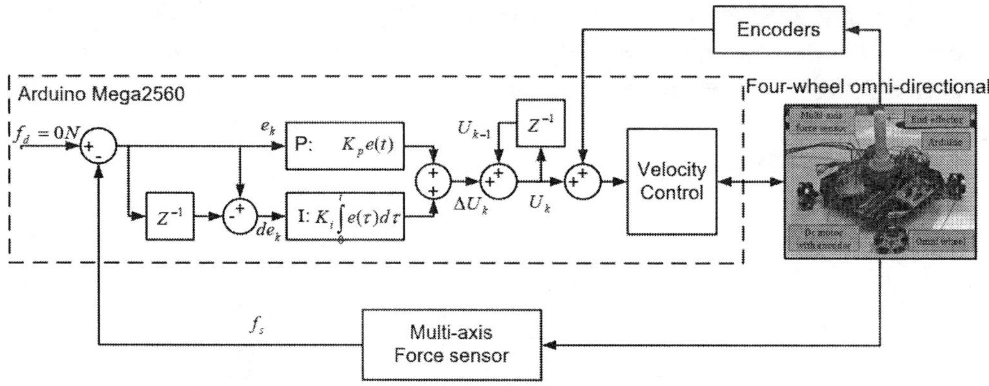

Figure 5. Diagram of the force control strategy based on PI control.

In the same way, the output of the PI velocity control implemented was assumed to be the incremental position $\Delta u(k)$, where increment PI velocity control algorithm can be calculated by applying the equation as shown below.

The incremental robot PI velocity control is defined by

$$u(k) = u(k-1) + k_p[e(k) - e(k-1)] + k_i e(k) \quad (16)$$

$$e(k) = V_d - V_s \quad (17)$$

The input e of PI control was generated as the difference in magnitude between the desired (V_d) and actual (V_s) velocities, and the input de is the change in error (e) where V_d is desired velocity, and V_s is the actual velocity (measured by the encoder sensor), respectively.

5. External Implementation

In order to improve the stability and performance of the autonomous robotic trainer designed to guide a stroke patient through an upper-limb rehabilitation task, the proportional integral (PI) force/velocity control technique has been clearly addressed in the last section. However, one of the most principal challenges in the control development is to design a set of appropriate PI gains to achieve an effective system response. Various techniques for PI gain tuning have been developed, such as the Ziegler-Nichols, Cohen-Coon, Chien-Hrones-Reswick or manual (trial and error) techniques. In this paper, Ziegler-Nichols tuning method was adopted to establish appropriate PI gains for the robot's velocity control algorithm. However, as complicated robot force control while the robot interacts with the human environment, a trial-and-error tuning method based on a circular tracking experiment was used in the robot's external force control.

5.1. Experiment 1: Gain tuning of the PI velocity control

Ziegler-Nichols method was applied to simplify the determination of PI velocity control's gain parameters to meet the desired performance specifications, in which the system overshoot response is kept to minimum but at the same time the response should be as fast as possible. However, in this paper this tuning technique still needs operator's experience for a fine tuning until the system specification being completely achieved. As the systematic approach based on the method of Ziegler and Nichols rules was applied to simplify PI gains, its performance was analyzed based on the root mean square error (*E_RMS*) of the velocity. It can be said that the performance of the system response can be evaluated in terms of variation in the radial forces, in which the lower the variation in radial forces, the better the performance of the system. The equation used to calculate the magnitude of error deviations of *E_RMS* is expressed as:

$$E\_RMS = \sqrt{\frac{\sum_{i=1}^{n}(X_d - X_s)^2}{n}}, \tag{18}$$

where n is the number of evaluated values, X_d is desired value, and X_s is the actual value (measured by the encoder sensor).

Properly tuning the gain parameters for the PI velocity control algorithm can be a challenge because, if the time constants in the process are large, the time to do the optimization can be excessively long. This following process presents how the appropriate gain tuning was achieved. Firstly, the autonomous rehabilitation robot was pulled by constant forces using a mass of 5N (simulating a human pulling force). The integral term (K_i) was turned off and only the proportional gain (K_p) was initially set to zero. The K_p component subsequently increased slowly while the robot velocity was simultaneously measured and monitored until it can be noted that the velocity profile exhibited sustained oscillations. At this point, the experiment finalizes the critical gain, and the K_u, T_u unknown parameters were calculated. Thus, using Ziegler and Nichols technique gave the proportional gain (K_p) of 0.45 and integral gain (K_i) of 0.10. Nevertheless, as mentioned earlier, to adopt the set of proper PI gains this method still needs operator's experience for a fine tuning. A set of the experiments were undertaken to examine the relationship between the adjusted proportional gain and root mean square error (*E_RMS*) of the velocity where V_d is desired velocity, and V_s is the actual velocity. The test results are illustrated in Table 1.

5.1.1. Experimental results: K_P gain fine tuning (robot velocity control algorithm)

Table 1. The results of preliminary tests to evaluate the gain K_p

K_p *Gain*	E_{RMS} *of velocity* [*RPM*]
0.25	14.07
0.35	8.17
0.45	4.86
0.55	5.44
0.65	5.93
0.75	8.13
0.85	9.61

After using Ziegler-Nichols technique to roughly estimate the proportional and integral gains, the operator's experience was needed for a fine tuning to delivery an optimized control system. A set of the experiments was proposed and undertaken to examine the relationship between the adjusted proportional gain and root mean square error (*E_RMS*) of the velocity where V_d is desired velocity, and V_s is the actual velocity. The gain tuning for the proportional velocity control was successfully implemented. Based on the results of the experiments, it can be concluded that the best performance as specified by the lowest *E_RMS* of the velocity is observed with the proportional $K_p = 0.45$, in which this value was subsequently employed in the robot velocity control in order to finalize the appropriate integral gain K_i.

IOP Conf. Series: Materials Science and Engineering **297** (2018) 012051 doi:10.1088/1757-899X/297/1/012051

5.1.2. Experimental results: K_i gain fine tuning (robot velocity control algorithm)

Table 2. The results of preliminary tests to evaluate the gain K_i

K_i *Gain*	E_{RMS} *of velocity* [RPM]
0.09	6.49
0.10	6.20
0.11	4.86
0.12	5.38
0.13	5.63

In the same way, Table 2 shows the results of preliminary tests to evaluate the gain K_i presented in the performance results of the tests in which the range of the integral gain (K_i) were between 0.09 and 0.13. The system performance can be also identified based on the *E_RMS* of velocity as in the previous case. The best performance of this test is represented by minimum *E_RMS* of velocity, and was achieved at a gain K_i of 0.11.

5.2. Experiment 2: Gain tuning of the PI force control

As suggested by De Schutter and Van Brussel [9], proportional integral (PI) control is appropriate for robot force/position control in order to provide the smallest possible force control error. The journal of Nerano and Bicker [10] also stated that according to the Ziegler-Nichols method, K_u, T_u and the critical gain can be calculated once the K_p component is slowly increased until the robot motion profile exhibited sustained oscillations; however, at this point if the robot system has very high unstable oscillation, this could damage the robot. Therefore, the proper gains of PI external force control were experimentally tuned using the trial-and-error method based on a set of circular tracking experiments which allow a human participant to move the robot along the constrained circular path whilst attempting to minimize the radial force. The designed circle was drawn on a green template mounted on the fixed platform where the circular tracking test in an X-Y plane required the operator to track a circle radius selected of approximately 25 cm. The robot was positioned such that its pointer was at the home position as shown in Figure 6. The performance was analyzed based on the root mean square error (*E_RMS*) of the radial forces, in which the lower the variation in radial forces, the better the performance of the PI control system.

Ten participants were adopted and required to perform a set of tests in order to become familiar the test rig before completing five repetitions of the substantive experiments. Prior to running the real-time circular tracking test, the human was requested to sit down in comfortable position in front of the test rig and naturally grasp the supporter provided using only one hand. When the timer trigger is activated, the human participant is asked to start moving the robot from the home positions towards the constrained circular path. In the meantime, the robot position and velocity components are simultaneously detected and monitored in real time. The human operator is then required to continually track the constrained circular path in a clockwise direction until the test is totally completed when a full revolution has been achieved and the timer is stopped.

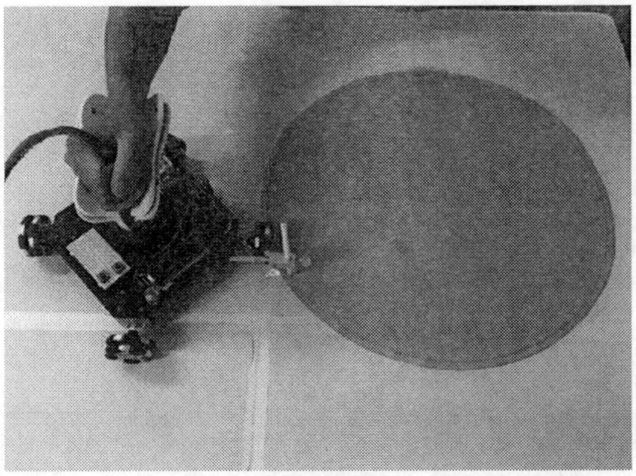

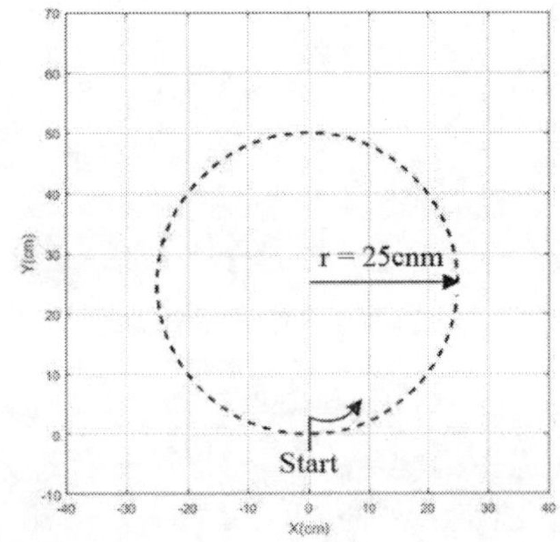

Figure 6. experimental set-up.

5.2.1. Experimental results: K_p gain tuning (robot force control algorithm)

As a trial-and-error tuning based on a circular tracking experiments was adopted. Firstly, the integral term (K_i) was initially turned off and only the proportional gain (K_p) ranging between 0.2 – 0.8 with 0.1 resolution was set. Choosing the optimum empirical control gains is a cautious exercise if a high proportional gain is preferable as its implementation will increase the transparency of the operation; nevertheless, the gain can only be increased to a certain upper limit; if the gain is set too large the system becomes unstable which could easily damage the mobile robot. The following table and figures demonstrate a representative of results of the relationship between the proportional gain and the root mean square error (*E_RMS*) of the radial force applied to the robot.

Table 3. The results of preliminary tests to evaluate the gain K_p

K_p *Gain*	E_{RMS} *of force* [N]
0.2	4.33
0.3	1.71
0.4	1.92
0.5	1.45
0.6	1.39
0.7	0.96
0.8	0.98
0.9	1.29

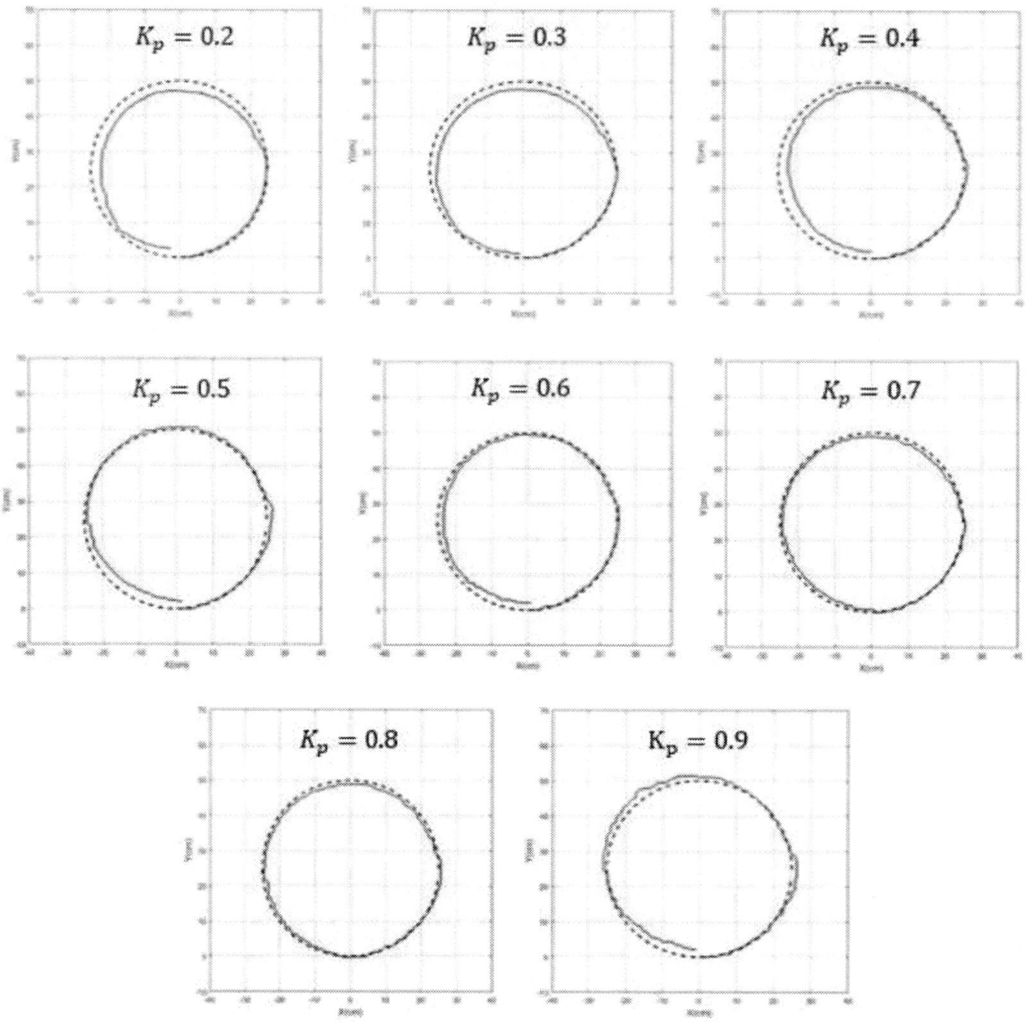

Figure 7. Circular tracking tests to evaluate the gain K_p.

Figure 7 presents the examples of the performance results with the different values of proportional gain K_p ranging between 0.2 and 0.9 in which the system performance can be identified based on the *E_RMS* of force. The highest performance of this test is represented by the minimum *E_RMS* of the radial force and was achieved at a gain K_p of 0.7, where the minimum *E_RMS* value is 0.96 N. Additionally, as expected the tangential force applied to the robot decreases when the gain K_p increases, and is approximately inversely proportional to the K_p gain value. Hence, the optimized proportional gain proposed was then utilized in K_i gain tuning of the robot force control algorithm expressed in the following section.

5.2.2. Experimental results: K_i gain tuning (robot force control algorithm)

To optimize the integral gain (K_i), the gain K_p was initially set at 0.7, and then tuning of the integral gain is marked by increasing K_i until the best performance indicating by the minimum *E_RMS* of the radial force is achieved. The same procedure for the first circular tracking test developed for evaluating the performance of the gain K_p was used. The same group of the participants was employed to perform the assigned tests in which a range of integral gains varying from 0.25 to 1.25 with 0.25 resolution.

Table 4. The results of preliminary tests to evaluate the gain K_i

K_i *Gain*	E_{RMS} *of force* [N]
0.25	2.44
0.50	1.26
0.75	0.91
1.00	1.03
1.25	1.27

Table 4 presents the performance results of test different values of integral gain K_i ranging between 0.25 and 1.25. The system performance can be identified based on the *E_RMS* of force. The best performance of this test is represented by minimum *E_RMS* of force, and was achieved at a gain K_i of 0.75.

The performance of the circular tracking test for different values of integral gain (K_i) is illustrated above. It can be noted that the best performance of the K_i tuning test is 0.75, in which the lowest *E_RMS* value is 0.9. To sum up, based on a set of experiments which allow a human participant to move the robot along the constrained circular path whilst attempting to minimize the radial force, the performance was analyzed based on the root mean square error (*E_RMS*) of the radial forces, in which the lower the variation in radial forces, the better the performance of the system. The outstanding performance results to optimized proportional (K_p) and integral (K_i) gains are observed with 0.7 and 0.75, respectively.

6. Conclusion

This paper has presented a new development of the force/velocity control for an autonomous rehabilitation robot system. This new robotic system has its emphasis on real-time control, programmability and repeatability in which the robot provides as advantages for guiding a stroke patient through an upper-limb rehabilitation. A well tuned proportional plus integral (PI) control was successful developed and it allows the robot to meet the desired response and satisfactorily stable performance. Based on the experimental results, it was confirmed that the best performance of the robotic force/velocity control as specified by the *E_RMS* of radial force has been observed with proportional and integral gains of the velocity control equal to 0.45 and 0.11, respectively. And proportional and integral gains of force control are 0.7 and 0.75, respectively. Therefore the development of an autonomous robotic trainer designed to guide a stroke patient through an upper-limb rehabilitation task has been carried out successfully.

IOP Conf. Series: Materials Science and Engineering **297** (2018) 012051 doi:10.1088/1757-899X/297/1/012051

Reference

[1] Krebs H I, *et al., J. Neuroengineering Rehabil.* **1**(1) p 5

[2] Krebs H I, Hogan N, Aisen M L and Volpe B T 1998 *IEEE Trans. Rehabil. Eng* **6**(1) pp 75–87

[3] Lum P S, Burgar C G, Shor P C, Majmundar M and Van der Loos M 2002 *Arch. Phys. Med. Rehabil.* **83** (7) pp 952–959

[4] Loureiro R, Amirabdollahian F, Topping M, Driessen B and Harwin W 2003 *Auton. Robots* **15**(1) pp 35–51

[5] Reinkensmeyer D J, Kahn L E, Averbuch M, McKenna-Cole A, Schmit B D and Rymer W Z 2014 *J. Rehabil. Res. Dev.* **37**(6)

[6] Reinkensmeyer D J, Dewald J P A and Rymer W Z 1990 *IEEE Trans. Rehabil. Eng.* **7**(1) pp 1–11

[7] Fazekas G, Horvath M and Toth A 2006 *Int. J. Rehabil. Res.* **29**(3) pp 251–254

[8] Oliveira H P, Sousa A J, Moreira A P and Costa P J 2009 *Modeling and Assessing of Omni-Directional Robots with Three and Four Wheels*

[9] De Schutter J and Van Brussel H 1988 *Int. J. Robot. Res.* **7**(4) pp 18–33

[10] Neranon P and Bicker R 2016 *Therm. Sci.* **20**(2) pp 537–548

[11] Lum P S, Burgar C G, der Loos M V, Shor P C, Majmundar M and Yap R 2005 *9th Int. Conf. on Rehabilitation Robotics* pp 511–514

[12] Housman S J, Le V, Rahman T, Sanchez R J and Reinkensmeyer D J 2007 *2007 IEEE 10th Int. Conf. on Rehabilitation Robotics* pp 562–568

[13] Hashemi E, Jadidi M G and Babarsad O B 2009 *Computational Intelligence in Robotics and Automation (CIRA), 2009 IEEE Int. Symp.* pp 272–277

[14] http://world-stroke.org/ [4 January 2016].

IOP Conf. Series: Materials Science and Engineering **297** (2018) 012052 doi:10.1088/1757-899X/297/1/012052

Multiple products management system with sensors array in automated storage and retrieval systems

Supachai Vongbunyong*, Perawat Roengritronnachai, Savanut Kongsanit, Chawisa Chanok-owat and Pongsakorn Polchankajorn

Innovation and Advanced Manufacturing Research Group, Institute of Field Robotics, King Mongkut's University of Technology Thonburi.
126 Pracha Uthit Rd, Bang Mot, Thung Khru, Bangkok 10140, Thailand

* Corresponding Author: supachai.von@kmutt.ac.th, Tel: +662-47-9339,
Fax: _662-470-9703

Abstract. Automated Storage and Retrieval Systems (AS/RS) have now been widely used in a number of industries due to its capability to automatically manage the storage of products in effective ways. One of the key features of AS/RS is that each rack is not assigned for a specific product resulting in the benefit of space utilization and logistics related issues. In this research, sensor arrays are equipped at each rack in order to enhance this feature. As a result, various products can be identified and mixed in each rack, so that the space utilization efficiency can be increased. To prove the concept, a prototype system consisting of a Cartesian robot that manages the storage and retrieval of products with 9 variations based on size and color. The concept of Cyber-Physical System and self-awareness of the system are also implemented in this concept prototype.

1. Introduction

The warehouse is one of the most common and important sections in factories. It is used for storing goods which can vary from raw materials to finished products. Traditionally, in order to prevent confusion, raw materials and finished products need to be completely separated. In addition, each type of product needs to be in their own slot and rack without mixing with others. As a consequence, the space may not be utilized efficiently due to the dead space that needs to be reserved for particular types of goods. To resolve this problem, the concept of a "Smart Warehouse" or "Automated Storage and Retrieval Systems (AS/RS)" have been introduced and used in a number of industries these days.

AS/RS is an automated system that fully utilizes the storage space by using information system in order to systematically organize the products to achieve certain optimal conditions. It is generally employed in the industries that deal with a high volume of products that need to be stored densely in a certain location with space constraints. It is also employed in applications where accuracy is a primary concern. The system consists of robots and automation used for physically loading and unloading the products to and from the particular storage locations thus eliminatingthe uncertainties due to human error. Products will then be able to be stored and retrieved accurately and safely with minimal possibility of damage.

Much research was conducted in order to improve the level of intelligence of the warehouse. For the AS/RS, [1] proposed a smart Warehouse Management System (WMS) and Warehouse Control System (WCS) that will increase the efficiency of warehouse operations. However, it is not limited to only AS/RS but also incorporates human operators. For instance, [2] used smart sensors and wearable device to assist the user to efficiently search the products and organize the inventory. [3] study the effects of smart glasses on the operators' activities in warehouses.

In Industrie 4.0, "Cyber Physical System"(CPS) is one of the key concepts that enhances the capability to monitor and control systems. Sensors network, control system, and Internet-of-Things (IoT) are integrated in order to enable the smart interaction between machines and/or human that is capable of achieving specific purposes. This concept has been implemented in a number of industrial applications [4-6]. CPS is also known as "digital twins", where items in reality can be visualized, monitored, and controlled in the virtual world. The interaction can be performed in a smarter way in comparison to the traditional monitoring system. CPS is implemented in AS/RS to enable the full control of the system. To reveal the information of each product, technology of RFID and barcode are normally used, e.g. [7, 8].

In this research, a prototype AS/RS that is able to store products with various types and sizes is developed. The products can be mixed and the space utilization can be optimized according to the desired goal. Sensor arrays are equipped at each slot in order to enable the concept of CPS and self-awareness of the system. As a result, the accuracy and security of the system will be improved.

2. Methodology

2.1 *Cyber-Physical System*

The concept of CPS is implemented in this project in order to enhance traceability of the products and the condition monitoring of the inventory. In general, RFID or barcode tags are tagged on each product so that the product can be identified by using a reader [7, 8]. However, extra cost and process are needed due to the tags that must be on every product.

In this project, the tagging system is replaced by an array of sensors and a camera that are permanently installed at the inventory. Therefore, none of the extra process and cost is needed for individual products. The system can identify the *size, type,* and *existence of the products* by using (a) an array of proximity sensors installed at each slot of the storage and (b) a camera equipped at the load-in dock. The sensor arrays, in this case: contact switches, are used to identify the size and existence while the camera with vision system is used to identify the type of products. The data obtained from these sensors will be aligned with the data in the database in order to update the current condition of the storage.

2.2 *Storage slot and product variations*

In this conceptual prototype, one storage rack can contain a maximum of 6 unit-size products. Therefore, 6 contact switches are placed according to this storage capacity, meaning that each contact switch is able to sense the condition of one unit. The arrangement is shown in figure 1. The contact switch located at the back of the slot will be on if the product exists.

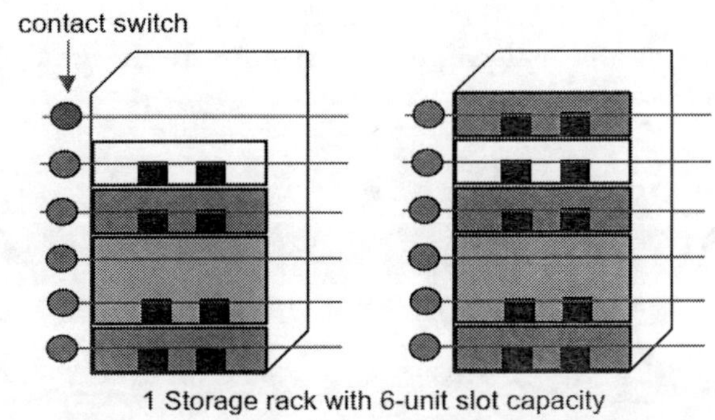

Figure 1. Product detection with sensor array

According to the product samples, 9 variations of products are used in order to demonstrate that the type and size of the products can be identified. The nine variations are the combination of 3 sizes (1-unit, 2-unit, and 3-unit) and 3 types (red, green, and blue). The summary is shown in figure 2. As a result, the sample of grouping methods – by size and by type – can be demonstrated as shown in figure 3.

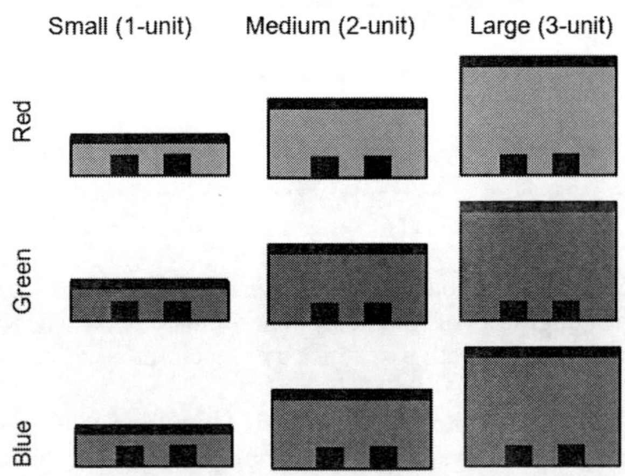

Figure 2. Product variation summary

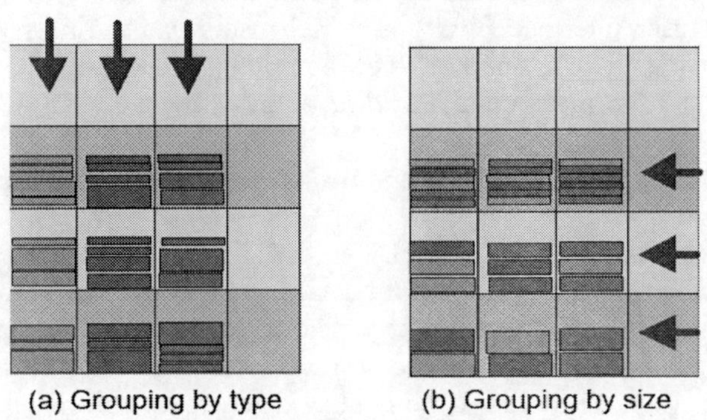

Figure 3. Grouping method

2.3 *System overview*
The prototype in this research is an automatic system consisting of (a) an inventory warehouse with sensors, (b) a Cartesian robot arm with a gripper, (c) software for the control system with database. The system overview is shown in figure 4.

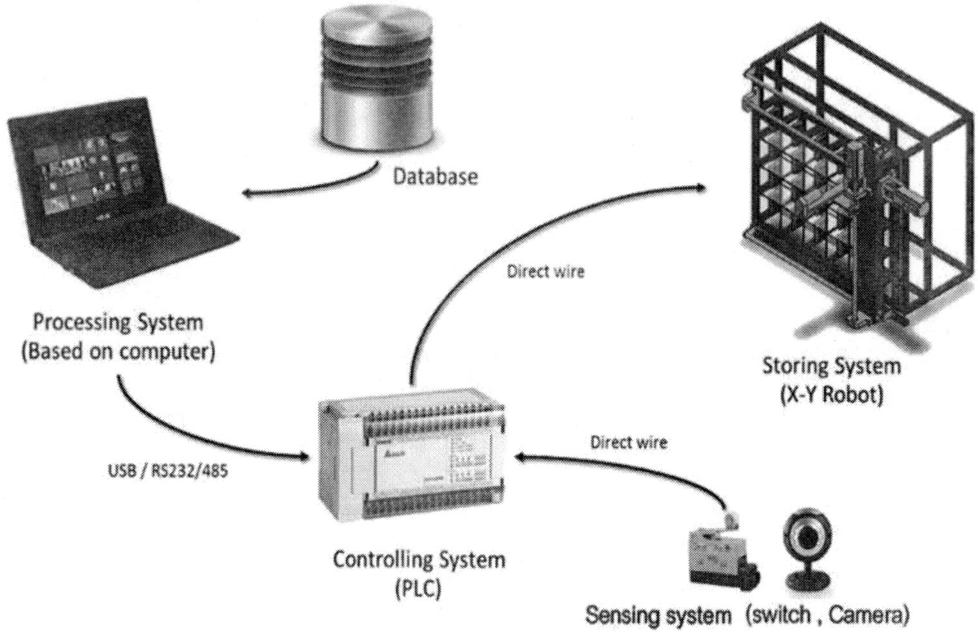

Figure 4. System overview

2.4 *System architecture*
As a n overview, this system is a PC-based control system. The system consists of 3 levels of control layer: *high-level, mid-level*, and *low-level*. The system architecture is shown in figure 5. The software and hardware in three layers work together seamlessly via a number of communication protocols.

2.4.1 High-level control layer. The Intelligent Agent (IA) is responsible for sequence planning of the products and storage arrangement. The sequence planning deals with the sequence of actions that the robot gripper will pick the products to insert or remove from the slots. The IA works based on the information regarding the current statuses of the storage which is in the database. The system also consists of a graphical user interface (GUI) in which users can monitor and control the operation. These are operated on the main computer. The computer is connected to a Programmable Logic Controller (PLC), DVP-SV2, which is at the *Mid-level* layer, by using RS-485 protocol.

2.4.2 Mid-level control layer. Two primary modules are operated which are *Signal controller module* and *Vision system module*.

(a) *Signal controller module* consists of a PLC with digital input expansion. The PLC is responsible for signal acquisition and processing. The input signal is mainly obtained from the sensor array, where the acquisition is through the digital input expansion module. This signal passes to IA for being processed in order to obtain the status of the storage slots. This digital input expansion is directly connected to the sensor arrays at the low-level control layer.

Regarding the robot's movement, the locations (x,y,z) of each slot, products, and its corresponding position due to the actions of product manipulation are predetermined and stored in PLC's program.

(b) *Vision system module* is developed based on EmguCV (OpenCV in .NET) [9] which is operated on the computer. The main function is to detect the type of the product according to their color. The vision system module is connected to a webcam via a USB which is at the low-level control layer.

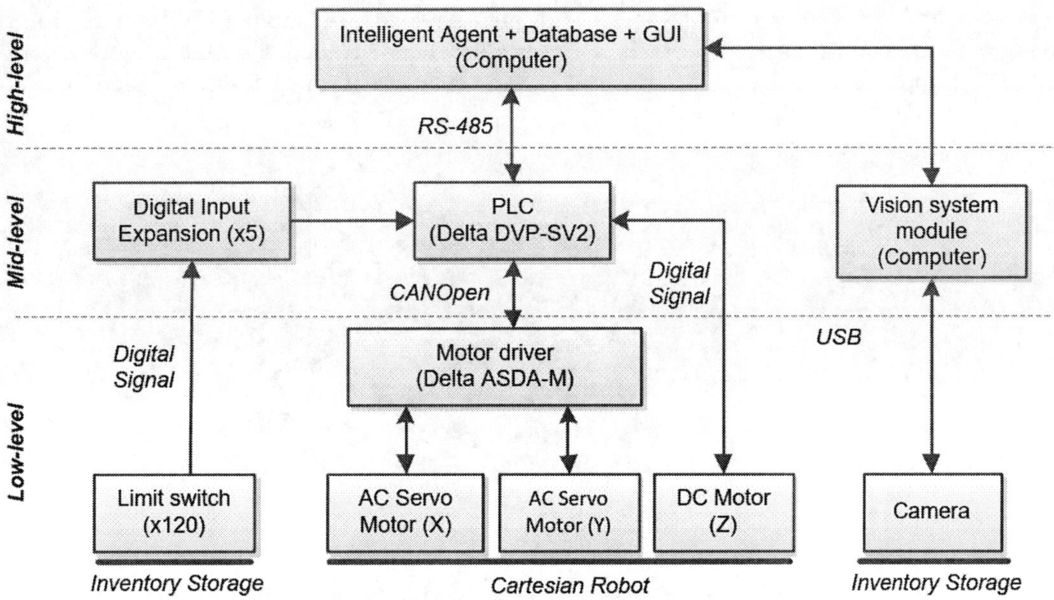

Figure. 5 System architecture

2.4.3 Low-level control layer. Hardware is considered in this level. The hardware consists of 3 motors, sensor arrays of 120 contact switches (only 112 switches are used for the demonstration of this prototype), and a webcam.

The positions of the slots and products in *xyz*-space will be used to command the motors. For x-axis and y-axis, the motor driver ASDA-M is used to perform motion control of two AC servo motors to drive the robot to particular locations. CANOpen is used to connect between ASDA-M and the PLC. For z-axis, DC linear motor is used and controlled directly from the PLC. Besides, the limit switches connect directly to the digital input expansion module. The webcam is plugged into the computer with USB.

3. Prototype development

3.1 Mechanical design

3.1.1 Inventory storage. The inventory storage of the prototype consists of 14 racks (12 storage racks, 1 load-in rack, and 1 load-out rack). For each rack, 6 contact switches (representing 6 slots) are mounted at the back of each slot. The dimension of the slot is designed based on the dimension of the product required. In this case, the dimension of the sample product of 1-unit size is 80mm x 120mm x 10mm*. The height (*) is 20mm or 30mm, in case of 2 and 3-unit size product. The sample product is inspired by wellplates that are normally used in clinical labs as this project will be applied to be used in that field.

The main structure of the system is constructed with Aluminium profiles. The dimension of the entire structure is 500mm x 500mm x 300mm. The inventory storage prototype is shown in figures 6 and 7.

3.1.2 Cartesian robot. The driving mechanism for x-axis and y-axis is based on power screws in order to convert rotary motion to linear motion. The ϕ16mm lead screws with linear guide rails are used due to the firm structure and able to the object with high payload. AC servo 400W motors are used for driving both axes. With respect to z-axis, a DC linear actuator equipped with a gripper at one end is used. The gripper is designed to be passive similar to the platform of forklifts. Therefore, no extra actuator is needed for operating the gripper. The robot prototype is shown in figures 6 and 7.

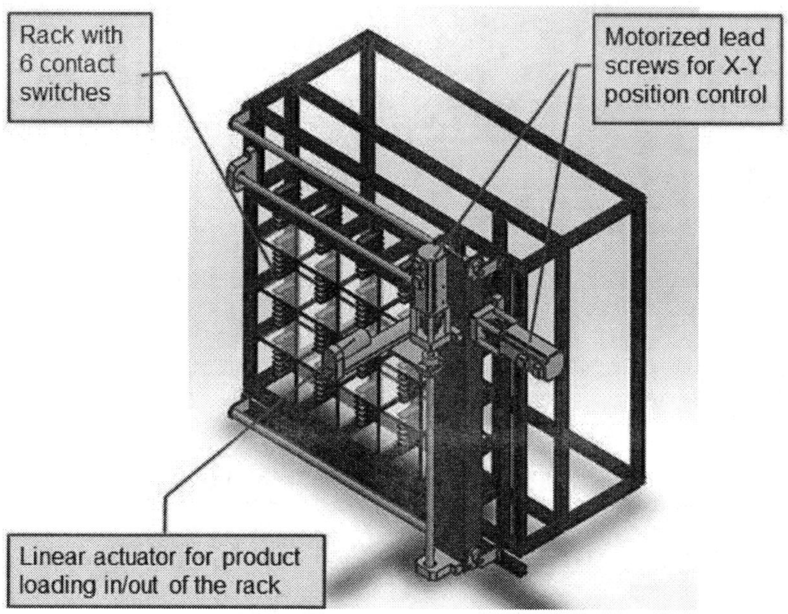

Figure. 6 Prototype design - Isometric view

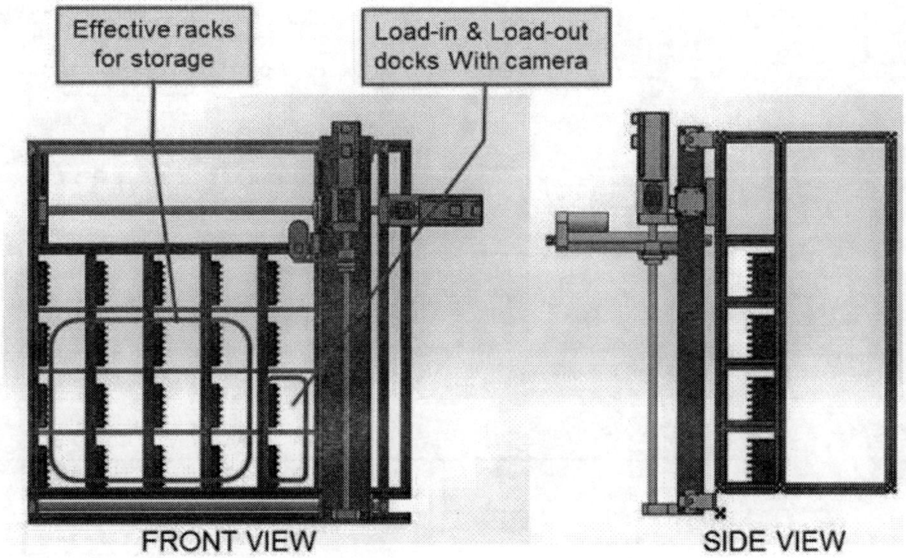

Figure. 7 Prototype design - Front and side views

3.2 Electrical Design

The PLC is the primary component that controls all the signals of commands transferred between high-level and low-level. The PLC DVP consists of the 3 primary modules and 5 digital input expansion.

(a) DVP-28SV is a main controller PLC which is connected to the computer with RS-485. The linear actuator is connected through 2 relays. (b) DVP-PS02 is a 24VDC power supply unit which was converted from 220VAC input. (c) The digital input expansion module connects to the terminal board with sensor array of 120 contact switches. (d) DVP-COPIN is a communication module for connecting to the motor driver ASDA-M by using CANOpen. The complete wiring diagram is shown in figure 8.

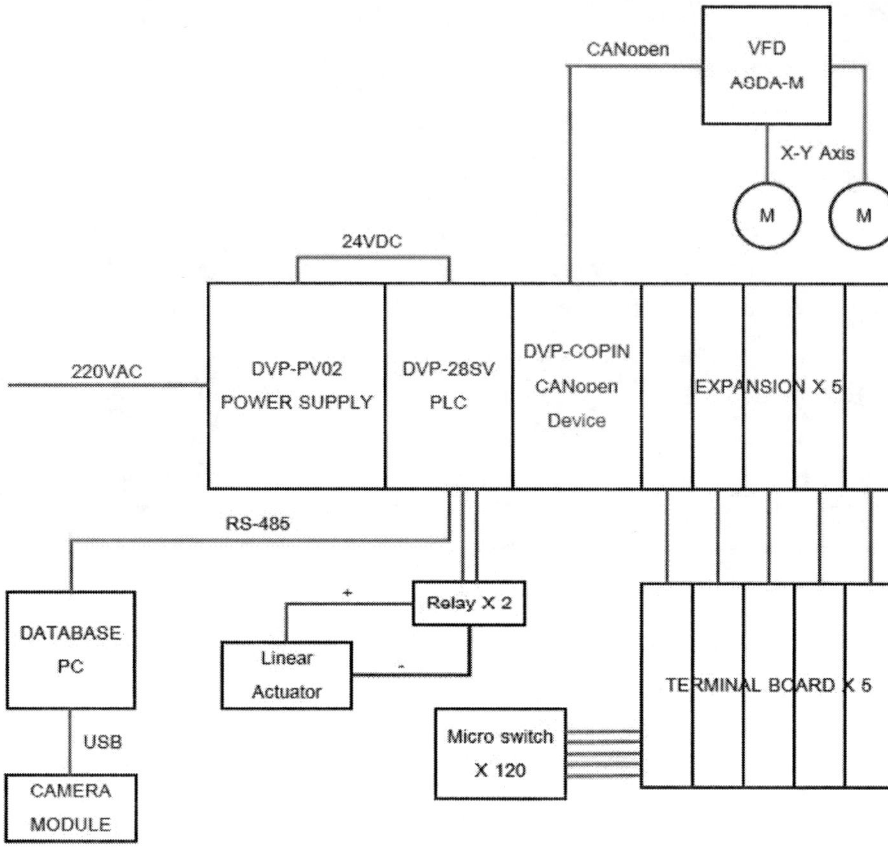

Figure. 8 Wiring diagram in detail

3.3 Programming

3.3.1 Identity checking. The product's identity - size and type - is firstly checked at the load-in dock. To ensure the accuracy, it will be checked again at the retrieval process.

The *size* of the product is checked according to the number of the occupied slots, which are checked by the status (0 = free, 1 = occupied) of the switch array (see figure 1). The *type* of the product is identified with the color detection from the camera. Hue-Saturation-Value (HSV) color space is used as it is robust to the change of illumination from the ambient [10]. As the detected color is sensitive to the ambient light temperature, color balance correction is needed. In summary, for the variables: *size* = 1,2, or 3 and *type* = 'r', 'g', or 'b'.

3.3.2 Database. The structure of database is shown in figure 9. The location of the storage rack is indexed with *(i,j)* where $i =$ 1,2,3,4 and $j =$ 1,2,3. In each rack, there are 6 slots which are represented with $k = (1,2,\ldots 6)$. In summary, the location of each slot is represented with *slot(i,j,k)*.

The IA will check and update the information with the database and the sensor array in every operation cycle. In case that the products have been removed from the storage by someone (not the robot), the system will acknowledge the absence of the product. Hence, this behavior improves self-awareness of the system resulting in an increase of security level.

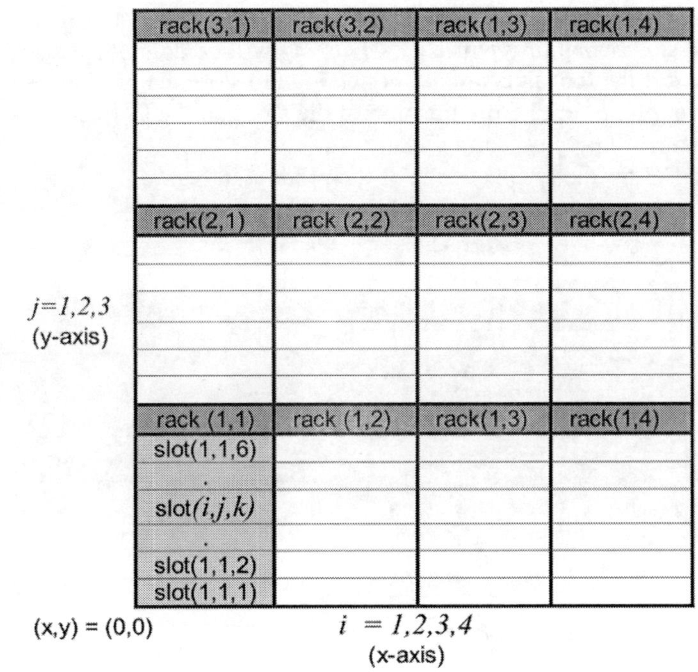

Figure. 9 Data structure in database

3.3.3 Robot movement. For each operation cycle, the robot is responsible for picking product from one rack and placing to the target rack. The picking and placing locations are accurately referred to the reference position *(x,y)* corresponding to *slot(i,j,k)* that is described in the previous section.

For the movement in z-axis, the sequence of robot's movement is predefined for picking and placing procedure. The predefined movement is illustrated in figure 10 where the sequence for picking is *a-b-c* while the sequence for placing is reversed.

In summary, the action for picking and placing a product in the storage are *pick(slot(i,j,k))* and *place(slot(i,j,k))*. The term *slot* is substituted by *LoadInDock* and *LoadOutDock* for the load-in and load-out process.

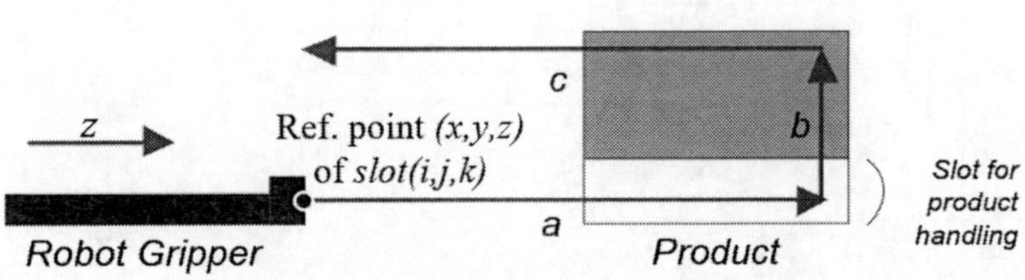

Figure. 10 Predefine movement path for pick and place

3.3.4 Program flow and communication. The IA will gather the information from each component including database in order to generate the command for operating the robot. The operation can be divided into 2 modes: load-in and load-out. The procedure for loading-in is shown in the flowchart in figure 11 and briefly described as follows:

(1) User selects the grouping method. In this case, 3 grouping methods are available which are group by type, group by size, and optimal space. The product with the desired qualifications will be grouped, e.g. place all red product in the one column-i.

(2) Product is loaded at the load-in dock

IOP Conf. Series: Materials Science and Engineering **297** (2018) 012052 doi:10.1088/1757-899X/297/1/012052

(3) Identify the loaded product type and size with the camera and sensor array at the load-in dock. This is performed by *checkSize()* and *checkType()*

(4) IA checks the slot availability with DB. Checking will be done for each rack. It will move on to another available rack if the free slots are not big enough to store the product.

(5) IA sends a string representing the location of the slot that the Cartesian robot will go to store the product . This is performed by *pick()* and *place()*.

(6) IA update the DB for the type and size of the product in regard to the slot occupied.

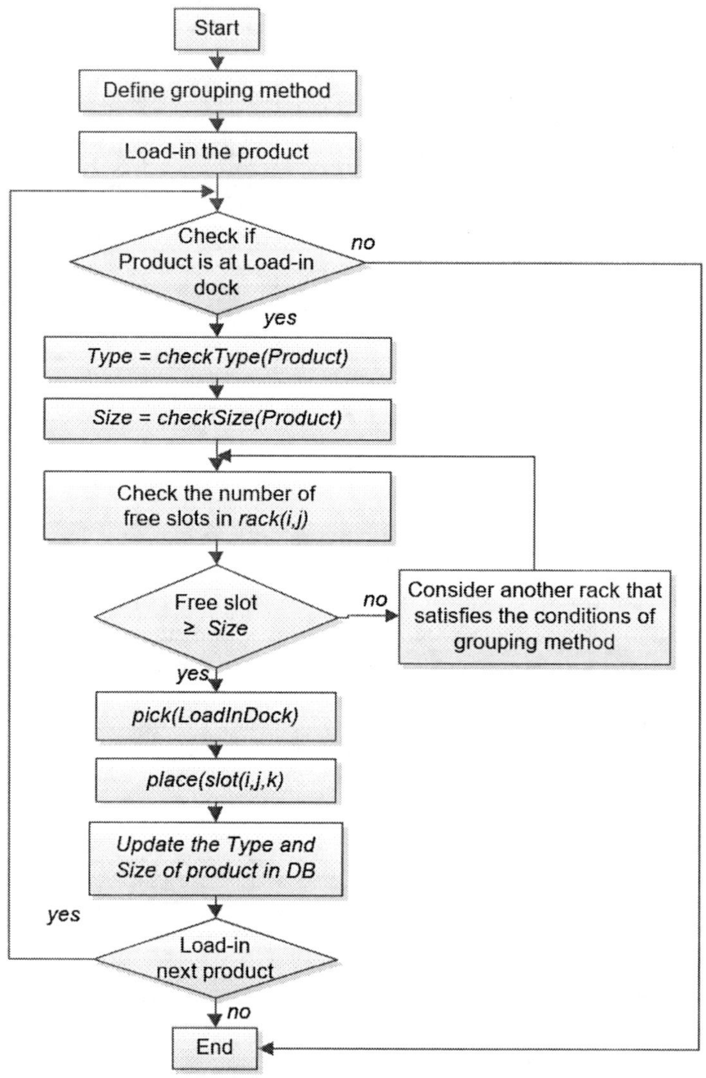

Figure. 11 Flowchart of load-in procedure

Regarding the Load-out process, the user needs to select the product to be loaded-out from the GUI. Then the robot will go to that slot to pick the product and place it on the loaded-out rack. In case there are other products stacked over the desired one, the products on the top will be temporarily moved to a free slot. Then they will be transferred back to the original rack after the desired product has been loaded out. The IA will update the current status with the DB.

4. Experiment and discussion

The goal of the experiment was to demonstrate the proposed functions which are (a) load-in and load-out the products, (b) to group the products according to the type, and (c) to group the product according to the size.

4.1 Load-in/out experiment

The actual prototype as shown in figure 12 was used. First, the robot is able to pick-and-place the products according to the command generated automatically by the system. The movement to the reference position of the gripper was accurate. However, error occurred due to the imprecise mechanical parts and assembly resulting in the failure of products handling. Second, the sensing from contact switches sometimes was inaccurate due to the installation that lead to a small gap between the contact point and a product. Non-contact sensors are recommended to resolve this problem.

Figure. 12 Prototype

The grouping experiments (b) and (c) were tested in the mathematic simulation software. The system was able to organize the products according to the locations that satisfied the goals. The utilization of storage capacity is more efficient than a conventional warehouse that the product with various sizes and types cannot be mixed in each rack.

5. Conclusions

A prototype AS/RS that is capable of storing a variety of products has been developed. When the products in a rack can be mixed, the storage space can be utilized with higher efficiency in comparison to conventional AS/RS that the products in each rack are fixed. The proposed sensor array together with IA enables the system to be able to organize the storage with mixed products efficiently. In addition, the function of CPS is enabled by using the sensors array. As a result, the security and accuracy of the system has been improved. The concept has been proved with the current version of prototype.

For future work, the mechanical and sensors system of the concept prototype will be improved to achieve better reliability. The design will be more specific according to the application. In this case, it is expected to be used for personalized bio-samples storage where a high level of security is needed.

References

[1] Son DW, Chang YS, Kim NU and Kim WR. 2016 *Proc. Proceedings - 2016 5th IIAI Int.Congress on Advanced Applied Informatics, IIAI-AAI* pp 980-984
[2] Diete A, Sztyler T, Weiland L and Stuckenschmidt H. 2016 *Proc. UbiComp 2016 Adjunct – Proc. of the 2016 ACM Int. Joint Conf. on Pervasive and Ubiquitous Computing* pp 1755-1758
[3] Friemert D, Ellegast R, Hartmann U. 2016. Data glasses for picking workplaces: Impact on physical workloads. In *Lecture Notes in Computer Science (including subseries Lecture Notes in Artificial Intelligence and Lecture Notes in Bioinformatics)* pp 281-289
[4] Rho S, Vasilakos AV and Chen W 2016. *Future Generation Computer Systems* **56** pp 436-437
[5] Monostori L, Kádár B, Bauernhansl T, Kondoh S, Kumara S, et al. 2016. *CIRP Ann. - Manuf. Technol.* **65** pp 621-641
[6] Ochoa SF, Fortino G and Di Fatta G. 2017 *Future Generation Computer Systems* **75** pp 82-84
[7] Zhou W, Piramuthu S, Chu F and Chu C 2016 *Decision Support Systems*
[8] Huang S, Gan OP, Hwang ZQ, Song H and Xie L 2015 *Proc. IECON 2015 - 41st Annual Conf. of the IEEE Industrial Electronics Society* pp 5026-5031
[9] Emgu CV 2017. *Emgu CV*. http://www.emgu.com/wiki/index.php/Main_Page
[10] Vongbunyong S, Pagnucco M and Kara S 2016 *Int, J. Autom. Tech.* **10** pp 708 - 716

IOP Conf. Series: Materials Science and Engineering **297** (2018) 012053 doi:10.1088/1757-899X/297/1/012053

Transferring experience labs for production engineering students to universities in newly industrialized countries

A. Leiden[1,*], G. Posselt[1], V. Bhakar[2], R. Singh[2], K. S. Sangwan[2] and C. Herrmann[1]

[1] Institute of Machine Tools and Production Technology, Chair of Sustainable Manufacturing and Life Cycle Engineering, TU Braunschweig, Langer Kamp 19b, Braunschweig 38106, Germany

[2] Birla Institute of Technology and Science, Pilani, Rajasthan, India

* Corresponding Author: a.leiden@tu-braunschweig.de

Abstract. The Indian economy is one of the fastest growing economies in the world and the demand for the skilled engineers is increasing. Subsequently the Indian education sector is growing to provide the necessary number of skilled engineers. Current Indian engineering graduates have broad theoretical background but lack in methodological, soft and practical skills. To bridge this gap, the experience lab ideas from the engineering education at "Die Lernfabrik" (learning factory) of the Technische Universität Braunschweig (TU Braunschweig) is transferred to the Birla Institute of Technology and Science in Pilani (BITS Pilani), India. This Lernfabrik successfully strengthened the methodological, soft and practical skills of the TU Braunschweig production-engineering graduates. The target group is discrete manufacturing education with focusing on energy and resource efficiency as well as cyber physical production systems. As the requirements of industry and academia in India differs from Germany, the transfer of the experience lab to the Indian education system needs special attention to realize a successful transfer project. This publication provides a unique approach to systematically transfer the educational concept in Learning Factory from a specific university environment to a different environment in a newly industrialized country. The help of a bilateral university driven practice partnership between the two universities creates a lighthouse for the Indian university environment.

1. Introduction

In newly industrialized countries, such as India, the demand for skilled engineers is increasing massively. Especially the manufacturing sector is growing and skilled professionals are required to foster this development. Manufacturing is the motor to increase the wealth of an emerging economy against the background of sustainable development. To increase the number of local skilled professionals, governments increase the number of graduates by increasing the number of university placements to skill more young people. For example, in India the number of enrolled students increased from 3.6 million in 1985/86 to 19.0 million in 2015/16 [1]. However, most university graduates have a broad theoretical background but lack in methodological, social and personal skills. Therefore, many of them are not ready to be employed just after graduation and need further training to become ready for employment. For example in India 60 % of the graduates do not get a direct job placement after graduation but have to go to finishing schools organized by the Indian industrial sectors [2].

To address this gap in teaching university engineering students, learning factories were introduced in many industrialized countries during the last 25 years. Learning factories can help to increase the employability of graduates by developing their methodological, social and personal competencies [3]. The first learning factory was founded by the national science foundation at the Penn State University in the USA in 1994[4]. Tisch and Metternich identified three waves of learning factory development. The first wave lasts from the end of the 1980s to 2005, which was the beginning of the development of learning factories. During the second wave from 2005 to 2010, local learning factories were building up in Europe. The third and currently ongoing wave is dominated by networking and scientific consideration. [5] In this context, TU Braunschweig and BITS Pilani developed a concept that allows the Indian partner to build on the experience in developing and operating a learning factory at a leading German university. To date, no relevant learning factory exists in India, which bases on the learning factories concept from industrialized countries. The introduction of learning factories at Indian universities could help to overcome the lack in the demanded competence portfolio of engineering graduates, which is mainly responsible for the low employability of Indian graduates in the different industry sectors.

Transferring the learning factory concept to universities in a newly industrialized country will not only help these universities but the universities in industrialized countries also can benefit from such cooperation. Suppliers all over the world including newly industrialized countries produce most products from the manufacturing sector globally. Further, many important challenges, for example global warming, must be solved globally. The integration of these aspects into the learning factories of industrialized countries will enable their students to get a better understanding of working in a globalized working environment and deal with global challenges.

Therefore, successfully transferring the learning factory concept to a newly industrialized country could be the starting point for the next learning factory wave, which extends the existing network and includes newly industrialized countries supporting the capacity development. (see figure 1).

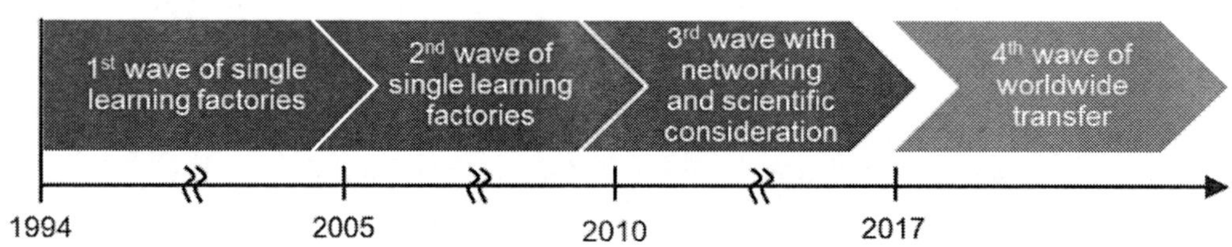

Figure 1. Waves in development of learning factories.

Simply copying the concept of learning factories to newly industrialized countries would not have the maximum effect of learning factories. As shown in figure 2, the setting of a learning factory is influenced by many factors. The educational system with its characteristics as well as the industry environment has a major impact on the required learning factory concept. Beside these quantifiable factors, also the culture, in particular the teaching and learning culture must be considered.

Therefore, the concept must be adapted to the local university and industry context to ensure that a maximum of competence portfolio can be effectively addressed. Against this background, this paper will present an approach to transfer single conceptual building blocks from existing learning factories from industrialized countries into learning factories of newly industrialized countries. Afterwards the approach is applied within a case study of the "Joint Indo-German Experience Lab" (JInGEL) project to evaluate the transfer of the experience lab from the learning factory of TU Braunschweig to BITS Pilani.

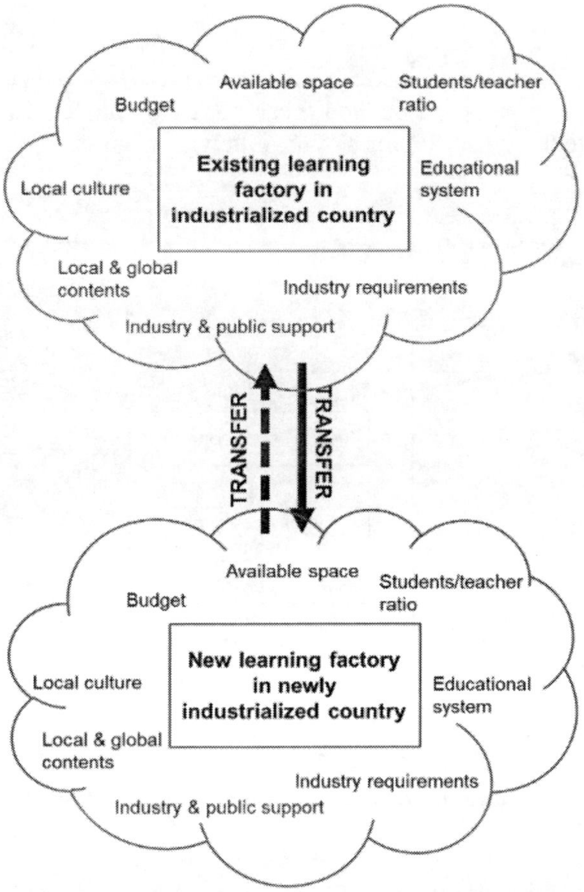

Figure 2. Transfer to newly industrialized country.

2. State of research

Most of the researches in the field of learning factories are focused on the competence requirements of young employees in industrialized countries. Most of these learning factories have gained a higher degree of maturity as they were developed over a longer time span with significant support from public funding organizations. To measure the maturity of these learning factories Enke et al. [6] developed a maturity model, which is still in the validation phase. It is expected to estimate the maturity state of existing learning factories before they are to be transferred to other environments.

To foster the research in the field of learning factories and increase the maturity of the learning factories, cross-national networks were established to bring researchers working on learning factories together. An example for such a cross-national network is the network of innovative learning factories, which brings European learning factories together. The goals are to support international mobility, establish innovative educational programs as well as enhance the quality of existing and future learning factories.[7] However, this network only considers developments in industrialized countries.

As the learning factories in industrialized countries gained a higher degree of maturity, learning factories are nearly unknown in newly industrialized countries. Learning factories could foster the industrial development of those countries significantly by increasing the employability of engineering graduates and make it possible to bring more practice relevant contents into university teaching. For this cross continental network or collaborations with newly industrialized countries should be established to make a knowledge transfer to the newly industrialized country possible. To the best of author's knowledge, no such a collaboration exists to date.

IOP Conf. Series: Materials Science and Engineering **297** (2018) 012053 doi:10.1088/1757-899X/297/1/012053

3. Method

This section introduces a method to transfer learning factories from industrialized countries to newly industrialized countries. The proposed method focuses on the transfer between two universities with within one or more engineering disciplines. The following figure 3 gives a brief overview of the proposed method.

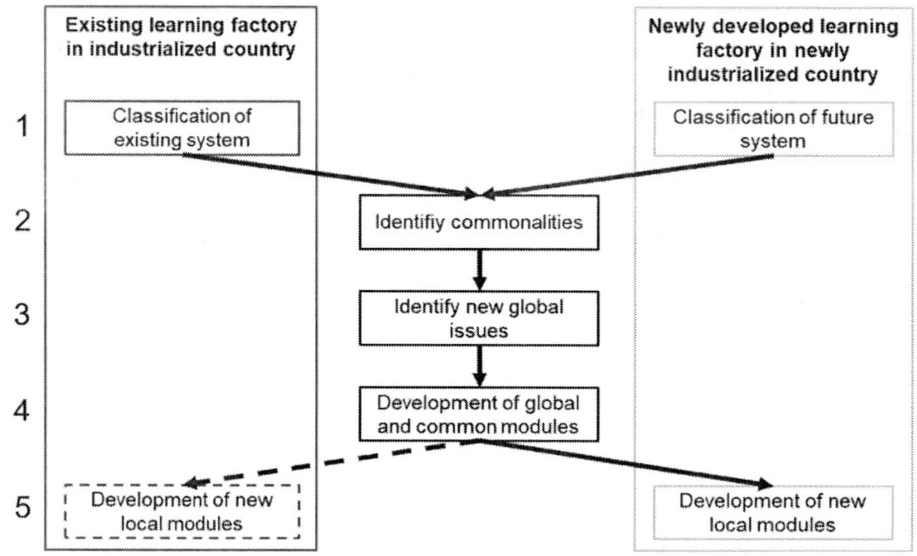

Figure 3. Structure of the proposed method.

The proposed method consists of five steps as shown in figure 3:

1. Classify existing and future systems
2. Identify commonalities
3. Identify new global issues
4. Develop global and common modules
5. Develop new local modules

This five step method fosters the transfer of an existing learning factory from an industrialized country to a university in a newly industrialized country and ensures that the new learning factory fits to the demand of the university in the newly industrialized country.

First, the existing learning factory has to be classified and in parallel, the new, yet to develop learning factory in the newly industrialized country has to be classified. For this, the Learning Factory Morphology from Tisch et al. [8] is applied. The morphology is divided into the seven dimensions of learning factories:

- Operating Model
- Purpose and Targets
- Process
- Setting
- Product
- Didactics
- Metrics

Each dimension is further specified with six to 12 subcategories. At this step, it is assumed that no learning factory exists in the specific engineering education environment of the newly industrialized country. Therefore, experts must be consulted to define the future learning factory, involving

IOP Conf. Series: Materials Science and Engineering **297** (2018) 012053 doi:10.1088/1757-899X/297/1/012053

stakeholders from industry, academia and governmental institutions. This classification step will help to point out the real demand in the newly industrialized country and ensures that the system from the industrialized country will not simply be copied without considering the specific setting and boundary conditions, including economic, technical and most importantly cultural dimensions.

In the second step, both classifications will be compared to identify commonalities. In few cases, a simple transfer to the new learning factory might be possible. For example in the process dimension both learning factories could be favored to have the same hardware setup as it enables a fast implementation and future bilateral support culture. However, there can be operational reasons, which avoid such a simple transfer, especially in terms of the operating model dimensions such as available budgets for the learning factories.

Care has to be taken if there is no commonality or a little commonality. In the worst case, an expensive learning factory will not be used and maintaining as it is not adapted to the contextual, didactic or cultural demands of the teaching and learning system in the newly industrialized country. Hence, the employability of the engineering graduates will not increase.

Apart from commonalities, general global issues at the contents side (processes) should be identified. One possible global topic could be sustainable development. For this, new common modules can be developed and used at both universities. An example is the global warming challenge. As measures to reduce the rate of global warming have to be adapted by all industries in the world, the teaching modules for both leaning factories can be the similar. In these cases, a long-term collaboration, for example with virtually connected modules, could be an option to be sought. The virtual linking of learning factories of two universities especially makes sense when a global supply chain connects the industries at both locations. Under such circumstances, linking the learning factories can show the impact of local decision on the global supply chain. This helps to optimize the whole supply chain instead of only the local processes.

Finally, in the last phase new modules must be developed to ensure that local requirements are considered especially for the new learning factory. The experience from the learning factory in the industrialized countries can help to build these new modules, but the modules must fit to the local requirements in the newly industrialized country. Also, new modules can be developed in the industrialized country after broadening the horizon of the involved experts. The view on different processes and structures can lead to changes in the own learning factory environment.

4. Case study JInGEL

The proposed method is applied in this case study within the cooperation between the TU Braunschweig in Germany and BITS Pilani in India. A learning factory was established at the Institute of Machine Tools and Production Technology of TU Braunschweig, Germany in 2013. The learning factory (called "Die Lernfabrik") consists of three major pillars: the research lab, the experience lab and the education lab (see figure 4). For all three labs, the focus topics are energy and resource efficient manufacturing, industry 4.0 as well as urban production. The research lab represents a real-sized manufacturing environment consisting machine tools, technical building services and information technology. The main objective is to investigate new concepts and principles of prior mentioned focus topics at a real life scale. Although, students work in the research lab, the focus is clearly on research topics and a guided learning concept. In the experience lab, education becomes a more dominating topic. Industry processes are scaled down to small modular stations and many practical lab sessions are held here. Although the focus is set on practice and research-oriented learning, it is also possible to investigate new technical and didactic concepts within this infrastructure. The third pillar consists of education lab, which is used to educate the apprentices of TU Braunschweig, like industrial mechanics or precision mechanics. In this learning environment, real life scaled machine tools slimed down to educational demands are used for practice-oriented learning under supervision.

Figure 4. "Die Lernfabrik" at TU Braunschweig.

For the transfer process to BITS Pilani, in the initial stage, only the experience lab is taken into consideration (see Figure 5). It consists of customized adapted Modular Productions System (MPS) from FESTO Didactic SE. Each station is part of a fixed process chain and represents a scaled-down production process. Over the period of five years, the use of the experience lab at TU Braunschweig was synergetically extended and used by four different engineering lectures. The most important topics are energy and resource efficiency in manufacturing and cyber physical production systems. The specific developments and adaptions are results of educational and research projects, which are well documented and evaluated in the literature. [9], [10]

Figure 5. Experience Lab at TU Braunschweig.

Since 2009, the TU Braunschweig maintains a close collaboration with BITS Pilani (examples for project results: [11], [12] or [13]). Büth et al. [2] showed already that there is a qualification gap between academia and industry in India. The department of mechanical engineering of BITS Pilani invested into a new workshop building with space for an experience lab.

First, the experience lab of "Die Lernfabrik" was classified with the learning factory morphology. Afterwards, through collaborative work with BITS Pilani researchers, the learning factory morphology is used to investigate the demand of BITS Pilani for a learning factory.

Next, the results of the classifications are compared. Table 1 summarizes the results of the comparison. In general, both experience labs are comparable in most dimensions. The main goal of both universities is to train their production-engineering students in the topics of sustainable manufacturing and cyber physical production systems. The industrial processes are scaled down to ensure that the students can work self-guided in the experience labs without any safety risks. However, within the process dimension some differences are identified for which specific modules are to be developed. While the TU Braunschweig is focusing on cutting, additive manufacturing and changing material properties in their learning factory, BITS Pilani is focusing on cutting and joining. Therefore, modules with focus on joining should be developed to ensure that this topic would be addressed within the experience lab. Another relevant difference from the metrics perspective is the number of participant per year. While at TU Braunschweig 200 to 500 students use the experience lab, at BITS Pilani only 50 to 200 participants use the infrastructure. One of the reasons for this is the size of the universities. At TU Braunschweig 20,000 students are enrolled and at BITS Pilani, Pilani campus only 5,000. Also, the available space at BITS Pilani is 70 sqm which significant smaller than at TU Braunschweig.

The universities in India particularly technical universities/institutes are not big. Therefore, in India, the learning factory should be designed to use less space and bring different topics together within the same hardware. This not only reduces the necessary amount of space, but decreasing the demand for available space can reduce the cost for a learning factory.

Table 1. Summary of results of classification.

Dimension	Degree of accordance	Similarities	Differences
1. Operating Model	Very high	Similar funding structure	Involvement of professors at BITS
2. Purpose and Targets	High	Focus on trainings for students	Students and employee trainings at TU BS
3. Process	High	General industrial process are the same	Differences in manufacturing methods
4. Setting	Very high	Scaled down processes in factory environment	None
5. Product	Very high	Same product	None
6. Didactics	High	Self-guided trainings for students	Focus on standarized trainings at BITS
7. Metrics	Medium	Number of participants per training	Size and number of trainings per year

TU BS = TU Braunschweig BITS = BITS Pilani

Global issues, which are addressed by both universities, are sustainable manufacturing and global warming. Energy and resource efficiency are relevant at both locations and should be taken into consideration. At TU Braunschweig these aspects are integrated into each MPS station. As the use of electrical energy drives CO_2 emissions, it could make sense to integrate renewable energies into the experience labs. At TU Braunschweig, solar panels are already installed close to the experience lab. Also BITS Pilani has a high potential for solar power. Thus it makes sense to integrate the topic of renewable energies with a commonly developed MPS station.

Recently a setup of five MPS stations has been installed at BITS Pilani (see figure 6). These stations were carefully selected considering the boundary conditions at BITS Pilani. Currently new lecture units, which will take place in the new experience lab, are in development and will be held by BITS Pilani lecturers. In the next steps, new common modules and specific modules for each learning factory will be developed. In addition, psychologists are tracking the success of this project by testing the employability of the BITS Pilani graduates before and after the introduction of the experience lab.

Figure 6. New experience lab at BITS Pilani.

5. Conclusion

The learning factories can be used economically and effectively to enhance the employability of young engineering graduates in newly industrialized countries. This paper applies the generic five step approach to transfer learning factory to other environment. A study has been within the cooperation between the TU Braunschweig and BITS Pilani. Proper care has been taken to develop learning factory at BITS Pilani to take care of the differences in the size and education methods of the two universities. Also, the cultural dimension, for example the learning and teaching culture, has been synchronized in the transfer. Currently, the learning factory is build up at BITS Pilani and the development of local modules has just started.

6. Outlook

The proposed approach is very generic and it could be used for other institutions and must not necessarily be applied for collaboration between industrialized and newly industrialized countries. Further, during the transfer of the experience lab to BITS Pilani many other universities and institutes showed high interest in the project JInGEL. As BITS Pilani is one of the highest ranked universities in India, many small universities and institutes in India, which train the majority of Indian graduates can also benefit from the learning factory at BITS Pilani. The new learning system should become ready to be transferred to these universities and institutes to ensure that a large number of graduates can benefit from the learning factory concept.

Acknowledgement

The mentioned practice partnership "JInGEL – Joint Indo-German Experience Lab" is supported by the DAAD – German Academic Exchange Service (Project ID 57219215) and FESTO Didactic SE. It aims to transfer the positive experience from the TU Braunschweig experience lab MPS stations to the teaching environment of BITS Pilani.

References

[1] University Grant Comission 2016 Annual Report 2015-16, Delhi, India, URL: https://www.ugc.ac.in/pdfnews/3710331_Annual-Report-2015-16.pdf, accessed on 16/10/2017.
[2] Büth, L. et al. 2017 *7th Conf. on Learning Factories* (Darmstadt, Germany)
[3] Müller-Frommeye L, et al. 2017 *7th Conf. on Learning Factories* (Darmstadt, Germany)
[4] Abele E. et al. 2015 *5th Conf. on Learning Factories* (Bochum, Germany)
[5] Tisch M and Metternich J 2017 *7th Conf. on Learning Factories* (Darmstadt, Germany)
[6] Enke J, et al. 2017 *7th Conf. on Learning Factories* (Darmstadt, Germany)

[7] ESB Business School, Hochschule Reutlingen 2017 NIL Network Innovative Learning Factories, Reutlingen, Germany, URL: http://www.esb-business-school.de/en/research/research-activities/nil-network-innovative-learning-factories/, accessed on 14/08/2017.
[8] Tisch M, et al. 2015 *The Turkish Online J of Educational Technology* **14**(2) pp 356-363
[9] Thiede S, et al. 2015 *6th Conf. on Learning Factories* (Gjøvik, Norway)
[10] Blume S, et al. 2015 *6th Conf. on Learning Factories* (Gjøvik, Norway)
[11] Juraschek et al. 2016 *Indo German Conference on Sustainability 2016* (Chennai, India)
[12] Bhakar V, et al. 2015 *The 22nd CIRP Int. Conf. on Life Cycle Engineering* (Sydney, Australia)
[13] Mittal V, et al. 2012 *The 19th CIRP Int. Conf. on Life Cycle Engineering* (Berkeley, USA

IOP Conf. Series: Materials Science and Engineering **297** (2018) 012054 doi:10.1088/1757-899X/297/1/012054

Mind map learning for advanced engineering study: case study in system dynamics

Denchai Woradechjumroen*

Department of Mechanical System Engineering and Industrial Innovation, School of Engineering, Sripatum University, Bangkok, 10900, Thailand

* Corresponding Author: denchai.wora@gmail.com

Abstract. System Dynamics (SD) is one of the subjects that were use in learning Automatic Control Systems in dynamic and control field. Mathematical modelling and solving skills of students for engineering systems are expecting outcomes of the course which can be further used to efficiently study control systems and mechanical vibration; however, the fundamental of the SD includes strong backgrounds in Dynamics and Differential Equations, which are appropriate to the students in governmental universities that have strong skills in Mathematics and Scientifics. For private universities, students are weak in the above subjects since they obtained high vocational certificate from Technical College or Polytechnic School, which emphasize the learning contents in practice. To enhance their learning for improving their backgrounds, this paper applies mind maps based problem based learning to relate the essential relations of mathematical and physical equations. With the advantages of mind maps, each student is assigned to design individual mind maps for self-leaning development after they attend the class and learn overall picture of each chapter from the class instructor. Four problems based mind maps learning are assigned to each student. Each assignment is evaluated via mid-term and final examinations, which are issued in terms of learning concepts and applications. In the method testing, thirty students are tested and evaluated via student learning backgrounds in the past. The result shows that well-design mind maps can improve learning performance based on outcome evaluation. Especially, mind maps can reduce time-consuming and reviewing for Mathematics and Physics in SD significantly.

1. Introduction

System Dynamics (SD) is one of the most significant fundamental subjects, which can be used to virtually model, applied engineering applications in the real world. To be more specific for significances in modern engineering areas, SD can be applied in a special modeling technology, especially, for reducing cost-prohibitive issue called "virtual sensor". The concepts of this low-cost and effective sensor can be categorized in three criteria [1]: measurement characteristics-based criterion (transient-state and steady-state data-based approach); modeling method-based criterion (white-box, gray-box and black-box models) and application purposes-based criterion (replacement and observing) [2]. For first criterion, steady-state based approach is typically applied in quick modeled process from changed inputs or slowly changed inputs of a system, so it is suitable for intelligent diagnostic module called "automated fault detection and diagnosis (AFDD)". Virtual sensor examples can be found in the recent literature such as virtual wall surface temperature [3] and refrigerant charge [4]. In contrast to steady-state method, the transient-state based approach using a transient model to predict an immeasurable variable is typically appropriate for real-time feedback control or transient document used in FDD. With these behaviors, fast system responses such as chemical process control [5, 6] have applied transient virtual sensors to detect fast system changes due to existing faults and measure cost-effective values for a feedback control system.

In developed counties, beyond engineering program, several educational programs leverage this course in terms of educational programs [7-9] because this program can be used to initially test an innovative concept in terms of simulation-based modeling in any engineering areas before conducting a prototype [10]. However, since students have learned how to model any engineering system in terms of mathematical modeling, strong backgrounds in mathematics and physics are required for qualified students. For now, SD has been taught with lecture-based learning. The problems are: students in private schools lack of strong backgrounds in Applied Mathematics (e.g. Differential Equations) and Engineering Dynamics; as a result, these students cannot learn SD effectively and also cannot further applied SD in the continuous course – feedback control system for selecting optimal controllers to control the dynamic models.

To mainly enhance student learning with weak engineering and mathematical backgrounds, Woradechjumroen and Sinjindawong [10] applied project-based learning (PBL) as the idea of "learning by doing" [7]; each student group selected the project of a system modeling in the real world based on their backgrounds as the semester project. In addition, since PBL mainly relies on subjective rubrics based on instructor experiences for student satisfaction results in attending the course, outcome-based learning (OBE) method had been designed to evaluate how students can improve their performance compared with related student backgrounds before joining the class. Although the prior research can leverage some student performances, some students still cannot apply the accomplished project to similar applications because they cannot effectively link or create the relations between similar applications.

To reduce the aforementioned limitations, this paper proposes a mind map engineering learning for SD course. Firstly, student backgrounds are evaluated based on the designed course content relations. Then, the instructor will review and conclude the concept in terms of overview pictures. Then, the student learning is assigned via four projects for designing their mind maps for improving their studies. Mid-term and final exams are used to evaluate how students improve themselves efficiently.

2. Backgrounds and limitations

2.1 Course contents relation

In order to potentially evaluate the student study backgrounds, the course content relation relating to SD was designed as depicted in figure 1 [10]. In the figure, it illustrates the correlations of each related course in control engineering areas. As directly related course backgrounds to SD, the fundamental engineering background explains the phenomena engineering movement – Engineering Dynamics is significantly utilized to derive simplified mechanical systems in terms of rotational and transitional systems, whereas other systems are thermal, fluid and electrical system modeling do not associate with Engineering Dynamics. This is so challenging to develop potential mind map to link the three systems in one system for reducing required backgrounds in Heat Transfer, Fluid Mechanics and Electrical Engineering course, respectively.

In advanced engineering applications, especially relating to building energy systems, Building Control, Energy Modeling and Smart Building Solutions can be further assigned to interested students in building energy systems as cooperative educational students [11].

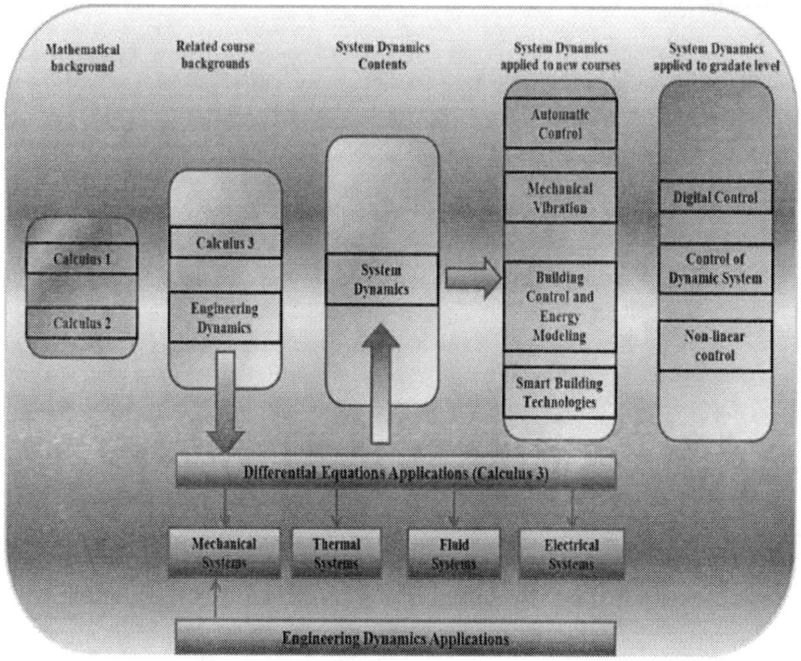

Figure 1. The relation illustration in control engineering courses [10]

2.2 Student Backgrounds

According to designed course syllabus, the objectives of main map engineering method are that students who pass SD should be able to: 1) understand the phenomena of engineering systems; 2) know how to build mathematical modeling of an engineering system; 3) analyze system response via characteristics equation of the system modeling; and 4) understand similarity of thermal, fluid and electrical systems and the equivalence equation between mechanical and electrical systems. According to table 1, it is evident that the class majority has weakness in mathematical solving problems. As a result, at least one mind map of each student were designed to improve and illustrate how to obtain the system response by using a mind map method.

Table 1 student study performance backgrounds

Related course backgrounds	Grade ≥ B	C ≤Grade < B	D ≤ Grade < C	Do not enrol the course
Engineering Dynamics	0	7	29	0
Calculus 3 or Differential Equation	1	6	20	9
Calculus 2	12	12	11	1
Calculus 1	8	9	19	0

3. Mind Maps Development Procedures

This section provides the general steps for constructing mind maps efficiently. At first, mind map theory and method are briefly explained to students with social science pattern. Then, the system overview of the course should be illustrated concisely composing of the general form of engineering system dynamics in terms of input, output and element. With the general form, system equivalence between mechanical engineering and electrical system is demonstrated via a mind map. Then, each student will be evaluated by understanding the system similarities via transferring the designed mind map to associate with thermal, fluid and real world system. A further step, simplified mathematical solving procedures are designed to link the previous mind maps for obtaining system response. Then, the evaluation will be systematically conducted step by step to well notice student that performance improvement and any limitations of mind map utilization.

3.1. Basic mind map learning

Mind map was mainly designed to use the advantage of a right brain to memory any contents in system configuration instead of using only left brain which emphasize to use logics and theories [12]. As shown in figure 2, the related contents can be grouped together from the main concept or main feature in which action plan will link between the main system and each subsystem. With the good side, the instructor will teach students first and then will draw the conclusion in terms of mind map for each class. However, mind map has been intensively used in social science course rather than engineering courses due to the content complexity. Attending students are required to study basic general mind map concept for further applying in engineering education to improve their study performance.

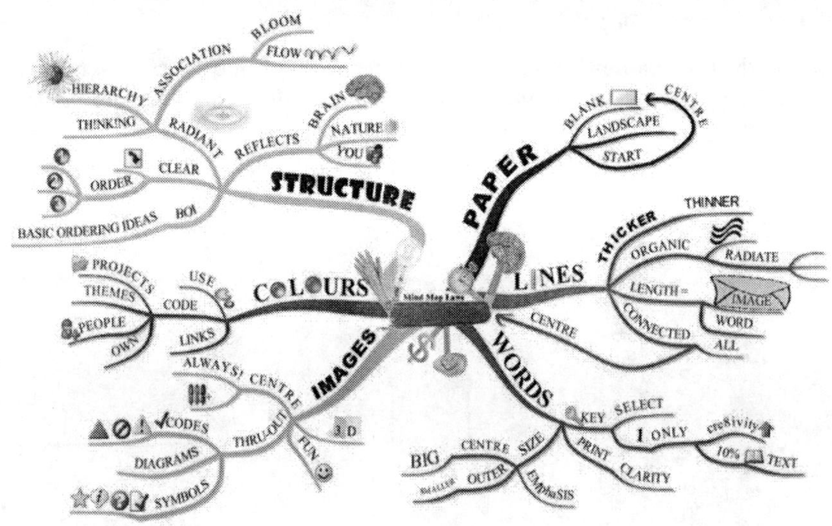

Figure 2. General mind map illustration [12]

3.2. System Overview

In this step, the main system of SD includes: input, output and system element in terms of resistance (R), capacitance (C) and Inductance (L). The three elements are used to model mechanical and electrical systems. Also, force or current is a driving force condition or input for mechanical and electrical system, respectively. Meanwhile, velocity is equivalent to voltage as output of electrical system as demonstrated in figure 3. At this step, the instructor has to conclude all initial and fundamental concepts for the system modelling procedures.

IOP Conf. Series: Materials Science and Engineering **297** (2018) 012054 doi:10.1088/1757-899X/297/1/012054

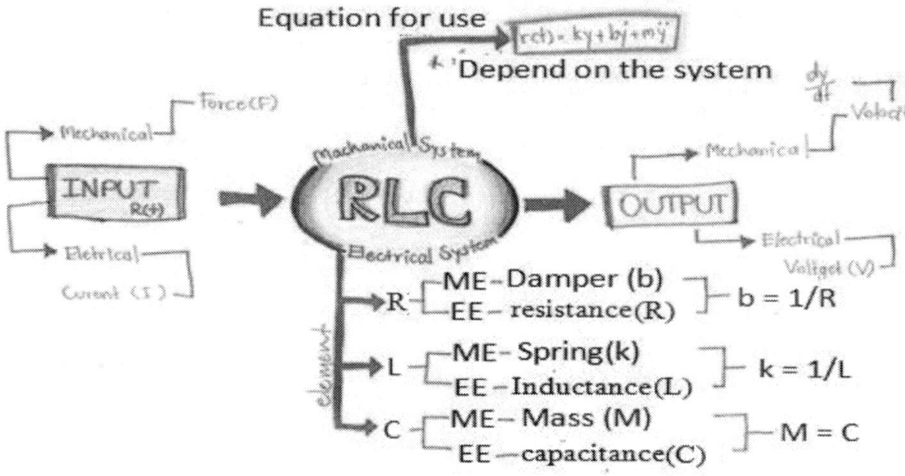

Figure 3. System overview between input, output and element of mechanical and electrical system

3.3. System modelling equivalence

This section is to design mathematical models in terms of mind map 2, which this mind map is initially conducted with free body diagram with Newton's law, and then is converted to electrical system without using Kirchhoff law, using system equivalence compensation (figure 4). With the system similarities, students who were poor in electrical system modelling can understand more processes. In addition, this method could be used for the learning exchange between mechanical and electrical engineering students.

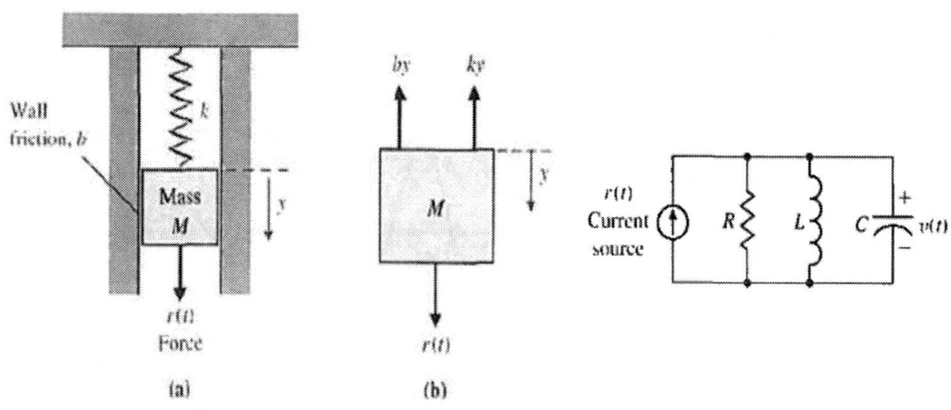

Mechanical system is equivalent to the electrical circuit $(\therefore R \equiv b, L \equiv k, C \equiv M)$

Mechanical system: $M\frac{d^2y(t)}{dt^2} + b\frac{dy(t)}{dt} + ky(t) = r(t),$

Electrical system: $\frac{v(t)}{R} + C\frac{dv(t)}{dt} + \frac{1}{L}\int_0^t v(t)\,dt = r(t).$

Figure 4. Mathematical modelling mind map

3.4. Mathematical solving

After any students can well model the engineering applications in mechanical systems, the next step is to learn how to solve the system without strong mathematical backgrounds. In this class, trial solution in figure 5 is mapped in order to teach students how to solve transient response without using Laplace transform in case of 1 degree of freedom (DOF) or 2 DOF with constant input or step function excitation.

However, in case of 2-DOF with more complex inputs such as harmonic or periodic functions, trial solution technique is not suitable to solve non-homogenous equations. To tackle this issue, Laplace transform table with simple examples is used to build potential mind map for solving homogenous equations as shown in figure 6.

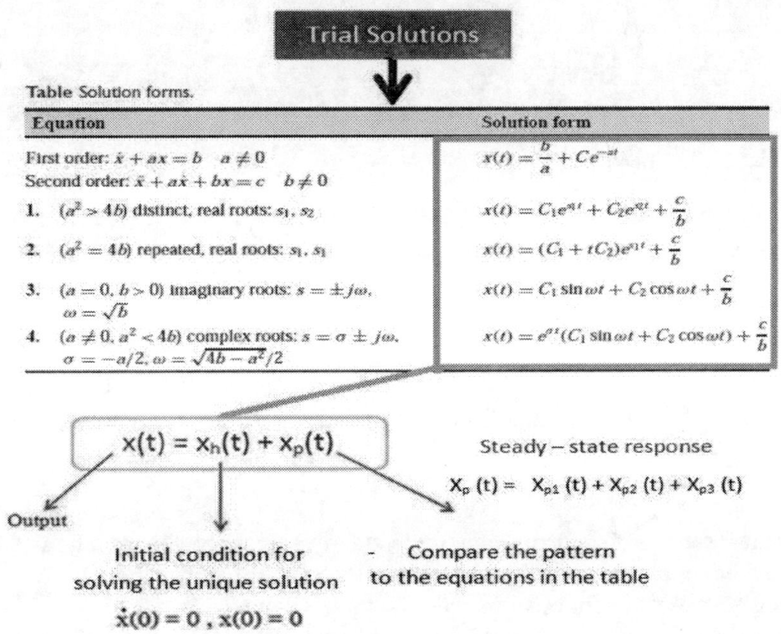

Figure 5. Mathematical solving mind map using trail solution

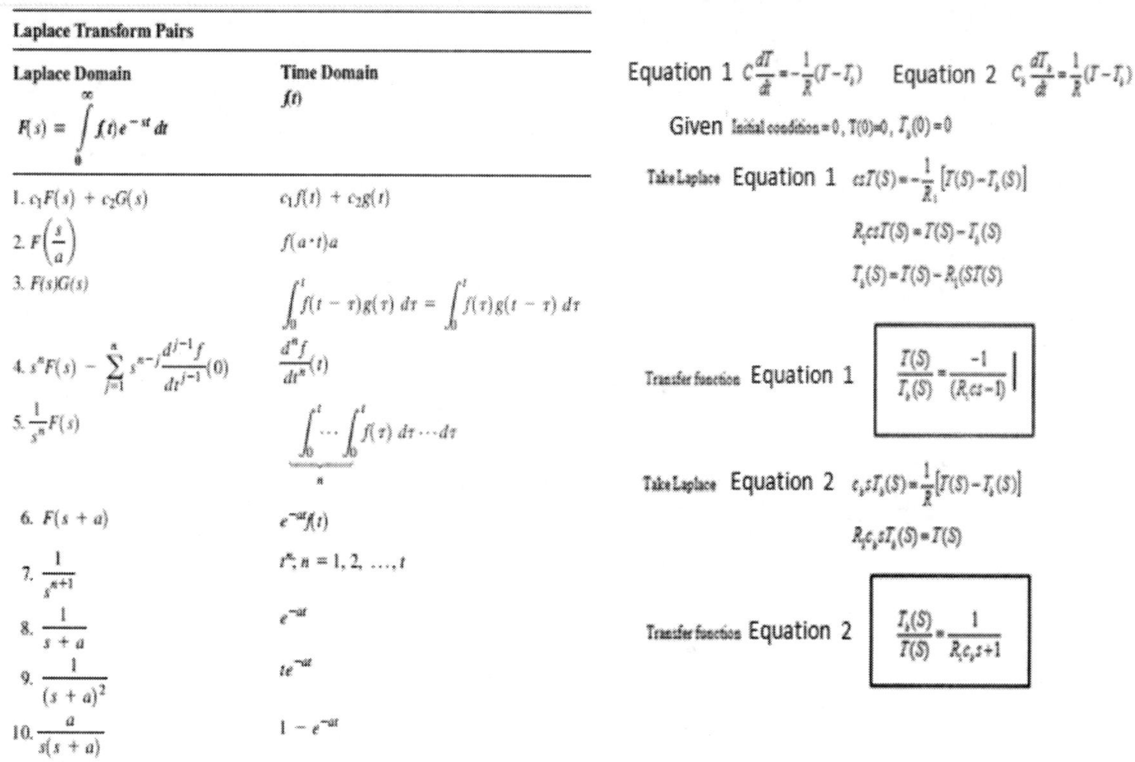

Figure 6. Mathematical solving mind map using Laplace transform

4. Student Performance Evaluation

4.1. System relation

To test and evaluate trained students can understand and apply the mind map concept for other engineering areas beyond mechanical and electrical systems. One of the five reports called lab 4 is designed to evaluate these student outcomes as shown in figure 7.

In figure 7, the mind map example demonstrates the similarities between mechanical, electrical, fluid and thermal systems. Especially, the electrical system can be further used to model thermal network used for a thermal system, and can be used for flow diagram in a fluid system as well. After that, Laplace transform main map in figure 6 can be applied to accomplish transfer function and transient response based on inverse Laplace transforms.

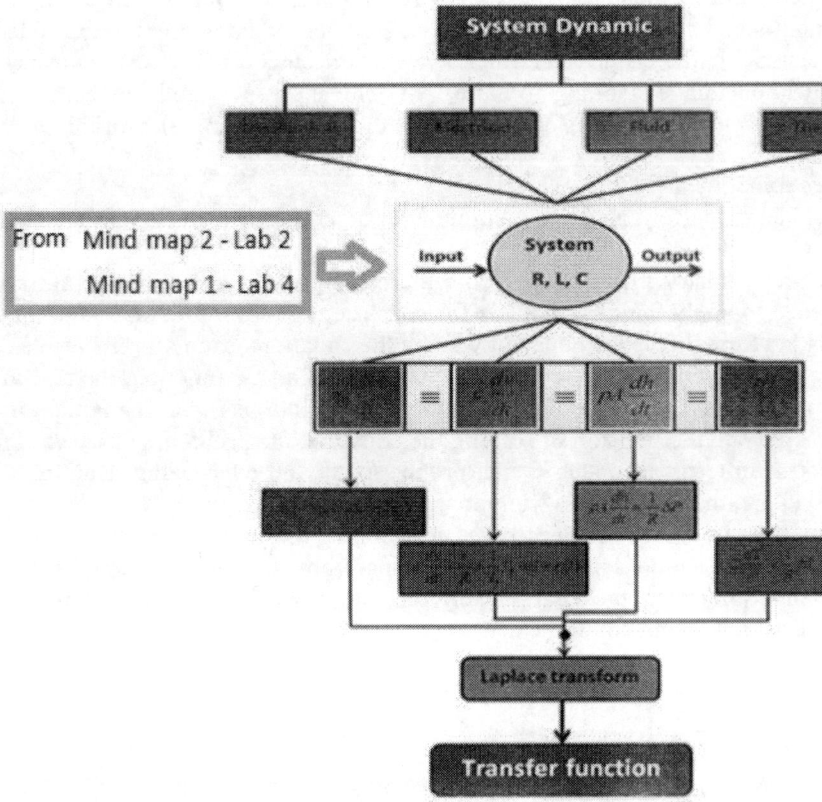

Figure 7. Multiple system mind map relation for final examination

4.2. Extending mind map for more complex systems

To test and evaluate how well trained students understand and can apply the mind map concept for other engineering areas beyond mechanical and electrical systems. One of the five reports called lab 4 is designed to evaluate these student outcomes as shown in figure 8.

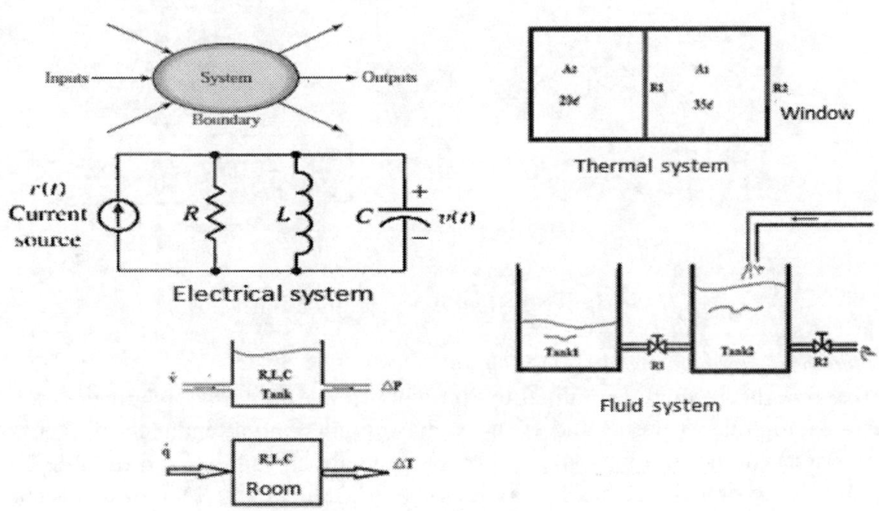

Figure 8. System similarity for final examination

In the figure, the similarities of electrical, fluid and thermal systems are first to review in terms of generic components and elements for a system. After that, students should explain how complex systems in the right-side hand of figure 8 are related together by extending the left-side hand concept. This extension can tackle the original learning styles of the students who always remember problems and solutions without understanding of the physical meanings and concept. Eventually, the mathematical mind map will be also used to solve solutions of each complex problem.

4.3. Real Application Solving

To further test students' performance for continuous courses and real world applications, one problem in a thermal system is designed by relating the mind map concept as depicted in figure 9 as lab 5. In this lab, the lecturer primary introduces a room thermal network with a heat pump application. All students are needed to apply their mind maps for constructing a room thermal equation with the heat pump. Then, Matlab Simulink is introduced by using the equations from aforementioned steps.

Due to using artificial outdoor air temperature in terms of a sine function, the add-in mathematical solver of Matlab is so convenient to obtain the continuous solutions rather using the original solutions from the mind map. With the revision of solving the equation, the problem is used to simulate how the heat pump performs in terms of under-sizing, proper-sizing and over-sizing. Runtime fraction (RTF) [13] is a simplified parameter for identifying the heat pump performance.

After learning the real world problem-based learning as the first mind map of lab 5, students are required to achieve the second activity for converting from the room heating system to a cooling system application. Similar to the heating solution, students must have an answer based on the simulation what the right-sizing cooling capacity of the room is.

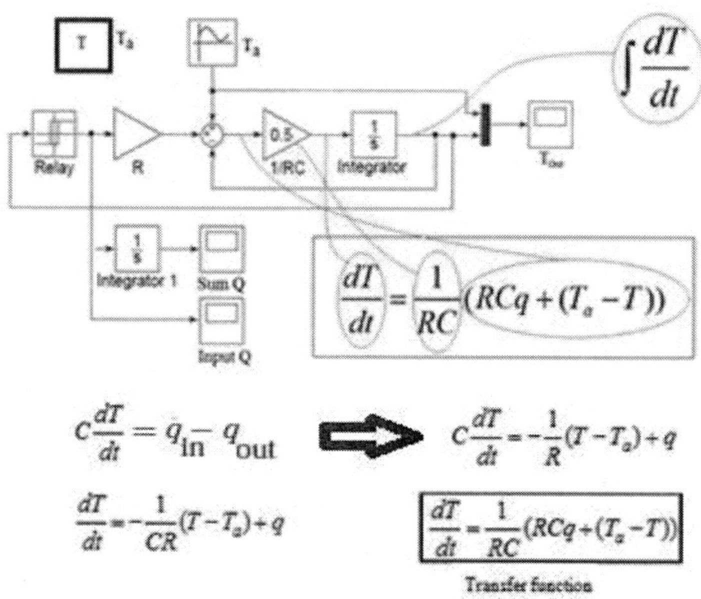

Figure 9. Real world application

4.4. Post student performance based on mind map

Eventually, five reports are totally summed to 30 points from 100 points, whereas other 70 points are from mid-term exam both 15 points and final exam 50 points and attendance of 5 points. The study performance was based on mind map application is summarized as tabulated in table 2 by comparing with the student backgrounds in table 1. The interesting findings are: 1) 7 students obtaining higher than grade B who do not have strong backgrounds in Engineering Dynamics and Calculus 3 (only one obtained B); 2) 9 students obtaining between C and B can improve themselves and studying the behaviors in terms of solution memorization since 6 students from the total remember the solutions when they attended Calculus 3 and the rest 3 students can leverage the study performance; and 3) the rest would obtained D+ and D because they are still concern the improved study method, and do not pay attention to apply mind maps for improving their study styles and performances.

Table 2 student-learning results

Student backgrounds	Grade $\geq$ B	C $\leq$Grade < B	D $\leq$ Grade < C	Do not enrol the course
Engineering Dynamics	0	7	29	0
Calculus 3 or Differential Equation	1	6	20	9
SD student grade	Grade $\geq$ B	C $\leq$Grade < B	D $\leq$ Grade < C	D $\leq$ Grade < C and do not learn Calculus 3
	7	9	11	9

5. Conclusions

SD is a significant and fundamental engineering course to develop system modeling for testing engineering concept before producing a prototype. To study SD well, students are required to have strong backgrounds in Mathematics and Physics, which are not inherent behaviors of engineering students in private universities. To overcome these issues, the mind map application is proposed for SD course. The instructor briefly explains the main concept in each class by using mind maps. To evaluate how well students understand, they are required to accomplish 5 reports are designed for a new students study method in the course. The first two reports are designed for study familiarity of mind map application. Then, lab 3 and lab 4 are designed to extend the application for more complex systems. Finally, one of the real world applications is used as lab 5 to test how student can apply mind maps and improve their study behaviors and performances. We found that mind map engineering applications can reduce the complexity of mathematics and physics by mapping the similarity of each system and by using solving pattern of system similarity. Their study performances can be improved if users systematically design potential mind maps.

Acknowledgement

The author is deeply grateful to Teaching and Learning Support and Development Center (TLC), Sripatum University for financial support, and gratefully acknowledges the recommendations from Dr. Sirinthorn Sinjindawong, Director of TLC.

References

[1] Li H, Yu D and Braun JE 2011 *HVAC&R Res* **17**(5) pp 619–645
[2] Woradechjumroen D, Yu Y and Li H 2016 *Appl. Energy*. **162** pp 1323-1335
[3] Woradechjumroen D, Li H and Yu Y 2014 *Int .J Build Urban Interior Landsc Technol*. **3** pp 5-21
[4] Li H and Braun JE 2007 *Int. J. of Refrig.* **30**(3) pp 546-64
[5] Bonne D and Jorgensen SB 2004 *Proc. Int. Conf on ADCHEM* (Hong Kong)
[6] James S, Legge R and Budman H 2002 *J. of Process Control* **12**(1) pp 113-121
[7] Pavlov OV, Doyle JK, Saeed K, Lyneis JM and Radzicki MJ 2014 *Department of Social Science and Policy Studies Working Papers* (No. 001)
[8] Nuhoglu N and Nuhoglu M 2007 *J. Turkish Science Education*. **4** (2)
[9] Barlas Y 1993 *Proc. Int. on System Dynamics Conf.* (Cancun: Mexico)

[10] Woradechjumroen D and Sinjindawong S 2016 *Proc. Int. STEM Education Conf.* (Pattaya: Thailand)
[11] Woradechjumroen D and Jansa P 2016 *Proc. Int. Conf on Mechanical Engineering* (Chaingmai: Thailand)
[12] Buzan Centre Australia <http://www.buzan.com/
[13] Woradechjumroen D, Yu Y, Li H,Yang H and Yu D 2014 *Energy Build.* **69** pp 131-143

Comparing the engineering program feeders from SiF and convention models

Warawaran Roongruangsri[1*], Niwat Moonpa[2], Janyawat Vuthijumnonk[1] and Kampanart Sangsuwan[1]

[1]College of Integrated Science and Technology, Rajamangala University of Technology Lanna, 128 Huaykaew Road, Changpuek, Muang, Chiangmai, 50300, Thailand
[2]Mechanics, Materials and Engineering Design Research Group, Rajamangala University of Technology Lanna, 128 Huaykaew Road, Changpuek, Muang, Chiangmai, 50300, Thailand*

* Corresponding Author: warawaran@rmutl.ac.th

Abstract. This research aims to compare the relationship between two types of engineering program feeder models within the technical education systems of Rajamangala University of Technology Lanna (RMUTL), Chiangmai, Thailand. To illustrate, the paper refers to two typologies of feeder models, which are the convention and the school in factory (SiF) models. The new SiF model is developed through a collaborative educational process between the sectors of industry, government and academia, using work-integrated learning. The research methodology were use to compared features of the the SiF model with conventional models in terms of learning outcome, funding budget for the study, the advantages and disadvantages from the point of view of students, professors, the university, government and industrial partners. The results of this research indicate that the developed SiF feeder model is the most pertinent ones as it meet the requirements of the university, the government and the industry. The SiF feeder model showed the ability to yield positive learning outcomes with low expenditures per student for both the family and the university. In parallel, the sharing of knowledge between university and industry became increasingly important in the process, which resulted in the improvement of industrial skills for professors and an increase in industrial based research for the university. The SiF feeder model meets its demand of public policy in supporting a skilled workforce for the industry, which could be an effective tool for the triple helix educational model of Thailand.

1. Introduction

Innovation in education has led to the development of life skills acquisition, using both integrated learning and contextual teaching/ learning models. Both learning models have a similar aim of developing students' life skills [1]. In Thailand, a study by the National Economic and Social Development Board or NESDB [2] found that a disparity between demand and supply in science and technology (S&T) manpower, especially in terms of quality. The main reason was discontinuity in knowledge, skills and attributes between the expectation of the industry and the capacities of educational institutions, particularly when it came to the chasm between practical and theoretical modes of learning. The study went on to suggest that overcoming such demand and supply disparity in the quality of manpower. There was a need for every college and university to established in-depth cooperation with industrial enterprises [3]. Therefore, Rajamangala University of Technology Lanna (RMUTL), in Chiangmai, Thailand decided to implement a project focused on the "School in factory" (SiF) models through an effective collaboration including three parties (triple helix), namely industrial partner (Siam Michelin Co., Ltd), the academic sector (RMUTL) and government agency (National Science Technology and Innovation Policy Office, STI). The objectives were to solve the problem of shortage in qualified technicians, reduce the turnover rate of employees and increase the country's productivity.

Within the SiF project, Siam Michelin Co. Ltd was the main stakeholder, offering the work venue, and "in cash investment" such as: students' school fees, monthly allowances, accommodation costs, and payment for lecturers, facilitator and on-site trainers. The RMUTL provided an "in kind investment" such as the contributions to lecturers, facilitators and school management. The STI, as the government agency, fosters and supports this form of participatory educational management. Students graduating from this program receive a diploma certificate from Rajamangala University of Technology Lanna. The pilot project was first launched in 2012. It is a two-year educational program. Each year, students spend three months studying at the university and nine months at the factory. The intake of students on the SIF project are students who graduated from higher secondary school (general education) or vocational certificate programs. In view of this diversity in students' background, the university assisted them with two months training program in basic and higher professional skills before attending the factory [4].

This paper addresses two typologies of feeder models, which are the conventional and the school in factory (SiF) models. The new SiF model was developed as a collaborative educational project between the industry, government and academic sectors using work-integrated learning. The present research discussed and compared the features of the SiF model and the conventional models in terms of learning outcomes, budget for study, and advantages and disadvantages from the point of view of students, professors, the university, government and industry.

2. Methods

This research purposed to survey and make a comparison between the two types of engineering program feeder models in the technical education systems. The target group was the first year students (30 students) in the bachelor degree in industrial engineering of the Faculty of Engineering, at Rajamangala University of Technology Lanna (RMUTL), in Chiangmai, Thailand. These students would either (i) followed the conventional model, or (ii) participated in the school in factory (SiF) model which was an innovative model developed in collaboration between three parties: industry (Siam Michelin Co., Ltd), government (National Science Technology and Innovation Policy Office) and the academic sector (RMUTL), using work-integrated learning.

The information was collected in relation due to (i) learning outcomes – grade point average (GPA) and grades in the main subjects of the students in the first year of the 2016 academic year (semester 1 and 2), (ii) the budget involved in studying within the two types of engineering program feeder models, and (iii) the advantages and disadvantages for students, professors, the university, government and industry – acquired by personal communication.

3. Results and Discussion of Research

The research establish a result of the feeder students from the SiF models that showed positive learning outcomes with a GPA above average, as well as greater skills and ability to study as demonstrated by the grades in the main subjects displayed in table 1.

Table 1. Learning outcomes (GPA and grades) in the main subjects for feeder student for the SiF and the conventional models

List of Grades	SiF Models (10 students)	Conventional Model (20 students)	
GPA	2.81^{a}	2.67^{b}	
Calculus 1	2.96	2.87	ns
Physic 1	2.83^{a}	2.69^{b}	
Calculus 2	3.65^{a}	3.19^{b}	
Physic 2	2.92	2.88	ns
English in everyday use	2.46	2.46	ns
Chemistry for engineer	3.08	3.00	ns
Production process	2.46^{a}	2.10^{b}	
Engineering Drawing	2.79^{a}	2.60^{b}	
Engineering Material	3.38^{a}	3.22^{b}	

Data are expressed as mean values.
Mean values with different superscripts in the same column differ significantly at $P \leq 0.05$.
The symbol ns means that the mean values are not significantly different.
Grades are weighted as follows: A (Excellent), 4.0; B+, 3.5; B (Good), 3.0;C+, 2.5; C (Satisfactory), 2.0; D+, 1.5; D (Barely Passing), 1.0; F (Failing), 0.

The project as developing the technical education SiF models of high vocational certificate in industrial technique (2 years), was funded by Siam Michelin Co. Ltd, in the amount of approximately 3,808,400 bath for 11 students in the program. Each student gain full financial support for attendance as follow: free tuition and fees for the whole course and living expenses, notably room - 2,500 bath per month, personal expenses - 7,600 bath per month, books and supplies - 2,000 bath per year, student uniform - 4,000 bath per year, welfare - 2,000 bath per year and health insurance - 2,500 bath per year. The conventional model students, on the other hand, had to pay for tuition and fees: approximately 7,500 per semester and also to support themselves with regards to living and personal expenses or obtain student government loans.

The SiF models provided an opportunity for total immersion in the industrial environment and skill setting. On such excursions, the student practices their industrial skills and critical thinking abilities outside of the controlled class setting. Being able to ask experts about their particular area of study on the spot requiring thinking. They were able to learn real-world lessons. This model offered activities and programs that integrate academic learning in its application in the workplace. The practice was real and occurred in the workplace.

This model can have several advantages for students; the course was designed to put theory into practice in order to enhance students' skills in terms of reflection on future practice. This was the opportunity for students to see their subject area in practice. The SiF model also helps to develop ‘soft' skills such as communication, teamwork, leadership and career development. Teachers in this project found satisfaction in seeing students develop and mature. Most of them accepted their role as a participant in a learning process and took genuine responsibility for their learning whilst undertaking the SiF project.

The RMUTL University' s strategic interest in the project focused on work integrated learning (WiL) and this was an important project for them due to the potential that WiL has in enhancing graduate competencies. So the university decided that one of the main priorities was to incorporate an opportunity for work based, experiential learning opportunities in undergraduate programs. The intention was to enhance and extend work-integrated learning strategy, to enhance the quality of teaching and corporate/community engagement. During the SiF project, the university addressed the professional development requirements and also ensured that the university' s wide procedures and regulations were amended in order to enable a more flexible integration into the programs of study. One of the outputs from the SIF project was to develop a curriculum based on industrial expertise in assessment methods, by working with employers who have experience in assessing employability skills. This will continue the Work Integrated Learning (WiL) curriculum used by other universities in Thailand. Moreover, it also sought to identify new and innovative research opportunities linked to immediately practical contexts.

In Thailand, the government and the Ministry of Education use WiL as a key-approach and have launched in 2009 a documen "Guidelines for the Development of WIL Programs". The aim was to encourage universities to implement WiL efficiently, and in 2007 roughly 50% of all institutions from higher education had adopted cooperative education [5]. On the industry side, the SiF project was aimed at improving the employability of graduates by giving them valuable practical experience that is directly related to thecourses offered by universities. TheSiF project also improves the transition from university to work and increase productivity outcomes for the employer and the economy. In the project, students can understand, adapt and apply skills in the workplace. The industry helps ensuring the students are able to plan, investigate and navigate careers in an environment where conceptual, adaptive, personal, technical and vocational skills are essential. On the government side, WiL programs help establish competencies that reduce the personal and community risk of economic downturn. Beyond that, it helps produce a highly skilled workforce to meet industry and community needs.

For Rajabhat Universities, the development of WiL can be a developmental process that creates new tools, models, and methods for the administration of education in the service industry field. The researchers found that WiL provided opportunities for students to acquired real-life work experience and offered opportunities to develop their social skills. They can be a tool for problem solving, as it was an experience learning method with which students had opportunities to adopt knowledge, working skills, and other specific skills in relation to their occupation. From the instructors' perspective, instructors gained real experiences and knowledge. From the perspective of the educational institutions, WiL brought about learning exchanges, acceptance, and trust in the quality of graduates among entrepreneurs. Finally, the advantages for the industry or entrepreneurs, was that they acquired competent employees able to meet their demands, and did not waste time with new employee; this boosted the industrial partners' image [6].

Promwong and Pittayasophon reported that in Thailand, the adoption of work-integrated learning (WiL) needed government support to achieve effective collaboration in order to reach specific social objectives. Moreover, in order to attain an effective triple helix model, good governance and a culture of integrity seemed to be basic challenges [3].

Gellerstedt et al. reported, about WiL and higher education, that it is important that higher education takes responsibility for increasing the professional readiness of students, and for finding a systematically strategy to support WiL education. People were indeed must prepare for a profession that demands lifelong learning. These were therefore the qualities needed to become an efficient student in higher education; they are very much the same abilities required for effective professional practice [5].

4. Conclusion

The results of this study indicate that the developed SiF feeder model is the most pertinent ones to meet the needs of the university, the government and the industry. The SiF feeder model showed positive learning outcomes with low expenditures per student by family and university were alike. In parallel, the sharing of knowledge between university and industry has become increasingly important, which resulted in an improvement in the industrial skills of professors and an increase in industrial based research for university. The SiF feeder model meets the demands of public policy in supporting

IOP Conf. Series: Materials Science and Engineering **297** (2018) 012055 doi:10.1088/1757-899X/297/1/012055

a skilled workforce for the industry, which could be an effective tool for the triple helix educational model of Thailand.

References

[1] Martawijaya DH 2012 *J. of Technical Education and Training* **4** p 2229
[2] NESDB 2005 Human Resources Development Strategy for Building Up Competitiveness in Key Industrial Sector, Bangkok: National Economic and Social Development Board, Publish in Thai
[3] Promwong K and Pittayasophon S 2010 *Proc. Int. Conf. on Triple Helix (Madrid)* vol 1 (Madrid: Spain/LASALLE parquet innovation) p 19
[4] Phalasoon S 2017 *TVET@Asia* **9** 1 Online: http://www.tvetonline.assia/issue9/phalasoon_tvet9.pdf (retrieved 30.06.2017)
[5] Gellerstedt M, Johansson K and Winman T *Systemics, Cybernetics and Informatics* **13** p 1690
[6] Snguanyat O, Boonchart J, Moonlapat K 2016 *Mediterranean J. of Social Sciences* **8** p 216

Performance characteristics of the Savonius turbine

Supakit Worasinchai[1,*], Kiatkong Suwannakij[2]

[1] Renewable Energy Laboratory, Materials for Energy Research Unit, National Metal and Materials Technology Center, 114 Thailand Science Park, Phahonyothin Road Khlong Nueng, Khlong Luang, Pathum Thani 12120, Thailand
[2] Electrochemical Materials and System Laboratory, Materials for Energy Research Unit, National Metal and Materials Technology Center, 114 Thailand Science Park, Phahonyothin Road Khlong Nueng, Khlong Luang, Pathum Thani 12120, Thailand

* Corresponding Author: supakitw@mtec.or.th

Abstract. The paper presents performance investigation of the Savonius turbine both in terms of starting and power extraction characteristics. Effects of the number of blade and the overlap ratio were investigated experimentally using a wind tunnel at a wind speed of 4.45 m/s. Results show that the number of blades and the overlap ratio have a great impact on both performances. In terms of starting, the two-bladed rotor was found to be able to produce higher peak torque than the three-bladed case. However, the three-bladed rotor has a greater starting capability due to its more continuous torque production. The overlap ratio was found to reduce the amount of torque and hence their starting performance. With regard to power extraction characteristic, the three-bladed rotor without the overlap ratio was found to exhibits the highest performance.

1. Introduction

The Savonius turbine was invented by the Finnish Engineer, Sigurd Johannes Savonius, in 1922 [1]. It is categorized as a drag type machine as the incoming wind will 'drag' the blade to spin around its vertical axis (figure 1). The drag force caused by the incoming wind actually acts on both sides of the rotor blade. The force on the convex side is however higher than that of the concave side and this causes a net force that drive the rotor (figure 2).

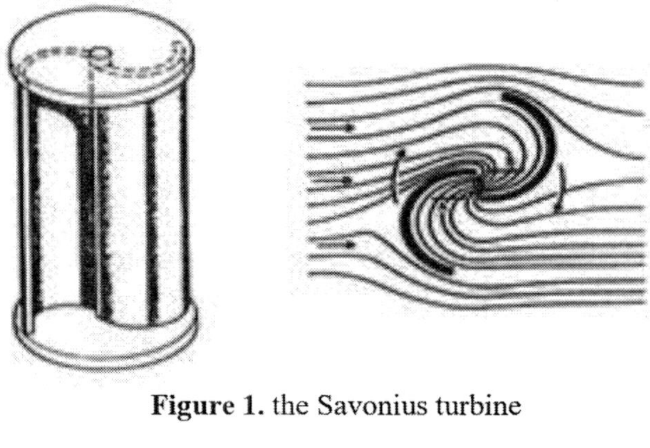

Figure 1. the Savonius turbine

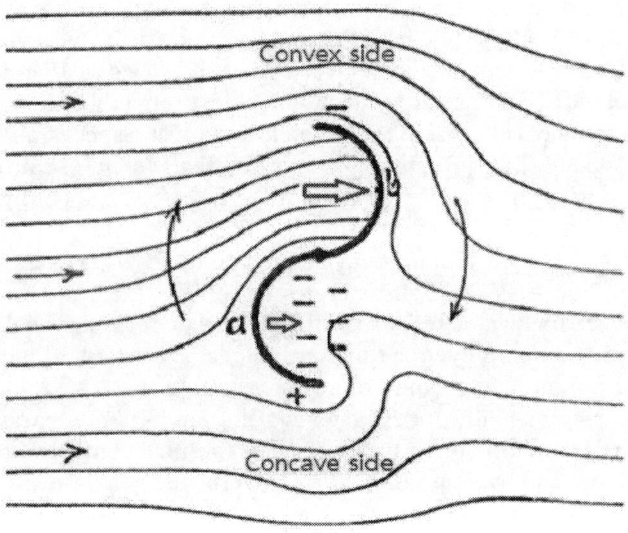

Figure 2. Net force acting on the Savonius blade

Many tests were conducted focusing on power extraction performance [2-4] and little work had been done to investigate its starting characteristics [5, 6]. For example, Saha et al [2] conducted some tests to evaluate aerodynamic performance of Savonius rotors with different stages. Tests conducted by Ali [3] showed that two-bladed rotor has a higher power coefficient than the three-bladed rotor and the author reasoned that the three-bladed configuration has a higher drag at the operating condition. Typically, the Savonius turbine will operate at the tip speed ratio of around 1 and has a power coefficient of about 0.15 (figure 3).

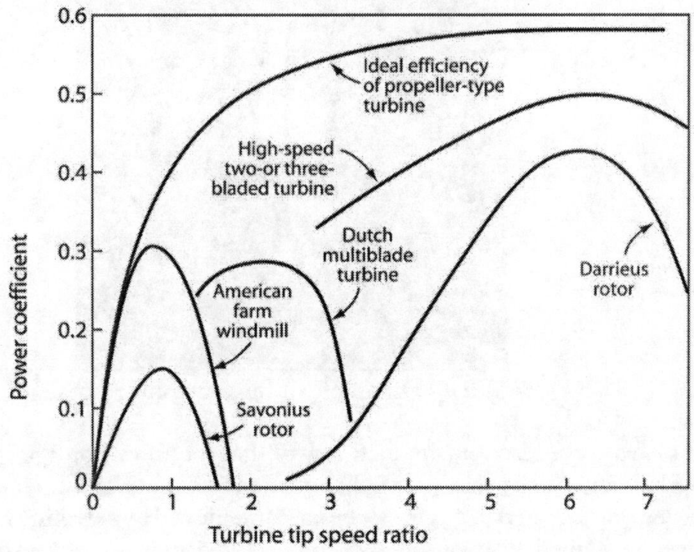

Figure 3. Wind turbine generic performance [7]

Starting characteristic of this type of turbine is however not investigated in greater detail. Measurement of static torque of semi cylindrical blades together with flow visualization were conducted by Sawada et al [5]. The measurement had shown that the static torque varies continuously over the azimuthal angle but had not shown that how that static torque affects the starting behavior of the turbine. In addition, the tests were also conducted at a very low wind speed of 15 cm/s. A more detail study was made by Ushiyama et al [6]. In this study, wind tunnel tests on a model which was made of steel sheet was conducted. A number of design parameters such as aspect ratio and overlap ratio were systematically investigated. The investigation revealed that, for the semi cylindrical blade, the two-bladed rotor with zero gaps and the overlap ratio of 0.2 is the optimal configuration.

It is worth noting that the rotor investigated in the previous research was made of steel, which is heavier than other materials such as aluminum and carbon fiber. With an application of lighter materials, the optimal configuration may be significantly changed as the rotor inertia can be reduced. The understanding of this starting characteristics with appropriate rotor materials is of importance as recent research had shown that the improvement of starting characteristic will increase the turbine operating range and can lead to a significant amount of power output [8].

This paper then experimentally investigates starting and energy-extraction characteristics of the aluminum-made Savonius rotor in order to figure out an optimal configuration. A number of parameters such as number of blades and overlap ratio, which governs the generation of torque, were investigated.

2. The Savonius Rotor

The Savonius rotor employed in this study is configured to be 36 cm width and 45 cm height. It consists of rotor blades that are covered by end plates at the top and the bottom (figure 4). The blades are semi-circular of 15 cm diameter and made of aluminum. The end plates are laser-cut and also made of aluminum. The whole rotor set is attached to a vertical metal shaft which is held by a tower that sits on a stand.

Figure 4. Example of the Savonius rotors

The rotor end plates were laser-cut in such a way that the rotor configuration such as number of blades and the overlap ratio can be easily varied. For example, they were laser-cut radially with a series of holes and the change of the overlap angle can be easily achieved by selecting different radial positions. This study however is limited to two and three blades and one set of the overlap ratio. All tested configurations are presented schematically in figure 5 and table 1 summarizes specification of all rotors.

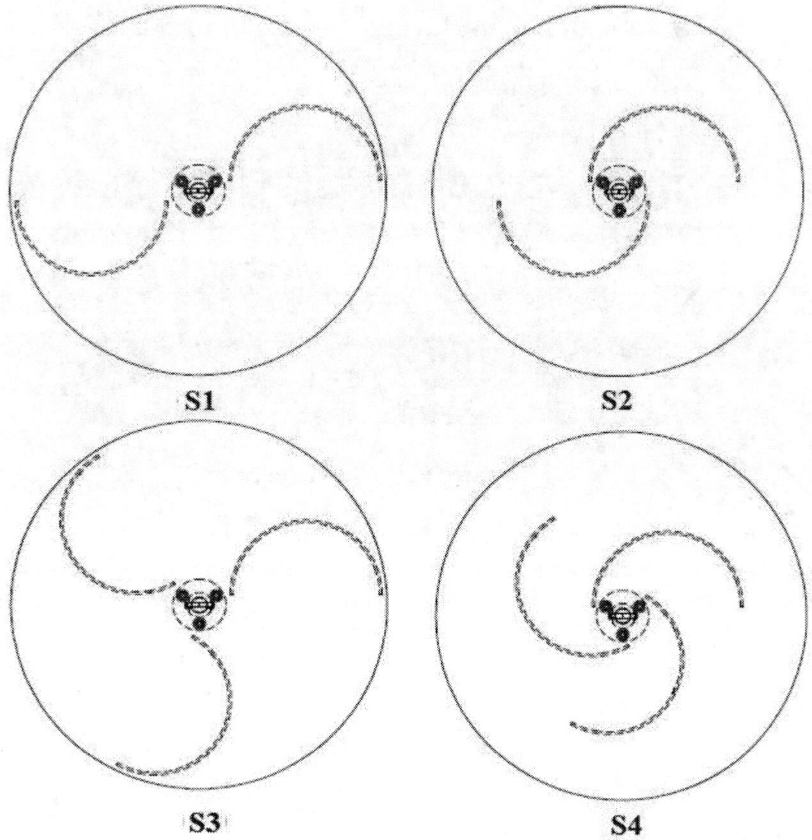

Figure 5. Two and three bladed rotors with and without the overlap angle

Table 1. Rotor Configurations.

Rotors	No. of Blades	Overlap Ratio (%)	Blade Width (cm)	Diameter (cm)	Height (cm)
S1	2	0	15	36	45
S2	2	10	15	24	45
S3	3	0	15	36	45
S4	3	10	15	24	45

3. Testing Facility and Procedure

The wind tunnel used for the experiments was an open-jet type which discharges directly to the atmosphere (figure 6). It has a working section of 0.8 x 0.7 m. Turbulence intensity was found to be 3.5% at the maximum wind speed of 4.45 m/s. The turbine was installed on a stand in front of the tunnel exit (figure 6).

Figure 6. Wind tunnel setup

Starting torque measurement was conducted using a Futek torque sensor. In every measurement, the rotor will be set at a reference point (the azimuthal angle of zero as shown in figure 7). After the measurement completed, the rotor will be rotated counterclockwise by 10 degree to measure starting torque at the next azimuthal angle. This process continues until it completes the revolution. This test was conducted at a wind speed of 4.45 m/s

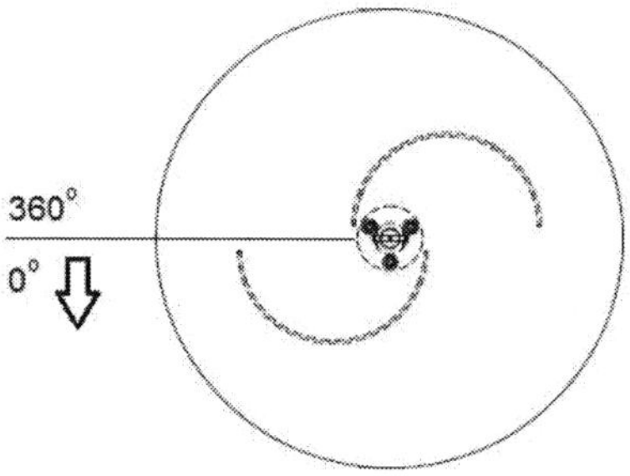

Figure 7. Sign convention

Starting behavior of the rotor also performed at the wind speed of 4.45. In this kind of test, the tunnel will be turned on and the rotor was held stationary. After the wind speed stabilized, the rotor was released and its rotational speed was recorded at a sampling rate of 1 Hz.

Power measurement was also conducted using the Futek sensor by applying a known load to the other end of the torque sensor (figure 8). With this load, the strain gauge senses a net torque that the rotor has to exert to spin the load and a rotational speed that the whole turbine spins at the same time. Power of turbine is the product of the two measured quantities.

Figure 8. Power measurement

4. Results

4.1 Starting Torque

The starting torque of the two-bladed rotors as a function of the azimuthal angle are presented in figure 9.

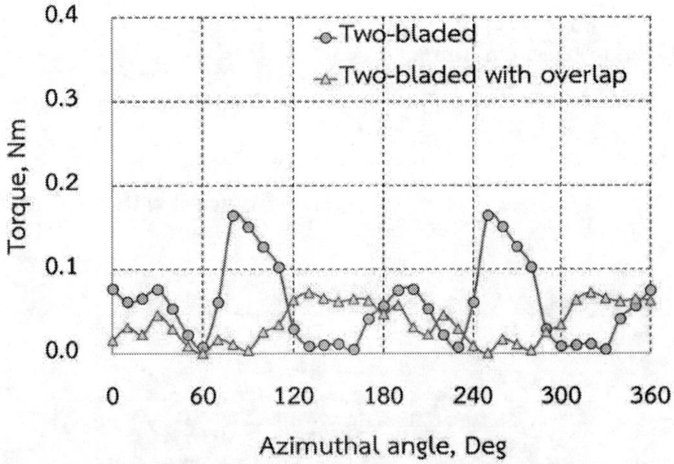

Figure 9. Starting torque: two-bladed

It can be seen that starting torque of the two-bladed rotor is around 0.077 Nm at the reference angle. The torque roughly decreases with the increasing azimuthal angle (as the rotor moves away from the incoming wind) until reaches its lowest value of 0.008 Nm at the azimuthal angle of 60. After that, it increases to the peak of 0.15 Nm at the azimuthal angle of 80 and then reduces as the rotor moves to 180 degree at which the same pattern is repeated. The azimuthally averaged torque of the rotor was found to be 0.083 Nm.

The torque distribution over the azimuthal angle is significantly affected by the incorporation of the overlap ratio, as seen in figure 9. The amount of torque at the reference angle is reduced to 0.015 Nm, compared to 0.077 Nm and this is anticipated to be the result of reduced swept area that shorten the effective area of the rotor blade (figure 10). This results in a reduction of torque over the azimuthal range (figure 9). The average torque is then decreased to 0.039 Nm.

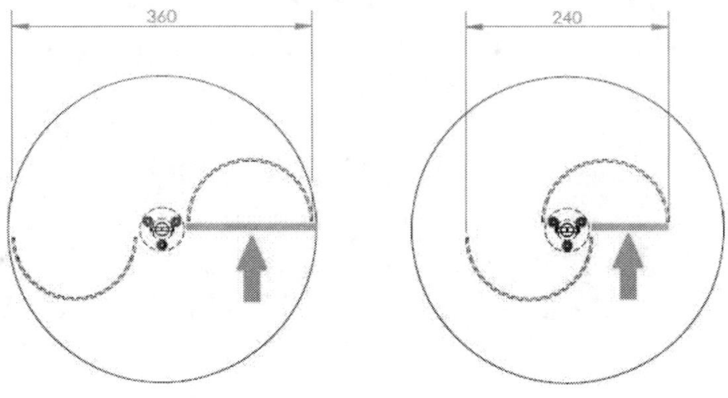

Figure 10. Effect of overlap angle on swept area and net force

The peak of the torque is however less observed in the three-bladed configurations (figure 11). For three-bladed rotor without the overlap angle, the torque value is about 0.027 Nm at the reference point. It is worth noting that this stating torque is lower than that of the two-bladed one and this is thought to be the effect of interference between blades. The maximum torque is reduced to 0.1 Nm and the average torque over the revolution is 0.076 Nm. In this scenario, the airflow will be diverted by the nearby blade. This effect increases when the blade further move away from the incoming wind as shown in figure 12. Although the two-bladed configuration can generate higher torque, it is subjected to a periodic pulsation which will lead to bearing failure.

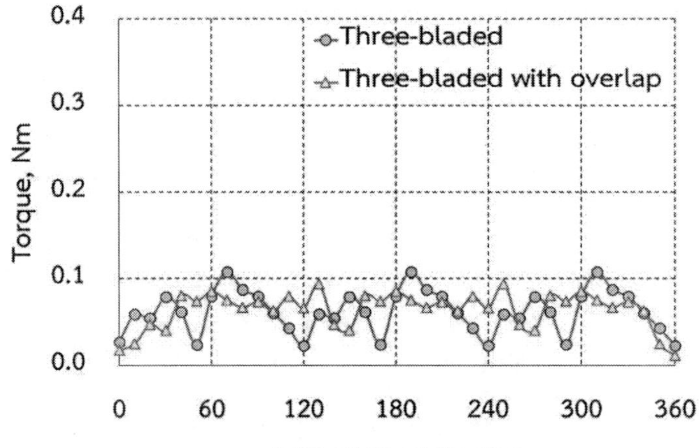

Figure 11. Starting torque: three-bladed

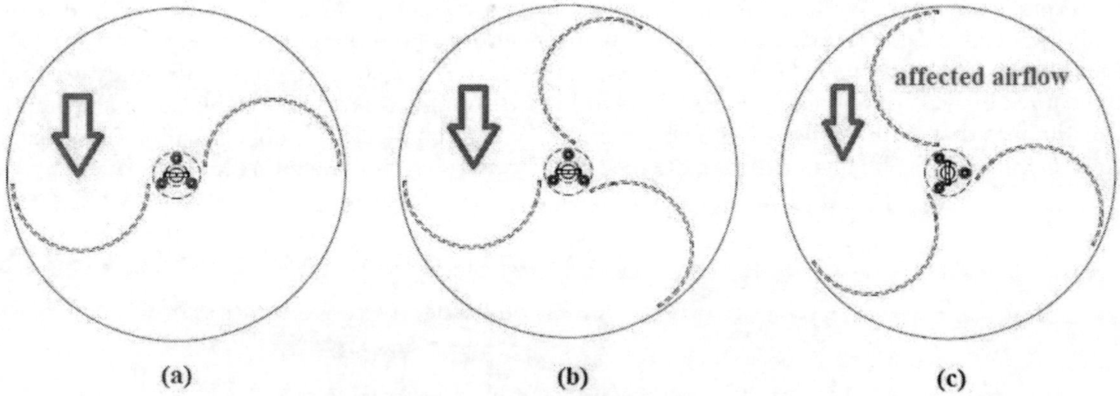

Figure 12. Effect of blade interference

Comparison of torque distributions of the two- and three-bladed rotors in figure 9 and 11 shows that, although the maximum torque of the three-bladed rotor is lower than the two-bladed case, its torque distribution is comparatively flat and does not experience a large drop to a nearly zero value at the azimuthal angles between 130-180, as presented in the two-bladed cases. The incorporation of the overlap ratio again reduces the overall torque generation of the rotor.

4.2. Starting Behavior

Figure 13 compares starting characteristic of the two-bladed rotors and it can be observed that the two-bladed rotor can accelerate to its final rotational speed of around 150 RPM. The incorporate of the overlap ratio, however, decreases the rotor ability to start and the rotor can accelerate to around 110 RPM and takes about 120 seconds to reach its final rotational speed. This reduced performance is the effect of the overall change of the swept area of the rotor that reduced the amount of incoming wind that strike on the rotor blade (figure 10). In addition, the overlap ratio also change the radius of the resultant force on the blade, resulting in a significant reduction of the torque and the degradation of starting capability.

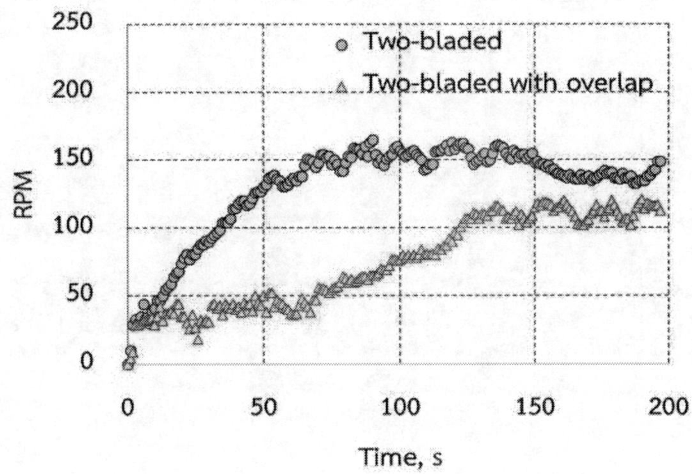

Figure 13. starting characteristic: two-bladed rotors

Similar behavior was found in the three- bladed case (figure 14). The incorporation of the overlap ratio significantly affect the turbine ability to start, especially the time that the rotor takes to reach the final rotational speed. While the three-bladed rotor takes 70 seconds to reach the final RPM, the rotor with the overlap ratio takes up to 200 seconds to reach its equilibrium. The final, operating RPM is reduced from 190 to around 150. It is worth noting that the optimal configuration in this study was found to be three-bladed configuration rather than two-bladed configuration as shown by Ushiyama et al [6]. This implies that, with the use of lighter material, the addition of the third blade will not increase the rotor inertia significantly and can promote the ability to start despite generating lower torque.

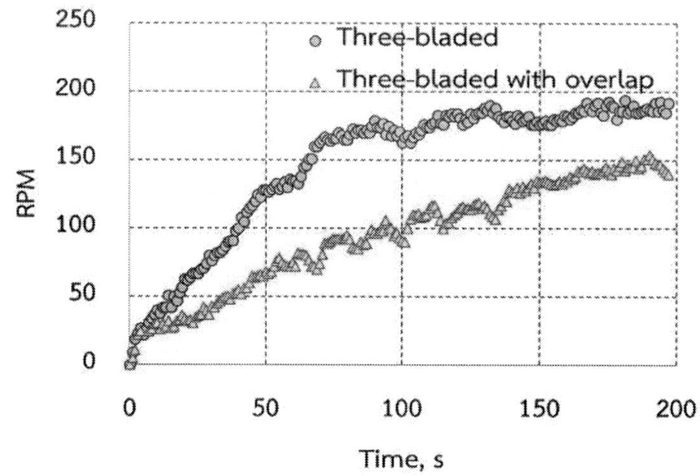

Figure 14. starting characteristic: three-bladed rotors

4.3. Cut-In Speed

With the starting behavior characteristics in the previous section, it is interesting to explore what cut-in speed that each rotor begins to start spinning. This investigation was conducted by increasing the inverter frequency by 5 Hz and rerun the test until the rotor begins spinning and be able to sustain that spinning to a particular rotational speed. It is however worth noting that the torque sensor used in this study has a significant amount of friction and this inevitably hinders the starting of the turbine. The results then compare the effect of rotor configuration on the cut-in speed qualitatively. The real cut-in speeds were not obtained from this kind of measurement. Figure 15 shows the results.

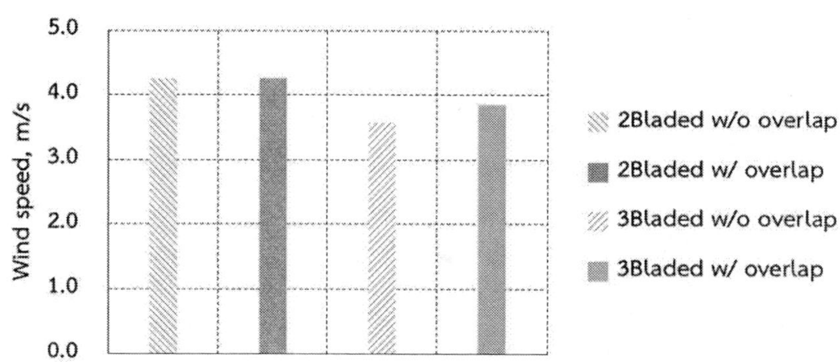

Figure 15. Cut-in wind speed

It can be observed from the figure that the 3 bladed rotor without the overlap ratio has the lowest cut-in wind speed of 3.5 m/s. This characteristic is closely related to the starting torque characteristics of the rotor that the 3 bladed rotor manage to produce a more continuous torque. As expected, the incorporation of overlap ratio, which reduces the rotor swept area and the radius of the net force, causes the rotor to be more difficult to spin and needs a higher wind speed to be able to start.

4.4. Power

Since the three bladed rotor without the overlap ratio has the best performance characteristics, only power coefficient of this turbine is shown. Figure 16 presents its power coefficient as a function of tip speed ratio and

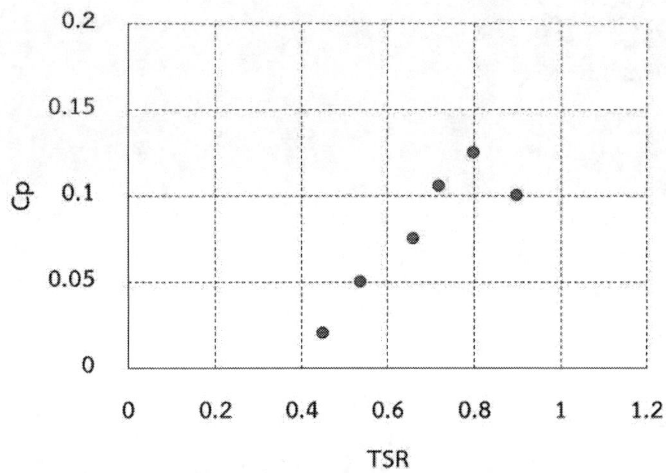

Figure 16. Power coefficient

It can be seen from the figure that the turbine has a maximum power coefficient of 0.125 at a tip speed ratio of 0.79. This maximum power coefficient is lower than the generic performance presented in figure 3 and occurs at a lower tip speed ratio (0.79 compared to around 1 in figure 3). This is anticipated to be the effect of the tested wind speed which is comparatively low. The low wind speed typically lowers the tip speed ratio that the turbine operates and also its power coefficient.

5. Summary

The performance characteristic of the Savonius turbine was experimentally examined and the following conclusions are found:

- The number of blades has a great impact on the starting characteristic. Three bladed rotor will always produce a more continuous torque, on both with and without the overlap ratio
- The overlap ratio generally reduces the torque generated by the turbine and hence its starting capability, as the swept area is smaller.
- The use of lightweight materials is beneficial and the optimal configuration might be different if different materials are used.
- Of investigated, the three-bladed rotor without the overlap ratio exhibits the highest starting capability and power extraction performance.

References

[1] Savonius SJ 1925 *The Wing Rotor in Theory and Practice* (Finland: Savonius and Co, Helsingfor)
[2] Saha UK, Thotla S and Maity D 2008 *J. of Wind Engineering and Industrial Aerodynamics* **96** pp 1359-1375
[3] Ali MH 2013 *Int. J. of Modern Engineering Research (IJMER)* **3**(5) pp 2978-2986
[4] Wenehenubun F, Saputra A and Sutanto H 2015 *Energy Procedia* **68** pp 297-304
[5] Sawada T, Nakamura M and Kamada S 1986 *Bulletin of JSME* **29**(253) pp 2095-2099
[6] Ushiyama I, Nagai H and Shinoda J 1986 *Bulletin of JSME* **29**(258) pp 4130-4138
[7] Eldridge FR 1980 *Wind Machines* 2nd Ed., the MITRE Energy Resources and Environmental Series, (Van Nostrand Reinhold Company)
[8] Worasinchai S, Ingram GL and Dominy RG 2012 *ASME J. of Engineering for Gas Turbines and Power* **134**(4) 042603

A parametric simulation of solar chimney power plant

Lim Beng Hooi, and Saravana Kannan Thangavelu*
Faculty of Engineering, Computing and Science, Swinburne University of Technology, 93300 Kuching, Sarawak, Malaysia

* Corresponding Author: sthangavelu@swinburne.edu.my

Abstract. The strong solar radiation, continuous supplies of sunlight and environmental friendly factors have made the solar chimney power plant becoming highly feasible to build in Malaysia. Solar chimney power plant produces upward buoyancy force through the greenhouse effect. Numerical simulation was performed on the model of a solar chimney power plant using the ANSYS Fluent software by applying standard k-epsilon turbulence model and discrete ordinates (DO) radiation model to solve the relevant equations. A parametric study was carried out to evaluate the performance of solar chimney power plant, which focused on the temperature rise in the collector, air velocity at the chimney base, and pressure drop inside the chimney were based on the results of temperature, velocity, and static pressure distributions. The results demonstrate reliability by comparing a model with the experimental data of Manzanares Spanish prototype. Based on the numerical results, power capacity and efficiency were analysed theoretically. Results indicate that a stronger solar radiation and larger prototype will improve the performance of solar chimney power plant.

1. Introduction

Malaysia is situated on the South China Sea that lies between 1° and 7° in North latitude and 100° and 120° in East longitude. The annual climate condition of Malaysia does not have a distinct seasonal variation which results in receiving almost equal amount of sunshine daily throughout the year. Theoretically, Malaysia should have high potential for harvesting solar energy. However, the solar radiation received is generally low due to the high presence of cloud [1].

Malaysia is an ideal country for running large-scale solar power system due to its geographic position being located at the equator of earth. The strong solar radiation, continuous supplies of sunlight, lower carbon emissions and environmental friendly factors have made the solar chimney power plant become highly feasible to build in Malaysia. Research shows that the measured monthly average daily solar radiation of Malaysia falls in the range of 600 W/m^2 and 1000 W/m^2 [2]. The average daily sunshine duration was found to be in the range of 4 to 8 hours per day with an average daily temperature of 26.5°C [3]. Hence, solar chimney power plant is very practicable to be developed in the climatic conditions of Malaysia.

Solar chimney produces electrical power from solar radiation. The system combines three components, which consists of a huge greenhouse collector, a chimney that placed in the center, and one or multiple wind turbines that is connected to the generators below the chimney (figure 1). Generally, the principle of solar chimney is based on the idea of natural convection that using hot air produced by sun lights to drive the turbines. The conversion between solar radiations to electricity involves two stages. During the first stage, thermal energy is produced by means of greenhouse effect.

This can be done by placing a collector above the ground to trap the radiant energy. In the second stage, the hot air generated by the radiant energy rise up in the chimney due to buoyancy effect. Thus, the movement of air drives the turbines at the base of chimney to generate the electricity [4-9].

In 1998, Sherif and Pasumarthi have carried out several experiments to examine the velocity and temperature distributions of the airflow inside the chimney [4]. The method of setting up the solar chimney is by modifying the shape and the radius of collector. All experimental results were found to be agreed within the calculation of mathematical model. On the other hand, Schlaich claimed that the output power of solar chimney is proportional to the height of chimney and the area of collector [5]. Besides, it was mentioned that solar chimney has poor efficiency because only a small amount of solar radiation can be converted into useful energy [5].

Nonetheless, many researchers have performed evaluation on the mathematical models of solar chimney power plant based on both theoretical and numerical analysis [6]. Moreover, Ming et al. [7] have developed different types of mathematical models for the solar collector, solar chimney and energy storage layer. The research has analyzed the characteristics of air flow in the solar chimney with energy storage layer. As a result, the effects of several parameters such as static pressure, driving force, output power, and efficiency were evaluated. Results show that the relative static pressure decreased, whereas the velocity increased significantly inside the chimney with the increase of solar radiation.

In recent years, Pastohr et al. [8] have performed study on solar chimney with two dimensional steady state simulations. The research has analyzed the energy flow on the collector, turbine, and the chimney. In the research, the profile of temperature, pressure, and velocity in the collector have been identified. Results show that when the collector radius is constant, the increase of solar radiation will increase the air velocity. Furthermore, the results of analytical calculations have shown a good agreement with the fluent results. Similarly, Sangi et al. [9] have performed two-dimensional axisymmetric numerical simulation of solar chimney using standard k-ε model and standard wall mode to describe the air flow. A good agreement was obtained between numerical results and the experimental data of Spanish prototype. However, further investigation is still needed for different sizes of solar chimney power plant.

In fact, Cao et al. [10] have presented the simulation of sloped solar chimney with respect to different chimney height using the validated numerical model. A conventional Spanish prototype was first simulated to observe the agreement between experimental data and the Fluent result. Accordingly, the numerical model is then used to simulate solar chimney in other conditions.

In this research, parametric study was conducted on solar chimney of different geometrical size to determine the performance of solar chimney power plant under different conditions of solar radiations, so that the power capacity could be evaluated. This study was performed using ANSYS Fluent software.

2. Methodology

As the size of solar chimney power plant will result in significant effect on the velocity and temperature rise, three different dimensions of design have been proposed to observe the difference. It should be noted that the dimensions of solar chimney used in second design (Case 2) were based on the parameters of Spanish prototype [8]. Meanwhile, other designs of solar chimney (Case 1 & 3) have been drawn based on the ratio of Spanish prototype which results in three different geometrical dimensions of solar chimney. The main dimensions of solar chimney prototype are given in table 1. Case 1 & 3 prototypes were drawn to consider minimum and maximum optimum size, respectively.

Table 1. Main dimensions of solar chimney prototype

Solar Chimney	Chimney Height (m)	Chimney Radius (m)	Collector Radius (m)
Case 1	100	3	75
Case 2	195	5	120
Case 3	500	15	400

The solar chimney power plant was divided into three parts, which consists of the collector, transition section in the middle, and the chimney tube. In ANSYS Fluent, the meshing process on different part was generated separately and independently. Since the transition section is the most sensitive part in simulation, it requires a finer mesh, which tetrahedral meshes were used. Meanwhile, the collector and the chimney tube were meshed using hexahedral meshes. The grid generation of solar chimney power plant is shown in figure 1. Boussinesq model was selected to describe the air flow. The change of density was only considered in the momentum equation. Besides, the pressure boundary condition was used to simulate the natural convection flow in the chimney. Since both convection and radiation exist in the transparent covering of collector, discrete ordinates (DO) radiation model and convection heat transfer boundary were applied. The mean daily solar radiation in Malaysia was found to be around 800 W/m^2 with the minimum solar radiation around 600 W/m^2 and the maximum solar radiation around 1000 W/m^2. Hence, these values were used to apply in the simulations. On the other hand, the ground is treated as an inner heat source. The setting has considered the transmission absorptivity because the ground surface absorbs solar energy and the energy also converts to the cover of collector. For the turbulent flow conditions, the standard k–ε model and standard wall mode were selected to describe the fluid flow. All the numerical calculations are selected to be performed with double precision by the solver. Table 2 shows the boundary conditions setting for simulation.

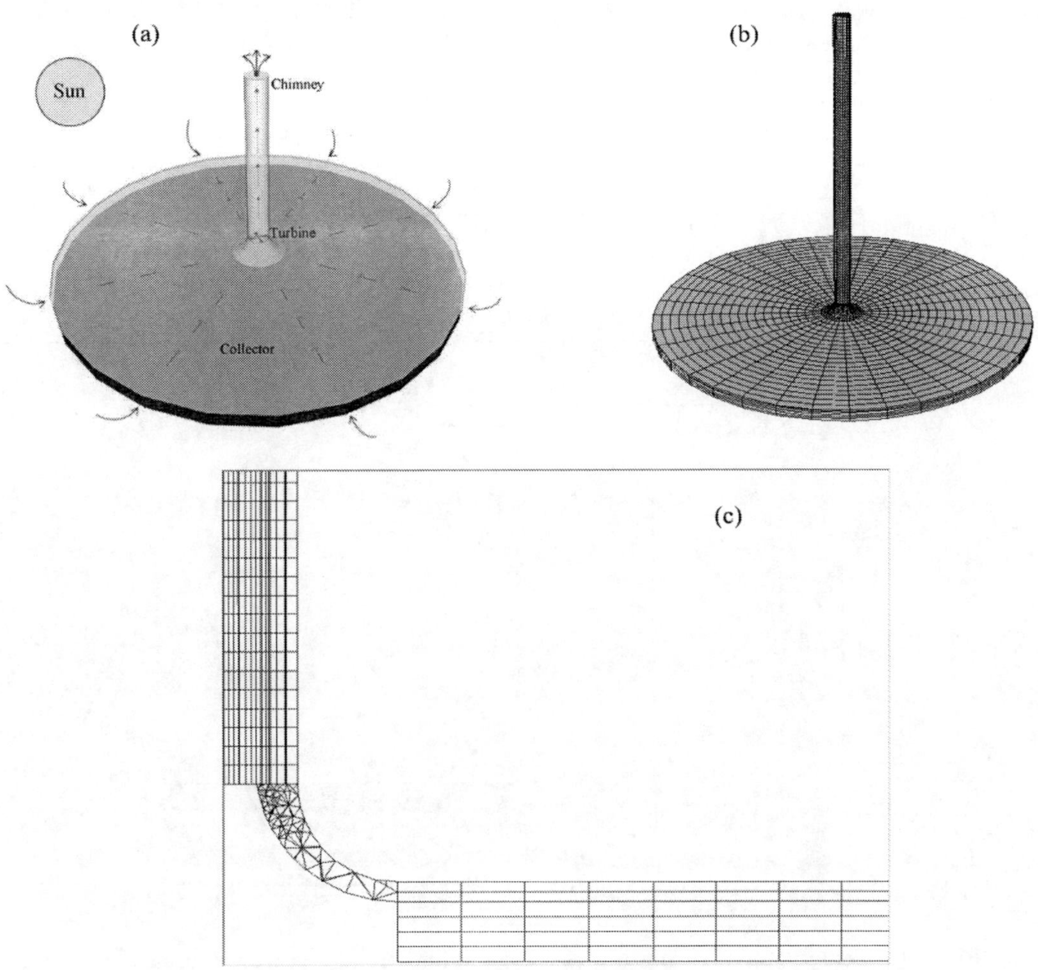

Figure 1. Solar chimney power plant (a) schematic illustration [9] (b) grid generation overall view (c) grid generation section view

IOP Conf. Series: Materials Science and Engineering **297** (2018) 012057 doi:10.1088/1757-899X/297/1/012057

Table 2. Setting of main boundary condition

Surface	Type	Value
Collector inlet	Pressure inlet	$\Delta P = 0$ Pa & $T = 299.5$ K
Chimney outlet	Pressure outlet	$\Delta P = 0$ Pa
Chimney wall	Adiabatic wall	Heat flux = 0 W/m^2
Transition section	Adiabatic wall	Heat flux = 0 W/m^2
Collector	Wall (Semi-transparent)	Heat transfer coefficient = 8 W/m^{2}K
Ground surface	Wall	$T = 299.5$ K

3. Results and discussion

To validate the numerical model, the temperature rises in the collector and the upwind velocity at the chimney inlet of Case 2 prototype are compared with the experimental data of Spanish prototype. Experimental data of Spanish prototype shows that when the solar radiation is 1000 W/m^2, the temperature increase through the collector reaches 20 K, and the upwind velocity at the base of chimney have a value of 15 m/s under no load condition [9]. The temperature, velocity, and pressure distribution of solar chimney power plant are shown in figure 2, figure 3 and figure 4, respectively. Based on the test of validation in table 3, a good quantitative agreement was obtained between the experimental data and the result of numerical simulations. In addition, benchmarking of this simulation results were conducted by comparing to previous work [10] and [11]. It can be concluded that the reliability of the simulation results in this research project is proven. Therefore, the developed method including boundary conditions is feasible to simulate the other prototypes.

Table 3. Comparison between the experimental data and numerical result

Results	Temperature Increase (K)	Upwind Velocity (m/s)
Experimental	20	15
This work	21.9	15.7
Asnaghi and Ladjevardi [11]	21.8	16.2
Cao et al. [10]	21.5	14.33

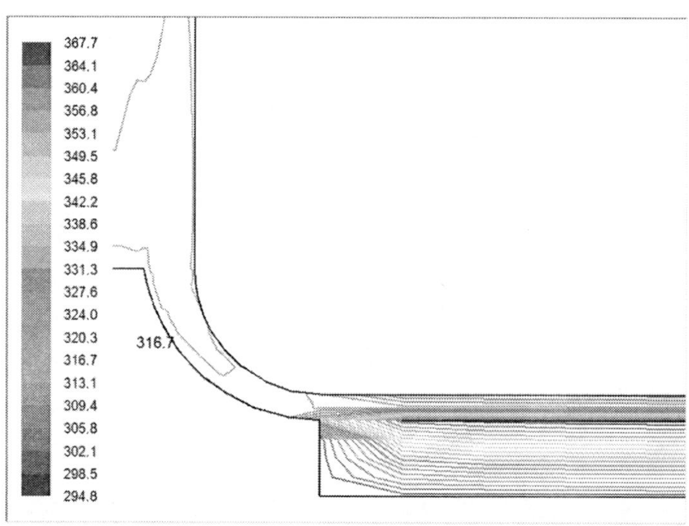

Figure 2. Temperature distribution of Case 2 solar chimney power plant at 1000 W/m^2

Figure 2 shows the temperature distribution of Case 2 power plant when the solar radiation is 1000 W/m^2. The figure shows that the temperature of air rise through the collector reaches about 21.9 K. Figure 3 shows the velocity distribution of the solar chimney. When the solar radiation is 1000 W/m^2, the maximum velocity inside the solar chimney is about 15.7 m/s. It is illustrated that the velocity of air increases through the collector and the maximum velocity is at the chimney base. The static pressure distribution of the solar chimney is shown in figure 4. Results show that the static pressure decreases through the collector up to -122.3 Pa and starts increasing through the chimney up to 0 Pa. It is observed that the minimum static pressure lies at the bottom of the chimney.

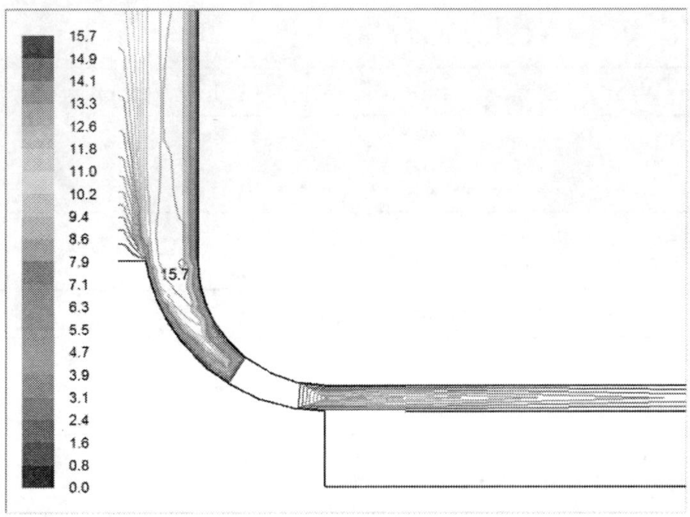

Figure 3. Velocity distribution of Case 2 solar chimney power plant at 1000 W/m^2

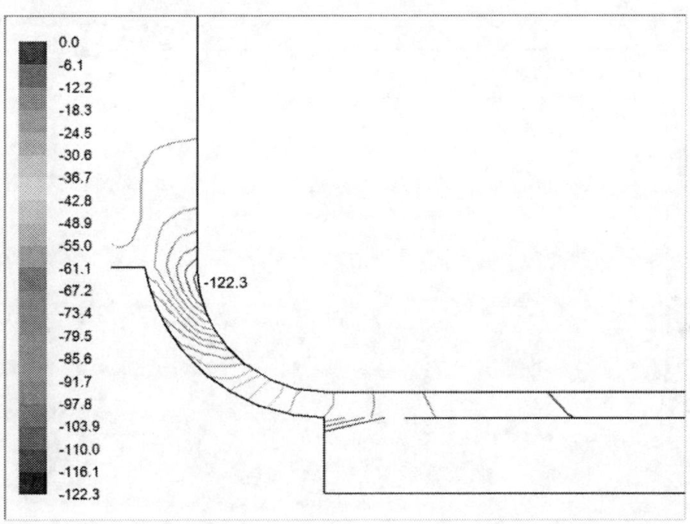

Figure 4. Static pressure distribution of Case 2 solar chimney power plant at 1000 W/m^2

Table 4 to 6 shows the influence of different solar radiation and geometrical dimensions of solar chimney on the numerical results (Case 1 to 3). Based on the numerical results of temperature rise in collector, airflow velocity at chimney, and pressure drop inside the chimney as function of the solar radiation ranging from 600 to 1000 W/m^2, the results show that when the geometrical dimensions are constant, the raising of solar radiation increases the air temperature and velocity while decreasing the static pressure in the chimney. This is because more heat is absorbed and the increase of the heat quantity causing the air temperature to rise while dropping the air density at the same time. The decrease of density increases the pressure difference which results in more air flowing through the chimney. Similarly, when the solar radiation is constant, the increase of geometrical dimensions causes the air temperature to increase which results in the decrease of air density. Hence, the buoyancy differences increases which produces greater driving force and thus the pressure drop. Thereby, the increase of driving forces results in a higher velocity of air inside the chimney. Figure 5 shows the velocity distribution of Case 3 prototype solar chimney power plant (maximum of 27 m/s).

Table 4. Simulation results of Case 1 prototype

Solar Radiation (W/m^2)	Temperature Rise (K)	Airflow Velocity (m/s)	Pressure Drop (Pa)
600	14.5	9.8	51.1
800	16.5	11.1	60.2
1000	18.5	12.1	68.7

Table 5. Simulation results of Case 2 prototype

Solar Radiation (W/m^2)	Temperature Rise (K)	Airflow Velocity (m/s)	Pressure Drop (Pa)
600	16.2	12.4	86.6
800	18.4	14.3	105.7
1000	21.9	15.7	122.3

Table 6. Simulation results of Case 3 prototype

Solar Radiation (W/m^2)	Temperature Rise (K)	Airflow Velocity (m/s)	Pressure Drop (Pa)
600	22.2	22.0	250.2
800	24.0	24.7	301.3
1000	26.6	27.0	350.4

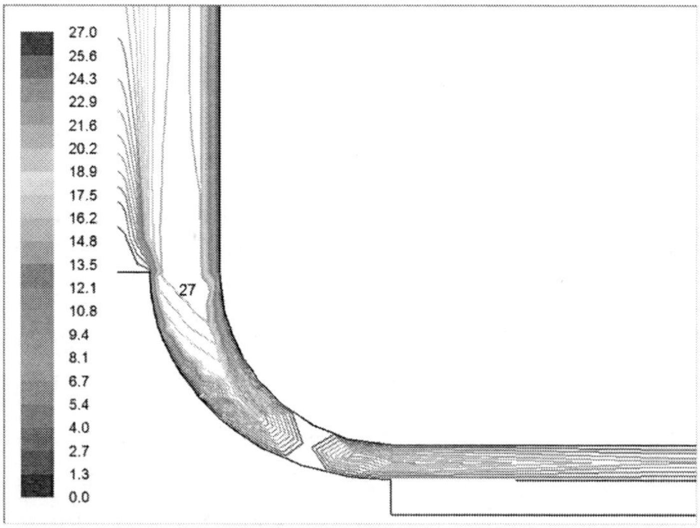

Figure 5. Velocity distribution of Case 3 prototype at 1000 W/m^2

Table 7 to 9 shows the power capacity and efficiency of the solar chimney from Case 1 to 3, respectively. The power generated by turbine and the overall system efficiency are analyzed by,

$$P_{out} = \eta_t \Delta P_t V_{ch} A_{ch} \tag{1}$$

$$\eta = \frac{P_{out}}{A_{gro} q} \tag{2}$$

Where, ΔP_t is the driving forces to impel the air flow, and A_{gro} is the area of ground [10].

Table 7. Performance evaluation of Case 1 prototype

Solar Radiation (W/m^2)	Power Capacity (kW)	Efficiency (%)
600	32.64	0.12
800	43.56	0.12
1000	54.19	0.12

The mentioned equation was used to analyze the performance of solar chimneys theoretically. Results show that the power capacity of solar chimney increases with higher solar radiations and larger geometrical dimensions. When the chimney height reaches 500 m with collector radius of 400 m, the efficiency surpasses 1%. The solar chimney power plant was estimated to produce power of about 600 kW, under the solar radiation of 1000 W/m^2. Power generation could further be increased if the chimney height increases, which results in improving the efficiency.

Table 8. Performance evaluation of Case 2 prototype

Solar Radiation (W/m^2)	Power Capacity (kW)	Efficiency (%)
600	70.00	0.26
800	98.53	0.27
1000	125.17	0.28

Table 9. Performance evaluation of Case 3 prototype

Solar Radiation (W/m^2)	Power Capacity (kW)	Efficiency (%)
600	358.82	1.32
800	485.14	1.34
1000	616.73	1.36

The main reason that causing the low efficiencies of solar chimney is due to the heat loss in the turbine, whereby the heat energy generated from the solar energy could not transfer well into work energy during the isentropic expansion process. Despite the efficiency of solar chimney is very low, it still able to generate over 100 kW when the efficiency remains less than 1%. Considering that solar chimney power plant does not consume fossil energy, the materials used to build the plant can be easily assessed which initiated the investment for commercial application. From an economic viewpoint, it is well known that these power plants were only considered to be constructing in desert regions where the land price is very cheap. Nonetheless, the power plant is not only used in power production, on the other hand the space under the collector roof can also be planned for agricultural purposes (drying).

4. Conclusion

This paper presents a description of the temperature rise, air velocity, pressure drop, power output and efficiency as dependent on the effects of solar radiations and geometrical dimensions of solar chimney power plant. The distributions of the temperature, velocity and static pressure were illustrated for a solar chimney power plant model, which reported reasonable agreements between the numerical results and experimental data of temperature difference and upwind velocity. Other models result should be validated in future work as well, while more parameters can be explored without time constraint. Based on the results from parametric study, it can be concluded that the performance of solar chimney depends significantly on its geometric dimensions and the intensity of solar radiation to some extent. Further investigations are required to improve the efficiency of the solar chimney power plant. The effectiveness of geometry such as the ratio between chimney height, chimney radius, and collector radius of the solar chimney has to be investigated. The optimization and chimney modification of the solar prototype models are also required. Additionally, the effectiveness of material for the solar chimney need to be examined to improve the airflow velocity and temperature.

References

[1] Baharin K A, Abd Rahman H, Hassan, M Y and Kim GC 2013 *Proc. IEEE Student Conf. on Research and Development (SCOReD)* p 191

[2] Sopian K and Othman M Y H 1992 *Renewable Energy*, **2** p 319

[3] Mekhilef S, Safari A, Mustaffa W E S, Saidur R, Omar R and Younis M A A 2012 *Renewable and Sustainable Energy Reviews* **16** p 386

[4] Pasumarthi N and Sherif S A 1998 *Int. J. of Energy Research* **22** p 443.

[5] Schlaich J 1995 *The solar chimney: electricity from the sun, Axel Menges*

[6] Zhou X, Yang J, Xiao B and Hou G 2007 *Applied Thermal Engineering* **27** p 2044

[7] Tingzhen M, Wei L and Guoliang X 2006 *Int. J. of Energy Research* **30** p 861

[8] Pastohr H, Kornadt O and Gürlebeck K 2004 *Int. J. of Energy Research* **28** p 495

[9] Sangi R, Amidpour M and Hosseinizadeh B 2011 *Solar Energy* **85 p** 829

[10] Cao F, Li H, Zhang Y and Zhao L 2013 *The Scientific World Journal* **2013** p 8

[11] Asnaghi A and Ladjevardi S M 2012 *Renewable and Sustainable Energy Reviews* **16** p 3383

[12] Sangi R 2012 *Renewable and Sustainable Energy Reviews* **16** p 704

IOP Conf. Series: Materials Science and Engineering **297** (2018) 012058 doi:10.1088/1757-899X/297/1/012058

A novel direct-fired porous-medium boiler

Boonrit Prasartkaew*

Combustion and Solar Energy Laboratory (CASE Lab.), Faculty of Engineering, Rajamangala University of Technology Thanyaburi, Klong 6, Thanyaburi, Patumthani 12110, Thailand

* Corresponding Author: boonrit.p@en.rmutt.ac.th

Abstract. Nowadays, power and heat generation systems pay an important role in all economic sectors. These systems are mainly based on combustion reaction and operated under the second law of thermodynamics. A conventional boilers, a main component of heat and power generators, have thermal efficiency in the range of 70 to 85%, mainly owing to they have flue gas heat loss. This paper proposes a novel type of boiler, called a Direct-fired Porous-medium Boiler (DPB). Due to being operated without flue gas heat loss, its thermal efficiency cloud be approximately close to 100%. The steam produced from the proposed boiler; however, is not pure water steam. It is the composite gases of steam and combustion-product-gases. This paper aims at presenting the working concept and reporting the experimental results on the performance of the proposed boiler. The experiments of various operating parameters were performed and collected data were used for the performance analysis. The experimental results demonstrated that the proposed boiler can be operated as well as the conceptual design and then it is promising. It can be possibly further developed to be a high efficiency boiler by means of reducing or suppressing the surface heat loss with better insulator and/or refractory lined.

1. Introduction

The energy and environmental issues become the most serious problem of all lives on the earth and it also keeps on dramatically increasing and will not be remedied easily. One of the best ways to overcome aforesaid problems is utilization of renewable energy via the high efficiency systems with environmental friendly processes [1, 2].

Up to the present day, there are three main types of typical boiler designs: fire tube, water tube and tubeless boilers. These can be deeply classified as: fire tube, vertical tubeless, electric, water tube, cast iron and copper fin types. Currently, boilers are manufactured in many different sizes and configurations depending on the fuel used, specified steam output, and required emissions levels. Inside these boiler (excludes electric one), the combustion or oxidation process is taken place, oxygen (from supplied air) reacts with carbon, hydrogen, and other elements in the fuel to generate a flame and hot combustion gases. These hot gases are drawn for one pass or multi-pass through the boiler. Regarding all of the aforesaid boilers, heat is transferred indirectly to water via conduction through surface or wall. Eventually the hot combustion gases flow through a stack and into the atmosphere. To obtain higher efficiency, the components of boiler are designed to promote efficient combustion, heat transfer and control emission. Literatures revealed that a substantial amount of energy is wasted through high temperature flue gas of the boiler. In typical boilers, an economizer might have conducted for a large size boiler with higher investment cost. At their flue gas temperature ranges from 150 to 250 °C, their thermal efficiency is in the range of 70 to 85% [3, 4]. In addition, nowadays, the environmental issues:

global warming, climate changing and serious disasters are mainly attributed to the combustion of such systems and these serious problems must be overcame as soon as possible.

From the literature, there are many works regarding the industrial, commercial, and institutional boilers (ICI boilers). It can be summarized that no one has proposed about this novel concept of boiler, called a 'Direct-fired Porous-medium Boiler' (DPB), as review reported in [5]. It should be noted that, there are some differences between the proposed boiler and the conventional boilers (includes tubeless boiler), such as: no flue gas, no stack, direct heating process without effective wall/surface between hot combusted gases and water, and porous medium (PM) is used for enhancing the heat transfer and recovery. Regarding the PM application in boiler system, there are very few researches about this, as in [6] and still studied on a laboratory-scale test. It can be expected that this study will full-fill the aforesaid research gap.

With the rational hypothesis and expected performance, the proposed DPB might outperforms the conventional one with many expected advantages, such as:

- High thermal efficiency, the flue gas loss is suppressed
- It is compact in construction, directly heating and mixing without barrier surface
- It requires short time for rising steam, the required residual water for operation is quite small
- Low surface heat loss, with water jacket and lesser outer surface area
- Low emission and particulate matter, more uniform temperature and mixed with water vapor
- Higher superheated temperature, with a compact PM super-heater
- Safety, low amount of hot residual liquid water and no flame to wall contact problem
- High dryness fraction, direct heating with controlled water feed rate
- No stack and economizer, combustion gases mixes and leaves as useful high temperature mixture
- Can be possibly developed for a novel internal combustion steam engine
- Low cost, not complicate, etc.

This current paper aims to presents the working concept of DPB and reports the experimental investigation on its performance to assessing the development feasibility of this new porous burner boiler concept, as the above expected features.

2. Working concept of a novel DPB

The proposed DPB is a double-walled reactor with water jacket, as shown in figure 1. The main different between DPB and conventional boiler is that the hot combustion gasses and the produced steam are mixed and heated before leaving as the useful thermal fluid without flue gas loss. To achieve this novel concept, the porous medium and pressurized flame are conducted.

The working concept of the DPB can be explained as the following 4 steps:

1) There are two exit pipes with two stop valves: bigger stop-valve (V1), for flue-gas or wet steam and smaller stop-valve (V2), for superheated composite-gas. During the cold starting step, these valves must be opened to permit the flue gas easily exit the confined vessel. After the fuel and air are mixed and supplied to the pressurized combustor, the fuel-air mixture will be ignited by a spark plug and the pressurized flame will be taken place inside the combustion chamber. The combustion chamber is partitioned or allocated by the porous medium matrix. This porous medium matrix will act as an 'optical thermal wall' to keep the proper condition for a sustainable combustion under higher pressure and moisture condition than usual.
2) After the porous medium is heated as much as possible (depends on the melting point of porous medium used), this heat will be transferred from flame to water inside the water jacket effectively via porous medium. After the water temperature inside the water jacket (T3 as shown in figure 2) reached the saturated temperature at any required steam pressure, then open a vapor feed control valve (V3) while a liquid feed control valve (V4) must be closed. The feed water pump will be turned on whenever the water level is lower than the set lower level, to maintain the constant water level. During this step, V4 may be opened to increase the steam generation rate. This wet steam (at about 1atm) will mix with the combustion hot gases, form more dry steam,

and exit through V1. This composite gases (CG) of steam and combustion hot gasses can be utilized as the useful thermal fluid suitable for indirect heating processes.

3) To get the high pressure superheated CG, V3 and V4 have to be closed. The water-jacket pressure and temperature will be increased. When this pressure reach at about the required steam pressure, which can be seen at the water-jacket pressure gage (P1), then V1 must be closed and V3 must be opened. The wet steam and hot gases are mixed together at the upper region of boiler, far from the optical thermal wall, therefore the steam or moisture will not disturb the flame stability. Finally, the CG are forced to flow through the counter-flow helical coil super-heater and exit the boiler through V2 as the useful superheated thermal fluid suitable for indirect heating processes or power generation at high steam pressure.
4) The steam pressure depends upon the steam consumption rate and fuel-air mixture feed rate or combustion rate. For the steam consumption rate, V2 can be adjusted for the individual boiler test. The combustion rate will be controlled to conserve the constant steam pressure. This combustion rate control may be achieved by two methods: a) a single combustor with motorized flow rate control valves of fuel and air, and b) multi-combustor with pressurize pilot flame and combustor on-off control.

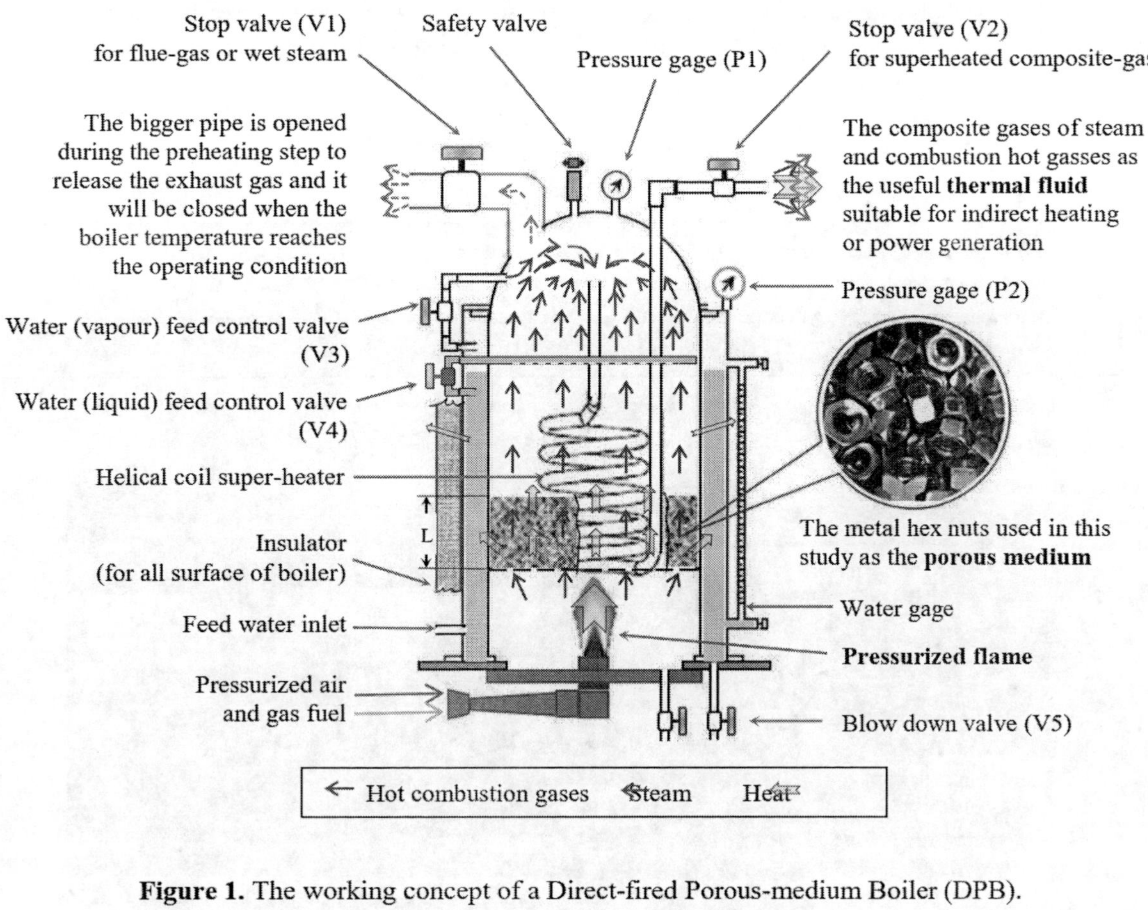

Figure 1. The working concept of a Direct-fired Porous-medium Boiler (DPB).

3. Experimental setup

To assess the feasibility of the proposed system practicality, the prototype of DPB was fabricated and tested. In this study, the 25kg metal hex nuts were used as the porous medium, with the porous medium thickness ('L' as shown in figure 1) of 7 cm. Figures 2 shows a schematic diagram and photo of the experimental setup used in this study. The experimental setup was designed, concerning the three-phase heat transfer mechanism, to allow the gas and liquid flow in and out of the system to be controlled and measured. In this feasibility test, the flow rate and pressure were controlled manually so that operation over a range of power heat input can be explored. To maintain the constant combustion rate, the LPG and air flow rates were manually controlled for all over the experiments. All temperatures were recorded every ten-second intervals via data acquisition (DAQ) system, comprised of a data logger (Datataker, DT605 model) equipped with K-type thermocouples, and computer software for system monitoring. While the flow rate, pressure and feed water consumption rate were manually measured and recorded.

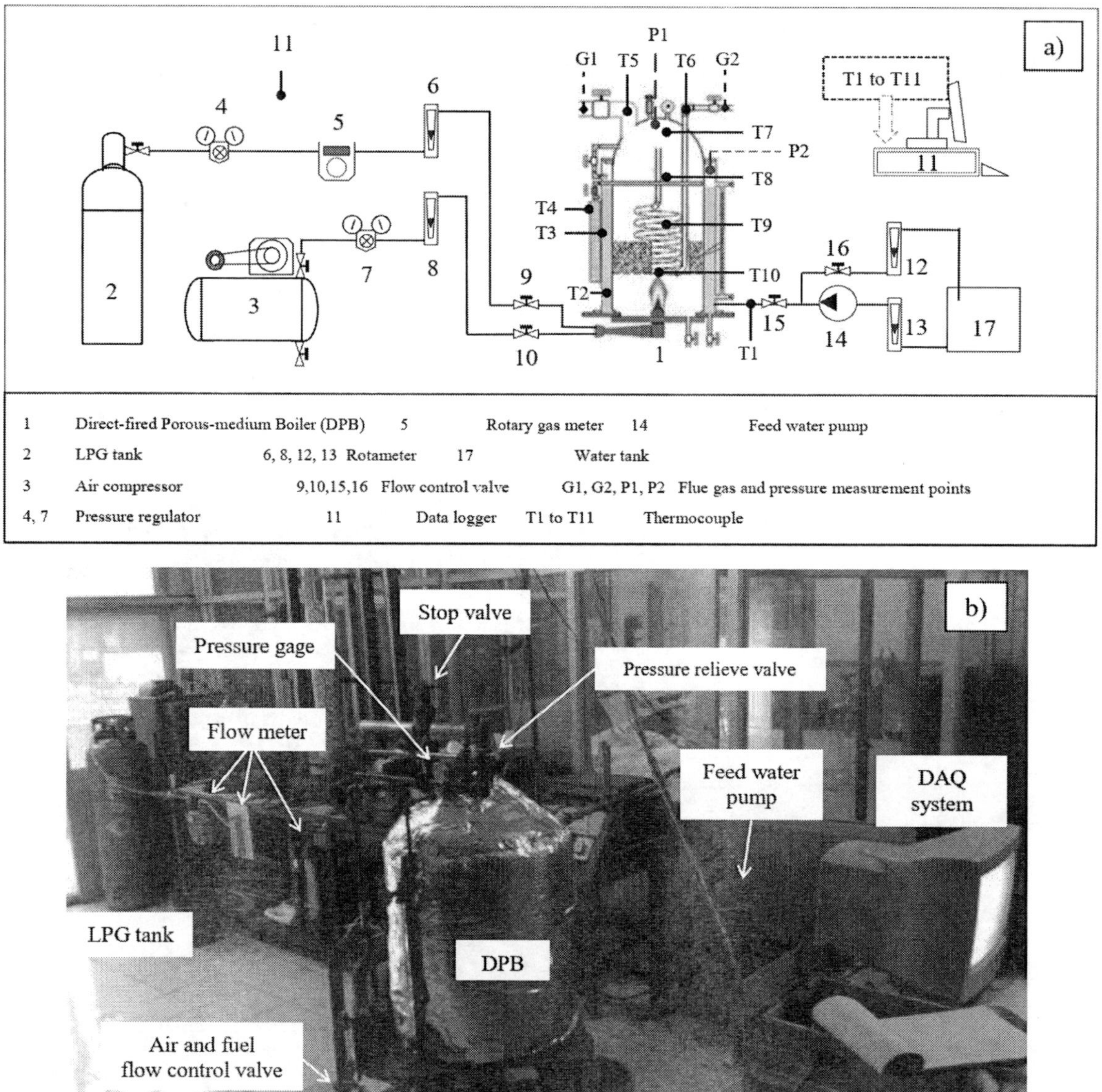

Figure 2. The experimental setup: a) schematic diagram and b) photo of the proposed boiler system.

IOP Conf. Series: Materials Science and Engineering **297** (2018) 012058 doi:10.1088/1757-899X/297/1/012058

4. Performance evaluation

For the boiler performance analysis, let us consider the DPB system as shown in figure 1. To simplify the analysis, the following assumptions are made:

1) Performance is in steady state
2) There is only surface heat loss with the natural convection only
3) The insulator surface is at uniform temperature
4) Properties of all working mediums are independent of temperature
5) The fuel is completely combusted inside the boiler
6) No radiating loss occurs from low temperature insulator surface.

To know the efficiency of a general boiler, the ratio of useful heat exported by outlet steam ($\dot{Q}_u$) to the total input heat supplied by fuel ($\dot{Q}_i$) was calculated as the following equations:

$$\eta_{boiler} = (\dot{Q}_u/\dot{Q}_i) \times 100\% \tag{1}$$

As same as the other thermal systems, the overall energy balance equation at steady state can be expressed as:

$$\dot{Q}_u = \dot{Q}_i - \dot{Q}_l \tag{2}$$

where $\dot{Q}_u$ = useful heat output [kW]
$\dot{Q}_i$ = input heat supplied by fuel [kW]
$\dot{Q}_l$ = heat losses [kW]

Owing to the useful heat of the proposed boiler is not exported by the pure steam, but it is the gas mixtures or composite gas of steam and combustion hot gases. Therefore, between the useful heat and heat loss, the later quantity with the above assumptions can then be estimated easier than the former. Then the equation (1) can be rewritten as:

$$\eta_{boiler} = \left[(\dot{Q}_i - \dot{Q}_l)/\dot{Q}_i\right] \times 100\% \tag{3}$$

5. Experimental results and discussions

In all experiments, the total input heat supplied by fuel ($\dot{Q}_i$) was fixed with the fuel and air volume flow rates of 0.000146 m^3/s and 0.00431 m^3/s, respectively. To maintain this input heat, these flow rates were controled by manually adjusting the flow control valves every half an hour.

5.1. The proposed boiler without porous medium

The proposed boiler was firstly tested without porous medium. To know whether it can work without porous medium or not. The experimental results show that, during the preheating step, the water temperature increased slowly. After the preheating temperature reached the saturated temperature at the lowest pressure (1atm), the water (vapour) feed control valve (V3) was then be opened. When the steam was fed into the mixing chamber, the pressurized flame cannot be sustainable or it was extinguished by steam. It can be summarized that, without the thermal wall and heat transfer enhancement of the porous medium, the direct fired steam boiler cannot be actually operated.

5.2. The proposed boiler with porous medium

In this first practicality test, the influences of steam pressure (P1 and P2) on the performance of the proposed DPB were mainly taken into account. There are three steam pressures (P2): 1, 1.5 and 2 bar, were used in the experiments. Refer to the graph shown in figure 3 to 8, the water feed control valve

IOP Conf. Series: Materials Science and Engineering **297** (2018) 012058 doi:10.1088/1757-899X/297/1/012058

(V3) was opened from the seventeenth minute to thirty-seventh minute. Therefore, the DPB was tested for 20 min until it reached around the steady state condition. It was assumed that the input heat supplied by fuel was constant at 13.72 kW. The efficiency of DPB (during the steam generation as shown in figure 4, 6 and 8) was then be estimated using the aforesaid assumptions and equation (3).

Figure 3 shows the temperature profile of the working fluid (T1 to T11) at feed water pressure (P2) of 1 bar (gage). After the preheating step (V1 and V2 were opened), when the water temperature (T3) of water jacket reached more than 123°C, then V1 was closed and V3 was opened. During the steady state condition (from the twenty-seventh minute to thirty-seventh minute), the CG can then be generated from the DPB with the properties of: T6=217°C, P1=0.4bar (gage). From the graph, it can be demonstrated that the wet steam did not substantially affect on the flame front, the temperature at the porous medium inlet (T10) was constant at about 850°C. At this condition, the steam generation rate of about 18 kg/hr was obtained. Regarding the thermal efficiency, figure 4 shows the efficiency of DPB with respected to input heat of combustion and surface heat loss. With the low cost 10cm-thickness fiberglass insulator and approximated convective heat transfer coefficient (h) of 2.5 W/m^{2}K, the average surface heat loss during the steady state was about 2.24 kW. Finally, the thermal efficiency during the steady state of the proposed DPB was calculated and the average thermal efficiency of 84% was achieved.

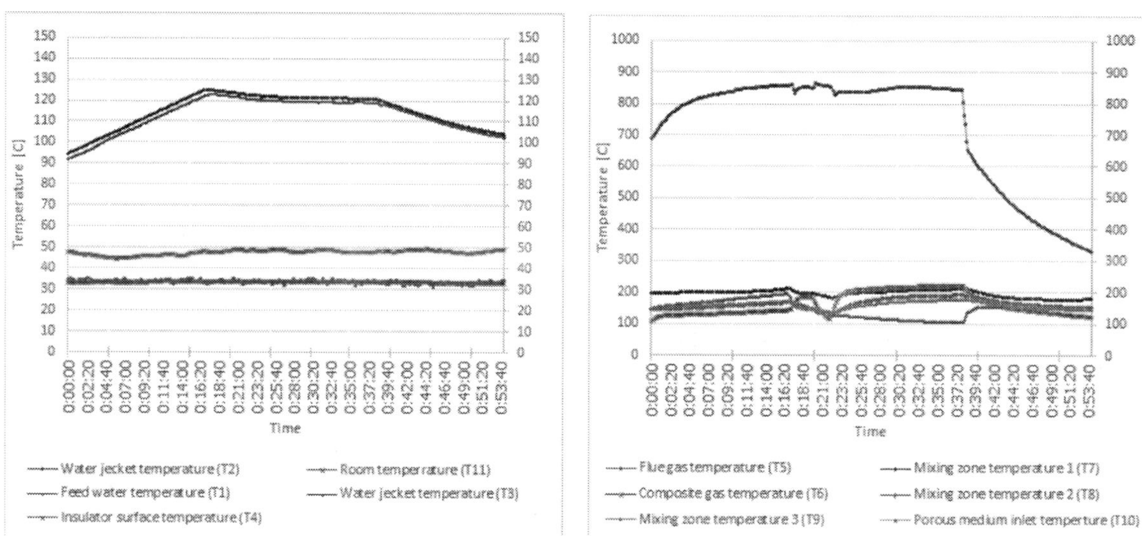

Figure 3. Temperature of working fluid at feed water pressure of 1 bar (gage).

Figure 5 shows the temperature profile of T1 to T11 at P2 of 1.5 bar (gage). After the preheating step, when T3 of water jacket reached more than 125°C, then V1 was closed and V3 was opened. During the steady state condition, the CG can then be generated from the DPB with the properties of: T6=215°C, P1=0.6bar (gage). From the graph, it can be demonstrated that the wet steam not substantially affected on the flame front, the temperature at the porous medium inlet (T10) was constant at about 873°C. At this condition, the steam generation rate of about 19.5 kg/hr was obtained. Regarding the thermal efficiency, figure 6 shows the efficiency of DPB with respected to input heat of combustion and surface heat loss. The average surface heat loss during the steady state was about 0.5 kW. Finally, the average thermal efficiency during the steady state of the proposed DPB was about 97%.

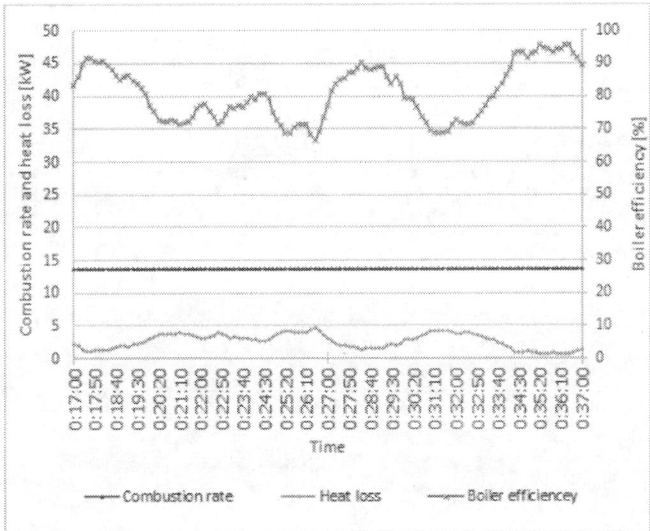

Figure 4. Heat rate and thermal efficiency at feed water pressure of 1 bar (gage).

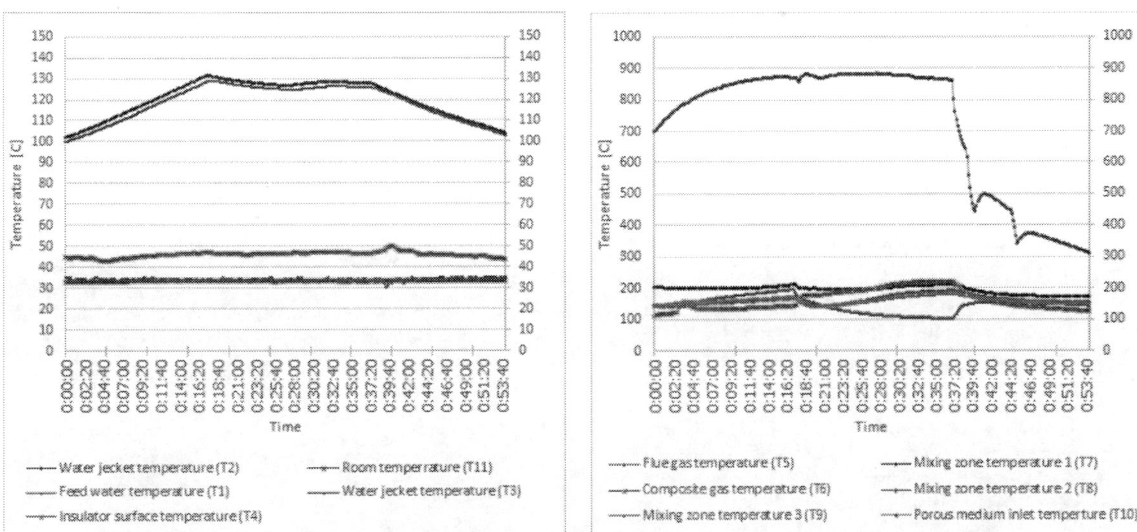

Figure 5. Temperature of working fluid at feed water pressure of 1.5 bar (gage).

Figure 7 shows the temperature profile of T1 to T11 at P2 of 2 bar (gage). After the preheating step, when T3 of water jacket reached more than 132°C, then V1 was closed and V3 was opened. During the steady state condition, the CG can then be generated from the DPB with the properties of: T6=237°C, P1=1bar (gage). From the graph, it can be demonstrated that the wet steam not substantially affected on the flame front, the temperature at the porous medium inlet (T10) was constant at about 867°C. At this condition, the steam generation rate of about 21 kg/hr was obtained. Regarding the thermal efficiency, figure 8 shows the efficiency of DPB with respected to input heat of combustion and surface heat loss. The average surface heat loss during the steady state was about 4.1 kW. Finally, the average thermal efficiency during the steady state of the proposed DPB was about 70%.

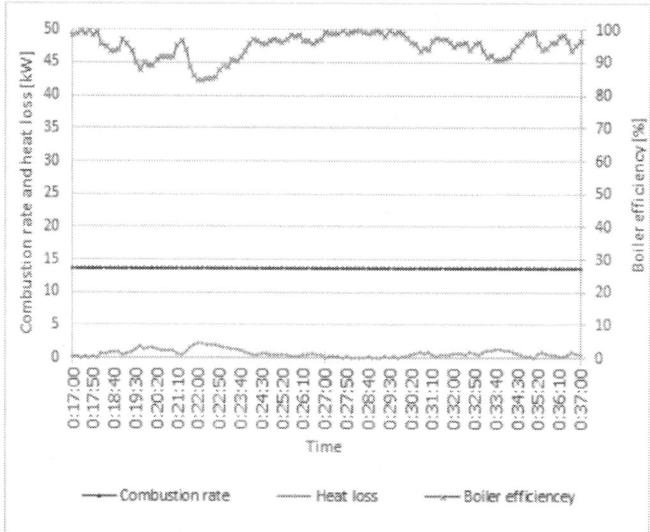

Figure 6. Heat rate and thermal efficiency at feed water pressure of 1.5 bar (gage).

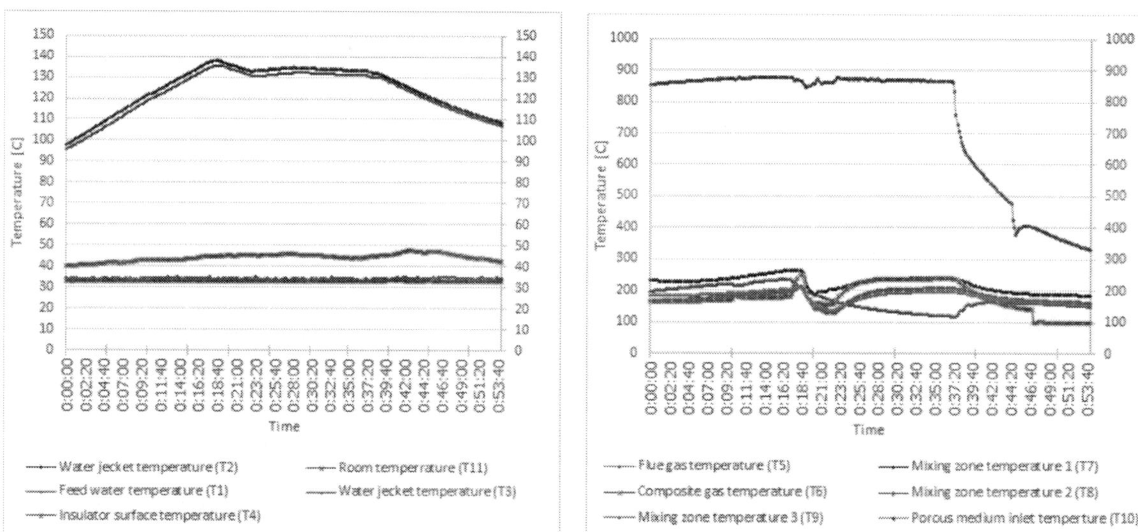

Figure 7. Temperature of working fluid at feed water pressure of 2 bar (gage).

6. Conclusions

A prototype of DPB was fabricated and tested to assess the feasibility of this new type boiler, as the above expected features. The experimental results show that the proposed DPB can be practically operated with high expected performance. It can be said that this innovation can be achieved by using a thermal wall of porous medium. This thermal wall not only maintain the condition for sustainable combustion but it also enhance the heat transfer from the flame to the fed water effectively. Among these three pressures, with the existing components, the proposed DPB should be operated at 1.5 bar (gage) to get the high thermal efficiency of about 97%. With this perspective, it can be concluded that the proposed DPB is promising and can possibly developed to be a high performance actual system for both heat and power generation systems.

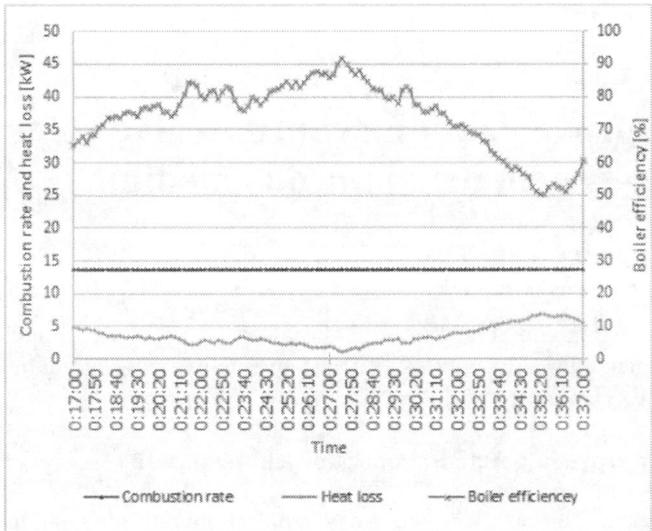

Figure 8. Heat rate and thermal efficiency at feed water pressure of 2 bar (gage).

Acknowledgments
The author wishes to thank Rajamangala University of Technology Thanyaburi (RMUTT) for providing financial support for his research and publication.

References

[1] Prasartkaew B and Kumar S 2010 *Energy and Buildings* **42** pp 1453–1462
[2] Prasartkaew B and Kumar S 2013 System *Renewable Energy* **57** pp 86-93
[3] Saidur R, Ahamed JU and Masjuki HH 2010 *Energy Policy* **38(5)** pp 2188–97
[4] Regulagadda P, Dincer I and Naterer GF 2010 *Applied Thermal Engineering* **30(8–9)** pp 970–6
[5] Barma MC, Saidur R, Rahman SMA, Allouhi A, Akash BA and Sait SMA 2017 *Renewable and Sustainable Energy Reviews* **79** pp 970–983
[6] Barcellos WM, Carlos L, Souza EO, Saveliev AV and Kennedy LA 2011 *Experimental Thermal and Fluid Science* **35** pp 570–580

IOP Conf. Series: Materials Science and Engineering **297** (2018) 012059 doi:10.1088/1757-899X/297/1/012059

Efficiency improvement of a concentrated solar receiver for water heating system using porous medium

Boonrit Prasartkaew*

Combustion and Solar Energy Laboratory (CASE Lab.), Faculty of Engineering, Rajamangala University of Technology Thanyaburi, Klong 6, Thanyaburi, Patumthani 12110, Thailand

* Corresponding Author: boonrit.p@en.rmutt.ac.th

Abstract. This experimental study aims at investigating on the performance of a high temperature solar water heating system. To approach the high temperature, a porous-medium concentrated solar collector equipped with a focused solar heliostat were proposed. The proposed system comprised of two parts: a 0.7x0.7-m^2 porous medium receiver, was installed on a 3-m tower, and a focused multi-flat-mirror solar heliostat with 25-m^2 aperture area. The porous medium used in this study was the metal swarf or metal waste from lathing process. To know how the system efficiency could be improved by using such porous medium, the proposed system with- and without-porous medium were tested and the comparative study was performed. The experimental results show that, using porous medium for enhancing the heat transfer mechanism, the system thermal efficiency was increased about 25%. It can be concluded that the efficiency of the proposed system can be substantially improved by using the porous medium.

1. Introduction

Nowadays, solar energy plays a tremendous role as the most powerful renewable energy source for a sustainable developments. It is the biggest energy resources and can be said that every hour, solar energy reaches the earth is enough for the world's energy demand for a whole year. Among the applications of solar thermal energy, solar water heating and solar air heating systems (SWHS and SAHS) are impressive in their beneficial manner [1]. Owing to it is a renewable energy based system and can be possibly developed for a high efficiency environmental friendly system, such system is vital and should be encouraged for overcoming the energy and environmental issues [2, 3].

The literatures reported that the total energy consumption for heating water in many countries around the world is ranging from 18 to 23% [1]. There are many applications of SWHS, e.g.: air conditioning (heating and cooling), washing, drying, and distillation, etc. Almost all of the SWHSs are comprised of a solar thermal collector. In terms of the solar collector of SWHS, it is designed to fulfill the required temperature as per the local climatic conditions. Actually, when the high temperature is required, the concentrated solar collector is taken into consideration. There are many types of concentrating collector, e.g.: flat plate collector equipped with tilted plate reflectors, parabolic trough, parabolic dish and central receiver, etc.

The thermal performance of the solar collectors mainly depend on the useful energy produced from the collector, this mainly depends on the collector's heat removal factor (F_R) and overall heat loss coefficient (U_L). To improve the efficiency of solar collector, by enhancing the heat removal factor and reducing heat loss, the application of porous medium (PM) was proposed for this current study.

From the literature and previous works, it can be summarized that no one performed in which porous medium (PM) is used for improving the performance of the SWHS [1, 4]. Moreover, the literatures also reveal that there is no one proposed in which the solar heliostat and tower receiver are used as a SWHS. Therefore, it was proposed in this study with the aim to investigate on the performance of a PM concentrated solar collector (PMCSC) of a small scale high temperature water heating system.

2. Experimental setup

Figure 1 shows the schematic diagram of the forced circulation PMCSC. It was setup at latitude of 14.03N and longitude of 100.72E. This system consists of two main components, a PM tower receiver and a focused solar-heliostat as shown in figure 2 and 3, respectively. The tower receiver was located in the south of heliostat with the distance of 20 m. The center of tower receiver and heliostat are on the same horizontal level at 3 m from the ground level.

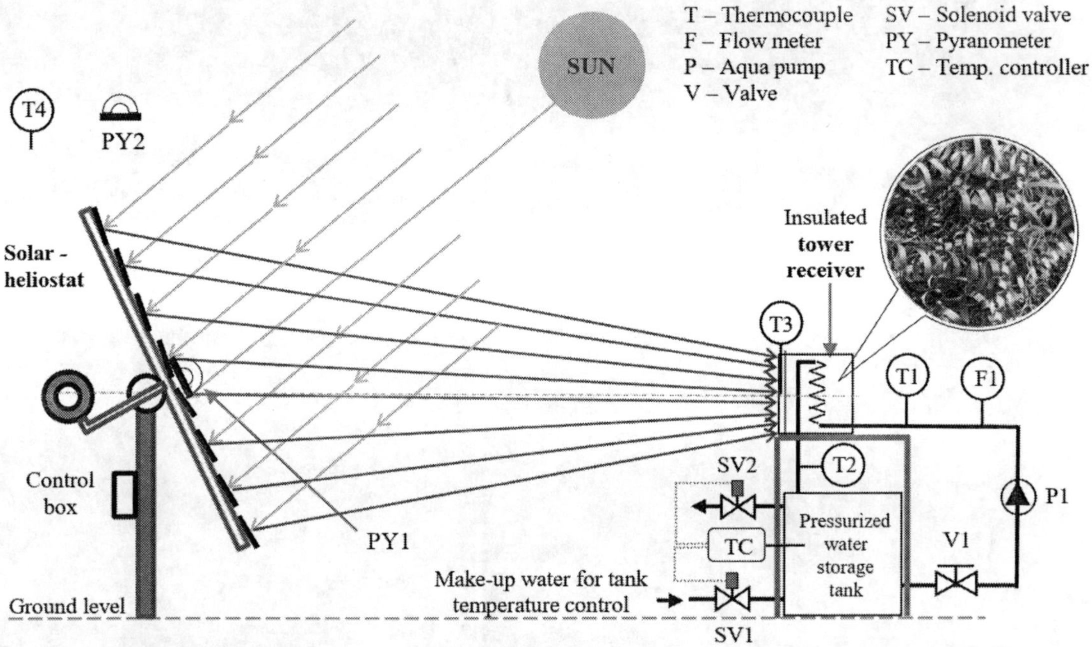

Figure 1. Schematic diagram of the forced circulation PMCSC system used in this study.

Figure 2 shows a 5x5-m^2 focused solar heliostat used in this study. It was fabricated by using the local low-cost materials. Regarding this low investment cost, it may be attractive or suitable for some small enterprises of heat process applications, such as: cooking, cooling (via absorption or adsorption refrigeration cycles), heating, drying, etc. The working principle and control system of this heliostat were described as in [5-7]. The aperture area of this current heliostat was bigger than two times of the previous one. The solar radiation strike on this heliostat would be redirected to the tower receiver via 289-plate of flat fixed mirrors. Each plate has 0.3x0.3-m^2 and was individually attached on an adjustable mounting set, so the desired tilting angle can be achieved with more flexibility. With this configuration, this tracking solar thermal system cloud be fabricated by using lighter and simpler structure, compared with the parabolic dish system. The focus of the heliostat was adjusted under the real solar radiation during 1 to 2 hours before and after solar noon.

Figure 3 shows a 0.7x0.7x0.7-m^3 insulated solar tower receiver. It has multi U-tube bundle with common inlet and outlet headers. A single glass cover was used to reduce the losses of the solar collector. Water was supplied to the receiver via an aqua pump (P1). The porous medium (as shown in figure 1) was fully filled in the receiver box.

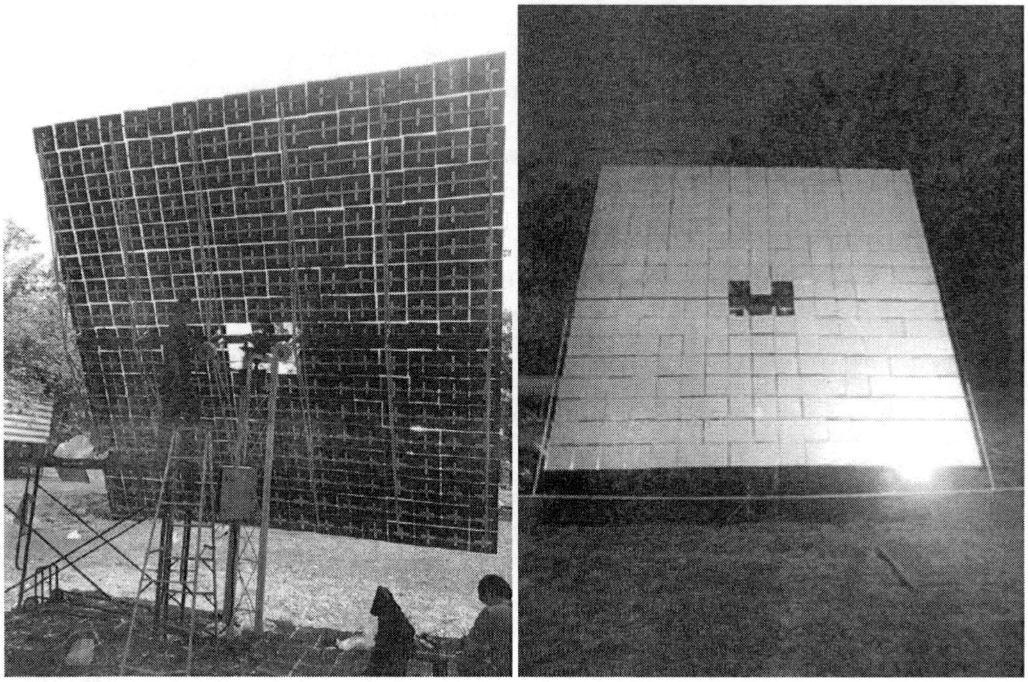

Figure 2. The focused multi-flat-mirror solar heliostat used in this study.

Figure 3. Photos of insulated solar tower receiver used in this study.

The temperatures of water and air at different locations (as shown in figure 1) were measured with K-type thermocouples. The flow rate was recorded on hourly basis. Radiation data included global horizontal radiation (PY2) and global horizontal radiation on tilted surface (on heliostat, PY1) were also measured and recorded. All temperature and solar radiation measurements were recorded using a data logger (Datataker model DT605) at 1-minute intervals.

The flow rate of water was manually controlled by a flow control valve (V1). During the test, the hot water was drawn off at the bottom of the pressurized water storage tank. A pump circulated water through the U-tube bundle of the tower receiver. To avoid the water evaporation under atmospheric pressure, the water temperature at the middle level of the tank was controlled at about 60 °C by means of draining hot water at the top and filling the make-up water at the bottom of the tank.

The comparative study was conducted to compare two different cases of study, with- and without-PM in order to find the most efficient one. The PM used in this study was the waste metal scrap or swarf from the lathing process. The experiments were conducted at the time from 9am to 4pm. During the test, the water mass flow rate was fixed at 0.04 kg/s. Thus, to maintain this value, the water flow rate was measured and manually adjusted every half an hour.

3. Evaluation of system performance

The steady conditions of the ASHRAE 93-1986 solar flat plate collector test standard [9] was conducted for this current study with steady state conditions, at steady radiation conditions, steady fluid flow, constant inlet temperature, and constant ambient conditions, are chosen for the analysis. The collector outlet water temperature is also constant. The useful energy gain ($\dot{Q}_u$) from the solar collector and collector efficiency (η_c) during the steady state condition were then estimated as [8]:

$$\dot{Q}_u = \dot{m}c_p\left(T_{c,o} - T_{c,i}\right) \quad (1)$$

and

$$\eta_c = \dot{Q}_u / A_c G_T \quad (2)$$

The tests were conducted on clear sky days when high quantity of the solar beam radiation and the collector surface is closely normal to the direct ray. The expression for determine the useful energy could be written as [8],

$$\dot{Q}_u = A_c F_R\left[G_T(\tau\alpha) - U_L\left(T_{c,i} - T_a\right)\right] \quad (3)$$

Substituting equation (3) into equation (2), gives,

$$\eta_c = F_R(\tau\alpha) - \left[F_R U_L\left(T_{c,i} - T_a\right)\right] / G_T \quad (4)$$

where, F_R is the collector heat removal factor
$(\tau\alpha)$ is the collector transmittance-absorptance product
U_L is the overall heat loss coefficient
C_p is the specific heat of water.

4. Experimental results and discussions

The experimental system of forced circulation PMCSC system was operated during winter season (18 February to 11 March 2017). The water volume flow rate was fixed at 0.04 l/s, this value was controled by manually adjusting the inlet air damper actuator of the blower every half an hour. The weather data of testing days and experimental data are shown in figure 4. The daily incident solar radiation during these days varied in the range of 18 to 21 MJ/m^2-day.

It should be noted that only data that consistent to the test standard was choosen for the performance anlaysis in the next sections. The steady state conditions during 5 minutes interval must be [9]:

- Solar radiation : >790 W/m²; ±32 W/m²
- Inlet water temperature : ±1 °C
- Water flow rate : ±1.1 lite/hr
- Ambient temperature : ±1.5 °C

4.1 The tower receiver without-porous medium

Figure 4 shows the average ambient temperature and total instantaneous solar radiation when the tower receiver without porous medium was conducted.

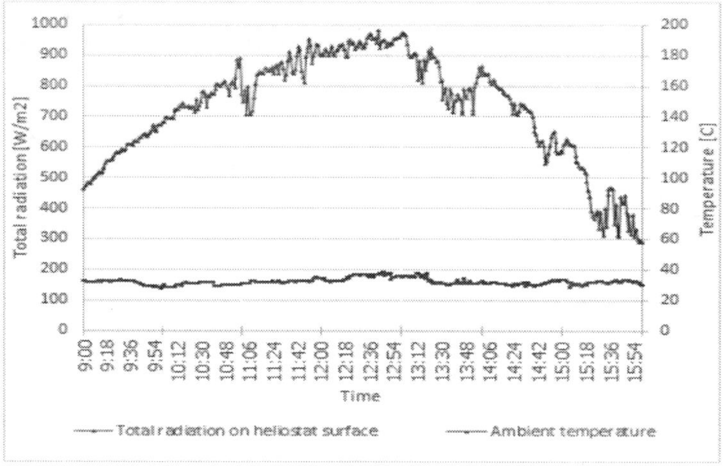

Figure 4. Ambient temperature and total solar radiation (case: without PM).

Figure 5 shows the hot water temperature as a function of solar irradiance during the steady state condition, when the tower receiver without porous medium was tested. The graph shows that the temperature of the hot water produced from the tower collector during the steady state condition varied in the range about 80 to 130°C with the average of about 114°C. This temperature can be reached this level when the total solar irradiance incident on the heliostat should be in the range of 800 to 950 W/m^2 with pressurized circulation system.

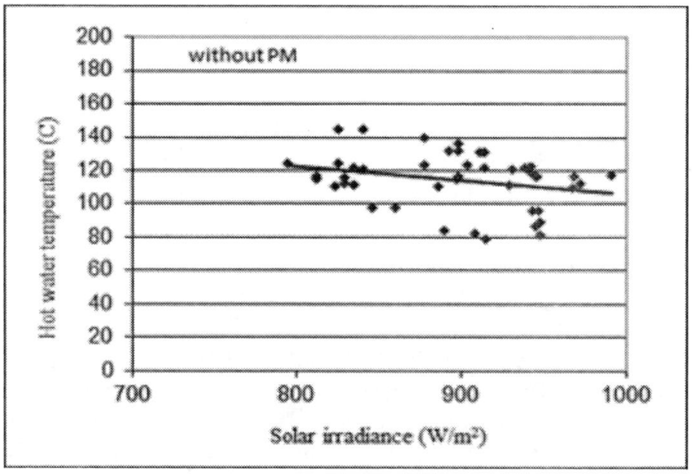

Figure 5. Hot water temperature as a function of solar irradiance (case: without PM).

Figure 6 shows the influence of solar irradiance on the useful heat output (calculated as equation (1)) during the steady state condition. The experimental results show that with the 25m^2 heliostat, the proposed system without porous medium can produce the hot water with heat output rate about 9 kW when the solar irradiance is about 800 to 950 W/m^2.

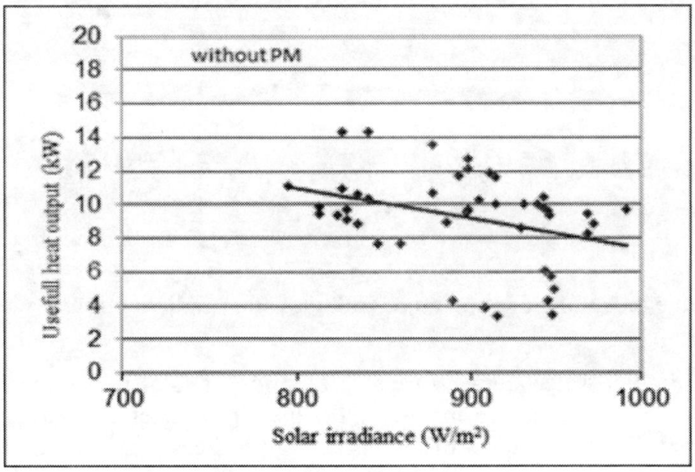

Figure 6. Useful heat output as a function of solar irradiance (case: without PM).

The most important performance index is overall system efficiency. Figure 7 shows the efficiency data of the proposed system without the porous medium. The vertical axis shows efficiencies during steady state condition, determined from equation (2). It can be seen that the average efficiency of about 42% can be obtained.

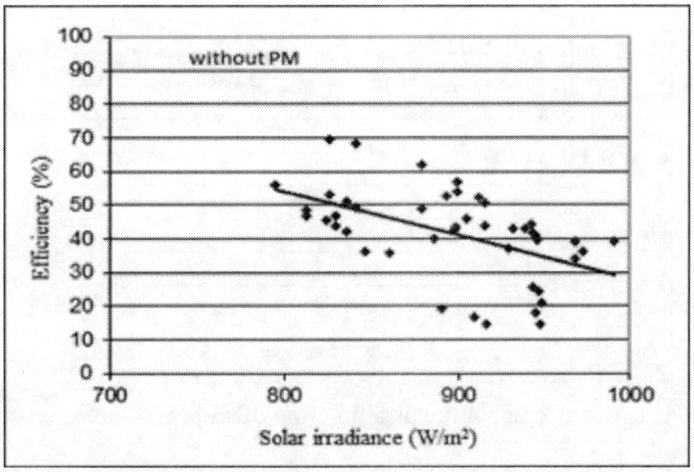

Figure 7. Overall efficiency as a function of solar irradiance (case: without PM).

4.2 The tower receiver with-porous medium

Figure 8 shows the average ambient temperature and total instantaneous solar radiation when the porous medium was used for performance improvement of the proposed PMCSC.

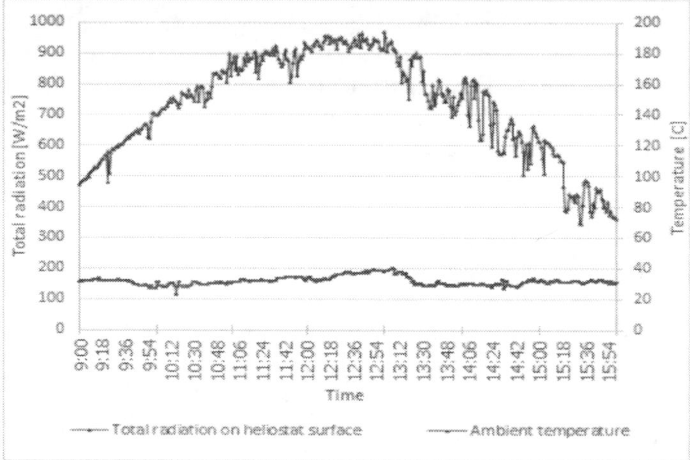

Figure 8. Ambient temperature and total solar radiation (case: with PM).

Figure 9 shows the hot water temperature as a function of solar irradiance during the steady state condition, when the PM was used for enhancing the heat transfer effectiveness of the tower receiver. The graph shows that the temperatures of the hot water produced from the tower collector during the steady state condition were higher than the case of without-PM and they varied in the range about 100 to 140 °C. This temperature can be reached this range when the total solar irradiance incident on the heliostat should be in the range of 800 to 950 W/m^2 with pressurized circulation system.

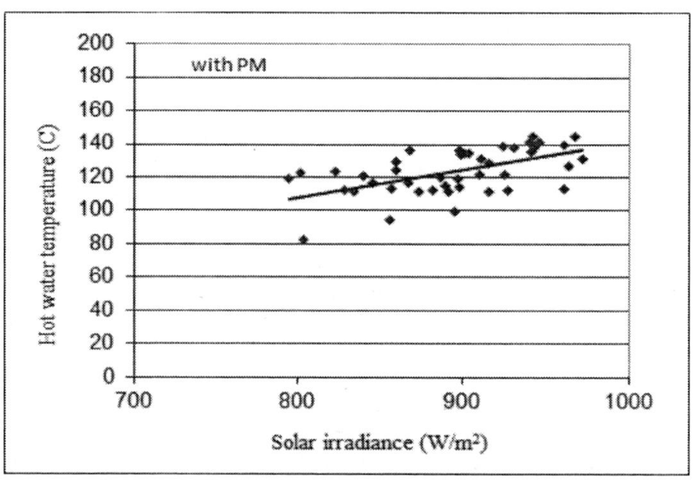

Figure 9. Hot water temperature as a function of solar irradiance (case: with PM).

Figure 10 shows the influence of solar irradiance on the useful heat output of PMCSC (calculated as equation (1)) during the steady state condition. The experimental results show that with the 25m^2 heliostat, the proposed system with porous medium can produce the hot water with heat output rate about 11 kW when the solar irradiance is about 800 to 950 W/m^2.

Figure 11 shows the efficiency data of the proposed PMCSC system. The vertical axis shows efficiencies during steady state condition, determined from equation (2). It can be seen that the average efficiency of about 52% can be obtained.

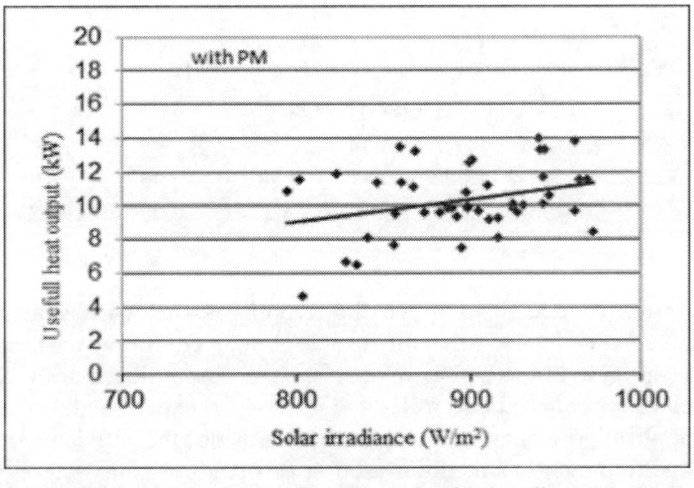

Figure 10. Useful heat output as a function of solar irradiance (case: with PM).

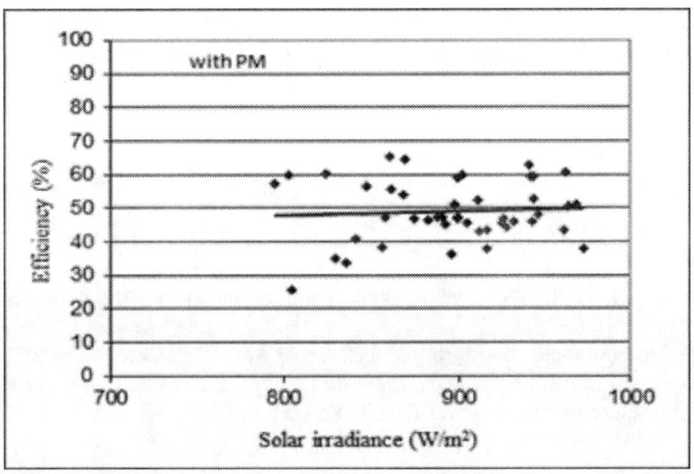

Figure 11. Overall efficiency as a function of solar irradiance (case: with PM).

4.3 Comparison between the receiver with- and without-porous medium

From the experimental results analysis of chosen steady state data, it was found that the performance of the proposed system can be improved significantly by using the porous medium. The different of two cases of study: with- and without-porous medium was summarized as shown in table 1. At the constant water flow rate, the temperature of hot water produced from the proposed system was increased from 114 to 126°C (increased by about 10%).The overall system efficiency can be increased from 42% to 52% (increased by about 25%).

In addition, as aforesaid about two important parameters, $F_R(\tau\alpha)$ and $F_R U_L$ (as in equation (4)) which are used to describe how the flat plate solar collector efficiently works, are also taken into account for this study. Due to the $F_R(\tau\alpha)$ is an indication of how energy is absorbed and $F_R(\tau\alpha)$ is an indication of how energy is loss. From the experimental result and analysis, in can be summarized that the values of $F_R(\tau\alpha)$ and $F_R U_L$ cannot be used for describing the behaviour of this proposed system, owing to the plot between $\left[\left(T_{c,i} - T_a\right)\right]/G_T$ and the efficiency (η_c) resulted in negative slope.

Table 1. The differences of average values between two cases of study.

Average data (Figure 4 – 11)	$T_{f,i}$ [°C]	$T_{f,o}$ [°C]	G_T [W/m²]	Ta [°C]	$T_{f,o}$ - $T_{f,i}$ [°C]	Flow rate [kg/s]	Qu [kW]	($T_{f,o}$ - T_a)/G [°C m²/W]	Eff. [%]
Case: w/o PM	59.21	114.43	895.64	33.93	55.23	0.04	9.26	0.09	41.74
Case: w/ PM	63.26	126.37	894.64	33.74	63.11	0.04	11.08	0.1	52.36
Different	**4.05**	**11.94**	**-1.00**	**-0.19**	**7.88**	**0.00**	**1.82**	**0.01**	**10.62**
[%]	**6.84**	**10.43**	**-0.11**	**-0.56**	**14.27**	**0.00**	**19.65**	**11.11**	**25.44**

5. Conclusions

Experimental study was conducted to prove the hypothesis that porous medium can improve the performance of a concentrated solar collector for a high temperature water heating system. Hot water temperature and overall system thermal efficiency were compared between with- and without-porous medium tower receivers at constant controlled variables. The experimental results reveal that, for the proposed system, application of porous medium is promising, the efficiency can be increased about 25%. In addition, due to the porous medium used in this study was the waste material with very low cost. It can be summarized that the substantial efficiency improvement with very low investment cost of the proposed system seem to be a vital measures for addressing the energy issues.

Acknowledgments

The author wishs to thank Rajamangala University of Technology Thanyaburi (RMUTT) for providing financial support for his research and publication.

References

[1] Gautam A, Chamoli S, Kumar A and Singh S 2017 *Renewable and Sustainable Energy Reviews* **68** pp 541–562
[2] Prasartkaew B and Kumar S 2010 *Energy and Buildings* **42** pp 1453–1462
[3] Prasartkaew B and Kumar S 2013 System *Renewable Energy* **57** pp 86-93
[4] Jaisankar S, Ananth J, Thulasi S, Jayasuthakar ST and Sheeba KN 2011 *Renew Sustain Energy Rev* **15**(6) pp 3045–50
[5] Prasartkaew B, Ngernplabpla A and Poowadin T 2015 *RMUTI Journal* **Special Issue 1** pp 55-62
[6] Ngernplabpla A and Prasartkaew B 2013 *Proc. Int. Conf. on 2013 Alternative Energy in Developing Countries and Emerging Economies (AEDCEE)* (Bangkok: Thailand)
[7] Prasartkaew B and Ngernplabpla 2013 *Proc. Int. Conf. on 2013 Int. Conf. on Frontiers of Environment, Energy and Bioscience (ICFEEB 2013)* (Beijing: China)
[8] Duffie JA and Beckman WA 1991 *Solar Engineering of Thermal Processes* second ed. (New York: *John Wiley*)
[9] ASHRAE standard 1986 *American Society of Heating, Refrigeration, and Air Conditioning Engineers* (New York: ASHRAE)

IOP Conf. Series: Materials Science and Engineering **297** (2018) 012060 doi:10.1088/1757-899X/297/1/012060

Electric motorcycle charging station powered by solar energy

Akarawat Siriwattanapong*and Chawin Chantharasenawong

Department of Mechanical Engineering, Faculty of Engineering King's Mongkut's University of Technology Thonburi
126 Pracha-uthit Rd., Bang Mod, Thung Khru, Bangkok 10140, Thailand

* Corresponding Author: pipe_pepper@hotmail.com

Abstract. This research proposes a design and verification of an off-grid photovoltaic system (PVS) for electric motorcycle charging station to be located in King's Mongkut's University of Technology Thonburi, Bangkok, Thailand. The system is designed to work independently (off-grid) and it must be able to fully charge the batteries of a typical passenger electric motorcycle every evening. A 1,000W Toyotron electric motorcycle is chosen for this study. It carries five units of 12.8V 20Ah batteries in series; hence its maximum energy requirement per day is 1,200Wh. An assessment of solar irradiation data and the Generation Factor in Bangkok, Thailand suggests that the charging system consists of one 500W PV panel, an MPPT charge controller, 48V 150Ah battery, a 1,000W DC to AC inverter and other safety devices such as fuses and breakers. An experiment is conducted to verify the viability of the off-grid PVS charging station by collecting the total daily energy generation data in the raining season and winter. The data suggests that the designed off-grid solar power charging station for electric motorcycle is able to supply sufficient energy for daily charging requirements.

1. Introduction

Motorcycles are by far the most popular vehicles in Thailand. There are over 20 million motorcycle registrations in Thailand in 2016, which are approximately 4 times of 4-wheel vehicles [1] and the number of motorcycles in Thailand has grown steadily every year as shown in figure 1. The huge number of vehicles in Thailand caused an increase in carbon dioxide (CO_2) emission and resulting in global warming. The CO_2 emission data from transportation in 2016 collected by Thailand greenhouse gas management organization (TGO) stands at 71.866 million tons-CO_2 as shown in figure 2 [2].

Renewable energy is a way to reduce CO_2 emission, because it is generated from clean resources. Current renewable energy technologies provide a variety of options such as solar, wind, geothermal heat, biomass, biofuels and hydroelectricity. Global trends in renewable energy investment [3] have also seen a steady increase since 2004 in both developed and developing countries as shown in figure 3.

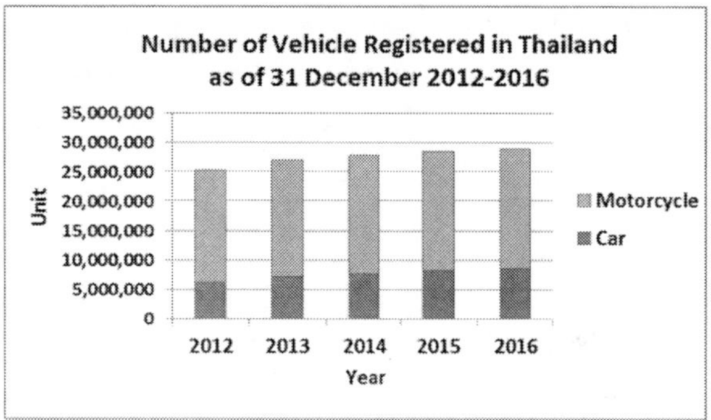

Figure 1. Number of vehicle registered in Thailand 2012-2016.

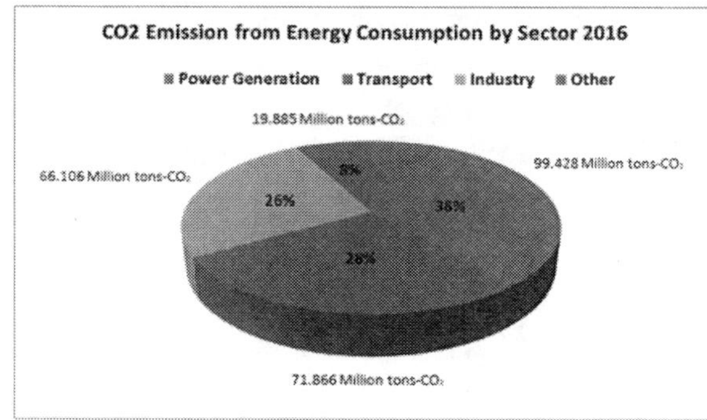

Figure 2. CO_2 emission from energy consumption by sector in 2016

The global trends in new investment in renewable energy [3] show that Thailand is in the pole position in Non-Organization for Economic Co-operation and Development (non-OECD) in the Asia-Oceania countries with the renewable energy investment totaling USD 1.4billion as shown in figure 4.

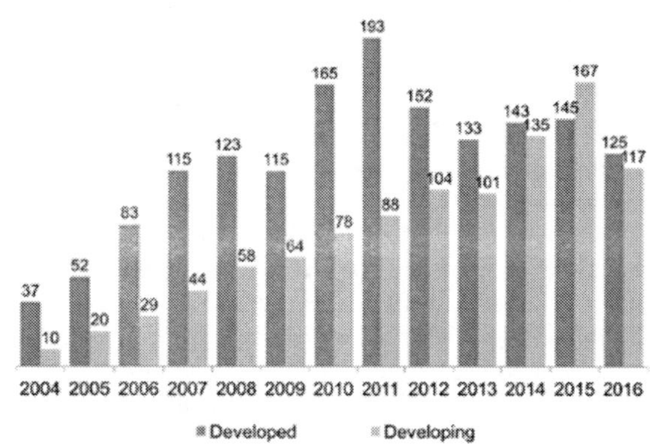

Figure 3. Global new investment in renewable energy developed and developing contries, 2004-2016.

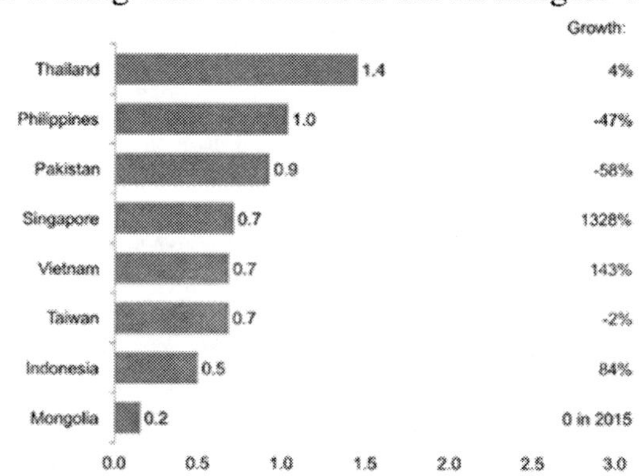

Figure 4. Renewable energy investment in non-OECD Asia (excluding China and India) in 2016 and growth in comparison to 2015.

Given the large number of motorcycles in Thailand its likely a continuation to remain high due to the compactness, low cost and convenience using for both in the cities and rural areas, there is an opportunity to create an impact in reducing the carbon footprint related to their use by promoting the use of electric motorcycles. Moreover, the carbon dioxide emission reduction could achieve during the charging process by utilizing renewable energy.

This research focuses on the design of a photovoltaic system (PVS) to charge an electric motorcycle on a daily basis without using electricity from the grid (off grid system). Solar energy is selected to power the charging station because of the falling costs of PV panels, inverter and installation to below half of the average costs in 2009 as shown in figure 5 [4]. The proposed solution explored in this work is similar to the successful work in [5] which designs a rooftop grid-connected PVS charging station for electric vehicle with 10kW PV panels in the Netherland. This work shows that a 10kW PVS is able to reduce grid energy exchange by 25%.

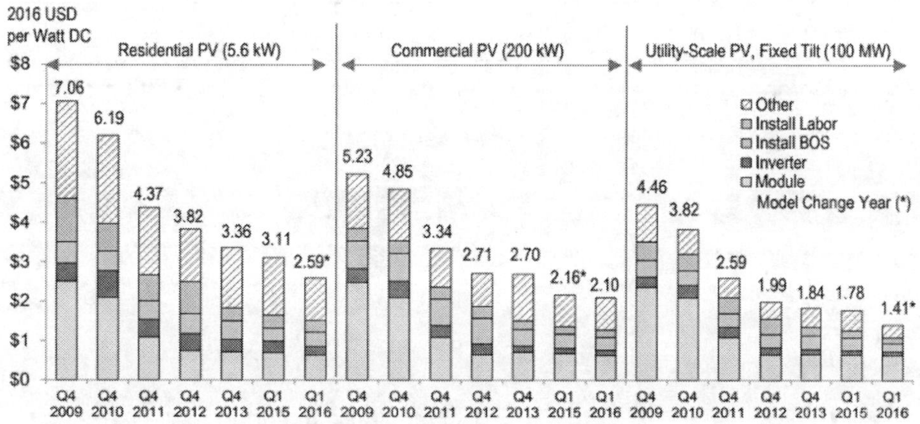

Figure 5. Falling cost of energy of photovoltaic system (PVS) between 2009-2016.

2. Methodology

This research attempts an off-grid PVS design for an electric motorcycle charging station, which must be able to fully charge the batteries of an electric motorcycle on a daily basis. The designed PVS contains 4 main devices: PV panels, charge controller, storage battery and inverter [6, 7] as shown in figure 6.

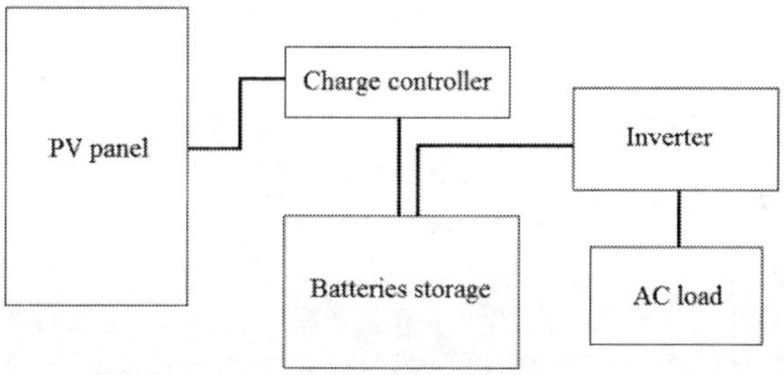

Figure 6. Illustration of PVS diagram with the main devices

2.1. Total demand energy requirement

This section describes the methodology to determine the energy requirement by an electric motorcycle in this study. The electric motorcycle model is Toyotron TX-1 as shown in figure7 and its specifications as shown on table 1.

Figure 7. Toyotron TX-1 electric motorcycle

IOP Conf. Series: Materials Science and Engineering **297** (2018) 012060 doi:10.1088/1757-899X/297/1/012060

Table 1. Toyotron TX-1 specifications.

Item	Specification
Motor	Integrated rear hub 2,000 Watt high torque
Battery	60 Volt (12V x 5)
Wheels	12 inches Alloy
Climbing	20-25 degrees
Speed	50-60 km/h
Range	1 person: 50-80 km 2 person: 40-60 km
Brakes	180 mm. disk and drum rear
Dimensions WxLxH (mm.)	460 x 1,920 x 1,150
Road clearance	150-190 mm.
Weight	95 kg

The Toyotron TX-1 electric motorcycle contains 5 $LiFeMnPO_4$ Prismatic 12.8V 20Ah batteries in series to make a 60V 20Ah battery. The battery specifications are shown on table 2.

Table 2. Battery specifications.

Item	Specification
Voltage	12.8V (working) /15.2V (peak) /10V (cut-off)
Capacity	20 Ah
Cycle life	> 1000 cycles (80% of initial capacity at 0.2C rate, IEC Standard)
Energy density	85.33 Wh/kg
Charging rate	1C rate (20A) recommended/ 3C rate (60A) Maximum
Max discharge rate	Constant discharge current 40A (2C rate) Impulse discharge current (<10 sec) 200A (10C rate)
Dimensions WxLxH (mm.)	78.5 x 180 x 168
Weight	3 kg.

The total energy requirement is equal to maximum energy storage of battery of electric motorcycle. Ohm's law (1) states that

$$V \times A = W \tag{1}$$

where V is voltage, A is current and W is power in Watts. The maximum energy storage of the Toyotron electric motorcycle battery is 1,200 Wh. However, there is loss of energy in the PVS from the devices, hence a need to allow for extra energy production. A safety factor of 1.3 (assume 30% greater than

IOP Conf. Series: Materials Science and Engineering **297** (2018) 012060 doi:10.1088/1757-899X/297/1/012060

requirement because losses in system) is selected in the design [8]. Therefore, the maximum possible total energy requirement is 1,560 Wh/day.

In actual use, the electric motorcycle is rarely utilized until its battery is completely drained. Daily usage of electric motorcycle using data [9] shows a data of energy consumption in each day for 20 days by volunteers using the electric motorcycle in their regular routine, i.e. traveling between home and work. The average daily energy consumption is 842.25 Wh. The data is shown on table 3.

Table 3. Daily energy consumption of an electric motorcycle in real life usage over 20 days

Day	Energy (Wh)	Day	Energy (Wh)
1	687	11	1,013
2	859	12	771
3	488	13	1,153
4	1,000	14	778
5	1,000	15	879
6	644	16	849
7	780	17	848
8	767	18	967
9	878	19	984
10	876	20	624

2.2. PV sizing

This section provides a PV panel selection methodology. Different sizes of PV panels will produce different amount of power. The total peak Watt (*Wp*) produced depends on the size of the PV panel and climate of each location. The important factor for calculating the appropriate PV panel size is the peak sun hour. Peak sun hour (PSH) is an average daily solar insolation. It is the maximum value for a certain number of hours at 1 kW/m^2 of solar irradiation (peak solar radiation). In Thailand, average monthly PSH is approximately 4-6 hours per day, but in July, 2015 has the lowest average PSH of 3.5 hours per day [6]. The *peak Watt* of the PV panel can be calculated by equation (2)

$$Wp = \frac{Load(Wh/day)}{PSH(h/day)} \tag{2}$$

In this case, the *Load* is the maximum possible total energy requirement of 1,560 Wh/day and the worst case of *PSH* is 3.5 h/day. The *Wp* of PV is 445.714≈500 W, therefore the selected size of PV panel for this study is 500 Wp. This research has received an industrial support from Stars Microelectronics (Thailand) Public Company Limited who has provided TenKSOLAR Apex 500W PV module for use in this research. The PV panel specifications are given in table 4.

Table 4. TenKSOLAR Apex 500W PV module specifications

Item	Specification
Peak power (P_{max})	500 W
Module size	2,018 x 1,298 mm.
Cell type	Mono-Crystalline (PERC)
Power tolerance	±3%
Module efficiency	19.1%
Max current output (I_{MP})	9.1 A
Operating DC voltage (V_{MP})	35-59 V
Operating temperature range	- 40 to 85 °C
Module NOCT	46 °C
Temperature coefficient (P_{MP})	-0.42% / °C

2.3. Storage Battery sizing
This section describes the method to determine the suitable size of storage battery. The suitable battery type for use in the PVS is a deep cycle battery. Deep cycle battery is designed for being discharged to low energy and be rapidly recharged more cycles than standard lead acid batteries. The storage battery should be large enough to store sufficient energy to operate over 3 days, even at night and on cloudy days. Battery sizing can be calculated by equation (3)

$$Battery\,capacity(Ah) = \frac{Total\ appliances\ use}{battery\ efficiency \times DOD \times system\ voltage} \times Days\ of\ autonomy \quad (3)$$

where total appliances use in this case is the requirement from Toyotron electric motorcycle battery of 1,200 Wh/day. The battery efficiency is approximately 0.80-0.95. The depth of discharge (DOD) is used to describe how deeply the battery is discharged. Fully charged battery is 0% of DOD and fully discharged battery is 100% of DOD. DOD is a factor to be considered for sizing the battery because the DOD can effect to battery life span.

In this case, a conservative DOD value of 50% is chosen. There are three possible values of system voltage: 12V, 24V and 48V depending on peak power voltage of PV. The 48V system is chosen because it must be compatible with the PV panel operating DC voltage, which in this case is between 35-59V. Days of autonomy represent the number of days where no additional energy is supplied from the PV panel (completely no sun). In this case, the system is designed to have 3 days of autonomy. Therefore, the suitable battery size from is 48V 187.5Ah or 48V 200Ah.

2.4 Charge controller sizing
This section describes the methodology to determine the appropriate charge controller size selection. The charge controller is typically rated against current and voltage capacities. The solar charge controller must match the voltage of PV panel and the storage batteries. The size of charge controller can calculate by equation (4).

$$I_{Charge\,controller} = \frac{Wp}{V_{system}} \tag{4}$$

where Wp of the PV panel is 500W and V_{system} is 48V. Therefore, the charge controller size is 10.42A≈ or 11A. The design should include a 30% safety margin, hence, the charge controller size should be greater than 14.3A.

2.5. Inverter sizing

This section describes the inverter sizing selection. The inverter is a device to convert DC power input to AC power output. The inverter must have the same nominal voltage as the battery it is connected to. For off-grid systems, the inverter must be large enough to handle the maximum amount of power usage when all appliances are in use at the same time. The inverter size should be 25-30% higher than the total combined power of all appliances for safety. The size of the inverter can calculate by equation (5).

$$W_{Inverter} = Total\ watt\ of\ all\ appliances\ use \times additional\ safety \tag{5}$$

where the total combined power of appliances is from the Toyotron electric motorcycle charging adapter. The adapter requires 220V AC 3.5A and an additional safety factor of 1.3 (30%). The suitable inverter size is 1,000W.

2.6. Collecting data

Solar irradiation data with 14 degree tilt angle from horizontal and facing South as the latitude of Bangkok, Thailand is 13.73617 North [6] is used in this study. The solar irradiation data is collected with the use of Kipp&Zonen CM11 pyranometer whose specifications are shown in table 5. The records show solar irradiation in 30 second time steps and average data to solar irradiation per day. Equation (6) shows the energy production from PV enough for energy demand can use solar irradiation data and calculate by equation (6)

$$E = SI \times A \times \eta_{PV} \times \eta_{system} \tag{6}$$

where E denotes the energy produced from PV, SI is solar irradiation, A is PV panel area (m^2), η_{PV} is PV efficiency and η_{system} is system efficiency.

Collecting daily 14 degree tilt to horizontal solar irradiation data by Kipp&Zonen CM11 pyranometer from 1 July 2016 to 31 December 2016. The solar irradiation data in figure 8 shows a variation over the 6 months because of the seasonal changes between the raining season and winter in Thailand.

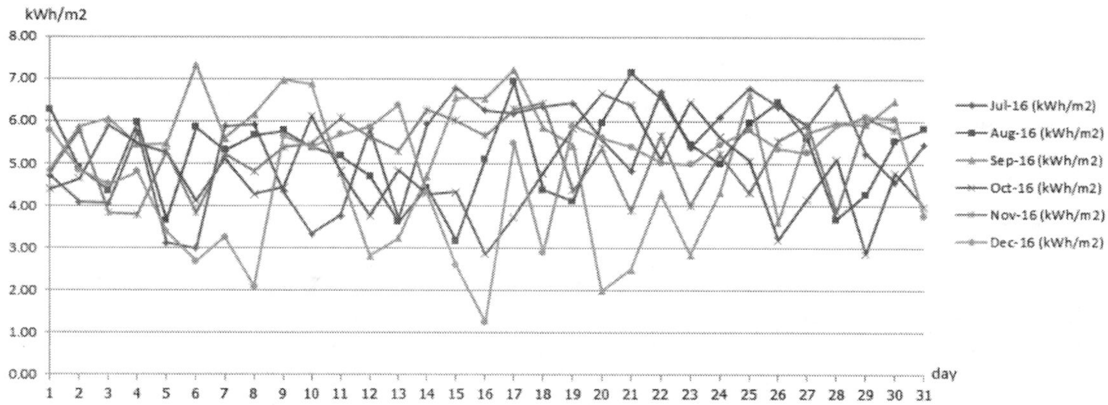

Figure 8. Solar irradiation per day from 1 July to 31 December 2016.

Table 5. Kipp&Zonen CM11 specifications

Item	Specification
Spectral range	310-2,800 nm (50% points) 340-2,200 nm (95% points)
Sensitivity	Between 4 and 6 $\mu V/Wm^{-2}$
Impedance	700-1,500 Ohm
Response time	< 15 s (95% response) < 24 s (99% response)
Non-linearity	< 0.6% (< 1,000 W/m^2)
Temperature dependence of sensitivity	< 1% (-10 °C to 40 °C)
Directional error	< 10 W/m^2 (beam 1,000 W/m^2)
Tilt error	< 0.25% (beam 1,000 W/m^2)
Operating temperature	-40 °C to 80 °C
Zero- offset due to FIR (Ventilated with CV 2)	< 7 W/m^2 at 200 W/m^2 net thermal radiation
Zero-offset due to temp. changes	< 2 W/m^2 at 5 K/h temp. change
Viewing angle	2π sr

3. Results and discussions

From collected daily solar irradiation data can be calculated to find produced energy of PVS by equation (6), where *SI* is solar irradiation, *A* is 2.619364 m^2 (from PV specifications), η_{PV} is 19.1% (from PV specifications) and η_{system} is 70% (30% loss from assumption). The results of produced energy calculation as shown in figure 9.

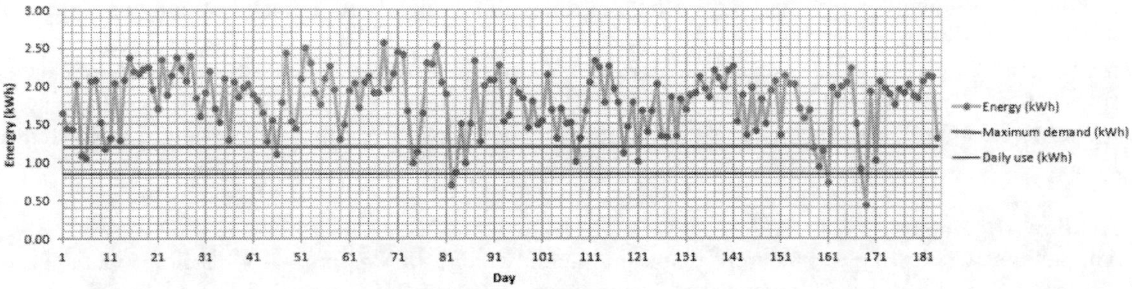

Figure 9. Comparison between energy from PVS, maximum demand (1.2 kWh) and average energy in daily use (0.84225 kWh).

Figure 9 shows that over 90% of the records, the calculated daily energy produced by the PVS has exceeded the maximum demand. The maximum record daily energy production is 2.57 kWh on 6 September, while the minimum energy production is 0.44 kWh, which is observed on 16 December. The average value of daily energy production in this period is 1.79 kWh. There are only 18 days from 184 collecting data days where the produced energy is lower than the maximum demand.

Considering days of insufficient energy production (lower than 1.2 kWh), the maximum number of days with insufficient energy that has been observed is 3 days in a row between 6-8 December. This is the only instance. When considering the average energy in daily use, there are only 3 separate days where insufficient energy is observed.

Considering accumulative energy of storage battery, this research design for 3 days of autonomy. The possible maximum storage energy of 48V 200Ah battery is 9.6 kWh. The battery is charged in daytime and discharged for supply to electric motorcycle in nighttime. The accumulative energy of storage battery was shown in figure 10.

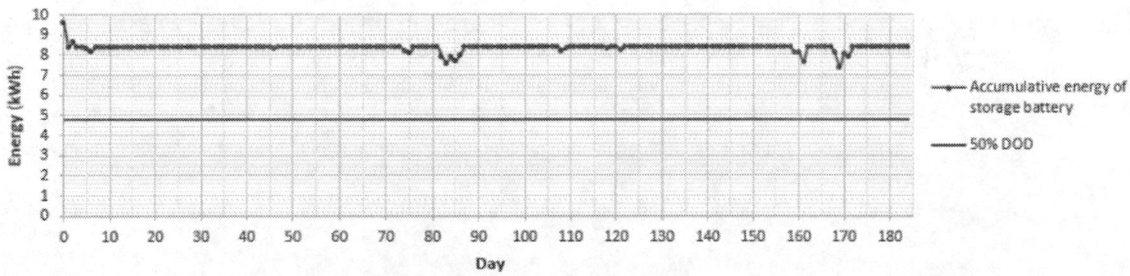

Figure 10. Accumulative energy of storage battery over 184 days.

From figure 10, the system starts with a fully charged storage battery of 9.6 kWh on the first day and is discharged to electric motorcycle of 1.2 kWh every night. The system is charged again in daytime from the PVS. The lowest accumulative energy of the battery of 7.34 kWh is observed on day 169 (16 December) which is 23.54% of DOD when compare with 50% DOD designing.

With this system, the energy can supply 1.2 kWh to an electric motorcycle without charging storage battery for 3 days at 50% of DOD and can supply storage battery for 6 days at 100% of DOD (including

IOP Conf. Series: Materials Science and Engineering **297** (2018) 012060 doi:10.1088/1757-899X/297/1/012060

battery loss from battery sizing calculation). However, higher percent of DOD can decrease battery life span.

4. Conclusion

The proposed design of the off-grid PVS provides a sustainable energy storage for charging an electric motorcycle from solar energy. The most challenging seasons for PVS to provide sufficient energy are the raining season and winter. Even during these seasons, the system is still able to produce more energy than the electric motorcycle requirements over 90% of the days. The system has shown that it is capable of holding sufficient energy for 3 days of autonomy with the maximum demand of daily use. However, this research still assumes 30% for total loss of PVS which not accurate for real application. There are many cause of loss in system such as loss from PV, loss from cable, loss from charge controller, loss from batteries and loss from inverter which depends on specification, type or market brand. Some system in real application has other device such as meter, diode, breaker or other safety device to cause more loss.

Acknowledgements

The authors would like to sincerely thank and acknowledge the support of CES Solar Cells Testing Center of King Mongkut's University of Technology Thonburi (CSSC-MKUTT) for providing the test facilities and solar irradiation data and Stars Microelectronics (Thailand) Public Company Limited for the support of PV panels.

References

[1] Ministry of Transport (Thailand) *Registered vehicle 2012-2016 in Thailand*
[2] Thailand Greenhouse Gas Management Organization 2016 *CO_2 Emission from Energy Consumption in 2016*
[3] Frankfurt School FS-UNEP Collaborating Centre *Global Trends in Renewable Energy Investment 2017* Chapter 1 pp 15-31
[4] National Renewable Energy Laboratory *U.S. Solar Photovoltaic System Cost Benchmark Q1 2016* p 7
[5] G.R. Chandra Mouli, P. Bauer, M. Zeman 2016 *System design for a solar powered electric vehicle charging station for workplaces*
[6] CES Solar Cells Testing Center of King Mongkut's University of Technology Thonburi (CSSC-MKUTT) *Basic of Photovoltaic System Design*
[7] Gunn Panprayun, Nipon Ketjoy, Wattanapong Rakwichian and Prapita Thanarak 2011 *Int. J. of Renewable Energy* **6**(1) pp 59-61
[8] Baltus1 CWA, Eikelboom J A and van Zolingen R J C 1997 *14th European Photovoltaic Solar Energy Conference* (Barcelona)
[9] Piyatida T, Amornrat K, Pawinee I, Thaned R, Thana R and Yossapong L *Evaluation of Electric Motorcycle Driving Behaviors Comparing with Conventional Engine Motorcycle in Thailand*

IOP Conf. Series: Materials Science and Engineering **297** (2018) 012061 doi:10.1088/1757-899X/297/1/012061

Experimental study on the drying of natural latex medical gloves

Mano Chankrachang[1], Pisit Yongyingsakthavorn[1], Atitaya Tohsan[2] and Udomkiat Nontakaew[1,*]

1 Department of Mechanical & Aerospace Engineering, Faculty of Engineering, King Mongkut's University of Technology North Bangkok, Bangkok, 10800, Thailand

2 Department of Materials and Production Technology Engineering, Faculty of Engineering, King Mongkut's University of Technology North Bangkok Bangkok, 10800, Thailand

* Corresponding Author: udomkiat.n@eng.kmutnb.ac.th

Abstract. The purpose of this research was to study latex film drying at 70 °C using a laboratory drying oven. Two different total solid content (TSC) latex compounds, which 45% TSC and 35% TSC were used. The undried latex films were prepared according to the common procedures used in latex gloves manufacturers, that is, by dry coagulant dipping process. The experimental results such as initial moisture content, the amount of moisture and drying time of latex films in each latex compound formula were determined. After that, the results were projected to calculate on the production capacity expand by 1 million piece/day of natural latex medical gloves. Finally, the rate of moisture entering the latex drying oven and the energy consumption of the drying oven were estimated. The results indicated that when the 35% TSC of latex compound was used. The initial moisture content of latex film was higher than 45% TSC of latex compound about 7%. The drying time of 35% TSC was longer than 45% TSC for 2.5 min and consume more energy about 10%. As a result, the 45% TSC latex compound was the better way to saving energy and managing humidity in the production line. Therefore, it was found to very useful to an approximate design length and size of actual of latex drying oven and the rate of moisture entering the oven as well.

1. Introduction

Nowadays, natural latex medical gloves industry are widely used in the clinical and medical setting and for various objectives, which helps to protect the disease infection or contamination. The main reasons of increment for using the medical gloves is due to people are now becoming more hygiene awareness and want to be prevented them from impurity and chemicals substance [1]. In this way, the demand for latex rubber gloves industry is viewed to increase steadily. It is reported that a demand of medical gloves is increasing around 5 - 10% per year [2].

Concentrated natural rubber (NR) latex is the main raw material to produce the dipping products such as medical glove, nipple, hot/cold bag, balloon, sporting goods and so on. Thailand is the world largest producer and exporter of the concentrated latex, accounting for one-third of the total exports volume in the

Nomenclature

$\dot{P}$	Production rate of latex films (g/s)	$\overline{W}_{m,t}$	Weight of moisture on ceramic plate at time = t (g)
W_t	Weight of undried latex film from each former (g/piece)	$\dot{Q}$	Total heat energy requirement (kW)
W_d	Weight of dry matter from each former (g/piece)	$\dot{Q}_d$	Sensible heat of dry matter (kW)
W_m	Weight of moisture from each former (g/piece)	$\dot{Q}_m$	Sensible heat of moisture (kW)
$\dot{W}_d$	The amount of dry matter entering the oven (g/s)	$\dot{Q}_e$	Latent heat of vaporization of water (kW)
$\dot{W}_m$	The amount of moisture entering the oven (g/s)	T_i	Initial Temperature of latex films (°C)
n	Number of formers entering the oven (piece/s)	T_f	Final Temperature of latex films (°C)
$\dot{M}_r$	The amount of moisture to be removed (g/s)	C_{pw}	Specific heat capacity of water (kJ/kg K)
M_i	Initial moisture content of latex films (% wb)	C_{ps}	Specific heat capacity of rubber (kJ/kg K)
M_f	Final moisture content of latex films (% wb)	h_v	Latent heat of vaporization of water (kJ/kg)
$\overline{W}d$	Weight of dry matter on ceramic plate (g)		

world [2]. Although, Thailand is the number one producer and exporter of concentrated NR latex but in term of the products of concentrated NR latex, for examples, medical gloves, Thailand has the proportion only 22 percent of the world market. The main reason of the slow progress in the development of rubber product industry of Thailand is the lack of continuous improvement in research and development.

The methods of making medical gloves via dipping process are consisted of the following steps [2,3]: first, a former is dipped into the latex compound, withdrawing a former, drying and vulcanizing the latex film, respectively [4]. In all actuality, the drying process is involved in the several production lines, for example: drying process was applied to cleaning, dipping coagulant and latex compound, leaching and chlorination section. The hot air was supplied to each drying oven in order to remove moisture and also maintain the temperature of formers and latex films properly. Thus, the drying process not only takes into account the dipping process production line but also consumes amount of heat energy in this process more than 80% of total energy consumption [3].

The objective of this work was to study latex films drying using a laboratory drying oven by measuring amount of moisture of latex film on the ceramic plates. The latex films were prepared according to the common procedures used in latex gloves manufacturers, that is, by dry coagulant dipping process. The experimental data were projected to calculate the rate of moisture entering the actual latex drying oven in the factory, and also determine drying kinetics of the undried latex film to anticipate how long is the undried latex films keep stay in the actual drying oven. Last, the heat energy required for actual oven also calculated when the production capacity expand by 1 million piece/day.

Consequently, by conducting this experiment, we wish to elucidate the energy of drying characteristic of medical rubber gloves, which can be a great advantage for development of the gloves industry in Thailand.

IOP Conf. Series: Materials Science and Engineering **297** (2018) 012061 doi:10.1088/1757-899X/297/1/012061

2. Drying Kinetics of Latex Film

The hot air was supplied by various heat sources such as thermal hot oil and fired burner that transferring heat from a heat exchanger to fresh air. The undried latex film on the formers were entering into the latex drying oven and the moisture content in the latex films were removed to hot air. The most important thing has to concern, which determines the flow rate of hot air supply in the oven that could be maintained the drying kinetics of the latex film firmly. Therefore, mass and heat balance were applied to this study.

2.1. Mass Balance

In order to increase the productive capacity to 1 million piece/day, it means that about 12 pieces of glove must be pass through the latex drying oven per second. Each of undried latex film at initial moisture content and temperature needs to be dried to a final moisture content at a temperature, thus the latex gloves production rate into the latex drying oven can be calculated as follows: [5]

$$\dot{P} = nW_t \tag{1}$$

The weight of undried latex film from each former consists of weight of dry matter and weight of moisture in latex film is calculated as follows:

$$W_t = W_d + W_m \tag{2}$$

The rate of moisture entering the oven is calculated as follows:

$$\dot{W}_m = \frac{M_i \dot{P}}{100} \tag{3}$$

The rate of dry matter that passes through the oven is calculated as follows:

$$\dot{W}_d = \dot{P}\left(1 - \frac{M_i}{100}\right) \tag{4}$$

Hence, the rate of moisture to be removed is calculated as follows:

$$\dot{M}_r = \dot{P}\left(\frac{M_i - M_f}{100}\right) \tag{5}$$

2.2. Heat Balance

In the classical concepts for drying process the total heat requirement comes from 2 types of heat: [6] that is, the combination of sensible heat of moisture and dry matter, in this case stand for dry rubber and latent heat of vaporization of moisture. For this study, the rate of required heat calculation was focused on the rate of moisture and dry matter entering the latex drying oven. Other mechanical parts such as former holder, conveyor chain, drying oven wall were neglected. Therefore, the rate of required heat for drying latex films can be calculated as follows:

$$\text{Heat required} = \begin{bmatrix}\text{Sensible heat of dry}\\ \text{matter to increase}\\ \text{temperature}\\ \text{from } T_i \text{ to } T_f\end{bmatrix} + \begin{bmatrix}\text{Sensible heat of}\\ \text{moistureto increase}\\ \text{temperature from}\\ T_i \text{ to } T_f\end{bmatrix} + \begin{bmatrix}\text{Latent Heat required}\\ \text{to evaporate the}\\ \text{moisture at}\\ \text{temperature } T_f\end{bmatrix} \tag{6}$$

For this model, e.g. 6 can be written as follows:

$$\dot{Q} = \dot{P}\left[C_{pd}\left(1-\frac{M_i}{100}\right)+C_{pm}\frac{M_i}{100}\right](T_f - T_i) + \dot{P}h_v\left(\frac{M_i - M_f}{100}\right) \tag{7}$$

The final drying temperature of the latex films was assumed to be 70 °C referring to the manufacturing procedure. A summary of the parameters used in the estimation is shown in table 1.

Table 1. Calculation Parameters used for the drying process. [6, 7]

Parameter	Parameter Value	
M_f	5.00	% wb
T_i	25	°C
T_f	70	°C
C_{pm}	4.19	kJ/kg K
C_{pd}	3.20	kJ/kg K
h_v	2333	kJ/kg
n	12	piece/s

3. Materials and Methods

3.1. Material

NR latex was an ammoniacal latex containing 0.7 % NH_4OH with a dry rubber content (DRC) of 60 %, which was purchased from Thai rubber latex public Co., Ltd., Thailand. Potassium hydroxide dispersion, potassium laurate dispersion, sulfur dispersion, zinc dibutyl dithiocarbamate dispersion (ZDBC), wingstay-L dispersion and ZnO dispersion and calcium nitrate were prepared from the commercial grade reagents.

3.2. Latex Compounding Preparation

The NR latex was stirred at room temperature using stirring speed of 300 rpm, then it was firstly mixed with 20% Potassium laurate solution and 10% Potassium hydroxide solution which were functioned as stabilizer. Next, 50% sulfur dispersion, 50% Zinc dibutyl dithio carbamate dispersion, 50% Wingstay–L dispersion, 50% ZnO dispersion were respectively added into the latex mixture. Finally, the DI water was added to obtain TSC = 45% and 35% of latex compound, respectively. The amount of the chemicals were listed in table 2. After the mixing was completed, the latex compound was stirred at 30 rpm for 24 hrs maturation time.

Table 2. The amount of chemicals in 45% and 35% TSC latex compound

Chemicals	Undried weight (g)	
	45% TSC	35% TSC
60% Concentrated Latex	167.0	167.0
20% K-laurate solution	1.0	1.0
10% KOH solution	3.0	3.0
50% Sulfur dispersion	1.0	1.0
50% ZDBC dispersion	2.5	2.5
50% Wingstay – L dispersion	1.0	1.0
50% ZnO dispersion	0.5	0.5
DI water	52.2	117.4

3.3. *Film Preparation*

The undried latex films were prepared according to the common procedures used in latex gloves manufacturers, that is, by dry coagulant dipping process. The ceramic plates that used in this experiment were cut from the palm and the back of the hand area of medical gloves ceramic dipping formers. The sizes were set on a 30 mm x 40 mm. It is important to clean up and dry in an oven before staring the experiment. First, the clean ceramic plates were heated up at 60°C and dipped by using 10% calcium nitrate as the coagulant and dried in an oven at 120°C for 5 min then cooled down in desiccator until the temperature reached 60°C. Next, the ceramic plates were dipped into the latex compound for 8 seconds. After reaching a dwell time, the ceramic plates were slowly withdrawn and instantly weighted by an analytical balance in order to record the initial mass of undried latex film. After that, the undried latex film were dried in the oven with a time step equal to 2.5 minute and weighted the latex film again also repeat the drying process until the mass of latex film was constant to complete the experiment.

Thus, the moisture content of the latex film at each time step is calculated by ratio between amount of moisture and the total weight of undried latex film at each time step as follows:

$$M_t = \frac{\overline{W}_{m,t}}{W_t} \tag{8}$$

4. Results and Discussion

4.1 The Amount of Moisture

In this study, two different total solid content (TSC) latex compounds, which 45% TSC and 35% TSC were used. From table 2, it is shown that DI water in 35% TSC latex compound was larger than 45% TSC latex compound 65.2 g or 125%.

The surface area of ceramic plate was 0.0012 m^2. To find the total surface area of the medium size medical gloves former, the 3D scanner was used for measuring the surface area so that the dipped surface area of the former is about 0.0635 m^2. After calculating the experimental data shown in figure 1. The weight of moisture of 45% TSC and 35% TSC latex compound was about 10.32 and 11.33 g/piece and initial moisture content of latex films is about 50.37% and 56.91%, respectively. As a result, the rate of moisture to be removed which was calculated by using e.g. (5) were about 111.46 and 124.11 g/s for 45% TSC and 35% TSC, respectively.

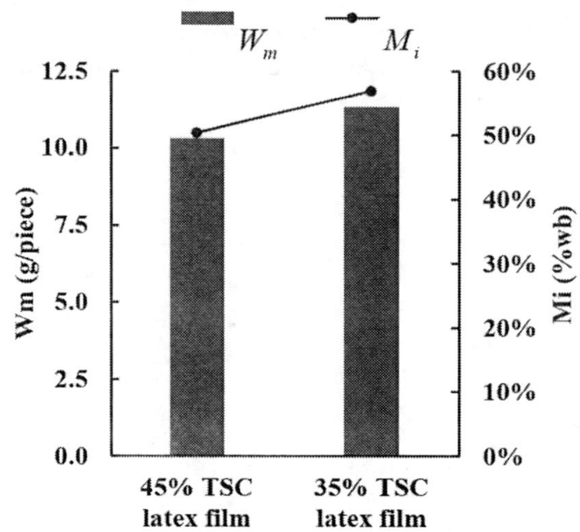

Figure 1. W_m and M_i value of each former dipped by 45% and 35% TSC latex compound

4.2 Film Formation and Drying Time

Some previous research [8–10], They indicated that there are 3 steps of the latex films formation. Step 1, a linear cumulative moisture loss occurs overtime and ends when irreversible contact between latex particles is achieved. During this moment, the evaporation rate is similar to the evaporation of pure water. The mass transfer limitation is localized in the gaseous phase [8, 11]. Step 2, Latex particles deform and come into close contact with so-called honeycomb structure, which is associated with a sharp decrease in the water evaporation rate. The overall rate of evaporation decreases considerably [8]. Step 3, a homogenous film is formed and the evaporation rate strongly decreases and the mass transfer limitation is also clearly within the film. The remaining moisture leaves the film by diffusion through the air/dried latex layer interphase [8, 11]

In this experiment, the weight of undried latex films on the ceramic plates and drying time were recorded. The result of change in moisture content of the latex film dipped by 45% TSC and 35% TSC compound were represented in figure 2. Each of moisture content value was calculated by e.g. (8), whereas the cumulative moisture loss of the latex film dipped by 45% TSC and 35% TSC compound were shown in figure 3. The amount of moisture loss at each time step were recorded and counted to plot the cumulative moisture loss.

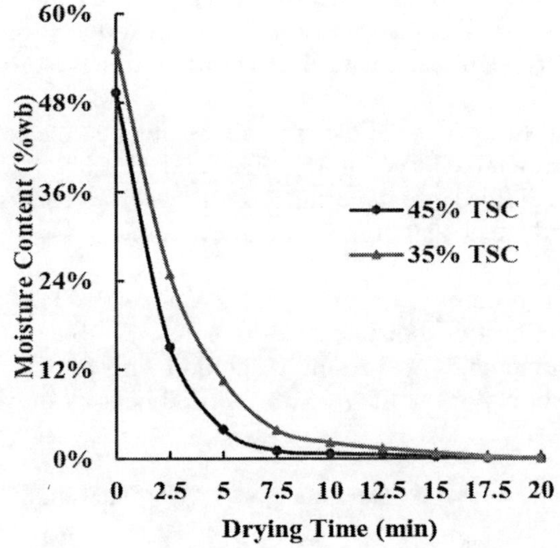

Figure 2. Change in the moisture content of the latex film dipped by 45% TSC and 35% TSC latex compound

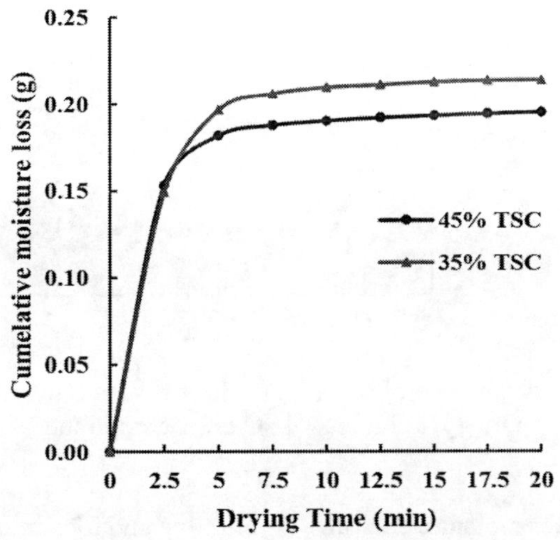

Figure 3. Change in the cumulative moisture loss from latex film dipped by 45% TSC and 35% TSC latex compound

From the experimental data, it was observed that, when the drying time increased from 0 - 2.5 min, the moisture content and cumulative moisture loss are dramatically decreased and increased, respectively. It can be seen that, when drying time is reached 5 min for 45% TSC and 7.5 min for 35% TSC, a constant weight of latex film was detected. These results can be explained that the moisture content of latex film and the humidity ratio of hot air were balanced. In term of the film formation, the honeycomb structure was completely formed and become to a homogenous film. Consequently, the 5 min and 7.5 min drying time from dipped latex film by 45% TSC and 35% TSC latex compound was applied in this study.

4.3 Energy Requirement

As the previous section, the rate of required heat for drying latex films was focused on the latex film. Other mechanical parts were not included. In order to increase the production capacity expand by 1 million piece/day, the big problem that has to deal with the massive of moisture in the undried latex film.

The rate of required heat in e.g. (7) was calculated by using parameters value from various research [6,7,11] about 300.94 kW and 330.03 kW for 45% TSC and 35% TSC, respectively. For more details, the rate of required heat was divided into three parts as follows from e.g. (6), sensible heat of dry matter, sensible heat of moisture and latent heat of moisture to evaporate.

Sensible heat of dry matter was about 17.56 kW for 45% TSC and 14.83 kW for 35% TSC compound. Similarly, sensible heat of moisture was about 23.33 kW for 45% TSC and 25.65 kW for 35% TSC compound. Meanwhile, latent heat of moisture to evaporate was about 260.05 kW for 45% TSC and 289.54 kW for 35% TSC compound. As a result, it should be noted that the latent heat of evaporation takes on an important role about 86% of the rate of required heat as shown in figure 4.

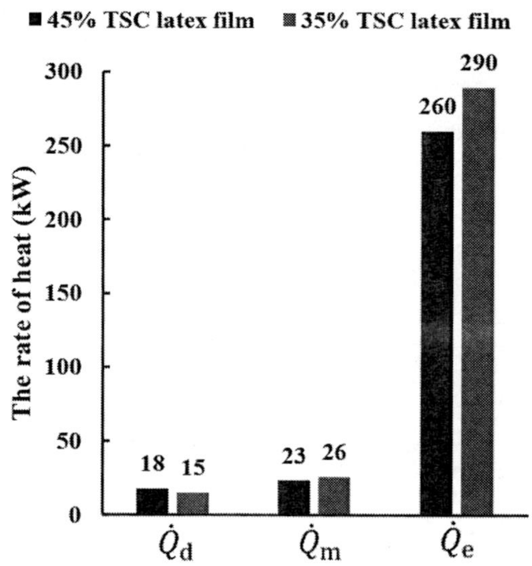

Figure 4 .The rate of required heat for drying latex film dipped by 45% and 35% TSC latex compound

5. Conclusion

For this study, the amount for moisture entering, drying time and total energy usage of latex drying oven were determined. The effect of different TSC of latex compound causes impact to the increase of energy consumption about 10% and drying time for 2.5 min. It also affects to expand the whole length of production line as well. At the present, the most of latex compound formula try to reduce operating cost by reducing TSC of latex compound. The energy supply for drying oven also has to increase inevitably. Therefore, it was found to be very useful to an approximate design length and size of actual of latex drying oven and the amount of moisture entering as well.

Acknowledgement

This research was funded by and the Thailand Research Fund (TRF) (Project No. RDG 5950036).

References

[1] Ken Research 2017 *Global Natural Latex Gloves Industry In-depth Investigation and Analysis Report* (India)
[2] Thaiplastic (Plastics Institute of Thailand) 2014 *Rubber intelligence unit* (Bangkok)
[3] DEDE (Department of Alternative Energy Devolpment and Efficiency) 2007 Project on Studying of Energy Effciency Index in Rubber Industry (Thailand)
[4] Saidur R and Mekhilef S 2010 *Appl. Energy* **87** pp 2746–58
[5] Ghaly A E and Macdonald K N 2012 *J. Eng. Appl. Sci. Publ. Online* **5** pp 136–50
[6] Cengel Y A and Boles M A 2006 *Thermodynamics: An Engineering Approach* (McGraw-Hill)
[7] Kiil S 2006 *Prog. Org. Coatings* **57** 236–50
[8] Steward P a., Hearn J and Wilkinson M C 2000 *Colloid Interface Sci.* **86** pp 195–267
[9] Keddie J L and Routh A F 2010 *Fundamentals of Latex Film Formation Processes and Properties* (New York: Springer Dor)
[10] Ludwig I 2008 *Drying, Film Formation and Open Time of Aqueous Polymer Dispersions* (Demand: Universitätsverlag Karlsruhe)
[11] Bouyer D, Philippe K, Wisunthorn S, Pochat-Bohatier C and Dupuy C 2009 *Technol.* **27** 59–70

Possibility study of gasifier with axial circulating flue gas for reducing Tar

T. Poowadin*, **M. Polsongkram and P. Khantikomol**

Renewable Energy Technology Research Laboratory (RTER)
Department of Mechanical Engineering, Faculty of Engineering and Architecture,
Rajamangala University of Technology Isan,
744 Suranarai Rd., Muang, Nakhonratchasima 30000, Thailand

* Corresponding Author: poowadin@yahoo.com

Abstract. This present research article aims to study the possibility of gasification by axial core flue gas circulating kiln and find the efficiency of syngas production. An axial core flue gas circulating tube was installed in the center of the updraft gasifier in purposing of tar reducing. In the present study, the eucalyptus wood chip 4, 8, and 10 kg with the moisture content 16% were examined. Several type-K thermocouples were employed to measure the temperatures at preheat, combustion, reduction, pyrolysis, drying, and gas outlet zone. The results showed that the temperatures in the combustion and the reduction zone of the kiln with the axial core flue gas recirculating were lower than the kiln without the core owing to installing the core would reduce the combustion zone area in biomass burning. Obviously, the temperature in the pyrolysis and drying zone were nearly the same as both with and without the core. In consideration of syngas components, it was found that CO production from the gasifier with the core was higher than the gasifier without the core about 25%. Other gases, however, were almost same. The syngas production efficiency obtained from the gasifier with the core decreased with increasing the mass of biomass. It showed that the highest efficiency was 30% at 4 kg supplying biomass. In comparison, the efficiencies of both the kilns with and without the core were not different. For liquid product, the amount of liquid decreased about 47.23% comparing with the gasifier without the core.

1. Introduction

Gasification is an attractive technology for convert fuel from biomass through incomplete combustion at high temperatures to syngas. This technology is environmentally friendly that is to reduce carbon dioxide (CO_2) from the burning of biomass fuels. The processes of gasification include drying, pyrolysis, oxidation and reduction. The first stage of gasification is drying. Usually the moisture content of biomass in this zone is removed by the radiation heat from oxidation zone. The temperature in this zone remains less than 120 °C. Then pyrolysis process starts to destroy structure of biomass at temperature about 200 – 400°C. Char, water vapour, methanol, acetic acid are the main products of pyrolysis zone. The moisture vaporised in the drying zone and others in the pyrolysis zone travels down towards oxidation zone. Temperature in this zone rises to about 900 – 1200°C. The main product of oxidation process is carbon dioxide (CO_2)[1]. The product of oxidation zone then passes through the reduction zone. The temperature in this zone is about 600 – 900°C. Gasifier is the reactor for producing syngas. Syngas normally consist of carbon monoxide (CO), carbon dioxide (CO_2), hydrogen (H_2), methane (CH_4) and tar. There are four types of gasifiers classified according to application: updraft or counter current gasifier, downdraft or co-current gasifier, crossdraft gasifier and fluidized bed gasifier. Other types of gasifiers are currently under development. In all types of gasifiers, the carbon dioxide (CO_2) and water vapour (H_2O) are converted (reduced) as much as possible to carbon monoxide, hydrogen and methane, which are the main combustible components of syngas. The updraft gasifier achieves the highest efficiency but the disadvantage is the excessive amount of tar in syngas. Tar is the seriously problem to operation of internal combustion engine. This problem is reduced in the downdraft gasifier. Crossdraft gasifier has certain advantages over updraft and downdraft gasifiers but the disadvantages are high exit gas temperature and poor CO_2 reduction. The information from previous study about "Tar" levels from gasifiers shows that general agreement about the relative of tar production, with updraft gasifiers being the dirtiest, downdraft the cleanest and fluidized bed intermediate [2, 3]. There are many attempts to develop of gasifier that expects of tar formation and tar conversion or removal during gasification. However it is not easy to find new technic or design that can fulfill all requirements. Installing an axial core flue gas circulating tube in the center line of the kiln (reactor) to change the direction of syngas to reduce "Tar" is the idea of this research.

2. Materials and methods

2.1 Characteristics of fuel

As shown in figure 1, Eucalyptus wood chip is biomass that can be processed into energy and it has plentiful in Thailand. Chemical formula for eucalyptus is $C_{4.1}H_{5.7}O_{2.8}$ Eucalyptus wood chip was used as fuel for producing syngas by updraft gasifier with axial core flue gas circulating kiln. The consumption of biomass was varied at 4, 8, and 10 kg with the size was 5-10 mm thick and the moisture content 16%.

2.2 Experimental Setup and Measuring Devices Used

Figure 2 shows the schematic of the updraft gasifier with axial core flue gas circulating kiln consisted of an axial core flue gas circulating tube installed in the center line of the kiln (reactor). The kiln's capacity was 0.41 m^3. For combustion zone, the throat size diameter was 40 cm and diameter of the syngas outlet tube was 6 cm.The temperature of the syngas was cooled down by the condenser. Several type-K thermocouples were employed to measure the temperatures at preheat, combustion, reduction, pyrolysis, drying, and gas outlet zone.

Figure 1. Eucalyptus wood chip

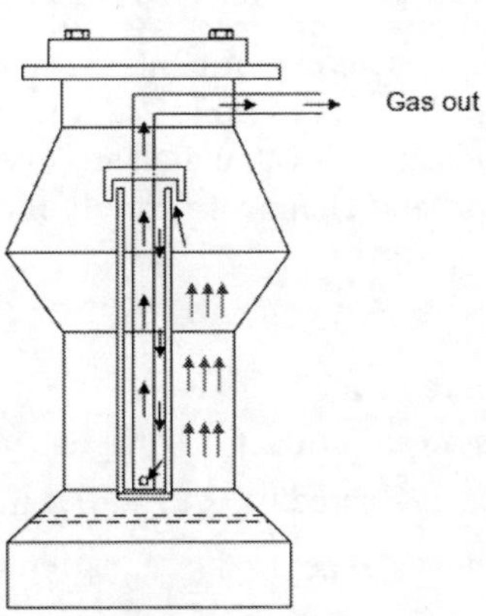

Figure 2. Schematic of updraft gasifier with axial core flue gas circulating kiln

Production of syngas from eucalyptus wood chips by updraft gasifier with axial core flue gas circulating kiln for reducing "Tar" is the target of this research. The experimental apparatus diagram was shown in figure 3. The experimental apparatus was made of the steel with 140 cm high and 60 cm base diameter. The kiln was insulated by thermal insulator outside the combustion zone. An air was used to be a working fluid blew by the blower through the flow meter into the combustion chamber at bottom of the gasifier. The air temperature in front of the gasifier was 30-32 °C and volume flow rate 0.0033 m^3/s. Type K-thermocouples were used to measure the temperatures of the flowing air at inlet of the gasifier (T1), inlet of the combustion zone (T2), inlet of the reduction zone (T3), inlet of the pyrolysis zone (T4), inlet of the drying zone (T5) and inlet of the gas outlet zone (T6). Moreover, the ambient temperature was also measured during the experiment.

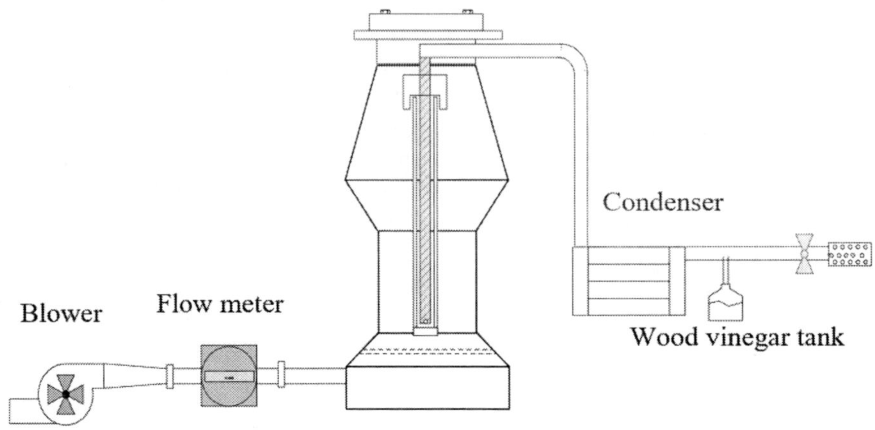

Figure 3. The schematic diagram of the experimental apparatus

3. Data processing

In the present study, Eucalyptus wood chip was used as a biomass to produced syngas by the process of gasification. Eucalyptus was chopped into pieces the length of 3-5 cm and 5-10 mm thick with the moisture content 16%. Moisture content indicates the amount of water contained in the material compared to the mass of material. Moisture content can be determined by using the equation as follow:

$$M_w = \frac{w-d}{w} \times 100 \quad (1)$$

where, M_w is the moisture content (%), w is the mass of material (kg), d is the mass of dry material (kg)

In the present experiment, the mass balance principle is used for calculating the Air to Fuel ratio (A/F ratio). The first variable is the amount of air used for incomplete combustion, which is between 20-30%. This research is calculated at a maximum percentage of 30%. Then air consumption 30% for gasification process and chemical formula for eucalyptus $C_{4.1}H_{5.7}O_{2.8}$ are used to analyse. Using the relationship is as follow.[4]

$$C_{4.1}H_{5.7}O_{2.8} + 0.3(O_2+3.76N_2) \Longrightarrow xCO+yH_2+zCH_4+1.128N_2 \quad (2)$$

The syngas production efficiency (η_s) of the updraft gasifier with axial core flue gas circulating kiln can be evaluated by the ratio of mass of syngas production (m_g) and the consumption of biomass(m_f) which was expressed in equation (3).

$$\eta_s = \frac{m_g}{m_f} \times 100 \quad (3)$$

where, η_s is syngas production efficiency (%), m_g is mass of syngas production (kg), m_f is the consumption of biomass (kg)

IOP Conf. Series: Materials Science and Engineering **297** (2018) 012062 doi:10.1088/1757-899X/297/1/012062

4. Results and Discussion

4.1. Temperature distribution in the gasifier

The updraft gasifier with axial core flue gas circulating kiln used in the present work have the same element dimensions of width and capacity. Therefore, the density of fuel depends on the amount (mass) of biomass. In the present work, the eucalyptus wood chip 4, 8, and 12 kg with the moisture content 16% were examined. The relationship of biomass time and temperature distribution in combustion zone was shown in figure 4. The results indicated that at a biomass of 8 kg, the temperature in the combustion zone was higher than the other because amount of biomass is not too tight, resulting in a sufficient space for combustion. For a biomass of 4 kg, the heat from the combustion is low, so the temperature in the combustion zone tends to increase slowly. The temperature in this zone is about 400°C. Furthermore, the results showed that the temperatures in the combustion and the reduction zone of the kiln with core were lower than the kiln without core owing to installing the core would reduce the combustion zone area in biomass burning. Obviously, the temperature in the pyrolysis and drying zone were nearly the same as both with and without the core

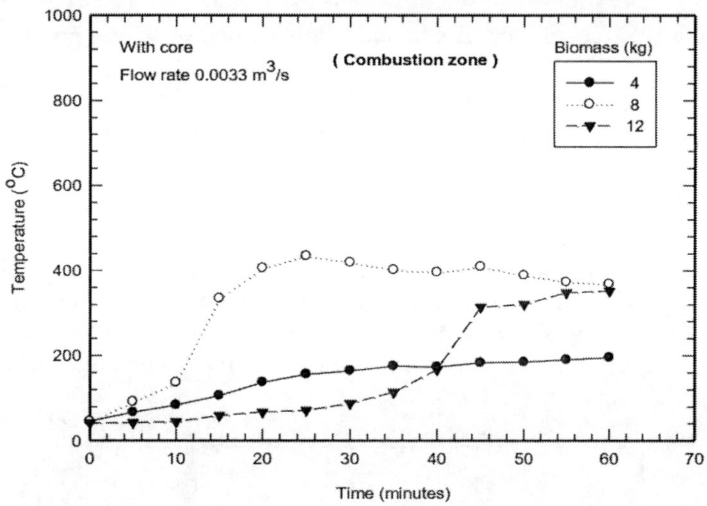

Figure 4. Temperature distribution in the combustion zone for kiln with core.

4.2. Heating value of syngas production.

For the result of heating value of syngas production, figure. 5 shows the effect of type of kiln and heating value of syngas production that the heating value of syngas received from gasifier with axial core flue gas circulating kiln was higher than the case of without the core. This result corresponds to the increased carbon monoxide (CO) content. It was found that the CO content produced by the gasifier with axial core flue gas circulating kiln was increased by 25% compared to the gasifier without the core. In installation of the core, the gas has been heated for a long time to produce more CO. Some of tar may was destroyed become CO. However other gases, carbon dioxide (CO_2), hydrogen (H_2), and methane (CH_4), were almost the same.

IOP Conf. Series: Materials Science and Engineering **297** (2018) 012062 doi:10.1088/1757-899X/297/1/012062

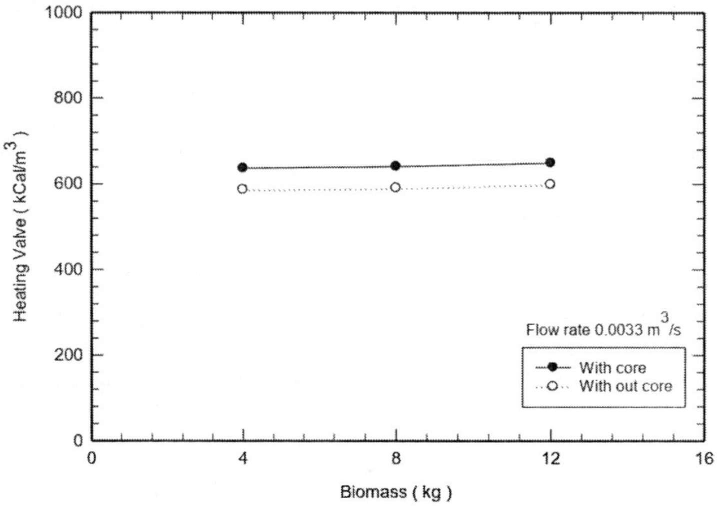

Figure 5. Effect of type of kiln and heating value of syngas production.

4.3. Syngas production efficiency

The syngas production efficiency was calculated by the ratio of mass of syngas production and the consumption of biomass. As shown in figure. 6, the graph shows the effect of type of kiln and syngas production efficiency. It was found that increasing of fuel consumption affects to decrease the syngas production efficiency. Syngas production efficiency obtained from the gasifier with axial core flue gas circulating kiln was higher than the gasifier without core by 10% for fuel consumption 8 and 12 kg. However, the experimental result shows that the highest efficiency was 30% at 4 kg supplying biomass. In comparison, the efficiencies of both kilns with and without the core were not different.

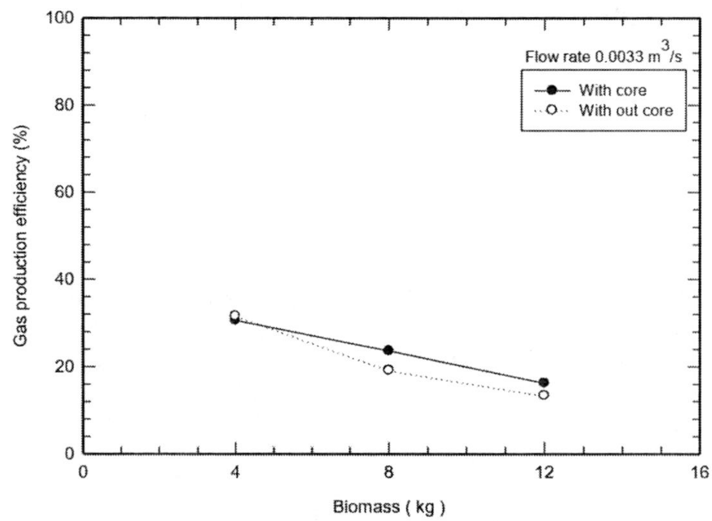

Figure 6. Effect of type of kiln and syngas production efficiency.

IOP Conf. Series: Materials Science and Engineering **297** (2018) 012062 doi:10.1088/1757-899X/297/1/012062

4.4. Liquid product
When moisture, gas and smoke condensed at the condenser, it became a liquid product. As shown in figure 7, the experimental result showed the effect of type of kiln to the liquid product quantity. The results show that a liquid product obtained from the gasifier without core was higher than the gasifier with core by 30-40% for all supplying biomass. This result corresponds to the increased of the syngas production efficiency. Due to the installation of the core was increasing time to reduce and changed a liquid product to be syngas. Expecting some of tar may was reduced in this process. Due to the axial flow system, syngas flows back to the combustion zone again before it is sent to the outlet pipe. The liquid product is destroyed become more gas.

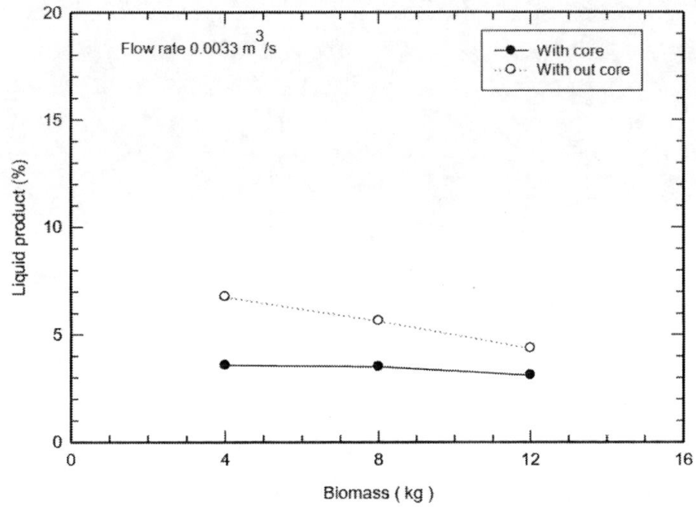

Figure 7. Effect of type of kiln and liquid product.

5. Conclusion
The present study conducted about the possibility for reducing tar in syngas. The updraft gasifier was modified by installing an axial core flue gas circulating tube in the center line of the kiln. From the experimental results and discussion of experimental data, the issue can be summarized as follows:

1. The temperatures in the combustion and the reduction zone of the kiln with core were lower than the kiln without core owing to installing the core would reduce the combustion zone area in biomass burning. For other zones were almost the same.
2. The syngas production efficiency of both the kilns with and without the core was not different. The highest efficiency was 30% at 4 kg supplying biomass.
3. The heating value of syngas received from the kiln with core was higher than the kiln without core that corresponds to the increased of carbon monoxide (CO). Carbon monoxide produced the kiln with core was increased by 25%.
4. The liquid product obtained from the kiln without core was higher than the kiln with core by 30-40% for all supplying biomass. The experimental result points that some of tar may was reduced and changed to be more gas.

References
[1] Food and Agriculture Organization of the United Nations 1986 *Wood gas as engine fuel* (Rome, Italy: FAO)
[2] Milne T A, Evans R J and Abatzoglou N 1998 *Biomass Gasifier "Tars": Their Nature, Formation, and Conversion* (Colorado: NREL/TP-570-25357)
[3] Kivisaari T, Bjornbom P and Sylwan C 2002 *J. of Power Sources.* **104** pp 115-124

[4] Tinnakorn P and Sangwan B 2016 *13th Eco-Energy and Materials Science and Engineering Symposium* (Udonthani: Thailand) p 204-207
[5] Anil K. R 1986 *Alternative Energy in Agriculture* (Maharashtra: CRC Press)
[6] Yan C, Yang W, John T. R and Wei-Ping P 2006 *Fuel Processing Technology* **87** pp 343 – 353
[7] M. Dogru, C.R. Howarth, G. Akay, B. Keskinler and A.A. Malik 2002 *Energy* **27** pp 415–427
[8] Murat D, Adnan M and Colin R. H. 2002 *Fuel Processing Technology* **75** pp 55–82
[9] Adnan M, Murat D, Colin R. H, Mike J. L and Teoman A 2001 *Energy Conversion & Management* **42** pp 157-172
[10] Philippe M and Raphael D 2002 *Energy Conversion and Management* **43** pp 1291–1299

Study on light and thermal energy of illumination device for plant factory design

A. Yoshida[1,*], K. Moriuchi[2], Y. Ueda[2] and S. Kinoshita[1]

[1] Department of Mechanical Engineering, Osaka Prefecture University, 1-1 Gakuen-cho, Naka-ku, Sakai 599-8531, Japan

[2] Technology and Development Department, Seiken Co., Ltd, 1-2-18 Katsube, Toyonaka 561-0894, Japan

* Corresponding Author: ayoshida@me.osakafu-u.ac.jp

Abstract. To investigate the effect of illumination devices on the yield of crops cultivated in a plant factory, it is necessary to measure the actual cultivation environmental factors related to the plant growth and understand the distribution ratio of light and thermal energy to the electrical energy injected into the illumination device. Based on cultivation results, we found that light intensity greatly affected the growth of plant weight. Regarding the selection of illumination device, its spectral components also affected the morphological change. Lighting experiments using a high frequency (Hf) fluorescent lamp and a light emitting diode (LED) bulb were performed. A certain difference was found in the distribution ratio of light energy to electrical energy between Hf and LED. It was showed that by placing the safety equipment or internal circuits outside the cultivated site, the air conditioning load could be reduced.

1. Introduction

There is increasing awareness of the destabilizing effects of abnormal weather on crop production, agricultural businesses, and food safety. Therefore, the focus is on plant factories, which allow for stable crop production, and efficient and safe vegetable production. However, plants produced artificially in factories are more expensive than those produced in open air. To lower production costs, the cultivation environment needs to be appropriately controlled, and the growth rate of plants needs to be improved [1]. The following environmental factors are associated with growth: air temperature, humidity, wind speed, CO_2 concentration, light strength, and nutrient components. The decrease in productivity, owing to inconsistency in cultivation quality, is the effect of inconsistent environmental factors in these factories. To optimize the cultivation environment and air-conditioning in plant factories, it is important to evaluate first their air temperature, humidity, wind speed, and CO_2 concentration distribution with a numerical simulation [2]. The relation among the harvest, the light strength and the light quality were discussed, and the required lighting energy consumption was estimated [3]. The effects of lighting on the cultivation state are large. It is required to offer comprehensive information about the lighting to design the plant factory.

In this study, we measured the actual cultivation environmental factors and fresh weights of lettuce in a plant factory, and analyzed the effects of these factors on its growth. Based on the results, we investigated the effects of environmental factors, especially light intensity, on the fresh weights of plants by measuring the distribution of factors and fresh weights of plants cultivated in a plant factory. With respect to the factor related to the selection of illumination device, the effect of not only light intensity but also wavelength distribution on growth of plants was examined [4-7].

In a fully artificial-light type of plant factory, which uses illumination devices in place of sunlight, there is little cooling load caused by ventilation or wall heat transmission, and approximately 80 % or more of the cooling load comes from heat generated by illumination devices [8, 9]. Therefore, the illumination device has a significant effect on the air conditioning load. Further, it is considered that the photosynthetic rate is also affected by the lighting in the cultivation room. The distribution of the various types of energy by the illumination devices has been investigated in the past [10], but LED has not been measured and the other lighting devices have gradually become more efficient. The heating value of the illumination devices, including LED [11], is clarified, but the energy distribution ratio of light in relation to normal rated power has not been investigated.

In this study, in order to comprehend the effect of the latest illumination devices on air conditioning load, lighting experiments using high frequency (Hf) fluorescent lamps and LED are performed, and the distribution rate of light to electrical energy from each illumination device is investigated. The results of these experiments are incorporated into an air conditioning simulation, which included transpiration and photosynthesis of the cultivated plants [2], and its effect on crop yields is investigated. Based on these results, methods for reduction of air conditioning load in plant factories are proposed.

2. Effect of light intensity on plant growth

2.1. Effect of light intensity on plant growth weight

2.1.1. Cultivation room and measurement method. We examined the effect on frill lettuce (Lactuca sativa L.) growth of light intensity, air temperature, and wind speed within the measured cultivation room. In addition, by comparing them against the results of cultivation experiments performed in an environment-simulator room, in which air temperature, humidity, and wind speed were regulated in turns, we evaluated the regulated conditions of these environmental factors in the cultivation room.

Figure 1 shows plane and cross-section views of the actually measured cultivation room for this study, and the measurement locations of each cultivation shelf. The cultivation room had six four-tier cultivation shelves. We measured the air temperature, humidity, and wind speed, and light intensity from the top to the third tier of one shelf.

The regulated air temperature of the cultivation room was 22 °C, the light period (daytime) was 16 hours, the dark period (nighttime) was 8 hours, and the CO_2 concentration was kept at 600 ppm. An Hf fluorescent lamp (NEC, FHF32EX-D-HX-S, 32 W, 1200 mm) was used as lighting equipment. The number of lamps was four per shelf. The lettuce grew on the cultivation shelf three weeks after seeding, and therefore it was harvested three weeks after cultivation. Each tier of the cultivation shelf was divided into six sections, and 4 x 9 plants of lettuce were cultivated per section. The air temperature, humidity, wind speed, and CO_2 concentration in each section were constantly measured at the positions shown in figure 1, at five-minute intervals from planting to harvest. Furthermore, the light intensity affecting photosynthesis was evaluated in terms of photosynthetic photon flux density (PPFD). PPFD is a value expressed in number of light quanta in the wavelength range of 400 nm to 700 nm, which is the effective for photosynthesis. We measured the PPFD per cultivation position in the state prior to lettuce planting. The lettuce was harvested over three days, and then winnowed according to the cultivation day (21 to 23 days). We measured not only the weight of the lettuce after harvesting, but also the fresh weight of seedlings before planting.

2.1.2. Experimental results and discussion. The harvest took 3 days (cultivation period of 21 to 23 days) owing to the very large number of lettuces. A lettuce with 23 days of cultivation was larger than that harvested in a different day or that had been cultivated in the third stage of the shelf. Therefore, using the fresh weight after 23 days cultivation, the effect of each environmental factor on plant growth was investigated. figure 2 shows the relationship between light intensity and fresh weight of the harvested lettuce. The light intensity distribution within the shelf was determined. The tendency that light intensity was lower around the shelf than at the center was confirmed. The fresh weight of lettuce increased with increasing light intensity. It is known that the photosynthetic rate of lettuce increases proportionally with increasing light intensity, and converges to a constant value when it reaches 350 μmol $m^{-2}s^{-1}$ [12]. In this cultivation room, it was found a range that increased with increasing light intensity, as shown in figure 2. There was a high correlation of produced weight with light intensity, but a certain deviation was recognized. Besides light intensity, the effects of other environmental factors, such as wind speed, and the individual differences of cultivated plants should be examined [1].

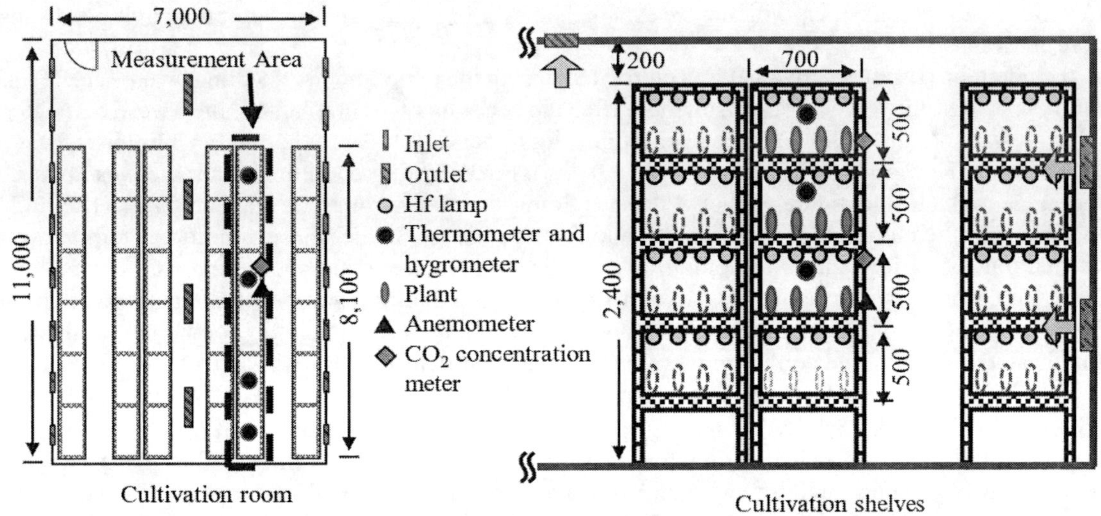

Figure 1. Actual cultivation room and measurement points (dimensions in mm).

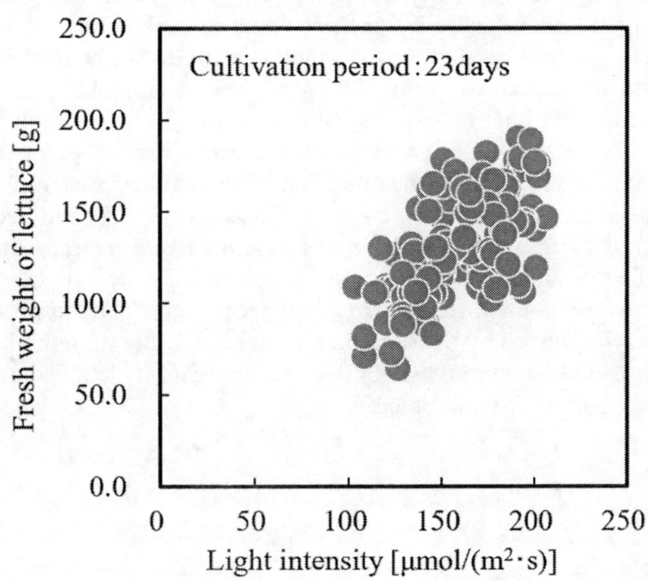

Figure 2. Relationship between light intensity (PPFD) and fresh weight.

IOP Conf. Series: Materials Science and Engineering **297** (2018) 012063 doi:10.1088/1757-899X/297/1/012063

2.2. Effect of light quality on plant growth rate

2.2.1. Cultivation equipment and conditions. The experiments were carried out in the cultivation environment simulation room at the R&D Center of the Plant Factory, Osaka Prefecture University. In this facility, there is a ventilator located on each cultivation rack, and environmental factors such as air temperature and wind speed can be controlled by manipulating the position of the ventilation duct, wind direction, and wind speed. The facility has Hf fluorescent lamps (Panasonic, FHF16EX-D-H, 23 W, 600 mm) and LED lighting (Shibasaki, DP-10344-01, 17 W, 600 mm). Because LED is small in size, has long life and strong intensity, and the wavelength of emitted light can be freely changeable, its introduction to the plant factory has been an advance [8]. The experimental cultivation was carried out in various cultivation zones, which were established by altering the air temperature, wind speed and CO_2 concentration.

Frill lettuce (Lactuca sativa L.) was cultivated during the experiments. Seedlings were selected based on their weight three weeks after sowing. After the seedlings were replanted, they were cultivated for two weeks under the following basic conditions, light: dark cycle 14:10 h, relative humidity: 70 %, and feed: Otsuka House 1/2A (EC 1.3–1.5, pH 5.0–6.5). Four lighting condition zones were established, as listed in Table 1, and figure 3 shows these lighting condition zones in practice. Hf3, Hf4, and Hf7 indicate that the number of lamps is 3, 4, and 7, respectively. Under these conditions, cultivation was carried out for various air-conditioning factors (air temperature, wind speed, and CO_2 concentration). Nine plants, as shown in figure 4, were cultivated in each condition zone; five of them (positions A, C, E, G, and I) were used for growth analysis, and the other four were utilized for component analyses [13]. In this study, the adopted condition was 20 °C in air temperature, 1000 ppm in CO_2 concentration, and 0.55–1.12 ms^{-1} in wind speed. Air temperature, relative humidity, and CO_2 concentration were controlled within ±0.3 °C, ±5 %, and ±80 ppm, respectively. We measured not only the fresh and dry weight of the lettuce after harvesting, but also the fresh weight of seedlings before planting. The leaf total area was also measured by separating all leaves from a stem, and displaying them planarly. The net assimilation rate (NAR) $gcm^{-2}day^{-1}$ was obtained by dividing the average dry weight change rate during the cultivated period by leaf total area.

2.2.2. Experimental results and discussion. Regarding lighting conditions, the light intensities of LED and Hf3 were the same, as presented in Table 1. For these conditions, the effect of light quality on plant growth was investigated. The spectral distributions are shown in figure 5. The relative values of lighting energy per unit wavelength are indicated, with 100 % for each maximum value. A peak value is recognized at one wavelength for Hf, but there are three peak values in wavelength for LED.

The NAR of plants cultivated with LED was around 2 times that of plants cultivated with Hf fluorescent lamps, as shown in figure 6(a). On the other hand, the total leaf area of plants cultivated with Hf was larger than that with LED, as shown in figure 6(b). Finally, with a similar level of light intensity, the fresh weight increase, i.e., the growth rate, was similar with LED and Hf, as shown in figure 6(c). As shown in figure 6(d), the dry weight of plants cultivated with LED was slightly larger than that of plants cultivated with Hf, because of the difference in the water content. The morphological change was caused by the quality of light expressed by the spectral distribution of the illumination device [4-7, 14]. The lettuce cultivated with LED had a large weight per leaf area and its leaf became thick, while that with Hf had a large leaf area, and its volume became large.

Table 1. Light intensity at cultivation point.

No.	LED	Hf3	Hf4	Hf7
			[μ mol m^{-2} s^{-1})]	
A	202.3	206.0	288.0	567.7
C	183.0	182.0	254.3	465.3
E	237.7	236.0	315.7	628.0
G	196.7	199.0	293.0	539.0
I	184.0	185.7	265.0	465.3
Average	200.7	201.7	283.2	533.1

Figure 3. Lighting condition zones in cultivation environment simulation room.

IOP Conf. Series: Materials Science and Engineering **297** (2018) 012063 doi:10.1088/1757-899X/297/1/012063

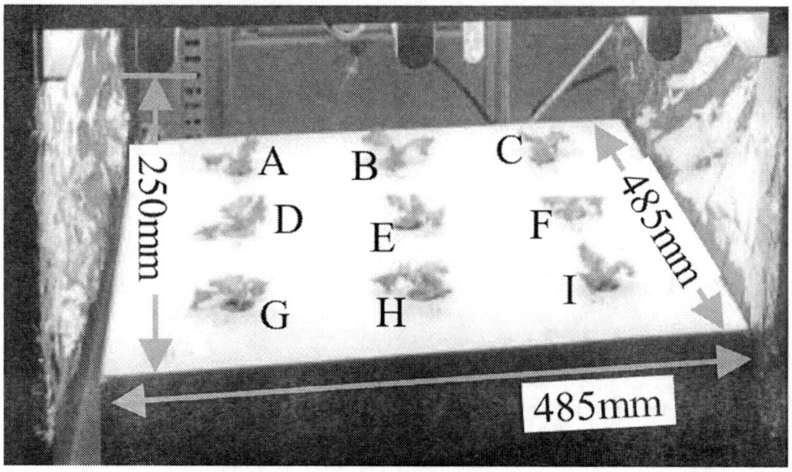

Figure 4. Cultivation situation.

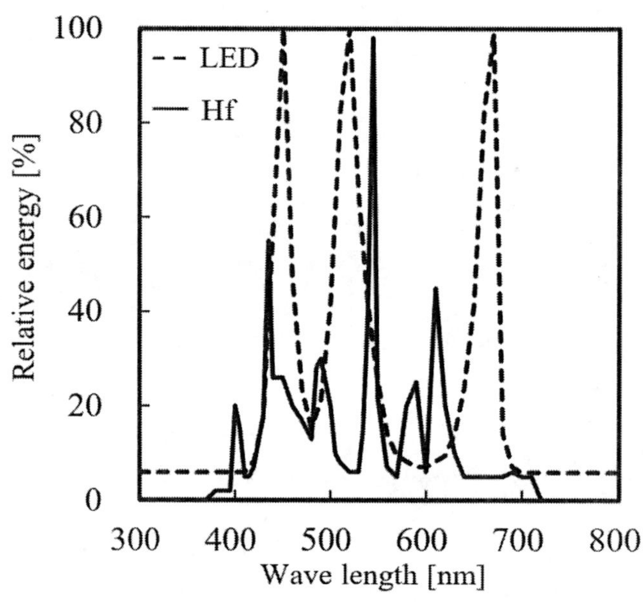

Figure 5. Spectral distribution of Hf and LED.

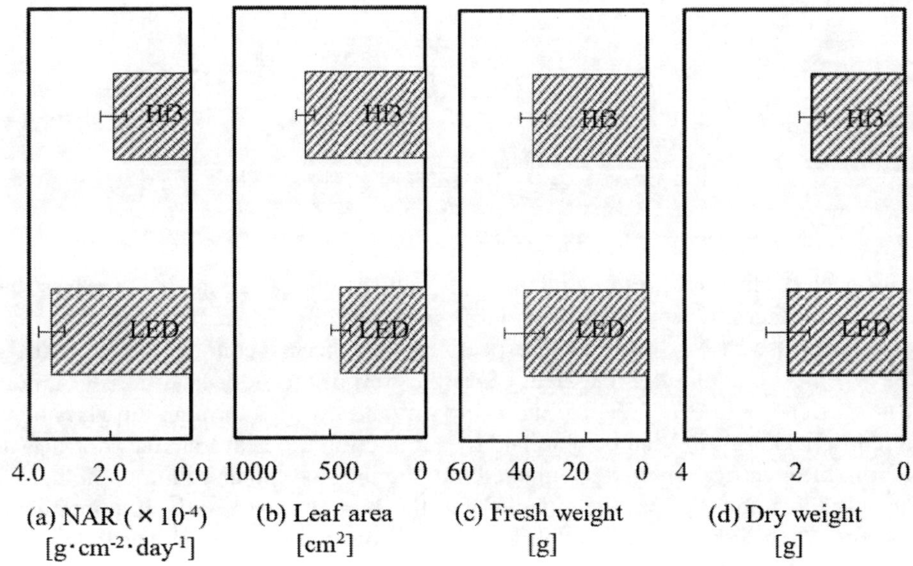

Figure 6. Effect of light quality on growth rate.

3. Energy distribution of illumination device

3.1. Experimental equipment and method

The experimental equipment is shown in figure 7. There are insulation materials (polystyrene foam 30 mm thick) on areas without ceiling, and the ceiling section has glass with a thickness of 5 mm (float board glass). The illumination device is placed in the center so that it shines on the ceiling area. Two small fans are placed inside, which operate during the experiment to maintain internal uniformity of air. The inside of the illumination device and experimental equipment, including the fan, are tightly sealed. The internal air temperature and wall, floor, and ceiling internal and external surface temperatures are measured using thermocouples (type T, wire diameter of 0.65 mm). Further, the small fan and illumination device voltage and current are measured using a clamp wattmeter. The measurement positions for the air temperature, and floor, wall, and glass surface temperatures are shown in figure 8.

Figure 9 shows the internal air temperature over time during the experiment. The experiment first runs the fan internally, in a lights-off state. After operation of the internal fan, the air temperature becomes constant in about 2 hours and reaches a steady state. At this time, the differences between internal and external temperatures at the measurement locations are constant, in the same way as the internal temperature. After reaching a steady state, the light is turned on and measurement continues until the differences between internal and external temperatures at each measured location become constant. As in the case of the illumination device, the temperature increased immediately after turning on the light, but approximately 2 hours after, it became constant. The temperature external to the experimental equipment was controlled to become constant, and maintained at approximately 24 °C ±2 °C. Illumination of the lighting device was not controlled. Based on the above measurement results, the energy distribution ratio ω can be obtained according to the following procedure.

The electrical energy is converted to heat and light in the illumination device. The thermal energy heats the illumination device, and the heat generated by the illumination device heats the surrounding air through convective heat transfer, and this is finally transmitted to the wall, floor, and ceiling surfaces. This amount of energy is calculated as heat transmission from the differences between internal and external temperatures of the wall, floor, and ceiling. The heat transfer amount Q_W for the floor and wall

IOP Conf. Series: Materials Science and Engineering **297** (2018) 012063 doi:10.1088/1757-899X/297/1/012063

can be obtained from equation (1). The ceiling heat transmission amount Q_G can be obtained in the same way as equation (2).

$$Q_W = k_W \frac{(T_{Win} - T_{Wout})}{d_W} A_W \tag{1}$$

$$Q_G = k_G \frac{(T_{Gin} - T_{Gout})}{d_G} A_G \tag{2}$$

where, for each of the variables in equations (1) and (2), the subscript W expresses the values for the floor and wall surfaces, G represents the values for the ceiling surface, and the subscripts of internal and external are expressed as "in" and "out," respectively. k is thermal conductivity, T is surface temperature, d is material thickness, and A is the area. Q_W is the sum of the floor heat transfer amount Q_{W1} and the wall heat transfer amount Q_{w2}. Visible lights are driven outwards through the glass of the ceiling. This transmitted light energy $E_L{}'$ can be obtained by subtracting the heat transfer Q_{W1}, Q_{W2} and Q_G for each surface from the electrical energy Q_E injected into the internal fan and illumination device. Further, the heat transmission amount Q_G can be thought of as the sum of Q_{G1} from the illumination device and Q_{G2}, which occurs when light passes through the glass. From the above, the distribution ratio ω is found, according to the following equation:

$$\omega = \frac{E_L}{Q_E} = \frac{E_L{}'/C_G}{Q_E} = \frac{(Q_E - Q_{W1} - Q_{w2} - Q_G)/C_G}{Q_E} \tag{3}$$

where C_G in equation (3) is the glass board transmission rate. When the light energy E_L obtained from equation (3) is multiplied by the glass board absorption, the value of Q_{G2} is calculated. The value of Q_{G1} is given by subtraction of Q_{G2} from Q_G.

In this experiment, 32 W type Hf fluorescent lamps (Hitachi Appliances, HNK4205V-MEN14) and 20 W type LED bulbs (Hitachi Appliances, NHE4205-JI14D) are employed. The number of lamps and bulbs is two, as shown in figure 7. The illumination of both devices is essentially the same, and it is that usually required in plant factories. Further, it is known that heat is also generated from areas without fluorescent lamps, such as the ballasts in Hf or internal circuits in LED. It was considered possible to reduce the internal load by locating the ballast or internal circuit outside of the cultivation room; thus, the experiments are performed with the respective ballast and internal circuit located outside of the cultivation room.

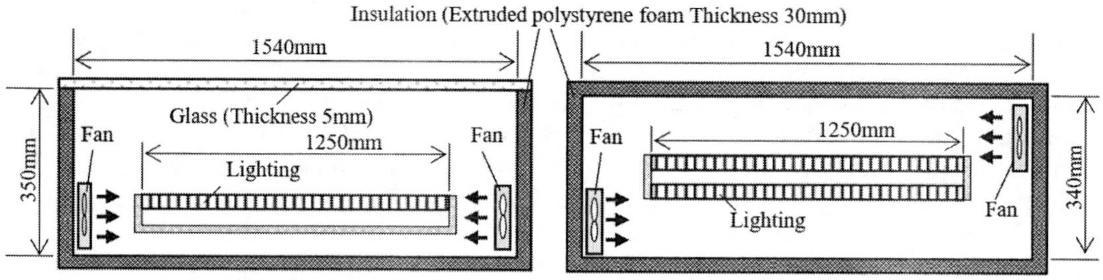

Figure 7. Experimental equipment of lighting experiment.

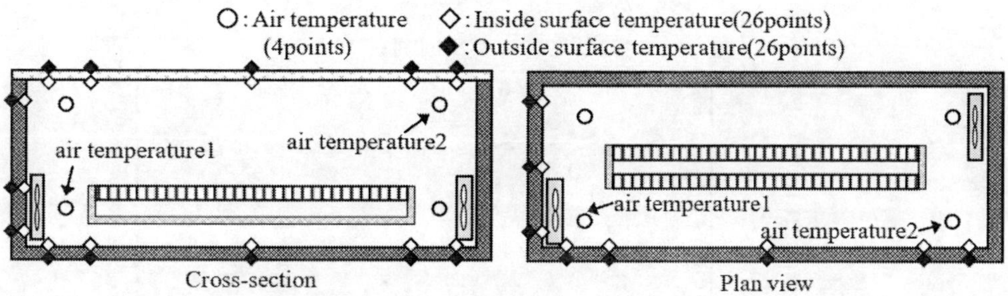

Figure 8. Measurement points.

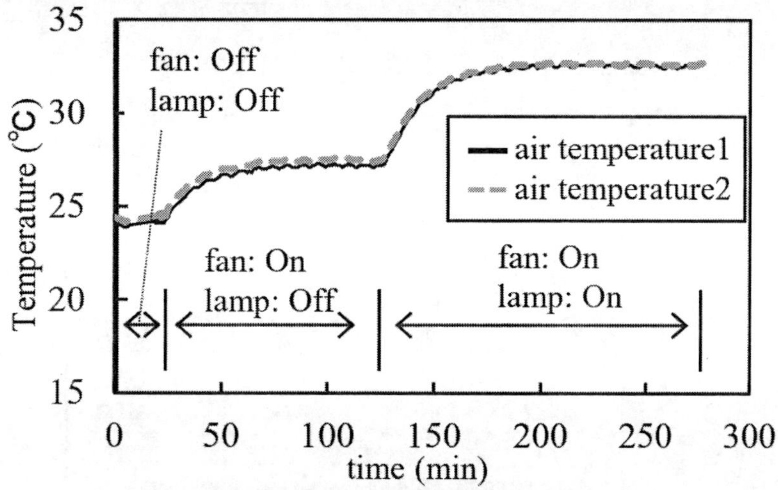

Figure 9. Internal air temperature during experiment.

3.2. *Experimental results and discussion*

Figure 10 shows the internal surface temperature T_{in} of the floor surface and the difference between external and internal surface temperatures on the floor ΔT after turning on the lights. The internal surface temperature on the floor surface was higher with Hf than with LED. The difference between the internal and external temperatures of the floor surface was smaller with LED than with Hf. The temperature differences of glass and wall also showed the same tendencies. The temperatures with both Hf and LED were slightly higher around the center, but this is considered to be influenced by the heat generated by the ballast and internal circuit.

The difference between the internal and external surface temperatures on the floor surface in the experiment with the Hf without ballast and with LED without internal circuit is shown in figure 11. The ballast or internal circuit was placed outside of the experimental box as shown in figure 7. By removing the ballast and internal circuit, the difference between the internal and external temperatures decreased. The amount of this decrease was more significant in the case with LED than in that with Hf.

The values used for the calculation are shown in Table 2 [15, 16]. The results of the analysis of energy distribution by equations (1) and (2) with the experimental results are shown in figure 12. The

IOP Conf. Series: Materials Science and Engineering **297** (2018) 012063 doi:10.1088/1757-899X/297/1/012063

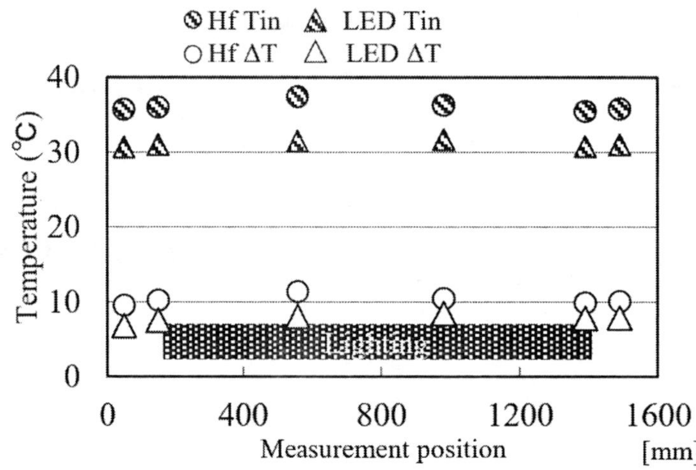

Figure 10. Temperature distribution at floor.

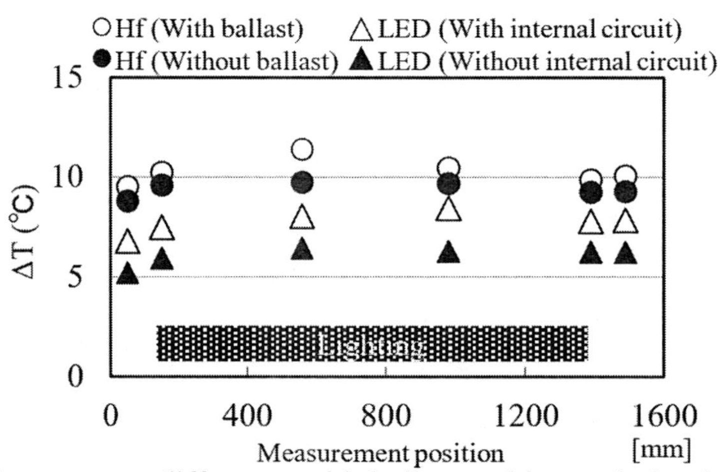

Figure 11. Temperature difference with ballast and internal circuit at floor.
Figure 11. Temperature difference with ballast and internal circuit at floor.

number of lighting experiments was three for Hf and two for LED, respectively. The heat transfer amount of the wall Q_{w2} and glass $Q_G(=Q_{G1}+ Q_{G2})$ are virtually the same, and the heat transmission amount of the glass is approximately 2.5 times that of the floor Q_{W1}. However, no difference in the distribution ratio was seen between Hf and LED. As the power consumption of the illumination device measured with the clamp wattmeter showed a difference of approximately ±1 W from the normal rated power, Q_E in equation (3) is treated as the normal rated power. Further, the float glass transmission and absorption ratios were 0.83 and 0.10, respectively [16]. Based on the above assumption, the energy distribution rate ω can be obtained from equation (3). The result for LED, 0.56, was slightly larger than that for Hf, 0.52.

Figure 13 shows the results of the energy distribution with Hf without the ballast and with LED without the internal circuit. The number of lighting experiments was two for LED and Hf, respectively.

The values of Q_{W1}, Q_{W2}, and Q_G were calculated from the measured data. As the value of E_L', the averaged value shown in figure 12 was used. From the diagram, it can be seen that when the ballast is removed from the Hf, the amount of heat generated by the illumination device is reduced by approximately 2 W. In the same way, the amount of heat generated by LED is also reduced by 3 W. The ω of Hf is 0.52 and the generated heat amount is 31.2 W. Therefore, this is a reduction in the lighting load of approximately 6 %. The heat amount generated by LED is 19.8 W, representing a decrease in the load of about 15 %. From the above, it is thought that placing the ballasts or internal circuits outside of the cultivation room can reduce air conditioning load. It is considered that the reduction is more effective in the case of LED than that of Hf.

Table 2. Physical values of experimental equipment.

	Thermal conductivity	Thickness	Area		
			Floor	Wall	Roof
	[W m^{-1} °C^{-1}]	[m]	[m^2]		
Extruded polystyrene foam	0.038*	0.030	0.524	1.316	-
Glass	0.030**	0.005	-	-	0.524

* and ** refer to [15] and [16], respectively.

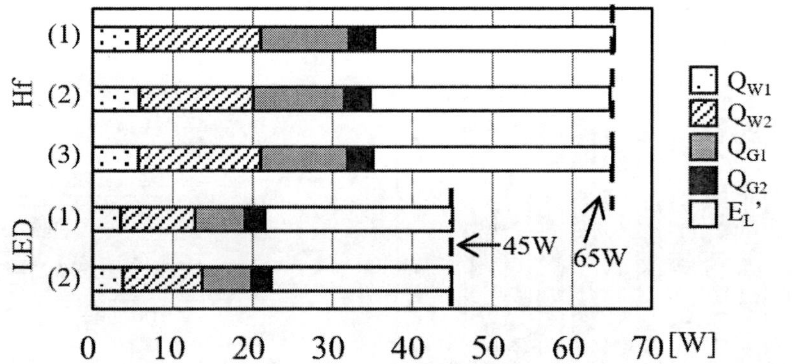

Figure 12. Heat and light energy distribution of illumination device.

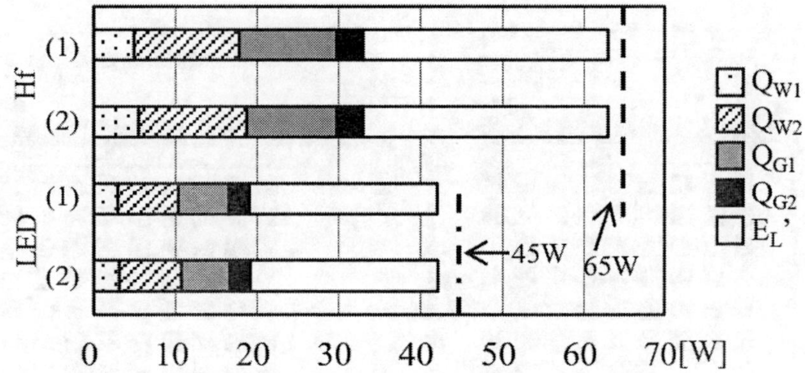

Figure 13. Effect of ballast on Hf and internal circuit on LED.

The distribution ratio of light energy against electrical energy obtained in the lighting experiments was incorporated into the air conditioning simulation and the effect of different illumination devices on the temperature distribution within the cultivation room was investigated. An analytical model based on the cultivation room of a plant factory is shown in figure 14. Regarding the analysis conditions, air flow velocity is 0.1 ms^{-1}, supply air temperature is 23 °C, and relative humidity is 70 %. In the lighting experiment, simulations of both Hf and LED lighting were conducted by applying the temperature distributions obtained in the experiments. The cultivated plants were assumed as leaf lettuces, for which the characteristic values had been investigated in previous research [2]. The analytical results of the temperature distribution in the cultivation room for Hf and LED are shown in the thermal distribution diagram in figure 15. The amount of heat generated by the illumination device increased the air temperature surrounding it. Further, it was recognized that the temperature surrounding the plants was higher in the case of Hf than with LED. From the above, it is considered that Hf has a greater effect on air temperature distribution within the cultivation room than LED. Thus, it is suggested that the cooling load can be reduced by using LED instead of Hf.

The photosynthetic rate is estimated under the condition of 22.5 °C in air temperature, 75.5 % in relative humidity, 0.8 ms^{-1} in wind speed, and 1000 ppm in CO_2 concentration [2]. The electric power consumption of LED was assumed the same as HF, with 65 W. The heat amount generated from both illumination devices is almost the same. The value of PPFD of LED is larger than that of Hf. The results are shown in figure 16 as a function of the distribution ratio of light energy for each illumination device. It was found that there is a difference in photosynthetic rate, i.e., product amount, between the illumination devices, even though the electric power consumption of the devices and their cooling loads are the same. If an illumination device with higher conversion efficiency is developed, a decrease in air conditioning load and an increase in harvest can be expected.

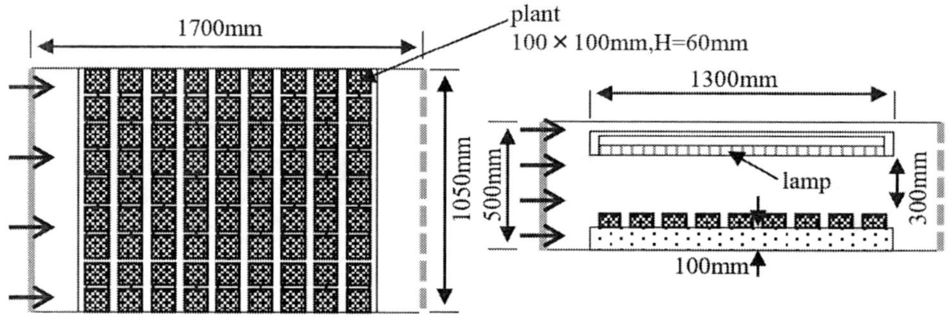

Figure 14. Simulation model.

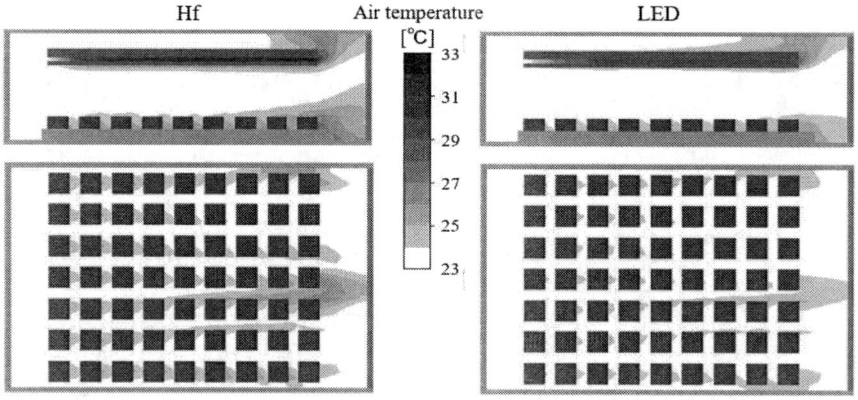

Figure 15. Calculated result of air temperature.

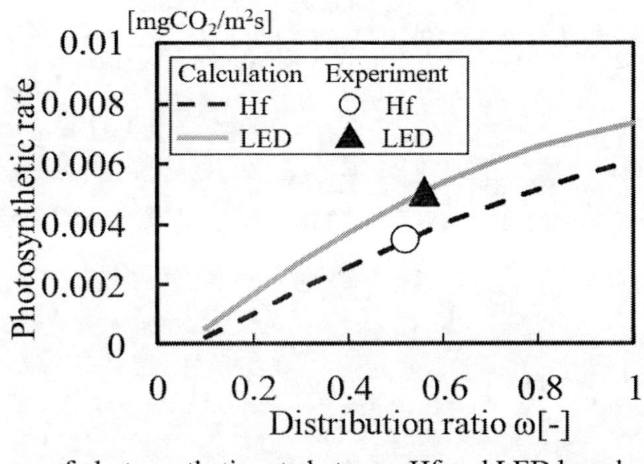

Figure 16. Difference of photosynthetic rate between Hf and LED based on lighting experiment.

4. Conclusion

To study the effect of lighting devices on air conditioning load in plant factories, lighting experiments were performed, and by incorporating those results into an air conditioning simulation with plant activity, the following results were obtained.

1) The light intensity distribution within the actual cultivation shelf was recognized. The fresh weight of lettuce increased with increasing light intensity.
2) The light spectral components affected the morphological state.
3) From the lighting experiment, the distribution rate ω of light to electrical energy was 0.52 for Hf and 0.56 for LED, and no major differences were seen.
4) By removing the ballasts and internal circuits, there was a reduction in lighting load of 6 % for Hf and 15 % for LED. Therefore, by placing the ballasts or internal circuits outside the cultivation room, air conditioning load can be reduced.
5) The result of integrating the experiment results in the simulation showed that Hf had a greater effect on increasing temperature in the vicinity of the plants than LED. By using LED, therefore, it is possible to suppress temperature increases in the area around plants.
6) The photosynthetic rate is higher for LED than for Hf under the same power consumption. Therefore, a greater yield can be expected by using LED.

References

[1] Moriuchi K, Ueda Y, Okamura N, Yoshida A, and Kinoshita S 2016 *Proc. 4th Int. Forum on Heat Transfer* (Sendai, Japan) 6 pages
[2] Komatsu S, Ueda Y, Okamura N, Inou T, Yoshida A, and Kinoshita S 2014 *Trans. of the JSRAE* **31**(3) p 345
[3] Furuyama S, Ishigami Y, Hikosaka S and Goto E 2017 *J. Soc. High Tech. in Agriculture* **29**(2) p 60
[4] Yorio N C, Goins G D, Kagie H R, Wheeler R M, and Sager J C 2001 *Hort Sci.* **36**(2) p 380
[5] Folta K M, and Childers K S 2008 *Hort Sci.* **47**-7 1957
[6] Fuyuyama T, Ohashi K, Ono E, and Watanabe H 2013 *J. Soc. High Tech. in Agriculture* **25**(4) p 175
[7] Yamamoto D, Fukumura K, Matsui M, and Saito T 2015 *Eco-Eng.* **27**(1) p 1
[8] Kozai T 2012 *Plant factory with artificial light* (Tokyo: Ohmsha)
[9] Ohyama K, Yoshinaga K, and Kozai T 2000 *J. Soc. High Tech. in Agriculture* **12**(3) p 160
[10] Morita M 1977 *National Technical Report* **23**(5) p 674
[11] Miyaoka Y, Nakayama H, Teranishi Y, Yoshizawa N, and Hirota M 2016 *Trans. of the JSRAE* **33**(2) p 125
[12] Tatsumi M and Hori Y 1969 *Bulletin of the Horticultural Research Station.* Series **A**-8 p 127

[13] Yoshida A, Okamura N, Furuakawa H, Myojin C, Moriuchi K, and Kinoshita S 2017 *AIP Conf. Proc.* **1851** 020087-1 doi:10.1063/1.4984716
[14] Mortensen L M, and Stromme E 1987 *Scientia Horticulture* **33** p 27
[15] SHASE 2002 *SHASE Handbook* **13** (Tokyo: Maruzen) p 130
[16] Asahi Glass Co Ltd 2015 *Architectural Glass General Catalog Technical Appendices*

IOP Conf. Series: Materials Science and Engineering **297** (2018) 012064 doi:10.1088/1757-899X/297/1/012064

Improvement for enhancing effectiveness of universal power system (UPS) continuous testing process

Lerdlekha Sriratana*

Faculty of Engineering, Ramkhamhaeng University, Huamark, Bangkapi, Bangkok, Thailand 10240

*Corresponding Author: lerdlekha_s@ru.ac.th

Abstract. This experiment aims to enhance the effectiveness of the Universal Power System (UPS) continuous testing process of the Electrical and Electronic Institute by applying work scheduling and time study methods. Initially, the standard time of testing process has not been considered that results of unaccurate testing target and also time wasting has been observed. As monitoring and reducing waste time for improving the efficiency of testing process, Yamazumi chart and job scheduling theory (North West Corner Rule) were applied to develop new work process. After the improvements, the overall efficiency of the process possibly increased from 52.8% to 65.6% or 12.7%. Moreover, the waste time could reduce from 828.3 minutes to 653.6 minutes or 21% , while testing units per batch could increase from 3 to 4 units. Therefore, the number of testing units would increase from 12 units up to 20 units per month that also contribute to increase of net income of UPS testing process by 72%.

1. Introduction

In recent global business, industrial processes need to be improved in order to ensure the quality in both national and international levels as well as to reduce waste and excessive costs of production for price competitiveness. Therefore, several methods have been applied to encourage the success of process improvement such as work study, time study, and work scheduling [1, 2]. Electrical and electronic industry is one of the major industrial businesses in Thailand, which has played an important role for years. Similar to other countries, the electrical appliance industries in Thailand have been controlled by the government as adopting international and national standards to ensure the quality and safety of the products. Normally, there are several tasks testing required in order to achieve the industrial standards before distributing to the customers, which would take a long time to complete the production process and lead to excessive costs in addition to the products [3, 4]. Electrical and Electronic Institute is the independent organization that is responsible for testing electrical appliances, that includes Universal Power System (UPS) regarding to TIS 1291-3-2555 standard. Initially, there was only one operator in UPS continuous testing process with the average work efficiency was about 52.8%. Therefore, waste time can be observed in this testing process and work target could not meet its requirements.

This study aims to enhance the effectiveness of UPS continuous testing process of Electrical and Electronic Institute by improving the overall efficiency while reducing waste time in testing process. Yamazumi chart and job scheduling theory (North West Corner Rule) were adopted by determining the standard time and developing new testing process. After the improvement, the developed work process would enhance the effectiveness of the operator who contributes to the well-organize work plan of the institute.

2. Methodology

For Electrical and Electronic Institute, the major problem of UPS testing process was the testing target could not met its requirement on time. Therefore, time Study was applied to analyse the standard time of well-trained operator with normal pace. Work study and time study are related to the work ability per unit of time (second or minute) regarding to work standard. Time study can be further used to develop work schedule and work plan.

2.1. Yamazumi Chart

Comparison of cycle time (the time from the beginning to the end of process) and takt time (the average time between the start of production of one unit and the start of production of the next unit) were shown in Yamazumi Chart. Both cycle time and takt time were used to rearrange and develop the appropriate work plan to improve the effectiveness and also reduce wastes in the production line [5]. Yamazumi chart is therefore commonly and effectively used for time study in several production processes such as assembly line, forming process and also for production flow analysis [6-8].

2.2. North West Corner Rule (NCR)

Job scheduling method namely the North West Corner Rule (NCR) was also applied for evaluating a basic feasible solution to rearrange new operation process of UPS testing. Theoretically, NCR is a heuristic approach which is used to solve the problem by permutation in order to control the daily-used resources based on the demand and managing the demand, which does not override the supply such as the optimal delivery plan (from factories to warehouses) in supply chain model [9-11].

2.3. UPS testing process

UPS testing process is responsible for issuing the certificate of TIS 1291-3-2555 [12] to manufacturers in order to guarantee the quality of the products before distributing them to the customers. All UPS products with requirement of this standard should be performed by testing process according to the specification and label provided by the manufacturer. Normally, there were 3 units of UPS per batch required for testing. The average values of testing time spending in each task per unit and per batch were assessed. Details of testing are as given in Table 1 [13] and Appendix A.

To improve the effectiveness of UPS testing process, details of research methodology are as follows:

- Analyse initial work process and develop new one by rearranging daily tasks with consideration of standard time as well as feasible and optimal solution (NCR method) and limitation of each task. Moreover, new testing process should not influence on the results of UPS testing and should be applicable to all UPS products with requirement of TIS 1291-3-2555.
- Consider the possibility to increase testing unit from 3 units to 4 units per batch.
- Assess standard time of each testing task before and after improvement by measuring time of work (10 times), averaging time of work per unit and per batch, estimating standard time and cycle time with consideration of overall allowance.
- Comparison of effectiveness in terms of efficiency and waste time before and after improvement.

Table 1. UPS testing process before improvement.

Testing Title (TIS1291-3-2555)	Testing Process	Details	Testing Time (min)		
			Per unit		Per batch
			Average	STD time[a]	
Day 1					
6.3.2	Input testing	On current	30.5	33.3	91.6
5.2.2h	Harmonic		6.20	6.8	18.6
6.3.3	Inrush current		21.9	23.9	65.8
6.3.4.1	Output Testing 1		9.5	10.4	28.6
6.3.4.3	Output Testing 3	Off current	10.7	11.6	32.1
6.3.4.2	Output Testing 2	On current	8.9	9.7	26.6
6.3.4.4	Output Testing 4	Off current, increasing load[b]	8.0	8.7	24.0
		Cycle time			**287.2**
Day 2					
6.3.5.1	Overload 1	On current	11.4	12.5	34.3
6.3.5.2	Overload 2	Off current	9.7	10.6	29.2
6.3.6.1	Linear load 1	On current	8.5	9.2	25.4
6.3.6.2	Linear load 2		4.9	5.4	14.8
6.3.6.4/6.3.5.1	Overload 3		6.6	7.2	19.9
6.3.6.4/6.3.5.2	Overload 4		6.7	7.3	20.0
6.3.8.1	Non-linear load 1		17.5	19.0	52.5
6.3.8.2	Non-linear load 2	Off current, increasing load[b]	13.3	14.5	40.0
		Cycle time			**235.9**
Day 3					
6.3.8.1/6.3.8.3	Linear/Non-linear	On current	15.7	17.1	47.1
6.3.7.1	Step load 1		6.7	7.3	20.1
6.3.8.4	Step load 2		15.3	16.7	45.9
6.3.8.6	Step load 3		16.1	17.6	48.4
6.3.9.1	Energy storage	Off current, increasing load[b]	12.3	13.4	37.0
		Cycle time			**198.4**
Day 4					
6.3.10	Efficiency 1	On current	16.8	18.3	50.3
6.3.10	Efficiency 2		14.7	16.0	44.0
6.3.10	Efficiency 3		21.8	23.8	65.5
6.3.10	Efficiency 4	Off current, increasing load[b]	10.2	11.1	30.6
		Cycle time			**190.3**
Day 5					
6.3.5.3	Short circuit 1	Off current, final tasks	7.5	8.1	22.4
6.3.5.4	Short circuit 2		7.5	8.2	22.4
		Cycle time			**44.8**

[a] STD time is the standard time calculated by adding 9% of overall allowance (5% of personal allowance and 4% of fatigue allowance) to average time [5].
[b] Battery power is exhausted after testing and needs to be charged up to 8 hours before performing next day testing process.

3. Process Improvement

Daily work period of the operator is at 8 am to 5 pm or 540 minutes per day with no overtime required. Excluding the breaks, the network time or actual takt time is 435 minute per day. The cycle time of each work day is much lower than net time of work that causes waste time as shown by Yamazumi Chart in Figure 1. For Day 5, short circuit testing should be moved to other work days due to low cycle time in order to complete all testing tasks within 4 days if possible. However, rearranging this testing should be well considered as it could cause damage of UPS and also battery power exhausted after testing.

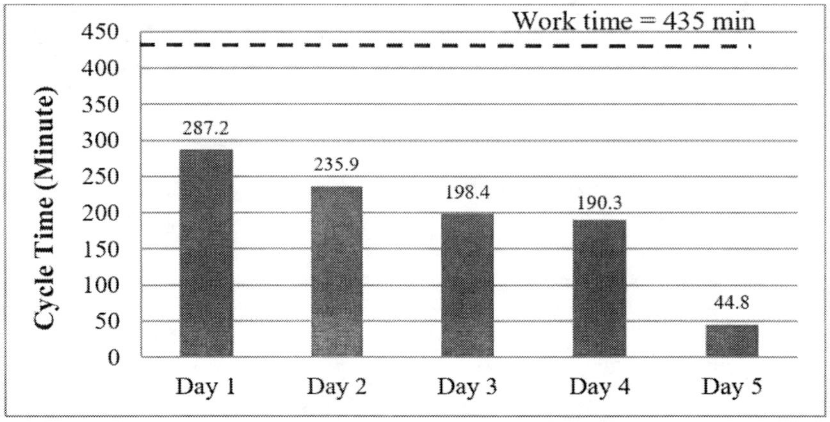

Figure 1. Comparison of Cycle time and net time of work on each work day.

From the net time of work, it can be assumed that the total work time of the operator is 1740 minute per batch. As there are 3 units per batch, the unit time is therefore 580 minute which is also used as the standard time (the time required by an average skilled operator). To evaluate the efficiency of the testing process, it can be calculated by [6]:

$$\text{Efficiency} = \frac{\text{Cycle Time(per day)}}{\text{Net time of work(per day)}} \tag{1}$$

The efficiency of Day 1, Day 2, Day 3, and Day 4 are 66%, 54.2%, 45.6%, and 43.8%, respectively. Thus, the average efficiency of the process is about 52.8% which is relatively low. From analysis, it can be noted that low efficiency and waste times (about 4.2 hours per day on average) would be caused by the limitations of individual testing task and also the specification of battery used.

Initially, the testing process has been performed regarding to the testing category based on TIS 1291-3-2555. Waste time could then caused by several movements and battery power exhausted. To improve the efficiency of the process and to increase the testing units per batch, work scheduling method was applied by using North West Corner Rule (NCR). Standard time was applied in NCR table where Demand was the standard time of each task and Supply was the actual work time of the operator. Daily demand or total standard time from several testing tasks of each work day were rearranged to not exceed the supply or actual work time to control overtime wage. Moreover, each task rearranged should not affect other tasks or overall process. The conditions applied are as follows:

1. Rearrange testing tasks to complete within 4 work days
2. Rearrange testing tasks with the same load to perform on the same day
3. Rearrange similar testing tasks to perform continuously
4. Rearrange testing tasks that cause exhausted battery power to perform last
5. Rearrange short circuit testing to perform in the last step if possible due to risk of damage

From NCR method, the appropriate work schedule was developed by considering 4 units per batch. Details are as shown in Table 2 and comparison of cycle time before and after improvement is shown by Yamazumi chart in figure 2.

Table 2. UPS testing process after improvement.

Testing Process	Details	Cycle time (min)	
		Estimated[a]	Actual[b]
Day 1			
Input testing		133.1	122.7
Harmonic		27.0	18.9
Inrush current	On current	95.6	46.3
Output Testing 1		41.6	37.2
Output Testing 3		46.6	41.7
Output Testing 2	Off current	38.7	43.2
Output Testing 4[c]		34.9	40.4
	Cycle time	**417.4**	**350.3**
Day 2			
Overload 2	Off current	42.4	25.6
Overload 1		49.8	85.3
Linear load 1		36.8	24.5
Linear load 2	On current	21.5	27.0
Step load 1		29.1	27.0
Efficiency 1		73.1	60.8
Efficiency 2		63.9	46.6
Efficiency 4[c]	Off current	44.5	27.1
	Cycle time	**361.1**	**324.2**
Day 3			
Linear/Non-linear		68.36	49.8
Step load 2		66.71	49.2
Step load 3	On current	70.33	49.0
Efficiency 3		95.14	74.9
Non-linear load 1		76.21	73.5
Non-linear load 2[c]	Off current	58.08	77.2
	Cycle time	**434.8**	**373.5**
Day 4			
Overload 3	On current	28.91	2.9
Overload 4		28.99	10.2
Short circuit 1		32.48	25.3
Short circuit 2	Off current	32.79	25.2
Energy storage[c]		53.76	27.1
	Cycle time	**176.9**	**90.7**

[a] $\text{Estimated cycle time} = \text{STD time (Table 1)} \times 4 \text{ units}$.

[b] Actual cycle time was evaluated from average cycle time measured from each task.

[c] Battery would be exhausted after testing and required charging up to 8 hours before next day testing process.

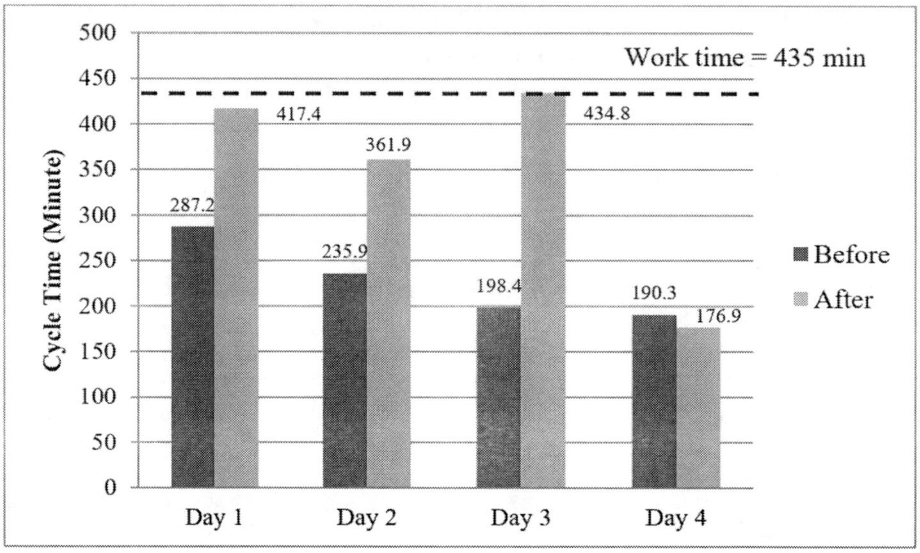

Figure 2. Comparison of estimated cycle time before and after improvement.

After improvement, the operator could perform well in all testing tasks with less waste time. From equation (1) using actual cycle time, the efficiency after improvement of Day 1, Day 2, Day 3, and Day 4 are 80.5%, 74.5%, 85.9%, and 20.8%, respectively. The overall efficiency after improvement is 65.6% or increasing 12.7% on average. For Day 4, the efficiency would decrease after improvement as several tasks could be shifted to other workdays. Therefore, the operator can manage the paperwork to do on this day.

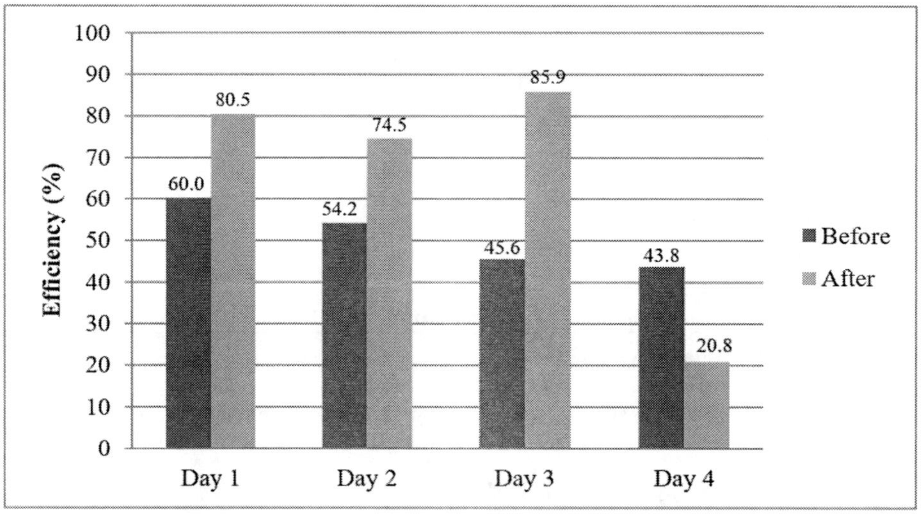

Figure 3. Comparison of efficiency before and after improvement.

To estimate waste time, net time of work and actual cycle time per batch were compared. Details of waste time before and after improving process are as illustrated in figure 4. From study, it can be noted that overall waste time of the process can be decreased from 828.3 minutes to 653.6 minutes or about 21%.

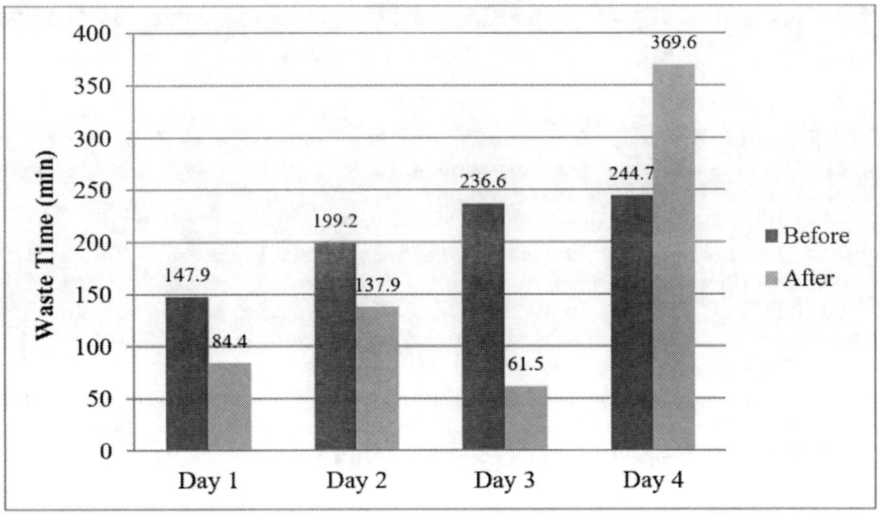

Figure 4. Comparison of waste time before and after improvement.

4. Discussion
Improvement of UPS continuous testing process can enhance the overall efficiency of the process and work day can be reduced from 5 days to 4 days. Moreover, the units of testing per patch could increases from 3 units to 4 units. Normally, the operator works about 20 day per month and receives wage about 9,000 Baht per month. Before improving process, work day per batch would take about 5 days. Therefore, work ability of this process should be 4 batches or 12 units per month. Assuming the institute earns 10,000 Bath per units of testing, the income from UPS testing should be 120,000 Baht per month. Excluding operator wage, the net income from UPS testing process would be 111,000 Baht. After improving process, work day per batch would take about 4 days and work ability should be 5 batches or 20 units per month. The total income from UPS testing process would be 200,000 Baht per month and the net income would be 191,000 Baht per month that increases 80,000 Baht or 72%.

5. Conclusion
Electrical and Electronic Institute provides UPS testing service for electrical appliance manufacture in order to obtain TIS 1291-3-2555 standard before distributing to the customers. However, the UPS continuous testing process has been observed low efficiency as the cycle time is much lower than work time of the operator. Initially, testing process would take about 5 days and there were 3 units of testing per batch. This study aims to enhance the effectiveness of the UPS testing process by reducing waste time and increases the overall efficiency of the operation process. Yamazumi chart and job scheduling theory namely North West Corner Rule were adopted by determining the standard time and then developing the new testing process with consideration of applicability to all UPS products with requirement of TIS 1291-3-2555. The developed testing process was performed by rearranging all tasks to be completed within 4 work days, rearranging tasks with the same load to perform on the same day, rearranging similar tasks to perform continuously, rearranging tasks that cause exhausted battery power to perform last and rearranging short circuit testing to perform in the last step.

Moreover, unit of testing per batch could be increased from 3 units to 4 units. After improvement, the overall efficiency would increase from 52.8% to 65.6% or increasing 12.7%. Also, waste time of the process could be decreased by 21% after improvement. Moreover, the institute could earn more net income from UPS testing process due to increasing testing unit by about 72% and the delivery target could be met.

Acknowledgement
Lerdlekha S. wishes to thank Electrical and Electronic Institute for valuable information to accomplish this study.

References

[1] Sharma K L S 2017 *Overview of Industrial Process Automation (Second edition)* pp 321-357
[2] Anders H 2015 *Int. J. Industrial Ergonomics* **50** pp 170-177
[3] Burhan B 2016 *World Patent Information* **45** pp 33-46
[4] Dong-hyu K, Heejin L and Jooyoung K 2017 *Research Policy* **46** pp 1234–1254
[5] Joseph N 2014 *The Lean Expert: Educating and Elevating Lean Practitioners Throughout Your Organization* (Florida: CRC Press)
[6] Mahmoud N, Frank C and Hung-Da W 2017 *Procedia Manufacturing* **11** pp 593-601
[7] Noor A, Mohd. S, Salmiah K and Ahmed J 2012 *Procedia Engineering* **41** pp 1702-1707
[8] Rahani A R and Al-Ashraf M 2012 *Procedia Engineering* **41** pp 1727-1734
[9] Natha M R C 2007 *Industrial Engineering and Management* (New Delhi: New Age International)
[10] Bettina K and Gerhard J W 2011 *Discrete Applied Mathematics* **159**(12) pp 1284-1289
[11] Lee Y C E, Chan C K, Langevin A, and Lee H W J 2016 *Transportation Research Part E: Logistics and Transportation Review* **91** pp 68-89
[12] Ministry of Industry (Thailand) 2012 *TIS 1291-3-2555* (Bangkok)
[13] Ekpraween S, Ubolrat S and Apinat J 2015 *Increasing Effectiveness of Universal Power Systems testing process using Work scheduling method: A Case study of Electrical and Electronic Institute* (Bangkok: Ramkhamhaeng University)

Appendix

A. Details of Testing Tasks

Electrical and Electronic Institute provides UPS electric testing service mainly in Chapter 6.3 of TIS 1291-3-2555 standard (Test of UPS operation according to Manufacturer's Specifications) [12]. Details of testing title (Table 1 and 2) are briefly described as follows:

Testing Title	Details
5.2.2h	Test of total input harmonic distortion
6.3.2	Test of input voltage and frequency tolerance band
6.3.3	Test of UPS inrush current
6.3.4.1	Test of static output characteristic (normal mode/no load)
6.3.4.3	Test of static output characteristic (stored energy mode/no load)
6.3.4.2	Test of static output characteristic (normal mode/full load)
6.3.4.4	Test of static output characteristic (stored energy mode/full load)
6.3.5.1	Test of static output characteristic (normal mode/overload)
6.3.5.2	Test of static output characteristic (stored energy mode/overload)
6.3.5.3	Test of static output characteristic (normal mode/short circuit)
6.3.5.4	Test of static output characteristic (stored energy mode/short circuit)
6.3.6.1	Test of dynamic output characteristic (normal to stored energy mode/resistance linear load)
6.3.6.2	Test of dynamic output characteristic (stored energy to normal mode/resistance linear load)

Testing Title	Details
6.3.6.4	Test of operation mode change (normal to bypass mode if applicable)
6.3.7.1	Test of output dynamic load characteristic (linear step load)
6.3.8.1	Test of UPS output (output tolerance/non-liner load/normal mode)
6.3.8.2	Test of UPS output (output tolerance/non-liner load/stored energy mode)
6.3.8.3	Test of UPS output (linear load /normal to stored energy mode)
6.3.8.4	Test of UPS output (non-linear step load/normal mode/rated ≤ 4.0 kVA)
6.3.8.6	Test of output (non-linear step load/stored energy mode)
6.3.9.1	Test of stored energy time
6.3.10	Test of input efficiency and power factor

Experimental study on the heat transfer enhancement of the effect of teardrop protrusion-to-protrusion spacing

O. Yemin, M. Wae-Hayee*, and C. Nuntadusit

Energy Technology Research Center and Department of Mechanical Engineering, Faculty of Engineering, Prince of Songkla University, Hat Yai, Songkhla 90112, Thailand

* Corresponding Author: wmakatar@eng.psu.ac.th

Abstract. An experimental analysis was presented for heat transfer on the surface of teardrop protrusions with an inline arrangement. The air flow was passed through perpendicularly over the single row of 5 protrusions in the rectangular wind tunnel. The protrusion height from the bottom of wind tunnel was H/D=0.2 and the spacing of teardrop protrusion-to-protrusion was adjusted by four different spacing cases such as S/D= 1.125, 1.25, 1.5 and 2. The spacing can affect heat transfer enhancement and flow characteristics. Therefore, the spacing which was analysed in this study considered four cases because of experimental limitations. The temperature distributions of surface with protrusions was detected by using the thermochromic liquid crystal sheet (TLC) at Re_H=20,000. Image processing was used to evaluate distribution of Nusselt number. The result shows that the highest average Nusselt number was found in S/D=1.125D case.

1. Introduction

Using passive heat transfer enhancement techniques such as ribs, pin fins, vortex generators, dimpled surface and surface with arrays of protrusions can enhance heat transfer in internal flow passages and can increase secondary flow and turbulence level for mixing enhancement. These kinds of flow can remove heat away from surfaces. Therefore, this passive heat transfer enhancement techniques were used in many industrial applications such as cooling of turbine airfoil, combustion chamber, electronics cooling devices and heat exchangers. Dimples and protrusions are an attractive method for internal cooling due to significant enhancement of turbulence level, enhancement in heat transfer convection with minimal pressure drop penalties and form multiple vortex pair that enhance Nusselt number distributions.

Acharya et al. [1] investigated experimentally and numerically on heat transfer and flow structure for four types of dimple shape (square, triangular, circular, and teardrop) in square internal passage. They showed that the teardrop dimple was the highest heat/mass transfer among four dimples by comparing with both experimental and numerical results. And then, the triangular dimple was the lowest heat transfer enhancement. For flow pattern, the circular and teardrop geometries have single vortex roll that was formed in wake region, but square dimple showed no noticeable vortex roll.

The flow and heat transfer characteristics of teardrop dimple and protrusion was investigated and compared with hemispherical dimple/protrusion by Xie et al. [1]. They showed the thermal performance for teardrop dimple and protrusion is higher than hemispherical dimple and protrusion for lower Reynold number and thermal performance increased gradually as the centre moves downwards for the teardrop dimple/protrusion.

Rao et al. [4] investigated an experiment and a numerical study on the heat transfer of turbulent flow in channel with staggered arrangement of spherical and teardrop dimples in 11 rows. They found that heat transfer performance of teardrop dimple was distinctively higher than spherical dimple. Although many researchers have studied about heat transfer and flow characteristics with many dimple and protrusion row on the surface, it is still difficult to describe the phenomenon of flow and heat transfer characteristics. Therefore, the heat transfer characteristics with single protrusion row over the surface was researched in this work.

Many literatures have been analysed about the different arrangement of dimpled and protruded surface. The aim of this work was to investigate experimentally the effect of spacing between protrusion-to-protrusion on heat transfer characteristics with a single row of 5 protrusions. The spacing can affect the different value of heat transfer enhancement. The different four spacings in this work was considered according to experimental dimensions.

1.1. Model of protrusion

Figure 1 shows the model of 5 protrusions which was made by plastic and installed on the internal surface of rectangular wind tunnel. The air with fully developed flow passes through the tear drop protrusions. The origin of Cartesian coordinate system was existed at the centre of middle dimple such as X-axis is the along the flow direction of wind tunnel, Y-axis is the rectangular wind tunnel height direction and Z-axis is normal direction to the flow.

The details of protrusion are shown in figure 2. The spherical with the radius of r =0.725D and the projected diameter on the surface was D=26.4mm. The height of protrusion from the surface was 0.2D. The spacing between protrusion-to-protrusion was varied with S/D= 1.125, 1.25, 1.5 and 2. The Reynolds number of mainstream air inside the wind tunnel based on the hydraulic diameter of wind tunnel was fixed at Re_H=20,000 at the middle of wind tunnel by using Pitot tube to measure velocity. The position of Pitot tube is shown in figure 4.

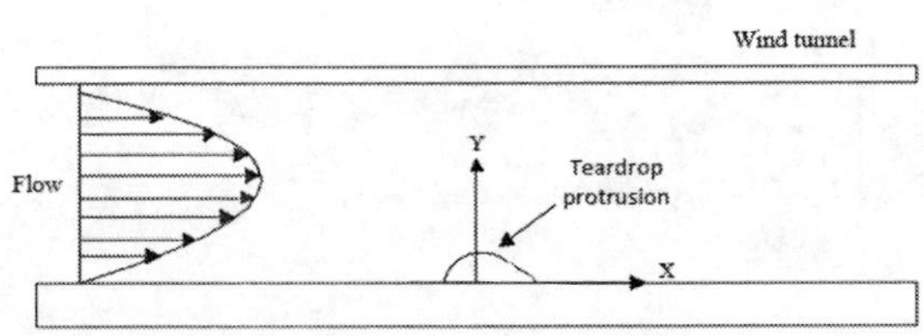

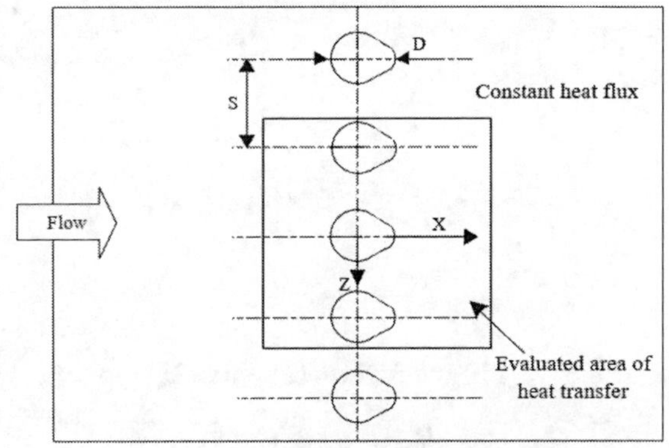

(a) Side view **(b)** Top View

Figure 1. The model of investigation for teardrop test section.

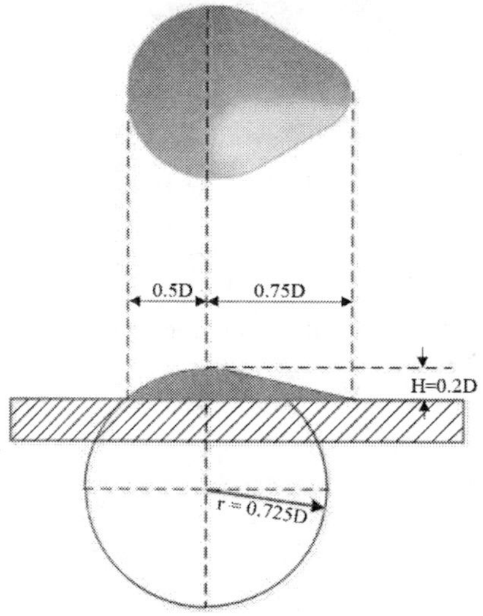

Figure 2. The detail of protrusion.

1.2. Wind tunnel

Figure 3 shows the details of wind tunnel in the experiment. The rectangular wind tunnel was designed with three parts: the upstream of test section (1700 *mm*) that was sufficient distance to get fully developed flow, test section (280*mm*) where was formed teardrop protrusions and downstream of test section where the air was leaved out from the wind tunnel. The wind tunnel height was 26.4 *mm* (1*D*).

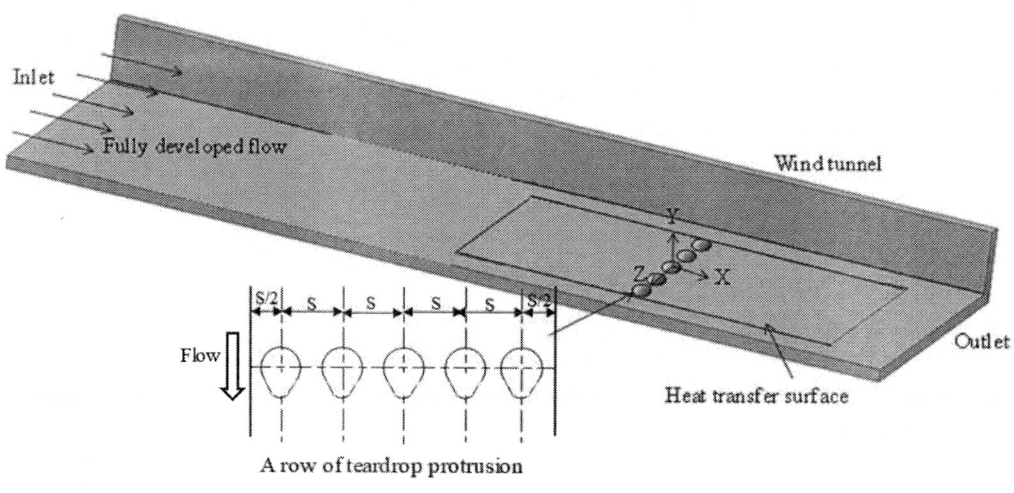

Figure 3. Experimental 3D model of wind tunnel with single teardrop protrusions row.

1.3. Measurement of heat transfer

The test section for heat transfer measurement is shown in figure 4. The heat transfer surface was made of a stainless-steel foil with thickness of 0.03mm. The foil was attached with Thermochromics liquid crystals sheet (TLC_S) on the side of the wall. The stainless-steel foil was stretched between a couple of copper bus bars. The heat transfer surface was heated by DC power source that can supply current through the copper bus bars. A digital camera was used to capture color on the TLC sheet. Images of color pattern on the TLC sheet were converted from a red, green, and blue (RGB) system to a hue, saturation, and intensity (HSI) system. The air was introduced through the inlet chamber, flow straightener, two layer of mesh plates, test section and chamber outlet by using blower.

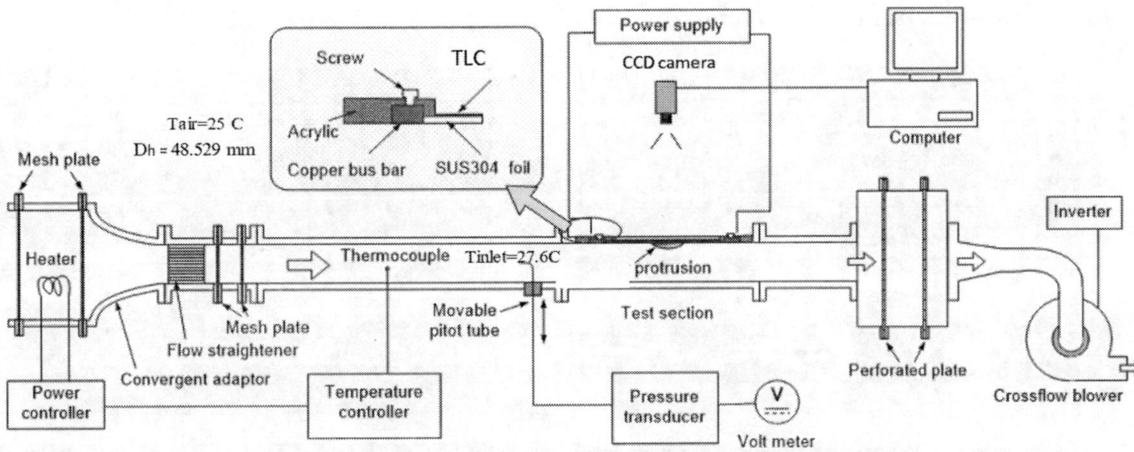

Figure 4. Schematic diagram of experimental set-up.

Electrical energy dissipated in the stainless-steel foil can be calculated from:

$$\dot{Q}_{input} = I^2 \cdot R \tag{1}$$

where, I is the supplied electrical current and R is the electrical resistance of stainless-steel foil.

The air temperature was controlled at 27.6°C and flow on the surface of stainless-steel sheet. The local heat transfer coefficient (h) can be calculated from this equation:

$$h = \frac{\dot{Q}_{input} - \dot{Q}_{losses}}{A(T_{LC} - T_a)} \tag{2}$$

where, $\dot{Q}_{input}$ is the rate of heat generation in the stainless plate, $\dot{Q}_{losses}$ is the rate of heat loss for convection and radiation, A is the area of heat transfer surface, T_{LC} is the temperature of the colour that appears on the TLC plate and T_a is air temperature.

The local Nusselt number can be calculated from

$$Nu = \frac{hD_H}{k} \tag{3}$$

where, D_H is hydraulic diameter and k is thermal conductivity of air.

3. Results and discussion

The contours and the values of Nusselt number on the surface of only three protrusions are described in figure 5. The solid part area of teardrop was not considered about Nusselt number. At the front edge of teardrop protrusion, low Nusselt number value was found because flow impingement became strong and this effect caused the small circulation flow region. In addition, the flow separation mainly occurred over the downstream half of teardrop protrusion surface, and the turbulent wake reattached immediately downstream of the protrusion. Low-pressure region formed in the rear section of protrusion and, the flow coming from two sides entered this region. Therefore, the distribution of high Nusselt number could be seen at the downstream of protrusions due to the flow separation and attachment flow immediately appeared along the downstream of protrusion [1, 5] and this effect can enhance heat transfer augmentation. Increasing or decreasing Nusselt number on the dimple and protrusion surface can occurr because of both attachment and circulation flow effects that was studied in literature [1, 4]. Furthermore, the results were found that the Nusselt number increased when the protrusion-to-protrusion spacing decreased because of the flow tendency to the lateral side of longitudinal vortex pair between protrusion-to-protrusion spacing and turbulent flow.

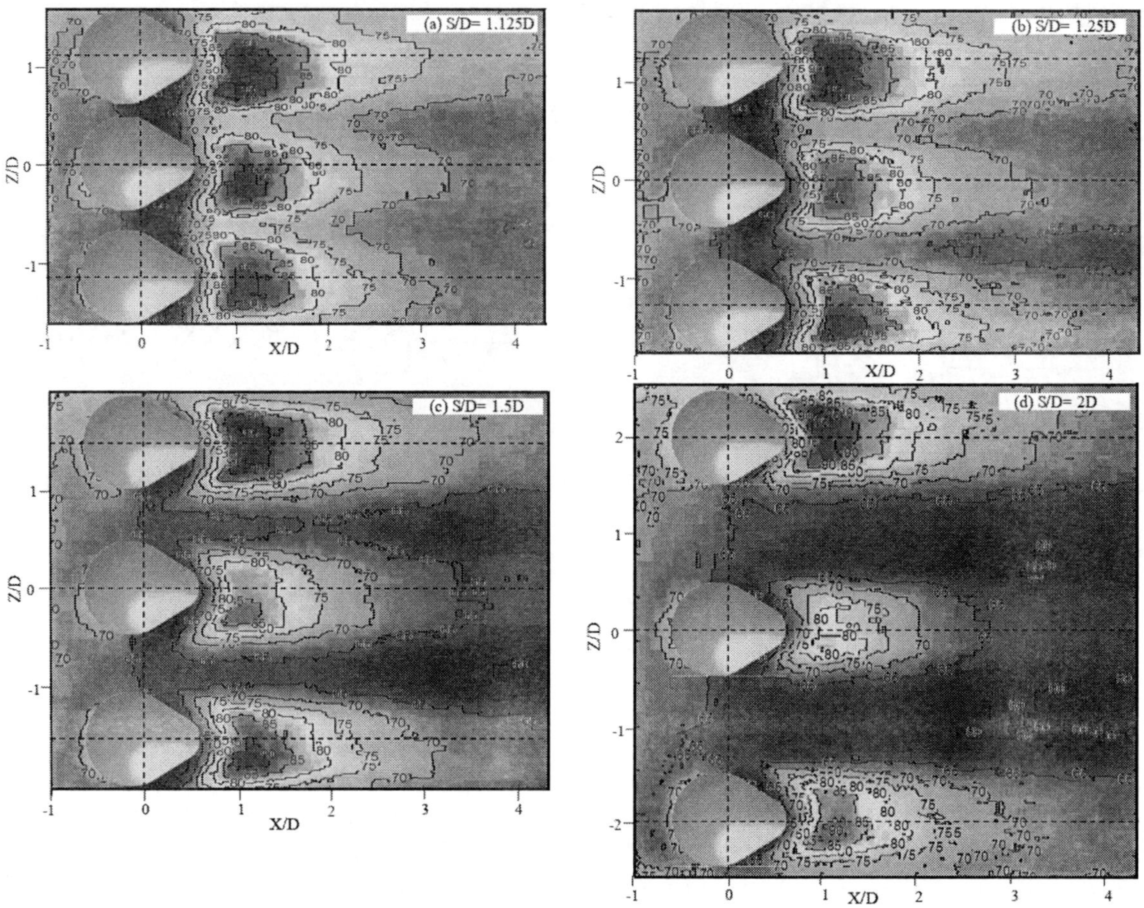

Figure 5. Contour of Nusselt number on teardrop protrusion surface.

The distribution of Nusselt number value along the protrusion centre *Z/D*=0 for all *S/D* cases was shown in figure 6. The area of the range (-0.5≤*X/D*≤0.75) was not considered because it is a protruded teardrop It was found that the maximum peak trend occurred at *S/D*=1.125 and this trend was higher than the other S/D cases.

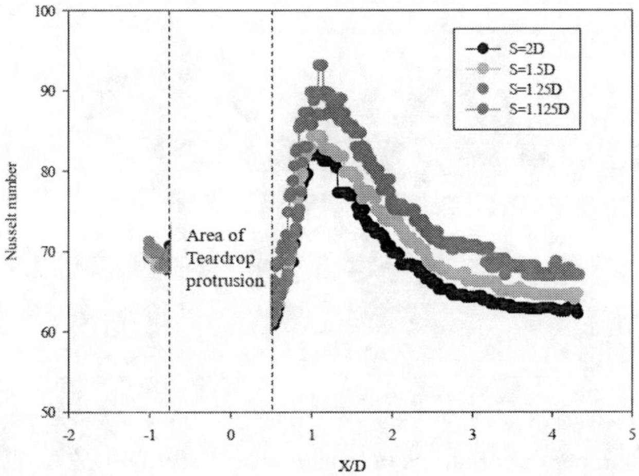

Figure 6. Nusselt number distributions in stream wise direction (*X/D*) (*Z/D*=0, Re_H=20,000)

Figures 7, 8 and 9 show the span wise Nusselt number distributions at the downstream portion of protrusion (*X/D*=1, 2 and 3) for all *S/D* cases. Three peak regions of maximum Nusselt number cases occurred behind the protruded teardrop smooth surface. When the distance downstream behind the protrusion became large, the value of Nusselt number became small because the reattachment flow was occurred at the position just behind the teardrop protrusion. It can be clearly seen that S/D=1.125 case was the highest Nusselt number distribution trend for every X/D cases among different spacing cases. In figure 7, the value of Nusselt number values for S=1.125D case was 75<Nu<100. However, the value of Nusselt number values for S=2D case was 60<Nu<95. In addition, the smallest spacing case (S=1.125D) can develop more turbulent wake and longitudinal flow compared to other cases.

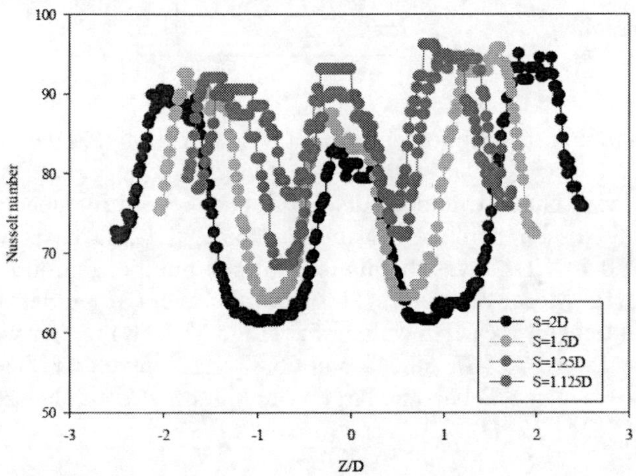

Figure 7. Nusselt number distributions in span wise direction (*Z/D*)) (*X/D*=1, Re_H=20,000).

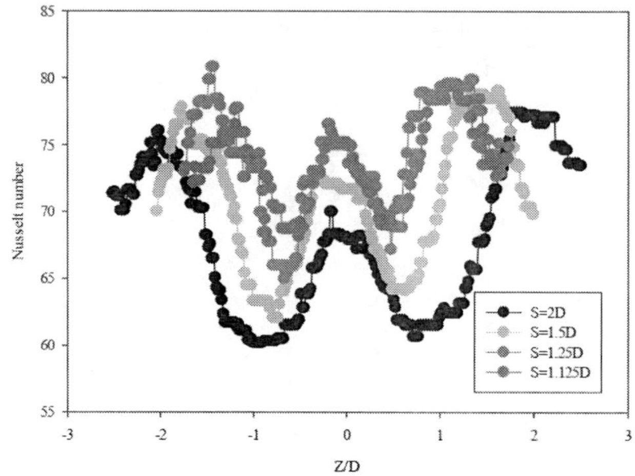

Figure 8. Nusselt number distributions in stream wise direction (*Z/D*) (*X/D*=2, Re_H=20,000).

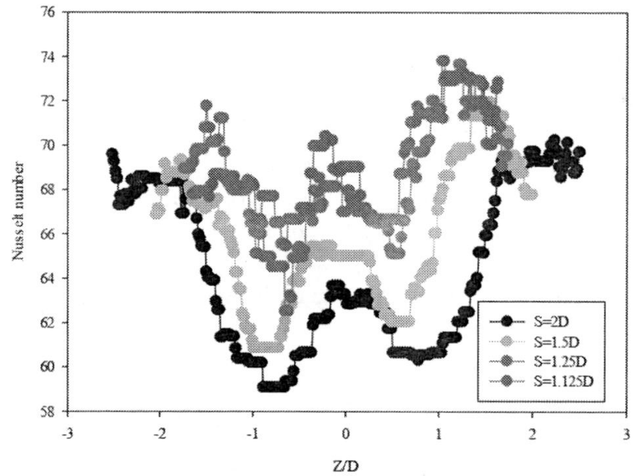

Figure 9. Nusselt number distributions in stream wise direction (*Z/D*) (*X/D*=3, Re_H=20,000).

The average stream wise Nusselt number distribution of respective spacing case is illustrated in figure 10. The trend of Nusselt number distribution for S=1.125D case was the highest Nusselt number. The position of 0.5<X/D<2 was the highest Nusselt number region because the reattachment flow happened in this area. Moreover, figure 11 shows the average Nusselt number over protrusion surface of evaluated heat transfer area. The value of average Nusselt number was 71.287 for S=1.125D case while the other cases were 70.467, 68.617 and 66.837. The result can be shown that the trend of average value become low when *S/D* become large. The highest average Nusselt number case was the narrowest spacing case (*S/D*=1.125*D*).

IOP Conf. Series: Materials Science and Engineering **297** (2018) 012065 doi:10.1088/1757-899X/297/1/012065

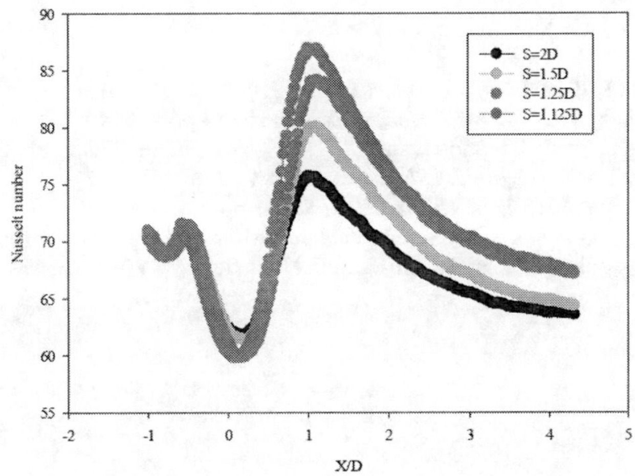

Figure 10. Nusselt number distributions in streamwise direction (*X/D*) (-2.4<*Z/D*<2.4 for *S*=2*D*, -2<*Z/D*<2 for *S*=1.5*D*, -1.7<*Z/D*<1.7 for *S*=1.25*D*, -1.6<*Z/D*<1.6 for *S*=1.125*D*, Re_H=20,000).

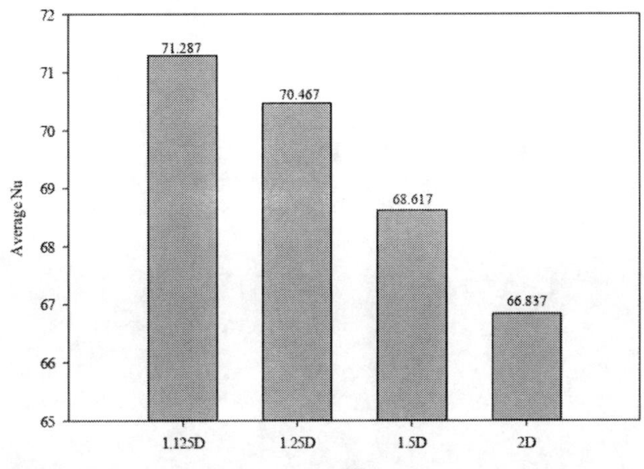

Figure 11. Average Nusselt number of teardrop protruded test section surface (Re_H=20,000).

4. Conclusions

In this work, the investigation of heat transfer characteristics over the teardrop protrusion surface of evaluated heat transfer area in wind tunnel was studied. The spacing was varied with different spacing at *S/D*=1.125, 1.25, 1.5 and 2. The summery of this work are as follows:

1. The value of average Nusselt number was increased when protrusion-to-protrusion spacing was decreased. The highest Nusselt number average value can be found in *S/D*=1.125 case.
2. The heat transfer was low at just in front of the protrusion due to circulation flow that appeared in this region. The region downstream of protrusion, the heat transfer was high because of reattachment flow.
3. The high Nusselt number can be observed at the position for the downstream of protrusion (*X/D*=1) because of reattachment flow after that Nusselt number values became low along the downstream of smooth surface.

Acknowledgement
The research grant was supported by the Research and Development Office (RDO), Prince of Songkla University, grant No. ENG590725S.

References

[1] Acharya S and Zhou F 2012 *J. Turbomach.* **134** pp 0610281– 06102813
[2] Xie Y, Qu H and Zhang D 2015 *Int. J. Heat Mass Transf.* **84** pp 486–496
[3] Chang S W, Chiang F, Yang T L and Huang C C 2008 *Exp. Therm. Fluid Sci* **33** pp 23–40
[4] Li P, Zhang D and Xie Y 2014 *Int. J. Heat Mass Transf.* **73** pp 456–467
[5] Rao Y, Li B and Feng Y 2015 *Exp. Therm. Fluid Sci* **61** pp 201–209
[6] Mohammad A E and Danesh K T 2008 *J. Turbomach.* **130** pp 0410161–0410169
[7] Wae-Hayee M, Tekasakul P, Eiamsa-ard S and Nuntadusit C 2015 *Experiment of Heat Transfer* **28** pp 511–530

IOP Conf. Series: Materials Science and Engineering **297** (2018) 012066 doi:10.1088/1757-899X/297/1/012066

Flows developing around a finite rotating disk enclosed by a cylindrical casing

Takashi Watanabe* **and Soma Endo**

Graduate School of Informatics, Nagoya University, Nagoya, 464-8601, Japan

* Corresponding Author: watanabe@i.nagoya-u.ac.jp

Abstract. Numerical investigation has been performed on the flows around a rotating disk in a cylindrical casing. The sizes of the disk and the casing are finite, and the flow field has a radial clearance and an axial clearance. Taylor-Couette type flow develops in the radial clearance while the rotor-stator cavity flow appears in the axial clearance, and the interaction of the flows in these clearances makes complex flow developments. The flow structures are classified into some types including normal and anomalous modes in the radial clearance and circular and polygonal vortices in the axial clearance. The effect of the acceleration rate of the disk is estimated and it is found that the gradual increase of the disk angular velocity tends to form circular vortices in the normal mode.

1. Introduction

For the flow around a rotating disk, the three-dimensional shear layer with a cross flow has been studied [1-3]. In the present study, the flow around a finite rotating disk in a finite stationary cylindrical casing is numerically investigated. The flow field has a radial clearance between the disk rim and the side wall of the casing and an axial clearance between the surface of the disk and the end wall of the casing, and it can be regarded as a simple model of fluid machineries. One interest is the interaction of the flows in the clearances and it is expected that some flow structures appear.

The flow in the cylindrical cavity where the rotating and stationary coaxial disks are confined by a stationary cylindrical side wall has been studied in the experimental and numerical approaches. Schouveiler et al. [4, 5] used the visualization method and the ultrasonic Doppler velocimeter and experimentally showed that circular vortices and spiral vortices appeared in the case of wider axial clearance and turbulent spots and solitary waves emerged in the narrower axial clearance. They also offered their comment that the radial clearance of their experimental apparatus had some effects on the entire flows [5]. Poncet et al. [6] pointed out that the wave number of the spiral vortices was sensitive to the experimental condition. The annular cavity flow is the flow in the cylindrical cavity with a central axial hub. In this flow, the stationary hub suppressed the disturbances which grew in the Bödewadt layer on the stationary disk and propagated to the Ekman layer on the rotating disk [7]. On the other hand, the hub rotating with the disk enhanced the disturbances and made the flow turbulent [8-10]. Some reviews of the rotor-stator cavity flow with or without the hub have been reported [11, 12]. The flow around two or more disks connected by the hub can be found in the electric devices and chemical reactors. Hendriks [13] numerically predicted the flow in the hard disk drive with two platters and found that Taylor vortices appeared between the disk rim and the shroud of the casing. Washizu et al. [14] showed that the rib mounted on the shroud suppressed the pressure-based vibration of the hard disk drive. Meeuwse et al. [15] and de Beer et al. [16] examined the characteristics of the multi-phase and single-phase flow in the multi-stage reactor, respectively, and Huang et al. [17] and Uenishi et al. [18] classified the polygonal flow structures found in the flow between two co-rotating disks shrouded by a cylindrical wall.

When the disk rotating in the casing has a finite thickness, centrifugal-force driven vortices emerge between the disk rim and the side wall of the casing [13]. Original Taylor vortices between the coaxial rotating cylinders have been studied in many areas [19-21]. When the cylinders have finite lengths and the stationary walls on their ends, the Ekman pumping mechanism forms the secondary flows on the end walls at the Reynolds number lower than that for the onset of Taylor vortices and the bifurcation phenomena become subcritical [22, 23]. With this shift of the physical situations, Taylor vortex flows show various structures including circular, wavy and modulated flows in normal and anomalous modes and the change of the incremental rate of the cylinders' angular velocities may bring multiple flow structures even though the Reynolds number and the geometrical configurations are fixed [24-26]. Recently, the stability of the quasi-Keplerian flow, in which the angular velocity radially decreases and the angular momentum radially increases, has been attracted researchers' interests [27-29], and some conclusions might be given [30].

We have carried out experimental and numerical investigations of the flow around a finite rotating disk in a cylindrical casing. In [31, 32], the numerical method was established and some experiments were performed. The numerical result showed good agreements with the experimental result of the circular vortices, spiral vortices and the phase velocity of the vortex structures. In [33], the effect of the width of the radial clearance on the flow was estimated. In the present study, the effect of the axial clearance width is numerically considered. The flows in the radial and axial clearances are classified and the diagrams of the flow structures in the space of the Reynolds number and the axial clearance width are presented. Two acceleration rates of the disk are used and their effect on the final flow structures is shown.

2. Formulation

The schematic construction of the flow field is represented in figure 1. The cylindrical casing is stationary and its inner radius and the length are r_c and h_c, respectively. The rotating disk with a driving shaft is set at the center of the casing. The radius and the thickness of the disk are r_d and h_d, respectively, and the radius of the driving shaft is r_s. The disk rotates with the driving shaft at the angular velocity ω. The coordinate system is the cylindrical system (r, θ, z). Its origin is at the center of the lower end wall of the casing and the z axis aligns with the central axis of the flow field. In numerical method, the governing equations are made nondimensional by the reference length of the disk radius, the reference circumferential velocity at the disk rim and the reference time given by the reciprocal of the angular velocity. The size of the casing is fixed and $r_c = 1.118$ and $h_c = 0.3150$. The shaft radius r_s is 0.0787 and it is small enough not to have any influence on the flow field [6]. Eight disks with h_d from 0.1575 to 0.2677 are introduced to see the effect of the axial clearance width. The thickness ratios γ defined by the fraction h_d / h_c are 0.50, 0.55, 0.60, 0.65, 0.70, 0.75, 0.80 and 0.85.

In order to estimate the dimensional phenomena from the nondimensional results, the dimensional disk radius D of 127 [mm], the dimensional angular velocity Ω [1/s] and the kinematic viscosity ν of 4.0 [mm^2/s] of glycerin aqueous solution are assumed. The Reynolds number Re is given by $D^2\Omega / \nu$ and the dimensional time is estimated from the nondimensional time t by $4032 \times t / Re$.

The governing equations are the equation of continuity and the unsteady three-dimensional incompressible Navier-Stokes equations:

$$\nabla \cdot \boldsymbol{u} = 0, \tag{1}$$

$$\frac{\partial \boldsymbol{u}}{\partial t} + \nabla(\boldsymbol{u}\boldsymbol{u}^T) = -\nabla p + \frac{1}{Re}\nabla^2 \boldsymbol{u}, \tag{2}$$

where $\boldsymbol{u}$ is the velocity vector with its radial, circumferential and axial components u, v and w, and p is the pressure. The boundary condition is the no-slip condition on the solid walls. Initially, the flow is at rest in the entire region. The discretization method is based on the finite difference method [34] and the whole circumferential region of θ from 0 to 2π is adopted. In the case that the acceleration time of the disk is zero, the disk is rotated suddenly at the angular velocity giving a prescribed Reynolds number. When the acceleration time is finite, the angular velocity is linearly increased. The acceleration rate α is defined by the incremental of the Reynolds number in one dimensional second.

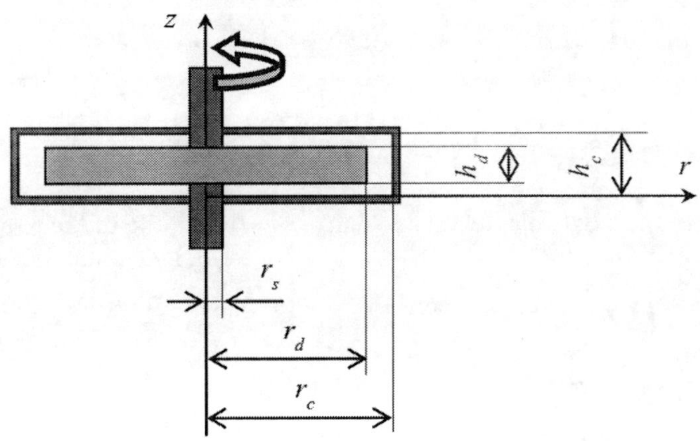

Figure 1. Schematic drawing of the flow field.

3. Results and Discussion

3.1. Flow structures in the radial clearance

The modes of Taylor vortex flows between rotating cylinders with finite lengths have been determined by the number of vortices in the meridional section and the flow direction on the stationary end wall [35]. In the flows around a rotating disk in a cylindrical casing, flow modes similar to those of Taylor vortex flows can be found in the radial clearance between the disk rim and the side wall of the casing. In the present study, the flow structures are classified in four modes. Figure 2 shows the flow in the normal 2-cell mode. The thickness ratio γ is 0.5, the Reynolds number *Re* is 3,000 and the acceleration rate α is 200. Figure 2 (a) depicts the contour of the axial velocity component *w* in the meridional section (*r*-*z* plane) near the radial clearance, and the disk rim is on the left side and the side wall of the casing is on the right side. In the upper half of the figure, a pair of a cold color region on the left and a warm color region on the right represents a toroidal vortex rotating in the counterclockwise direction. The lower half denotes the vortex rotating in the clockwise direction. These two vortices have radially inward flows on the upper and lower end walls of the casing. The mode of Taylor vortex with a radially inward flow on the end wall is normal [35]. The flow in figure 2 (a) has two vortices (two cells) in the normal mode and the flow mode is the normal 2-cell mode. The flow between two vortices is radially outward near the midplane in the axial direction. Figure 2 (b) shows the profile of the circumferential velocity component *v* (abscissa) along the axial direction (ordinate) at midsection of the radial clearance (r = 1.06). The cusp at the axial midplane results from the convection of the circumferentially accelerated flow near the rotating disk rim by the radially outward flow.

Figure 3 shows the normal 4-cell mode in the radial clearance. In figure 3 (a), four pairs of the warm and cold regions, from upper to lower, indicate the vortices rotating in the counterclockwise, clockwise, counterclockwise and clockwise directions, respectively. The flows of the vortices on the end walls are radially inward and they are normal. The flow has three boundaries between the four vortices. The flows near the upper-most and the lower-most boundaries are radially outward and the flow near the mid-boundary is inward. The radially outward and inward flows carry accelerated and decelerated circumferential flows, respectively, and the profile of the circumferential velocity component has two maxima and one minimum as shown in figure 3 (b).

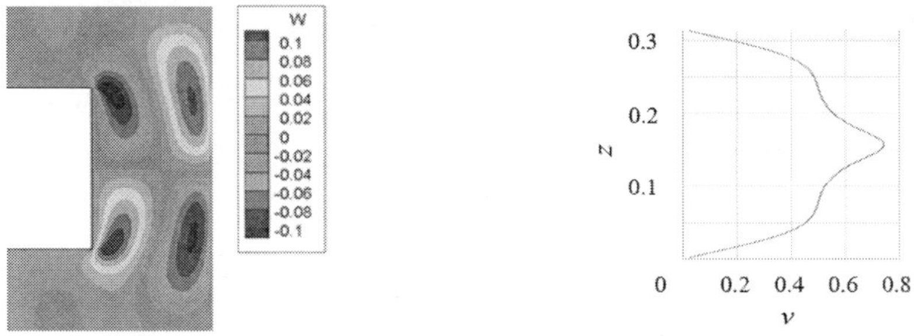

(a) Contour of the axial velocity component near the radial clearance.

(b) Profile of the circumferential velocity component at $r = 1.06$.

Figure 2. Normal 2-cell mode. $\gamma = 0.5$, $Re = 3000$, $\alpha = 200$.

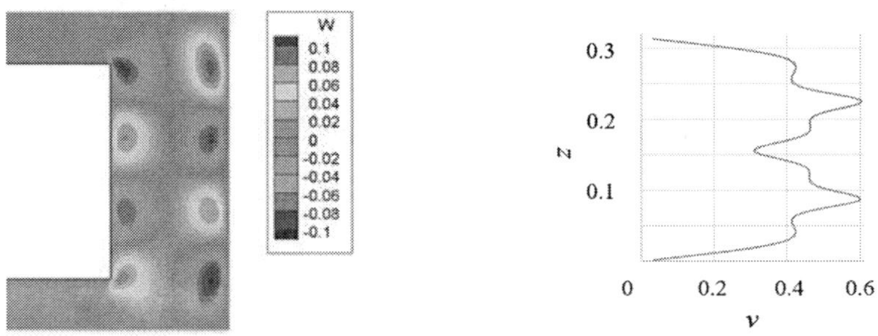

(a) Contour of the axial velocity component near the radial clearance.

(b) Profile of the circumferential velocity component at $r = 1.06$.

Figure 3. Normal 4-cell mode. $\gamma = 0.7$, $Re = 5000$, $\alpha = 1000$.

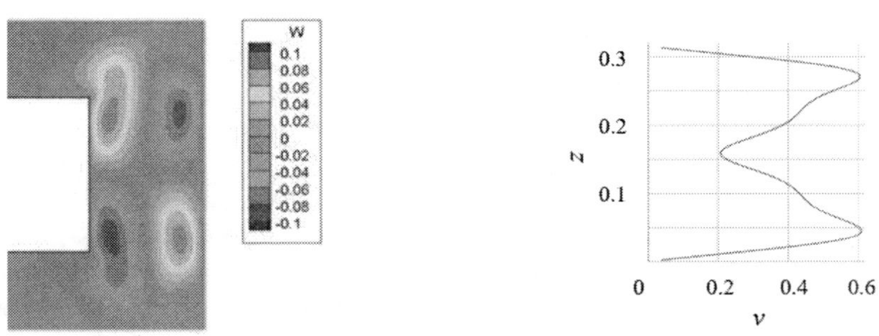

(a) Contour of the axial velocity component near the radial clearance.

(b) Profile of the circumferential velocity component at $r = 1.06$.

Figure 4. Anomalous 2-cell mode. $\gamma = 0.5$, $Re = 3000$, $\alpha = 1000$.

The flow in figure 4 has two vortices in the radial clearance. The upper one and the lower one in figure 4 (a) rotate in the clockwise and counterclockwise directions, respectively. This means that the flows on the end walls are radially outward and the mode of the vortices is anomalous [35]. The flow mode is the anomalous 2-cell mode. The radial flow near the axial midplane is inward and the profile of the circumferential velocity component in figure 4 (b) has a minimum at the midplane, while it has peaks near the end walls.

Figure 5 shows an asymmetric flow pattern with three vortices. In figure 5 (a), the lowest vortex rotates in the counterclockwise direction and it has a radially outward flow on the lower end wall. The lowest vortex is anomalous, and the flow mode is the anomalous 3-cell mode. The three vortices have two boundaries and the profile in figure 5 (b) has its maximum and minimum near these boundaries.

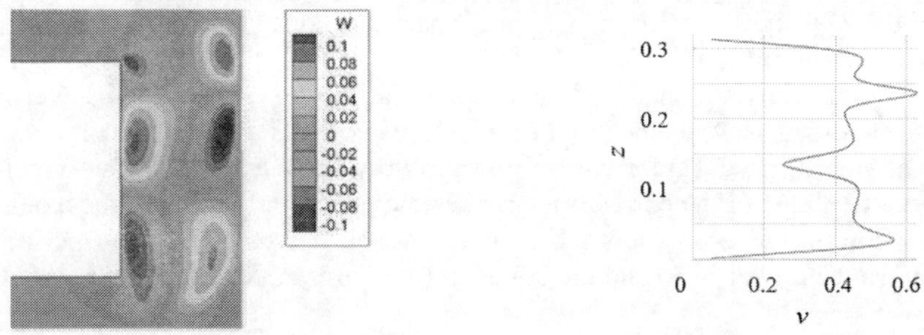

(a) Contour of the axial velocity component near the radial clearance.

(b) Profile of the circumferential velocity component at $r = 1.06$.

Figure 5. Anomalous 3-cell mode. $\gamma = 0.7$, $Re = 7000$, $\alpha = 500$.

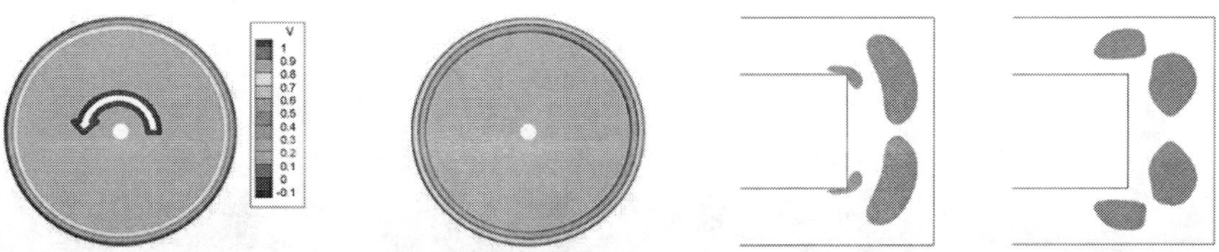

(a) Top view of the normal 2-cell mode.

(b) Top view of the anomalous 2-cell mode.

(c) Vortex regions of the normal 2-cell mode.

(d) Vortex regions of the anomalous 2-cell mode.

Figure 6. Steady state with circular vortices in the normal and anomalous 2-cell modes.
(a) and (c): $\gamma = 0.5$, $Re = 3000$, $\alpha = 1000$. (b) and (d): $\gamma = 0.5$, $Re = 3000$, $\alpha = 200$.
(a) and (b): Contours of $Q = 1.0$. (c) and (d): Vortex regions at $0.0 \leq Q \leq 1.0$.

3.2. Flow structures in the axial clearance

In this subsection, the three-dimensional flow structures in the axial clearance are visualized by the second invariant of the velocity gradient tensor (Q value) formulated in the cylindrical coordinate system [36]. Figure 6 represents the flow structures as the contours of $Q = 1.0$, which are in the steady state and in the normal 2-cell mode. The rotating direction of the disk is counterclockwise. The thickness ratio γ is 0.5 and the Reynolds number Re is 3,000. Figures 6 (a) and (b) show the normal mode flow at α = 1,000 and the anomalous mode flow at $\alpha = 200$, respectively. The color on the contours corresponds to the value of the circumferential velocity component v shown in the legend. The vortex structures are confined in the radial clearance and they are axisymmetric and circular. Figures 6 (c) and (d) depict the vortex regions indicated by the value of Q from 0.0 to 1.0 in the meridional sections near the radial clearance of the flows in figures 6 (a) and (b), respectively. The flow in the normal mode in figure 6 (c) has small vortex regions at the edges of the disk, and these small structures correspond to the inner yellow ring with higher circumferential velocity component in figure 6 (a). In figure 6 (d) of the anomalous mode flow, vortex regions appear above and below the disk and two vortex regions emerge in the radial clearance. The latter two regions have higher circumferential velocity component near the end walls and they can be seen in the outer yellow region in figure 6 (b).

Figure 7 shows the flow which appears at the moderate Reynolds number and has polygonal vortices. The mode is the anomalous 2-cell mode. Figures 7 (a) and (b) represent the contours of $Q = 1.0$ and the legend of the color map is the same as that in figure 6. The vortex structure in figure 7 (a) is a nonagon with nine almost periodic patterns and the circumferential wave number is 9. Figure 7 (b) is the bird's eye view. Intermitted and elongated patterns emerge at the top and the bottom of the figure. The two vortices in the radial clearance are in the wavy state. In order to see the details of the flow pattern, the vortex regions at Q from 0.0 to 1.0 in the meridional sections denoted by the numbers from (1) to (5) in figure 7 (b) are shown in figure 7 (c). Other than the four main vortex regions similar to those in figure 6 (d), small vortex regions appear and/or disappear at the upper-right and lower-right corners of the casing. These small vortices attached to the anomalous vortices were called extra vortices by Nakamura et al. [24] who investigated Taylor vortex flows in very short cylinders. The circumferential variation of the size and the shape of the extra vortex makes the polygonal vortex in the axial clearance. Figure 8 describes the flow in the anomalous 2-cell mode, which has more frequent periodicity and bead-like patterns. The wave number is 30 and the coherent structures appear at the corners of the casing.

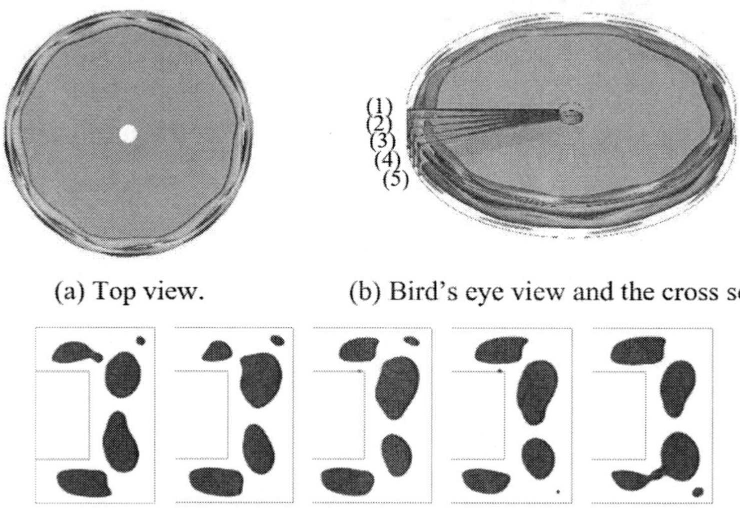

(a) Top view. (b) Bird's eye view and the cross sections.

(1) (2) (3) (4) (5)

(c) Vortex regions ($0.0 \leq Q \leq 1.0$) near the radial clearance in the cross sections.

Figure 7. Wavy state with polygonal vortices in the anomalous 2-cell mode and the wave number of 9. $\gamma = 0.5$, $Re = 4000$, $\alpha = 200$ and $Q = 1.0$.

Figure 8. Wavy state with polygonal vortices in the anomalous 2-cell mode and the wave number of 30 (bead-like pattern). $\gamma = 0.7$, $Re = 8000$, $\alpha = 1000$ and $Q = 1.0$.

The flow structures at higher Reynolds number are shown in figure 9, where the flow mode is the anomalous 2-cell mode. In the flow in figure 9 (a) at $\gamma = 0.50$, the disturbances generated in the radial clearance propagate onto the inner region of the disk and form spiral vortices with some periodicities. Figure 9 (b) is the flow at $\gamma = 0.65$. Only a few spiral vortices are found in the axial clearance and the flow structure is unconstructed. Though the state of the flows in the radial clearances in figures 9 (a) and (b) is a turbulent state, the flows show some coherence similar to that which has been called herring-bone-like streaks [37].

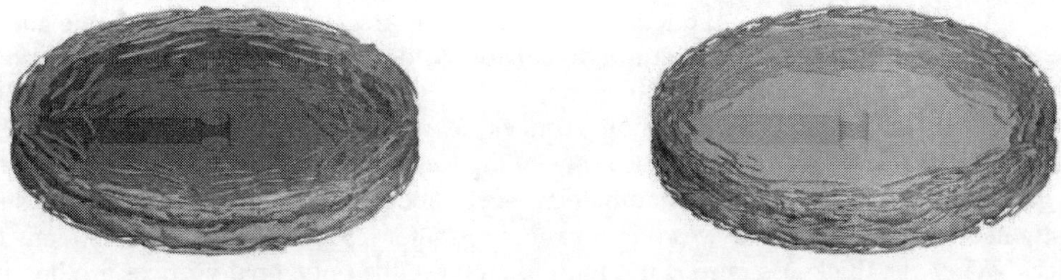

(a) Spiral vortices in the axial clearance. $\gamma = 0.50$. (b) Unconstructed flow structure. $\gamma = 0.65$.

Figure 9. Turbulent state. $Re = 20{,}000$, $\alpha = 10{,}000$ and $Q = 1.0$.

Table 1. Final flow structures [a].

(a) Sudden acceleration.

	γ							
	0.50	0.55	0.60	0.65	0.70	0.75	0.80	0.85
3000	C	C	C	C	C	C	C	C
4000	P	C	C	C	C	C	C	C
5000	P	P	C	C	C	C	C	C
6000	P	P	C	C	C	C,S	C	C
7000	P	P	S	C	C	C,S	C	C
8000	P	P	S	S	S	S	C	C
9000	P	P	S	S	S	S	C	C
10000	P	P	P	S	S	S	C	C
11000	P	P	P	S	S	S	C	C
12000	P	P	P	S	S	S	C	C
13000	P	P	P	S	S	S	C	C
14000	P	P	P	S	S	S	C	C
15000	P	P	P	S	S	S	C,S	C
16000	P	P	P	S	S	S	C,S	C
17000	P	P	P	S	S	S	S	C
18000	P	P	P	S	S	S	S	C
19000	P	P	P	S	S	S	S	C
20000	P	P	P	S	S	S	S	C

(b) Gradual acceleration at $\alpha = 1{,}000$.

	γ							
	0.50	0.55	0.60	0.65	0.70	0.75	0.80	0.85
3000	C	C	C	C	C	C	C	C
4000	P	C	C	C	C	C	C	C
5000	P	P	C	C	C	C	C	C
6000	P	P	C	C	C	C	C	C
7000	P	P	S	C	C,S	C	C	C
8000	P	P	S	S	S	C	C	C
9000	P	P	S	S	S	C	C	C
10000	P	P	S	S	S	C,S	C	C
11000	P	P	P	S	S	C,S	C	C
12000	P	P	P	S	S	S	C	C
13000	P	P	P	S	S	S	C	C
14000	P	P	P	S	S	S	C	C
15000	P	P	P	S	S	S	C	C
16000	P	P	P	S	S	S	C	C
17000	P	P	P	U	S	S	C	C
18000	P	P	P	U	S	S	C,S	C
19000	P	P	P	U	S	S	C,S	C
20000	P	P	P	U	S	S	C,S	C

[a] yellow: normal 2-cell mode, orange: normal 4-cell mode, green: anomalous 2-cell mode, blue: anomalous 3-cell mode. C: circular vortices, P: polygonal vortices, S: spiral vortices, U: unconstructed flow.

3.3. *Effect of the acceleration rate on the flow structures*

The final flow structures with the sudden acceleration and the gradual acceleration of α = 1,000 are presented in tables 1 (a) and (b). In each table, the rows are the Reynolds number *Re* and the columns are the thickness ratio γ. The colors distinguish the flow structures in the radial clearance, which are the normal 2-cell mode (yellow), the normal 4-cell mode (orange), the anomalous 2-cell mode (green) and the anomalous 3-cell mode (blue). The characters denote the axial clearance flow structures with the circular vortices (C), polygonal vortices (P), spiral vortices (S) and unconstructed flow (U). In the condition of the sudden acceleration in table 1 (a), when the thickness ratio is 0.85 and the axial clearance is the narrowest, the flow mode in the radial clearance is the normal 2-cell mode and the circular vortices appear in the axial clearance at the Reynolds numbers up to 20,000. At γ = 0.80, the anomalous 2-cell mode and spiral vortices are found at the higher Reynolds numbers. The flow in the anomalous 3-cell mode appears at *Re* = 15,000 and 16,000 and this flow has the circular vortices near the end wall where the normal vortex appears and the spiral vortices near the end wall on which the anomalous vortex emerges. At smaller thickness ratios, the range of the Reynolds number at which the normal 2-cell mode emerges is limited. At $\gamma \leq 0.55$, the anomalous 2-cell mode occurs in the entire Reynolds number range investigated in the present study. At the thickness ratio not less than 0.65, the flow tends to have spiral vortices. When the thickness ratio is not more than 0.60, the polygonal vortices are dominant. Table 1 (b) in the case of the gradual acceleration shows that the normal 4-cell mode with circular vortices tends to appear more often at the moderate thickness ratios, while the anomalous mode seems to occur at the higher Reynolds numbers. At γ = 0.65 and the higher Reynolds numbers, the flow has no clear spiral vortices and it is unconstructed. The range of the Reynolds number at which the flow in the anomalous 2-cell mode has the polygonal or spiral vortices is almost fixed in the flows with the sudden and gradual rotations at $\gamma \leq 0.65$. Regardless of the acceleration conditions of the disk, the normal mode flows have circular vortices.

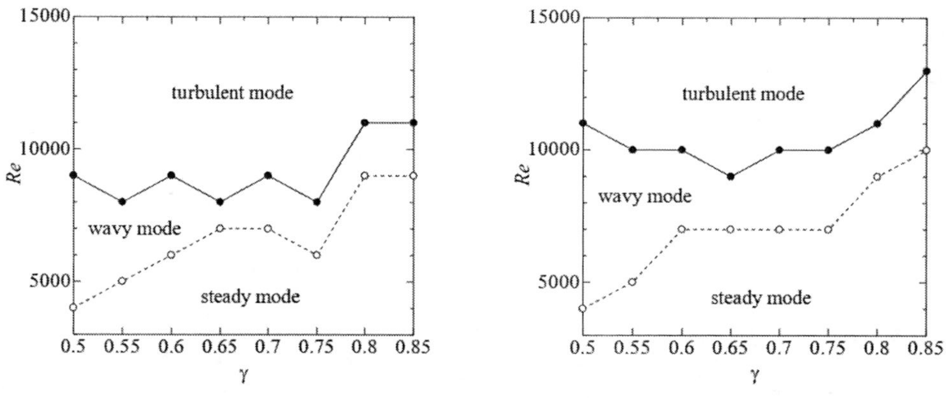

(a) Sudden acceleration. (b) Gradual acceleration at α = 1,000.

Figure 10. Flow state diagram.

The diagrams for the onset of the steady, wavy and turbulent states mentioned in subsection 3.2 are shown in figure 10. The abscissa is the thickness ratio and the ordinate is the Reynolds number. The open circles and the filled circles denote the lower limits of the appearances of the wavy state and the turbulent state, respectively. The sudden start case is in figure 10 (a) and the gradual start case at α = 1,000 is in figure 10 (b). In both acceleration cases, the critical Reynolds number for the wavy state tends to increase as the thickness ratio becomes larger and the axial clearance turns narrower, and the effect of the acceleration rate on the critical values is not large. The diagram at the sudden acceleration shows the minima of the critical values at γ = 0.75. When γ is 0.75, the fraction of the radial clearance

IOP Conf. Series: Materials Science and Engineering **297** (2018) 012066 doi:10.1088/1757-899X/297/1/012066

width to the thickness of the disk is 2.0 and the fraction of the axial clearance width to the width of the radial clearance is 0.333. This geometrical condition may contribute to make the minima and it is expected that this can be clarified by the investigation of the scenario of the flow development from the stationary state. The Reynolds number for the onset of the turbulent state is almost constant at $\gamma \leq 0.75$, and it increases at $0.80 \leq \gamma$. This means that the narrower axial clearance suppresses the growth of the disturbances. The critical value of the turbulent state in the case of the gradual acceleration is about 100 larger than the value at the sudden acceleration.

4. Conclusion

The flow around a rotating disk in a cylindrical casing is numerically investigated. The Reynolds number based on the disk radius and the circumferential velocity at the disk rim is from 3,000 to 20,000. Eight disks are considered and the ratio of the disk thickness to the length of the casing is from 0.50 to 0.85. The sudden acceleration and the gradual acceleration of the disk rotation are adopted. By following the categorization of Taylor vortex flows, the flow structures in the radial clearance are classified into the normal and anomalous modes with 2, 3 and 4 cells. The flows in the axial clearance show the circular vortices, polygonal vortices with various wave numbers, spiral vortices and the unconstructed flow. When the disk is thick and the axial clearance is narrow, the normal mode easily appears in the radial clearance and the flow has circular vortices. As the thickness of the disk becomes thinner, the range of the Reynolds number at which the anomalous mode appears widens at the higher Reynolds numbers. The flow includes the spiral vortices at the moderate thickness ratios. At lower thickness ratios, the flow in the anomalous mode has the polygonal vortices resulted from the circumferential shape change of the extra vortices associated with the anomalous vortices. When the acceleration of the disk is gradual, the normal mode likely emerges and the critical Reynolds number between the wavy state and the turbulent state increases. The Hopf bifurcation from the steady state to the wavy state is not largely affected by the acceleration rate.

Acknowledgement

This work is partly supported by the JSPS Grants-in-Aid for Scientific Research 15K05791.

References

[1] Jasmine H A and Gajjar T S B 2005 *Phil. Trans. R. Soc. A* **363** pp 1131-44
[2] van Eeten K M P, van der Schaaf J, Schouten J C and van Heijst G J F 2012 *Phys. Fluids* **24** 033601
[3] Lingwood R J and Alfredsson P H 2015 *Trans. ASME App. Mech. Rev.* **67** 030803
[4] Schouveiler L, Le Gal P and Chauve M P 1998 *Phys. Fluids* **10** pp 2695-7
[5] Schouveiler L, Le Gal P and Chauve M P 2001 *J. Fluid Mech.* **443** pp 329-50
[6] Poncet S, Serre É and Le Gal P 2009 *Phys. Fluids* **21** 064106
[7] Serre E, Crespo del Arco E and Bontoux P 2001 *J. Fluid Mech.* **434** pp 65-100
[8] Séverac É Poncet S, Serre É and Chauve M-P 2007 *Phys. Fluids* **19** 085113
[9] Viazzo S, Poncet S, Serre E, Randriamampianina A and Bontoux P 2012 *Flow Turbulence Combust.* **88** pp 63-75
[10] Makino S, Inagaki M and Nakagawa M 2015 *Flow Turbulence Combust.* **95** pp 399-413
[11] Saric W S, Reed H L and White E B 2003 *Annu. Rev. Fluid Mech.* **35** pp 413-40
[12] Launder B, Poncet S and Serre E 2010 *Annu. Rev. Fluid Mech.* **42** pp 229-48
[13] Hendriks F 2010 *Microsyst. Technol.* **16** pp 93-101
[14] Washizu T, Lubisch F and Obi S 2013 *Flow Turbulence Combust.* **91** pp 607-21
[15] Meeuwse M, van der Schaaf J and Schouten J C 2012 *AIChE J.* **56** pp 247-55
[16] de Beer M M, Keurentjes J T F Schouten J C and van der Schaaf J 2014 *Chem. Engng. J.* **242** pp 53-61
[17] Huang R F and Hsieh M K 2011 *Exp. Fluids* **51** pp 1529-47
[18] Uenishi R, Tsugawa S, Maeda T, Uesaka M, Watanabe T, Tanigawa H and Hirata K 2016 *Proc. the 7th TSME Int. Conf. on Mech. Engng. (Chiang Mai)* TSF0025
[19] Tagg R 1994 *Nonlinear Science Today* **4** pp 1-24
[20] Fardin M A, Perge C and Taberlet N 2014 *Soft Matter* **10** pp 3523-35
[21] Grossmann S, Lohse D and Sun C 2016 *Annu. Rev. Fluid Mech.* **48** pp 53-80

[22] Benjamin T B 1978 *Proc. Royal Soc. Lnd. A* **359** pp 1-26
[23] Benjamin T B 1978 *Proc. Royal Soc. Lnd. A* **359** pp 27-43
[24] Nakamura I and Toya Y 1996 *Acta Mechanica* **117** pp 33-46
[25] Watanabe T, Furukawa H and Nakamura I 2002 *Phys. Fluids* **14** pp 333-41
[26] Furukawa H, Watanabe T, Toya Y and Nakamura I 2002 *Phys. Rev. E* **65** 036306
[27] Ji H, Burin M, Schartman E and Goodman J 2006 *nature 444* 343-6
[28] Avila 2012 *Phys. Rev. Lett.* **108** 124501
[29] Edlund E M and Ji H 2015 *Phys. Rev. E* **92** 043005
[30] Lopez J M and Avil M 2017 *J. Fluid Mech.* **817** pp 21-34
[31] Watanabe T and Furukawa H 2010 *Exp. Fluids* **48** pp 631-6
[32] Watanabe T and Furukawa H 2010 *Phys. Fluids* **22** 114107
[33] Hara S, Watanabe T, Furukawa H and Endo S 2015 *J. Visualization* **18** pp 501-10
[34] Lygren M and Andresson H I 2001 *J. Fluid Mech.* **426** pp 297-326
[35] Nakamura I, Toya Y, Yamashita S and Ueki Y 1990 *JSME Int. J. Ser. II* **33** pp 685-91
[36] Jeong J and Hussain F 1995 *J. Fluid Mech.* **285** pp 69-94
[37] Dong S 2007 *J. Fluid Mech.* **587** pp 373-93

IOP Conf. Series: Materials Science and Engineering **297** (2018) 012067 doi:10.1088/1757-899X/297/1/012067

Validation of CFD simulation of recoilless EOD water cannon by firing experiments with high speed camera

Tonkid Chantrasmi*, Premsiri Hongthong and Manop Kongkaniti

Department of Mechanical and Aerospace Engineering, King Mongkut's University of Technology North Bangkok, Bangkok 10800, Thailand

* Corresponding Author: tonkid.c@eng.kmutnb.ac.th

Abstract. Water cannon used by Explosive Ordnance Disposal (EOD) were designed to propel a burst of water jet moving at high speed to target and disrupt an improvised explosive device (IED). The cannon could be mounted on a remotely controlled robot, so it is highly desirable for the cannon to be recoilless in order not to damage the robot after firing. In the previous work, a nonconventional design of the water cannon was conceived. The recoil was greatly reduced by backward sprays of water through a ring of slotted holes around the muzzle. This minimizes the need to manufacture new parts by utilizing all off-the-shelf components except the tailor-made muzzle. The design was then investigated numerically by a series of Computational Fluid Dynamics (CFD) simulations. In this work, high speed camera was employed in firing experiments to capture the motion of the water jet and the backward sprays. It was found that the experimental data agreed well with the simulation results in term of averaged exit velocities.

1. Introduction

Improvised Explosive Devices (IEDs) are a growing concern in many parts of the world including the southern part of Thailand. Many research and engineering works have been done to develop water cannon capable of defusing them by ejecting a high-pressure water jet to disrupt electronic circuits of the explosives before they can detonate. For additional safety, the water cannon can be mounted on a remotely controlled robot so that human operators are as far away from the dangerous areas as possible. It is highly desirable for the firing of the water cannon to be recoilless so as not to damage or immobilize the carrying robots. This allows the robots to be controlled back to the operation base, thus avoiding the need for the human operators to retrieve them from the dangerous area.

In our previous works [1-3], the authors proposed a design methodology to create a recoilless muzzle when other components of the water cannon were given. The method involved both firing experiments to quantify the given bullet driving force as well as a series of Computational Fluid Dynamics (CFD) simulations to find the operating point – i.e. the driving pressure and the water mass flow rate – of an actual firing [1-2]. In essence, the proposed methodology separated effects of the driving bullet from those of the muzzle design. In this way, a smallest number of actual firing tests were required and most work could be done through simulations by iterating on the designs of the muzzle alone.

The proposed methodology had been applied to a problem posed by Explosive Ordnance Disposal (EOD) division of the Royal Thai Army. From the existing off-the-shelves components, the .50 BMG bullets and water cannon barrel were given by EOD to the researchers who were asked to design a new muzzle that would reduce the recoil force. At the beginning of this research, a series of muzzle designs were tested and found that the recoil force could be reduced significantly [2]. Eventually, a recoilless

IOP Conf. Series: Materials Science and Engineering **297** (2018) 012067 doi:10.1088/1757-899X/297/1/012067

design was found experimentally in a firing experiment (using some trend information from the previous simulations of older designs) [3]. Figure 1 shows this recoilless design, which featured 24 slotted holes with a certain shape around the muzzle. When fired, the cannon released backward sprays through these slotted holes to counterbalance the forward destructive water jet. The holes were at 45° angle with the main water jet velocity.

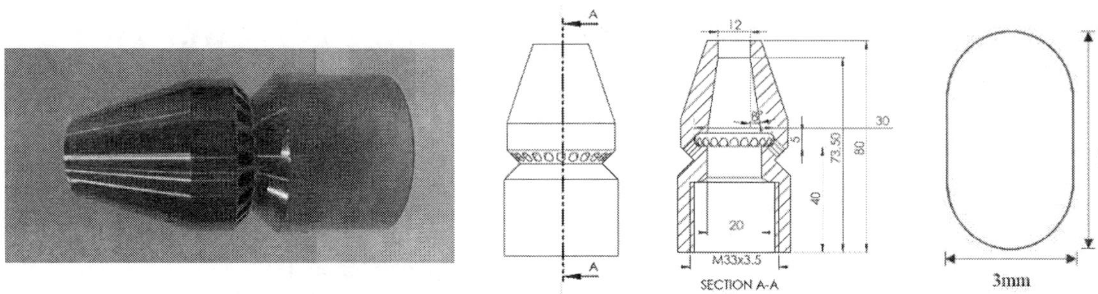

Figure 1. The current recoilless muzzle design.

Afterward, a series of CFD simulations of this recoilless design (to find the load curve) were carried out along with a detailed numerical study to find a set of optimal numerical parameters [4]. The simulations predicted that the water velocity at the main outlet was 501.5 m/s (using a steady-state simulation).

In this work, the authors describe a new firing experiment of the same water cannon and its result. A high-speed camera was deployed to capture the development of the water jet and the backward water sprays. In the next section, the setup of the firing experiment was described along with the remark on the recoil force or the lack thereof. Then, in the following section, the penetrated metal plates were shown and the destructive capability of the water cannon was discussed. Next, the results from high-speed recordings were analyzed and compared with the simulation results from the previous work in order to validate it. Finally, all aspects of the experimental results were summarized as the conclusion.

2. Experimental Setup and a Remark on the Recoil Force

Figure 2 shows the experimental setup with a high-speed camera in a military test-firing tunnel. The water cannon and the metal plate target were mounted on a metal frame with a wide base for stability. Both the cannon and the plate were at 45° angle with the leveled ground as shown. The main outlet of the muzzle and the middle of the metal plate were aligned along the direction of the water jet and they were 10 cm apart.

The target plate was firmly affixed to a mount at the end of a hollow rectangular tube, which is a part of the base, to allow the water to go through after penetrating the plate. Six hexagonal screws and a 1-cm-thick metal gasket with 10 cm inner diameter were used to hold the plate in place. On the other hand, the water cannon was free to slide upward along its rail since no fastening of any kind was used. The only forces kept it from doing so were its own weight and friction.

On the side of the base assembly, a high-speed camera was set at the same horizontal level as the muzzle with two additional stage lights for high-speed recordings. A clear acrylic screen was placed between the base assembly and the camera equipment for safety. The high-speed camera and the firing of the water cannon could be controlled remotely so that the researchers could operate them from outside the tunnel. It is worth noting that the acrylic screen turned out to be unnecessary in this case because the backward sprays dispersed as very small particles well before reaching the camera.

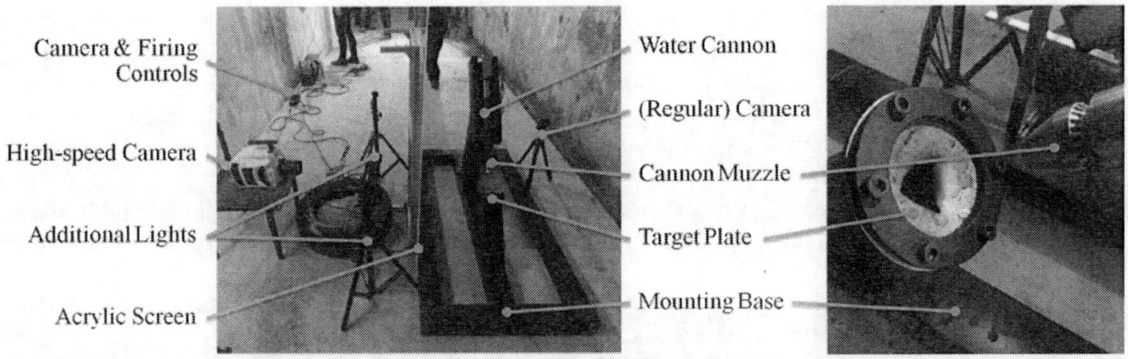

Figure 2. The experimental setup with a high-speed camera

The high-speed camera used was Phantom v2512 and the video recordings were done at 31 kfps for the first two firings and 39 kfps for the third firing.

In all video recordings, no movement of the muzzle could be detected at all. This validated that the firing of this water cannon was indeed recoilless or at least that the recoil was less than the amount of force required to slide the water cannon up (which could be easily done by an average person with one arm and could be estimated to be less than 200N).

3. Destructive Capability

Four metal (ST-41) plates were used as targets for the firings as described in the section above. The thickness ranged from 0.8 to 2.0 mm. Figure 3 shows the aftermath of each metal plate. The water jet was clearly capable of penetrating all the plates in the experiment. The size of the (water) bullet hole decreases as the plate thickness increases as expected.

(a) 0.8 mm thick

(b) 1.2 mm thick

(c) 1.6 mm thick

(d) 2.0 mm thick

Figure 3. The results of firing on target plates with different thicknesses.

4. High-speed Camera Results

Figure 4 shows four key frames from the video recording of the first firing. These four key frames were chronologically (a) when water started coming out of the backward holes, (b) when water started coming out of main outlet, (c) when the water impacted the metal plate target, and (d) when water inside the cannon was depleted. In the other firings, similar behaviors were observed in the same order but with variations in some quantitative measurements.

Before the evidence from high-speed recordings, it was unclear to the authors whether the water would be out of the main outlet or the backward holes first. A rubber stopper inside the muzzle was forcefully pressed against the main outlet on the inside as the high velocity water mass slammed into it. This caused the water to deflect backward to go out of the backward holes instead; however, eventually the rubber stopper had to be pushed out of the muzzle by the water pressure. It was unclear whether this would happen before or after some water came out of the backward holes. As it turned out in all firings in this experiment, the water came out of the backward holes approximately 0.2-0.3 ms before it did the main outlet. This small delay guaranteed that the recoil force was zero at all times during the firing as opposed to being damped over time.

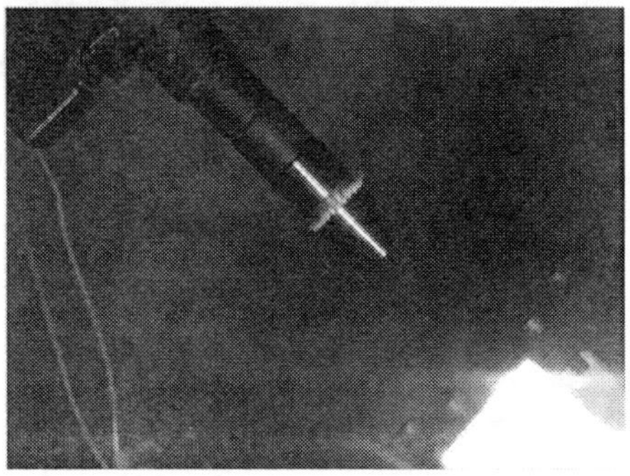

(a) water out from the backward holes

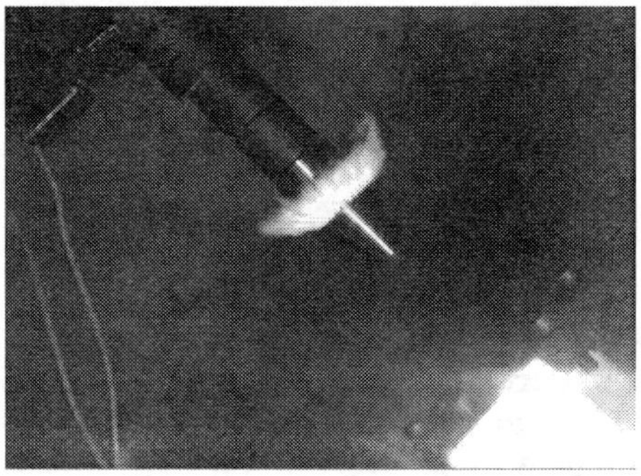

(b) water out from the main outlet

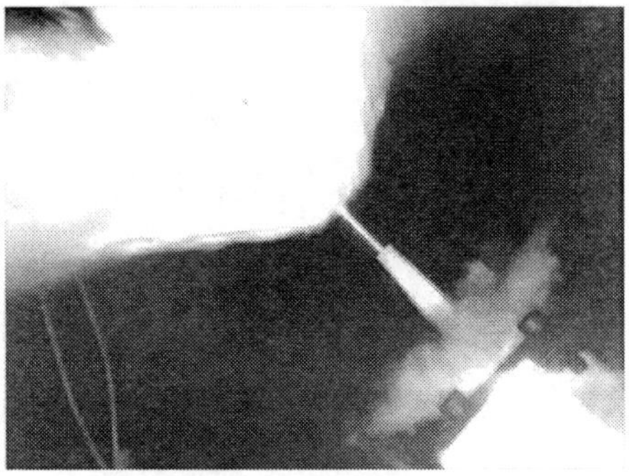

(c) water impacting the target

(d) out of water

Figure 4. Key frames in development of the water flow during a firing.

All the video recordings were processed frame by frame to find the position of the leading edge of the main water jet in each frame. This was done by counting the pixels (via graphical software) and converted it into a physical length by using the known length of 8 cm of the muzzle as a reference. Figure 5 (left) shows the displacement of the leading edge of the water jet as a function of time from all the video recordings. The time was defined to be zero on the frame that some water was seen exiting the muzzle through the main outlet.

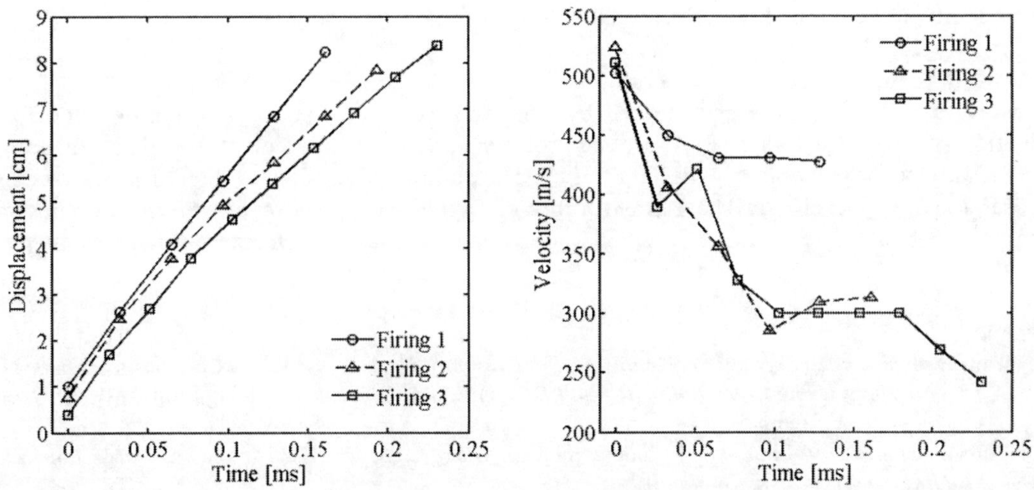

Figure 5. (Left) displacement and (right) velocity of the leading front of the water jet from the main outlet as a function of time from all three firings.

All firings show the same trend in the displacement, especially near the beginning of each firing. The discrepancy could be mainly due to the discrete time frames – the actual time when the water started to exit the main outlet could be anywhere between the two consecutive captured frames, but due to the limited captured frame rate it was locked down to the closest frame.

The first-order finite-difference approximation was used to estimate the velocity of the leading edge of the water jet from displacements of two consecutive frames and the known time interval between them. Figure 5 (right) shows this velocity approximation from all of the recordings. The authors would like to emphasize that this velocity by definition was the velocity of the leading edge of the water jet and *not* the velocity *at* the muzzle outlet. The two velocities were innately different because the aerodynamic drag force in the air slowed down the water jet that was out from the muzzle (and the cross-sectional area of the jet midair would be larger than the main outlet area). From the results shown in the above figure, the water could be reduced to as much as 50% of its speed at the muzzle outlet when it reached the target plate 10 mm away.

Fortunately, at the initial time as defined, the leading edge of the water was at the muzzle outlet. Therefore, the calculated velocity from the high-speed recordings at this point in time could be used to validate the velocity at the muzzle main outlet. Additionally, note that the initial velocities from firings were quite consistent. The average of these three velocities was 512.3 m/s, which was very close to the predicted result from the simulation of 501.5 m/s with a 2.1% relative error.

5. Conclusions

In summary, a firing experiment was setup and carried out to validate various aspects of the recoilless water cannon designed and manufactured in the previous works [3-4]. Metal plates of various thicknesses were used as a target to visually gauge the destructive force from the firing. A high-speed camera was setup to capture the development of the main water jet and the backward sprays. In all the firings, the water was shown to come out of the backward slots slightly before it did out of the main outlet. This transient behaviour ensured that the firings did not generate a recoil force from the very beginning as opposing to being able to damp it as time passed. The video recordings from 3 firings were processed to obtain an average water velocity at the main outlet.

From the experiment, three conclusions can be drawn: (1) the water cannon was confirmed to be *objectively* recoilless since the high-speed camera could not detect any movement of the cannon itself at all; (2) the firing of this water cannon was capable of penetrating a 2-mm-thick metal plate showing a significant destructive force; and (3) the experiment validated the simulation results in term of the water velocity at the main outlet with a relative error of 2.1%.

Acknowledgment

This work is part of the research granted by Thailand Research Fund (TRF) entitled "Performance Evaluation of End Effectors of EOD Robot Prototype and Development of Training Curriculum for EOD Robot Operators," grant no. RDG5850117. The authors would like to thank Explosive Ordnance Disposal division of the Royal Thai Army for providing the testing facility as well as a number of water cannon components. The authors also greatly appreciate technical help from Hi-Tech Resources (Thai) Co.,Ltd. in setting up the high-speed recordings for this experiment.

References

[1] Tansuwannarat S 2013 *Numerical Analysis and Design of High-Speed Water Cannons to Reduce the Recoil Force Using Computational Fluid Dynamics and Experiments* (King Mongkut's University of Technology North Bangkok)

[2] Tansuwannarat S, Vallikul P and Chantrasmi T 2013 *Proc. The 4th TSME Int. Conf. on Mechanical Engineering (Pattaya, Thailand)*

[3] Chantrasmi T, Tansuwannarat S and Vallikul P 2015 *Proc. The 1st Asian Conf. on Dedence Technology (Hua Hin, Thailand)* (New York: IEEE)

[4] Hongthong P and Chantrasmi T 2017 *Proc. The 3rd Asian Conf. on Dedence Technology (Phuket, Thailand)* (New York: IEEE)

IOP Conf. Series: Materials Science and Engineering **297** (2018) 012068 doi:10.1088/1757-899X/297/1/012068

Performance improvement in a tubular heat exchanger by punched delta-winglet vortex generators

C. Khanoknaiyakarn[1], P. Promvonge[2], C. Thianpong[2] and S. Skullong[3,*]

[1] Faculty of Engineering and Technology, King Mongkut's University of Technology North Bangkok, Rayong Campus, Nonglalok, Bankhai, Rayong 21120, Thailand
[2] Department of Mechanical Engineering, Faculty of Engineering, King Mongkut's Institute of Technology Ladkrabang, Bangkok 10520, Thailand
[3] Energy Systems Research Group, Department of Mechanical Engineering, Faculty of Engineering at Sriracha, Kasetsart University Sriracha Campus, 199 M.6, Sukhumvit Rd., Sriracha, Chonburi 20230, Thailand

* Corresponding Author: sfengsps@src.ku.ac.th, sompol@eng.src.ku.ac.th

Abstract. A novel tubular heat exchanger incorporated with punched delta-winglet vortex generators (called perforated delta-winglet vortex generator, P-DWVG) is proposed for improving its thermal performance and energy saving. The P-DWVG elements are punched out from a straight tape having its width nearly equal to the tube diameter before insertion. The main aim at employing the P-DWVG insert is to produce counter-rotating vortices along the tube to promote turbulence intensity inside as well as to transport the cold fluid at the central core to the near-wall regions. The experiment was performed to study thermal behaviors in a uniform heat-fluxed tube inserted with P-DWVGs. The P-DWVGs with the attack angle of 45° were mounted periodically with three different blockage ratios (B_R = 0.1, 0.2 and 0.3) and two pitch ratios (P_R = 2 and 3). Air as the test fluid was varied to obtain turbulent airflow for Reynolds number (Re) in a range of 4,150-25,500. The experimental results show that the P-DWVG provides a considerable increase in the rate of heat transfer around 3.1-4.01 times whereas friction factor increases around 11.44- 34.23 times higher than the plain tube. To assess the real benefits of P-DWVGs, thermal performance factor (TEF) is examined and in the range of 1.39-1.48 where its maximum is at B_R = 0.1 and P_R = 2.

1. Introduction

Heat exchangers are widely used in air conditioning, ventilation, refrigeration and thermal systems. Improving the thermal performance of such systems is crucial to meet energy costs and environment impact. Many techniques for heat transfer augmentation have been widely adopted to obtain a compact and small size of the systems with lower operating costs. For years, many investigations on the utilization of vortex generators (VGs) in improvement of heat exchanger tubes have been performed experimentally and numerically. Several kinds of VGs were used in enhanced tubes such as coiled square wires [1]; snail entry/twisted tape and coiled-wire [2, 3]; ribbed tube with double twisted tapes [4]; conical-nozzle [5]; regularly-spaced twisted tapes [6]; perforated twisted-tapes with parallel wings [7]; and perforated VGs [8]. The modifications of all twisted tapes mentioned earlier were found to yield the improvement of thermal systems around 10-35% higher than the typical twisted ones.

In another type of VGs in the form of conical/circular/vortex/V-shaped rings, Promvonge [9] studied fluid friction and thermal behaviors in a round tube contained with three arrangements of conical ring/nozzle geometries. Also, the effect of combined conical-nozzles and snail-type vortex generator mounted at the entrance on heat transfer behaviors was examined by Promvonge and Eiamsa-ard [10]. Promvonge et al. [11] again reported thermal characteristics in a circular tube inserted with 30° inclined vortex rings (VR) and found that the VR gives higher thermal performance than other twisted tapes, coiled wires and transverse rings. Chingtuaythong et al. [12] investigated heat transfer behaviors in a uniform heat-fluxed tube with V-shaped rings and showed that the V-ring is superior to other rings.

Appling winglet-type vortex generators (WVGs) mounted or punched on the heat transfer surface, is an innovation strategy to augment heat transfer with small increases of pressure drop penalty. Several investigations on the influence of wing/winglet parameters on the rate of heat transfer and pressure loss have been carried out. Promvonge [13] proposed that the use of V-winglet vortex generators was much better than the ribs or WVGs. Skullong et al. [14, 15] studied thermal and flow friction behaviors in a solar air heater channel with combined wavy-groove and perforated-delta wing; and winglet vortex generators. They showed that the combined devices yielded considerably higher thermal performance than the single device acting alone.

In the literature review mentioned above, the WVGs were used in flat surfaces/channels to generate longitudinal vortex flows along the channel. The investigation on thermal characteristics from using a straight tape with double-sided perforated delta-winglet vortex generator (P-DWVG) inserted into a circular tube has hardly been seen. In the current work, a novel 45° P-DWVG inserted into a circular tube is presented and designed to enhance heat transfer with appropriate pressure loss penalty. Thermal performance characteristics of the tube with P-DWVG inserts are studied experimentally for airflows in the Reynolds number range of 4,150 to 25,500.

2. Experimental setup

Experimental approach has been adopted to generate the data in the dimensionless form of Nusselt number (Nu) and friction factor (f) for a tubular heat exchanger with 45° P-DWVG to explore the effect of B_R, P_R and Re on Nu and f. The arrangement of the experimental apparatus is sketched in figure 1. In the apparatus settings, the copper test tube is 50.8 mm in inner diameter (D), 2 mm thick (t) and 3,000 mm long which included the test section length (L) of 1,200 mm. To obtain a uniform heat-flux condition, the test section was heated using a flexible electrical wire wound continually around it where the output power of the electric heater was controlled by a variac transformer. The outermost test tube was wrapped by insulations to prevent the convection heat loss to the surrounding. The room air from a 2.3 kW high pressure blower flowed toward the orifice meter to measure the volumetric airflow rate through the test section where the inclined manometer was used to measure the pressure drop across the orifice meter. The airflow rates in the test tube were varied by controlling the motor speed via an inverter. The fluid temperatures at the inlet and the outlet were measured by RTD thermocouples, Pt-100 type, while the temperatures along the tube surface were measured using 24 thermocouples, type-T, positioned equally on the upper and the sidewall of the tube. Once a steady-state condition was reached, all the temperature values were recorded through a data acquisition system (Fluke 2680A). The measurement of pressure drop of the tube was done by using a digital manometer. The complete details of experimental setup, method and uncertainty analysis are the same as another work of the authors [16, 17].

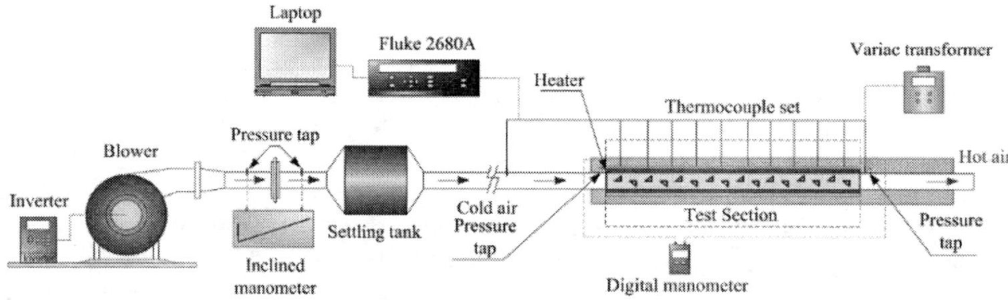

Figure 1. Experimental set-up.

Figure 2 presents details of the 45° perforated delta-winglet tape mounted inside the tube. The perforated delta-winglet tape in the present investigation was made of aluminum sheet (straight tape) with 1.0 mm thickness (t) and length of 1,200 mm. For forming delta-winglet, the tape was partially cut and extruded to become the delta winglet as seen in figure 2. There were two relative winglet pitches called pitch ratios ($P_R = P/D = 2$ and 3), and three winglet blockage ratios ($B_R = b/D = 0.1$, 0.2 and 0.3) with a single punched hole or pore diameter on the winglet ($d = 2$ mm).

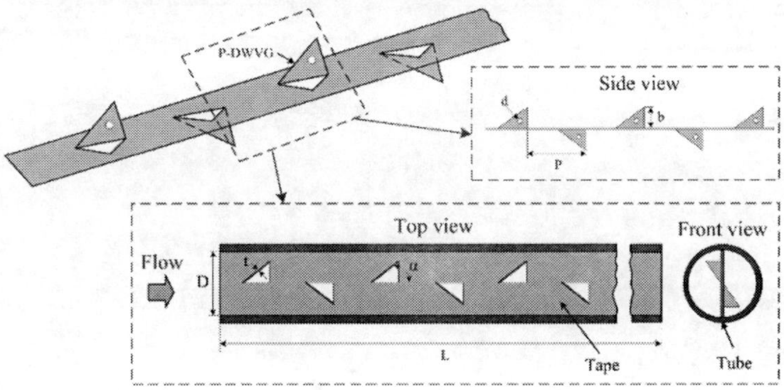

Figure 2. Test tube with 45° P-DWVG insert.

3. Data processing

The objective of the current work is to explore the heat transfer, friction factor and thermal enhancement factor in a tubular heat exchanger inserted with P-DWVGs. The independent parameters include Reynolds number (Re), pitch ratio (P_R) and blockage ratio (B_R). The Re is written as

$$\mathrm{Re} = UD/\nu \tag{1}$$

The calculation of friction factor (f) is performed using pressure drop (ΔP) across the tube as given by

$$f = \frac{2}{(L/D)}\frac{\Delta P}{\rho U^2} \tag{2}$$

in which ρ, ν and U are the density, kinematic viscosity and mean velocity of air, respectively.
In this work, the rate of heat transfer on the surface is identical to the heat absorbed by the air. Thus, the evaluation of the mean heat transfer coefficient (h) is conducted via

$$h = mC_{p,a}(T_o - T_i)/A(\widetilde{T}_w - T_b) \tag{3}$$

where

$$\widetilde{T}_w = \sum T_w / 24 \tag{4}$$

The dimensionless heat transfer coefficient or Nusselt number (Nu) is defined by

$$\mathrm{Nu} = hD/k \tag{5}$$

At a similar pumping power, thermal enhancement factor (TEF) defined by Webb [18] is given as,

$$\mathrm{TEF} = \left.\frac{h}{h_p}\right|_{pp} = \left.\frac{\mathrm{Nu}}{\mathrm{Nu}_p}\right|_{pp} = \left(\frac{\mathrm{Nu}}{\mathrm{Nu}_p}\right)\left(\frac{f}{f_p}\right)^{-1/3} \tag{6}$$

where h_p and h stand for the heat transfer coefficients for a plain tube and inserted tube, respectively.

IOP Conf. Series: Materials Science and Engineering **297** (2018) 012068 doi:10.1088/1757-899X/297/1/012068

4. Results and discussion

4.1. Validation test

The validation test was conducted to verify the reliability of the experimental setup. Nu and *f* data from the current plain tube are, respectively, compared with data from correlations of Gnielinski and Petukhov [19]. Verification tests as displayed in figure 3 are in good agreement between experimental and correlation data. The average deviations are within 4% for Nu and 5% for *f*.

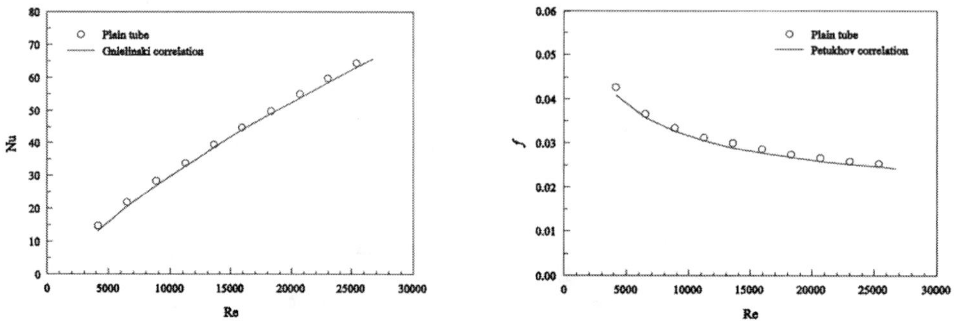

(a) (b)

Figure 3. Verifications of (a) Nu and (b) *f* for plain tube.

4.2. Effect of P-DWVGs on heat transfer

The effects of extending the flow and the winglet parameter on the heat transfer are clearly depicted in figures 4a and b, respectively. The relationship between Nu and Re of the inserted tube is demonstrated in figure 4a. Obviously, P-DWVGs give higher Nu than the plain tube and the Nu shows the increasing trend with the rise in Re but with decreasing P_R. This comes from the fact that the P-DWVG insert produces longitudinal vortex flows assisting to promote the flow turbulence intensity and to move the central core fluid to the near-wall areas. Regarding to its parameters, the P-DWVG with larger B_R can generate more flow obstruction resulting in stronger flow disturbance than the one with smaller B_R.

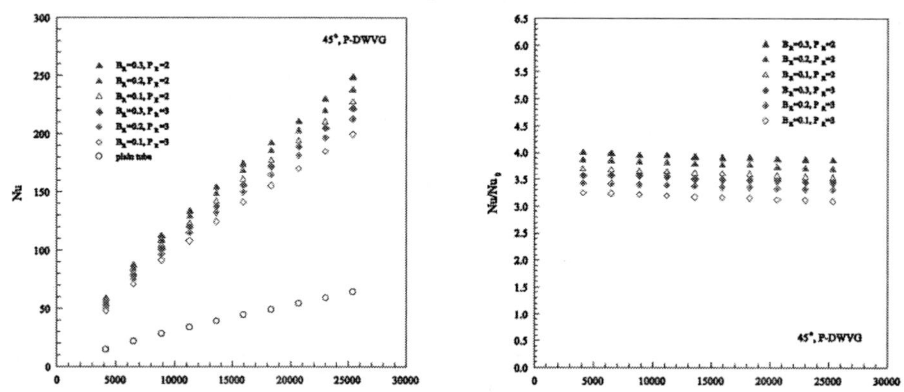

(a) (b)

Figure 4. Variations of (a) Nu and (b) Nu/Nu_0 with Re for P-DWVGs.

Nusselt number ratio (Nu/Nu_0) defined as Nu of the inserted tube divided by Nu of the plain tube is plotted against Re as depicted in figure 4b. It is noted that Nu/Nu_0 displays the increasing tendency with the increment of B_R but with the reduction of P_R due to higher turbulence degree level among the P-DWVGs. The quantitative results of the mean Nu/Nu_0 at P_R=2 and P_R=3 are, respectively, about 3.92, 3.78 and 3.61; and 3.51, 3.37 and 3.17 times for B_R=0.3, 0.2 and 0.1. This means that the employ of larger B_R and smaller P_R results in a considerable increase in the rate of heat transfer.

4.3. Effect of P-DWVGs on friction loss

The isothermal pressure drops for using the P-DWVG displayed in terms of f and friction factor ratio, (f/f_0) are, respectively, plotted against Re as shown in figures 5a and b. It is observed in figure 5a that f of the P-DWVG found to be much higher than that of the plain tube, shows the downtrend with the rise of Re and P_R.

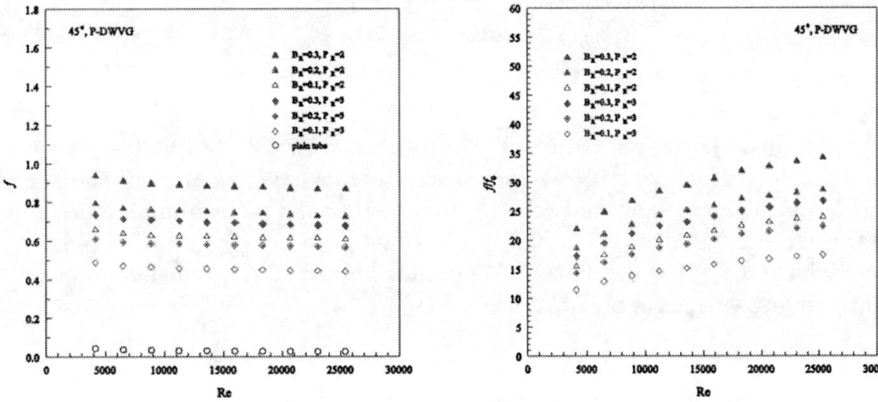

(a) (b)

Figure 5. Variations of (a) f and (b) f/f_0 with Re for P-DWVGs.

In figure 5b, f/f_0 tends to increase with the increment of Re. It is seen that in general, f/f_0 increases with increasing B_R but decreases with rising P_R. The mean values of f/f_0 at P_R=2 and P_R=3 are, respectively, about 29.47, 24.92 and 20.72; and 23.19, 19.33 and 15.17 times for B_R=0.3, 0.2 and 0.1.

4.4. Effect of P-DWVGs on thermal performance

To assess the potential of P-DWVG insert for real practical, thermal enhancement factor (TEF) is offered and is plotted against Re as depicted in figure 7. It is found that, in general, values of TEF are higher than unity, suggesting that the application of P-DWVG insert is superior to the plain tube. TEF displays the downtrend with increasing Re for all the P-DWVG cases and its peak is 1.48 at B_R=0.1 and P_R=2. At P_R=2, TEF values are, respectively, around 1.22-1.48, 1.20-1.46 and 1.19-1.43 for B_R=0.1, 0.2 and 0.3.

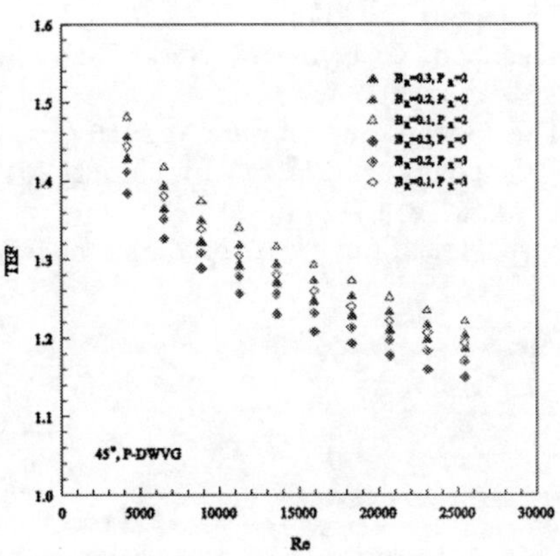

Figure 6. Variation of TEF with Re for P-DWVGs.

5. Correlations for Nu and *f*

Correlations generally are fruitful for use in the prediction of thermal performance or to determine the optimum geometric parameters for a particular application. The empirical correlations for Nu and *f* of using the 45° P-DWVG inserts are correlated as shown below. Nu and *f* correlations provided are in the function of Re, B_R and P_R. The discrepancies of Nu and *f* obtained from equation (7) and equation (8) are within ±6% and ±7%, respectively.

$$Nu = 0.148\, Re^{0.778}\, Pr^{0.4} B_R^{\ 0.083} P_R^{\ -0.29} \tag{7}$$

$$f = 3.203\, Re^{-0.043}\, B_R^{\ 0.348} P_R^{\ -0.662} \tag{8}$$

6. Conclusion

Thermal behaviors in a constant heat-fluxed tube inserted with 45° P-DWVGs for various B_R and P_R values in the turbulent regime, Re = 4,150-25,500 are experimentally examined. The rate of heat transfer of the tube insert is augmented around 3.1- 4.0 times above the plain/smooth tube alone whereas the friction factor is enlarged at about 15.2-29.5 times. However, the highest TEF around 1.48 is achieved for the delta-winglet tape at B_R=0.1 and P_R=2, pointing out that this optimal condition is regarded as a superior energy saving device for practical use.

References

[1] Promvonge P 2008 *Energy Convers. Manag.* **49** pp 980–987
[2] Promvonge P 2008 *Int. Commun. Heat Mass Transf.* **35** pp 623–629
[3] Promvonge P 2008 *Energy Convers. Manag.* **49** pp 2949–2955
[4] Promvonge P, Pethkool S, Pimsarn M and Thianpong C 2012 *Int. Commun. Heat Mass Transf.* **39** pp 953–959
[5] Promvonge P and Eiamsa-ard S 2007 *Int. Commun. Heat Mass Transf.* **34** pp 72–82
[6] Eiamsa-ard S, Thianpong C and Promvonge P 2006 *Int. Commun. Heat Mass Transf.* **33** pp 1225–1233
[7] Thianpong C, Eiamsa-ard P, Promvonge P and Eiamsa-ard S 2012 *Energy Procedia* **14** pp 1117–1123
[8] Skullong S, Promvonge P, Jayranaiwachira N and Thianpong C 2016 *Chem. Eng. Process. Process Intensif.* **109** pp 164–177
[9] Promvonge P 2008 *Energy Convers. Manag.* **49** pp 8–15
[10] Promvonge P and Eiamsa-ard S 2007 *Int. Commun. Heat Mass Transf.* **34** pp 838–848
[11] Promvonge P, Koolnapadol N, Pimsarn M and Thianpong C 2014 *Appl. Therm. Eng.* **62** pp 285–292
[12] Chingtuaythong W, Promvonge P, Thianpong C and Pimsarn M 2017 *Appl. Therm. Eng.* **110** pp 1164–1171.
[13] Promvonge P 2010 *Int. Commun. Heat Mass Transf.* **37** pp 835–840
[14] Skullong S, Promvonge P, Thianpong C and Pimsarn M 2016 *Appl. Therm. Eng.* **100** pp 611–620
[15] Skullong S, Promvonge P, Thianpong C, Jayranaiwachira N and Pimsarn M 2017 *Appl. Therm. Eng.* **122** pp 268–284.
[16] Skullong S, Promvonge P, Thianpong C and Pimsarn M 2016 *Int. J. Heat Mass Transf.* **95** pp 230–242
[17] Skullong S, Promvonge P, Thianpong C and Jayranaiwachira N 2017 *Appl. Therm. Eng.* **115** pp 229–243
[18] Webb RL 1981 P *Int. J. Heat Mass Transf.* **24** pp 715– 726.
[19] Incropera FP Witt PD Bergman TL and Lavine A. 2006 *Fundamentals of Heat and Mass Transfer.* (John-Wiley & Sons)

IOP Conf. Series: Materials Science and Engineering **297** (2018) 012069 doi:10.1088/1757-899X/297/1/012069

Effects of seawater flow rate and evaporation temperature on performance of Sherbet type ice making machine

C. H. Son[1], J. I. Yoon[1], K. H. Choi[1], H. K. Lee[2], K. S. Lee[2,*], C. G. Moon[3] and S. H. Seol[4]

[1]Pukyong National University, Busan, 48547, South Korea
[2]Pukyong National University Graduate School, Busan, 48547, South Korea
[3]Daeil Co., Busan, 46027, South Korea
[4]Hokkaido University Graduate School, Sapporo, 0608628, Japan

* Corresponding Author: suwe1862@gmail.com

Abstract. This study analyzes performance of the sherbet type ice making machine using seawater with respect to seawater volumetric flow rate, evaporation temperature, cooling water inlet and seawater inlet temperature as variables. Cooling water inlet and seawater inlet temperature are set considering average temperature of South Korea and the equator regions. Volumetric flow rate of seawater range is 0.75-1.75 LPM in this experiment. The results obtained from the experiment are as follows. As the seawater volumetric flow rate increases, or seawater inlet temperature increases, evaporation capacity tends to increase. At the point of seawater inlet temperature of 27°C and volumetric flow rate of 1.0LPM, evaporation capacity is over 2kW. On the other hand, results of COP change tendency are different from that of evaporation capacity. It appears to increase until volumetric flow rate of 1.0LPM, and decrease gradually from volumetric flow rate of 1.5LPM. This is due to the increase of compressor work to keep the evaporation pressure in accordance with the temperature of heat source. As the evaporation temperature decreases from -8 to -15°C, the evaporation capacity increases, but the COP decreases.

1. Introduction

Recently as well-being food is getting attention through the world, high quality fish as well as meat and fowl are in high demand. For fish, high quality is related to freshness. So, low temperature storage of fish as soon as they are caught is necessary in order to keep the freshness. Normally, fishery boats need ice to maintain freshness of fish, but littoral fishery boats in South Korea are small crafts less than 30-100 tons. It is difficult for them to be equipped with ice making machine, individually. These small sized fishery boats without ice making machines load ice at port before departure, and should return to port before exhausting the ice. For this reason, duration of voyage for littoral fishery boats in South Korea is usually within 7 days, and even shorter when a lot of fish are caught. To extend the duration and increase the number of catches, researches on developing compact ice making machines small enough for being installed on small vessels have been conducted actively. Ice made from the ground is usually angled cube in its shape and made of freshwater. Using freshwater to make ice gives merits of getting water easily and being produced on ground anytime. On the other hand, its demerits are insufficient cooling capacity and slow cooling speed when it is used for maintaining freshness of catches. Large and angled shape of the ice has the less surface to exchange heat between the ice and fish, so fish take longer time to cool. Also, the ice should be poured into fish storages with salt to prevent from osmotic pressure resulting from salinity difference between the ice and fish. Aside from this, using freshwater ice might cause the following problems: difficulty of adjusting salinity, damages on fish due to its crystalline shape. Damaged fish can rot quickly and are hard to be commercialized. Therefore, researches on developing machines making solid-liquid two-phase sherbet type ice have been being conducted throughout the world. The related researches are as follows.

Stamatiou et al. [1] examined basic crystallization mechanism of a fluidized bed ice slurry generator and a scraped surface ice slurry generator, which are commonly used.

Mouneer et al. [2] showed experimental comparison between super-cooled water jet type and scraped surface type ice making machines. The super-cooled water jet type machine has the lower initial cost and energy consumption than those of the scraped surface type. Additionally, the study showed results about enhanced heat transfer characteristics.

Lakhdar et al. [3] studied heat transfer resistances in scraped surface heat exchanger. Variables in the study were blades rotating speed and solute concentration.

Li et al. [4] investigated optimal operating conditions of seawater ice making machines. The study proposed effects of rotating speed of drum type ice making machines on temperature of produced seawater ice, amount of ice and refrigeration performance.

Yoon et al. [5] analyzed performance of seawater ice making machines that consist of two stages vapor-compression cycles. The study presented that R22, R600a and R717 have higher COP in the system. Besides, some variables such as condensing temperature, compression efficiency, super-heating degree and sub-cooling degree were considered important for enhancing COP.

Sherbet or slurry type ice, usually produced by double-pipe heat exchanger, has fine particles, resulting in less damage to the fish. Also, this kind of system is easy to adjust salinity in fish storage since it freezes seawater directly. That is why it has been already commercialized abroad. In spite of those advantages, studies related to ice making machines for small sized vessels are still desired in South Korea. Thus, this research conducts study on developing compact ice making machine using double-pipe heat exchanger and analyzes its performance characteristics.

2. Sherbet type ice making machine using seawater

Figure 1 shows schematic diagram of a sherbet type ice making machine using double-pipe heat exchanger. The sizes of machine are 1,400 mm wide, 650 mm long and 1,500 mm high. The machine weighs approximately 400 kg. These sizes are only for experimental equipment, and the machine which will be installed on boats might be different for different boat sizes. The machine consists of a compressor, an evaporator, a condenser, a receiver, an expansion valve, an oil separator and filter dryer. The evaporator is a double-pipe type with screw shaped scrapers. The scrapers are rotated by a motor installed on upper side of the evaporator. Its working principle is as following.

Vapor state working fluid with high pressure and temperature from the compressor outlet flows into the double-pipe type condenser after passing the oil separator. The working fluid is liquefied in the condenser by exchanging heat with cooling water, and flows to the expansion valve after passing the receiver, the filter dryer and a solenoid valve sequentially. In the expansion valve, the liquid state working fluid is expanded by wire drawing effect to the evaporation pressure. Expanded working fluid evaporates in the evaporator by exchanging heat with seawater thereby the seawater will be frozen. Then, the working fluid after getting heat from seawater flows to the compressor, completing the cycle. The compressor in this experiment is an inverter-type compressor, and the expansion valve is an electronic expansion valve. The expansion valve regulates evaporation pressure by loading super-heating degree from evaporator outlet. The evaporator in the experiment is semi-flooded type having superiority of cooling efficiency. But it has demerit of difficulty in recovering oil perfectly. In order to make oil circulation well, oil separator is installed between the compressor and the condenser. The condenser used in the experiment is a double-pipe heat exchanger made of titanium. Temperature of cooling water used to condensate the vapor state working fluid is maintained by a constant temperature bath. Likewise, another constant temperature bath controls temperature of seawater at the evaporator inlet.

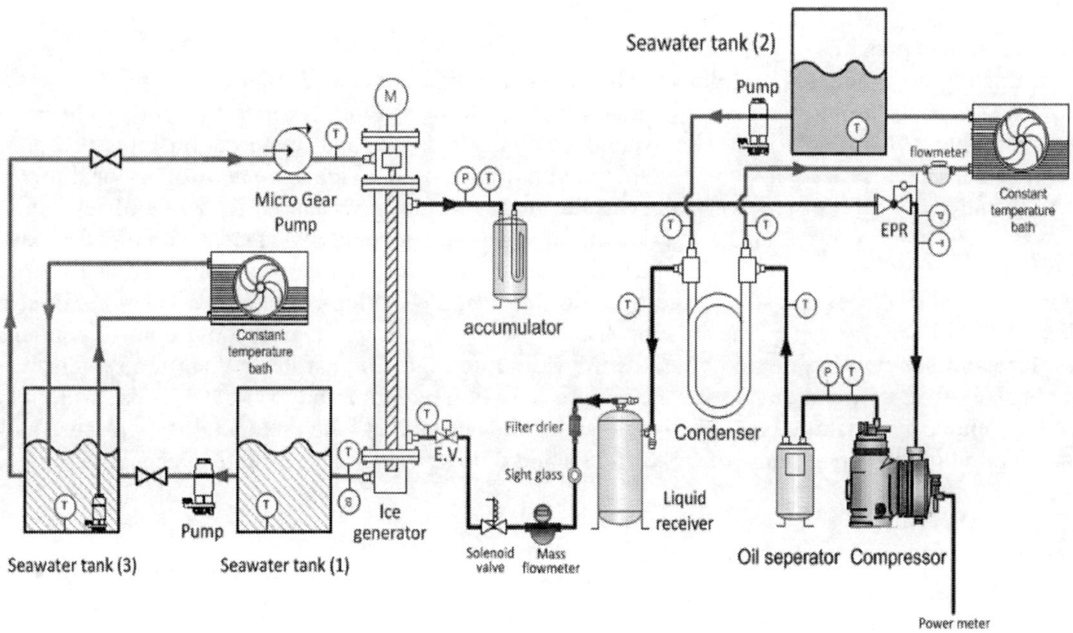

Figure 1. Schematic diagram of sherbet slurry type ice making machine.

3. Experimental conditions and methods

3.1 Experimental conditions

Table 1 shows operating conditions of the experiment. Initial seawater salinity was determined by common coastal salinity of 32% in South Korea. Based on related researches regarding slurry ice, freezing occurs too fast if evaporation temperature is too low, which may stop the scraper rotating. In this experiment, not only energy aspects such as COP but also finding sustainable operating conditions were considered important. That was why evaporation temperature condition of -10°C was set as standard condition. Standard inlet temperature of cooling water was 27°C based on general regulation of refrigeration equipment for the ocean industries. Working fluid was R22, considered high performance and having general use. Scraper rotation speed was 350 rpm.

Table 1. Experimental conditions.

Parameter	Value	Unit
Initial salinity	32.0*	‰
Flow rate of seawater	0.75*, 1.0,1.25,1.50, 1.75	LPM
Inlet temperature of seawater	10*, 15, 20, 27	℃
Evaporation temperature	-8, -10*, -15	℃
Inlet temperature of cooling water	27.0*, 32.0	℃
Refrigerant	R22	-
Scraper rotation speed	350	rpm
Super-heating degree	8	℃

*Standard condition

IOP Conf. Series: Materials Science and Engineering **297** (2018) 012069 doi:10.1088/1757-899X/297/1/012069

3.2 Experiment methods
Experiment was conducted as follows. The constant temperature bath maintained the temperature of seawater as both cooling water and heat source. The temperature was measured precisely by T-type thermocouple. In order to measure condensation pressure and evaporation pressure, pressure measurement devices were attached on inlet and outlet of both condenser and evaporator. Flow meters were installed at outlet of receiver and seawater tanks, so that it measured flow rate of refrigerant and seawater. Once the compressor started operating, super-heating degree was controlled by the expansion valve, and evaporation pressure was adjusted in accordance with compressor rotating speed. A pump to supply seawater into evaporator is micro motion gear pump, which could adjust flow rate of seawater close to 0.1LPM. Another pump to supply cooling water into condenser could control condensation pressure and sub-cooling degree by adjusting flow rate of cooling water. When the system was in a stable state, measurement devices measured pressure, temperature and flow rate. Those were recorded on a computer by a data logger. The stable state was defined as when temperature, pressure and flow rate were within the error range of 0.5°C, 5 kPa and 0.05 kg/min for 15 minutes, respectively.

3.3 Data analysis
Evaporation capacity can be calculated by equation (1) as following.

$$Q_{eva} = m_{ref} \times \Delta h \tag{1}$$

The m_{ref} means mass flow rate of working fluid measured by a flow meter, and Δh means difference of enthalpy at inlet and outlet of the evaporator. System COP can be calculated by equation (2) below. In equation (2), Q_{eva} means the evaporation capacity, and W is compressor consumption power from a power meter.

$$\text{COP} = \frac{Q_{eva}}{W} \tag{2}$$

Table 2 shows uncertainty analysis of experiment data. The uncertainty is obtained by using Moffat's method [6]. The P_H and P_L denote condensation and evaporation pressures, respectively. Also, T_C and T_E represent condensation and evaporation temperatures, respectively.

Table 2. Uncertainty analysis of experimental data.

Parameter	Systematic error	Random error	Total error
P_H	1.825083 kPa	0.948 kPa	0.185 %
P_L	0.334425 kPa	0.768 kPa	0.60 %
T_C	0.47775 °C	0.054 °C	1.377 %
T_E	0.26175 °C	0.037 °C	3.552 %
m_{ref}	0.07787 kg/min	0.001 kg/min	10.004 %

4. Results and discussions

4.1. Effect of seawater flow rate and inlet temperature
Figures 2 and 3 present evaporation capacity and COP with respect to seawater flow rate and seawater temperature to evaporator, respectively. The evaporation capacity (Q_{eva}) grew when volumetric flow rate of seawater increased. Also, the higher temperature of seawater resulted in the higher evaporation capacity. This is caused by two factors. The first factor is the increase of temperature difference between seawater and refrigerant, and the second factor is the rise of flow rate. Those factors affect heat exchange duty, so increase of those two factors contributes to rise of evaporation capacity. While evaporation capacity presented direct proportion value with changing volumetric flow rate of seawater, the tendency of COP was inversely proportional. That is, it reached peak value and then decreased gradually. In case

IOP Conf. Series: Materials Science and Engineering **297** (2018) 012069 doi:10.1088/1757-899X/297/1/012069

that the inlet temperature of seawater was 10°C and 15°C, COP showed peak value at volumetric flow rate of seawater of 1.0LPM. Evaporation capacity had maximum value of 2.1kW when seawater inlet temperature and volumetric flow rate of seawater were 27°C and 1.0LPM. On the other hand, COP had minimum value of 2.78 at same working condition. If inlet temperature or volumetric flow rate of seawater increases, the evaporation temperature and pressure rise. Then, rotating speed of the compressor increases so that evaporation pressure is maintained constantly, and it leads to increased compressor consumption power.

Since COP is a function of both evaporation capacity and compressor consumption power, each parameter regarding the working conditions should be considered simultaneously to analyze tendency of COP. When COP tends to decrease, it means increment of compressor consumption power has more effect on COP than that of evaporation capacity at the same working condition.

On the other hand, if increment of evaporation capacity takes the larger portion than that of compressor consumption power, COP tends to increase. For these reasons, COP shows inconstant tendency according to changing volumetric flow rate and inlet temperature of seawater, as shown in figure 3.

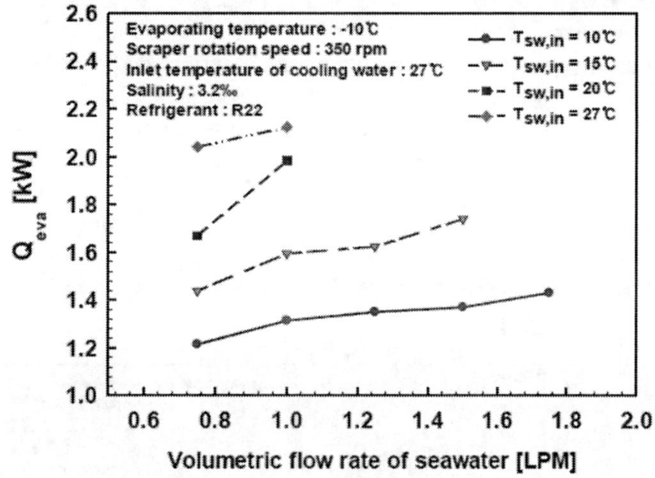

Figure 2. Evaporation capacities with respect to seawater flow rate.

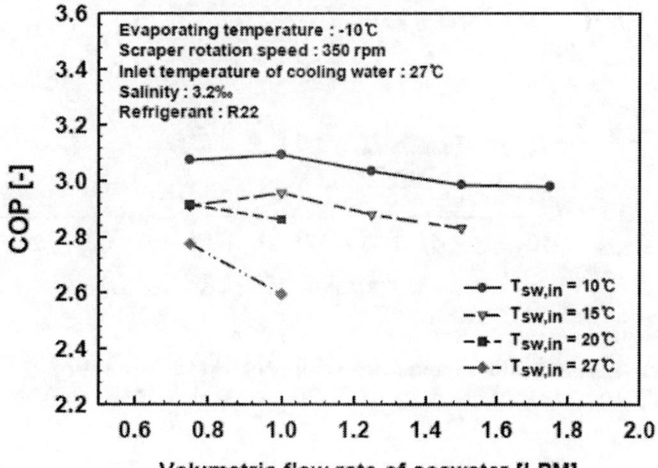

Figure 3. System COP with respect to seawater flow rate.

IOP Conf. Series: Materials Science and Engineering **297** (2018) 012069 doi:10.1088/1757-899X/297/1/012069

4.2. Effect of evaporation temperature and heat sink temperature

Figures 4- 6 show effects of evaporation temperature and heat sink temperature on evaporation capacity, COP and mass flow rate of refrigerant.

Evaporation capacity increased as evaporation temperature decreased due to the enlarged temperature difference between refrigerant and seawater. Additionally, it can be found that mass flow rate of refrigerant increased as the evaporation temperature decreased based on figure 6. The reason why the mass flow rate increased could be explained with analysis of COP.

COP had a different tendency from evaporation capacity. At the lower evaporation temperature, the lower COP was obtained. Increase of compressor consumption power was analyzed to be the main reason for this, and the followings are supporting reasons for increasing compressor consumption power. Firstly, decrease of evaporation temperature enlarged pressure difference between inlet and outlet of the compressor. Secondly, the larger pressure difference resulted in lower compressor efficiency. Lastly, the compressor should rotate with faster speed to get the lower evaporation, and simultaneously the expansion valve should reduce its opening. Here, the faster compressor rotating increases the mass flow rate of refrigerant, while reduced opening of the expansion valve decreases it. Considering both figure 4 and figure 5, increment of refrigerant flow rate by compressor is bigger than decrement by expansion valve, and the increase of the mass flow rate of refrigerant with the decrease of evaporation temperature was shown in figure 6 [7, 8].

Therefore, evaporation capacity in figure 4 increased as evaporation temperature decreases not only for the reason regarding enlarged temperature difference mentioned above, but also increased mass flow rate of working fluid.

Moreover, when heat sink temperature increased which refers to increase of condensation pressure, COP presented the same tendency with respect to changing evaporation temperature. The higher heat sink temperature also caused the larger pressure difference between compressor inlet and outlet. So, when seawater temperature flowing into the condenser was at 32°C, the COP revealed the lower value than that of the case with the seawater temperature of 27°C. Also, COP of ordinary compressor is shown in figure 5, which can get from BITZER software [9], and COP from the experiment is approximately 17% higher.

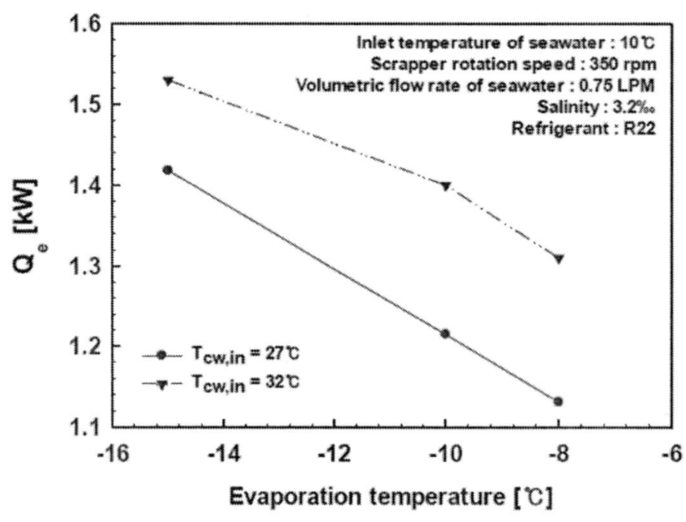

Figure 4. Evaporation capacities with regard to evaporation temperature

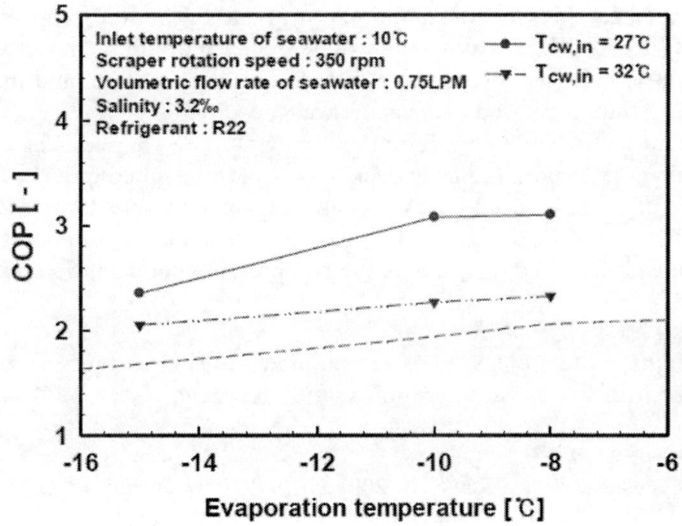

Figure 5. System COP with regard to evaporation temperature.

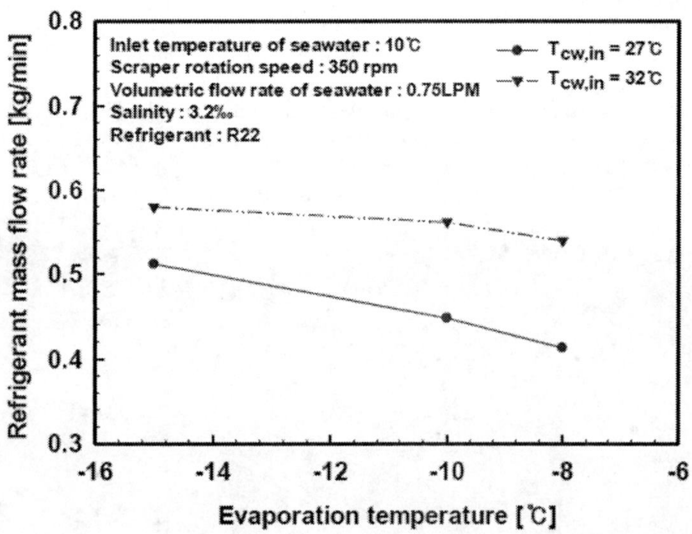

Figure 6. Refrigerant mass flow rate with respect to evaporation temperature

5. Conclusion

This study evaluated the performance of the compact seawater slurry ice making machine applying double-pipe heat exchanger for small vessels. The parameters used for experimental analysis were the volumetric flow rate and inlet temperature of seawater, the evaporation temperature and heat sink temperature. The summary of the experimental results is as follows.

(1) As the volumetric flow rate of seawater increased, the evaporation capacity also increased. This result was caused by two reasons as follows. The first reason is increase of the temperature difference between seawater and refrigerant. The second one is that the mass flow rate of refrigerant was increased by the increase of compressor rotating speed to maintain the evaporator pressure constantly.

(2) When the volumetric flow rate and the inlet temperature of seawater increased, COP presented inconstant tendency. The inconstant tendency of the COP was influenced by correlation of the changes between compressor consumption power and evaporation capacity.
(3) The evaporation capacity was affected by either the evaporation temperature or the condensation temperature. The enlarged temperature difference between them and the increased mass flow rate of working fluid increased the evaporation capacity.
(4) The COP decreased when the evaporation temperature decreased. The compressor rotated fast to keep the evaporator pressure from decreasing, and it resulted in increase of mass flow rate of refrigerant. While, the reduced opening of the expansion valve decreased the mass flow rate of refrigerant. Increment of the mass flow rate of refrigerant by compressor and decrement by expansion valve influenced on not only the compression consumption power but also the COP.

Acknowledgment
This research was a part of the project titled "Development of sherbet type ice making device using sea water," funded by the Ministry of Oceans and Fisheries, Korea.

References
[1] Stamatiou E, Meewisse J W and Kawaji M 2005 *Int. J. of Refrigeration* **28**(1) pp 60-72
[2] Mouneer T A, El-Morsi M Nosier S M and Mahmoud N A 2010 *Ain Shams Engineering J.l* **1**(2) pp 147-157
[3] Lakhdar M B, Cerecero R, Alvarez G, Guilpart J, Flick D and Lallemand A 2005 *Applied Thermal Engineering* **25**(1) pp 45-60
[4] Li H, Joo W J and Jeong S K 2010 *J.of the Korean Society for Power System Engineering* **14**(5) pp 76-82
[5] Yoon J I, Son C H, Heo S K, Jeon M J and Jeon T Y 2016 *J. of the Korean Society for Power System Engineering* **20**(2) pp 85-90
[6] Robert J Moffat 1988 *Experimental Thermal and Fluid Science* vol 1 (New York : Elsevier) pp 3-17
[7] Baek S M, Moon C G, Yoon J I, Jeong S K and Kim E 2011 *J. of the Korean Society for Power System Engineering* **15**(2) pp 37-41
[8] Baek S M, Moon C G, Kim H W, Jeong S K and Yoon J I 2010 *J. of the Korean Society of Marine Engineering* **34**(2) pp 282-288
[9] BITZER Software v6.7.0 rev1852

IOP Conf. Series: Materials Science and Engineering **297** (2018) 012070 doi:10.1088/1757-899X/297/1/012070

Performance characteristic of hybrid cooling system based on cooling pad and evaporator

J. I. Yoon[1], C. H. Son[1], K. H. Choi[1], Y. B. Kim[1],Y. H. Sung[2], S. J. Roh[2,*], Y. M. Kim[3] and S. H. Seol[4]

[1] Pukyong National University, Busan, 48547, South Korea
[2] Pukyong National University Graduate School, Busan, 48547, South Korea
[3] JUNG IN HI-TECH, Ulsan, 44716, South Korea
[4] Hokkaido University, N13-W8, Kita Ku, Sapporo 060-328, Japan

* Corresponding Author: nsj0821@naver.com

Abstract. In South Korea, most of domestic animals such as pigs and chickens might die due to thermal diseases if they are exposed to the high temperature consistently. In order to save them from the heat wave, numerous efforts have been carried out: installing a shade net, adjusting time of feeding, spraying mist and setting up a circulation fan. However, these methods have not shown significant improvements. Thus, this study proposes a hybrid cooling system combining evaporative cooler and air-conditioner in order to resolve the conventional problems caused by the high temperature in the livestock industry. The problem of cooling systems using evaporative cooling pads is that they are not effective for eliminating huge heat load due to their limited capacity. And, temperature of the supplied air cannot be low enough compared to conventional air-conditioning systems. On the other hand, conventional air-conditioning systems require relatively expensive installation cost, and high operating cost compared to evaporative cooling system. The hybrid cooling system makes up for the lack of cooling capacity of the evaporative cooler by employing the conventional air-conditioner. Additionally, temperature of supplied air can be lowered enough. In the hybrid cooling system, induced air by a fan is cooled by the evaporation of water in the cooling pad, and it is cooled again by an evaporator in the air-conditioner. Therefore, the more economical operation is possible due to additionally obtained cooling capacity from the cooling pads. Major results of experimental analysis of hybrid cooling system are as follows. The compressor power consumption of the hybrid cooling system is about 23% lower, and its COP is 17% higher than that of the conventional air-conditioners. Regarding the condition of changing ambient temperature, the total power consumption decreased by about 5% as the ambient temperature changed from 28.7°C to 31.7°C. Cooling capacity and COP also presented about 3% and 1% of minor difference at the same comparison condition.

1. Introduction

The hot and humid air of around 33 °C in the summer of Korea has a negative effect on manufacturing industries and livestock industries. In the livestock industry field, diseases such as reproductive disorders and dyspnea of domestic animals can be caused by the high temperature and humid climate of summer. Similarly, in the manufacturing industry, the production efficiency declines due to the increased labor's discomfort index. Especially if the production amount during the summer occupies a large portion from the entire production amount, it will have more fatal effect. Even though these symptoms are usually

IOP Conf. Series: Materials Science and Engineering **297** (2018) 012070 doi:10.1088/1757-899X/297/1/012070

caused by high air temperature, in fact, the influence of high humidity which is often seen in regions having hot climate is larger. In terms of characteristics of summer climate of the West United States, Mexico and Spain, their climate condition appears slightly different from Korea's climatic conditions; summer climate of those countries is usually high and dry, otherwise that of Korea appears high and humid. Therefore, problems coming from summer climate of Korea that mentioned above are hardly occurred in those countries. This is because the absorbed heat can be dissipated more rapidly and its amount is also larger if the humidity is low in the hot summer climate. Many researches on 'Evaporative coolers' which utilizes a cooling pad for space cooling have been conducted in countries with high and dry climate. The related researches are as follows.

Huang et al. [1] analyzed the heat and mass transfer between air and water film in an evaporative cooler. The flow speed of the air flowing into the cooling pad and the thickness of the pad have a large influence on the cooling effect of the evaporative, and the optimum air flow rate according to the experimental result is 2.5 m/s according to the author.

Ramin et al. [2] compared performance of the air-cooled condenser and the evaporative condenser, and found that the performance of the evaporative condenser was about 51% higher than that of the air-cooled condenser.

Abdollah et al. [3] conducted experimental analysis on pressure drop, humidity distribution, efficiency, etc. on the cellulosic evaporative cooling pad. The pressure drop and the amount of water evaporated are increased as the flow rate increases, and as the thickness of the cooling pad increases. On the other hand, efficiency and humidity tended to decrease as the flow rate increased.

Wasim et al. [4] listed types and features of evaporative coolers and recorded the difference of direct, indirect, and direct or indirect evaporative coolers.

Sarker et al. [5] analyzed performance improvement of cooling capacity by dry-air and moist-air in the case of the hybrid closed circuit cooling tower. The difference of evaporation capacity between the cases that dry air circulates and moist air circulates the cooling tower was about from 26% to 260%.

However, in the case of Korea, there is no research or development of the "evaporative coolers" at all. Therefore, in this paper, the performance characteristics of a hybrid cooling system combining evaporative cooler targeting Korea's hot and humid climate.

2. Evaporative cooler

As mentioned above, the structure of the evaporative cooler used in the country with high and dry climate is shown in figure 1. When outside air flows through the inlet part at the left and right side, the outside air is cooled by the cooling pad where water flows through the form of waves of it. Then, the cooled air is taken out by the blower. Sensible heat exchange, evaporative latent heat exchange and substance exchange occur in the process of contact between outside air and water.

Here, sensible heat means the exchanged heat amount caused by difference of a dry-bulb temperature between water and air flowing through the cooling pad. And the evaporative latent heat refers to the amount of heat delivered to the water from the air, occurred by the evaporation of water as it contacts the air. Lastly the mass transfer is a phenomenon which molecules of water move from water to air by evaporation of water. In short, the outside air passing through the cooling pad of the evaporative cooler is cooled by sensible heat exchange and evaporative latent heat exchange. However, due to the mass transfer phenomenon, its humidity at the outlet increases. Therefore, it is possible to improve the cooling capacity which is lowered by high humidity by combining evaporative cooler and general air conditioners. The combined system lowers the takeout temperature from the evaporative cooler and reduces the power consumption compared to a general air conditioner. Therefore, this paper attempts to demonstrate economical cooling performance through performance analysis of coupled systems under hot and humid operating conditions.

3. Hybrid cooling system

Figure 2 is a schematic diagram of an experimental apparatus of a hybrid cooling system. In order to combine the evaporative cooler and the general air conditioner, a fan and a water pump were used. And cooling pads were attached at the front part of the evaporator and the condenser, respectively. The refrigerant in the system absorbs the heat from the air, which is precooled at the cooling pad, in the evaporator, and then compressed by the compressor. After that, it releases heat to the cooling air, which is also precooled by the cooling pad, in the condenser, and then is expanded by the expansion valve. Finally, the refrigerant circulates the system to complete the cycle. In the hybrid cooling system, the cooling pad copes with a part of the cooling heat capacity of the air conditioner, so more economical cooling is possible compared to the general air conditioner with the same cooling capacity. In this study, the performance of the hybrid cooling system with respect to the ambient temperature is experimentally analyzed, and the effect of installation of the cooling pad at the condenser inlet is reviewed. The experimental working conditions of the system are shown in table 1.

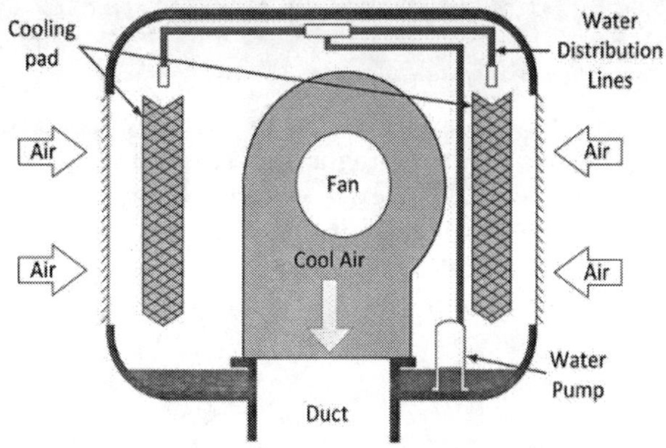

Figure 1. Schematic diagram of evaporative cooler.

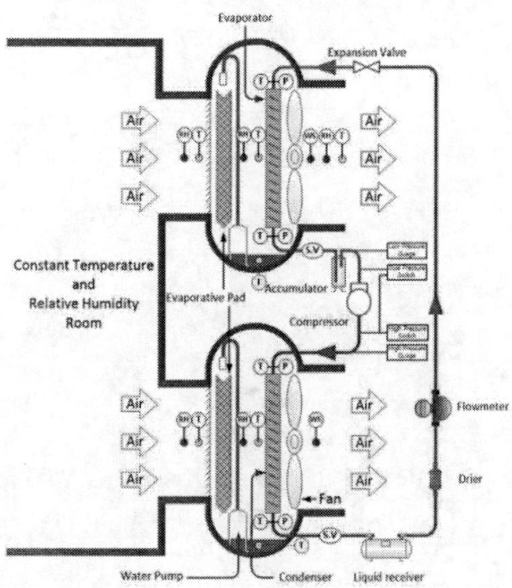

Figure 2. Schematic diagram of hybrid cooling system.

Table 1. Operating conditions of hybrid cooling system.

Parameter	Value	Unit
Ambient temperature	28~32	°C
Relative humidity of Ambient	65	%
Air velocity	2.42	m/s
Circulating water temperature	25	°C
Refrigerant	R-22	
Case	Case 1 : Without a cooling pad at the condenser inlet	Case 2 : With a cooling pad at the condenser inlet

※ In both cases, a cooling pad is installed at the evaporator inlet

4. Data reduction

The values of the physical property values of refrigerant and air applied in this study were calculated using the EES (Engineering Equation Solver) program, and the following equations were used to analyze the experimental data. First, in order to compare the hybrid cooling system with and without a cooling pad to a general air conditioner, mathematical equation of cooling capacity (Q_{TCC}), consumption power (W_{total}), COP of the whole system (COP_{total}) were used. And in terms of hybrid cooling system, the formula of cooling heat (Q_{HCS}), consumption power (W_{HCS}), COP (COP_{HCS}) were used. Each of equations are as followings.

$$Q_{TCC} = Q_{evap} + Q_{CP} \tag{1}$$

$$W_{total} = W_{comp} + W_{fan} \tag{2}$$

$$COP_{total} = \frac{Q_{TCC}}{W_{total}} \tag{3}$$

$$COP_{ref} = \frac{Q_{evap}}{W_{comp}} = \frac{m\cdot(h_{evap,outlet} - h_{evap,inlet})}{W_{comp}} \tag{4}$$

$$Q_{HCS} = Q_{evap} + Q_{CP}$$
$$= m_{R-22} \cdot (h_{evap,outlet} - h_{evap,inlet}) + G_{air} \cdot (h_{CP,inlet} - h_{CP,outlet}) \tag{5}$$

$$W_{HCS} = W_{comp} + W_{fan} \tag{6}$$

$$COP_{HCS} = \frac{Q_{HCS}}{W_{HCS}} \tag{7}$$

In equation (1), Q_{evap} and Q_{CP} denote the cooling capacity of the evaporator and the cooling pad, respectively. M_{R-22} and G_{air} represent the mass flow rate of the refrigerant and the supplied air. $h_{evap,inlet}$ and $h_{evap,outlet}$ in equation (4) represent the enthalpy of refrigerant at the inlet and outlet of the evaporator, respectively. $h_{CP,inlet}$ and $h_{CP,outlet}$ in equation (5) depict the enthalpy of air at the inlet and outlet of the cooling pad, respectively. Also, W_{comp} and W_{fan} in equation (6) show the power consumption of the compressor and that of the fan, respectively. In addition, uncertainty of data is shown in table 2.

IOP Conf. Series: Materials Science and Engineering **297** (2018) 012070 doi:10.1088/1757-899X/297/1/012070

Table 2. Uncertainty of analysis data.

Parameter	Deviation(-)	Uncertainty(%)
Cooling capacity	15.02±0.014	0.44
Power consumption of compressor	2.6±0.016	0.506
Power consumption of Fan	0.7±0.029	0.93

5. Experimental Results and Discussions.

When the cooling pad is installed at the front of the condenser, the temperature of the cooling air flowing into the condenser decreases, and the condensation temperature decreases due to its influence.

Therefore, it is necessary to analyze the performance of the hybrid cooling system with respect to the condensation temperature change. Two cases of working conditions are analyzed; when the cooling pad is not installed in front of the condenser (case 1), and when it is installed (case 2). In both case 1 and case 2, the cooling pad is installed at the inlet of the evaporator since the precooling process obviously increased the cooling capacity and improved the performance. Figure 3 presents a performance characteristic of the refrigerator of the hybrid cooling system with respect to the presence or absence of the cooling pad at the front of the condenser.

During the experiment, the evaporation capacity was kept relatively constant by controlling the mass flow rate of the refrigerant through the expansion valve to compare each case. In fact, when the condensation temperature decreased due to precooled cooling air from the cooling pad, the cooling capacity of the evaporator was constantly increased. That is because increased subcooling degree resulted in the enlarged enthalpy difference of the evaporator inlet and outlet. Thus, the mass flow rate of the refrigerant was decreased to maintain the constant cooling capacity. Compressor power consumption of case 2 was measured lower than case 1, due to not only the decreased amount of mass flow rate of the refrigerant, but also lowered condensation pressure. COP_{ref}, which means COP of only refrigerator, tended to increase in case 2. The reason is that the power consumption of the compressor decreased when the evaporation capacity is kept constant.

Figure 4 depicts performance characteristic of the hybrid cooling system with respect to the presence or absence of the cooling pad at the front of the condenser. The same parameters as in figure 3 were analyzed in figure 4, and all parameters presented the same tendency as figure 3 when comparing case 1 and case 2. Q_{HCS} showed about 54-62% higher value than Q_{evap} due to additional cooling capacity obtained from the cooling pad (Q_{CP}). W_{HCS} included power consumption of the fan (W_{fan}), thereby showing about 26-31% higher value than W_{comp} from figure 3. Therefore, COP_{HCS} was about 22-23% higher than COP_{ref}. In short, even though additional power consumption of the fan was needed in the hybrid cooling system, additionally obtained cooling capacity from the cooling pad occupied the higher portion in the hybrid cooling system.

Figure 5 represents the cooling capacity of the hybrid cooling system and the temperature of supplied air to the space, according to changing ambient temperature. As the ambient temperature rose, the temperature of supplied air increased due to the limited amount of obtainable cooling capacity at the cooling pad. Increase of the ambient temperature resulted in enlarged temperature difference between the ambient and the water that flows the cooling pad. Also, the one between the ambient temperature and evaporation temperature increased. Thus, the cooling capacity of the cooling pad (Q_{CP}) showed slight increase, and the cooling capacity of the evaporator (Q_{evap}) presented increase as well. As a result, the cooling capacity of the hybrid cooling system (Q_{HCS}) slightly increased as the ambient temperature rose.

Figure 6 depicts analysis about the total power consumption and the COP of the hybrid cooling system as ambient temperature changes. First, the total power consumption amount showed slight increase as the ambient temperature rose. This was because the mass flow rate of the refrigerant and the pressure difference at the compressor increased, resulted from risen condensation pressure. Based on the experimental data of figure 5 and figure 6, the cooling capacity of the hybrid cooling system at the ambient temperature of 28.7 °C was about 3% higher than when it was 31.7 °C. On the other hand, the total power consumption at the ambient temperature of 31.7 °C showed about 5% higher value compared to that of 28.7 °C. Considering these two values, the COP of the hybrid cooling system decreased about 1% as the ambient temperature changed from 28.7 °C to 31.7 °C.

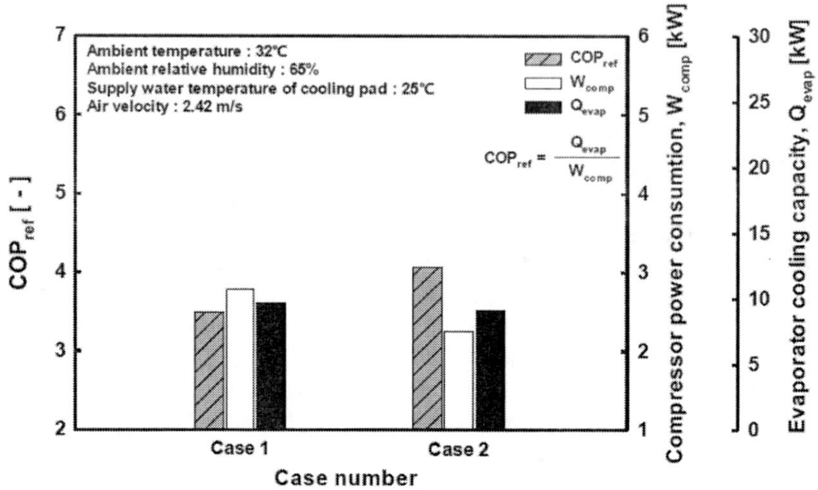

Figure 3. COP of refrigerator, compressor power consumption and cooling capacity of evaporator with respect to presence or absence of cooling pad in front of condenser.

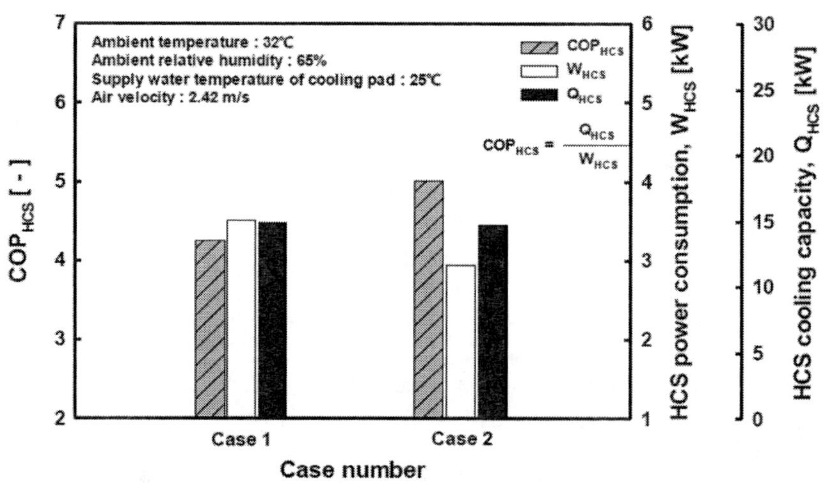

Figure 4. COP, power consumption and cooling capacity of hybrid cooling system with respect to presence or absence of cooling pad in front of condenser.

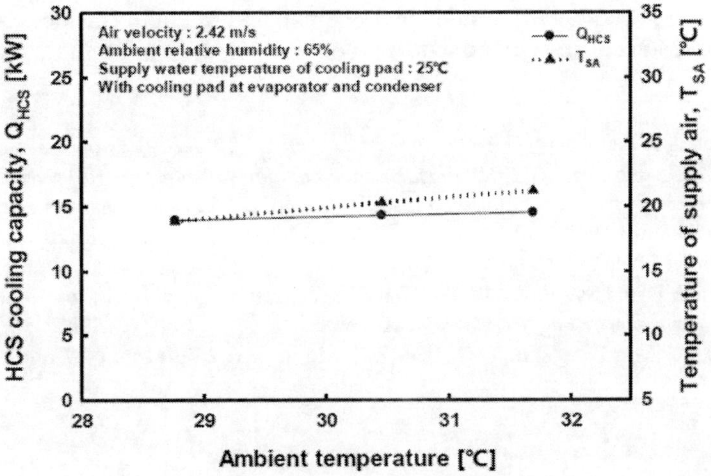

Figure 5. Cooling capacity of hybrid cooling system and temperature of supply air with respect to ambient temperature.

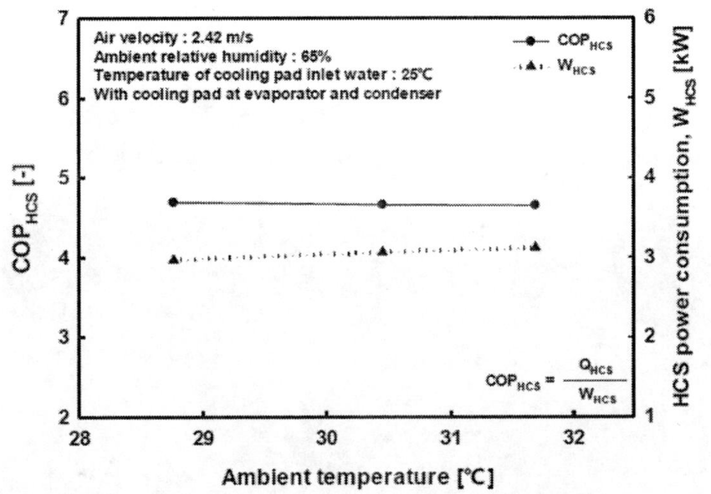

Figure 6. COP and power consumption of hybrid cooling system with respect to ambient temperature.

6. Conclusion

In this study, performance analysis on the hybrid cooling system was analyzed with respect to presence and absence of the cooling pad at the condenser inlet. Also, the effect of the ambient temperature on the hybrid cooling system was studied. As the result, applying the cooling pad at the condenser inlet made the compressor power consumption about 23% lower and COP of the refrigerator about 17% higher than that of the case without the cooling pad. On the other hand, effect of changing ambient temperature on the system was minor. The total power consumption decreased about 5% as the ambient temperature changed from 28.7 °C to 31.7 °C. Cooling capacity and COP also presented about 3% and 1% of minor difference at the same comparison condition. In summary, it was experimentally proved that the more economical space cooling is obtainable with the hybrid cooling system applied to the cooling pad, compared to a conventional cooling system.

Acknowledgement
1. This research was a part of the technology innovation development project titled 'Development of Hybrid cooling system'[C0549960], funded by the Small and Medium Business Administration, Korea.
2. This research reorganized and developed a part of research of 'Y. H Sung', coauthor of this paper.

References

[1] Wu J. M, Huang X and Zhang H 2005 *Applied Thermal enginnering* **29** pp.980-984
[2] Ramin F, John L and Sean G 2010 *Performance comparison of evaporatirvely-cooled condenser versus air-cooled condenser air conditioners, ACEEE* **1** pp.117-127
[3] Abdollah M, Hamid R.S, Mohammad L, Seyedmehdi S and Hamid B 2011 *Energy Conversion and Management* **52** pp.2598-2603
[4] Wasim S, Frank B and Steven T 2010 *Technical research on evaporative air conditioners and feasibility of rating their energy performance*, USA
[5] Yoon J. I, Sarker M. M. A, Shim G. J, Lee H. S, Moon C. G 2009 *Applied Thermal engineering* **29** pp.3328-3333

IOP Conf. Series: Materials Science and Engineering **297** (2018) 012071 doi:10.1088/1757-899X/297/1/012071

Effects of mesh type on a non-premixed model in a flameless combustion simulation

Seekharin Komonhirun[1,*], **Pisit Yongyingsakthavorn**[1], **and Udomkiat Nontakeaw**[1]

[1] Department of Mechanical and Aerospace Engineering,
King Mongkut's University of Technology North Bangkok, Bangkok, Thailand

*Corresponding author: k.seekharin@gmail.com

Abstract. Flameless combustion is a recently developed combustion system, which provides zero emission product. This phenomenon requires auto-ignition by supplying high-temperature air with low oxygen concentration. The flame is vanished and colorless. Temperature of the flameless combustion is less than that of a conventional case, where NOx reactions can be well suppressed. To design a flameless combustor, the computational fluid dynamics (CFD) is employed. The designed air-and-fuel injection method can be applied with the turbulent and non-premixed models. Due to the fact that nature of turbulent non-premixed combustion is based on molecular randomness, inappropriate mesh type can lead to significant numerical errors. Therefore, this research aims to numerically investigate the effects of mesh type on flameless combustion characteristics, which is a primary step of design process. Different meshes, i.e. tetrahedral, hexagonal are selected. Boundary conditions are 5% of oxygen and 900 K of air-inlet temperature for the flameless combustion, and 21% of oxygen and 300 K of air-inlet temperature for the conventional case. The results are finally presented and discussed in terms of velocity streamlines, and contours of turbulent kinetic energy and viscosity, temperature, and combustion products.

1. Introduction

Flameless combustion is a recently developed combustion system, which provides zero emission product. According to work done by Gupta and Arghode [1], flameless combustion is developed from a high temperature air combustion (HiTAC) in an industrial furnace. This technology is investigated to apply with a high thermal intensity gas turbine in a term of flameless oxidation or colorless distributed combustion (CDC). This phenomenon provides ultra-low NOx of about 1 ppm and reduces carbon-monoxide (CO) emission at the operation point. Flameless combustion is promoted by controlling hot product gas recirculation to provide fast fuel/oxidizer mixing at high temperature, which results in a distributed combustion condition, and avoid hot spot region formed in gas turbine combustor.

Flameless combustion is commonly designed in non-premixed combustion regime in order to discretely and directly inject oxidizer and fuel in a confined volume to avoid the direct mixing and reaction of fuel and combustion air prior to the operation point. Fast diluted of oxidizer and internal recirculation are turbulent flow phenomena which induced reactants to form a turbulent-combustion couple relation.

Abuelnuor et al. [2] mention that flameless oxidation reactions are combustion with oxidizer dilution inlet, and the oxygen concentration must be reduced to below 7% to suppress the peak temperature along with internal recirculation inside a combustion chamber to maintain stable combustion. At low oxygen concentration environment, it is difficult for the combustion to occur. To overcome this problem, the inlet air must be preheated until it is reached a fuel auto-ignition temperature in order to sustain the oxidation reaction. When these two conditions meet, the flame is suddenly vanished and the peak temperature is decreased to around 1,000-1,300 K, the reaction is distributed entirely over the combustion region and the emission is suppressed to zero emission.

Duwig et al. [3] used planar laser-induced fluorescence (PLIF) and Rayleigh scattering measurements to study turbulent combustion interactions in distributed reaction regimes of flameless combustion in a laboratory-scale jet burner. Oxygen depletion reactions are varied from equivalent ratio 0.8 to a very lean case at 0.4. All cases are observed by detecting CH_2O/OH signal. The results indicated a complex and intermittent interplay between turbulent flow and chemical reaction involving a coherent structure arising in a jet shear layer. They concluded that a stronger intermittent and the high dilution drive the invisible reaction in a particular case.

Hossieni et al. [4] investigate biogas flameless combustion by applying numerical simulation. They used standard k-ε model with the two-step combustion model developed by Westbrook and Dryer (WD model). This model consists of combustion reaction in equation (1) and oxidation of carbon-monoxide (CO) and carbon-dioxide (CO_2) reaction in equation (2) including with gas phase kinetic file from Arrhenius equation in equation (3) where Ai is pre-exponential factor βi is temperature coefficient and Ei is activation energy.

Combustion reaction:

$$CH_4 + \frac{3}{2}O_2 \rightarrow CO + 2H_2O \tag{1}$$

Carbon-dioxide oxidation reaction:

$$CO + 1/2O_2 \leftrightarrow CO_2 \tag{2}$$

Arrhenius equation:

$$k_{fi} = A_i T^{\beta i} exp(-\frac{E_i}{R_C T}) \tag{3}$$

Arrhenius equation describes a reaction rate in energy form. The fluctuation combustion can be solved from a standard k-ε model and energy equation. Arrhenius equation is weakly implied turbulent combustion phenomena present in a real flameless combustion. Mancini et al. [5] investigates a flameless burner with a strong oxidizer jet and two weak natural gas jets by simulation method. The Reynolds-Averaged Navier-Stokes equation (RANS) and the eddy break up sub-model with two step reaction scheme are applied to predict turbulent non-premixed behavior. They found that the error in predictions of the entrainment was caused by a combustion reaction that slightly occurs within the natural gas jets. The temperature rises within the fuel jets resulting from the entrainment of the hot combustion products. These results present the failure of RANS equation with the two-step reaction scheme.

Pope [6] introduced a PDF method for a turbulent reactive flow where turbulent and combustion are distributed over a probability density function (PDF) which developed from the Heaviside and Dirac delta function adaptation (4)-(6). Pope [6] also proposed that a mean and variance of any parameters such as velocity, temperature and mixture fraction can be described as in equation (7)-(11), respectively. Mean and variance can then be distributed in PDFs as defined in equations (12) and (13) called a joint-PDF equation. This PDF can be coupled solving with other parameters such as pressure, enthalpy and radical mixture which is more approached to a real turbulent combustion regime.

Heaviside function (Unit step function) is

$$H(y) = 0, y \leq 0, \\ = 1, y > 0, \tag{4}$$

Dirac delta function (Impulse function) is a derivative of Heaviside function as

$$\delta(y) = \frac{dH(y)}{dy} \tag{5}$$

The distribution of Dirac delta function is

$$\int_{-a}^{a} \delta(y)dy = H(a) - H(-a) = 1 \tag{6}$$

Distribution function of velocity as a probability density function (PDF) generating from the distribution of Dirac delta function is

$$F_u(V) \equiv P(U < V) \tag{7}$$

Derivative of the distribution function of velocity is

$$f_u(V) \equiv \frac{dF_u(V)}{dV} \tag{8}$$

Mean velocity distribution (U) is given by

$$< U > = \int_{-\infty}^{\infty} V f_u(V)dV \tag{9}$$

Variance of U can be written as

$$D(U) =< u^2 > = \int_{-\infty}^{\infty} (V-< U >)^2 f_u(V)dV \tag{10}$$

where u is a velocity fluctuation as

$$u \equiv U-< U > \tag{11}$$

PDF distribution function Q of mean velocity U and mean temperature ϕ (joint PDF equation) is

$$< Q(U, \emptyset) > = \iint_{-\infty}^{\infty} Q(V, \varphi) f_{u\emptyset}(V, \varphi)dVd\varphi \tag{12}$$

where $f_{u\emptyset}(V, \varphi)$ is a derivative of probability density function (PDF) $F_{u\emptyset}(V, \varphi)$ represent in

$$f_{u\emptyset}(V, \varphi) = \frac{\partial^2 F_{u\emptyset}(V, \varphi)}{\partial V \partial \varphi} \tag{13}$$

Ren et al. [7] applied the PDF method for solving turbulent non-premixed flame problem by comparing the method between an eddy dissipation concept (EDC) and PDF. The results demonstrate that EDC predicted a thinner flame with higher flame temperature than the PDF method. Compared with experiment data, EDC provides over-prediction for flame temperature and OH concentration while PDF model yields more accurate prediction.

The literature review above shows the relation between combustion and turbulence. Both are random base phenomena which required a proper type of mesh. Selecting an improper mesh can lead to numerical error caused by fluctuating parameters. Therefore, this research aims to identify suitable meshing in terms of mesh types, namely tetrahedral and hexagonal, with turbulent non-premixed combustion model (PDF model) for a flameless combustion simulation then compares and discusses results from each case.

2. Methodology

This study is a part of a flameless combustion design for industrial application. Flameless combustion is a new type of combustion with high efficiency with zero emission. This combustor could play a major role in a future technology for gas turbine or drying process heat source. The design uses computational fluid dynamics (CFD) to simulate phenomena inside a combustor. Mesh selection is the first major step before varying other parameters in a further calculation process to prevent inaccurate results. The study of mesh effects started by selecting a geometrical model (refer to Hossieni et al. [3]) as demonstrated in figure 1. The model has full length of 600 mm and a 5-mm single fuel inlet hole at the center surrounding by 8 holes of air inlet with the same diameter. The outlet diameter is 50 mm.

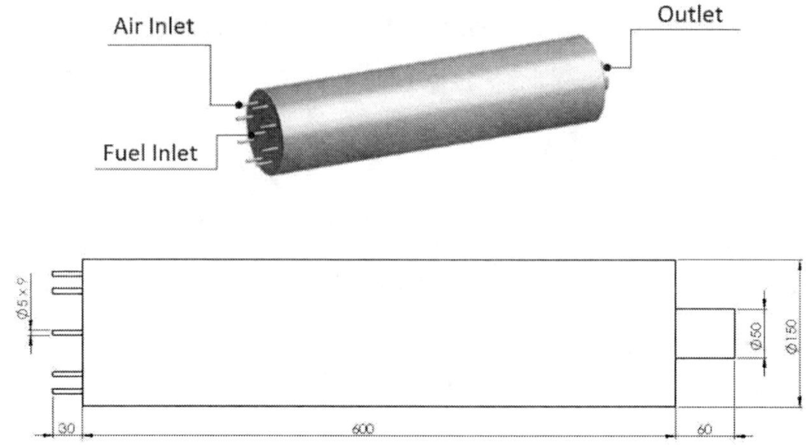

Figure 1. Fluid domain and its dimensions

Next mesh types are selected as tetrahedral and hexagonal in figures 2 and 3, respectively. Both mesh types are generated by ANSYS 16.2 meshing and note that hexagonal mesh is constructed by multi-zone meshing method. Then mesh quality is adjusted by advanced sizing function: proximity and curvature. The number of elements is 270,312 elements which is comparable to Hossieni et al. [3]. The mesh edge sizing effect is also included. The boundary condition at inlet for flameless mode and conventional mode are shown in table 1. Biogas is applied as fuel for both cases.

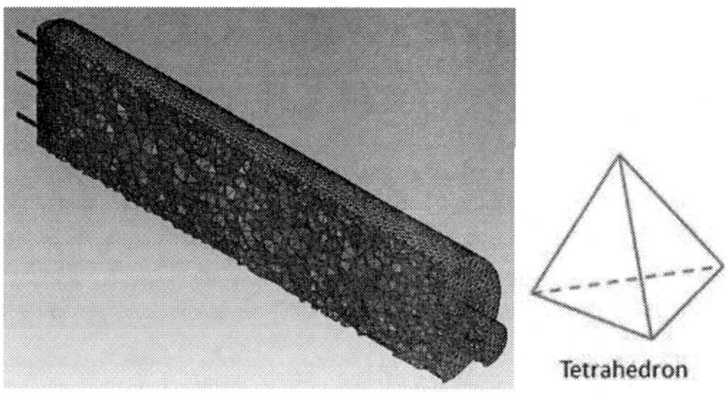

Figure 2. Tetrahedral mesh generated by ANSYS meshing

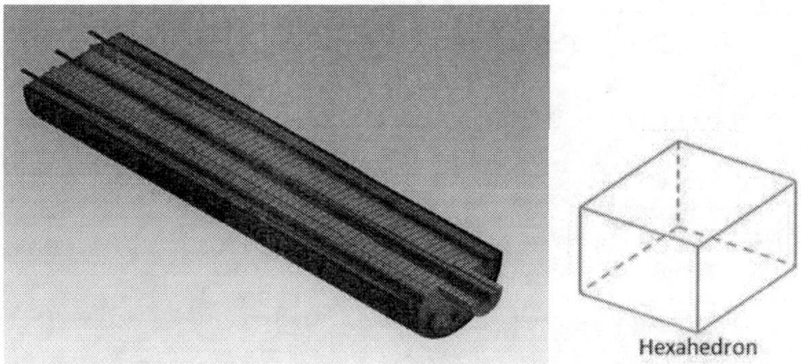

Figure 3. Hexagonal mesh generated by ANSYS meshing

Table 1. Boundary condition

Boundary Conditions	Value
Biogas Composition	CH_4 60% , CO_2 40%
Oxidizer Composition (Flameless mode)	N_2 95% ,O_2 5%
Oxidizer Composition (Conventional mode)	N_2 79% ,O_2 21%
Inlet Air Temperature (Flameless mode)	900K (627 ^{o}C)
Inlet Air Temperature (Conventional mode)	300K (27°C)
Inlet Fuel Temperature	300K (27°C)
Equivalent ratio ϕ	1
Pressure	1 atm
Inlet air speed	30 m/s

For turbulent model, the RNG k-ε model is chosen along with energy equation, as shown in equations (14)-(21), and a PDF non-premixed combustion. This PDF method of ANSYS FLUENT 16.2 is developed by Pope [6]. In each iteration, PDF method is solved by calculating mean mixture fraction and mixture fraction variance in equations (19) and (20), respectively. Then, it uses both parameters to create a beta function as in equations (22)-(24) (see figure 4), instead of Dirac delta function as in Pope [6]. PDF function is mapped into other parameters such as enthalpy, pressure, velocity by equation (12) which prior solving from turbulent model and energy equation. This represents the randomness and interconnection between turbulent and combustion phenomena. The results are compared with species transport methane-air 2 step in equations (1)-(3). Second-order upwind numerical scheme is applied in all cases. Finally, all results are compared and discussed.

Continuity equation:

$$\frac{\partial\rho}{\partial t} + \nabla(\rho\vec{v}) = 0 \tag{14}$$

RNG k- ε turbulent model – k equation:

$$\frac{\partial}{\partial t}(\rho k) + \frac{\partial}{\partial x_i}(\rho k u_i) = \frac{\partial}{\partial x_j}\left(\alpha_k \mu_{eff}\frac{\partial k}{\partial x_j}\right) + G_b + G_k - \rho\epsilon - Y_M \tag{15}$$

RNG k- ε turbulent model – ε equation:

$$\frac{\partial}{\partial t}(\rho\varepsilon) + \frac{\partial}{\partial x_i}(\rho\varepsilon u_i) = \frac{\partial}{\partial x_j}\left(\alpha_k \mu_{eff}\frac{\partial\varepsilon}{\partial x_j}\right) + C_{1\varepsilon}\frac{\varepsilon}{k}(G_k + C_{3\varepsilon}G_b) - C_{2\varepsilon}\rho\frac{\varepsilon^2}{k} - R_\varepsilon \tag{16}$$

Ideal gas equation:

$$P = \rho RT \tag{17}$$

Energy equation in term of enthalpy (H) of all species:

$$\frac{\partial}{\partial t}(\rho H) + \nabla \cdot (\rho \vec{v} H) = \nabla \cdot \left(\frac{k_t}{c_p} \nabla H\right) \tag{18}$$

Mean mixture fraction in term of atomic mass fraction:

$$f = \frac{Z_i - Z_{i,ox}}{Z_{i,fuel} - Z_{i,ox}} \tag{19}$$

Mean mixture fraction transport equation:

$$\frac{\partial}{\partial t}(\rho \bar{f}) + \nabla \cdot (\rho \vec{v} \bar{f}) = \nabla \cdot \left(\frac{\mu_t}{\sigma_t} \nabla \bar{f}\right) \tag{20}$$

Mixture fraction variance transport equation:

$$\frac{\partial}{\partial t}\left(\rho \overline{f'^2}\right) + \nabla \cdot \left(\rho \vec{v} \overline{f'^2}\right) = \nabla \cdot \left(\frac{\mu_t}{\sigma_t} \nabla \overline{f'^2}\right) + C_g \mu_t (\nabla \bar{f})^2 - C_d \rho \frac{\varepsilon}{k} \overline{f'^2} \tag{21}$$

Beta function PDF:

$$P(f) = \frac{f^{\alpha-1}(1-f)^{\beta-1}}{\int_0^1 f^{\alpha-1}(1-f)^{\beta-1} \, df} \tag{22}$$

where α and β are:

$$\alpha = \bar{f}\left[\frac{\bar{f}(1-\bar{f})}{\overline{f'^2}} - 1\right] \tag{23}$$

$$\beta = (1-\bar{f})\left[\frac{\bar{f}(1-\bar{f})}{\overline{f'^2}} - 1\right] \tag{24}$$

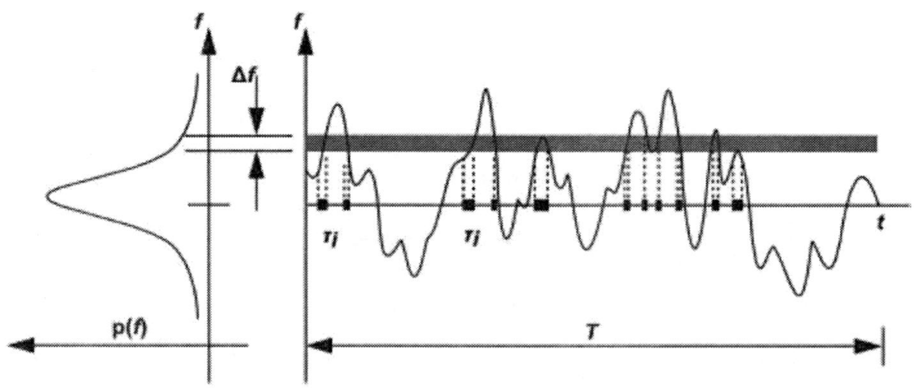

Figure 4. Beta function probability density function

The difference in each case is observed by contour and streamline plots of the following parameters: velocity, temperature and turbulent parameters. They are also compared with theoretical values from literature review mentioned above. The conclusion is expected to provide a guidance for further flameless simulation in term of mesh types and turbulent combustion model selection.

3. Results discussion

3.1 Mesh types comparison

The simulation results indicate that tetrahedral mesh is sensitive to the randomness of turbulent fluctuation than hexagonal mesh as seen in both the temperature distribution in figure 5 and streamlines in figure 6. Figure 5 demonstrates a rough result of flameless and conventional modes. In flameless mode, the peak temperature is suppressed by the effect of dilution oxidizer and temperature is rapidly dropped, but in conventional mode the peak temperature arises by a concentrated oxidizer and fuel. PDF method demonstrated the strongly turbulent interplay between streamlines and temperature distribution. The randomness jumping between tetrahedral mesh causes streamline and internal recirculation distort in both cases with severely induced temperature distribution that represent an unstable combustion zone. Flameless combustion is the most stable combustion phenomena, so the tetrahedral mesh prediction is substantially unacceptable, the turbulent randomness has extremely effected on this mesh type.

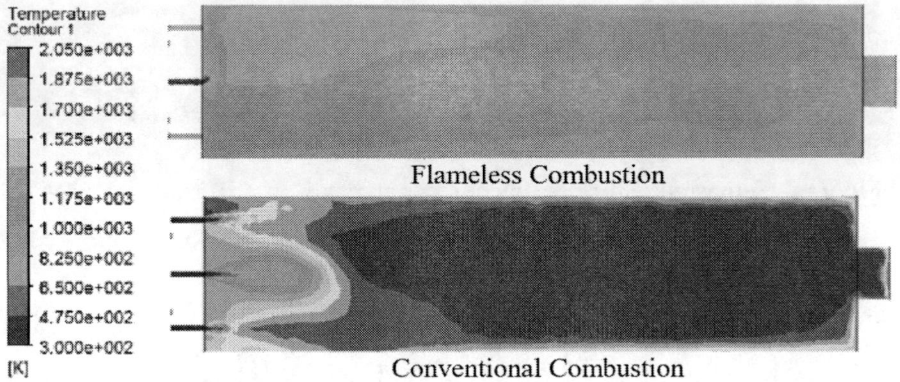

Figure 5. Temperature distribution contour comparison of tetrahedral mesh case for flameless mode and conventional mode

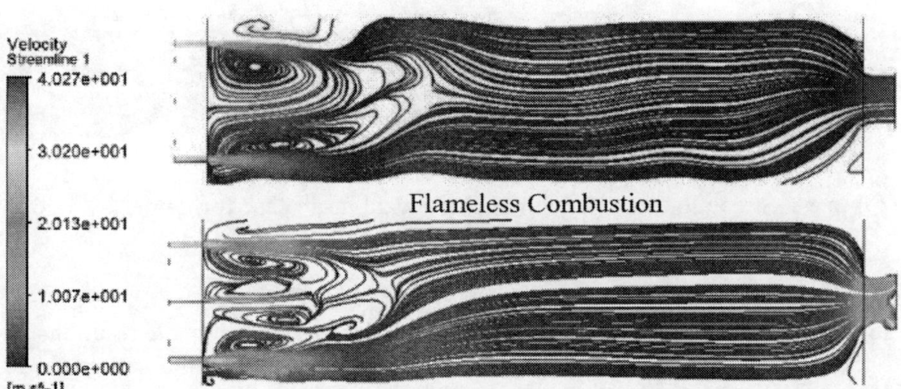

Figure 6. Streamline comparison of tetrahedral mesh for flameless mode and conventional mode.

The hexagonal mesh shows more stable results. In figure 7, the temperature distributions show the distinction between flameless and conventional modes. In flameless mode, temperature is distributed symmetrically across the entire combustor and temperature is drastically dropped compared to a conventional one which characterizes the flameless phenomena. The peak temperature in conventional mode is confined only in a small range of combustion regions. In figure 8, the streamlines is fully aligned along the combustion length. It is concluded that the randomness in turbulent scheme has less effect on this mesh type. The internal recirculation is not severely distorted as in a tetrahedral mesh. The turbulent combustion still has a major interplay by inducing the temperature distribution as a streamline-contour mapping in figure 8. The results shows a successful prediction of flameless combustion in agreement with literature review mentioned earlier.

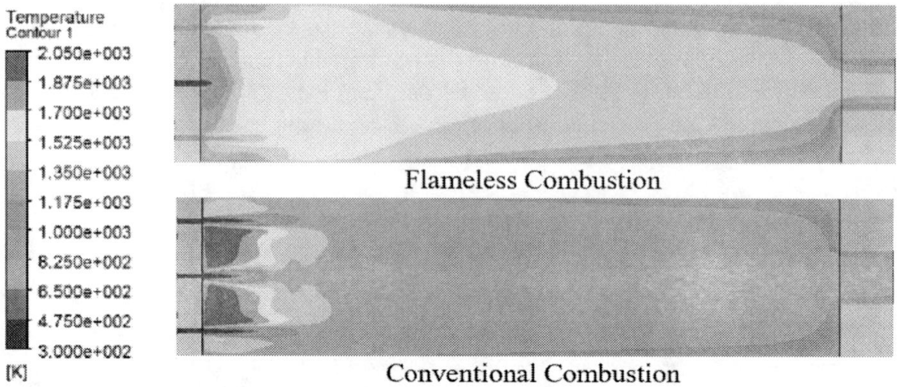

Figure 7. Temperature distribution contour comparison of Hexagonal mesh case for flameless mode and conventional mode

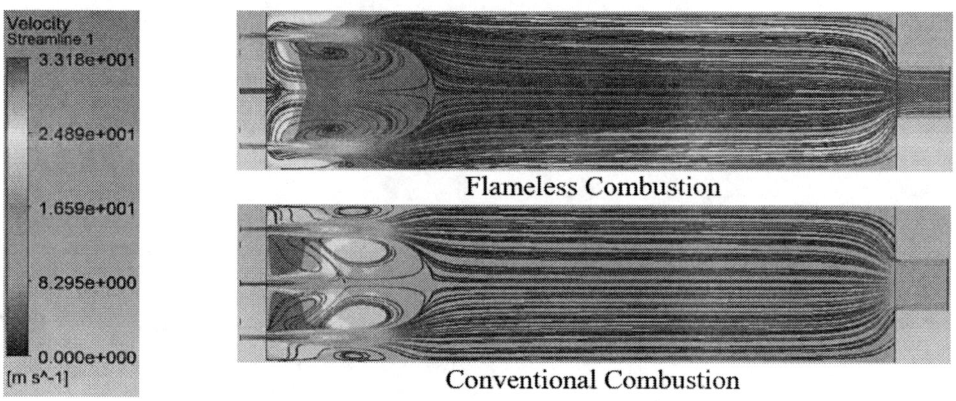

Figure 8. Streamline comparison of Hexagonal mesh case for flameless mode and conventional mode

3.2 The effect of edge sizing

In the last subsection, the results show that hexagonal mesh has a more reliable result than the tetrahedral type due to the less sensitivity to turbulent randomness. In this section, the hexagonal mesh is further improved by using edge-sizing functions to vary the details of edge sizing as demonstrated in figure 9. Edge sizing 1 sets the number of division around combustor circumference to 30. Edge sizing 2 is a biased division along combustion chamber length with a bias type shown in figure 9 with a bias factor 30. Finally, edge sizing 3 sets around all of inlet holes circumference with number of division 10. The boundary condition is similar to the latest cases with biogas fuel. The number of elements is 260,805 elements which is comparable to 270,312 elements in Hossieni et al. [3]. The PDF model and RNG turbulend3 scheme are also applied.

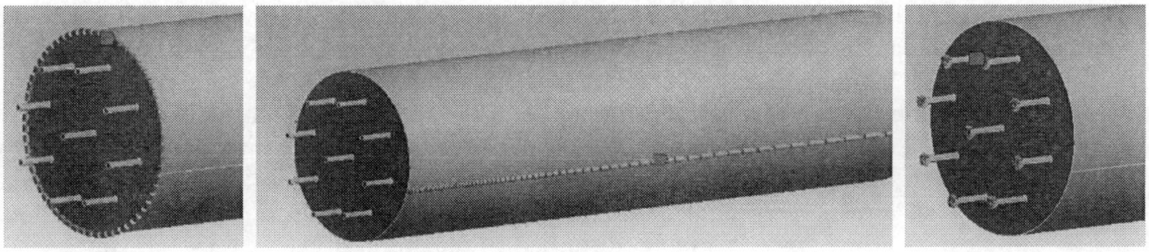

(A) Edge Sizing 1

Number of Divisions = 30

(B) Edge Sizing 2

- Number of Divisions = 60
- Bias Type: -------- ---- -- --
- Bias Factor: 30

(C) Edge Sizing 3

Number of Divisions = 10

Figure 9. Position of edge sizing

When applying an edge-sizing function, the temperature results represented more smoothly distribution in both flameless and conventional mode. Streamlines arrangement is more symmetrical than the case without edge sizing function. Now the turbulent kinetic energy is included in the results. In conventional mode, turbulent kinetic energy has a strong influence on flow pattern than flameless mode. In flameless mode, fuel stream flow is weaker causing the internal recirculation to form more symmetric flow patterns. In contrast, stronger fuel stream in conventional mode causes streamlines collision within reaction zone which increases turbulent kinetic energy.

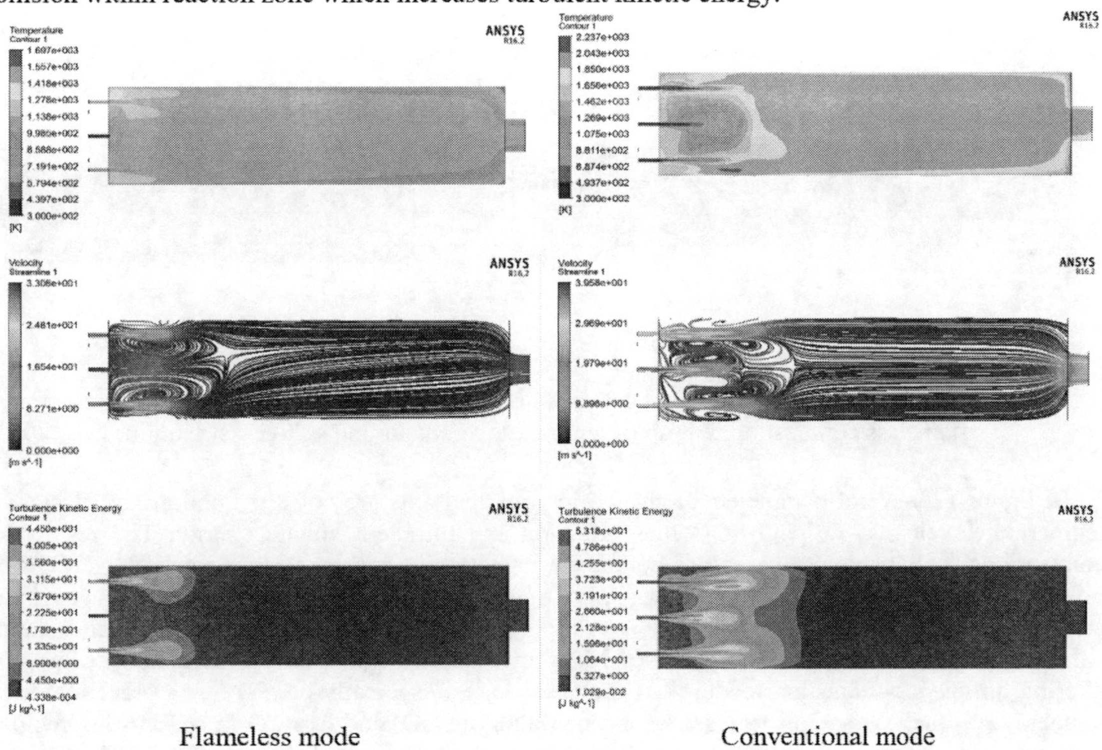

Figure 10. Temperature, Turbulent kinetic energy contour and streamline after using edge sizing functions.

3.3 The effect of turbulent combustion model

The combustion model is compared between two-step species transport model and PDF model for non-premixed combustion. Hexagonal mesh type and edge sizing function mentioned in subsection 3.2 are applied to each case. In figure 11, the results show that both models can be used to superficially estimate flameless combustion characteristics. The difference of these two models can be detected when observed deeply in a turbulent scheme. In flameless case, the two-step model has less influence to turbulent than PDF model. Two-step model represents the less large scale fluctuation when observed from streamlines which causes less disturbance to a temperature distribution. In contrast, the non-premixed model results in less turbulent kinetic energy but more disturbed in streamlines and excessively interplay with combustion reaction.

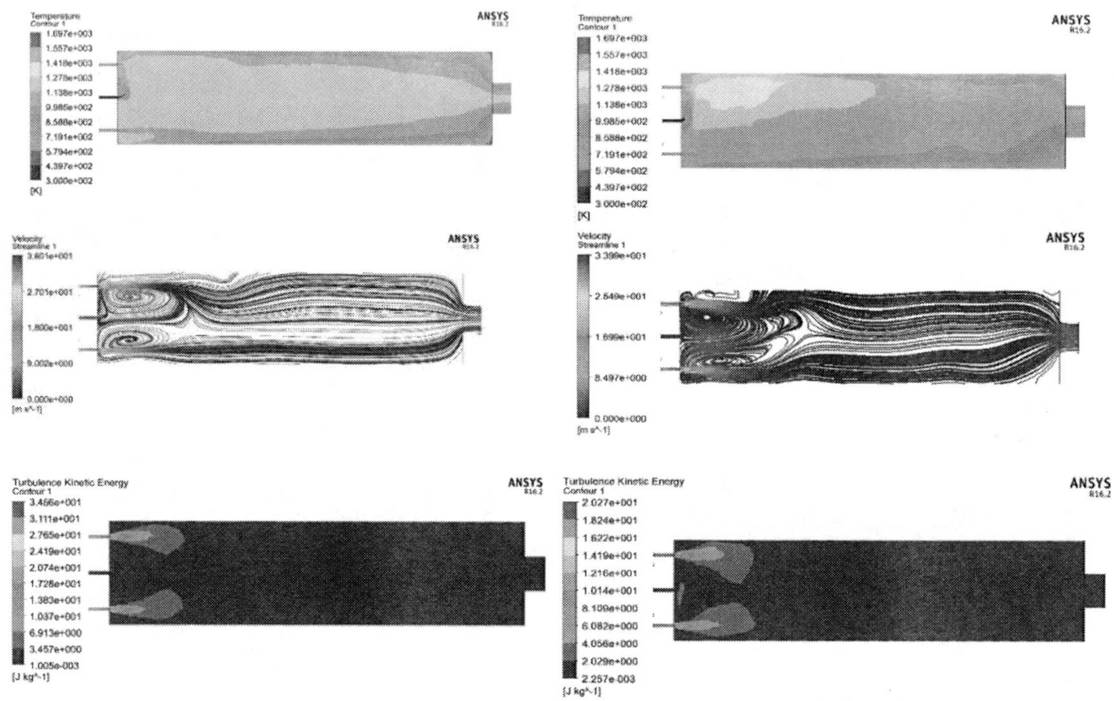

Two-step species transport with RNG k-ε PDF non-premixed model with RNG k-ε

Figure 11. Comparison of two-step species transport and PDF non-premixed combustion for flameless combustion in term of temperature contour and velocity streamlines.

In figure 12, several parameters along the longitudinal axis are collected and analysed including temperature, velocity, O_2, CO_2, CO mass fraction and turbulent kinetic energy. The results from conventional and flameless cases are compared in both two-step species transport and PDF non-premixed combustion models which show similar trend of velocity and temperature along an axis. Two-step species transport indicate a slightly over-predicted turbulent kinetic energy but still maintain a similar trend to the PDF model in both cases. Mass fraction of conventional cases has major effects from altering turbulent combustion model than flameless cases, especially on CO_2 – CO reaction in peak temperature zone. Two-step species transport predict more CO_2 and less CO in combustion region. In contrast, PDF model shown less CO_2 production at peak temperature but more CO which indicate an incomplete combustion. In fact, CO_2 – CO reaction is the slowest chemical reaction of all combustion mechanism which require more reaction time. Two-step model predicted reaction faster than expect, this could be a noticeable results for further calculation.

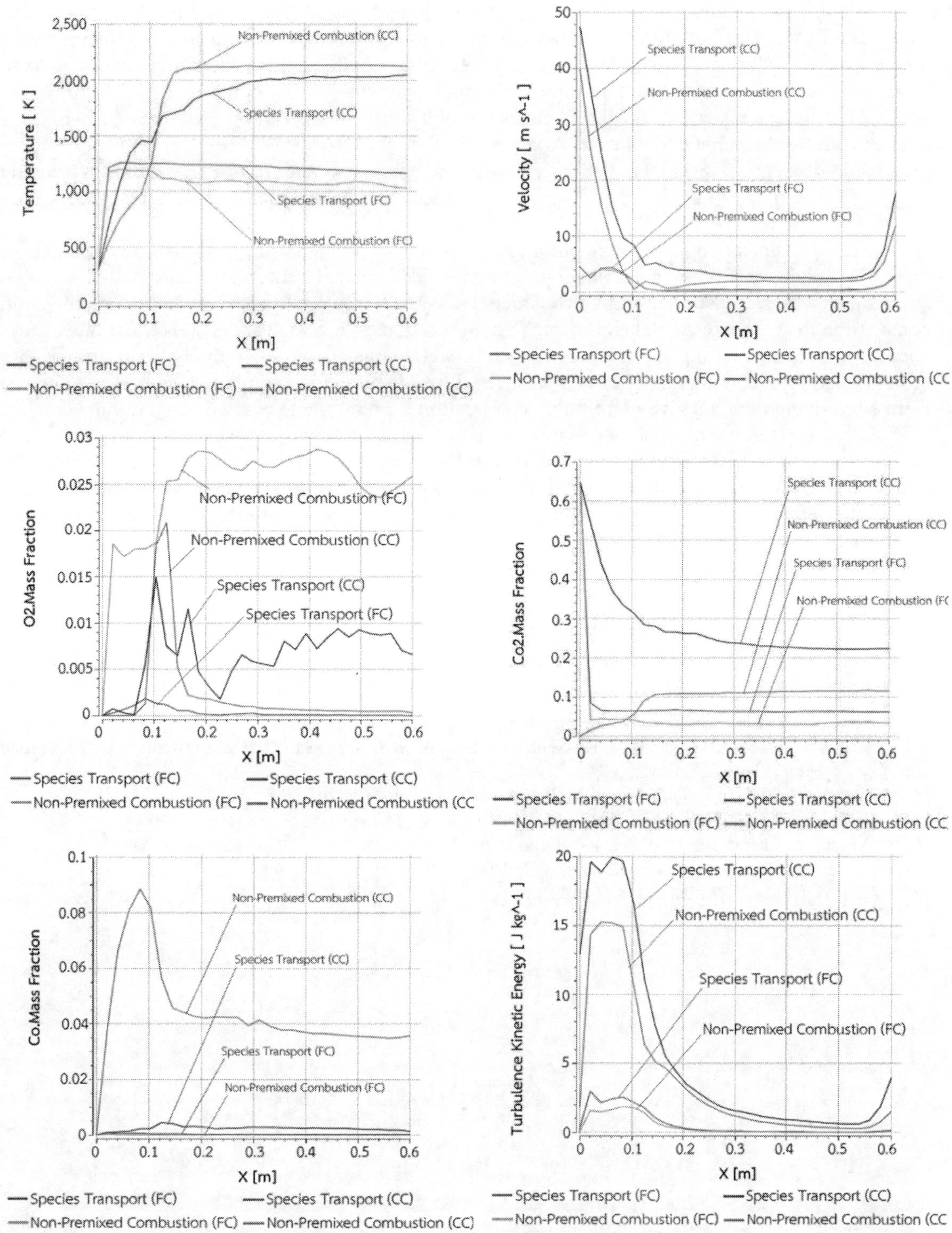

Figure 12. Comparison of two-step species transport and PDF non-premixed combustion in term of temperature, velocity, CH_4, CO_2, CO, O_2 mass fraction, turbulent kinetic energy and enthalpy of both flameless and conventional cases. CC = Conventional Case and FC = Flameless Case

4. Conclusion

This research studies the effect of mesh types for simulations of flameless turbulent non-premixed combustion. Mesh effect study is important to investigate prior to a further design process. An inappropriate mesh selection can lead to significantly inaccurate results. The study chooses two types of mesh which are tetrahedral and hexagonal to simulate with boundary condition of 5% of oxygen and 900 K of air-inlet temperature for the flameless combustion, and 21% of oxygen and 300 K of air-inlet temperature for the conventional case. Applying Probability Density Function (PDF) method for non-premixed turbulent combustion in both mesh types and boundary conditions. PDF method is a simulated randomness from beta function that distribute on others parameters such as velocity, temperature and mass fraction to represent turbulent flow behaviour.

The results concluded that tetrahedral mesh is an inappropriate mesh type for turbulent combustion due to its high sensitivity to turbulent randomness. Hexagonal represents more stable solution in streamlines with less disturbed from turbulence. However, turbulent randomness affected conventional combustion than flameless combustion due to collision of a high velocity fuel and oxidation stream in a combustion regime. Hexagonal mesh is chosen to study edge sizing by applying edge sizing function around combustor and inlet holes circumference with edge bias along combustor centreline. The results from edge sizing function give more stable combustion for both flameless and conventional cases

Finally, the turbulent combustion models are compared. Two-step species transport and PDF turbulent non-premixed model are applied to both flameless and conventional combustions. The results demonstrated the similarity of two model on predicting large scale phenomena. The results are slightly different due to turbulent interplay. Flameless cases again admit less effect from altering turbulent combustion model than the conventional one. Note that CO_2 – CO reaction has different prediction in both turbulent combustion model in conventional mode. Two-step species transport predict the formation of CO_2 significantly higher than PDF model in combustion zone around peak temperature region, indicating the difference in residence time prediction of reaction that could be noticeable in a further work.

References

[1] Arghode V K and Gupta A K 2013 *Applied Energy* **111** pp 930-956
[2] Abuelnuor A A A, Wahid M A, Hosseini S E, Saat A, Saqr K M, Sait H H and Osman M 2014 *Renew. Sustain. Energy Rev.* **33** pp 363–370
[3] Duwig C, Li B, Li Z S and Alden M 2012*Combustion and Flame* **159** pp 306-316
[4] Hosseini S E, Bagheri G and Wahid M A 2014 *Energy Convers. Manag* **81** pp 41–50
[5] Macini M, Schwoppe P, Weber R, Orsino S 2007 *Combustion and Flame* **150** pp 54-59
[6] Pope S B 1985 *Prog. Energy Combust. Sci.* **11** pp 119-192
[7] Ren Z, Goldin G M, Hiremath V and Pope S B 2013 *Fuel* **105** pp 636 – 644

IOP Conf. Series: Materials Science and Engineering **297** (2018) 012072 doi:10.1088/1757-899X/297/1/012072

Heat transfer enhancement of flow insulator by combined stainless steel fibrous and wire net porous materials

P. Khantikomol*, M. Polsongkram, W. Apisitpinyo and T. Poowadin

Renewable Energy Technology Research Laboratory (RTER)
Department of Mechanical Engineering, Faculty of Engineering and Architecture,
Rajamangala University of Technology Isan,
744 Suranarai Rd., Muang, Nakhonratchasima, 30000, Thailand

* Corresponding Author: preecha@rmuti.ac.th

Abstract. The present research article aims to propose the heat transfer enhancement of the flow insulator using combined fibrous and wire net stainless steel porous material. The stainless fibrous plate with porosity of 0.9292 was combined to the stainless steel wire net having pore per inch (PPI) of 16 and total thickness of 30 mm. Two models of the arranging porous plates were prepared, which were model BA and model AB. Each porous plate segment had the same thickness. The examined porous plate model have porosities of 0.8452. The porous plate was placed normal to the flow direction. The air was used as working fluid heated by 5 kW electric heater, which was controlled by the automatic temperature control. Type-K thermocouples were employed to measure the air temperatures. The temperature at front of the porous plate was varied to be 350, 450, and 550°C. The air flow rate was varied in the range of 4-12 m^3/hr. The experimental result showed that the temperature drop across the porous plate and the thermal efficiency increase with the inlet temperature. The air velocity slightly affects the temperature profile inside the test section at the upstream side of the porous plate but greatly affects temperature inside the porous plate. In consideration of the arranging porous plate, placing of the stainless steel wire net at the upstream side and placing the stainless steel fibrous at downstream side (model BA) results in the highest temperature drop and the highest thermal efficiency. At Re 733 and inlet temperature 550°C for model BA at 30 mm thickness, the thermal efficiency was 50%. It was shown that the combined stainless steel fibrous and stainless steel wire net porous material could be a good flow insulator.

1. Introduction

Over the past two decades, heat transfer in the porous media has been studied both experimentally and theoretically. Because the porous media features large areas for heat transfer, it has high heat transfer coefficient at the surface. Many high temperature facilities were done by highly porous material, for example, a gaseous core nuclear reactor, plasma, combustion burner [1-3] and high temperature heat exchanger [4]. These applications could be done by using a multiphase medium consisting of fluid phase (gas) and particulate phase (solid) [5-8] of the porous media. Owing to the advantage of the porous media on high efficiency heat transfer, the energy of the flowing fluid flow through the porous element was recovered leading to the temperature difference between the upstream and downstream sides. This so-called novel concept was called flow insulation because the porous could act as a thermal insulator.

The concept of flow insulation system was presented by Echigo [9] in 1982. He placed a high-porosity steel porous material normal to the gas flow direction in an exhaust duct. It was found that the temperature dropped more than 60% across 15-mm-thick porous plate. When the porous media receives the thermal energy, it emits the thermal radiation into both upstream and downstream sides. In 2009, Viskanta [10] proposed a high-porosity porous material and its application. He showed that high-porosity porous materials could emit, absorb, and radiate the thermal radiation due to its large surface area for heat transfer. Several researchers [11-18] studied this both numerically and experimentally. They illustrated the recovering thermal energy from flowing fluid using the porous materials which involve with the flow insulation concept. Khantikomol et al. [19-20] proposed the numerical and experimental study on the flow insulation system using open-cellular porous material. They indicated that the upstream radiation temperature strongly affected the quantity of the gas temperature drop across the porous plate.

As mentioned above, the high-porosity porous material has significant characteristics in heat transfer. Preecha et al. [21] reported that the stainless steel fibrous material could be a good heat transfer recovery. However, the fluid flow was quite difficult through the fibrous due to complicated structure. In the present work, therefore, the enhancement in heat transfer characteristics of the flow insulation system using combined stainless steel fibrous and wire net material have been investigated. Three average porosities and three thickness were examined. Heat transfer characteristics were presented in terms of heat recovery, temperature drop and the thermal recovery efficiency.

2. Experimental Setup

Figure 1 shows the experimental apparatus diagram. The experimental apparatus was made of the steel tube with 130-mm inner diameter. To make sure that it is perfectly insulated (adiabatic), the ceramic fiber was insulated on both inner and outer sides. Therefore, the experimental tube had the inner diameter of 120 mm. Air was used as a working fluid moved by the blower through the electric heater controlled by the PID temperature controller. The combined porous plates made of stainless steel fibrous and wire net were placed normal to the hot air flow at the top of the experimental tube at a distance of 300 mm from the electric heater. The air temperatures in front of the porous plate (inlet temperature) were varied from 350-550 °C and volume flow rate from 4-12 m^3/h. Several type-K thermocouples were used to measure the temperatures of the flowing air at inlet of the heater (T_1), outlet of the heater (T_2), inlet of the porous plates (T_3) and out of the porous plate (T_4). Moreover, the ambient temperature (T_∞) was also measured during the experiment.

In the present experimental study, the stainless steel fibrous and wire net material were combined in two models, called model AB and model BA as shown in figure 2, where A and B indicated the fibrous and wire net stainless steel, respectively. In model AB, the fibrous plate was placed upstream and the wire net plate was placed at downstream side. Here, the fibrous plate having porosity $\phi = 0.9292$ and wire net having pore per inch (PPI) 8, 12, and 16 were examined. The total combined porous plate thickness were 10, 20, and 30 mm, with each type having the same thickness.

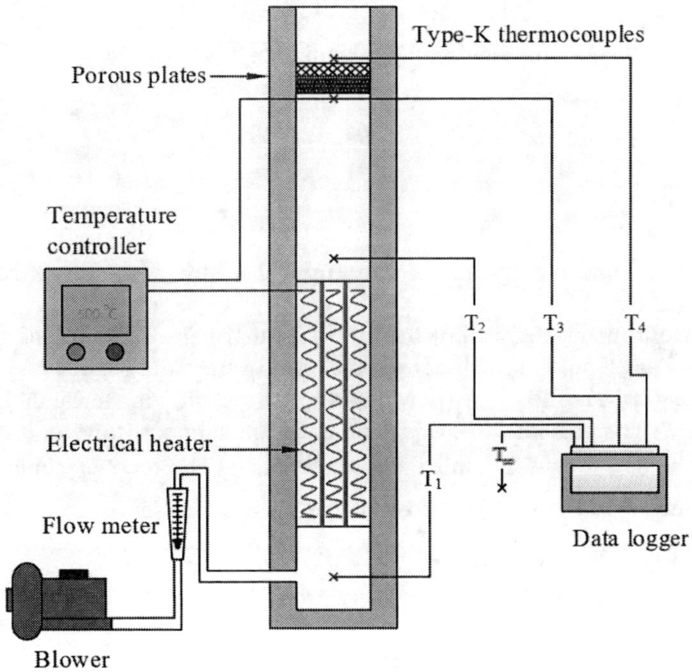

Figure 1. The schematic diagram of the experimental apparatus.

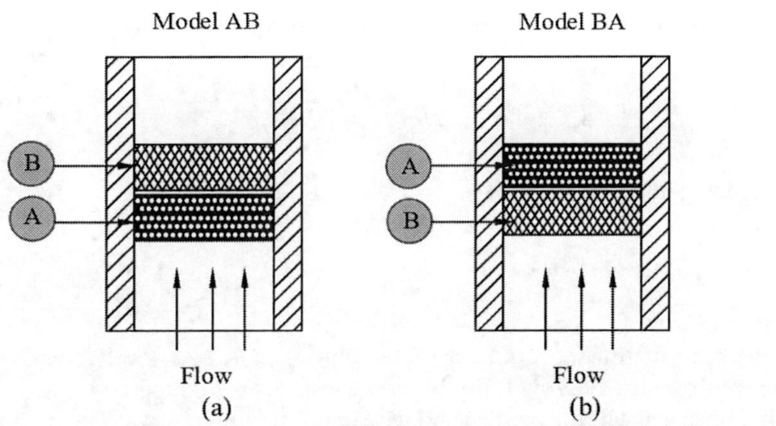

Figure 2. Combined porous plate model: (a) model AB and (b) model BA.

3. Data Analysis

When the fluid moes through the porous element, the heat is exchanged between the fluid and the solid element of the porous material by both convection and radiation. Some of energy is transferred by convection to the solid element and some is transferred by fluid element. Therefore, the energy transfer from the fluid to the solid element of the porous was called the heat recovery by the porous. For flowing fluid, the velocity would be presented in term of Reynolds number based on the tube diameter, Re_D. To investigate the heat transfer between the flowing fluid and the solid element of the porous plates, the heat recovery or heat transfer from the fluid to the porous plate was conducted. Here, the temperature drop (ΔT) and the thermal recovery efficiency of the flow insulator (η_r) were investigated. The flow condition (Reynolds number, Re_D), the inlet temperature of hot gas (T_{in}), the porosity (ϕ), porous thickness were considered as independent parameters.

The Reynolds number (Re_D) based on the tube diameter is given by

$$Re_D = \rho_\infty VD / \mu_\infty \quad (1)$$

where ρ_∞ is the ambient air density (1.225 kg/m^3), μ_∞ is the atmospheric absolute viscosity (1.789×10^{-5} Pa·s), V is the average air velocity (m/s) at upstream region of the porous plate, and D is the duct diameter (m).

In the present experiment, the wall of the tube with thermal insulator is assumed to be an adiabatic wall. Therefore, the gas temperature difference occurred across the porous plate is due to the exchange energy between the gas and solid phase (porous plate). The air flowing through the test tube was assumed to be a uniform flow. The present work was done under the laminar flow condition due to the fluid flow was rather low. The data was done under the steady state. The recovery energy rate ($\dot{Q}_{recovery}$) of the porous plate was equal to the rate heat loss of the air by convection to the solid element of porous material, which can be expressed as follows:

$$\dot{Q}_{recovery} = \dot{m}C_p\left(T_{in} - T_{out}\right), \quad (2)$$

where $\dot{m}$ is mass flow rate (kg/s), C_p is specific heat of the air (J/kg.K), T_{in} is inlet air temperature (K) and T_{out} is outlet air temperature (K).

The thermal recovery efficiency (η_r) of the porous plate can be evaluated by the ratio of recovery heat rate ($\dot{Q}_{porous}$) and the input energy rate ($\dot{Q}_{input}$), as in equation (3):

$$\eta_r = \frac{\dot{Q}_{porous}}{\dot{Q}_{input}} = \frac{T_3 - T_4}{T_4 - T_\infty} \quad (3)$$

where T_∞ is ambient temperature (30 °C).

4. Results and Discussions

4.1. *Temperature drop*

The temperature drop of flowing fluid across the porous plate (ΔT) indicates the amount of heat that the porous recover from the flowing fluid. Figures 3-6 show the temperature drops across the combined porous plates which consist of fibrous (A) and wire net (B) stainless steel. The model AB is the combined porous plates that places the fibrous at upstream side and places the wire net at downstream side. The experimental results reveal the effect of several parameters, including the inlet hot air temperature, Reynolds number, the porous plate thickness and the porosities, to the temperature drop due to the heat recovery by the porous media. Higher temperature drop of the hot air flowing through the porous plate depicted the high heat exchange between the hot air and the solid element of the porous. The experimental results revealed that the temperature drop across the porous plate increase with the inlet hot air temperature (figure 3) owing to the fact that the porous plate could act as the good radiation absorbing and emitting thermal radiation at high temperature as illustrated in figure 3. The temperature drop increases with the porous thickness (figure 4) due to the heat transfer rate increase with increasing heat transfer area by increasing porous thickness. While the temperature drop decreases with increasing Reynolds number ($\mathrm{Re}_D = \rho_\infty VD_{tube} / \mu_\infty$) as shown in figure 5 and increasing porosity (figure 6). At high

Reynolds number, the heat carried by the air element is higher than the heat transfer from the hot air to the porous element by convection leading to decrease the temperature drop of flowing gas. These reasons verify the results that the temperature drop across the porous plate could be high at low fluid velocity and high inlet hot air temperature.

In consideration of the effect of porosity, the temperature drop decreases with increasing porosity because increasing porosity decreases the heat transfer area. Since all of the wire net porous plate have the same diameter, increasing of porosity leads to decrease the heat transfer area.

In comparison of the combined porous plate models AB and BA, it was shown that the temperature drop across the combined porous plate model BA (figure 4a) was higher than that of model AB (figure 4b). This indicated that placing wire net porous at upstream side increased the heat transfer from the hot gas to the porous element. Placing the wire net at upstream side increased heat radiation to the upstream region leading to an increase in the heat exchange between the porous and the flowing fluid.

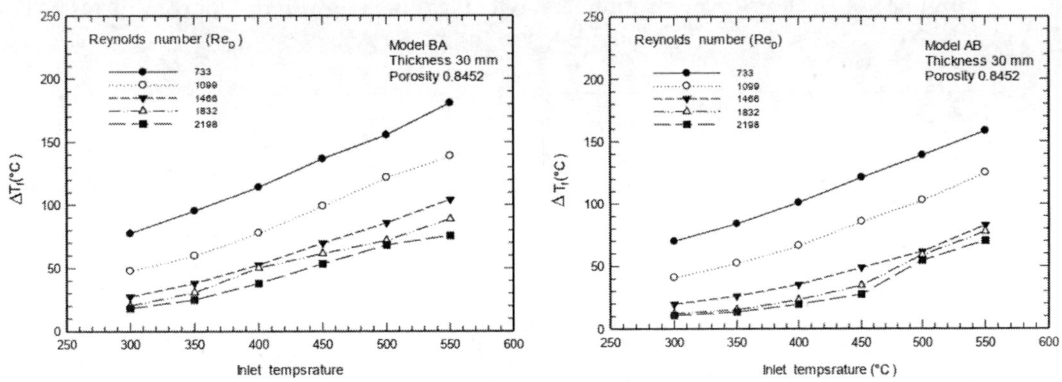

(a) (b)

Figure 3. Effect of inlet temperatures to temperature drop across porous plates: (a) Model BA and (b) model AB.

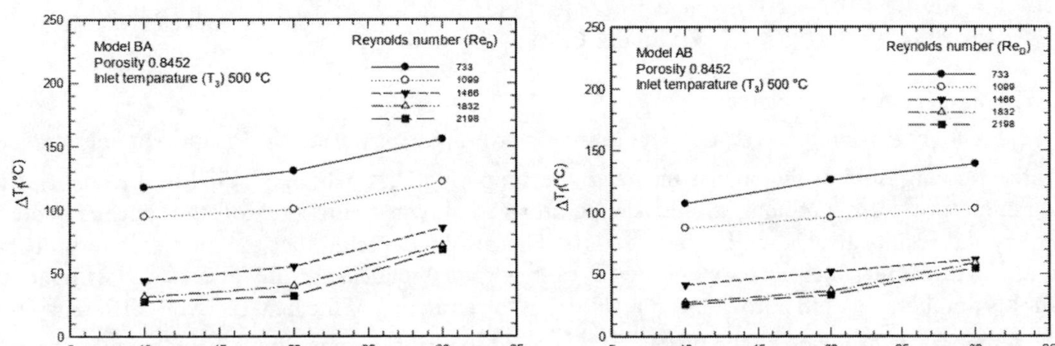

(a) (b)

Figure 4. Effect of porous thickness to temperature drop across porous plates: (a) Model BA and (b) model AB.

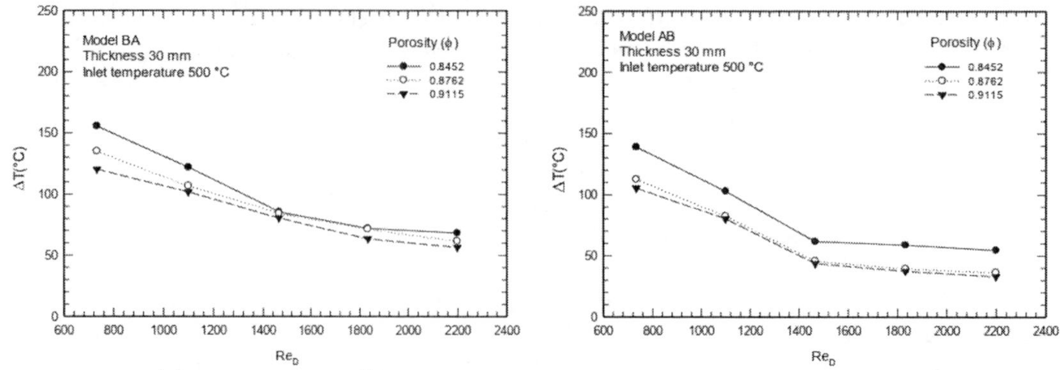

(a) (b)

Figure 5. Effect of Reynolds numbers to temperatures drop across porous plates: (a) Model BA and (b) model AB.

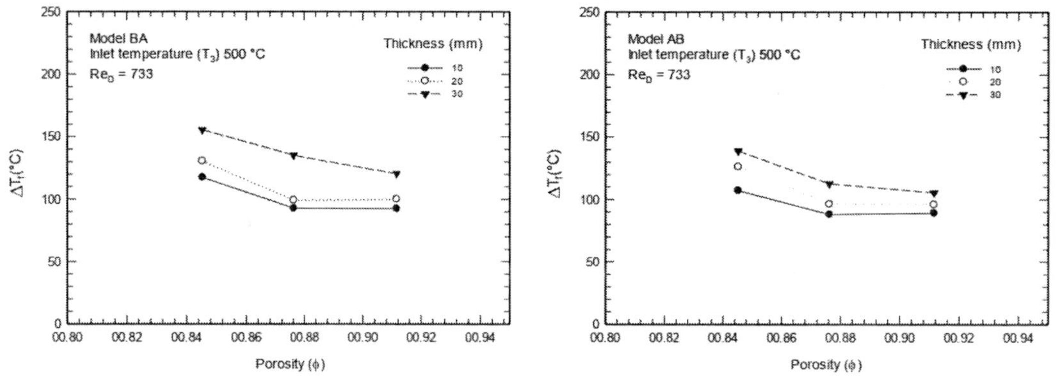

(a) (b)

Figure 6. Effect of average porosities to temperature drop across porous plates: (a) model BA and (b) model AB.

4.2. *Thermal recovery efficiency*

The heat recovery efficiency (η_r) of the combined porous plate is defined as the ratio of energy recovery from the flowing fluid of the porous media to the supplying heat, which is equal to the heat rise in the air temperature from ambient to the temperature at downstream side of the porous plate. The experimental results are shown in figures 7-10. The results depicted that the heat recovery efficiency increases with increasing inlet temperature (T_3) and porous thickness. While it decreases with increasing Reynolds number (Re) and porosities (ϕ). Note that the tendency of the heat recovery efficiency is same as the temperature drop across the porous plate due to the heat recovery have been calculated from the temperature drop. However, the heat recovery efficiency would indicate the portion of the heat that porous could recover from the flowing fluid. Clearly, placing the wire net porous plate at upstream, as opposed to the other way around, increases the amount of heat recovery significantly.

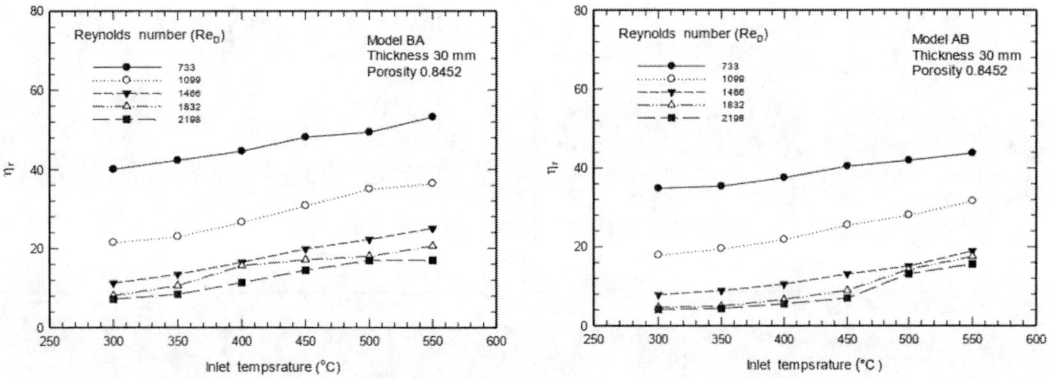

(a) (b)

Figure 7. Effect of inlet temperatures to thermal recovery efficiency: (a) model BA and (b) model AB.

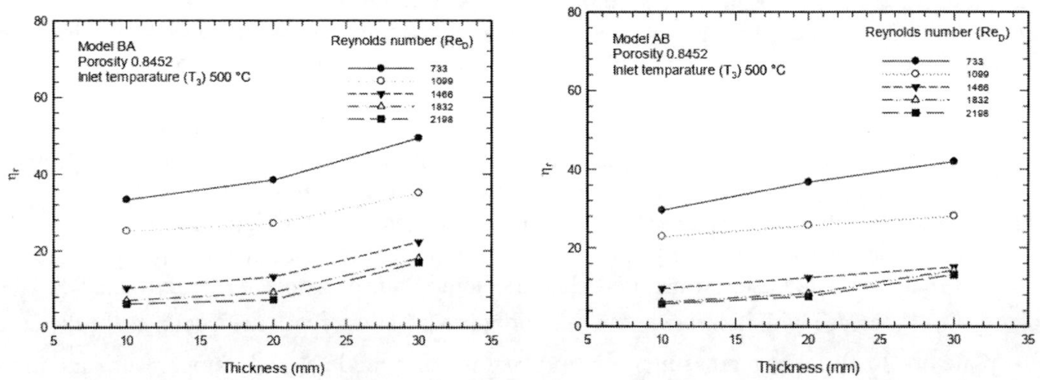

(a) (b)

Figure 8. Effect of Reynolds numbers and porosities to thermal recovery efficiency: (a) model BA and (b) model AB.

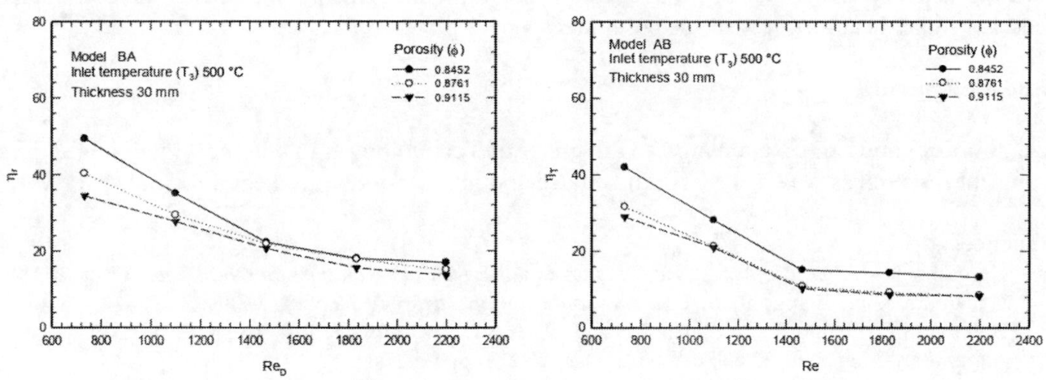

(a) (b)

Figure 9. Effect of Reynolds numbers to thermal recovery efficiency: (a) Model BA and (b) model AB.

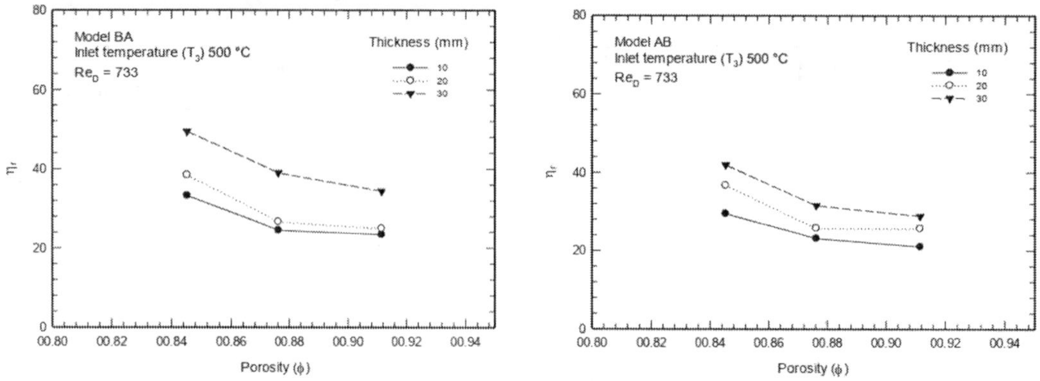

(a) (b)

Figure 10. Effect of Reynolds numbers and porosities to thermal recovery efficiency; (a) model BA and (b) model AB.

5. Conclusion

The present study conducted analysis of the heat transfer of flowing fluid moving through the combined porous plate made of the fibrous and wire net stainless steel porous plates. The experiment was done under the conditions according to the heat transfer characteristics of the porous media. The heat exchange between the hot air and the solid element of the porous was presented in term of temperature drop of the flowing air which indicates the amount of heat recovery. The heat recovery efficiency of the porous plate was also investigated to find conditions of heat transfer enhancement. The experimental results could conclude that the combined porous plate placed the wire net at the upstream side and placed the fibrous at downstream side rather effect to increase heat recovery of the porous plate from the flowing air. Combined porous material would enhance the heat transfer significantly.

In consideration of effect in many independent parameters to the temperature drop and the heat recovery efficiency. The result illustrated that the temperature drop (ΔT) and the heat recovery efficiency (η_r) increase with increasing the inlet hot air temperature and the thickness of the combined porous plate due to effect of increasing radiation heat transfer mode at high temperature and increasing of heat transfer area by increasing the porous thickness. While the temperature drop (ΔT) and the heat recovery efficiency (η_r) decrease with increasing the Reynolds number (Re_D) and the average porosity (ϕ) of the combined porous plate owing to the ratio of heat recovery to the heat carried by the flowing fluid decreasing with increasing velocity although the convection heat transfer mode increase with increasing velocity. In increasing the average porosity, the heat transfer area would decrease, leading to a decrease in the heat recovery of the combined porous plate.

Acknowledgment

The authors acknowledge Rajamangala University of Technology Isan who provided the funding for the present research as well as the working space and necessary research equipment.

References

[1] Golombock M, Prethero A, Shirvill L C and Small L M 1991 *Combust. Sci. and Tech.*77 pp 203-223
[2] Hanamura K and Echigo R 1991 *Warme-und Sttoffubertragung* **26** pp 377-383
[3] Hanamura K, Echigo R and Zhdanok S A 1993 *Int. J. Heat Mass Transf.* **16** pp 3201-3209
[4] Jugjai S and Chuenchit C 2001 A *Int. Energy J.* **2**(2) pp 95-104
[5] Echigo R, and Hasegawa S, 1972 *J. Heat Mass Transfer* **15** pp 2519-2534
[6] Echigo R, Hasegawa S and Tamehio H 1972 *J. Heat Mass Transfer* **15** pp 2595-2610
[7] Tamehiro H, Echigo R and Hasegawa S 1973 *Int. J. Heat Mass Transfer* **16** pp 1199-1213

[8] Hasegawa S, Echigo R, and Shimizu A 1986*Heat Transfer Operation* (Gulf Publishing Company) Vol 1 pp 523-558
[9] Echigo R 1982 *Proc.7th Int. Heat Transfer Conf.* Vol 6 pp 361-366
[10] Viskanta R 2009 *Proc. ASME 2009 Heat Transfer Summer Conf.* (San Francisco, CA)
[11] Wang K Y and Tien C L 1984 *ASME. J. Heat Transfer* **106** pp 453-459
[12] Khantikomol P, Saito S and Yokomine T 2009 *Proc. Int. Conf. Power Engineering (ICOPE-09*, vol 3 pp 37-42
[13] Khantikomol P and Kamiuto K 2008 *Proc. of Int. Conf. on Thermal and Fluids Engineering* (The 7th JSME-KSME)
[14] Kamiuto K 1977 S *JSME Int. J., Series B* **40** pp 577-582
[15] Kamiuto K 2000 *JSME Int. J., Series B* **43** pp 273-278
[16] Kamiuto K 1999 *Trends in Heat, Mass & Momentum Transfer* **5** pp 141-161
[17] Kamiuto K, Saito S and Ito K 1993 *Numerical Heat Transfer Part A* **23** pp 433-443
[18] Kamiuto K and Yee S S 2005 *Int. Communi. Heat Mass Transfer* **32**(7) pp 947-953
[19] Khantikomol P and Kamiuto K 2008 *J. Thermal Science and Technology* **3**(2) pp 319-329
[20] Khantikomol P, Saito S, Kamiuto K and Yokomine T 2001 *Int. Communications in Heat and Mass Transfer*, **37** pp 1015–1024
[21] Khantikomol P, Polsongkram M, and Poowadin T 2016 *J. of Research and Applications in Mechanical Engineering* (JRAME) **4**(2) pp 148-155

Institute of Physics
Dirac House, Temple Back
Bristol BS1 6BE UK

ISSN: 1757-8981
ISBN 978-1-5108-5836-7

9 781510 858367

CURRAN ASSOCIATES INC.
proceedings
.com